The Molecular Biology of Poliovirus

Friedrich Koch and Gebhard Koch

Springer-Verlag Wien New York

Dr. Friedrich Koch
Prof. Dr. Gebhard Koch

Abteilung Molekularbiologie
Universität Hamburg, Federal Republic of Germany

Softcover reprint of the hardcover 1st edition 1985

With 122 Figures

Library of Congress Cataloging in Publication Data. Koch, Friedrich, 1957–. The molecular biology of poliovirus. Includes index. 1. Poliovirus. 2. Molecular biology. I. Koch, Gebhard. II. Title. QR201. P73K58. 1985. 576'.6484. 85-4757.

ISBN-13:978-3-7091-7467-8 e-ISBN-13:978-3-7091-7000-7
DOI: 10.1007/978-3-7091-7000-7

Preface

Years ago when we were asked to write a book on the present-day knowledge of the molecular biology of poliovirus, we did not expect that such an apparently simple task could involve so much time and effort. Our writing was hampered by the fact that both of us are full time "workers", so that this monograph is mainly a spare time expedience.

The main attention of this book focuses on a detailed review of the molecular biology of poliovirus and especially on the advances of the last decade; medical and environmental aspects are only briefly mentioned. Observations from older studies are considered in view of more recent information. Some of the older observations provided fundamental insights and paved the way for present day research; too often such data has been neglected or independently rediscovered. Today, poliovirus research has again attracted considerable interest. High points gained within the last few years were the elucidation of the complete nucleotide sequences of the RNAs of the three poliovirus serotypes and the corresponding vaccine strains, the demonstration of genome evolution during transmission of poliovirus in an epidemic, further characterization of the antigenic sites on the virus particle and of the antigenic drift, characterization of alternate conformational states of the virion capsid, the development of monoclonal antibodies against some of the virus proteins, observations on the role of the plasma membrane, cytoskeleton, and cytoplasmic membranes as mediators in the virus-induced redirection of the synthetic machinery of the host cell, and characterization of proteins involved in RNA replication. The review is certainly not all-inclusive. We are grateful for any comments, corrections, additions, and criticism.

We wish to thank Kurt Bienz, John Bilello, Richard Crowell, Keith Dunker, Denise Egger, John Mapoles, Paolo La Colla, Roland Rueckert, and Douglas Scraba for many fruitful discussions and helpful comments. We also thank all those who provided us with material for illustrations.

The personal proceeds of this book will be donated to a preventive health care project in Nicaragua. As part of this project, the Nicaraguan health ministry has initiated mass vaccination programs against poliomyelitis, which will for the first time succeed in reaching most of the children in this country.

Hamburg, March 1985 *F. Koch* and *G. Koch*

Contents

Part I

The Poliovirus

1
History

I. Introduction

Within the first half of this century, poliomyelitis was a frightening epidemic infectious disease found throughout the world. Severe paralytic afflictions of the central nervous system characterized the disease, hence the name poliomyelitis (polio = grey; myelos = marrow, spinal chord).The vast majority of infections, however, took an abortive course with no or only minor gastrointestinal symptoms. Thus, even in the largest outbreaks, clinically apparent poliomyelitis was not very prevalent: less than 1% of the infected persons developed the severe paralytic form of the disease. Still, the sight of the usually young and often severely crippled survivors made even small epidemics terrifying.

The causative agent of the disease was identified as one of the smallest existing viruses, the poliomyelitis virus. The only natural hosts of the virus are humans and monkeys. Since the introduction and wide scale application of killed vaccines in the USA (1955) and of live vaccines in the USSR (1959), the incidence of poliomyelitis has drastically declined all over the world.

Today, in many parts of the world, the wild type polioviruses have been replaced by the attenuated vaccine strains. However, a complete eradication of the disease, as for small pox, has not yet been achieved. Especially in "developing" countries, the fight against the disease by large scale immunization is confronted with difficulties: other gastrointestinal viruses interfere with the replication of the attenuated polio vaccine viruses in the intestine of the vaccinees, immunization campaigns are often inadequate and do not reach the entire population. In many tropical and subtropical countries wild type polioviruses still circulate among unvaccinated or incompletely vaccinated children. In some small tropical countries with good health services and well-organized programs of annual mass vaccinations of almost all children under a certain age, paralytic disease caused by polioviruses has been elimated (Cuba) or promises to be eliminated in the near future (Nicaragua) (Sabin, 1982). But even in the "developed" countries, sporadic cases of poliomyelitis are still reported. The last natural epidemic in the United States occured in 1973. In 1978/79, a wildtype poliovirus was passed on from the Netherlands via Canada to the USA within a group that had refused to participate in vaccination programs on religious grounds (Nathanson, 1982).

Until the beginning of this century, poliomyelitis was primarily an occasional disease of infants (hence the German name "Kinderlähmung" = infantile paralysis), and this pattern is still seen today in communities with primitive sanitation where the disease is endemic. Prior to widespread immunization, improvement of sanitation paradoxically was accompanied by an increasing prominence of poliomyelitis epidemics. At the same time, a drift in the age distribution of this disease to include adults occurred and an increase in the severity of the disease in older persons was observed. These aspects of the disease are explicable today: Under poor sanitary conditions, nearly everyone became infected during childhood—the development of the paralytic form of the disease being rare—and thus usually acquired a life long immunity against infection by poliovirus. In contrast, under improved sanitary conditions, the first contact with the virus was postponed until a later age, at which point infection by the virus had more severe consequences (Sabin, 1949).

The poliomyelitis virus has been intensively studied during the last 30 years. Today the fundamental principles of poliovirus structure and replication are well understood. Poliovirus research parallels the advancements in our understanding of the fundamental biological concepts of life on a cellular level in general, and of viruses and their replications in particular. Indeed important impetuses to the general field of molecular biology have come from studies on polioviruses and tissue culture cell systems infectable by the virus.

Basic research on polioviruses is continued in laboratories all over the world even today (see Appendix I), and an (almost) complete understanding of this still interesting virus may be expected within the next 10 to 20 years.

II. Early History - Occasional Nonepidemic Cases of Poliomyelitis

The origin of poliomyelitis probably dates back before the recorded history of man (Paul, 1951, 1971). Representations of victims of crippling diseases in historic paintings show the characteristic muscular atrophy and deformities resulting from muscular paralysis in early life, resembling those of poliomyelitis. For example, a representation—possibly the first—of the disease can be found on an Egyptian relief from the 18th dynasty (1580–1350 B.C.), which shows the crippled leg of a young man from the Astarte Temple in Memphis (see Fig. 1; Fanconi *et al.*, 1945). Characteristic descriptions of poliomyelitis-like disease can be found in the literature of antiquity. Hippocrates described paralysis that afflicted patients predominantly in summer and autumn, *i.e.* the period which has been considered as the "polio season" (Armstrong, 1950). Biblical reports of persons with paralyzed or crippled extremities may also reflect affliction by poliomyelitis. In archaeological excavations in southern Greenland, 25 skeletons from the 15th century were dis-

Fig. 1. The first documented case of poliomyelitis?
This Egyptian stele shows a young man afflicted by a crippling disease reminiscent of poliomyelitis. The paralyzed right leg is shorter than the healthy left leg, indicating that the paralysis occurred in consequence of a disease during childhood. Notice also the marked atrophy of muscle in the afflicted leg typical for poliomyelitis. The stele is from the Astarte Temple in Memphis and was probably built between 1580 and 1350 B.C. This may be the first documentation of a poliomyelitis case in history. – Figure courtesy of the Carlsberg Glyptotek, Copenhagen

covered which showed bone deformities reminiscent of those typically associated with severe poliomyelitis.

These examples ascertain that occasional cases of poliomyelitis have occurred throughout the history of mankind. On the other hand, the scarcity of the reports indicates that the manifestations of poliomyelitis were rare, and that the disease did not often occur in an epidemic form. Numerous historical sources exist which vividly describe other diseases which from time to time broke out in devasting epidemics such as smallpox or plague. If there had been any early severe epidemic of poliomyelitis comparable to epidemics which occurred during the past centuries, we would surely have some corresponding historic records. Today, of course, we know that the switch in the nature of poliomyelitis from an endemic disease (with only occasional manifestations of the severe paralytic form) to the appearance of epidemics is related to the improvement of sanitary and hygienic conditions (see above).

III. The 18th and 19th Centuries: Polioepidemics, Poliomyelitis Is Described as a Clinical Entity

Recognition of poliomyelitis as a distinct clinical entity dates back only to the end of the 18th and to the beginning of the 19th century (Paul, 1951). The first description of poliomyelitis is generally accredited to Underwood (1789): he presented a general description of paralyses in the lower extremity, only some of which probably were poliomyelitis. The first exact publication that constituted poliomyelitis as a separate disease and differentiated it from other types of paralysis, appeared in 1840 by Heine (hence also the name Heine-Medin disease).

Near the end of the 19th century, severe epidemics of increasing magnitude occurred in Europe and North America. The first recorded polioepidemic occurred in 1889 in Stockholm, Sweden and afflicted 44 persons. The epidemic was described in detail by Medin (1891) who considered the disease to be an acute infection that could break out in epidemics. Seven years later, an epidemic afflicting 119 persons spread in Vermont, USA. Otherwise, until 1916, poliomyelitis became apparent only sporadically. Meanwhile, a description of the specific neuronal and inflammatory reaction in the central nervous system established poliomyelitis as a distinct clinical entity (Rissler, 1888). The typical clinical and epidemiological features of the disease were repeatedly described. (Caverly, 1896; Wickmann, 1905, 1907; Harbitz and Scheel, 1907; Peabody *et al.,* 1912). An alimentary route of infection in man was already suggested in 1899 (Bülow-Hansen and Harbitz, 1899). Nevertheless, the general assumption of an infectious route via the respiratory tract prevailed for a long time.

IV. Early 20th Century: Research on Polio Begins, Virology Is Still a Clinical Discipline

In 1916, while World War I was ravaging Europe, the worst polio-epidemic known in history spread throughout the USA, afflicting more than 27,000 persons with a fatal progress in 6,000 cases, in New York City alone. Panic broke out in New York: entire families fled from the city, desperate acts reminiscent of plague epidemics in the Middle Ages were reported.

Research on this devastating yet relatively rare disease had advanced onlv slowly to this date. Little was known at this time about the pathogenicity of poliomyelitis. Viruses as etiological agents of disease were first recognized by Iwanowski in 1892, who was able to transmit the tobacco mosaic disease to healthy plants by rubbing them with bacteria-free extracts from diseased leaves (virus = poison). In 1898, Loeffler and Frosch confirmed and extended this concept by showing that the foot and mouth disease of cattle could be similarly transmitted by a cell-free filtrate (foot and mouth disease virus is now known to belong to the same class of viruses as poliovirus, the picornaviruses). A virus as a causative agent for poliomyelitis was established by Landsteiner and Popper (1909) who provided the first evidence for a successful experimental transmission of the disease to monkeys by intracerebral inoculation of a bacteria-free filtrate from spinal cord material of a nine-year old poliomyelitis child, and by the subsequent work of Flexner and Lewis (1910) who passed the viral agent from monkey to monkey. It took another 30 years until poliovirus was adapted to cotton rats (Armstrong, 1939).

The 1916 USA epidemic gave great impetus to polio research, particularly in the USA. Additional psychological backing for polio research came from an episode which started only five years later: in 1921, Franklin D. Roosevelt contracted poliomyelitis during a vacation in Campobello, New Brunswick, Canada, directly following his defeat as candidate for vice president. Roosevelt was nearly completely paralyzed on both legs. A "small" epidemic with a few hundred cases also occurred that year in New York. With his increasing prestige and election to presidency (1932–1945), the psychological effect of Roosevelt's disease rose: he became a romantic symbol, under which the battle against poliomyelitis was intensified from the mid-thirties to the height of his political power, Roosevelt supported all measures that had hope of leading to a healing or prevention of the disease. In 1938, the National Foundation for Infantile Paralysis was founded in his name, perhaps one of the most famous and controversial private welfare programs of our time.

Yet even in the twenties and thirties, promising results from research were meager. A main portion of the energy was invested in theapeutic attempts: physical exercises to revitalize the paralyzed muscles, dietetic measurements to strengthen the overall health, operations involving transplantations of muscles, and all kinds of special apparatuses, social rehabilitation and integration of paralyzed patients etc. Such measurements, although very important, could only bring the damages of the disease to a reducible minimum. In any case, the only route towards a successful control of the disease had to be based on an elucidation of the cause and progress of the disease itself.

The difficulties and fruitlessness of early attempts to fight or prevent the disease by chemical agents are of course easily understandable today in light of the viral strategy for its replication: in contrast to bacteria, viruses do not have their own metabolism, instead they utilize the metabolic machinery of their host cell for replication. Chemical agents meant to destroy a virus by interfering with viral replication thus generally also have severe side effects on the host cell. Therefore, the early chemical therapeutic approaches against poliovirus were the cause of much frustration (in great historical contrast, *e.g.* to the enormous success in the therapy of bacterial infections with antibiotics following the discovery of penicillin by A. Flemming in 1928).

Two particular examples of claims of that time that a treatment for polio had been discovered are interesting and deserve mentioning. In 1936, Dr. E. Schultz, a bacteriologist of Stanford University in California–after having tested numerous chemicals–proposed that the application of a zinc-sulphate solution into the nose provided an efficient protection against infection by poliovirus (Schultz and Gebhard, 1933, 1938). This proposal was based on the then widespread belief–derived from experiments on monkeys–that the route of entry for poliovirus into the human body was the nose, and that the virus would proceed on its path of invasion from there via the olfactory nerve to the spinal cord. Dr. Schultz thought that zinc-sulfate promoted coagulation of proteins in the epithelial layers of the nose, thereby providing a protective barrier that prevented the adsorption of virus to the nerve cells. In 1936, a severe epidemic broke out in the USA, especially in the southern States. The zinc-sulfate solution was administered to more than 4,000 persons by local sanitary authorities, unfortunately, however, not precisely in the manner proposed by Dr. Schultz and without sufficient controls that would have permitted a clear evaluation of the efficiency of the mehod. One year later, during a poliomyelitis epidemic in Toronto, Canada, a sufficient test of the method was carried out under auspices of Dr. Schultz himself: 5,000 children were tested, another 5,000 served as controls. The spray was applied twice to each child within a period of 10–12 days. The efficiency of protection was said to depend on a temporary loss of the olfactory sense to ascertain that all nerves had been protected. The results were devastating: only some of the treated children showed a complete loss of the sense of smell, and of these, many still complained months after the spray application of having retained the loss of the olfactory sense. The evaluation of the results showed that the same number of treated as of untreated children had contracted poliomyelitis: the spray had had no therapeutic or preventive effect whatsoever.

Another claim for successful treatment of poliomyelitis during its early stages came from Dr. Retan, a pediatrician, who claimed to have developed a method for washing out the toxic products of the virus from the central nervous system. The method involved transfusion of large volumes of salt solution and puncture of the spinal cord to aspirate spinal fluid. Testing of the method on monkeys by the National Foundation for Infantile Paralysis, which at that time financed nearly all poliovirus research, again provided negative and frustrating results: the method did not show any therapeutic effect on infected monkeys, either prior to or after the onset of paralysis; worse yet, many monkeys, even in the uninfected controls, died as a direct consequence of the "treatment".

Many other drugs and chemicals were tested in the laboratories, none proved to be of any therapeutic value (Lo-Grippo *et al.*, 1949; Grulee *et al.*, 1950). Finally it became clear that the only hopeful method to conquer the threat of poliomyelitis would be the development of an efficient vaccine. Yet in the beginning of the 20th century, virology still evolved mainly as a clinical discipline. One important advance during that time was the elucidation of the actual route of infection: the alimentary tract. A successful alimentary tract infection in subhuman primates was demonstrated (Kling *et al.*, 1929; Sabin and Ward, 1941 b), and the regular presence of poliovirus in the stools of infected humans was discovered (Trask *et al.*, 1938).

V. The Development of Vaccines Against the Poliomyelitis Virus

The ancient knowledge passed on by Thucydides 2,500 years ago that humans were never attacked a second time by the same infectious disease led to deliberate attempts beginning in the Middle Ages to induce immunity by inoculating humans with material scraped from skin lesions of persons suffering from smallpox. The first safe procedure for immunization was established by Jenner in the late 18th century, who employed materials from pox lesions in cows rather than in humans. First attempts to immunize humans against poliomyelitis using infected monkey spinal cord suspensions were performed in 1935 and 1936 (Kolmer *et al.*, 1935; Brodie and Park, 1936). These, however, were unsuccessful and several vaccinees contracted the disease.

In order to prepare an effective vaccine, it was necessary to have a non-nervous but susceptible tissue in which sufficiently large quantities of virus could be grown. A landmark in the development of poliomyelitis vaccine came 30 years ago when Enders, Weller and Robbins (1949) showed that poliovirus could be isolated and readily propagated in cell cultures of non-neuronal human or monkey tissue (Nobel prize in 1954). Within three years, a formalin

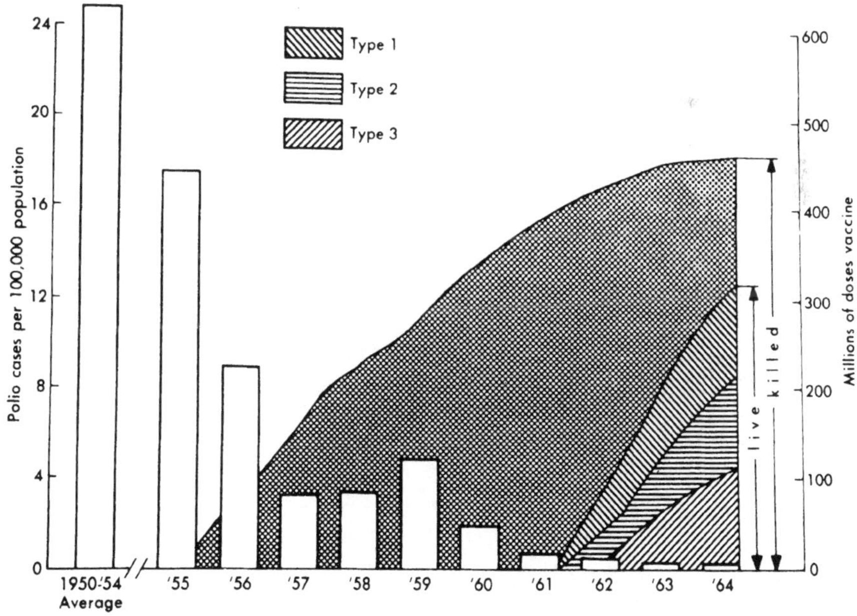

Fig. 2. The decline of poliomyelitis in the U.S.A. after the introduction of vaccines
The figure demonstrates the decline of poliomyelitis cases after the introduction of killed and live vaccines in the U.S.A. in 1955. The annual poliomyelitis case rate is correlated with the cumulative distribution of vaccine from 1950 to 1964. In the period from 1950–1954, the average incidence of reported poliomyelitis was 24.8 per 100,000 population; in 1964 it was 0.1 per 100,000 population. — Adapted from statistics of the Communicable Disease Center, Public Health Service, Atlanta, Ga., U.S.A. Figure from Davis *et al.* (eds.), 1973 [Microbiology, 3rd ed., p. 1295 (1973)]

inactivated vaccine was developed (Salk, 1953; Salk *et al.*, 1954). Large vaccination programs with inactivated virus were launched in many countries in the mid-fifties. Attenuated viruses were developed in tissue culture (Sabin, 1955; Koprowski *et al.*, 1956; Cox *et al.*, 1959) for the preparation of live vaccines. Such vaccines began to be employed on a large scale in 1959 in the USSR (Chumakov *et al.*, 1961). Figure 2 shows the decline of the incidence of poliomyelitis after the onset of vaccination programs in the USA.

A collaborative effort of many investigators (Committee on Typing of National Foundation of Infantile Paralysis, 1951) established that polioviruses belong to only three distinct serological types: type 1: Brunhilde (named after the female chimpanzee from which this virus type was first isolated), type 2: Lansing (named after the city in Michigan where the patient lived who was the host to this virus type), type 3: Leon (named after the patient from California from whom this virus type was isolated) (Bodian *et al.*, 1949; Kessel and Pait, 1949; Hahn, 1972).

VI. The Fifties and the Advent of Molecular Biology

The first progress towards determining the physical or chemical nature of viruses came in 1935, when Stanly succeeded in purifying and crystallizing tobacco mosaic virus (TMV) (Nobel prize 1946); allowing accurate chemical analysis which soon showed TMV to contain nucleic acid as well as protein. Poliovirus was crystallized 20 years later in 1955 (Schaffer and Schwerdt).

The successful cultivation of poliovirus in tissue culture (see above, Enders *et al.*, 1949; Enders, 1952) also paved way for detailed studies on the molecular biology of poliovirus. The introduction of this simple method to produce large amounts of virus resulted in an explosive development of animal virology in general and of poliovirology in particular. Of importance in this respect were also the development of a plaque assay for infectivity (Dulbecco, 1952, 1955; Dulbecco and Vogt, 1954, 1955), and the definition of the essential nutrients of cultured cells (Eagle, 1955; Eagle and Habel, 1956).

Work on DNA mediated transformation of bacteria (Avery *et al.*, 1944), and work on bacteriophages in the early fifties (Hershey and Chase, 1952) proved that the nucleic acid (DNA) of the phage alone was responsible for directing the replication of the phage. The infectious nature of the nucleic acid (RNA) of poliovirus was established in 1957: The isolated viral RNA was capable of producing an infection in mice (Colter *et al.*, 1957), and in tissue culture cells (Alexander *et al.*, 1958 a, b). The basic steps of the poliovirus replication cycle including the time course of synthesis of virus specific RNA and protein, and the formation and release of mature progeny viruses were described shortly thereafter (Darnell, 1958, 1962; Darnell and Levintow, 1960; Darnell *et al.*, 1961; Levy, 1961). The role of specific receptors in determining the susceptibility of a cell to viral infection was demonstrated, and the kinetics of poliovirus adsorption to cells were described (Holland and McLaren, 1959; Holland and Hoyer, 1962; Darnell and Sawyer, 1960). Early shut-off of the synthesis of host-cell RNA, DNA and protein was shown to be a characteristic feature of picornavirus infection (Salzman *et al.*, 1959; Holland, 1963). Other work in the fifties included analyses of the interaction of poliovirus with its antibody (Le Bouvier, 1955, 1957, 1959; Mandel, 1958; McBride, 1959), the discovery, characterization and differentiation of other human viruses with properties similar to polioviruses that lead to the establishment of the class of enteroviruses (polioviruses, coxsackieviruses, echoviruses) (Committee on Enteroviruses 1957, 1962), the induction and isolation of the first temperature sensitive mutants of poliovirus (Dubes and Chapin, 1958; Nomura and Takemori, 1960), the first X-ray crystallographic studies on poliovirus crystals (Finch and Klug, 1959), and the first electron microscopic analysis of poliovirus infected cells (Horne and Nagington, 1959).

VII. Advances in Polio Research During the Past Two Decades

Important advances in the sixties included the characterization of different RNA species in infected cells: dsRNA, replicative form (RF) (Montagnier and Sanders, 1963), and replicative intermediate (RI) (Erickson and Gordon, 1966; Girard, 1969), the description of a virus induced RNA dependent RNA-polymerase in the infected cell (Baltimore and Franklin, 1963 a), and of the kinetics of RNA synthesis (Baltimore, 1968), the description of virus specific protein synthesis under the direction of viral mRNA (Penman *et al.*, 1963), the development of the SDS-polyacrylamide gel systems for analyzing proteins with the demonstration of different viral capsid proteins, and other non-capsid virus induced proteins of infected cells (Summers *et al.*, 1965), the discovery of the polyprotein and the concept of protein precursor processing (Summers and Maizel, 1968; Holland and Kiehn, 1968; Jacobson and Baltimore, 1968 a, b), and the discovery of specific viral inhibitors, notably guanidine and 2-(α-hydroxybenzyl-benzamidazole) (HBB) (Rightsel *et al.*, 1961; Eggers and Tamm, 1962), the demonstration that poliovirus is able to replicate normally in the absence either of host DNA synthesis (Salzmann *et al.*, 1959) or of DNA directed RNA synthesis (Reich *et al.*, 1961), and electron-microscopic studies of poliovirus replication (Horne and Nagington, 1959; Mattern and Daniel, 1965; Dales *et al.*, 1965).

During the past decade, the basis for the infectivity of the three different viral RNA species were analyzed further (Koch, 1973), biochemical and genetic maps of the poliovirus genome were constructed (Rekosh, 1972; Saborio *et al.*, 1974; Cooper, 1969, 1977),the complete translation of the viral mRNA in vitro was achieved (Villa-Komaroff *et al.*, 1975), the entire nucleotide sequence of the genome was determined (Kitamura *et al.*, 1981; Racaniello and Baltimore, 1981 b; Toyada *et al.*, 1984), the genome associated protein VPg was discovered (Wimmer, 1979), important principles of viral structure and assembly were elucidated (Rueckert, 1976; Putnack and Phillips, 1981), the role of the accumulation of virus

Table 1. *Historic time chart*

Antiquity	representations of polio-like crippling diseases in paintings and literature
15th century	small epidemic in Greenland
Late 17th century	Jenner developes safe method of immunization against small pox
1789	treaties of paralyzing diseases of the legs by Unwood
1840	distinctive clinical features of poliomyelitis are described by Heine (Heine-Medin disease)
1887	first recorded polio-epidemic in Stockholm, Sweden (44 cases)
1892	Iwanowski discovers a subbacterial infectious agent (virus = poison) as a transmitter of the tabacco mosaic disease
1894	epidemic with 132 cases in Vermont, U.S.A.
1905	Wickman's book on Poliomyelitis acuata
1909	Landsteiner and Popper: A virus is demonstrated as the infectious agent of poliomyelitis by transmission of the disease from CNS material of a 9 year old child to monkeys
1909/10	Flexner and Lewis, Leiner and Wiesner: transmission of the disease from monkey to monkey
1916	(height of World War I) U.S.A. epidemic with more than 27,000 cases including 6,000 fatalities, panic breaks out in New York City where population is struck particularly severely
1910–1950	many fruitless attempts to prevent or treat poliomyelitis with drugs or chemical agents
1921	F. D. Roosevelt (U.S.-President 1932–1945) contracts the disease and is crippled in both legs
1928	penicillin is discovered to be a very potent antibacterial agent
1929	King: subhuman primates can be infected via the alimentary tract
1935/36	first (unsuccessful) attempts of immunization against poliomyelitis (Kolmer et al., Brodie and Park)
1936/37	the proposed protection against poliomyelitis by a zinc-sulfate spray proves to be a failure
1938	regular presence of poliovirus in the stool of patients observed by Trask; the National Foundation for Infantile Paralysis is established in the U.S.A.
1939	poliovirus is adapted to propagation in laboratory rodents (Armstrong)
1944	(height of World War II) Avery, McLeod and McCarthy: nucleic acids are proposed to be the carrier of genetic information: the advent of molecular biology
1949	Enders: cultivation of poliovirus in non-neuronal cell cultures derived from primates—the advent of tissue culture poliovirus is demonstrated in urban sewage: grouping of 14 strains into three basic immunological types (Bodian)
1950	diagnostic procedures: search for virus in the stool by tissue culture assays
1951	Committee on Typing of the National Foundation for Infantile Paralysis constitutes the existence of 3 sereologically different types of poliovirus: 1 = Brunhilde, 2 = Lansing, and 3 = Leon
1952	Dulbecco developes the plaque test for poliovirus Hershey and Chase demonstrate DNA to be the genetic material of bacteriophages
1950–1955	attenuation of wild poliovirus; introduction of the Sabine and Salk vaccines; neutralization of poliovirus; crystallization of poliovirus
1955–1960	phenotypic mixing, first *ts* mutants, infectious RNA, specific cellular receptors, the basic steps in the poliovirus replication cycle, cytopathic effect, differentiation from other human viruses: enteroviruses, x-ray crystallography, first electron microscopy of poliovirus, onset of eradications of poliomyelitis by world-wide vaccination programs with killed and live-vaccines

1960–1970	guanidine and HBB as inhibitors of viral replication, ds RNAs: replicative form (RF) and replicative intermediate (RI), RNA-dependent RNA polymerase, electronmicroscopy of infected cells, shut-off, intracellular membrane-proliferation, SDS gels, polyprotein-processing, pactamycin map
1970–today	biochemical and genetic maps of the poliogenome, *in vitro* translation of polio mRNA, nucleotide sequencing of the genome, discovery and characterization of non structural viral proteins: VPg, p22 protease, NCVPX, NCVP-4 replicase, principles of virus structure and assembly, changes in cytoskeleton and in kation concentrations during infection, cloning of the cDNA of the poliovirus genome, monoclonal antibodies

induced cytoplasmic membranes was elucidated (Caliguiri and Tamm, 1969, 1970 a, b), different viral RNA replication complexes were described (Caliguiri, 1974; Lundquist and Maizel, 1978 a, b; Etchinson and Ehrenfeld, 1981), the viral replicase was purified and characterized (Flanegan and Baltimore, 1979; Dasgupta *et al.*, 1979; Van Dyke *et al.*, 1982), and functions of several nonstructural viral proteins were defined.

Most recently, the application of recombinant DNA technology to poliovirus and the succesful cloning of its genome and defined fragments thereof have paved the way for the precise definition of poliovirus protein functions (Racaniello and Baltimore, 1981 b; van der Wef *et al.*, 1981). Monoclonal antibodies against several poliovirus proteins have been obtained during the past two years (Icenogle *et al.*, 1981; Emini *et al.*, 1982; Minor *et al.*, 1982). They promise to be of great aid in the further definition of viral protein functions.

The refinements of biochemical and immunological methods—oligonucleotide mapping, monoclonal antibodies—will permit the precise classification of poliovirus serotypes isolated from patients or from the environment (Minor *et al.*, 1982 a). It has become clear, that vaccine and wildtype viruses are genetically unstable, adapting rapidly to antibody pressure with variations in certain portions of the viral capsid proteins (Nottay *et al.*, 1981; Crainic *et al.*, 1983; Kew and Nottay, 1984). Modern genetic engineering technology may provide new means for the development of safe and more stable vaccines. Certain fragments of the capsid proteins which are essential for virus stability and/or virus adsorption to the host cell should be exposed on the surface of the virion and might be genetically more stable and constitute potential antigenic determinants. Corresponding peptides can be identified and produced in large amounts either by chemical synthesis or by cloning and expression of representative cDNA sequences in bacteria (Minor *et al.*, 1982 b). Alternatively, it should be possible to obtain stable, satisfactory attenuated mutants by cloning of viral DNA and then subjecting the cloned DNA to enzyme surgery to create viable deletion mutant strains (Chanock, 1982; Chanock and Lerner, 1984). The elucidation of the three-dimensional structure of the poliovirion by high resolution X-ray crystallography is expected to have important implications in understanding many fundamental properties of the virus including its assembly and uncoating, receptor recognition and neurovirulence, and of course its immunogenicity (Hogle, 1982, 1984).

Much emphasis of present day polio research is concerned with the occurrence of the virus in parts of the environment (especially water), and on methods for its elimination therefrom. This is of particular importance as there are plans for the recirculation of drinking water in some of the developing countries, and since the wild type virus has revealed some resistance to vaccination programs in these areas.

The history of poliomyelitis and the landmarks in the research on the poliovirus are summarized in Table 1.

Most of this monograph is devoted to a review of the more recent advances in the molecular biology of poliovirus.

2
Classification

I. General Considerations

Viruses are primarily classified according to the type of their nucleic acid (RNA, DNA), according to the expression of their genome (Baltimore, 1971 b), and other characteristics such as the nature of the envelope (Fig. 3 and 4, Table 2) (Wildy, 1971; Fenner, 1976).

A. Nature of the Genome

The genetic information of some viruses is encoded in single-stranded or double-stranded (two complementary strands) DNA (ssDNA, dsDNA), other viruses carry double or single stranded RNA (dsRNA or ssRNA). The RNA of ssRNA viruses may be of messenger RNA polarity or of polarity complementary to mRNA. The RNA of messenger polarity is termed the (+) strand, the messenger complement is the (–) strand. All viruses ultimately express themselves through the proteins which are synthesized under the direction of the viral messenger RNA, thus mRNA plays the central role in viral gene expression (Fig. 4). Some proteins are structural components of the viral capsid, other proteins are enzymes or factors (replicating enzymes = polymerases, processing enzymes = proteases or nucleases) required for viral replication.

Amongst the RNA viruses, some package their mRNA (poliovirus), others the (–) RNA, others yet dsRNA into a protein coat. Some viruses carry only one single strand of RNA (polio), others may carry several single or double stranded RNAs (influenza, reovirus), or two or more equivalent copies of one RNA (RNA tumor viruses). Since eukaryotic cells apparently lack enzymes which can (by themselves) copy viral RNA, those viruses whose genomic RNA is dsRNA or (–) RNA must carry a polymerase which copies the viral (–) RNA strand or the viral dsRNA after its entry into the host cell to produce mRNA, thereby permitting the synthesis of other viral proteins. Viruses which contain (+) RNA genomes, such as poliovirus,

Table 2. *Classification of RNA viruses*

Subphylum:	Nucleic acid	RNA								
	Symmetry of capsid or Core	Cubic					Helical or Pleomorphic			
Order:	Core	Naked		Enveloped			Enveloped			
	Number of capsomers	60	32	–	(32)	–	–	–	?	–
	Diameter of virion	25–30	60–80	50–300	40–70	80–100	80–120	150–300	70 × 200	75–165
	Mol. wt. of nucleic acid × 10^6	2.6	12–20	2–3	4	2–5	5	5–7	4	5.5–8
	Approximate number of genes	12	40	10	15	4	15	20	15	20
Family:	Virus family	Picorna-viridae	Reo-viridae	Arena-viridae	Toga-viridae	Retro-viridae	Orthomyxo-viridae	Paramyxo-viridae	Rhabdo-viridae	Corona-viridae

Table modified from Melnick, 1973 (in: Ultrastructure of Animal Viruses and Bacteriophages, 1973), Data from Matthews, 1979 (Intervirology 12, 129 to 296, 1979).

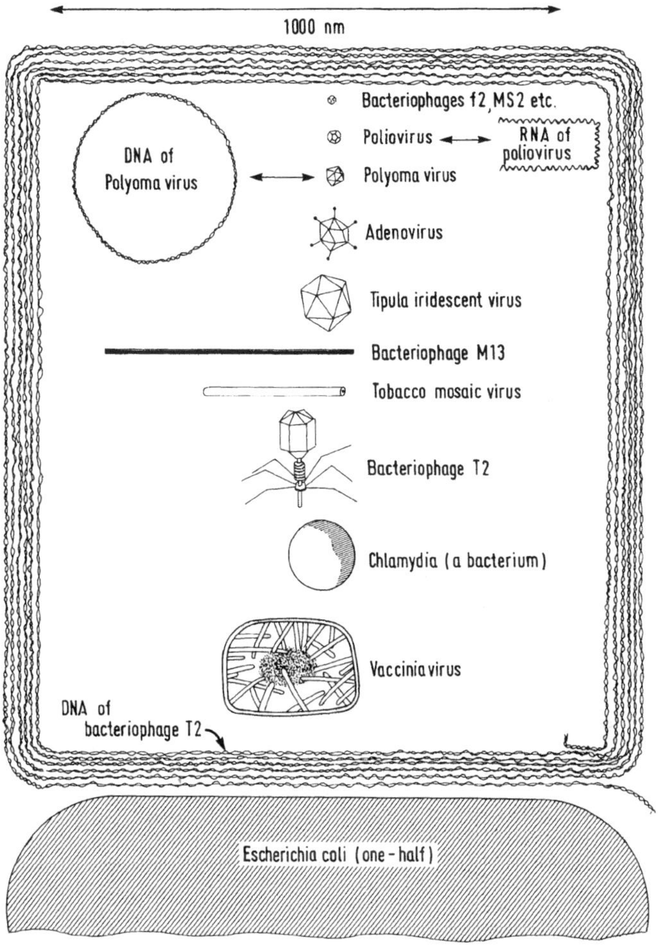

Fig. 3. The relative sizes of poliovirus and its RNA-genome
The figure illustrates the sizes of poliovirus and its genome in comparison to other viruses and the bacterium Escherichia coli. The viruses as well as the lengths of their nucleic acids are drawn on the same scale. – Figure modified from Dulbecco, 1980 [Davis *et al.* (eds.), Microbiology, p. 856 (1980)] and Horne, 1963 (Readings from Scientific American, The Molecular Basis of Life, 1963)

need not carry a polymerase (or other functional enzymes), since the incoming RNA can act as a message, thereby directing the synthesis of the proteins necessary for viral replication. Indeed, the isolated RNA of such viruses by itself may be capable of initiating productive infection, once it has penetrated into the host cell.

It is believed that the mRNAs of most animal viruses have only one initiation site for translation by cellular ribosomes, since eukaryotic ribosomes (unlike their prokaryotic counterparts) apparently cannot recognize "internal" initiation sites (*i.e.* those not located at or near the 5′ end of the RNA). For a more detailed discussion of this still controversial question see Chapter 8. Multicistronic RNA

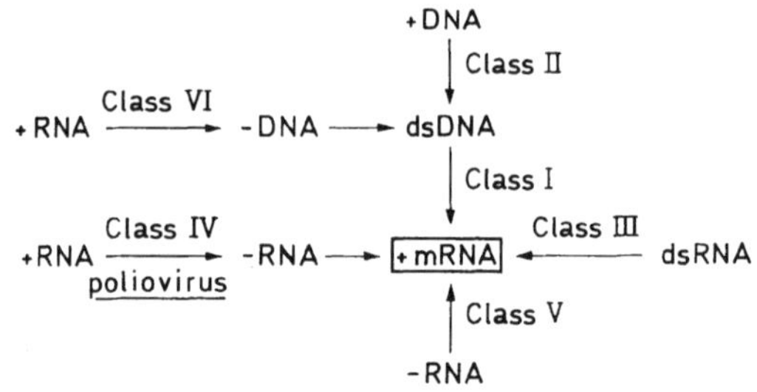

Examples:

I T 4 phage, vaccinia virus, adenovirus, herpesvirus
II ϕ X174
III reovirus
IV *picornaviruses,* tomato bushy stunt virus, Qβ phage
V influenza virus, vesicular stomatitis virus
VI RNA tumor viruses

Fig. 4. Classification of viruses according to the expression of their genomes
The figure illustrates the classification of viruses according to the expression of their genomes. Representative members of each class are listed below. — Figure modified from Baltimore, 1971 b [Bact. Rev. *35,* 235 (1971)]

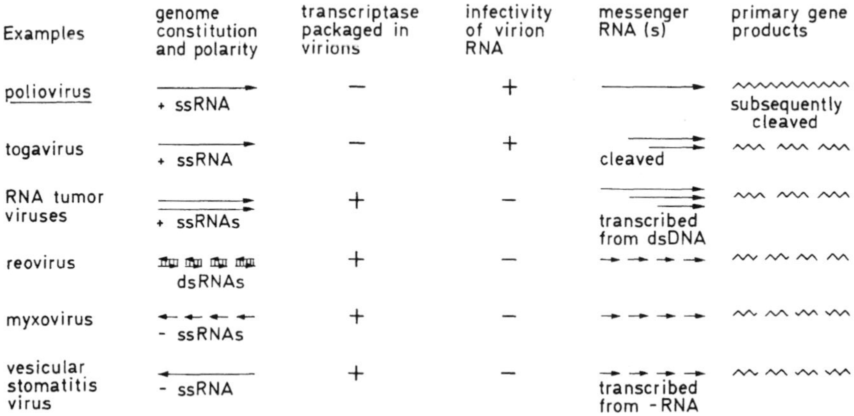

Examples
genome constitution and polarity
transcriptase packaged in virions
infectivity of virion RNA
messenger RNA (s)
primary gene products
poliovirus
+ ssRNA
−
+
subsequently cleaved
togavirus
+ ssRNA
−
+
cleaved
RNA tumor viruses
+ ssRNAs
+
−
transcribed from dsDNA
reovirus
dsRNAs
+
−
myxovirus
- ssRNAs
+
−
vesicular stomatitis virus
- ssRNA
+
−
transcribed from -RNA

Fig. 5. The genome strategies of some RNA viruses
The RNA viruses have been particularly useful in helping to understand how biological structures function in cells. All viruses ultimately express themselves through proteins which are synthesized under the direction of the viral messenger RNA. Some viruses package their mRNA (poliovirus), others package the c-RNA complementary to their mRNA ("antimessenger"), others still package double stranded RNA instead. If the genomic RNA of the virion is not mRNA, then the virion must carry a transcriptase which copies the incoming RNA strand and thereby permits mRNA to be synthesized, since cells evidently lack enzymes which can copy viral RNAs. If the genomic RNA is of mRNA polarity, the isolated virion RNAs are usually infectious (RNA tumor viruses may be an exception). In most eukaryotic viruses, the mRNA has only a single initiation site for translation by cellular ribosomes. In order to obtain functional protein products, multicistronic viruses solve this problem by cutting up either the RNA or the polyprotein. — Figure courtesy of R. Rueckert

viruses (having a genome containing information for several distinct polypeptides) may solve this problem either by cutting up a polyprotein (*e.g.* poliovirus) or by cutting up the RNA. Some of these aspects are illustrated in Figure 5. Other characteristics employed for virus classification are listed below.

B. The Cryptogram

A straightforward method of virus classification, based on the chief chemical, physical, and morphological characteristics has been devised (Gibbs *et al.*, 1966; Wildy, 1971). This information is assembled in cryptograms which are similar for similar viruses. The characteristics of any virus can be defined by a cryptogram which contains at least four sets of terms, a/b : c/d : e/f : g/h, where each term is a letter or a number related to a particular property of a virus particle. Each letter in the cryptogram is defined as follows: (a) describes the type of nucleic acid: D is written for DNA viruses, R for RNA viruses; (b) describes whether the nucleic acid is double (2) or single (1) stranded; (c) refers to the molecular weight of the nucleic acid (in millions); (d) refers to the percentage of nucleic acid in the virion; (e) describes the general shape of the virion: S = spherical, E = elongated with parallel sides, but the ends not rounded, U = elongated with parallel sides with rounded ends, X = complex; (f) describes the shape of the nucleocapsid as in (e); (g) describes the host organism: V = vertebrate, I = invertebrate; (h) describes how the virus is transmitted: O means that no vector is required, Ac and Si refer to insect vectors such as mites or fleas.

The cryptogram for poliovirus (and other picornaviruses) thus is R/1 : 2.6/30 : S/S : V/0, *i.e.* poliovirus is a single stranded RNA virus, the RNA has a molecular weight of 2.6×10^6, and it constitutes 30% of the mass of the virus particle. The particle is spherical in shape, has no envelope, and it infects vertebrates only.

II. The Distinguishing Features of Picornaviruses

Poliovirus is taxonomically classified as a member of the family Picornaviridae and subfamilial genus Enterovirus. Picornaviruses are very wide-spread in nature and many diseases are known to be caused by these viruses. They are among the smallest (pico) viruses currently known. Their sole genetic material is a single strand of RNA contained within a small capsid. Hence the name *pico RNA* virus. In Figure 3 above, the size and structure of poliovirus is compared to that of some other viruses.

A 'definition of picornaviruses' based on the presently known common properties of picornaviruses has been established by the Study Group on Picornaviridae of the International Committee on the Taxonomy of Viruses (ICTV) (Cooper *et al.*, 1978). The distinguishing features of picornaviruses are summarized in Tables 3 and 4, which are based on their report. The picornaviruses have the status of a virus family. Within this family four subfamilial groups (genera) of equal status have been clearly established to date, namely enterovirus (which includes the polioviruses), cardiovirus, rhinovirus, and aphthovirus (Tables 3 + 4). Recent information on the biological and physicochemical characteristics of the human hepatitis A virus (Siegl and Frösner, 1978 a, b; Siegl *et al.*, 1981) and the insect cricket paralysis virus (Moore and Tinsley, 1982) suggest that these viruses should also be classified under the family of picornaviruses. Further viruses may be added to the family when new information becomes available.

The distinct features that differentiate the picornaviruses into four genera are manifested in properties of pH stability, buoyant density in cesium salts, sensitivity to photoinactivation by dyes as well as some functional variations in genome

Table 3. *Features of picornaviruses*

Virion properties	
Type of coat	capsid (nonenveloped)
Diameter	22–30 nm
Conformation in electronmicrographs	spherical (slightly skewed)
Surface in electronmicrographs	finestructure barely detectable
Number of equivalent structural subunits	60 (approximately 10^5 daltons each)
Structural symmetry	5:3:2 fold cubic (icosahedral)
Dry molecular weight	$8–9 \times 10^6$ daltons
RNA in %	29–32
Protein in %	68–71
Buoyant density	1.3–1.45 g/cm^3
Sedimentation coefficient	150–160 S
UV absorption (260 nm/280 nm)	1.67–1.72
Virion proteins	
a) 58–60 copies	3 capsid proteins of 22–40 Kd (VP 1–3) and 1 capsid protein of 5–10 Kd (VP 4)
b) 0–2 copies	1 capsid protein of ca 42 Kd (VP 0)
c) 1 copy	genomic protein covalently linked to RNA of 2.4 Kd (VPg)
d) others?	phosphokinase activity
Virion genome	
Type of nucleic acid	ssRNA
Polarity	plus (mRNA equivalent)
Molecular weight	$2.5–2.6 \times 10^6$ daltons
Sedimentation coefficient	35 S
Viral proteins	
Posttranslational cleavage	very extensive
Non-structural proteins	numerous, derived from two thirds of primary translate: protease, RNA polymerase, others
Viral RNA	
Genome RNA	ssRNA (35 S) of + polarity, 5'end linked to VPg
mRNA	identical to genomic RNA, but lacking VPg
Replicative RNA	double (17 S) RF-RNA and multistranded RI-RNA (17–70 S)

Table 4. *Distinguishing features of picornavirus genera*

Genera	Enterovirus	Cardiovirus	Rhinovirus	Aphtovirus
Main members	polio, echo, and coxsackievirus	encephalomyocarditis virus group	human and animal rhinoviruses	foot and mouth disease viruses
Serotypes	70	1	110	7
Main hosts	human, vertebrate	human, rodent, vertebrate	human, vertebrate	most cloven-footed animals
Host range and tissue tropisms of individual strains	narrow to wide	wide	narrow	wide
Main habitat	gastrointestinal tract,	CNS, heart	respiratory tract	generalized
Characteristic diseases	diarrhea, paralysis	myocarditis	common cold	vesicular eruptions (mouth and extremities)
Optimal growth temperature	36–37	36–37	33–34	36–37
Virion properties				
% RNA	29	31	30	31.5
Diameter, nm	22–28	24–30	24–30	23–25
Infectivity at pH 3	stable	stable	labile	labile
Infectivity at pH 5–6	stable	labile in 0.1 M halide	(labile)	stable at high ionic strength
Sedimentation coefficient	155 S	155 S	155 S	145 S
Buoyant-density in CsCl, g/cm^3	1.33–1.35; and 1.44	1.34; and 1.44	1.38–1.41	1.43–1.45
Charge neutralization of RNA by	50 % capsid proteins 50 % ions (K^+)	50 % capsid proteins 50 % ions	25 % capsid proteins 75 % ions	90–100 % ions
Properties of RNA				
Nucleotide composition %				
C	20–24	24–26	20	28
A	27–30	25–27	34	26
G	23–28	23–24	19	24
U	23–25	24–27	27	22
Poly C tract near 5′ end	no	yes	no	yes
3′ end poly A tract	56–90	–60	90	10–90
5′ genome linked protein	yes	yes	yes	yes
size of untranslated regions				
– 5′ end	743	200	unknown	400
– 3′ end	72	120	42	92
Polyadenylation signal (AAUAAA)	no	yes	no	no
In-vivo production of empty capsids	yes	no	yes	yes

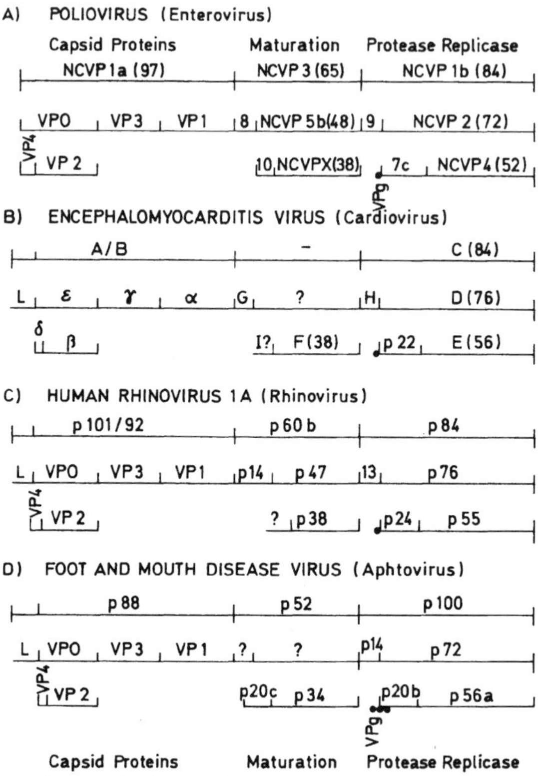

Fig. 6. Common features in the translational and processing maps of picornaviruses
Polypeptides with similar masses and mapping positions are evident in all viruses of the four picornavirus genera. The lateral positions represent the location of the corresponding gene locus on the viral RNA; the 5′ end of the RNA is to the left. The vertical positions represent precursor-product relationships. Line lengths are drawn roughly proportional to the corresponding lengths of the coding regions in the RNAs

structure and protein cleavage patterns (Fig. 6). Nonetheless, there are common principles of structure, assembly, and even of the overall genome infectious strategy amongst all these viruses (Table 3). They all have in common their small (pico) size (diameter approximately 30 nm, see Fig. 3), a single stranded RNA genome of mRNA polarity that is covalently linked to a small protein VPg, and a capsid composed of sixty equivalent sets of four major types of polypeptides arranged in an icosahedral lattice. One third of the genome's coding capacity is required for the coat proteins, the rest of the genome mainly codes for subunits of the polymerase required for genome replication, (a) protease(s) required for some of the complex processing of viral precursor proteins, and some smaller polypeptides. Besides the capsid proteins (and VPg) no other proteins are contained within the completed virus particle.

A uniform nomenclature of picornavirus proteins was adopted at the third meeting of the European study group of picornaviruses in September 1983. The nomenclature is shown in Figure 7; it illustrates the similarities in the genetic

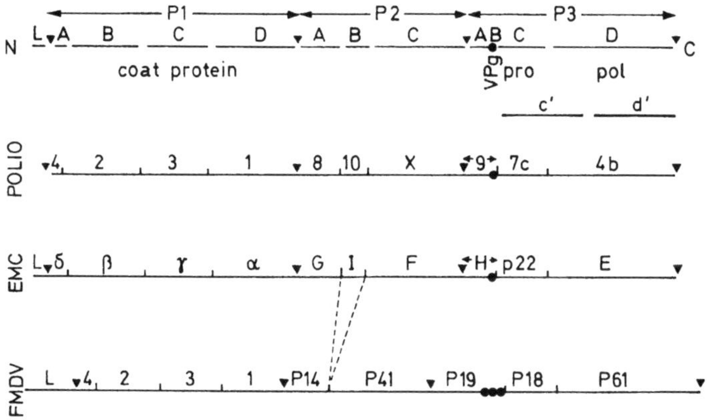

Fig. 7 A. The L434 diagram for picornaviral proteins

The polyproteins of poliovirus, EMC-virus, FMDV, and rhinovirus can be accommodated into the pattern L-ABCD-ABC-ABCD (L434), where L is a leader protein, and A, B, C, D represent the endproducts of a capsid piece (P1), a midpiece (P2), and a rightpiece (P3). Poliovirus lacks a leader while FMDV appears to lack a protein P2-B in the midpiece. Unusual proteins such as polio X/9 which overlaps the P2 and P3 regions is called 2C-3AB. The terms pro and pol are used to identify the genetic loci responsible for the synthesis of protease and polymerase, respectively. The L434 pattern provides a basis for describing the map location of all known picornavirus proteins

Map Coordinates	POLIO	EMC	FMDV (a)	FMDV (b)
L	–	L	p16	p16/20 a
L-1-2 A	–	A1	–	–
1-2 A	–	A	–	–
P1	NCVP1 a	B	P91	P88
1 A	VP_4		VP_4	VP_4
1 B	VP_2		VP_2	VP_2
1 C	VP_3		VP_3	VP_3
1 D	VP_1		VP_1	VP_1
1 A*B*	VP_0		VP_0	VP_0
1 ABC	NCVP3 a	–	–	–
1 CD	NCVP3 c	–	–	–
P2	NCVP3 b	–	p56	p52
2 A	NCVP8	G	deleted?	
2 B	NCVP10	I	p14	p20 c
2 C	NCVPX	F	p41	p34
2 AB	NCVP7 a	–	–	–
2 AC	–	–	= P2	= P2
P3	NCVP1 b	C	p102	p100
3 A	NCVP9 b	–	–	–
3 B*	VPg	VPg	VPg	VPg
3 C	NCVP7 c	p22	P18	P20 b
3 D	NCVP4 b	E	P61	P56 a
3 AB	NCVP9 a	H	P19	P14
3 CD	NCVP2	D	P81	P72
3 BCD	–	–	VPgP81	
2 C–3 AB	NCVPX/9	–	–	–
3 C′	NCVP6 a	–	–	–
3 D′	NCVP6 b	–	–	–

Fig. 7 B. Relationship of picornavirus proteins
(a) Plum Island nomenclature; (b) Pirbright nomenclature for FMDV

organization of the picornaviruses, and serves as common reference in order to facilitate comparison of the different picornavirus genera.

Indeed, the basic overall patterns of structure, genome organization, protein species synthesized, mode of RNA replication, and assembly, in short, the "genome strategy" is very similar amongst the viruses in the family, strongly suggesting that these viruses have evolved from some prototype picornavirus as a common ancestor. During the course of evolution a great number of mutations in the nucleotide sequences of the ancestor virion genome and subsequent picornavirus genomes must have occurred, with corresponding changes in the amino acid sequence and structure of the viral proteins, thus creating differences in the host range, *i.e.* the ability to infect and grow on certain types of cells, and in the susceptibility to influences from the environment. This could account for the present day multitude of picornaviruses and picornavirus induced diseases (Rekosh, 1977). (Today there exists a library of over 200 serologically distinct picornaviruses and respective reference antisera; Hahn, 1972.) Even such closely related viruses as the three subtypes of poliovirus were found to have mere 36–52% sequence homology of their RNA genomes by RNA hybridization techniques (Young, 1973 b). The sequence homology between the genomes of the different subgroups of the enteroviruses by hybridization are in the range of only 5% (Young, 1973 a).

Recent results from nucleotide sequencing of cloned cDNAs of the picornaviruses, however, have revealed substancial sequence homologies even in distantly related viruses. (Palmenberg, personal communication; Toyoda *et al.*, 1984). Within the replicase coding region sequence homologies of more than 60% occur even in the most distant relatives: poliovirus and foot and mouth disease virus. The amino acid sequence homology of the three subtypes of poliovirus is almost 90% (Stanway *et al.*, 1983; Toyoda *et al.*, 1984).

Other small RNA containing viruses with properties similar to those of the picornaviruses have from time to time been included in a larger grouping of picornaviruses. However, upon more detailed characterization some fundamental differences in composition or structure were detected (*i.e.* the calciviruses contain only one major capsid polypeptide), so that they should not be grouped in the same family.

III. Comparison of Different Picornavirus Genera

Originally the different viruses of the picornaviral family were distinguished on the basis of their host range and disease aspects. In the subsequent classification into subgroups the main distinction was made on the basis of density in cesium chloride and pH stability (see Table 4). To present a picture of the great variety of the poliovirus relatives in the picornavirus family, we shall briefly describe and compare the disease aspects and physicochemical properties of the major virus subgroups of this family.

A. Disease Aspects

The Enteroviruses primarily inhabit the alimentary tract where they normally cause little or no apparent illness (Committee on Enteroviruses, 1966; Melnick, 1976; Wenner, 1982). Occasionally, however, they spread from here to other parts of the body such as the heart (coxsackie), liver (hepatitis A), or central nervous system (polio), where they can cause severe, permanent damage, sometimes with fatal outcome. For all members of the group, however, infections with little or no apparent illness are far more common than clinically manifest diseases. The hepatitis A virus causes the well known epidemic hepatitis of man. In contrast to other picornaviruses, hepatitis A virus establishes a persistent infection, the replication cycle extends over at least one day, the progeny virions remain strictly cell associated, and viral antigen can be shown to accumulate over a period of weeks and even months (Siegl *et al.,* 1981 b).

The *rhinoviruses* are found in the nose or throat where they cause common colds.

The *cardioviruses,* the so-called 'murine viruses', are highly infectious for rodents where they typically induce fatal diseases. They are occasionallly involved in outbreaks of disease in higher animals and, rarely, in man, where they cause a mild febrile disease called the three-day-fever.

The *aphthoviruses* are the infectious agents of highly contagious diseases of livestock, typically manifested as fever and vesicular eruptions in the mouth, tongue and hooves (foot and mouth disease virus, FMDV). Although these diseases are rarely fatal, they have caused great economic losses, since the vesicular eruptions evidently are very painful to the inflicted animals, causing them to refrain from feeding.

B. Serotypes

Interestingly, there are great differences in the number of distinguishable serotypes of each of the picornavirus genera. While three serotypes of polioviruses and seven serotypes of foot and mouth disease viruses have been found, there exist more than 120 serotypes of rhinoviruses (Table 4). These differences may reflect adaptation of the viruses to different antibody-pressures in their hosts.

C. Physical Properties

Variations in the compositions and structure of the capsid proteins are reflected in characteristic buoyant densities and stabilities to pH and temperature (Rueckert, 1976). The buoyant densities in CsCl gradients are functions of the number of heavy cesium ions bound mainly to the RNA of the viron. The high buoyant densities of rhino- and aphthoviruses (1.40 and 1.43) appear to be due to internal binding of the cesium ions to the RNA in these viruses. No such binding appears to occur in the entero- and cardioviruses which exhibit a lower density of 1.34 g/ml. The capsids of rhino- and aphthoviruses thus appear to be more porous (*i.e.* readily penetrated by cesium ions) than those of entero- and cardioviruses, which are so compact as to prevent the cesium ions from entering. Another manifestation of the porous nature of aphthoviruses as compared to polioviruses is their sensitivity to inactivation by photodynamically active dyes (Brown, 1979). Infectivity of FMDV is rapidly inactivated by low concentrations (2 ug/ml) of proflavine in the presence of visible light (the RNA is inactivated), and the particles eventually disrupt into free RNA and 12S protein subunits (Brown and Steward, 1960). In contrast, poliovirus is relatively resistant to photodynamically active dyes, and any inactivation which occurs is due to oxidation of the protein coat (Brown, 1979).

The stabilities of virus particles below pH 7 can be correlated with their densities in caesium chloride. Thus, the enteroviruses and cardioviruses are stable at pH 3, whereas aphthoviruses are disrupted below pH 7. The rhinoviruses which have an intermediate buoyant density of 1.40 g/ml also occupy an intermediate position in pH stability, being stable between pH 7 and 5, but unstable below pH 5. The greater stability of enteroviruses to low pH than that of rhinoviruses is readily rationalized by the need of enteroviruses to pass through the highly acidic conditions of the stomach in order to reach the digestive tract, their typical habitat. It also explains why rhinoviruses are almost never found in the gut.

D. Relative Relatedness to Polioviruses

In evolutionary terms (reflected in physiochemical properties), the closest relatives of poliovirus probably are the other enteroviruses. The rest of the picornaviruses are of a more or less distant relationship as evidenced in different features such as the presence of a poly C tract in the RNA of cardioviruses and aphtho-

Table 5. *List of enteroviruses*

Human	poliovirus (3 serotypes) coxsackie A (23 serotypes) coxsackie B (6 serotypes) echovirus (32 serotypes) hepatitis A
Bovine and porcine	bovine enterovirus swine vesicular disease virus Teschen's disease virus porcine enteroviruses
Murine	murine encephalomyelitis virus Theiler's virus
Duck	duck hepatitis virus
Insect	Gonemeta virus Cricket paralysis virus

viruses, acid lability and the 'porous' capsid of rhino- and aphthoviruses. Cardio- and rhinoviruses are perhaps of an intermediate relatedness. FMDV appears to be the most distant relative of poliovirus.

The great similarity in the fundamental principles of structure and function amongst the different picornaviruses validates analogies made between the different picornavirus systems. In this review on poliovirus we shall rely primarily on experiments done on poliovirus itself. However, where corresponding experiments have not yet been performed on poliovirus, we shall from time to time "borrow" results obtained from other picornaviral systems, preferentially from other enterovirus systems. A certain scepticism about the conclusions drawn from such analogies is warrented nonetheless. Polioviruses and their closest relatives—*i.e.* the members of the enterovirus genus—are listed in Table 5.

3

Composition and Structure of the Virion

I. Introduction

The successful propagation of poliovirus in tissue culture cells in the early fifties permitted the production and isolation of large quantities of pure virus and provided sufficient material for successful investigations on virus structure. Crystallization of poliovirus particles was already achieved in the mid-fifties. Since then, detailed investigations on the compositional and structural features of the poliovirion have been carried out, including

1) studies on the overall virion architecture by electron microscopy and X-ray crystallography,

2) the isolation and characterization of the component parts – the genomic RNA and the different polypeptide building blocks,

3) the determination of the primary structures of RNA and protein,

4) discoveries and partial elucidation of stepwise processes of assembly and disassembly,

5) studies on the type of bonds involved in maintaining capsid integrity,

6) indirect examinations of the relative spatial localization and surface exposure of component parts, and

7) speculations on the correlation of spatial arrangements and observed functions.

X-ray crystallography of poliovirus crystals in the late fifties revealed a highly symmetric design inherent to poliovirus and shed some light on the fundamental construction principles of small isometric virus particles. A striking finding, upon similar investigations on small isometric plant viruses, was the apparent universal construction principle of small spheric virus particles: they appeared to have the same underlying symmetry relations and geometric design discovered for poliovirus. Theoretical evaluations of the possible design of spheric containers built from a large number of identical building units provided a plausible explanation: The geometric principle employed by poliovirus and other small viruses–the icosahedral lattice–is the most efficient design possible in terms of thermodynamic and geometric considerations.

The ensuing concepts of virus structure had their ups and downs. With the progress in physical and biochemical techniques, more detailed analyses of the building blocks were achieved. In contrast to the X-ray studies, the great variety of data from biochemical analyses of the proteinacious building blocks of different small viruses revealed great compositional diversity and complexity. As a result, the faith in a unifying concept of simple design and construction principles began to weaken. Comparisons of different icosahedral viruses illustrated and emphasized the differences. Drawing too close analogies, or the supposition of a unifying structural concept, seemed to be unwarranted.

Today, a close review of the literature and incorporation of a great mass of experimental data leads to a plausible and coherent model of poliovirus structure. A sympathetic aspect of the model is that it revitalizes some of the old unifying structural concepts for icosahedral viruses that were proposed more than 20 years ago on theoretical grounds, but had been partially abandoned with the apparent diversity of building block units. In particular, there are striking structural similarities to the small icosahedral RNA-plant viruses. The main aspects of the poliovirus model are consistent with nearly all hitherto reported data on picornavirus structure with one exception (Hordern *et al.*, 1979; Scraba, 1979) which is discussed below.

As we were writing this book, the complete nucleotide sequence of the poliovirus type 1 genome was elucidated (Kitamura *et al.*, 1981; Racaniello and Baltimore, 1981; Nomoto *et al.*, 1982 b). The amino acid sequence deduced from it for the capsid proteins correlates well with the reported amino acid composition in the literature. We have attempted to incorporate some of this information into our concepts of poliovirus structure. A far more thorough analysis of the amino acid sequence and its structural and functional implications with more sophisticated procedures is to be expected within the near future. The results of such analyses will certainly help to clarify some of the still controversial aspects of poliovirus morphology.

In the past ten years important advances in the field of protein structure and function have been made and new concepts have been developed. Relevant inferences from such studies can and should be made also for the proteinacious containers of viruses. Most important perhaps are the concepts of inherent protein dynamics. In a biological environment, proteins are not rigid structures but instead "living" masses with inherent dynamics (Gurd and Rothberg, 1979). Proteins "breathe" so to speak, reflecting constant minute changes of the local internal interactions between amino acid side chains in neighboring peptides. Although proteins in general are compact structures with little "empty space" inside, there may be flexible "pockets" or channels for water and ions. Even structural proteins such as the building block components of a seemingly rigid virus capsid probably have steady, albeit minor, inherent dynamics.

In the past few years the structural details at a resolution of 2–3 A^{O} of two small icosahedral RNA plant viruses were resolved in X-ray diffraction and neutron small angle scattering studies (Harrison *et al.*, 1978; Abad-Zapatero *et al.*, 1980), methods which so far, unfortunately, have not been applied with comparable success to any animal picornaviruses. These plant virus studies have revealed some beautifully intricate structural principles and features of virion design: On the construction of the capsid backbone and its surface projections, on protein-protein interactions and bonding, on the relative localizations of protein, RNA, and water, and on protein-RNA interactions. We shall discuss the important structural principles revealed by these studies, since we are convinced that some of these will be applicable to the picornaviruses as well. Briefly these are:

1) the construction of a compact 3–5 nm thick capsid with strict icosahedral symmetry from 60 copies each of 3 different types of protein (actually identical proteins in three different conformational states) arranged in 12 pentameric and 30 hexameric clusters;

2) localization of the RNA as a compact spherical coil or shell in the interior of the virion, rather than an RNA that is "threaded" through the proteinacious capsid as in the helical tobacco mosaic virus;

3) some interaction between the RNA and capsid protein–perhaps in a histone like manner–by means of polypeptide arms "dangling" into the interior of the virion, this internal ribonucleoprotein portion no longer exhibits icosahedral symmetry;

4) a high conservation of structural features in the capsid backbone contrasting to a high variability in capsid surface features, *i.e.*, in the external projections of the capsid proteins;

5) maintenance of capsid integrity by an intricate bonding system across the 3-fold symmetry axis;

6) a "built-in" capacity of the native virion capsid to undergo a concerted conformational transition to a more expanded stable state of the capsid. In the course of this transition, a previously internally located polypeptide may become exposed to the surface by passing through newly created "holes" in the capsid backbone.

The hypothesized structural analogies between the plant and animal picornaviruses, of course, can only be confirmed with an elucidation of the fine structure of picornaviruses by high resolution X-ray crystallography (Hogle, 1982, 1984). A comparison of the fine structure of these viruses should provide new clues towards an understanding of the origin and evolution of the small RNA viruses.

We have experienced that the construction of virus models greatly aids the visualization and comprehension of the building principles of poliovirus (and other icosahedral viruses). Two relatively simple construction plans for a poliovirus model are presented in Appendix II.

II. Composition and Physical Properties of the Virion

Poliovirus is a small, spherical particle, 28–30 nm in diameter. The general physical and biochemical properties of the native virion are summarized in Table 6. The virion has a dry molecular weight of 8.5×10^6 daltons, sediments at 156S (151–160S), and has a buoyant density of 1.34 g/cm^3 in CsCl. Approximately 30% by weight of the dry virion is RNA with its charge neutralizing cations (one

Table 6. *General physical and biochemical properties of the poliovirion*

A. Physical properties

Total dry molecular weight	8.25×10^6 daltons (dry particles)
Sedimentation coefficient (S°20, W)	150–160 S
Buoyant density in CsCl gradients	1.34 g/cm^3
Isoelectric point (pI): A-form	7.0
B-form	4.5
Diffusion coefficient (D°20, W 1×10^{-7})	1.44 cm^2/s
Partial specific volume ($\bar{v}$)	0.68 ml/g
E 260/280	1.69–1.74
Stability to	stable
– pH	– between pH 2.9 and 8.5
– heat	– up to 45 °C in isotonic salt
	– up to 60 °C in hypertonic salt
– lipid solvents	– in ether and chloroform
– desinfectants	– in chlorinated water
– denaturing agents	– in 1% SDS, up to 4 M urea or guanidiniumchloride

B. Biochemical properties

Composition	Copies per particle	Mol. weight in kd.	% of mass
Protein		5800	51
VP_0	0–2	37	0.7
VP_4	58–60	7	4
VP_2	58–60	30	15
VP_3	60	26	14
VP_1	60	34	18
VP_g	1 (covalently linked to the vRNA)	2	0.02
RNA	1 (7,500 nucleotides)	2600	23
H_2O	150,000	18	24
Ions	6,100		
K^+	4,900	39	2
Na^+	900	22	0.2
Mg^{++}	110	12	0.01
putrescine^{++} and spermidine^{+++}	50		
Lipid	not detectable		
Carbohydrates	not detectable		

single stranded molecule of RNA with a molecular weight of 2.5 x 10^6 daltons and 0.2 daltons of ions), the rest consists of protein with a total molecular weight of ca. 5.8 x 10^6 daltons (Schaffer and Frommhagen, 1965; Schwerdt and Schaffer, 1955, 1956; Rueckert, 1976; Cooper *et al.*, 1978). Purified poliovirus preparations do not contain detectable amounts of sugar or lipid (Drzeniek and Bilello, 1974). About 25–30% by weight of the native virion in an aqueous environment is water. Part of this water is probably located on the outer surface of the virion in the form of a hydration-shell that is responsible for balancing the charge distribution of the polar amino acid residues on the surface. Some H_2O is also expected in the interior of the virion in association with the RNA and in holes or pockets of the protein shell and in areas involved in RNA-protein interactions.

The protein component is used to construct the compact protective coat for the genomic RNA, the viral capsid. An exception is the small viral protein VPg—*v*iral *p*rotein *g*enomic—of 2,400 daltons, which is covalently bound to the 5' end of the RNA and is thus not an integral component of the capsid. The capsid is built from 60 equivalent (nearly identical) asymmetric structural protein units which are arranged in an icosahedral lattice. The subunits of the capsid are quite tightly packed: In its native form, the capsid is impermeable to phosphotungstic acid (used for staining in electron microscopy), to dyes which photosensitize RNA, and to a variety of ions and small molecules, such as Cs^+. Poliovirus traps photosensitizing dyes such as neutral red and proflavine or radioactive Cs^+ ions when the virus is propagated in their presence (Crowther and Melnick, 1961a; Mayor and Diwan, 1961; Schaffer, 1962; Mapoles *et al.*, 1978). All these substances remain tightly bound to the virion even during extensive incubation in dye or Cs^+ free buffer, implicating a tight and impermeable capsid.

The tight construction of the poliovirion with the resulting impermeability of the protein shell to hydrated ions has permitted the determination of the natural cations of poliovirus (Mapoles *et al.*, 1978; Mapoles, 1980). The virion contains approximately 4900 K^+ ions, 900 Na^+ ions, 110 Mg^{++} ions, 50 polyamine ions (putrescine and spermidine), and no significant amounts of Ca^{++}, glutamine, amonia, or free amino acids. These cations account for 6100 positive charges per virion, leaving 1400 phosphate residues of the RNA to be neutralized by capsid proteins or other sources. The divalent Mg^{++} ions cannot be removed from intact virions by the chelating agent EDTA, in contrast to surface bound Mg^{++} ions frequently observed in other viruses. Mg^{++} increases the stability of poliovirus, and provirions (see below) even require the presence of Mg^{++} for stability (Fernández-Tomas and Baltimore, 1973), suggesting that Mg^{++} may be involved in bonding forces between capsid units. The stoichiometry indicates that there may be 2 Mg^{++} ions per structural unit (Mapoles, 1980).

Table 7 summarizes the structural features of the poliovirion. Poliovirus particles prepared for electron microscopy by conventional procedures using phosphotungstic acid (PTA) negative staining, appear as small compact spheres with a diameter of 28—30 nm, only rarely showing a fine structure (Schaffer and Schwerdt, 1959; Brown and Hull, 1973). The electron microscopic studies on polioviruses indicate that the capsid is a relatively pliant structure. In electron microscopic pictures of the virus, when it is packed in a crystal array (intracellulary or in crystals of purified virus), the capsids appear to have a regular or somewhat skewed hexagonal contour (Fig. 8). On the other hand, when the virions are observed free of the constraint of the neighboring particles, they adapt a round shape (Fig. 9A, B), (Schaffer and Schwerdt, 1955; Sjöstrand and Polsen, 1958; Horne and Nagington, 1959; Hoyer *et al.*, 1959; Mayor, 1964; Agrawal, 1966; Boublik and Drzeniek, 1976, 1977). It is difficult to distinguish a fine structure of poliovirus particles by electron microscopy. The observation of a centrally located diamond with two 5-point radiating structures (Fig. 9C) was considered as indicative of a capsid constructed from 32 morphological clusters (Mayor, 1964) (see below). Poliovirus particles stained with PTA appear coreless since the PTA stain cannot penetrate through the tight viral capsid (Fig. 9A). The use of Cs^+ ions as a contrasting stain, however, revealed RNA-containing cores with a diameter of approximately 18 nm (Fig. 9D) (Boublik and Drzeniek, 1976). The protein capsid would thus appear to have a thickness of (28–18):2 = 5 nm.

There are some other interesting points that may be inferred from the X-ray and EM-studies on poliovirus crystals. Both methods reveal crystal lattices of virus

Table 7. *Structural features of the poliovirion*

Appearance	spherical, compact, dense (coreless)
Contour	a) individual particles: round b) particles in crystalline array: skewed-hexagonal
Surface fine structure	only observed in "favorably" oriented particles—here indicative of 30–40 morphological clusters, 10–12 nm in diameter each
Outside diameter of capsid	a) dry 28 nm b) wet 30 nm
Diameter of core	18 nm (visible only with special procedures)
Deduced thickness of capsid	5–6 nm
Arrangement of particles in crystals	a) body–centered cubic lattice (with a slight deviation from strictly cubic arrangement) (x-ray crystallography) b) close packed, face-centered cubic lattice (EM)
Geometric design of capsid	a) 5:3:2 fold cubic symmetry = icosahedral lattice with twelve equivalent 5-fold symmetry centers, twenty 3-fold symmetry centers, thirty 2-fold symmetry centers, all other points exist in groups of 60 equivalently situated sites on the icosahedron b) the capsid structure is not necessarily like that of the regular icosahedron, *i.e.*, with 12 prominent vertices and 20 flat sites. Depending on the localization of "corners" in, or projections from the building units, the number of apparent vertices rather may be a combination of the sum of numbers 12, 20, 30, n × 60
Arrangement of structural subunits within the capsid	a) 60 equivalent structure-units ($VP_{4,2,3,1}$)–protomers–occupy identical environments in the icosahedral lattice b) the relative orientation of subcomponents of neighboring structure-units may lead to a clustering of different subcomponents about defined geometric points in the icosahedron, forming 32 morphological clusters: 60 $VP_{2,4}$ = $(VP_2–VP_4)_5$ = twelve 5-fold vertices; 60 $VP_{3,1}$ = $(VP_3–VP_1)_3$ = twenty 3-fold vertices

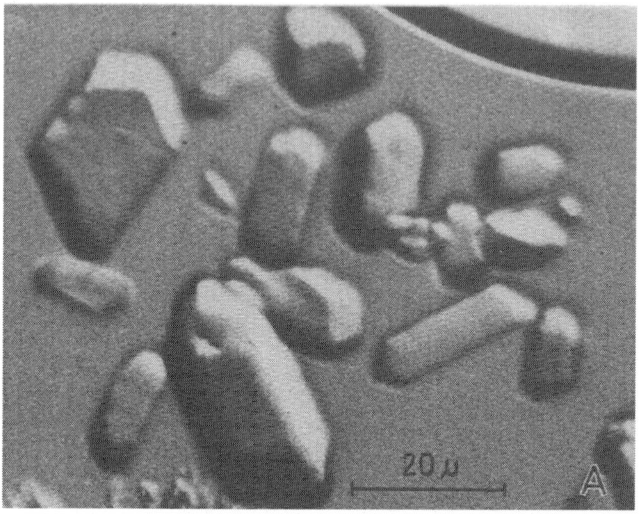

Fig. 8 A–C. Micrographs of poliovirus in crystalline arrays

Fig. 8 A. Light-microscopic images of poliovirus crystals

The figure shows the light microscopic appearance of purified type 1 poliovirus particles. Poliovirus was the first animal virus to be obtained in crystalline form. – Figure from Schaffer and Schwerdt, 1955 [Proc. Natl. Acad. Sci. *41,* 1020 (1955)]

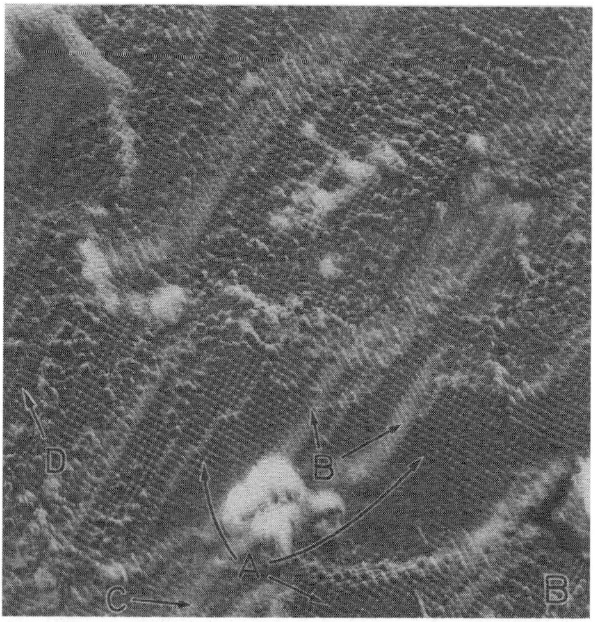

Fig. 8 B. Electron micrograph of poliovirus crystals

Electron micrograph of a replica of a fractured, frozen crystal of poliovirus type 1 showing various planes *(A, B, C)* within the crystal and a discontinuity *(D).* – Figure from Steere and Schaffer, 1958 [B.B.A. *28,* 245 (1958)]

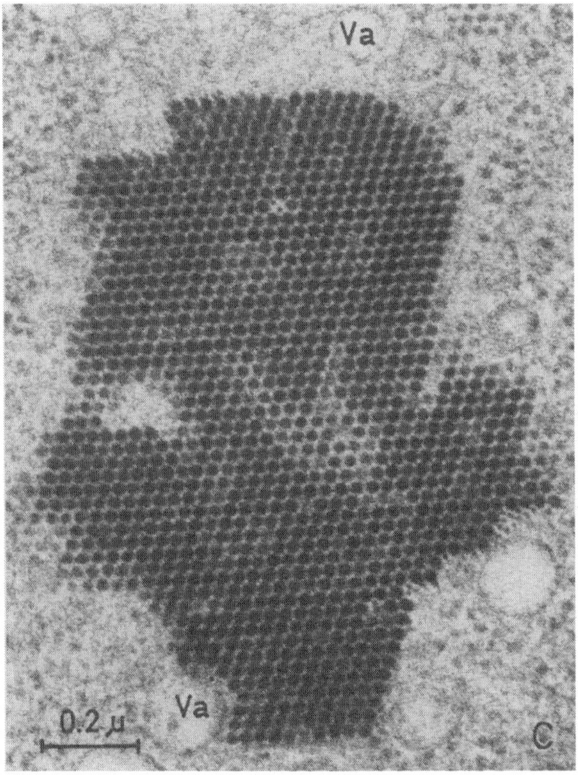

Fig. 8 C. Electron micrograph of intracellular crystals of poliovirus
The figure shows a large crystal of progeny poliovirus particles in a small area of cytoplasm in a poliovirus infected HeLa cell, 7 hours post infection. A small region within the crystal, presumably derived from the cytoplasm, is devoid of virus. – Figure from Dales *et al.*, 1965 [Virology *26*, 386 (1965)]

particles with cubic symmetry. However, electron micrographs invariably show the close face-centered cubic packing of particles, whereas the X-ray patterns of "wet" crystals imply a pseudo-cubic, body-centered packing. This is likely to reflect a distortability of the capsid under special conditions. Only a relatively small shear and/or torsion is required to convert a body-centered cubic structure to a close-packed (face-centered) one, and it seems quite reasonable that the more open structure of "wet" crystals that is held together by relatively weak forces, collapses under the special treatment necessary for electron microscopy of the specimen (drying, freezing, etc.), so that the particles are rearranged into the close-packed structure. The crystal lattice deduced from the X-ray diffraction patterns is not strictly cubic; there are relatively small departures from the body-centered cubic arrangement. This observation could reflect slight deviations from icosahedral symmetry in the individual virus particles that may be brought about by the presence of one or two "irregular" building blocks (containing the precursor VP_0 instead of VP_2 plus VP_4), or by the microheterogeneity of individual capsid polypeptides (Section III of this chapter).

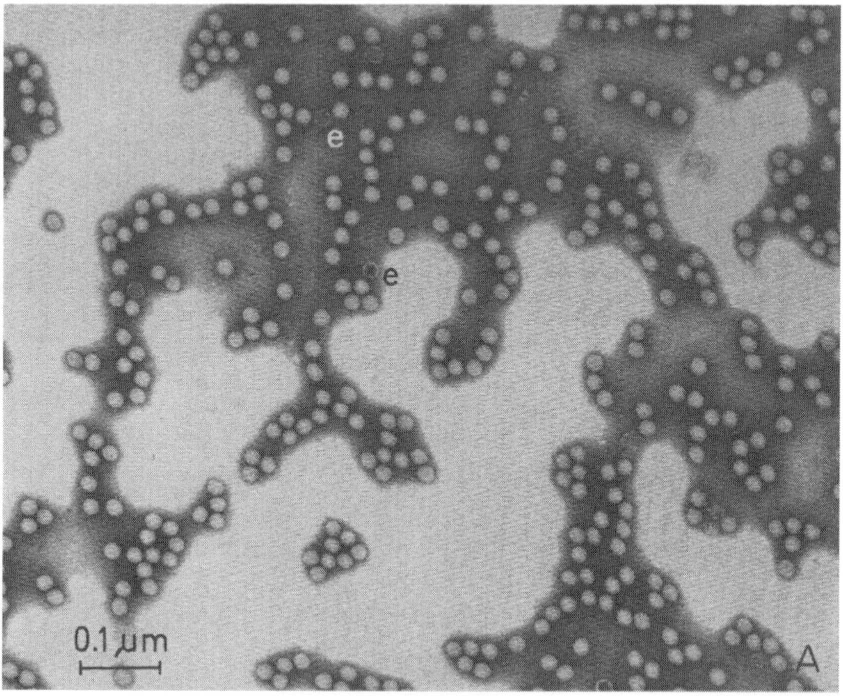

Fig. 9 A–D. Micrographs of individual poliovirus particles

Fig. 9 A. Electron micrograph of individual poliovirus particles
Poliovirus particles, stained with 1% PTA, adsorption technique; e = empty capsids. – Figure from Boublik and Drzeniek, 1977 [J. gen. Virol. *37,* 128 (1977)]

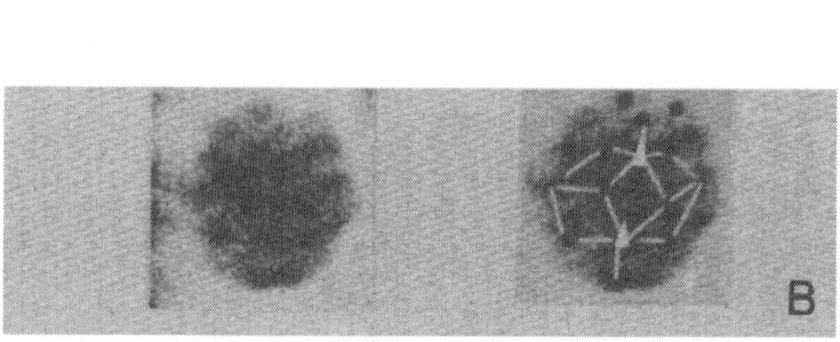

Fig. 9 B. High magnification of individual poliovirus particles
Left: High magnification electron micrograph of an isolated poliovirus particle printed in reverse contrast. Magnification: × 600,000. Right: The same particle as left marked to display the central rhombus and two clear axes of fivefold symmetry (white points). – Figure from Mayor, 1964 [Virology *22*, 158 (1964)]

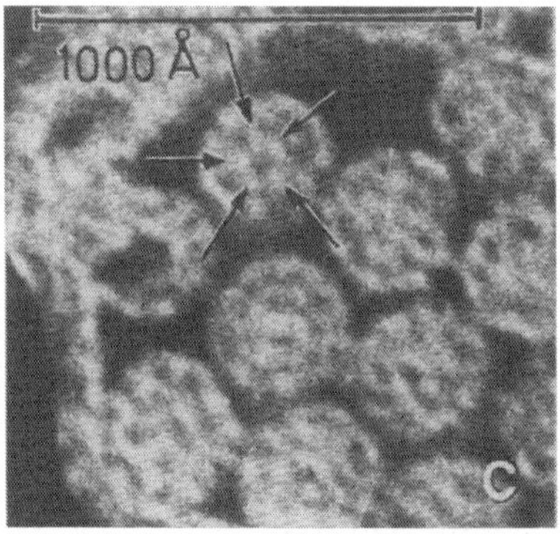

Fig. 9 C. High magnification of poliovirus particles in cytoplasmic vesicles from infected cells
This figure shows a high magnification electron micrograph of cell fragments from poliovirus infected cells at 4 hours p.i. Some virus particles are in suitable orientations to show possible 5-fold arrangement of the subunits (arrows). See also Fig. 102 A. – Figure from Horne and Nagington, 1959 [J. Mol. Biol. *1*, 333–338 (1959)]

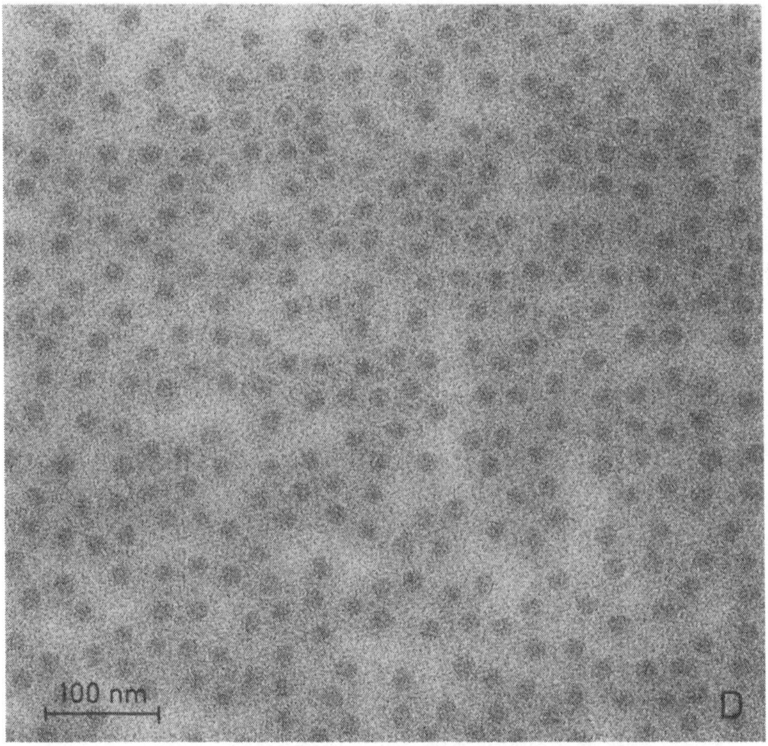

Fig. 9 D. Electron micrographic demonstration of a core in poliovirus particles
This figure shows an electron micrograph of unstained RNA-containing poliovirus particles in 3 M CsCl tris buffer. Suspension of poliovirus particles in 3 M CsCl during the preparation for electron microscopy is sufficient to visualize poliovirus particles without any additional staining procedures. A "core" is observed in the center of the particles surrounded by a lighter area. The diameter of the core is approximately 18 nm, the outer diameter of the virus particles approximately 28 nm. Empty capsids cannot be visualized by this procedure. The core presumably represents tightly packed virus RNA. – Figure from Boublik and Drzeniek, 1976 [J. gen. Virol. *31*, 448 (1976)]

III. Overall Architecture of the Viral Capsid

A. Building Blocks

1. Functional Requirements

The main function of the viral capsid is to provide a protective package for the genome. The essential infective agent of all viruses is their nucleic acid component; the isolated RNA of poliovirus, for example, by itself is capable of inducing a productive virus replication. The infectivity of a virus, however, must persist in a latent state outside of the host cell. Isolated nucleic acid molecules are very labile, particularly in an extracellular environment where nucleases are present in abundance. The viral capsid serves to protect and transmit this infectious agent in a functionally intact state through space and time to a susceptible host.

Properties required of a capsid to provide maximal protection are: maximal stability (strong bonds, low energy state), minimal susceptibility (minimal surface exposure, sufficient capsid thickness), and minimal permeability (tight, compact packaging).

2. Biological Arguments of Building Block Economy and Efficiency of Assembly

In 1956, Crick and Watson suggested that containers for genomes of small viruses are built from a larger number of identical protein subunits packaged together in a regular manner. The important biological arguments for this construction principle are: 1. The efficient use of limited genetic information contained in the virus nucleic acid: The coat proteins in form of small identical molecules require less nucleic acid space for their coding (Crick and Watson, 1956); 2. The repeated use of only a small number of different bonding types guarantees efficiency of assembly: When identical building blocks are used to construct a capsid, a minimum number of types of attractive forces between the subunits suffice to assemble the entire capsid. In the formation of the shell the identical subunits are packaged so that the same contacts and bonds between the subunits are used over and over again. As a consequence, a virus built in this fashion will have a uniform size and regular shape.

3. Thermodynamic Forces and Structural Consequences

The final structure of a virus particle is determined by thermodynamic forces tending to a minimal energy state and by geometric principles that determine the structural design with the optimal realization of these thermodynamic forces. The protein subunits and the nucleic acid chain readily associate into a virus particle because this is their lowest energy state. The assembly process is driven by the attractive forces between the structural subunits. The driving energy for shell-assembly is provided by the formation of the intersubunit bonds. Indeed, the attractive forces between the building blocks of the capsid are often sufficient for the structural units to self-assemble into a capsid in vitro under appropriate conditions in a process akin to crystallization. The lowest energy state of the final capsid structure will have the maximum number of most stable bonds.

The order in the final structure of a capsid built from many identical structural units is a necessary consequence of the compulsion to the lowest energy state. The units will be arranged in physically indistinguishable environments which necessarily produce a symmetric structure. Since the possible kinds of spherical symmetry are geometrically limited, there is also a geometric limitation on designs of viral capsids. Poliovirus employs the geometric design that is optimal for the construction of a capsid in accordance with these building principles: a lattice with icosahedral symmetry.

B. The Geometric Design: The Icosahedron

The key point concerning the structure of viruses was elaborated by Caspar and Klug (1962; Caspar, 1965): There are only a limited number of efficient geometric designs possible for a biological container which is constructed from a large number of identical subunits.

1. Helical Tubes Versus Spherical Shells

The two basic designs, each with different advantages, are helical tubes and spherical isometric shells. Poliovirus, like many other viruses, utilizes the latter design. The most stable conformation of a single-stranded RNA molecule is achieved by a high degree of intramolecular hydrogen bonding, *i.e.*, with a high degree of coiling into a compact globular structure (in contrast to a double-stranded nucleic acid, which is more stable in a linear conformation due to interchain hydrogen bonding). The coiled nucleic acid is contained most easily in a spherical capsid. In addition, the spherical shell has the advantages of using a minimum amount of protein for the packaging of a given quantity of nucleic acid. Also, a spherical container has a minimum of surface area exposed to the environment, whereas a helical structure leaves a larger area exposed.

2. Cubic Symmetries, the Platonic Polyhedra

Crick and Watson (1956) pointed out that there are only a few ways of building an isometric spherical shell from identical subunits so that the subunits are equivalently situated. The resulting symmetry relations are polyhedral.

Although the overall appearance of isometric structures built from distinct subunits may be nearly spherical, the underlying symmetry relations of the construction framework will always be one of the three types of cubic symmetry. This is a simple consequence of the fact, known to the ancient Greeks, that there exist only five perfectly regular, isometric polyhedra, and that all existing isometric polyhedra are related to one of these five principle polyhedra. Figure 10 shows representations of the Platonic polyhedra and illustrates their inherent symmetry relations. The three existing types of cubic symmetries are named after the corresponding Platonic polyhedra: tetrahedral (3- and 2-fold symmetries in the tetrahedron), octahedral (4:3:2 fold symmetries in the cube and octahedron), and icosahedral (5:3:2 symmetries in the dodecahedron and icosahedron).

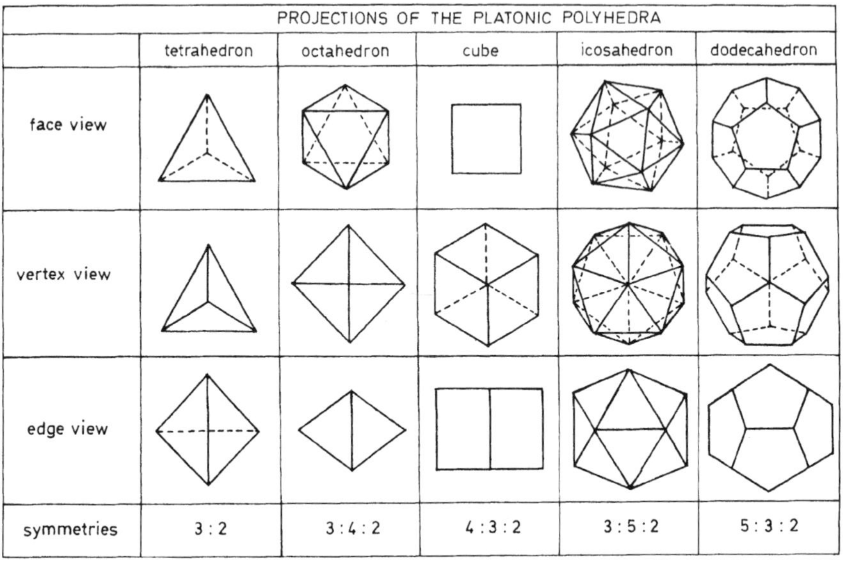

Fig. 10. The five platonic polyhedra. – From Pugh, 1976 [Polyhedra–A Visual Approach, p. 12 (1976)]

The Platonic polyhedra exhibit some beautiful geometric features and almost mystical interrelationship that have been studied and exploited not only by the ancient Greeks, renaissance and contemporary architects, mathematicians, but also by anthroposophists and, in the past thirty years also by virologists. We have summarized some of these features in Appendix III as they help to visualize some of the fascinating construction principles of the spherical viruses. More detailed descriptions and illustrations of these structures can be found in such monographs as "Polyhedra—a visual approach" by Pugh (1976) (a geometric approach) or in "Wandlungen – Freundschaft mit platonischen Körpern" by Keller von Asten (1980) (a philosophical approach).

3. *The Icosahedron*

a) Advantages of Icosahedral Symmetry

The icosahedral lattices offer a number of advantages over the tetrahedral and octahedral lattices for the construction of viral capsids in accord with the principles of economy and efficiency :

1) It allows the use of the greatest possible number, namely 60, of identical asymmetric structure units to build a spherical framework in which these units are identically packed (economy of genome space, efficiency of assembly).

2) The ratio of surface area covered to the number of subunits required is greatest (optimal) when icosahedral symmetry is employed (greater capsid stability, better protection).

The architectural advantages of an icosahedral lattice are described in a recent book on designs (Pearce and Pearce, 1980): "From the standpoint of stress distribution, the icosahedral subdivision results in an optimum structural framework from which to construct a sphere. Since a sphere has less surface relative to volume than any other geometric shape, this system can be quite efficient when a spherical form is particularly well suited to the objectives of a given architectural problem . . .".

The fact that the vast majority of independent evolutionary pathways for the capsid structures of spherical viruses have converged to icosahedral lattices (*e.g.*, picornaviruses, polyomavirus, adenovirus, the herpesvirus core, see also Figure 3 of Chapter 2)—only some bacteriophages employ octahedral symmetry—also illustrates the advantages of icosahedral lattices.

b) Limitations of the Icosahedral Skeleton as a Model for Virus Structure

Since the structural units of a virus particle have awkward shapes and a threedimensional extension, the resulting structure may bear little resemblance in gross appearance to the regular icosahedron. Thus, the icosahedral skeletons often used to represent viral structure (*e.g.*, Fig. 10, Appendix IIA) in fact do not properly reflect the actual conformation or shape of 60-subunit particles bearing icosahedral symmetry such as viruses, nor do the edges of the regular icosahedron necessarily correspond in any way to the actual borders of the viral structure units.

(Appendices II and III illustrate a number of different ways by which 60 identical structure units, whose edges do not correspond to those of the skeleton icosahedron, fit into the regular icosahedron). In general, icosahedral skeletons merely serve to illustrate the icosahedral symmetry relations and to define the relations of the subunits in corresponding spherical lattice.

In the case of poliovirus, the 60 identical structural units are sets of polypeptide chains that extend well above and below the imaginary intersphere of the icosahedral skeleton that passes through the subunits. In fact, the thickness of the structural unit is on the order of one half to one third of the radius of the imaginary sphere (the outside diameter of the poliovirion is 28 nm, capsid thickness 5 nm, diameter of imaginary sphere passing through the center of structure units thus is approximately 25–26 nm). Biological structure units usually do not have flat facets, sharp edges or corners: consequentially the vertices, facets and edges are "smoothed out" so that the appearance of an icosahedral virus particle may be spherical (see Fig. 9A, Appendix IIB).

Furthermore, the construction of an icosahedral virion capsid from 60 building blocks, each composed of a number of distinct polypeptide chains, allows for the clustering of polypeptides from adjacent building blocks—in particular around the symmetry centers. For example, the gross conformation of the structural units of poliovirus appears to be such, that certain morphological clusters are formed by the polypeptide subcomponents that stick out somewhat above the surface at the twelve 5-fold symmetry centers and the twenty 3-fold symmetry centers (represented by the facets of the icosahedral model) (see below). Thus, if one insists on talking about capsid vertices, the conformation of the poliovirus particle corresponds more to a thirty-two-apex structure rather than to the twelve-apex structure suggested by the icosahedral model.

c) Related Regular Polyhedra with Icosahedral Symmetry

In general, the appearance of an icosahedral viral capsid may bear more resemblance to regular, icosahedron-related polyhedra with different types of faces and vertices than to the strict regular icosahedron. These are described in Appendix IIIB since they are used occasionally for descriptions of poliovirus and other small icosahedral viruses, and since they illustrate the variety of shapes possible for isometric icosahedral viruses.

Depending on the point of view, one can chose either the vertices, edges or facets of an icosahedral lattice to represent the location of the viral icosahedral structure unit. When taking the facets as representing a structure unit, for example, the pentakis dodecahedron or trapezoidal hexacontrahedron would be particularly suited as a model for a 60 subunit virus such as poliovirus.

The capsids of the only viruses for which the structural arrangement has been determined in atomic detail—TBSV and SBMV—assume a pentakis dodecahedron or rhombic triacontrahedron-like shape: Each structural unit is constructed from 3 polypeptide chains, has a triangular conic shape with a near flat outer surface, and the surfaces of pairs of adjacent structure units lie in a similar plane (see Fig. 11 and Fig. 18 below).

In addition to the polyhedral structural lattice formed by the actual shell-forming part of the capsid proteins, there may be surface projections of capsid protein polypeptide chains at geometrically defined points. Thus the overall appearance of a viral surface may be more ruffled than is suggested by the smooth contours of the polyhedral models*.

d) The Bonding Pattern of an Icosahedral Lattice

Three principle types of bonds are associated with the icosahedral type of a lattice constructed from 60 structure units, namely a pentamer bond, a trimer bond, and a dimer bond corresponding to bonds across the three different symmetry axes. Figure 11 illustrates these three different types of bonds (see also Appendices II and III). A combination of at least two of these bond-types suffices to guarantee coherence of the structure. When considering the protein components of a viral capsid, it may be misleading to speak of "bonding types". The contact surfaces between the building blocks of a viral capsid are probably held together by a multitude of different types of interactions (hydrophobic-, hydrogen-, and ionic bonds) between the side chains of neighboring amino acid residues. The sum of many such interactions may represent a "bonding domain" that corresponds in principle to the bond types represented in Figure 11. Depending on the shapes of the subunits there may be additional bonding domains aside from the principle icosahedral bonding types (see, for example, the illustration of the pentagonal hexacontrahedron in Fig. 11C).

---

* The small plant RNA virus TBSV, for example, has 60 projections originating within the asymmetric structure unit, and 30 projections originating at the 2-fold symmetry centers; SBMV has 12 projections originating at the 5-fold symmetry centers, 30 projections at the 3-fold symmetry centers, and 60 at the asymmetric structure unit (see Sections C 4 and V A 2). Comparable projections have so far not been identified on poliovirus, and must await more detailed structural investigations.

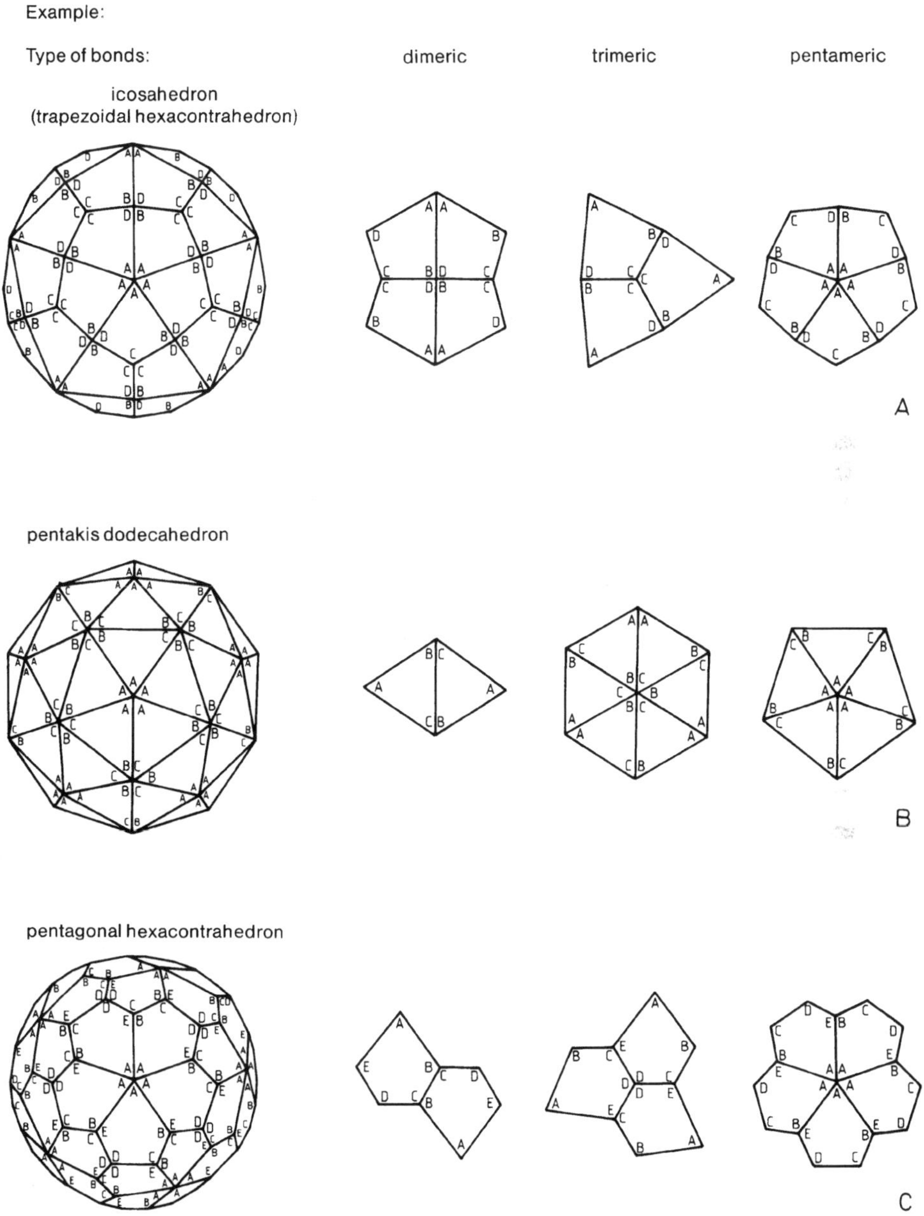

Fig. 11. The three bonding types associated with icosahedral lattices
This figure illustrates the types of bonds associated with icosahedral lattices. Three closely related polyhedra which are often used to represent viruses with icosahedral symmetry are used as examples (see Appendix III for further details). Dimeric bonds connect neighboring subunits across the two-fold axis, trimeric bonds connect subunits across the three-fold axis, and pentameric bonds connect subunits across the five-fold axis. The right hand panels illustrate the types of associations formed by the respective bonds. Note that the use of any two types of bonds between the subunits suffices to maintain the integrity of the polyhedron

For an icosahedral shell, whose 60 units in turn are constructed from different subcomponents (VP_{1-4}) as in the case of poliovirus, there must of course also be intrasubcomponent bonds in addition to the three icosahedral (and other) bonding types between the 60 structure units. The four protein subcomponents of the poliovirus structure unit are all derived from a common larger precursor protein that is cleaved at three specific sites before and during assembly. These cleavages do not affect the integrity of the structure unit, *i.e.,* the subcomponents always remain associated in a complex *in vivo* (Korant, 1973). As is usual for such protein aggregates, the subcomponents are probably held together by a multitude of interactions at their contact surfaces—each of the four subcomponents nonetheless probably occupies a discrete, globular domain rather than being extensively interlaced.

During the assembly of the picornavirus capsid, the icosahedral bonding domains symbolized by the bonding types in Figure 11, seem to be activated in sequence in order to guarantee an orderly assembly. The pentamer bonding type is employed first. The twelve symmetrical pentameric structures that are thus formed then associate, presumably as a consequence of the activation of the trimer bonding domain (see Chapter 10). The final capsid structure will be held together by multiple attractive interactions between the adjacent protein surfaces within each icosahedral structure unit and across all of the three icosahedral bonding types.

As will be discussed below, an interesting feature is built into the structural lattice of the picornavirus capsid; namely, the capacity to undertake a concerted conformational transition of all its structure units to a second stable icosahedral lattice state with a different set of intracapsid bonds and different surface features. The described cleavage of each poliovirus structure unit into four subcomponents may provide the required flexibility or rotational freedom for this conformational transition. In addition, cleavage seems to play a role in the described sequential activation of bonding domains.

C. Experimental Evidence that the Poliovirus Capsid Is an Icosahedral Lattice

With the accumulation of a large amount of evidence—which from time to time brought about some controversies—it is indeed clear today that the poliovirus capsid is constructed from 60 (nearly) identical structural units into an icosahedral lattice. The fundamental structural unit is termed the protomer. Each protomer in the native virion consists of a single copy of each of four individual proteins (VP_1, VP_2, VP_3, and VP_4 named in order of decreasing molecular weight), which are derived from a common precursor protein (NCVP1a). One or a few of the protomers may contain polypeptides VP_2 and VP_4 in the form of their uncleaved precursor protein VP_0. The twelve 5-vertices of the icosahedron are made from 5 such protomers. The pentamer of protomers about a 5-vertex appears to be the fundamental secondary structural unit of functional importance in viral assembly (see Fig. 111, Chapter 10): during virus assembly, the asymmetrical protomers are com-

combined to form the symmetric pentamer of protomers as assembly intermediates, 12 of which are subsequently assembled into a complete virion capsid.

1. X-Ray Diffraction Studies

Experimental studies on the structure of poliovirus were made possible by the development of methods for the purification and crystallization of the virus (Schaffer and Schwerdt, 1955, 1959; Steere and Schaffer, 1958). With the availability of such crystals, excellent X-ray diffraction patterns of crystallized poliovirus were obtained (Finch and Klug, 1959) (Fig. 12A). There are spikes of high intensity along certain directions which are related as the 5- and 3- and 2-fold axes of an icosahedron, indicating that the poliovirion possesses intrinsic 5 : 3 : 2 fold symmetry. Comparison of an optical diffraction pattern of 60 points on the surface of a sphere with icosahedral symmetry to the poliovirus pattern reveals that both show the same symmetry relations (Fig. 12B). Based on the theoretical grounds discussed above and by extrapolating from the X-ray crystallography data, it was concluded that the virion is indeed made up of 60 (or 60 x N) identical asymmetric "structure units" with a diameter of 6.0 to 6.5 nm each, arranged in an icosahedral lattice with 5 : 3 : 2 fold symmetry.

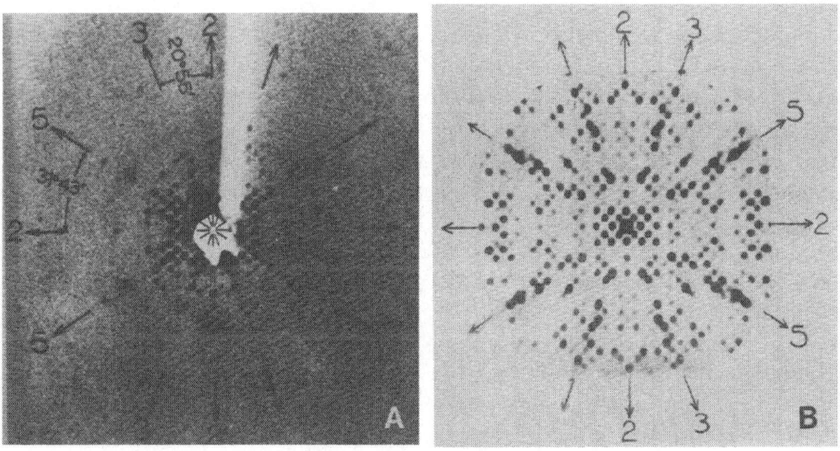

Fig. 12. Experimental evidence for the icosahedral lattice structure of poliovirus
A X-ray diffraction pattern of a poliovirus crystal. There are spikes of high intensity along certain directions which are related as the 5-, 3- and 2-fold axes of an icosahedron (indicated by the arrows) *B* Optical diffraction pattern of 60 points on the surface of a sphere with icosahedral symmetry. The intensity distribution of the poliovirus pattern shows the same symmetry relations as this optical analogue. – From Caspar and Klug, 1962 [Cold Spring Habor Symp. Quant. Biol. *27*, 9 (1962)]

2. Electron Microscopic Observations

The first electron microscopic observations reinforced the 60 subunit model: Particles with pentagonally arranged knobs about 5.0 nm in diameter were observed (see Fig. 9B, above) (Horne and Nagington, 1959). Electron microscopy studies on echovirus, an enterovirus-relative of poliovirus, however, favored a structure with 32 morphological subunits (Jamison, 1969). The data on the turnip yellow mosaic virus (TYMV), which at the time was considered a model for poliovirus (the X-ray diffraction pattern is very similar to that of poliovirus), also led to an erosion of confidence in the 60 equivalent subunit model. High resolution electron microscopy revealed that the 180 (identical) capsid proteins of TYMV were not organized into 60 but 32 morphological clusters composed of 12 clusters of 5 chains, and 20 clusters of 6 chains (Klug *et al.,* 1966; Finch and Klug, 1966). It seemed that a 5:3:2 fold symmetry pattern in X-ray crystallography provided insufficient evidence to conclude that a virion is constructed from 60 identical subunits (Rueckert, 1976). Actually, even these plant viruses are constructed from 60 identical structure units; in contrast to poliovirus, one such structure unit is composed of three identical chains in three different conformations. For both types of viruses, a 60-unit model can be superimposed on the 12-pentamer 20-hexamer model (see below).

A detailed analysis of poliovirus by electron microscopy (Mayor, 1964) indicated that the capsid had the structure of a polyhedron exhibiting icosahedral symmetry with 32 vertices (*i.e.,* with 32 distinguishable subunits in the EM). It was concluded that the capsomers of poliovirus are situated at the vertices of a rhombic triacontrahedron (*e.g.,* Mayor, 1964) (see Fig. 9C). The polyhedron with this curious name is simply related to a regular icosahedron: The long diagonals of the faces of the rhombic triacontrahedron form the edges of a regular icosahedron (see Fig. 122, Appendix III). To meet the mathematical requirement of 60N subunits, a clustering of subunits was proposed with 12 quasi equivalently arranged pentamers located at the 12 vertices of an icosahedron, the 20 triangular facets composed of hexamers (Mayor, 1964), corresponding to a triangulation number 3 (Caspar and Klug, 1962). The demonstration that picornaviruses are constructed from 4 individual polypeptides, (Maizel, 1963; Summers *et al.,* 1965) gave rise to more confusion as to whether the capsid is composed of 32 or 60 capsomers and whether it should be correctly classified as T = 1 or T = 3 in the theory of Caspar and Klug (1962) (Appendix III B).

Stoichiometry of the capsid proteins finally revealed that there were 60 copies of each of the 4 proteins. These observations in turn strongly supported the concept of 60 identical subunits. Degradation studies on different picornaviruses again yielded apparently conflicting results: Mild acid dissociation of cardioviruses yielded 60 identical oligomers of VP1, VP2, VP3 per virion supporting the 60 identical subunit models (Dunker and Rueckert, 1971). It was proposed that the organization of the capsid proteins within a single subunit species (termed protomer) ruled out 32 capsomer shells (which seemed to require construction from two different kinds of protomers, *i.e.* pentamers and hexamers) (Rueckert, 1971, 1976). Mild alkaline or urea treatment of enteroviruses (Johnston and Martin, 1971; Katagiri *et al.,* 1971; Philipson *et al.,* 1973), however, indeed yielded two types of oligomers, one composed of VP2 aggregates, the other of VP1–VP3 oligomers, leading to the proposal that the capsid proteins are arranged in two types of cluster: 12 VP2 (and VP4) pentamers and 20 hexamers of VP1 and VP3 with a triangulation of number 3 (Philipson *et al.,* 1973).

The two apparently inconsistent concepts are probably both valid, they may merely represent two different ways of looking at the same structure.

3. *Resolving the 32 or 60 Capsomers Controversy*

Today it is known that the capsids of the small icosahedral plant viruses (*e.g.*, TYMV or TBSV), which were once believed to be models for poliovirus, are composed of 180 identical single protein subunits, arranged in 12 clusters of pentamers and 20 hexamer clusters, at the twelve 5-vertices and twenty 3-vertices of an icosahedron (see Figs. 18 and 19 below). 60 clusters of trimers that occupy identical environments in an icosahedral arrangement can be superimposed on such a structure.

The poliovirus capsid evidently arrives at quite a similar structure with 60 identical subunits. Each subunit is composed of the 4 distinct capsid polypeptide chains VP1, VP2, VP3, and VP4. Since the 60 subunits are equivalent sets of polypeptides (VP4–VP2–VP3–VP1) and occupy identical environments on the surface of the spherical capsid, the capsid should be classified as having a triangulation number (T = 1). Packaging of the 60 protomers in equivalent environments nonetheless can lead to a clustering of the capsid polypeptides leading to a 32-subunit appearance (Rueckert *et al.*, 1969; Rueckert 1971; Dunker, 1974) (Fig. 13 and also Appendix II), for example, 12 subunits being formed by the pentamer of VP2VP4 about the twelve 5-vertices of the regular icosahedron, the other 20 by a clustering of trimers of VP1–VP3 about the twenty 3-vertices (20 faces) of the regular icosahedron. If the latter clusters stick out somewhat from the surface of the capsid (instead of the flat side or face of the regular icosahedron as seen in drawings of icosahedral lattices) (Fig. 10), they will appear as additional vertices, giving the impression of a 32-vertex rhombic triacontrahedron (Mayor, 1964) in the EM.

The combined molecular weight of a pentameric VP2–VP4 cluster of 187,000 would be quite similar to that of a hexagonal VP1–VP3 cluster of 180,000. This could explain the appearance of 32 similar morphological subunits in the EM. An alternative model would be the clustering of VP1 pentamers about the icosahedral apices (molecular weight of cluster 168,000) and VP2– VP3 hexamers about the icosahedral facets (molecular weight of cluster 170,000), with the localization of VP4 undefined. Evidence supporting the notion of clustering of VP2 in pentamers about the apices and the constitution of the capsid backbone by VP1 and VP3 in poliovirus is summarized below.

4. *Similar Construction Principles for Picornaviruses and the Small Plant Viruses*

The acquisition of similar structures with 180 identical subunits (small plant and insect viruses) or 60 identical subunits (picornaviruses), and the apparently awkward morphological clustering of the constituent proteins, may explain the confusion brought about from time to time by the accumulation of apparently controversial experimental data. It is also quite a fascinating example for the process of convergent evolution. TBSV and poliovirus have evolved quite similar principles of capsid design and structure by two independent evolutionary routes.

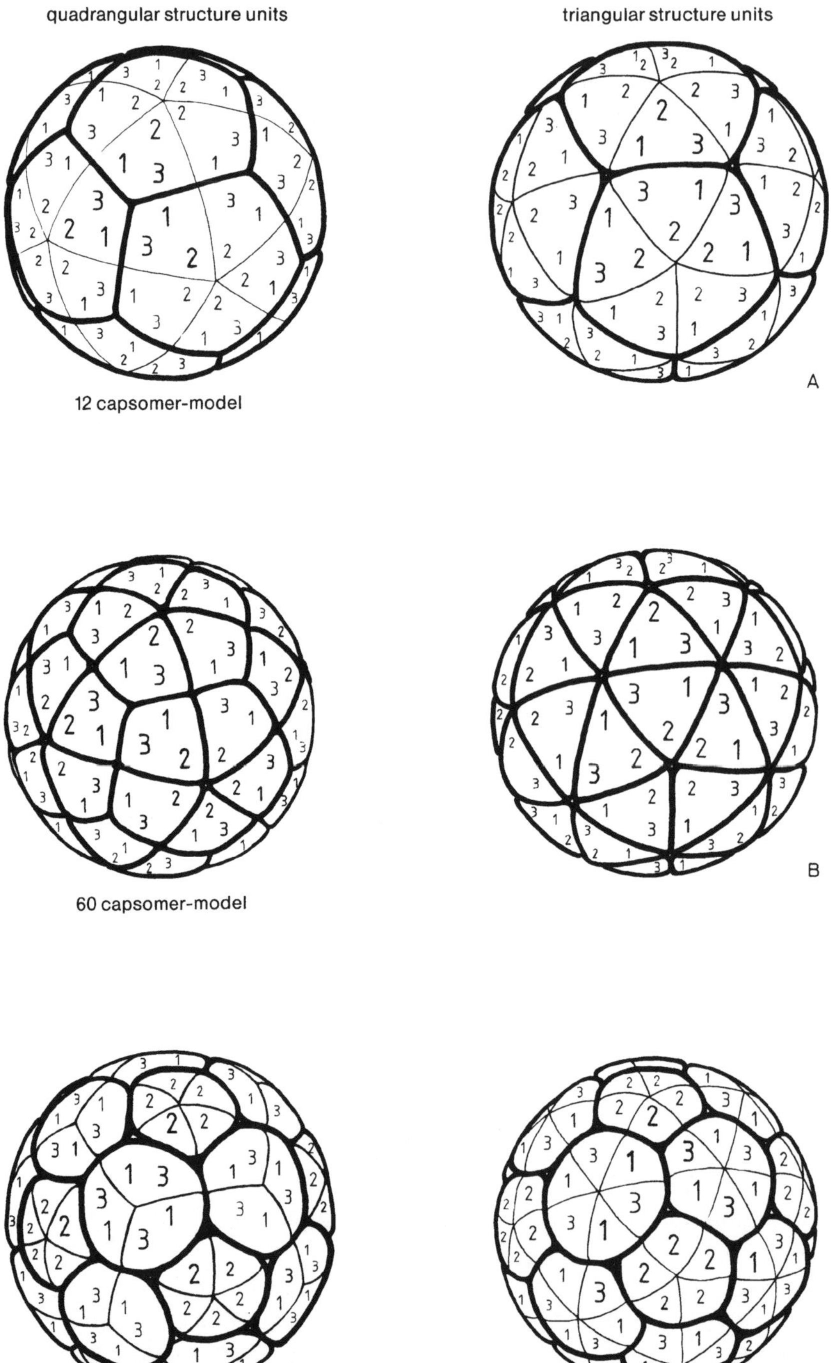
quadrangular structure units
triangular structure units
12 capsomer-model
A
60 capsomer-model
B
32 capsomer-model
C

The unifying concept is the construction of a structure with icosahedral symmetry from 60 identical structure units: In the case of TBSV, the structure unit is composed of 3 *identical* protein chains each, in the case of poliovirus of 3 *different* protein chains each, VP_1, VP_3 and VP_0 (the latter of which is cleaved finally into VP_2 and VP_4).

The definition of the number of capsomers is merely a matter of point of view. With the simplest compositionally identical structure unit, TBSV would have 180 capsomers, poliovirus 60. Taking the simplest asymmetric structure units that occupy identical environments within the completed capsid, both viruses would be composed of 60 capsomers (Fig. 13B). Considering clustering of polypeptides into morphologically identifiable units in the completed capsid, the number of capsomers for both viruses would be 32 (Fig. 13C) (20 of one kind, 12 of another kind). Other structure units that could be considered theoretically (identically situated, symmetric) are 30 dimers (corresponding to the icosahedral edges), 20 trimers (corresponding to the icosahedral facets), and 12 pentamers (corresponding to the icosahedral apices) (Fig. 13A). In the case of poliovirus the 12 pentameric structure units correspond to the fundamental second order bulding block during assembly (see Chapter 10). Figure 13 compares these clustering patterns and shows how the poliovirus capsid proteins may be arranged within the capsid.

The final confirmation of the icosahedron structure originally proposed in the late fifties was achieved only through the detailed characterization of the capsid proteins into 4 individual polypeptide chains derived from a common precursor and the determination of the capsid polypeptide stoichiometry as 60 copies per virion.

Fig. 13. Schematic models of the poliovirus capsid

This figure illustrates that the assignment of capsomer number to an icosahedral virus which is constructed from 60 copies of three similarly sized proteins is merely a matter of point of view. Starting with a triangular structure unit (right column) or a quadrangular structure unit (left column) composed of the three proteins, the appearance of the complete particle as containing 12, 32, or 60 subunits depends on the relative clustering of the three subunit proteins. Notice that the relative position of the proteins is the same in each of the three models: only very subtle shifts in the surface structure at the borders of the proteins are required to alter the appearance of the entire particle. The triangular (left side) and quadrangular (right side) shapes of the structure units were chosen, since these are the most uniform shapes which can be assumed by icosahedral structure untis (see Fig. 10). The actual structure units of the poliovirus may of course be shaped more awkwardly, the principle of clustering patterns will nevertheless be the same.

The relative positioning of the capsid proteins within the structure units is designated with the numbers 1, 2, and 3 in order to ease the visualization of the clustering patterns. We have placed VP_2 at the icosahedral 5-fold vertices, VP_{3-1} at the icosahedral 3-fold vertices, since we feel that this best fits the available data on poliovirus. Note, that a pentameric cluster of VP_{2-4} would be of the same size as a hexameric cluster of VP_{3-1} (180–185 kd molecular weight). VP_1 probably covers more of the surface area, VP_3 less than indicated. Part of VP_1 shows a high variation between different poliovirus strains and thus is not essential for capsid stability. It is likely that this portion of VP_1 projects from the backbone of the capsid and may also be responsible for the high antigenicity of this protein. The fine details of capsid structure, however, can only be ascertained when the data from high resolution X-ray crystallography becomes available

In the following section we will examine the individual capsid proteins in detail, describe their characteristics and the possible implication for the final structure of the capsid.

IV. Characterization of the Building Blocks: The Capsid Proteins

A. Separation and Identification

Early investigations performed to characterize the isolated capsid protein employed analytical ultracentrifugation and density gradient centrifugation. These methods revealed a relatively homogeneous protein component with a molecular weight of aproximately 30,000 daltons. The existence of several non-identical polypeptide components in the poliovirus capsid was discovered when their separation was first achieved by electrophoresis (Maizel, 1963, 1964). The electrophoretic separation techniques were improved, and with the introduction of the now classic polyacrylamide gel electrophoresis system containing sodium-dodecyl-sulfate (SDS-PAGE) in 1965, the presence of four distinct proteins in the capsid of the poliovirion was clearly demonstrated (Summers *et al.,* 1965). These were named VP1–VP4 (VP = viral protein) in order of increasing mobility in the SDS-PAGE system.

With the discovery that the electrophoretic mobility of a polypeptide in such SDS-PAGE systems bears a direct correlation to its molecular weight* (Shapiro *et al.,* 1967), it became possible to simultaneously separate and assign molecular weights to the capsid proteins: (VP1 33,000–35,000, VP2 28,000–31,000, VP3 24,000–27,000, and VP4 6,000–8,000) (*e.g.* Rückert, 1971, 1976). The bulk by weight and size of the capsid thus consists of VP1, VP2, and VP3, *i.e.,* more than 90%, VP4 making up less than 10% by weight. Figure 14 shows the typical patterns of poliovirus capsid proteins separated by SDS-PAGE. Today the SDS-PAGE is one of the most useful and widely applied techniques in protein characterization.

Soon thereafter, it was demonstrated that the four capsid proteins are all derived by cleavage from the larger precursor protein NCVP1a (Holland and Kiehn, 1968; Jacobson and Baltimore, 1968b). The stoichiometry of the capsid proteins in the virion capsid was calculated on different occasions (reviewed in Rueckert, 1976), the most proper estimates were of approximately 60 copies of each of the four viral proteins per virion.

Closer examination of the typical PAGE profiles of capsid proteins revealed the consistent finding of an additional protein component VP0 (MW 41,000) at a stoichiometry of approximately two (or more) copies per virion. Since VP0 is the uncleaved precursor to VP2 and VP4 (Jacobson *et al.,* 1970) it seems that one or two of the 60-structure units of the virion capsid may contain a copy of VP0 instead of VP4 plus VP2 (see also Table 6B above). Although the conformational differences between VP0 and its cleavage products VP2 and VP4 may be small, the

---

* The correlation is that the mobility of a polypeptide in the SDS-PAGE system is inversely proportional to the logarithm of its molecular weight.

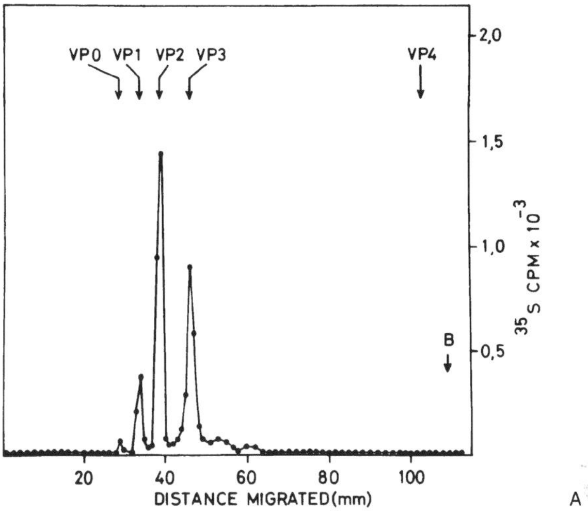

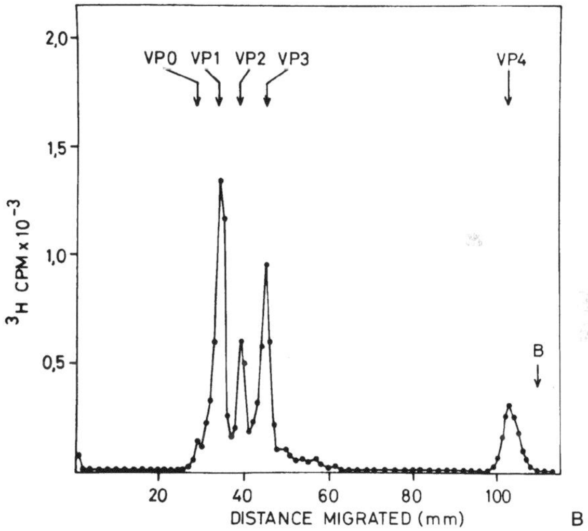

Fig. 14. SDS-polyacrylamid gel electrophoresis of the capsid proteins of poliovirus
The poliovirus polypeptides were labeled by incubation of infected cells with ^{35}S cysteine *(A)* or ^{3}H lysine *(B)*. The small polypeptide VP_4 does not contain any cysteine residues and is therefore not labeled in *(A)*. The direction of migration is from left to right. *B:* bromophenol blue. — From Wetz and Habermehl, 1979 [J. gen. Virol. *44*, 525–534 (1979)]

unusual VP0-containing capsomers are consistently observed in many different picornaviruses at approximately two copies per virion. The question naturally arises whether such capsomers are of any functional significance, for example, to confer a preferential direction to virions. The answer to this question is not known, although it is conceivable that the (slight) conformational incongruity conferred to the virion through a VP0-containing capsomer might play a role in the recognition of the virus by the specific virus receptors of the host cell during

Table 8. *Properties of the poliovirus capsid proteins*

Designation	Copies per virion	Molecular weight ($\times 10^3$)		Net[1] charge	Isoelectric points			Amino terminal				Carboxy terminal
		*	PAGE		Type I	Type II	Type III	Type I		Lsc2ab		
								*	a	*	b	*
VP_0	2	37.4	41	+3	6.6	n.d.	n.d.	.met	n.d.	met	n.d.	gln
VP_4	58	7.4	6	+2	7.3	7.2–7.5	7.2–7.5	met	blocked	met	—	asn
VP_2	58	30.0	28	+1	6.4	6.25	6.3	ser	asp	ser	ala (gly)	gln
VP_3	60	26.4	24	+0	6.0	6.0	6.2	gly	ser	gly	leu (gly)	gln
VP_1	60	33.5	35	+9	8.1	8.6–8.8	8.1	gly	gly	gly	gly (ala)	gln

* Deduced from the nucleotide sequence of poliovirus type 1 genomic RNA (see Fig. 44, p. 150).
[a] Dansylation.
[b] Edman degradation.
[1] All his residues protonated.

uncoating, or for its interaction with the immunological defense system of the host. The VP_0-containing capsomer(s) might also function in binding the two RNA ends or play some other role in assembly. The presence of one or two VP_0-containing capsomers in the virus particles may also explain the slight deviation from strict bodycentered cubic symmetry of poliovirus crystals that is observed in X-ray diffraction patterns (see above).

Recently new and refined techniques for the separation and characterization of the capsid proteins have been developed: SDS-pH gradient electrophoresis (Vrijsen and Boye, 1978), isoelectric focusing in urea-containing polyacrylamide-gels alone or in a 2D-analysis in combination with electrophoresis (Hamann *et al.*, 1977; Wiegers and Drzeniek, 1980), determination of amino acid composition (Wouters and Van der Kerckove, 1976), and mapping of the tryptic peptides of the capsid proteins (Kew *et al.*, 1980). Table 8 summarizes some of the properties of the individual capsid proteins.

B. Amino Acid Composition and Sequence

The amino acid composition of the virion capsid protein has been determined on different occasions (Levintow and Darnell, 1960; Munyon and Salzmann, 1962; Cooper and Bennett, 1973; Wouters and Van der Kerckove, 1976). Table 9 presents the amino acid composition of the capsid proteins as determined from their amino acid sequence deduced from the nucleotide sequence of poliovirus 1 (Kitamura *et al.*, 1981; Racaniello and Baltimore, 1981). Notable features are: a slight excess of basic (lys, arg, his) over acidic (asp, glu) residues, a low sulfur content (cys + met = 4–5 mol %), a substantial number of pro (6–8 mol %) and other helix-breaking residues (asn + tyr + gly = 15 mol %). Approximately 50% of the amino acid residues are apolar and 50% are polar. Optical rotatory dispersion and circular dichroism measurements of poliovirus indicate an α-helical content of only 5–10 % *in situ* in accordance with the large proportion of helix-disrupting and non-helix forming amino acid residues (Dernick, 1981). The capsid proteins carry the following net charge at pH7 (assuming that 50 % of the his-residues are charged at this pH): VP_1: +5.5, VP_3: –4.5, VP_2: –2, VP_4: +1.5 (VP_0: –0.5) (Table 10).

There have been conflicting reports as to the amino terminal amino acid of the capsid proteins. Dansylation of intact poliovirus revealed asp, ser, and gly as amino terminals (Burrel and Cooper, 1973). The amino terminals determined from the nucleotide sequence are gly for VP_1 and VP_3 and ser for VP_2 (Kitamura *et al.*, 1981). Automated Edman degradation of poliovirus capsid proteins revealed amino terminal sequence ambiguities for the capsid proteins, the amino acid terminals of the individual proteins being VP_1: gly/ala, VP_2: gly/ala, VP_3: gly/leu (Vrijsen *et al.*, 1978). The amino terminal sequence proposed by these authors is shown in Table 11. The VP_4 amino terminal is blocked, the nature of this block is still unknown. Of interest in this respect may be the fact that the amino terminal of VP_4 lies closest to the initiation point of protein synthesis and might be the

Table 9. *Amino acid compositions of the poliovirus capsid proteins*

	VP_4		VP_2		VP_3		VP_1		Protomer	
	a)	b)	a)	b)	a)	b)	a)	b)	a)	b)
Gly	4	5.7	20	4.8	13	5.4	17	5.4	54	5.3
Ala	7	11.4	21	8.0	13	7.1	23	7.3	64	7.8
Leu	2	2.8	26	7.6	22	9.2	20	6.4	70	7.2
Val	3	4.3	14	5.3	13	5.4	26	8.3	58	6.3
Ile	4	5.7	11	4.2	14	5.8	12	3.9	41	4.6
Pro	3	4.3	19	6.0	18	7.5	21	7.6	61	6.9
Phe	2	2.8	10	3.8	10	4.2	12	3.9	34	3.9
Trp	–	–	7	2.8	3	1.3	4	1.3	14	1.7
Met	2	2.8	6	3.1	11	4.6	5	1.6	24	2.9
Cys	–	–	8	3.1	5	2.1	2	0.7	15	3.1
Ser	9	12.8	18	8.4	19	8.8	24	6.7	70	6.3
Thr	5	7.2	23	8.9	21	10.0	32	10.3	82	9.5
Tyr	4	5.7	10	3.8	8	3.3	15	4.8	39	4.2
Asn	6	8.6	22	7.6	8	3.3	13	4.2	50	5.3
Gln	4	5.7	10	3.8	7	6.2	7	2.2	29	3.0
His (+)	1	1.4	6	2.3	5	2.1	7	2.2	9	2.2
Arg +	2	2.8	11	4.2	9	3.8	15	4.8	37	4.4
Lys +	5	7.2	5	1.9	10	4.2	15	4.8	35	4.0
Asp –	4	5.7	10	3.8	16	6.6	17	5.5	47	5.4
Glu –	2	2.8	11	4.2	10	4.2	11	3.6	34	4.2
Total:	69		271		238		302		880	

a) Number per peptide.
b) Mol %.

Table 10. *Some characteristic features of the poliovirus capsid proteins*

	VP 4	VP 2	VP 3	VP 1
	* **			
Basic amino acid residues	8–11	22– 8	24–10	37–12
Acidic amino acid residues	6– 9	21– 8	26–11	28– 9
Helix breakers (pro, gly, tyr, asn)	17–24	71–22	47–20	66–22
Net charge at pH 7.0[1]	+ 1.5	− 2	− 4.5	+ 5.5
Cys	0	8	5	2
Aromatic amino acids (trp, phe, tyr)	6	27	21	31
Trp	0	7	3	4

% α helix in viral capsid	10 %
SH groups per virus particle	900
SH groups on surface	0?
S-S bonds	0?

* Number of amino acids.
** Mol %.
[1] 50 % of his residues protonated.

Table 11. *Amino terminal sequence of capsid proteins*

VP_4	a) blocked- b) met-gly-ala-glu-val-ser-ser-glu-
VP_2	a) gly-pro-asn-ile-glu-ala ala-gly-thr-ile-leu-ala b) ser-pro-asn-ile-glu-ala-cys-gly-
VP_3	a) gly-leu-pro-val-Ø-asx- leu-leu-val-val-ala- b) gly-leu-pro-val-met-asn-thr-pro-
VP_1	a) gly-leu-gly-glx-met-leu- ala-leu-gly-ala-leu-leu- b) gly-leu-gly-gln-met-leu-glu-ser-

a) Automated Edman degradation.
b) Deduced from complete nucleotide sequence of genome.

only poliovirus protein amino terminal that does not arise by a cleavage mechanism.

The complete amino acid sequence of the four capsid proteins of poliovirus 1 as deduced from the nucleotide sequence of the genomic RNA is presented in Table 12. Figure 15 illustrates the degrees of hydrophobicity and hydrophilicity of consecutive 20 amino acid long stretches along the capsid proteins. The peaks of hydrophilicity indicate regions of the capsid proteins likely to be exposed on the surface of the virion. More hydrophobic regions probably constitute the capsid backbone. The distribution of positively and negatively charged residues along the capsid proteins is also indicated in Figure 15. Clusters of positively charged residues may be important for interaction with the viral RNA. Hydrophobic regions below the threshold line are also potential membrane insertion sites in the infected cell (von Heijne, 1981). The capsid proteins are indeed found in tight association with intracellular membranes in infected cells (Caliguiri and Tamm, 1970b; Korant, 1973), and may be secreted into the lumen of the endoplasmic reticulum for transport to internal locations in the virus-induced-perinuclear vesicles (Chapter 10).

The isoelectric points of the capsid proteins (poliovirus-type 1) are VP_1: 8.1, VP_2: 6.4, VP_3: 6.0, VP_4: 7.3 (Table 8) (Hamann *et al.*, 1977, 1978), *i.e.*, VP_1 is the most basic polypeptide of the capsid, and VP_3 the most acidic. The isoelectric point of VP_0 (isolated from empty capsid) lies between those of VP_2 and VP_4 at 6.6. The isoelectric points correlate well with the amino acid composition (Table 9) (*i.e.*, the isoelectric points decrease in the same order as the net charges).

The behavior of VP_3 and VP_1 in isoelectric focusing indicates a potential for strong attractive interactions between these two proteins. Of interest in this respect are the findings from crosslinking studies which indicate that VP_1 and VP_3 indeed lie in close proximity within the viral capsid (Wetz and Habermehl, 1979). Partial degradation of poliovirus also provides evidence for the role of a strong interaction between VP_1 and VP_3 in maintaining capsid integrity. In fact, the bonds between VP_1 and VP_3 apparently are sufficient to maintain a backbone structure resembling the skeleton of a virion particle (see below, Katagiri *et al.*, 1971). The two most basic proteins VP_1 and VP_4 have been implicated in extensive interaction with the negatively charged RNA. Clusters of basic amino acid residues, especially in VP_1 may provide binding sites for the RNA. VP_1 from urea degraded virion spontaneously binds to the negatively charged RNA, and VP_4 is crosslinked to the RNA upon UV irradiation of virions (see below). The largest of the capsid proteins VP_1 also appears to be the immunodominant surface peptide. Portions of this protein probably participate in the construction of prominent surface features which may function in the interaction with the host cell receptor.

Fig. 15. Relative hydrophobicity of 20 amino acid-long segments in the poliovirus capsid proteins This figure illustrates the relative distribution of hydrophobic and hydrophilic segments in the poliovirus capsid proteins. The plot is obtained by averaging the hydrophobic constants of the amino acid side chains over a stretch of 20 consecutive residues. For each consecutive stretch, the mean change in free energy required for insertion of the amino acid sequence into a hydrophobic environment was calculated. The "threshold" line at 5.9 kJ/mol indicated whether a certain segment of the protein could lie within a membrane (v. Heijne, 1981). Valleys in the plot indicate relatively hydrophobic segments, peaks indicate hydrophilic segments. Strongly hydrophobic segments are potential membrane insertion sites; the hydrophylic peaks are regions likely to be exposed on the capsid surface. The bottom of each figure illustrates the distribution of negatively (top row) and positively charged (bottom row) amino acid residues along the polypeptide sequence. Notice the clusters of positively charged residues, for example at the carboxy terminal end of VP_1: these regions may provide binding sites for the RNA. — Figures courtesy of J. Hoppe, Braunschweig

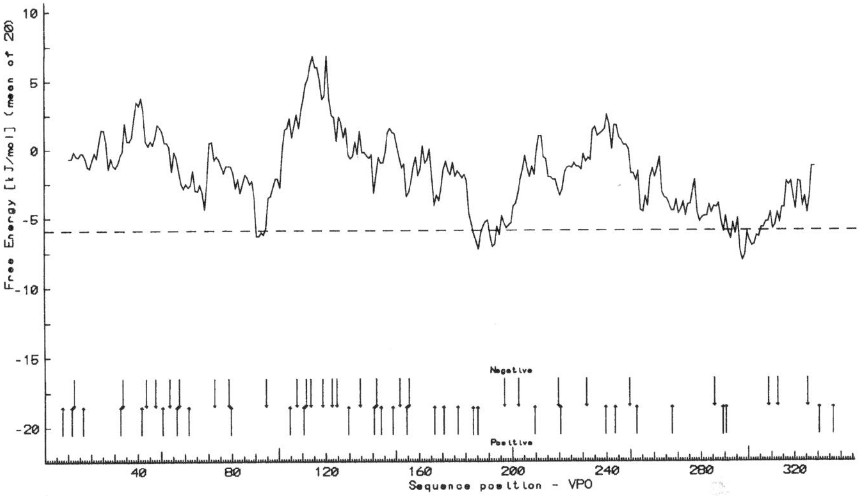

Free Energy [kJ/mol] (mean of 20)
10
5
0
-5
-10
-15
-20
Negative
Positive
40
80
120
160
200
240
280
320
Sequence position - VP0

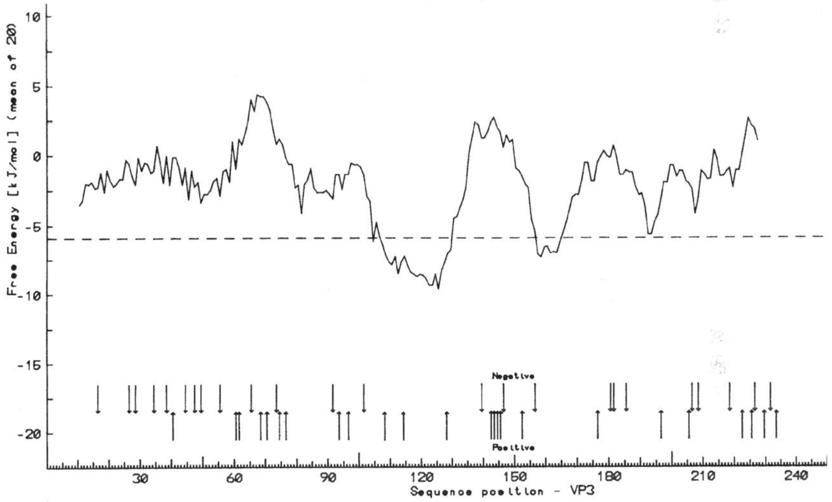

Free Energy [kJ/mol] (mean of 20)
10
5
0
-5
-10
-15
-20
Negative
Positive
30
60
90
120
150
180
210
240
Sequence position - VP3

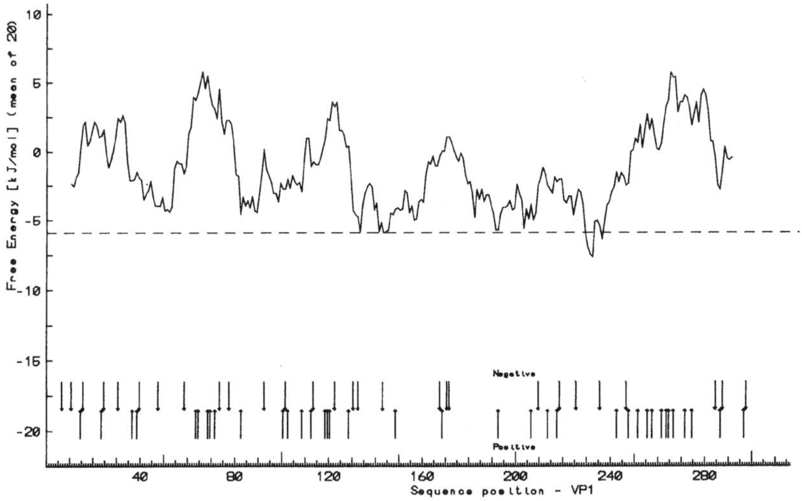

Free Energy [kJ/mol] (mean of 20)
10
5
0
-5
-10
-15
-20
Negative
Positive
40
80
120
160
200
240
280
Sequence position - VP1

Table 12. *The amino acid sequences of the poliovirus type I capsid proteins*

Sequence VP_4 (Length 69)

MET GLY ALA GLN VAL SER SER GLN LYS VAL GLY ALA HIS GLU ASN SER ASN ARG ALA TYR
GLY GLY SER THR ILE ASN TYR THR THR ILE ASN TYR TYR ARG ASP SER ALA SER ASN ALA
ALA SER LYS GLN ASP PHE SER GLN ASP PRO SER LYS PHE THR GLU PRO ILE LYS ASP VAL
LEU ILE LYS THR ALA PRO MET LEU ASN

Sequence VP_2 (Length 272)

SER PRO ASN ILE GLU ALA CYS GLY TYR SER ASP ARG VAL LEU GLN LEU THR LEU GLY ASN
SER THR ILE THR THR GLN GLU ALA ALA ASN SER VAL VAL ALA TYR GLY ARG TRP PRO GLU
TYR LEU ARG ASP SER GLU ALA ASN PRO VAL ASP GLN PRO THR GLU PRO ASP VAL ALA ALA
CYS ARG PHE TYR THR LEU ASP THR VAL SER TRP THR LYS GLU SER ARG GLY TRP TRP TRP
LYS LEU PRO ASP ALA LEU ARG ASP MET GLY LEU PHE GLY GLN ASN MET TYR TYR HIS TYR
LEU GLY ARG SER GLY TYR THR VAL HIS VAL GLN CYS ASN ALA SER LYS PHE HIS GLN GLY
ALA LEU GLY VAL PHE ALA VAL PRO GLU MET CYS LEU ALA GLY ASP SER ASN THR THR THR
MET HIS THR SER TYR GLN ASN ALA ASN PRO GLY GLU LYS GLY GLY THR PHE THR GLY THR
PHE THR PRO ASP ASN ASN GLN THR SER PRO ALA ARG ARG PHE CYS PRO VAL ASP TYR LEU
LEU GLY ASN GLY THR LEU LEU GLY ASN ALA PHE VAL PHE PRO HIS GLN ILE ILE ASN LEU
ARG THR ASN ASN CYS ALA THR LEU VAL LEU PRO TYR VAL ASN SER LEU SER LEU ASP SER
MET VAL LYS HYS ASN ASN TRP GLY ILE ALA ILE LEU PRO LEU ALA PRO LEU ASN PHE VAL
SER GLU SER SER PRO GLU ILE PRO ILE THR LEU THR ILE ALA PRO MET CYS CYS GLU PHE
ASN GLY LEU ARG ASN ILE THR LEU PRO ARG LEU GLN

Sequence VP_3 (Length 238)

GLY LEU PRO VAL MET ASN THR PRO GLY SER ASN GLN TYR LEU THR ALA ASP ASN PHE GLN
SER PRO CYS ALA LEU PRO GLU PHE ASP VAL THR PRO PRO ILE ASP ILE PRO GLY GLU VAL
LYS ASN MET MET GLU LEU ALA GLU ILE ASP THR MET ILE PRO PHE ASP LEU SER ALA THR
LYS LYS ASN THR MET GLU MET TYR ARG VAL ARG LEU SER ASP LYS PRO HIS THR ASP ASP
SER ILE LEU CYS LEU SER LEU SER PRO ALA SER ASP PRO ARG LEU SER HIS THR MET LEU
GLY GLU ILE LEU ASN TYR TYR THR HIS TRP ALA GLY SER LEU LYS PHE THR PHE LEU PHE
CYS GLY SER MET MET ALA THR GLY LYS LEU LEU VAL SER TYR ALA PRO PRO GLY ALA ASP
PRO PRO LYS LYS ARG LYS GLU ALA MET LEU GLY THR HIS VAL ILE TRP ASP ILE GLY LEU
GLN SER SER CYS THR MET VAL VAL PRO TRP ILE SER ASN SER THR TYR ARG GLN THR ILE
ASP ASP SER PHE THR GLU GLY GLY TYR ILE SER VAL PHE TYR GLN THR ARG ILE VAL VAL
PRO LEU SER THR PRO ARG GLU MET ASP ILE LEU GLY PHE VAL SER ALA CYS ASN ASP PHE
SER VAL ARG LEU LEU ARG ASP THR THR HIS ILE GLU GLN LYS ALA LEU ALA GLN

Sequence VP_1 (Length 302)

GLY LEU GLY GLN MET LEU GLU SER MET ILE ASP ASN THR VAL ARG GLU THR VAL GLY ALA
ALA THR SER ARG ASP ALA LEU PRO ASN THR GLU ALA SER GLY PRO THR HIS SER LYS GLU
ILE PRO ALA LEU THR ALA VAL GLU THR GLY ALA THR ASN PRO LEU VAL PRO SER ASP THR
VAL GLN THR ARG HIS VAL VAL GLN HIS ARG SER ARG SER GLU SER SER ILE GLU SER PHE
PHE ALA ARG GLY ALA CYS VAL THR ILE MET THR VAL ASP ASN PRO ALA SER THR THR ASN
LYS ASP LYS LEU PHE ALA VAL TRP LYS ILE THR TYR LYS ASP THR VAL GLN LEU ARG ARG
LYS LEU GLU PHE PHE THR TYR SER ARG PHE ASP MET GLU LEU THR PHE VAL VAL THR ALA
ASN PHE THR GLU THR ASN ASN GLY HIS ALA LEU ASN GLN VAL TYR GLN ILE MET TYR VAL
PRO PRO GLY ALA PRO VAL PRO GLU LYS TRP ASP ASP TYR THR TRP GLN THR SER SER ASN
PRO SER ILE PHE TYR THR TYR GLY THR ALA PRO ALA ARG ILE SER VAL PRO TYR VAL GLY
ILE SER ASN ALA TYR SER HIS PHE TYR ASP GLY PHE SER LYS VAL PRO LEU LYS ASP GLN
SER ALA ALA LEU GLY ASP SER LEU TYR GLY ALA ALA SER LEU ASN ASP PHE GLY ILE LEU
ALA VAL ARG VAL VAL ASN ASP HIS ASN PRO THR LYS VAL THR SER LYS ILE ARG VAL TYR
LEU LYS PRO LYS HIS ILE ARG VAL TRP CYS PRO ARG PRO PRO ARG GLN LEU ALA TYR TYR
GLY PRO GLY VAL ASP TYR LYS ASP GLY THR LEU THR PRO LEU SER THR LYS ASP LEU THR
THR TYR

Recently synthetic peptides corresponding to some of the hydrophilic regions in the three large capsid proteins of poliovirus type 1 have been constructed and tested for their immunogenicity and reactivity with antiviral antibodies (Emini *et al.*, 1983c, 1984). Results of these studies indicate that amino acid segments 132–143 of VP_2, 71–82 of VP_3, and 11–17, 70–75, and 93–103 of VP_1 are exposed on the capsid surface and are constituents of antigenic determinants. Segment 93–103 of VP_1 has also been implicated as an immunodominant antigenic determinant on the viral surface of poliovirus types 1 and 3 by genetic and immunological studies (Evans *et al.*, 1983; Minor *et al.*, 1983; Wychowski *et al.*, 1983; Van der Werf *et al.*, 1983).

C. Microheterogeneity

With isoelectric focusing as well as with refined SDS-PAGE methods, resolution of the major capsid proteins (VP_1–VP_3) into doublets have been reported (Van den Berghe and Boeye, 1972; Vrijsen *et al.*, 1978). Since these multiple bands were observed only in virus preparations that had been stored for longer periods and not in fresh preparations (Hamann *et al.*, 1977), it was concluded that they arise from modification of the "normal" capsid protein (by deamidation of glutamine or asparagine to the appopriate amino acid or by phosphorylation or dephosphorylation). Other authors have proposed that the microheterogeneity, *i.e.* doublets of VP_1, VP_2, and VP_3 originate in ambiguity during the post-translational processing of proteins (for example, due to multiplicity of cleavage sites) (Cooper *et al.*, 1970b; Fennell and Phillips, 1974; Beckman *et al.*, 1976; Vrijsen *et al.*, 1978). The report of amino terminal sequence ambiguity in the three major viral capsid proteins (VP_1–VP_3) is in concord with the latter proposal (Vrijsen *et al.*, 1978). However, the amino acid sequence deduced from the nucleotide sequence reveals no potential sites for such cleavage ambiguity (Kitamura *et al.*, 1981). Thus, the nature and significance of the observed microheterogeneity remain controversial. Slight variations in the composition of the viral capsids within a given virus population containing different mutant viruses may also provide a possible explanation for the heterogeneity of virus populations in respect to susceptibility to neutralization by antibodies and to degradation conditions (see below) or for the deviation from strict cubic symmetry in the arrangement of poliovirus particles within crystals (see Section II above).

V. Relative Localization of the Viral Proteins in the Capsid and Bonds Involved

A. General Reflections

Much indirect evidence on the relative configuration and spatial localizations of the individual poliovirus capsid proteins within a protomer, and their relative positions within the overall architecture of the icosahedral capsid, has accumulated over the years. Certainly it is of interest to know whether all (or which parts) of the 4 capsid proteins are exposed to the outer surface and thus are potential antigenic determinants or specific sites for recognition by the virus receptor on the host cell. Similarly, it is an open question whether all (or which parts) of the capsid proteins are exposed to the inner surface of the capsid, and thus are potentially involved in binding interactions with the virion RNA and in neutralization of the negative charges of the RNA. The relative localization of the capsid polypeptides will also determine the types and number of intracapsid bonds that are responsible for holding the capsid together. Lastly, knowledge of the relative localization of the capsid proteins may provide important insight into the evolution of capsid structure, the degradation or uncoating process of the viral capsid, the nature of the construction steps in assembly, and the mechanisms of the concerted conformational transitions of the overall capsid structure during poliovirus life (for example during assembly, or induced in the native virion by pH shifts, or by binding of the virion to the host cell receptor or antibody, see below).

Clearly, surface exposition of polypeptide chains or internal clustering is largely determined by thermodynamic driving forces. Since the virus is surrounded by an aqueous environment, there will be a tendency for a surface exposition of the more hydrophilic stretches of polypeptides and for most of the apolar chain-stretches to cluster in the wall of the capsid. As the RNA carries a great negative charge, stretches containing mainly basic amino acid residues will be likely to occupy internal locations.

The 7,500 negatively charged phosphate anions of the RNA need to be neutralized to allow a compact packaging of the RNA. The virion contains a number of cations required for the neutralization of the negative charge carried by the phosphate groups of the RNA (one negative charge for each of the 7500 nucleotides). Approximately 5800 monovalent cations (K^+ and Na^+ in a ratio of 5 : 1) contribute one positive charge each, and a few polyvalent cations (putrescine^{++} and spermidine^{+++}) together contribute another 110–120 positive charges. The remaining negative charges on virion RNA may be neutralized by basic amino acid residues on the capsid proteins. The latter interactions are probably important factors for the stabilization of the native virion capsid. They might play a role in the final steps of virion assembly: packaging of RNA and cleavage of VP_0. Divalent Mg^{++} ions seem to be important also for virion stability (Mapoles, 1980).

Likely candidates for the interaction with viral RNA are the most basic proteins VP_1 and VP_4, containing 38 and 8 basic amino acids and a net charge of +9 (+5.5) and of +2 (+1.5), respectively. The entire VP_1 component from urea dissociated poliovirus is bound spontaneously to the RNA upon dilution of the urea, indicating a strong affinity of VP_1 for the RNA (Wiegers *et al.*, 1976). UV irradiation of poliovirus particles induces the formation of covalent RNA-protein bonds. VP_4 is found to be most extensively bound to the RNA, followed by VP_1 and VP_2, while VP_3 apparently has little detectable interaction with the RNA (Wetz and Habermehl, 1981, 1982).

1. Geometric Restrictions

It is assumed that the basic structure of poliovirus in principle is that of a protein shell enclosing an RNA core. Neutron small-angle scattering should shed light on the precise relative proportions of protein, RNA and water in the concentric geometric shells of increasing radii from the particle center. Such studies so far have not been carried out on poliovirus. Application of the technique on the tomato bushy stunt virus (an icosahedral plant picornavirus) (Chauvin *et al.*, 1978) revealed four shells of different compositions (Fig. 16).

The best available estimates for the width of the poliovirus capsid come from electron micrographs of core-stained virions (see Fig. 9D above) (Boublik and Drzeniek, 1976): The backbone of the poliovirus capsid seems to be a spherical shell that extends from a radius of r = 9 nm to r = 14 nm. Polypeptide stretches or domains may protrude from the backbone to the external environment (receptor binding sites, antigenic determinants) or to the interior (interactions with RNA).

Extension of compartment (in term of radius of particle)	% composition in compartment		
Radius (nm)	% protein	% RNA	% H_2O
3– 5	0	38	62
5– 8	55	2	43
8–11	3	31	66
11–15.8	78	1	21
whole particle	34	48	18

Data from Chauvin *et al.*, 1978 [J. Mol. Biol. *124*, 641 (1978)]

Fig. 16 A. The relative contents of protein, RNA, and water in compartments of TBSV

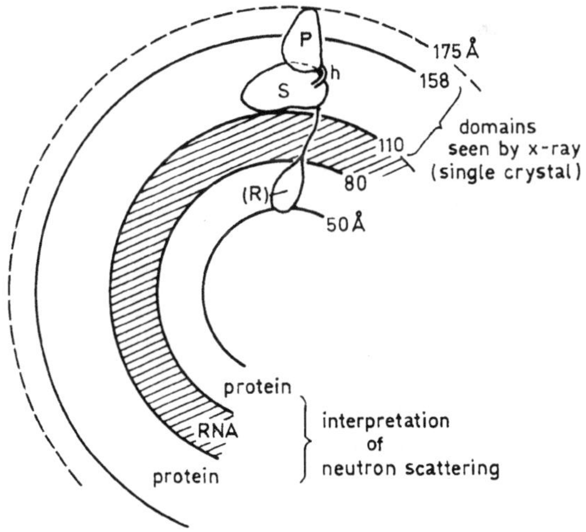

Fig. 16 B. Schematic representation of the distribution of protein and RNA in TBSV S, P, and R are components of the capsid protein, the shaded area indicates the major concentration of RNA. The "R domain" is inferred from an X-ray chain trace and the known amino acid chain length. Its location is implied by the results of neutron scattering (Fig. A above) and small angle X-ray scattering, although not all R-domains may lie in the shell indicated. The folding of the S (capsid backbone) and P (surface protrusions) domains is known precisely (see Figs. 18–20). It is not known whether the "R domain" is folded with comparable precision, although several lines of evidence suggest that it is. – Figure redrawn from S. Harrison, 1980 [Biophys. J. *10*, 140 (1980)]

A schematic representation of a prototype poliovirus model based on these measurements is presented in Figure 17, the corresponding values calculated for the surface areas, volumes, etc. of the icosahedral subcomponents are presented in Table 13. These calculations are rather abstract, but they serve to visualize some important features of poliovirus architecture (see Appendix II). A simple consequence of shell geometry is that a regular division of the shell results in subunits with the shape of truncated cones, *i.e.* with a larger external and smaller internal surface area. For a shell of 5 nm width, the external surface area of a structure unit is more than twice as large as the internal surface area.

Table 13. *Geometrical measurements for a prototype poliovirus*

	Virus-particle	Core	Shell	Pentamer	Cluster	Protomer	Capsid Proteins VP_4	VP_2	VP_3	VP_1
Diameter (nm)	28	18	–	–	–	–	–	–	–	–
Volume (nm^3)	11,494	3,053	8,441	703	264	141	11	44	38	48
Surface area (nm^2)										
External	2,463	1,020	2,463	205	77	41	5	12	11	13
Internal	–	–	1,020	85	32	17	2	5	4	6
Surface contour (nm)										
External	88	57	88	16.1	12.5	7.2	2.5	4	3.8	4.2
Internal	–	–	57	10.4	8.1	4.7	1.6	2.5	2.2	2.8
Thickness (nm)			5	5	5	5	5	5	5	5

The table presents some calculated measurements of the structural components of a prototype poliovirus. The calculations are based on an estimated outside diameter of intact particles of 28 nm, and a core diameter of 18 nm. For simplification of the calculations, the particles were assumed to be spherical and to contain a smooth surface as well as a smooth border between the capsid protein shell and the RNA core. The values for pentamer, cluster, and protomer are $1/20$, $1/32$, and $1/60$ times those of the shell. Values for the individual capsid proteins are obtained by assuming a regular cone-shaped penetration from core to capsid surface for each of the capsid proteins.

For electron microscopy, virions are usually thoroughly dehydrated. From the diffusion coefficient, a particle diameter of 30 nm is calculated for the hydrated virus particle, which correlates well also with measurements of virion density (see Table 14). Applying the measured density of poliovirus in CsCl gradients of 1.34 g/cm^3 to our 28 nm particle yields a mass of 15.4 x 10^{-18}g per particle. 14.2 x 10^{-18}g of these are protein and RNA salt (9.7 + 4.5), the rest must be 1.2 x 10^{-18}g H_2O corresponding to 37,000 molecules of water. Applying the same calculations to a particle of 30 nm diameter and taking the measured density of poliovirus in Cs_2SO_4 gradients of 1.32 g/cm^3 as a basis, one obtains a more realistic value of 4.4 × 10^{-18}g H_2O per particle or 150,000 molecules. Taking a density of 1.29 g/cm^3 for the hydrated protein shell then yields a density of 1.44 g/cm^3 for the RNA. This corresponds to a hydration of 0.2 g H_2O/g protein and 0.5g H_2O/g RNA. The thus hydrated RNA salt would occupy a volume of 4900 nm^3 corresponding to a core diameter of 21 nm, leaving a shell width of 4.5 nm from r = 10.5 to r = 15. The protein in the hydrated shell would have a density of

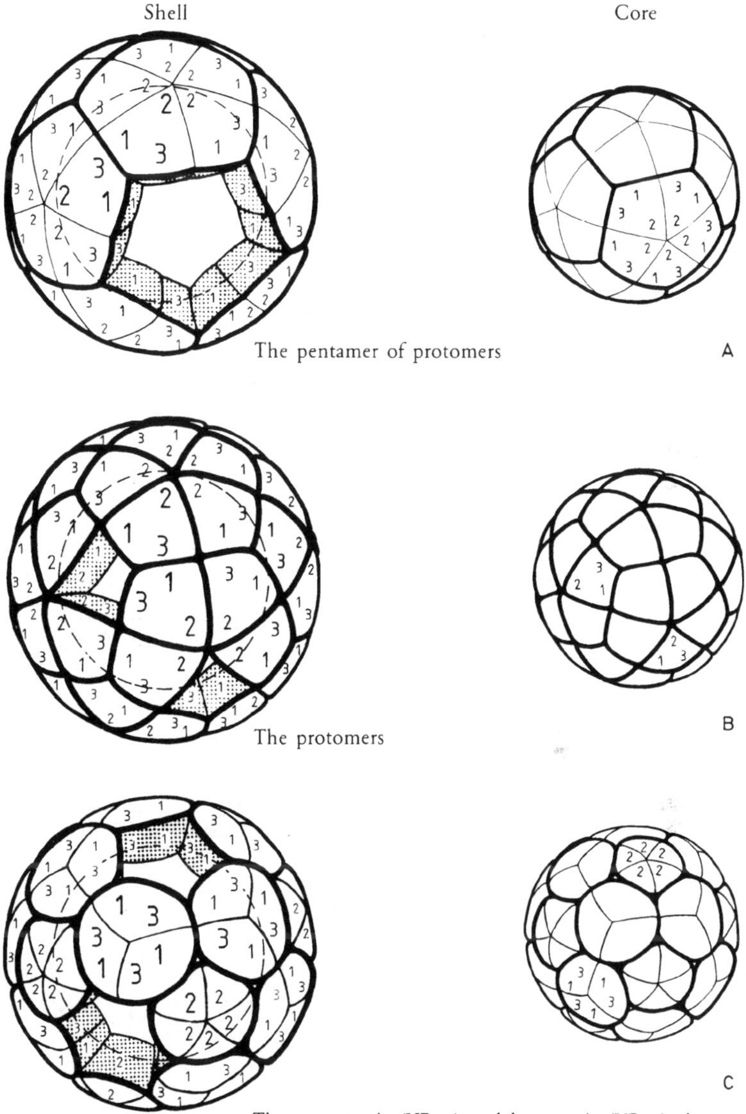

The pentameric (VP_{2-4}) and hexameric (VP_{3-1}) clusters

Fig. 17. The geometries of some subcomponents of a prototype poliovirus

This figure illustrates the relative sizes of some of the subcomponents of a prototype poliovirus. Figures on the left represent the shell of the capsid, figures on the right represent the core of the particle. In each figure, a subcomponent has been removed from the capsid in order to illustrate the thickness of the shell and its relationship to the RNA core: in *A* a pentamer of protomers, in *B* two single protomers, and in *C* a pentameric and a hexameric cluster of capsid proteins. The figures illustrate the relative areas of the capsid proteins which are exposed to the surface of the particle, as well as those available for interaction with adjacent capsid proteins, and for interaction with the RNA. Possible surface projections from the illustrated capsid backbone or regions of capsid proteins which may extend into the interior of the particle for more extensive interaction with RNA are not illustrated in these figures, since nothing is known about these regions for poliovirus

1.29 g/cm^3 corresponding to that measured for empty shells. Approximately 80,000 molecules of water would be near the RNA, corresponding to 10 molecules of water per nucleotide, still a relatively low value for ss RNA molecules. This leaves roughly 1,000 molecules of water per protomer of $VP_{4,2,3,1}$. Most of the latter water molecules are probably located in a hydration shell around the virion. Dehydration of the virion during the preparatory procedure for electron microscopy evidently removes a lot of water yielding the smaller diameters for core and particle listed above.

Table 14. *Molecular weights, masses and densities of some of the poliovirus components*

	Molecular weight kilodaltons	Mass $\times 10^{-18}$ g	Partial specific volume ml/g (g/ml) (dehydrated)	Density CsCl g/ml (hydrated)
Protein shell	5,850	9.7	0.73 (1.37)	1.29
RNA	2,500	4.2		
with Na^+/K^+	+ 200	0.3	0.53 (1.90)	1.89
with CsCl	+ 1,000	1.7	0.43 (2.33)	–
H_2O	2,700	4.4	–	–
Virion	11,200	18.6	0.68 (1.46)	1.34
Dense particles	12,000	20.0	–	1.45

The protein of the poliovirus shell was calculated to have a partial specific volume of 0.73 ml/g (Munyon and Salzman, 1962), corresponding to a density of 1.37 g/ml. This is a relatively common value for proteins, which usually have specific volumes ranging from 0.68–0.72 ml/g. The buoyant density of poliovirus empty capsids in CsCl gradients is 1.29 g/ml. The difference in densities indicates a hydration of the shell of 0.2 g H_2O/g protein (Mapoles, 1980). This is within the range of 0.05 to 0.35 found for other proteins in CsCl gradients (Kuntz and Kauzman, 1974). It is possible that the protein shell of empty capsids is slightly more expanded than that of intact virions, so that the density of the capsid proteins in native virions may even be a little higher than indicated by these calculations.

2. *Structural Principles Borrowed from the Plant Picornaviruses*

Some fascinating features of virion construction have been revealed with the elucidation of the capsid structure of two small icosahedral plant picornaviruses at 2.9 and 2.8 A^o resolutions (Fig. 18 and 19) (Harrison *et al.*, 1978; Abad-Zapatero *et al.* 1980). The tomato bushy stunt virus (TBSV) and the southern bean mosaic virus (SBMV) are constructed from 180 polypeptide chains (each with an approximate molecular weight of 38,000). The proteins within each virus have the same primary structure, yet they are present in three different conformations (named A, B and C). A triangular subunit of A, B, and C corresponds to the fundamental crystallographic unit, 60 of which are arranged equivalently in an icosahedral (Fig. 18A and 19A) lattice. These units are arranged so that pentamers of A are located at the twelve 5-fold vertices and hexamers of B and C at the twenty 3-fold vertices (Fig. 18B and 19B).

The two viruses are considered to be in chemically distinct groups of small spherical plant viruses (Hull, 1977) and are probably less related than the four animal picornavirus genera. Yet they exhibit some remarkable structural similarities: Those portions of the capsid proteins that are used to construct the "backbone" of the capsid (the outer 3–3.5 nm thick protein shell extending from r = 11 nm to r = 14.5 nm) are nearly exact homologous structures in the two virions (see Fig. 20A). The topographical equivalence is similar to that observed for the α and β chains of hemoglobin and is much better than that between the NAD-binding domains of lactate dehydrogenase and glyceraldehyde-3-phosphate dehydrogenase. These

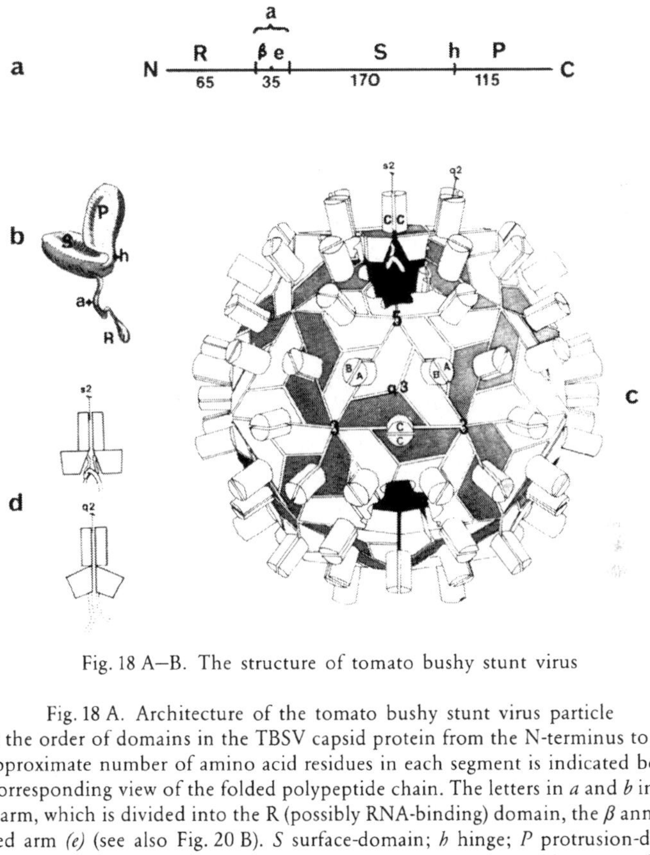

Fig. 18 A–B. The structure of tomato bushy stunt virus

Fig. 18 A. Architecture of the tomato bushy stunt virus particle

a illustrates the order of domains in the TBSV capsid protein from the N-terminus to the C-terminus. The approximate number of amino acid residues in each segment is indicated below the line
b shows a corresponding view of the folded polypeptide chain. The letters in *a* and *b* indicate: *a* the N-terminal arm, which is divided into the R (possibly RNA-binding) domain, the β annulus, and extended arm *(e)* (see also Fig. 20 B). *S* surface-domain; *h* hinge; *P* protrusion-domain
c illustrates the arrangement of subunits in the particle. A, B, and C denote distinct packing environments for the subunit; outer surfaces of C-subunit S domains are shaded. S domains of A subunits pack around fivefold axes; S domains of B and C alternate around threefolds. Examples of such axes are indicated by numerals 5 and 3. The local threefold axis *(q3)* relating S domains of an ABC trimer is nearly parallel to the adjacent strict twofold *(s2)*, across which P domains of C subunits are paired. Trimers present a rather flat surface across the strict dyad and a distinctly sharper dihedral angle (–40 °) across the quasi dyad *(q2)*. A and B P-domains, paired across q2, therefore have a hinge angle with respect to their S domain that differs by –20 ° from the angle on C
d shows the two principal states of the TBSV subunit viewed as dimers about s2 and q2. Subunits in C positions have the interdomain hinge "up" and a cleft between twofold related S domains, into which fold parts of the N-terminal arms. Subunits in quasi-twofold related A and B positions have hinge "down", domains abutting, and a disordered arm. – From Harrison, 1980 [Biophys. J. *10*, 141 (1980)]

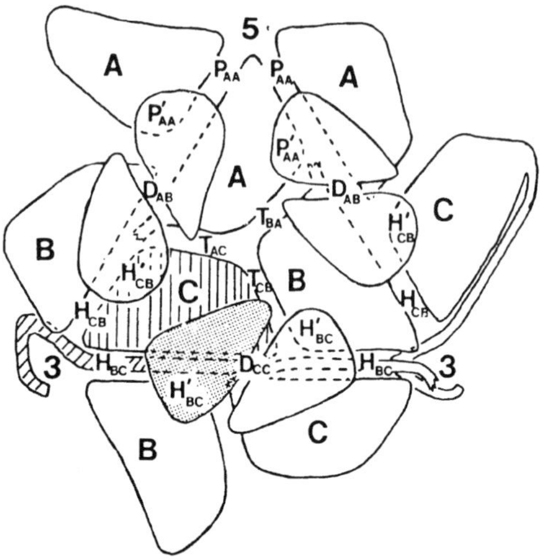

Fig. 18 B. The contacts of a capsid subunit in tomato bushy stunt virus
This figure illustrates the arrangements of subunit contacts and surface protrusions in TBSV. Domains are indicated in outline. Barrel-shaped P domains would protrude outward from the page, making only twofold contacts, while the more extensive S domains make contacts across all adjacent symmetry axes. The contacts are lettered by large letters with subscripts. The large letters indicate the class of contact (*D, T, P, H* for dimer, trimer, pentamer, hexamer, respectively), and the subscripts indicate participating subunits. — From Harrison, 1980 [Biophys. J. *10*, 143 (1980)]

regions are constructed largely from two ß-sheets and are shaped roughly like a triangular prism.

Internal protrusions of polypeptide stretches are also remarkably similar in the two viruses: The terminal arms of the three C subunits of a hexamer B/C cluster fold along grooves of the 2-fold symmetry axis below the borders of the adjacent B and C subunits; the three arms meet at the icosahedral 3-fold symmetry centers (below the center of the B/C hexameric clusters), where they form a β-anulus, a fascinating interdigitating structure (Fig. 19B and 20B). The N terminal arms of the A and B subunits project internally toward the RNA core, but they could not be traced in the crystallographic studies. Data from neutron small angle scattering of TBSV (Chauvin *et al.*, 1978) indicate that there is a region of extensive RNA-protein interaction down to r = 5 nm (see Fig. 16 above). The core of RNA and the protein arms do not exhibit any icosahedral symmetry relations.

Application of the secondary structure prediction technique (Argos *et al.*, 1976) to the corresponding N-terminal coat protein amino acid sequences of these viruses yielded helical predictions for these very basic N-terminal regions (Argos, 1981). The results suggested models of protein-nucleic acid interactions very similar to those proposed for DNA with histones and protamines (Argos, 1981).

In marked contrast to the homologies in the capsid backbones and in the intracapsid bonding mechanisms of TBSV and SBMV, the capsid portions constituting external projections are totally different. TBSV has two similar types of surface projections, each is composed of a dimer of equivalent polypeptide domains from two neighboring proteins. 60 dimeric A/B projections and 30 dimeric C/C

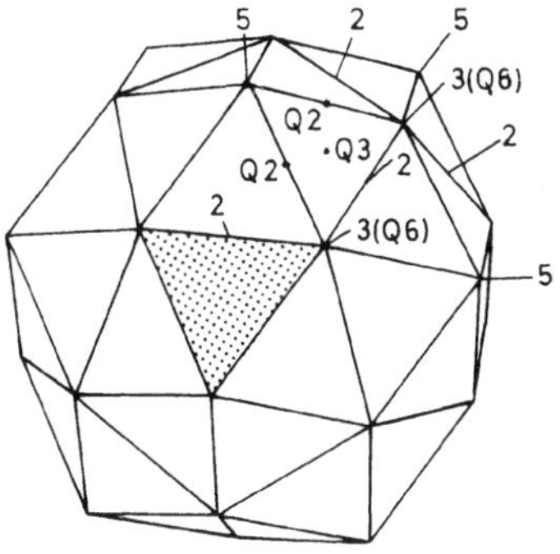

Fig. 19 A–B. The structure of southern bean mosaic virus

Fig. 19 A. The skeleton of the southern bean mosaic virus capsid
The complete viral shell is composed of 60 triangular icosahedral asymmetric structure units (shaded area). Each structure unit is composed of three covalently identical, but conformationally slightly different A, B and C subunits. As for TBSV (Fig. 18 A and B above), each icosahedral structure unit has a near planar surface, giving the entire particle the overall conformation of a pentacis dodecahedron (see Fig. 11). The particle has 32 prominent vertices–12 at the icosahedral 5-fold vertices (5) and 20 at the 3-fold symmetry centers (3, Q6). Q2, Q3 and Q6 represent the quasi-symmetry axes. – Figure redrawn from Abad-Zapatero *et al.*, 1980 [Nature *286*, 34 (1980)]

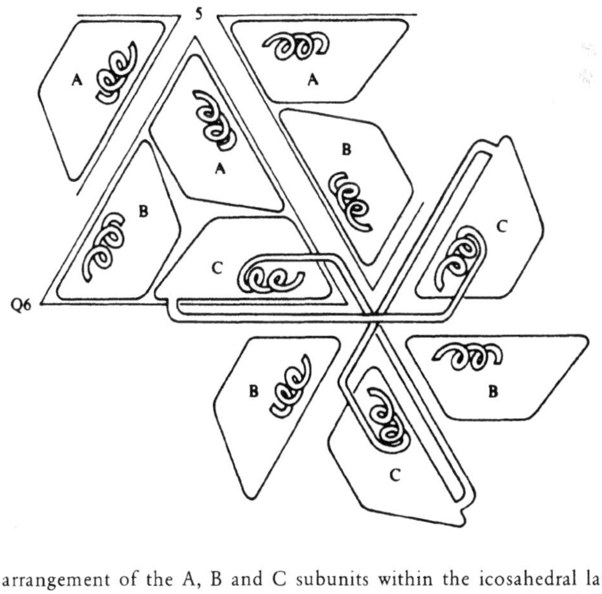

Fig. 19 B. The arrangement of the A, B and C subunits within the icosahedral lattice of SBMV
Arrangement of subunits *A*, *B*, and *C* within one icosahedral asymmetric unit as viewed from the inside of the virus, looking outward. This view permits a representation of the amino ends of the C subunits. – Figure from Abad-Zapatero *et al.*, 1980 [Nature *286*, 34 (1980)]

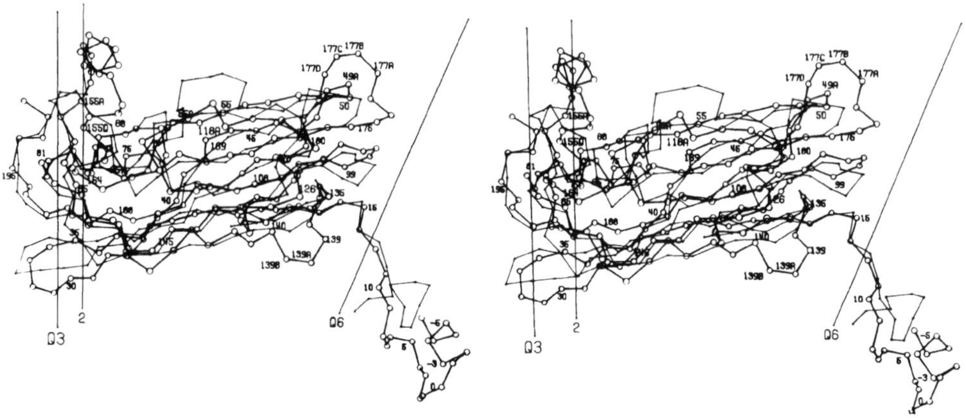

Fig. 20 A–B. Conserved features in the capsid backbones of the small plant picornaviruses

Fig. 20 A. Homologous structures in the capsid backbones of tomato bushy stunt virus and southern bean mosaic virus

Stereo diagram showing the Cα atoms in SBMV (large circles connected with thick bonds) superimposed on the TBSV shell domain (thin bonds). Both viruses are referred to the same icosahedral axial system. The amino acid numbering of SBMV is shown. It relates to a tentative system for TBSV. – Figure from Abad-Zapatero *et al.*, 1980 [Nature *286*, p. 36 (1980)]

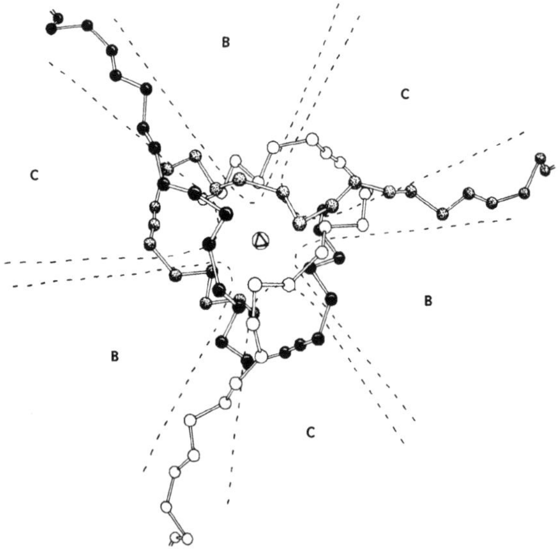

Fig. 20 B. Specific bonding contacts between the icosahedral structure untis at the 3-fold symmetry centers

Configuration of interdigitated arms of TBSV, the "β annulus", viewed down a threefold axis. The balls represent α-carbon positions. The residues here are the sequence "β" (Fig. 18 A) and a part of "e" extending into the intersubunit cleft. – Figure from S. Harrison, 1980 (Biophys. J. *10*, p. 149, 1980) projections are thus formed about local two-fold symmetry centers (see Figs. 18A and 18B). SBMV, on the other hand, has three different types of surface protrusions: a) twelve pentameric protrusions from adjacent A subunits about the 5-fold vertices, b) twenty hexameric protrusions from adjacent B and C subunits about the 3-fold vertices, and c) 60 prominent trimeric projections originating from the adjacent A, B and C subunits in the center of the 60 crystallographic units—here three tangential—helices form a keratin—like structure (see Fig. 21).

Important points to be taken from the plant-virus studies are:

1) The construction of the capsid backbone, the involved bonding mechanism, and aspects of protein-RNA interaction may be remarkably similar between distantly related viruses.

2) In contrast, the surface features, that are determined by different types of external protrusions from the capsid backbone, may be totally different. This is important, though not surprising, since it is the surface of a virion that determines the virion's host range and that must adapt to changes in the host cell receptor and to antibody pressures of the environment.

3) Theoretical considerations suggest that in an icosahedral shell that is constructed from three similarly sized capsid proteins, these capsid proteins will occupy discrete domains in space and that they will all penetrate through the width of the capsid backbones.

3. *Implications for the Capsid Features of Poliovirus: The Concept of a Rigid Capsid Backbone and Variable Surface Protrusions*

It is especially important that these points are kept in mind, when the results of studies on the accessibility of the capsid proteins in intact virions to antibodies, receptors, or chemical labels and crosslinking agents are interpreted in terms of the relative spatial localities of the individual capsid proteins. It is probable that the capsid components investigated with such techniques are mainly the surface protrusions rather than the capsid backbones. It probably does not make much sense to speak of "more external" or "more internal" capsid proteins (at least not for the three major poliovirus capsid proteins VP2/0, VP3 and VP1) as is sometimes done for the picornaviruses. An exception is the small polypeptide VP4 that is cleaved from VP0 only *after* the capsid has been assembled completely and the RNA encapsidated (see Chapter 10). VP4 is so small that it might not penetrate the entire width of the capsid backbone.

The great diversity of results on the capsid proteins from the different picornavirus genera in terms of "main antigenic site constituent", "receptor-binding protein", "surface protein", etc. probably reflects differences in the surface protrusions rather than any major differences in the basic construction of, and relative protein arrangement in, the capsid backbone. Within such a highly ordered capsid structure, it is certainly much easier to envision evolutionary changes in the conformation and constitution of surface protrusions than any fundamental shifts in the relative arrangement of protein chains.

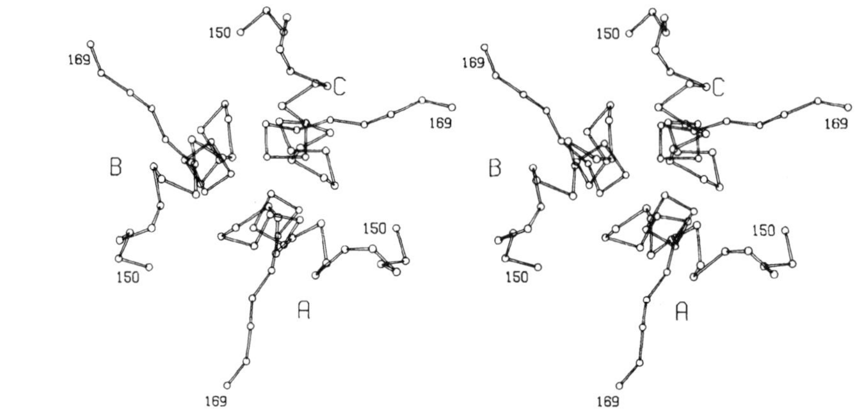
150
169
C
169
B
150
150
A
169
150
169
C
169
B
150
150
A
169

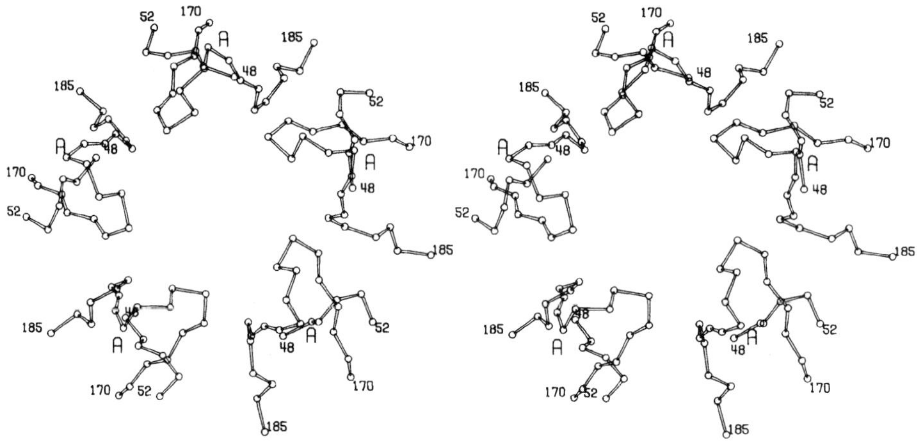
52
170
A
185
48
185
52
A
48
170
A
170
48
52
185
185
48
A
52
A
48
170
52
170
185
52
170
A
185
48
185
52
A
48
170
A
170
48
52
185
185
48
A
52
A
48
170
52
170
185

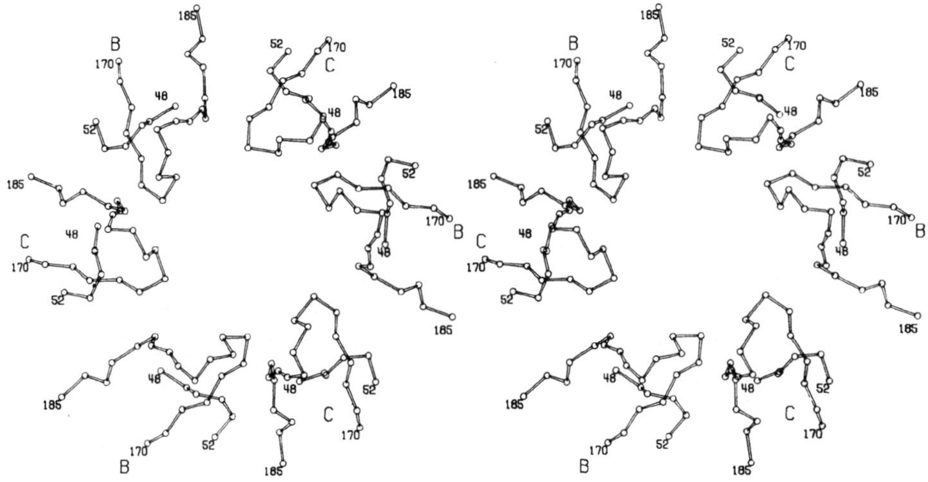
185
B
170
48
52
52
170
C
48
185
185
52
170
B
C
48
170
52
48
185
185
48
52
C
170
170
52
B
185
185
B
170
48
52
52
170
C
48
185
185
52
170
B
C
48
170
52
48
185
185
48
52
C
170
170
52
B
185

B. Experimental Results

For poliovirus (and other picornaviridae), only indirect experimental data are available to date on the destinct structural features of the viral capsid. The great variety of experimental approaches include:

1) Specific chemical modifications of capsid proteins in intact and disrupted virions,

2) chemical crosslinking of capsid proteins to one another and to the RNA,

3) step-wise destabilization and degradation of virions,

4) tracing of virus-antibody interactions.

Two principally different models have been proposed for the relative localization of the capsid proteins in picornaviruses. All data on poliovirus and other enteroviruses are consistent with the model which locates VP_2—VP_4 pentamers at the twelve 5-fold vertices and VP_3—VP_1 hexamers about the twenty 3-fold vertices (Martin and Johnston, 1972; Philipson *et al.*, 1973). A second model, proposed from electron microscopy of FMDV—antibody complexes and crosslinking studies on mengovirus (Hordern *et al.*, 1979) places VP_1 at the icosahedral vertices, VP_2 and VP_3 at the icosahedral facets. This model is more difficult to reconcile with the available results on poliovirus. Figure 22 shows the original illustrations of these models.

The final word on the fine details of the poliovirion architecture should be possible within the next 5 years, when detailed analyses of the elucidated amino acid sequences of the poliovirus capsid proteins and data from detailed X-ray diffraction analysis on the structure of poliovirus in crystals of virus become available. Such studies have recently been initiated and it should be possible to elucidate the poliovirus structure at 2.5 A^{o} resolution (Hogle, 1982, 1984). This information will provide a far sounder basis than the data discussed in the following sections for deducing the conformations of the individual capsid proteins and their positions relative to each other and to the external and internal faces.

As a supplement to the crystallographic studies, however, data from other investigations, like those described in the following sections, will be of value; since they also deal with dynamic features of poliovirus structure, an aspect that is not expressed within viral crystals.

1. Specific Chemical Modifications of Capsid Proteins

There are a number of possibilities to label exposed amino acid residues of proteins. For example, radioactive iodine (^{125}I) can be covalently bound to

---

Fig. 21. The surface protrusions of southern bean mosaic virus

The figure illustrates the surface protrusions of SBMV in views of the viral exterior surface

Top: Stereo diagram showing the protrusion around a quasi-3-fold axis produced by subunits *A, B,* and *C.* The tangential α-helix occurs only in SBMV. The radial helix forms a keratin-like structure about the quasi-3-fold axis

Center: Stereo diagram showing protrusion around an icosahedral 5-fold axis produced by adjacent A subunits

Bottom: Stereo diagram showing protrusion around a quasi-6-fold axis produced by three C subunits separated by three B subunits. — From Abad-Zapatero *et al.*, 1980 [Nature *286*, 37 (1980)]

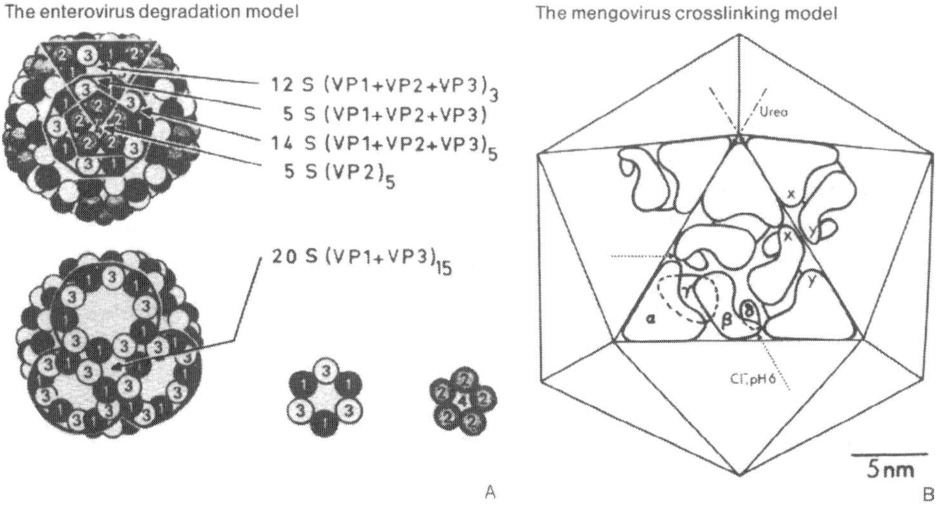

Fig. 22. Two alternative models of picornavirus capsids

A. The enterovirus-degradation model

The figure shows the first model for the relative localization of the picornavirus capsid proteins. This model is based on data from the dissociation pattern of enteroviruses to a variety of conditions (see text): release of VP_4 followed by release of the entire VP_2 component, leaving a skeleton capsid composed of VP_{3-1}. The figure on the top illustrates the cleavage planes of dissociation products from different picornaviruses. The 14 S pentamer and 5 S protomer were observed as dissociation products of Mengovirus (Dunker and Rueckert, 1971). The 12 S trimer was suggested as a breakdown product of FMDV (Talbot and Brown, 1972), more recent evidence, however, indicates that this is a pentamer corresponding to the 14 S product of mengovirus dissociation (Morrell and Brown, 1981)

The bottom figure shows a model of the skeleton capsid obtained after alkaline dissociation of poliovirus and bovine enterovirus (Katagiri *et al.*, 1971, Martin and Johnston, 1971). The proposed structure of the 20 S VP_{3-1} ring-like dissociation product of coxsackie virus is outlined. A schematic view of the proposed VP_{3-1} hexamers and VP_{2-4} pentamers are also presented (see also Fig. 17). – Figure from Philipson *et al.*, 1973 [Virology *54,* 78 (1973)]

B. The mengovirus crosslinking model

This model is based on data from an extensive study of the crosslinking of the capsid proteins in mengovirus. The shapes of individual polypeptides are drawn arbitrarily. This is a surface view. In contrast to the enterovirus model, VP_1 (= α) is located at the icosahedral vertices. $\beta = VP_2$, $\gamma = VP_3$. Arrows indicate the noncovalent interactions which usually are disrupted by chloride ions at pH6 in mengovirus and lead to the release of α-β-γ pentamers corresponding to the 14 S structure outlined in A. The interpretation of the crosslinking data is based on the assumption that virions crosslinked with DMA or DMS to crosslink individual capsid proteins dissociate in a fashion similar to untreated virions at pH6 (see Fig. 23). – Figure from Hordern *et al.*, 1979 [Virology *97,* 138 (1979)]

the hydroxyl group of tyrosine residues by catalysis with lactoperoxidase, iodoacetamide reacts with free SH-groups of cysteine residues, and labeled acetic anhydride can be covalently bound to free amino residues (*i.e.*, the amino group of lysine residues and the N-terminal amino group) (Bolton and Hunter, 1973; Montelaro and Rueckert, 1975; Lonberg-Holm and Butterworth, 1976). These methods were employed on intact and disrupted polioviruses to determine the accessibility of viral capsid proteins to such chemical labeling (Beneke *et al.*, 1977;

Wetz and Habermehl, 1979). The results are summarized in Table 15. It was found that VP_1 was readily labeled in several ways in both intact and disrupted virions, VP_2 and VP_3 were labeled less but still significantly in intact virions as compared to disrupted virions, and VP_4 was labeled only in disrupted virions. These observations were taken to suggest that VP_1 occupies most, VP_2 and VP_3 some, and VP_4 no part of the external capsid surface. Still it cannot be ruled out that some exposed parts of the capsid proteins may not be detectable by these labeling methods (*i.e.*, stretches containing no tyrosine residues or free amino residues could be exposed on the surface). The three amino terminal residues of VP_1–VP_3 of intact poliovirus were also labeled by dansylation of intact poliovirus (Burrel and Cooper, 1973).

Biological labelling of capsid proteins in intact virions can be achieved by treatment with proteases or specific antibodies. Most picornaviruses, including poliovirus, are resistant to proteases in their native configuration. Mild treatment of intact FMDV virions with trypsin cleaves VP_1 into two polypeptides (as determined by gel electrophoresis) (Wild and Brown, 1967), and thereby greatly decreases the ability of the virus to attach to and infect host cells (approximately 1,000-fold loss of infectivity), and also decreases the ability to elicit the production of neutralizing antibodies in guinea pigs (Rowlands *et al.*, 1971a; Cavanagh *et al.*, 1977). These and other studies on FMDV strongly suggest that at least the largest of the capsid proteins VP_1, is located on and occupies much of the external surface or surface protrusions of the native FMDV virion and plays an important role in binding to the host cell receptor, as well as in providing a major antigenic determinant. In mengovirus, antibody binding sites exist for both specific anti-α (VP_1) and anti-β (VP_2) antibodies of the virions, however, only anti-α-antibodies could block attachment of virions to cells (Lund *et al.*, 1977). For poliovirus, anti-VP_1, anti-VP_2, and anti-VP_3 antibodies have been found to interact with intact virion under certain conditions (see Section VIII). Care should be taken in drawing conclusions, or in comparing different picornaviruses, on the basis of such antibody-labelling studies. The main antigenic determinants on native virions are probably the most variable sites on the capsid surface, *i.e.*, the ones that vary the most between different viruses. The basic conclusion that can be drawn from the studies is that the three major capsid proteins VP_2, VP_3 and VP_1 are all exposed on the outer capsid surface, and that each picornavirus has different specific protein-regions that define the antigenic determinants.

There has been some controversy about the location of the smallest capsid protein, VP_4. While studies on the accessibility of capsid proteins to chemical labeling (see above) and crosslinking studies (see below) suggest that VP_4 polypeptides occupy internal locations in the intact virion, a detailed sereological study on VP_4 in polio virions suggested that VP_4 is located on the external surface (Breindl, 1971b) (for a discussion see Section VIII). Furthermore, since most of the cleavage of VP_0 to VP_2 and VP_4 occurs on already completed capsid structures, accessibility of the cleavage site to the responsible protease, which must act from the outside on the surface of the virion, requires that at least this portion of VP_4 near the cleavage site must lie on the capsid surface at the provirion stage. It would thus be of interest to examine whether VP_4 can be chemically

Table 15. *Chemical labelling of capsid proteins in intact virions and in empty capsids*

A. *Intact particles* *Labelling*/modification	Intact particles	VP_4	VP_2	VP_3	VP_1	References
Lysine and free amino groups		5*	5	10	15	
SIP, AA, Bolton-Hunter	+++	–	+	++	+++	a, b
DMA, DMS	++	–	–	+++	+++	a
Dansylation	++	–	+	+	+	c
Cysteine, free SH groups, and S-S bonds		0	10	5	12	
Na_2S_4, cystine	++	–	nd	nd	nd	d, e
IA	–	–	–	–	–	a, b
Tyrosine or free OH groups		4	10	8	15	
Lactoperoxidase iodination	++	–	(+)	+	+++	a, b
Proteinkinase phosphorylation	++	–	+++	(+)	+	f
Antigenic determinants	+	+	+	+	++	g
B. *Empty capsids*	Empty capsids	–	VP_0/VP_2	VP_3	VP_1	References
Lysine	++	–	++	+	++	b
Cysteine						
IA	(+)	–	nd	nd	nd	h
Tyrosine						
Lactoperoxidase	++	–	+++	+	+++	b
Proteinkinase	++	–	++	+++	++	f
Antigenic determinants	+	–	+	++	++	g

* Indicates the number of the specific amino acids per capsid protein.

Abbreviations: SIP ^3H succinimidyl *proprionate;* NEM N-ethyl maleimide; AA acetic anhydride; DMA dimethyl adipimidate; DMS dimethyl suberimidate; IA iodoacetamide.

References: a Wetz and Habermehl, 1979; Beneke *et al.*, 1977; b Lonberg-Holm and Butterworth, 1976; c Wouters and Van der Kerckhove, 1976; d Pons, 1964; e Pohjanpelto, 1958; f Schärli and Koch, 1984; g see section VIII; h Dernick, 1981.

labeled in provirions (*i.e.*, the VP4 containing part of VP0) and in the natural top component. If so, this would suggest that the conformational shift accompanying (following) the cleavage of VP0 to VP2–VP4 results in "burying" of VP4. The "burial" of VP4 must be readily reversible as revealed by the ease of the release of VP4 from the virion by various conditions, including binding to the host cell receptors (see below). In the poliovirus family relative bovine enterovirus, an antibody-mediated exposure of the hidden VP4 to the surface has been reported (Carthew and Martin, 1974; Carthew, 1976). In native bovine enterovirions, the three major capsid proteins VP1, VP2, VP3, but not VP4 were accessible to labeling by ^{125}I or pyridoxal phosphate-sodiumborohydride. After neutralization with homologous antisera, VP4 became accessible to labeling with ^{125}I.

2. *Chemical Crosslinking of Capsid Proteins*

Further interesting information on the spatial relationship of the capsid proteins involved in intra- and interprotomer binding has recently been obtained in studies on the chemically induced crosslinking of the capsid proteins (Scraba, 1979; Hordern *et al.*, 1979; Wetz and Habermehl, 1979). Neighboring proteins are linked with bifunctional crosslinking reagents, subsequently the polypeptide composition of the induced complexes is determined by gel electrophoresis procedures (Armstrong *et al.*, 1972; Peters and Richards, 1977). Crosslinking occurs via free amino groups by dimethyl suberimidate DMS (maximum spanning distance 1.1nm) or dimethyl adipimate DMA (maximum distance 0.8nm). Treatment of poliovirus with these agents leads to the formation of one new protein complex corresponding to the sum of VP1 and VP3 (Wetz and Habermehl, 1979). This indicates a direct neighborhood of VP1 and VP3 in the virus capsid. The finding is not surprising and the importance of a close VP1–VP3 neighborhood is underlined by the observation that stepwise degradation of poliovirus results in the release of VP4 and VP2 leaving a stable matrix of VP1 and VP3 (Katagiri *et al.*, 1971, see below).

A more detailed study of chemically-induced crosslinks of viral capsid proteins has been reported for mengovirus (a cardiovirus) (Scraba, 1979; Hordern *et al.*, 1979) (Fig. 23B). The picornavirus model derived from these studies differs in some fundamental aspects from the poliovirus model that we find to best fit the available data. The authors locate the (VP1) polypeptides in pentamers at the twelve icosahedral 5-vertices, since it is the only protein that was obtained in higher oligomers: $\alpha3$ and very small amounts of $\alpha4$ (but no $\alpha5$). The following bonds were proposed to be involved: 1) intraprotomer bonding arrangement seems to be α–β–γ (VP1–VP2–VP3) which is to be expected since these are the bonds which are theoretically required for holding together the protomer 2) bonding between pentamers involves α–β (VP1–VP2) links. The associated bonding-patterns appear somewhat awkward (Fig. 23B) (Dunker 1979). The conclusions are based on the finding that virions that had been incubated with the crosslinking reagents DMA or DMS, and then treated with 0.1–0.2 M NaCl at pH 6, resembled native virions in that they were dissociable into substructures, whereas DSP-treated virions were not (Mak *et al.*, 1974; Scraba, 1979).

Crosslinking reagent	Maximum linkage distance (nm)	Molecular weight of complex[a]	Probable composition	Relative occurrence	Composition confirmed[b]
Formaldehyde	0.3	54,000	$\beta\gamma$	++	
Dimethyladipimate (DMA)	0.8	58,000	$\alpha\gamma$	+++	
		65,000	α_2	+++	
Dimethylsuberimidate (DMS)	1.1	54,000	$\beta\gamma$	++	
		58,000	$\alpha\gamma$	++	
		65,000	α_2	++	Yes
		86,000	$\alpha\beta\gamma$	++	
		95,000	α_3	++	Yes
Dithiobis(succinimidyl propionate) (DSP)	1.5	54,000	$\beta\gamma$	++	Yes
		62,000	$\alpha\beta$	+	Yes
		86,000	$\alpha\beta\gamma$	++	Yes
		97,000	α_3	+++	Yes
		120,000	α_4	(+)	Yes

[a] Molecular weights estimated from migration during electrophoresis in SDS-polyacrylamide gels.
[b] Composition confirmed by chemical "reversal" of the crosslinks followed by electrophoresis in the second dimension.

Fig. 23 A. Chemical crosslinking of polypeptides in the mengovirion. – Data from Hordern *et al.*, 1979 [Virology *97*, 131 (1979)]

Fig. 23 B. Alternative interpretations of the crosslinking data

Ad Fig. 23 B. The model originally proposed by Hordern *et al.* on the basis of the above data is redrawn onto a spherical particle *(1)*. The interpretation is based on the assumption that DMA or DMS crosslinked virions dissociate as untreated virions into pentamers of α-β-γ *(3)*. The composition of the dissociation products was not confirmed by biochemical means. In electron micrographs, they appeared similar, albeit slightly smaller than the disruption products from untreated virions. An equally plausible interpretation *(2)* arises if one assumes the dissociation products of DMA or DMS treated virions to be trimers *(4)*. A summary of the interpretation of the crosslinks formed by bifunctional reagents would then be as follows:

DMA:
DMS:
1) links within protomers: β-γ, γ-α, β-γ-α
2) links within trimeric α-γ clusters: α_2, α_3, α-γ
3) links within pentameric β clusters: –

None of the DMA or DMS crosslinks (short bonds in *[3]* and *[4]*) can maintain capsid integrity as there are no crosslinks formed across the 5-fold axes. Degradation would result mainly in trimers.

DSP:
4) links between adjacent α-γ trimers: α_4
5) links between adjacent α-γ trimers and β-pentamers: α-β

DSP spans the greatest distance (long bonds in *[3]* and *[4]*) and is capable of crosslinking across larger "grooves" between adjacent clusters. Notice that the α-β and α_4 crosslinks are only formed after DSP treatment. All other DSP induced crosslinks can also be formed by DMA and DMS treatment. The alternative interpretation better explains why α_4 and α_5 are not seen after DMA or DMS treatment (A). If α was indeed located in pentameric clusters about the icosahedral apices, α_4 and α_5 bonds should be observed between the equally situated α s after DMA and DMS treatment. The individual subunits and the associated bonding pattern are also smoother in the enterovirus-like interpretation *(6)* than in the mengovirus interpretation *(5)*. Results from X-ray crystallography of picornaviruses should decide between the two models

1 mengovirus interpretation

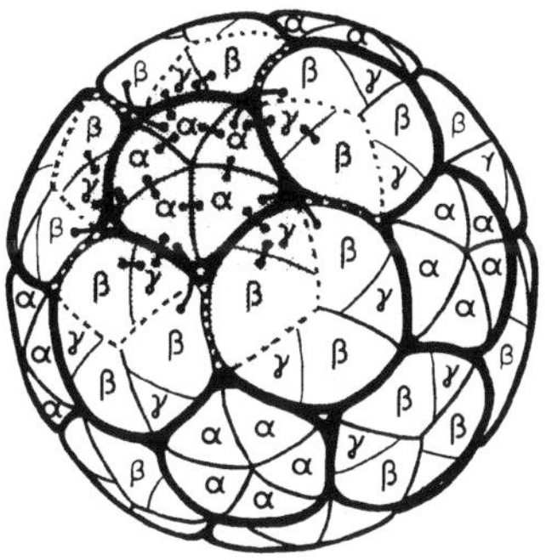

α at 5-fold vertices

2 enterovirus interpretation

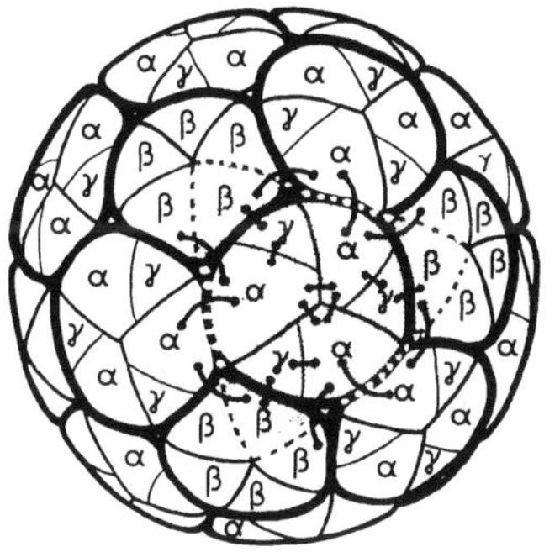

β at 5-fold vertices

3

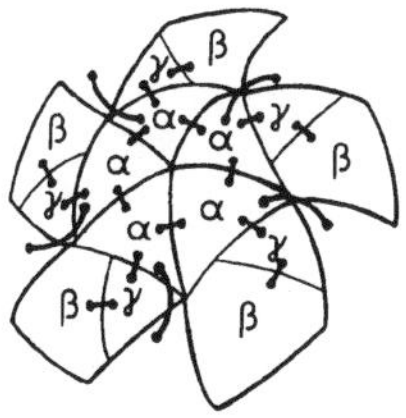

pentameric dissociation product

4

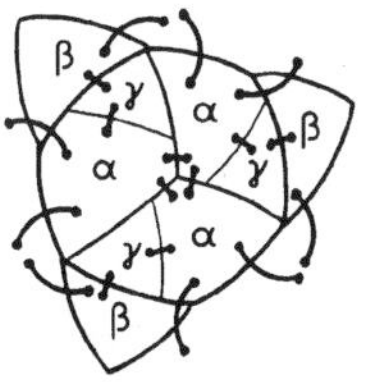

trimeric dissociation product

5

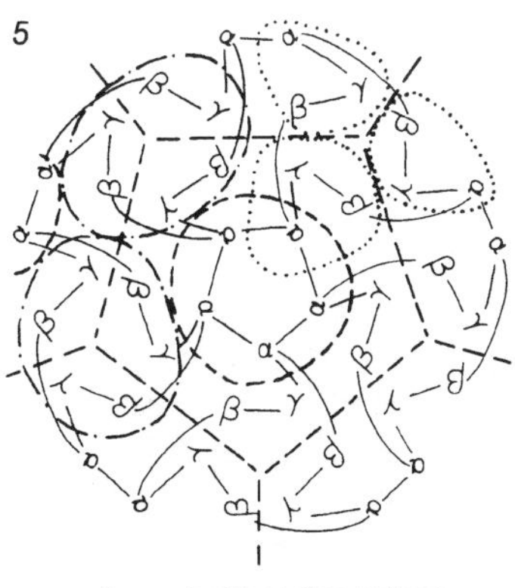

"complex" bonding pattern

6

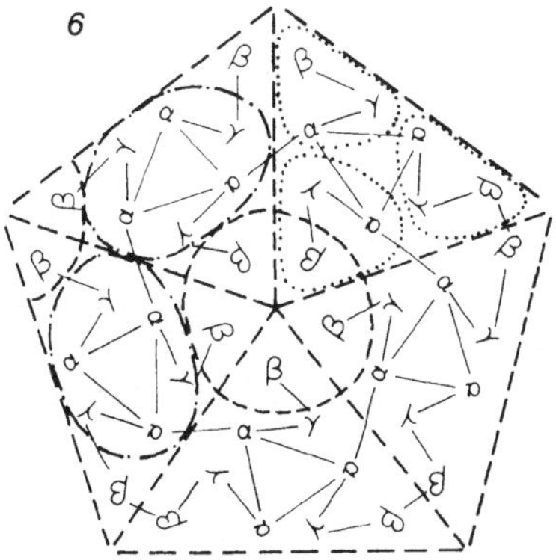

"smooth" bonding pattern

The substructures of dissociated untreated native virion are known to be composed of α, β, γ pentamers. Since the appearance of the substructures from dissociated DMA or DMS treated virions in the electron microscope had a similar appearance, it was concluded that the DMA/DMS substructures also represent α, β, γ pentamers and that, therefore, the DMA- or DMS-induced units are located within such pentamers. Since DSP induced crosslinks could prevent dissociation, it was predicted that the DSP-specific links are located between adjacent pentamers. Unfortunately, a physico-chemical analysis of the actual composition of the DMA/DMS substructures was never presented. An alternative interpretation of the DMA/DMS substructure compositions, namely trimers of α, β, γ leads to quite different, and in our opinion more plausible, structural implications, *i.e.*, the location of β pentamers about the 5-fold apices and hexamers of (VP3) and (VP1) about the 3-fold symmetry centers. The associated bonding pattern with such a structure is smoother (Dunker, 1979) (Fig. 23B). The substructures of DMA and DMS-treated, dissociated virions indeed appear slightly smaller than those of disrupted virions (Scraba 1979). Figure 23B illustrates the alternative enterovirus-like model for mengovirus.

The presented argumentation is a proposal to reconcile the structure of these different picornaviruses. Of course there is no *a priori* necessity for this. It is conceivable that mengovirus has a principally different localization of its capsid proteins than enteroviruses. For example, mengovirus does exhibit a fundamentally different susceptibility to acidic pH as enteroviruses (see Chapter 2), being degraded in a manner that has never been observed for enteroviruses under any degrading conditions. If the localization of mengovirus capsid proteins should indeed be fundamentally different from that of the polio capsid proteins, the interesting question arises, whether specific functions of the capsid—such as host-cell-receptor recognition—are correlated with geometric points of the virus structure (*i.e.*, 5-fold symmetry vertices or 3-fold symmetry vertices) or with specific viral proteins (*i.e.*, α and VP1). Based on the data from the plant picornaviruses described above, however, it seems unlikely that the localities of the capsid proteins shift significantly in the capsid backbone during evolution. It should be pointed out that the poliovirus model has never been proven directly so far, and that with some effort the available data on poliovirus and other enteroviruses can also be made to fit the mengovirus model. However, even after a stimulating dispute we feel more comfortable with the enterovirus model, while other scientists still prefer the mengovirus model.

3. *UV Irradiation of Poliovirus*

UV irradiation of polioviruses rapidly inactivates their infectivity (Katagiri *et al.*, 1967; Wetz and Habermehl, 1981; De Sena and Jarvis, 1982). Three types of crosslinks are induced by UV irradiation: intra chain RNA-RNA, RNA-protein and intermolecular protein-protein crosslinks. The rapid inactivation by UV irradiation is evidently due to alteration of the RNA. Minor conformational transitions of the viral capsid occur at lower UV levels, rendering the virion sensitive to RNase and leading to a buoyant density in CsCl of 1.45 g/ml, a value typical of dense particles. Major conformational changes of the virus capsid are observed after longer periods of irradiation with higher doses. After eight minutes, 50 % of the particles were converted to particles sedimenting at 104S, and after 16 minutes of irradiation the virus was completely converted to 104S. The 104S particles contain the four structural proteins and RNA, and the ratio of RNA to protein is unaltered. The reduced sedimentation coefficient implies a swollen particle. (A larger particle sediments slower than a small particle of identical composition.) The viral RNA, however, becomes sensitive to RNase. The 104S particles reveal a reduced stability, being readily converted into artifical 76S empty capsids deficient of VP4 and viral RNA. Unfortunately, it has not been determined whether these conformational alterations are accompanied by an alteration of antigenicity. A "swelling" of the viral capsid has also been observed as an early event in the stepwise degradation process induced by certain preparative procedures used for electron microscopy (see below) (Boublik and Drzeniek, 1977).

UV-induced intra chain RNA-RNA crosslinking is very extensive, indicating a close packaging of the RNA inside the capsid (Wetz and Habermehl, 1981, 1982). It is probable that the "swelling" of the virus particle upon heating (135S) (see below) or UV irradiation (104S) is a result of a breaking of bonds between the RNA and proteins and also between proteins themselves. A breaking of the ionic RNA protein interactions would permit the binding of a greater number of Cs ions during equilibrium density centrifugation (*i.e.*, maximum of 7,500 to neutralize all the RNA phosphate groups), explaining the increased buoyant density of 1.45 g/ml of UV irradiated virus. UV-induced protein—RNA crosslinks are formed predominantly to VP4, less extensively to VP1 and VP2, and only very little to VP3. The components of the products of intra-protein crosslinkages have not been identified.

VI. The Dissociation of Poliovirus

In the past 30 years, polioviruses have been subjected to a variety of chemical and physical treatments in order to disrupt their physical integrities and to inactivate their infectivity. Many reagents and conditions have been found to destabilize and to disrupt the virus particles including exposure to elevated temperatures, UV irradiation, exposure to extremes of pH, high salt, ionic and nonionic detergents, and a variety of chemical compounds, as well as staining procedures during preparation for electron microscopy and interaction with factors of the host cell membrane (see also Chapter 7). Other substances, in turn, have been discoverd that protect the virus particle against degrading conditions. Stabilizers include: high ionic strength, acidic pH, antibodies, hydrophobic compounds such as arildone or host cell membrane components, divalent cations (Mg^{++}), and sulfhydryl reducing substances. Figure 24 illustrates the effects of different degrading and stabilizing reagents and conditions.

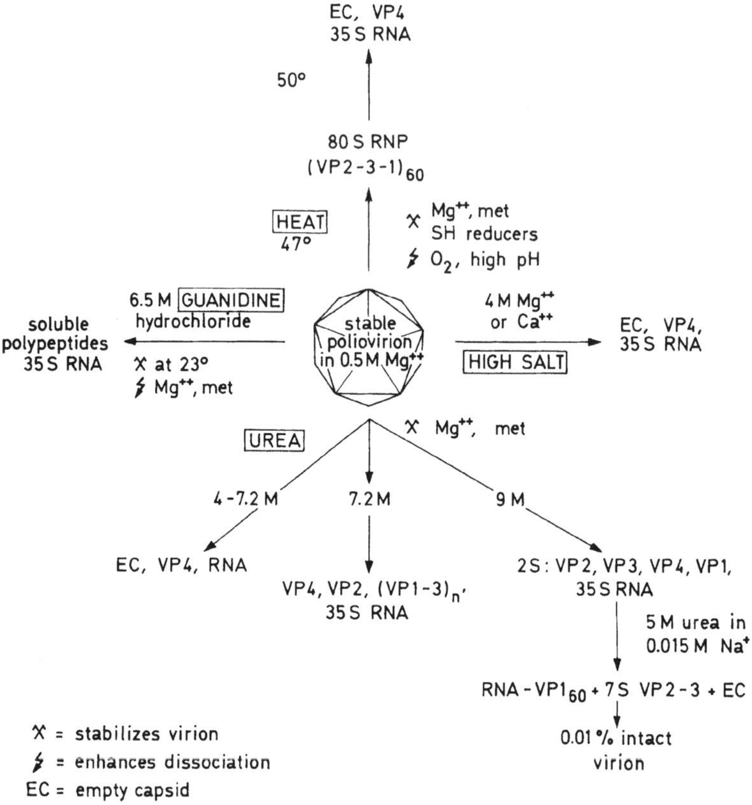

Fig. 24. Schematic overview of poliovirus degradation. References see text

A remarkable observation was the dissociation of the virus in a rather stereotype, stepwise fashion, irrespective of the particular degrading conditions employed. The basic pattern of virus dissociation is shown in Table 16. Typically, the virus particle first swells and snaps into a metastable alternate conformation. The bonds holding VP4 within the capsid are labilized significantly in the process and the surface of the virus particle becomes more negatively charged. Mg^{++} ions stabilize the virus in the native conformation suggesting that a release of Mg^{++} ions may be involved in the conversion of virions into the expanded more negatively charged state. VP4 is lost from the expanded virus particle unless the capsid is stabilized in the expanded state by antibody, arildone, or acid pH. If VP4 is lost, the step is irreversible. The release of VP4 is accompanied by a substantial alteration of capsid conformation, as evidenced by the exposure of a new set of antigenic determinants (C) and by the loss of attachment capacity. Either RNA or VP2 or both may be released in the following steps of dissociation, forming an empty capsid containing VP1,2,3 or a capsid skeleton composed only of VP1 and VP3. (Maizel *et al.*, 1967; Rueckert *et al.*, 1969; Katagiri *et al.*, 1971; Breindl, 1971b; Martin and Johnston, 1972; Philipson *et al.*, 1973; Talbot *et al.*, 1973). The surface

Table 16. *Typical steps in poliovirus dissociation*

Probable effects	Detected by
1. Destablilization of only a few bonds, "breathing" of capsid proteins	Exposition of new antigenic sites, shoulder in curves of kinetics of inactivation
2 a. Conversion to the alternate conformational state with low isoelectric point, loss of Mg^{++}	Isoelectric point of pH 4.5
b. Swelling of particles by an influx of water through holes in the capsid	Decreased sedimentation coefficient (150–140S)
3 a. Permeability to ions and dyes	Ethidium bromide, CsCl
b. Loss of the small capsid protein VP_4	PAGE, change to C-antigenicity, decreased sedimentation (135–90S), transition in fluorescence of tryptophan
4. Release of RNA	sucrose density gradients (80 S), RNase, double labeling
5. Loss of VP_2	PAGE
6. Breakdown of VP_1/VP_3 skeleton backbone	Sucrose density gradients

The selective consecutive release of individual capsid polypeptides suggests that the individual molecules occupy discrete localities within the capsid lattice.

charge of the capsid remains negative after the loss of VP4 and RNA. The C antigenic determinants are retained even in the capsid skeleton.

Interpretation of these phenomena are difficult and care is warranted. Nevertheless, the behavior of the virus particle to different conditions may provide some clues on the forces that hold the capsid together and on the relative distribution of the capsid components.

The individual peptides in the poliovirus capsid evidently interact with each other by hydrogen, ionic and hydrophobic bonds. S-S bridges apparently are not involved in intra-capsid protein bonds in polioviruses (Scharff *et al.*, 1964; Dernick, 1981). Ionic interactions are usually disrupted by high ionic strength, whereas hydrogen and hydrophobic bonds are stabilized under these conditions. Ionic and hydrogen bonds are broken by extremes of pH. Hydrophobic interactions and hydrogen bonds are disrupted by denaturing agents such as urea and guanidine. Lipophilic compounds can break (DPH) or stabilize (arildone) hydrophobic bonds.

From the stepwise character of the assembly process (see Chapter 10) and from a detailed study on the stepwise dissociation of other picornaviruses (mengo and cardioviruses) (Rueckert *et al.*, 1969; Dunker and Rueckert, 1971), it was proposed that two main bonding domains are responsible for holding together the capsid of these picornaviruses. It was suggested that the 5 protomers within a pentamer in mengovirus are held together by hydrophobic interactions disruptable with urea (Scraba, 1979), and that the 12 pentamers within a capsid are held together mainly by electrostatic interactions (disrupted in mengovirus by Cl^- ions at pH 6). Poliovirus—like all enteroviruses—is stable at acidic pH even down to pH 2. The polio capsid seems to be held together predominantly by hydrogen bonds and may be stabilized by ionic and hydrophobic bonds. Polioviruses and other picornaviruses are stabilized against heat inactivation in buffers of high ionic strength (Koch, 1960a; Wallis and Melnick, 1961), indicating that ionic interaction between capsid peptides is of secondary importance and might not be essential.

Table 17. *The characteristic dissociation products of disrupted poliovirions*

Sedimentation coefficient		Probable composition	Antigenicity
2 S	–	individual capsid proteins VP_4, VP_2, VP_3, VP_1 = (final dissociation products)	S
4 S–5 S	–	$(VP_2)_{2-5}$ or $(VP_2–VP_4)_{2-5}$ = icosahedral vertices (only for coxsackievirus)	D
13–14 S	–	$(VP_4–VP_2–VP_3–VP_1)_5$ (only for cardio- and aphto-viruses; never detected as dissociation product of polioviruses)	
15–20 S	–	$(VP_3–VP_1)$ aggregates = hexamer/icosahedron-facet / decamer/protomer ring	C
35 S	–	vRNA-VPg	
60 S	–	$(VP_3–VP_1)_{50-60}$ = capsid backbone	C
73–80 S	–	$(VP_2–VP_3–VP_1)_{50-60}$ = empty capsids	C
80–90 S	–	$(VP_2–VP_3–VP_1)_{50-60}$ + RNA = ribonucleoprotein particle	C
104 S	–	$(VP_4–VP_2–VP_3–VP_1)$ + RNA = UV induced swollen particle	C
135 S	–	$(VP_2–VP_3–VP_1)_{60}$ + RNA = A particle	C
145 S	–	$(VP_4–VP_2–VP_3–VP_1)_{60}$ + RNA = swollen virions (pI 4.5)	D
155 S	–	$(VP_4–VP_2–VP_3–VP_1)_{60}$ + RNA = native virions (pI 7.0)	D

Numerous structures of intermediate sedimentation properties are also reported in the literature, these generally represent products of degradation or aggregation with varying compositions.

In the following we will summarize the disruption procedures (for review see also Kaper, 1975; Rueckert, 1976). Some characteristic properties of individual breakdown products are listed in Table 17. An important point to be made from all these studies is that the poliovirus capsid should not be regarded as an absolutely rigid structure. Rather the capsid reveals an inherent potential for a variety of conformational modifications or structural rearrangements and disarrangements.

A. Breakdown During Preparatory Procedures for Electron Microscopy

Various conditions of specimen deposition and staining procedures for electron microscopy cause a stepwise breakdown of the poliovirus capsid. At the same time these procedures visualize the dissociation products (Boublik and Drzeniek, 1977). The dissociation of poliovirus here is initiated by physical shear or tensions on the capsid. Figure 25 shows the consecutive steps in the dissociation process of the viral capsid under these conditions: expansion (swelling) of the virus particle (about 15–20% increase in diameter); appearance of 1, 2 or more notches in the

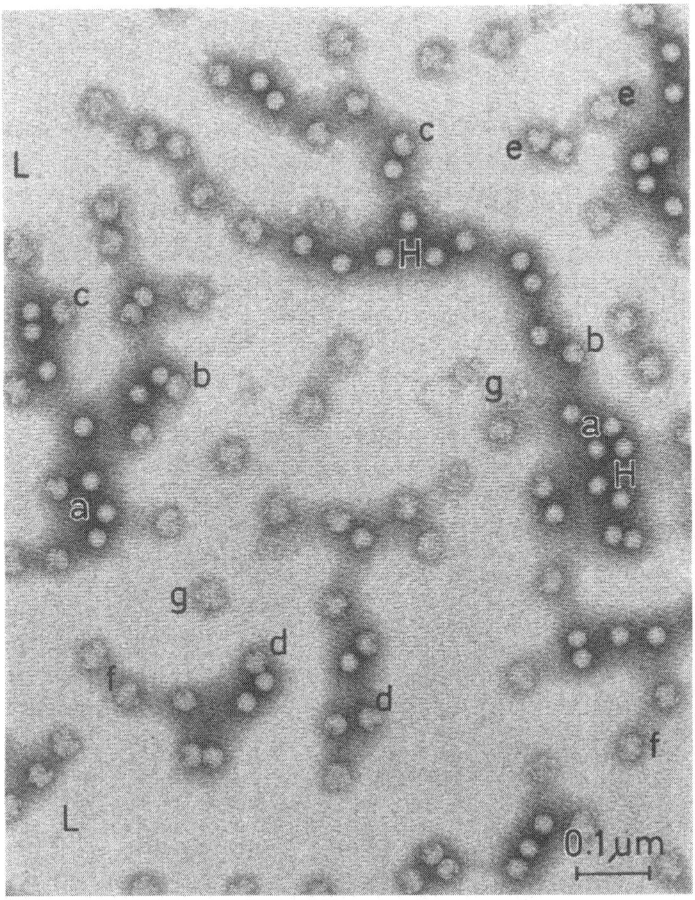

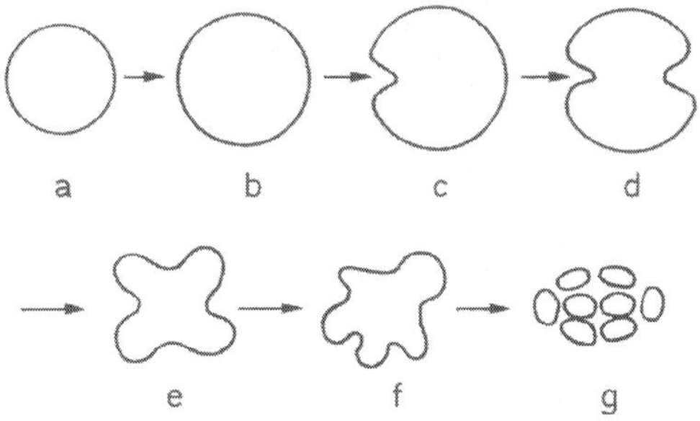

Fig. 25. Physical disruption of poliovirus during preparatory procedures for electron microscopy. – Top: Poliovirus particles, stained with 1% PTA, sandwich technique in a heavy *(H)* and light *(L)* stain deposit. *a* control poliovirus particles, *b* swollen particles, *c* and *d* swollen particles with notches, *e* and *f* intermediate forms of dissociated particles, and *g* sub-particles
Bottom: Scheme of the dissociation of poliovirus particles; symbols a to g correspond to that of the figure above
Figure from Boublik and Drzeniek, 1977 [J. gen. Virol. *37*, 130–131 (1977)]

protein coat; various intermediate stages of dissociation to a cluster of several (8 or more) equal subparticles approximately 10 nm in diameter. It was suggested that the equal sized subparticles correspond to the 14S pentamer of protomers (about the icosahedral vertices).

B. Alkaline Degradation

Alkaline pH causes destabilization of the virions and eventual breakdown of the viral capsid by disrupting hydrogen and ionic bonds (Van Elsen and Boeye, 1966; Katagiri *et al.*, 1971). The steps are shown in Figure 26. Increasing pH results in the consecutive dissociation of the VP4 and VP2 components from the capsids (Katagiri *et al.*, 1971). First VP4 and small amounts of the other capsid proteins are lost at pH 10, producing 80S empty particles composed of VP1, VP2, VP3 that exhibit a shift to H/C antigenicity and lack RNA (Boeye and van Elsen, 1967). VP4 probably interacts with some of its positive charges with the RNA and is otherwise held in the capsid to VP2 mainly by hydrogen and hydrophobic bonds. Fibrillar structures—probably representing ribonucleoprotein complexes (RNPs) between virion RNA and VP4—are seen in the EM after treatment at pH 10 (Van Elsen *et al.*, 1968). At pH 11 a further dissociation of the 80S particles to 60S particles and to slowly sedimenting components is observed. The 60S particle is a stable matrix of VP1 and VP3 and exhibits H/C antigenicity; the slow sedimenting component is mainly composed of aggregates of VP2 and some VP4. The 60S VP1—VP3 particles show that the capsid proteins VP1 and VP3 suffice to construct the backbone of a complete capsid. Assuming that the individual capsid proteins occupy discrete positions in the capsid, this is strong supporting evidence for locating pentamers of VP2 at the 12 icosahedral apices. This is the only structural cluster that can be removed entirely without destroying the integrity of the structure. We suggest that hydrophobic interactions stabilize a pentamer of VP2 (see Fig. 22A above and Chapter 10). Severe pH conditions result in complete dissociation of the viral capsid to the individual capsid proteins, which sediment at 2S.

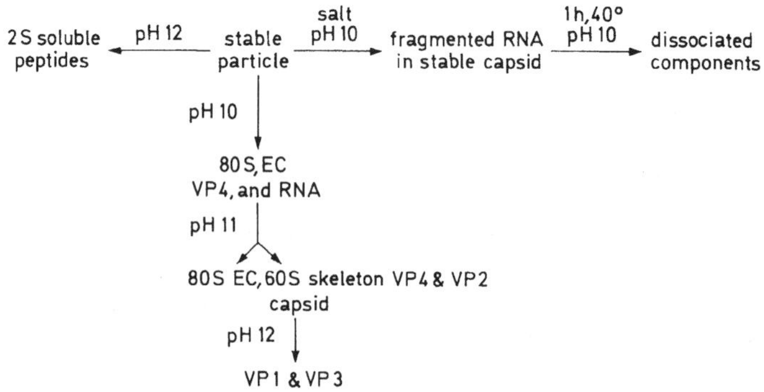

Fig. 26. Alkaline degradation of poliovirus. — Data from Katagiri *et al.*, 1971 [J. gen. Virol. *13*, 101–109 (1971)]

At mild alkaline treatment (pH 11 in the presence of 0.5 M NaCl—the latter stabilizes hydrogen bonds and opens ionic bonds), some destabilized particles are found with RNA still associated (van den Berghe and Boeye, 1973c). 90 % of the particle-associated RNA has a sedimentation coefficient of 11S (the rest is intact 35S RNA), *i.e.* RNA of about $^1/_{10}$–$^1/_{12}$ the size of intact polio RNA (Fenwick, 1968). The data show that the RNA within the partially degraded virus particle is accessible to the alkali and that there are approximately 10–12 points of access for alkaline hydrolysis of the RNA. This is consistent with the view that alkaline-induced removal of VP4 (and VP2) opens the capsid at its twelve icosahedral 5-vertices, leaving a backbone matrix of VP1 and VP3 located at the twenty 3-vertices (the icosahedral facets). An influx of water accompanying the loss of VP4 may be driven by the high internal cation concentration and could lead to a swelling of the RNA and its protrusion through the capsid at the "holes" formed by loss of VP4. These results are also consistent with the view that VP1 and perhaps also VP3 interact with the viral RNA by hydrogen and hydrophobic bonds holding most of the RNA within the capsid under these conditions.

The selective consecutive release of individual capsid polypeptides suggests that the individual molecules have discrete localities within the icosahedron. These observations also indicate that there is a considerable difference in strength of the bonds between VP1 and VP3 and the bonds between these proteins and VP2 (or VP0). The relative "ease" of dissociation of VP4 and VP2 is probably also a preparatory step to the release of RNA during the uncoating process early in infection (see Chapter 7).

Combined urea alkaline degradation studies of the enterovirus coxsackie virus B-3 (Philipson *et al.*, 1973) followed the same steps as that for poliovirus: release of VP4 and VP2—as a 5S component indicative of smaller protein aggregates—and larger VP1–VP3 structures (20S). In a more recent study, the 20S and 5S components were isolated and observed under the electron microscope (Beatrice *et al.*, 1980). The 5S component consists of amorphous dense structures, approximately 5 nm in diameter, having no clear morphological identity. The 20S component consists of ring-like structures averaging 9 -10 nm in diameter. The most likely interpretation is that the 5S component represents a trimeric or a pentameric aggregate of VP2 (about the 5-fold vertex) and the 20S component a hexameric (about the 3-fold vertex) or decameric (about the VP2–VP4 pentamer) ring of VP1 and VP3 (see Fig. 22 above).

Essentially the same observations were obtained for a third member of the enterovirus group, namely bovine enterovirus (Hoey and Martin, 1974). Alkaline treatment combined with short heating (30 seconds at 100° C) released VP4 and VP2, leaving an empty capsid-like structure of VP1 and VP3 which readily collapsed into fibrous-like material as seen in the EM. (Under these conditions the RNA is released in an intact state—in contrast to the breaking of RNA associated with modified particles of poliovirus under conditions of mild, prolonged (40 min) alkaline treatment in the presence of 0.5M NaCl, as described above.) For bovine enterovirus, dissociation of natural top component (NTC) was also studied: Here the precursor to VP4 and VP2–VP0 was released in an analogous fashion indicating that the locality and binding capacity of the VP0 protein component is not significantly disturbed by its cleavage into VP_2 and VP_4. In contrast, combined urea alkaline degradation of coxsackie B5 procapsids (VP_0, VP_1, VP_3) led to 40S components containing all of VP_0 and some of VP_1 and VP_3, and 20S components containing only VP_1 and VP_3, which is like the 20S substructure from degraded virions (Philipson *et al.*, 1973).

Alkaline treatment (pH 12) of poliovirion-neutralizing antibody precipitates (Breindl, 1971 b) results in the dissociation of the antibody-bound virion component—possibly the VP_2—VP_4 pentamers—from a component exhibiting C antigenicity (the VP_1—VP_3 backbone). Heating dissociates the antibody from the D reactive capsid structure concomitantly destroying the D reactive structure (see Section VIII below). When neutralizing antibodies were first bound to poliovirus, VP_4 was found to purify with the antibodies after disruption of the virions, indicating that VP_4 has a strong affinity for neutralizing antibodies (Breindl, 1971b). From this observation it was concluded that VP_4 is the antigenic determinant for poliovirus neutralizing antibodies. This conclusion, however, is quite unlikely in light of the recent strong evidence for an internal location of VP_4 (see above). Binding of the neutralizing antibody to the D-antigenic determinant(s), *e.g.* the VP_2 pentamers could distort the vertex-conformation and expose part of the small protein VP_4. For bovine enterovirus such a neutralizing antibody-mediated exposure of VP_4 was detected by acquired accessibility of VP_4 to chemical labelling after virion neutralization (Carthew, 1976). For coxsackie virus, alone the 5S VP_2 pentameric aggregate described above was capable of inducing neutralizing antibody production (Beatrice *et al.*, 1980).

C. Heat Degradation

For a long time it has been known that elevated temperatures destabilize the viral capsid and eventually cause its breakdown with the release of viral RNA (Koch, 1960a, 1962; Hinuma *et al.*, 1965; Dimmock, 1972; McGregor and Mayor, 1968, 1971a; Breindl, 1971a; Jordan and Mayor, 1974). Temperature induced disruption of viral capsids evidently occurs in a series of steps via a number of products in intermediate stages of breakdown, depending on temperature, duration of heat treatment, ionic strength of the medium, etc. Heat inactivation nonetheless seems to follow single hit kinetics (Hinuma *et al.*, 1965) (see Fig. 28A below).

The high temperature mediated increase in the kinetic energy of the capsid components should augment any inherent dynamics (protein "breathing") of the capsid structure and lead to a loosening of bonds. The slightest heat induced change in poliovirus is that to a 135S ribonucleoprotein particle (RNP), detectable after treatment at 56° C for 2 min. (Virions sediment at 155—160S.) This RNP particle is still infectious and appears virion-like in the EM (Jordan and Mayor, 1974). This change presumably represents a transition to the expanded state of the capsid by an influx of H_20 (see section VII below). Mg^{++} ions and low pH stabilize poliovirus against heat dissociation (Wallis and Melnick, 1961, 1962a; Melnick and Wallis, 1963).

Further incubation at elevated temperature induces the formation of a heterogeneous set of particles with sedimentation coefficients between 80 and 90S. These particles have lost their VP4 component (and possibly also varying amounts of other capsid proteins), but still contain infectious RNA (Koch, 1960a; Breindl, 1969, 1971a; Mietens and Koschel, 1971). The RNA is sensitive to RNase indicating that either the RNA protrudes from the capsid surface through some kind of "holes" or that the RNase can penetrate through "holes" into the virion. The particles are permeable to stain and appear as heterogeneous, partially empty particles in the EM, often with one or more small notches on one side of the capsid (see Fig. 27). This step is accompanied by a substantial rearrangement of the capsid structure as revealed by a shift of antigenicity from D (N) to C (H) antigenicity (Le Bouvier, 1955; Watanabe *et al.*, 1965; Breindl, 1971a), loss of infectivity and

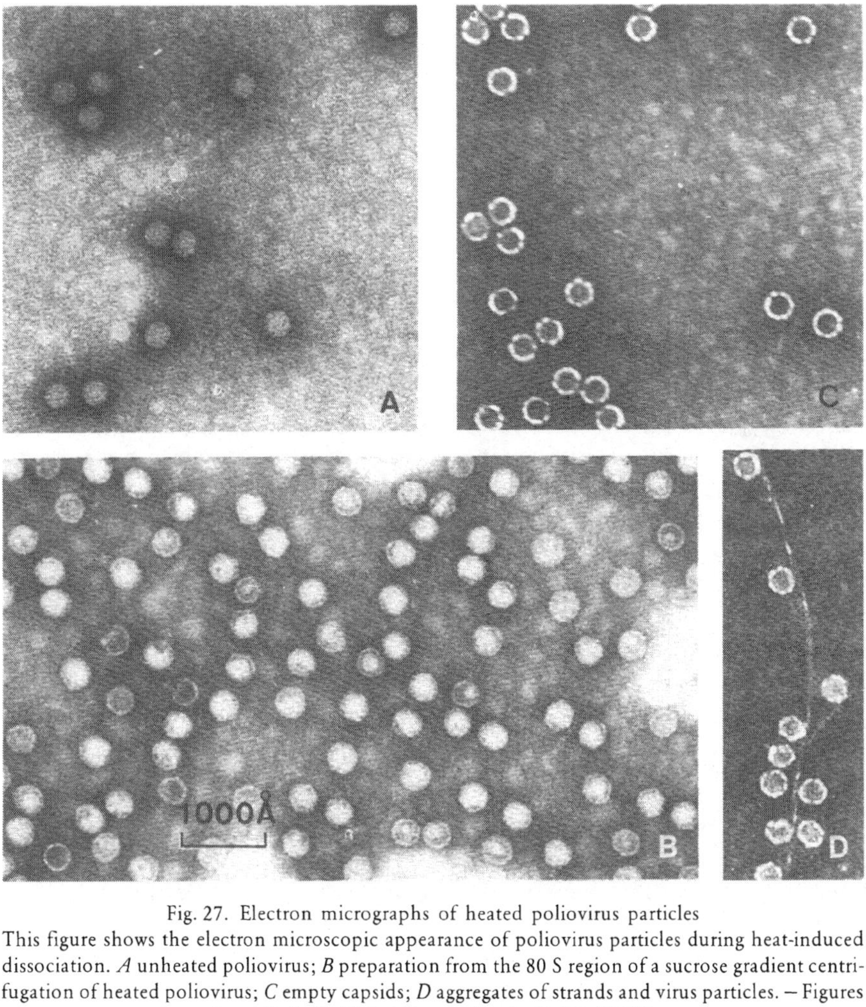

Fig. 27. Electron micrographs of heated poliovirus particles
This figure shows the electron microscopic appearance of poliovirus particles during heat-induced dissociation. *A* unheated poliovirus; *B* preparation from the 80 S region of a sucrose gradient centrifugation of heated poliovirus; *C* empty capsids; *D* aggregates of strands and virus particles. – Figures from Breindl, 1971 a [J. gen. Virol. *11*, 152 (1971)]

loss of ability to adsorb to cells (Graham, 1959; Katagiri *et al.*, 1968). Whereas 92% of intact virus particles adsorb to HeLa cells in 60 minutes at 5° C, only 7% of the 80–90S RNP were found to bind to HeLa cells under these conditions. The 80–90S RNP particles are dissociated into empty capsids and free RNA during CsCl equilibrium centrifugation (Breindl, 1971a) and into individual capsid proteins and free RNA by treatment with 1% SDS (virions are stable under these conditions). This step is followed by the release of intact infectious RNA (or broken RNA), leaving a 73–80S empty particle that does not contain any VP4 (Maizel *et al.*, 1967; Hinuma *et al.*, 1965; Drees and Borna, 1965). The EM reveals empty capsids, fibrillar structures, RNP strands and free RNA strands (van Elsen *et al.*, 1968; McGregor and Mayor, 1968; Wouters *et al.*, 1973b).

High salt concentrations stabilize poliovirus to heat degradation (Koch, 1960 a; Wallis and Melnick, 1961; Breindl, 1971 a). A mutant of poliovirus type 1, resistant to heat inactivation, differed from the parental strain in that it contained an altered VP3 capsid protein, as shown by a slightly slower migration in PAGE (Fennel and Phillips, 1974). The altered VP3 was detected also in cytoplasmic extracts of infected cells as well as in NTC and virions. Poliovirus Type 2 strains appear to be more heat and urea resistant than type 1 strains (Youngner, 1957; Fennell and Phillips, 1974) (Fig. 28A).

Arildone, an antiviral agent that reversibly blocks the uncoating of poliovirus (McSharry *et al.*, 1979) also stabilizes poliovirus against heat and alkaline induced inactivation (Caliguiri *et al.*, 1980). Arildone evidently interacts directly with the viral capsid via weak, nonpolar interactions. Arildone may be physically inserted into hydrophobic regions of the virus capsid. It prevents inactivation of poliovirus infectivity and the physical alterations typically induced by treatment at 47° C or pH 10.5 for 20 min. Even after 60 min at pH 10.5 there is still no significant loss of infectivity, yet the virion must be destabilized since it is converted to 80S particles lacking VP4 during centrifugation in sucrose gradients. The stabilizing effects of arildone are reminiscent of those induced by antibody binding, whereas the latter, however, also induces a significant reduction of infectivity. The studies with arildone indicate that there are differences in the mechanisms of destabilization of poliovirus by heat and alkaline treatment.

L-cystine interacts in a highly specific manner with heat sensitive strains of poliovirus and markedly increases the thermal stability at temperatures below 50° C (Pohjanpelto, 1958). Kinetic analyses showed that approximately 10 molecules of cystine must become attached to the virus to confer maximal stabilization. The binding of cystine to virus is slow. Dilution does not reduce the stabilzation effect of cystine indicating that it is tightly bound to the virus, presumably by S-S bonds to cysteine residues of the capsid proteins. The interaction of L-cystine with the virus is highly specific, D-cystine, glutathione and homocysteine and 17 other amino acids are ineffective. Cystine acts only at neutral pH, it does not stabilize virions at acidic pH indicating that the pI 4.5 form of the virion cannot be stabilized by cystine.

The thermal stability of virus is increased considerably in the absence of oxygen, indicating that oxidation plays an important role in thermal inactivation of virus. Reducing agents such as thioglycolate increase the thermal stability of poliovirus presumably by preventing oxidation (Pohjanpelto, 1958; Hirst; 1961). Poliovirus is also stabilized against heat inactivation by ^{35}S-sodiumtetrasulfide where ^{35}S is bound reversibly to the viral capsid (Pons, 1964a). During these procedures mild heat was required for interaction of virus and compound resulting in a fraction of inactivated viruses and a fraction of viruses with increased stability.

The results of these old studies on the interaction of cystine, Na_2S_4 and a number of reducing agents with poliovirus suggest that the abundant cysteine residues in the capsid proteins VP_2 (10), VP_3 (5) and VP_1 (12) may be more important to capsid stability and virion conformation than is generally assumed. Since neither iodoacetamide nor N-ethyl-maleimide react with intact virions (see Table 15 above), it appears that there are no free accessible -SH groups on the viral surface. Gentle disruption of intracapsid bonds, however, seems to permit substances such as cystine and Na_2S_4 to interact with S-S bonds within the particle, thereby increasing its stability. It would certainly be interesting to reinvestigate the role of SH groups and S-S bonds for virus structure in more detail.

D. Guanidine Degradation

Treatment of poliovirus with guanidine (6.5 M) degrades the virus into polypeptide units (Scharff *et al.*, 1964) (10^{-3} M guanidine inhibits poliovirus replication—see Chapter 10, Section VI). Molar concentrations of divalent cations (Mg^{2+}) sensitize the virion to inactivation by guanidine (Fujioka *et al.*, 1969; Fujioka and Ackermann, 1975b). The concentration of guanidine required for inactivation is in the range known to denature proteins (in contrast to urea, the concentration of which is usually much lower than that required for protein denaturation).

E. Urea Degradation

The dissociation of poliovirus mediated by concentrated urea induces a more complex series of steps than that induced by heat or alkaline pH (Fig. 28B). Urea is known to break noncovalent bonds and to denature the tertiary structure of proteins. Perhaps the relative "disorderliness" of the urea-mediated capsid breakdown reflects the concomitant breakage of different functional bonds (*e.g.* inter- and intraprotomer bonds, interpentamer bonds). Also the only claim of a successful complete reassociation of degraded poliovirus has come from attempts employing the protein components obtained by urea treatment of the virus particle (Drzeniek and Bilello, 1972 a+b, see below).

Like other degradation procedures, treatment of poliovirus with 7—8 M urea can result in the release of intact RNA with a concomitant formation of empty capsids (Cooper, 1962) that lack VP_4 (Maizel *et al.*, 1967) and exhibit altered antigenicity. Further degradation of the empty capsids is temperature dependent, occuring only from 32—40° C (Vanden Berghe and Boeye, 1973 b). The capsids

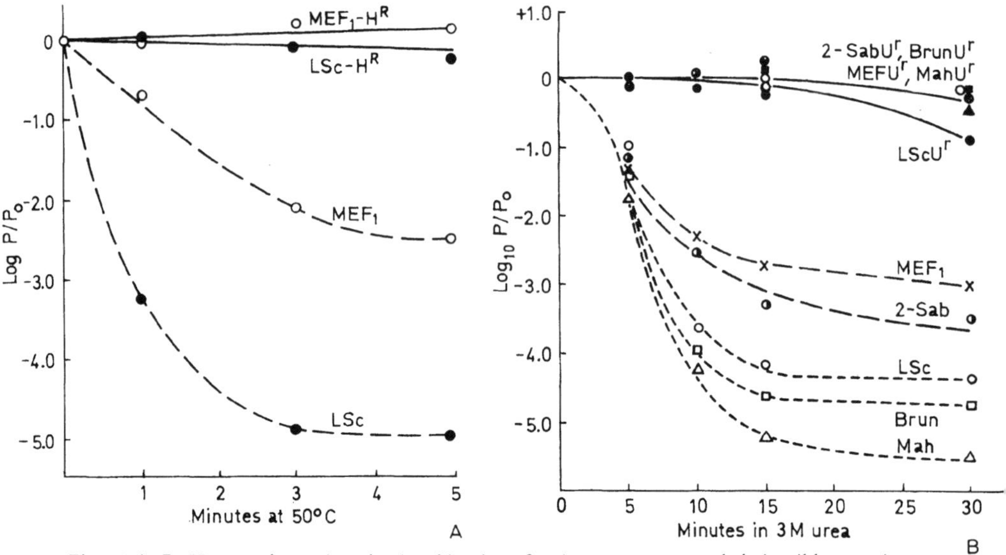

Fig. 28 A–B: Heat- and urea-inactivation kinetics of resistant mutants and their wild type viruses

A Heat inactivation kinetics of heat-resistant mutants and their wild type viruses
1 ml amounts of virus at 22 °C were added to 9.0 ml of phosphate buffer (0.02 M, pH 7.0) at 50 °C. At the times indicated, samples were diluted and assayed for infectivity of the surviving virus. Inactivation is expressed as log reduction in virus titer (PFU/ml)

B Urea inactivation kinetics of urea resistant mutants and their wild type viruses
Virus in phosphate buffer was added to equal volumes of 6 M urea in phosphate buffer and incubated at 37 °C for the times indicated
LSc wildtype poliovirus 1, LSc strain; *LScU*^{r} urea-resistant mutant of LSc; *Mah* wildtype poliovirus 1, Mahoney strain; *MahU*^{r} urea-resistant mutant of Mah; *Brun* wildtype poliovirus 1, Brunhilde strain; *BrunU*^{r} urea-resistant mutant of Brun; *MEF* wildtype poliovirus 2, MEF_1 strain; *MEFU*^{r} urea-resistant mutant of MEF; *2-Sab* Sabin vaccine strain of poliovirus type 2; *2-SabU*^{r} urea-resistant mutant of Sab; *LScH*^{r} heat-resistant mutant of poliovirus type 1 LSc strain; *MEFH*^{r} heat-resistant mutant of poliovirus type 2 MEF strain. – Figures from Fennel and Phillips, 1974 [J. Virol. *14*, 823, 824 (1974)]

exhibit a greater stability at higher temperatures, indicating an important entropic factor in the stablization of the capsid. (Empty capsids formed in concentrated urea at 50° C were resistant to further breakdown at temperatures above 50° C.)

Degradation of the empty capsid proceeds in a stepwise fashion: loss of most of the VP_2 component is predominant at first, followed and then accompanied by loss of VP_3 and also some VP_1. After treatment of poliovirus with 7.2 M urea for 15 min at 40° C, the sedimentable structures that remained were depleted of VP_2 and VP_4. Sucrose density gradient sedimentation analyses of the degradation products upon treatment of poliovirus with 7.2M urea at 37° C for 6–20 min revealed the stepwise disintegration via 70S particles (the predominant form after 6 min treatment, composed of VP_1, VP_3 and most of VP_2) into several species of intermediate sized particles. Up to 9 distinct species of particles were observed, the sedimentation coefficient of which ranged from 70S to 10S, suggesting a succession of stepwise decreases in particle size such as may be caused by the repeated loss of similar substructures from the capsid (Vanden Berghe and Boeye, 1973b).

Polypeptide analyses of the various intermediates revealed predominantly a successive release first of VP_2 (the 20–50S range being nearly devoid of VP_2), then of a smaller amount of VP_3, and VP_1. Electron microscopy of different fractions of the degradation products revealed recognizable particles only down to the 30–50S sedimentation range. After 9 min. of urea treatment mainly empty capsids are seen, often with notches or defects of 1/5 to 1/3 of the capsid circumference. The 33S–48S particle range obtained after 20 min. of urea treatment (composition VP_1–VP_3 at a ratio of 2 : 1), showed heterologous structures, similar in diameter to empty capsids, but flatter and lacking the core typical of empty capsids. Apparently these structures represent the more or less collapsed backbone of VP_1VP_3, after the additional loss of up to half of the VP_3 component.

Treatment of poliovirus particles with concentrated urea (9 M for 60 min at 25° C) results in the complete breakdown of the particles into free 35S (intact) RNA and protein components sedimenting at 2S, containing the individual capsid proteins.

In contrast to complete virus particles and empty capsids, stable 80S particles obtained from infected cells (NTC) were completely resistant to 7.2 M urea at 37° C but were disrupted at 62° C (Vanden Berghe and Boeye, 1973 b). In the presence of small amounts of Mg^{++} (0.0005 M $MgCl_2$), two distinct products sedimenting at 50S and 10S were found. The 50S contained most of the VP_0 component and some VP_1 and VP_3, the 10S and intermediate components were more enriched in VP_1 and VP_3. This may indicate that Mg^{++} ions can aggregate the VP_0 components of procapsids. The entropic factor preventing urea-mediated dissociation of the virion capsid at high temperatures is missing in the procapsid state.

The urea inactivation curve of poliovirus is characterized by three distinct slopes: an initial delay or shoulder, followed by a rapid exponential decrease in virus infectivity, and a final slow exponential decrease (Cooper, 1962; Vanden Berghe and Boeye, 1973 b). The initial shoulder of this inactivation curve indicates that the mechanism of urea inactivation of poliovirus is due to a process of cumulative damage or in its simplest form, a two-step reaction (see Fig. 28B, 29A). The rapid inactivation slope can be decreased (*i.e.*, the virus particle stabilized) by moderate concentrations of Mg^{++} without affecting the duration of the initial shoulder; higher Mg^{++} concentrations completely stabilize the virus particle to urea degradation (Fujioka and Ackermann, 1975a) (see Fig. 29B). Apparently treatment of poliovirus particles with urea first breaks the more susceptible hydrogen bonds converting the native virion into an intermediate, sensitized form that is still infectious (corresponding to the initial shoulder in the inactivation curve). The inactivation that ensues upon further treatment with urea apparently results from the breakage of the ionic bonds holding the VP_4 polypeptide on to the virion (these bonds are stabilized by Mg^{++}). It is tempting to speculate that the Mg^{++} stabilized bonds holding VP_4 in the capsid are equivalent to the bonds that cause aggregation of VP_0 in the presence of Mg^{++}.

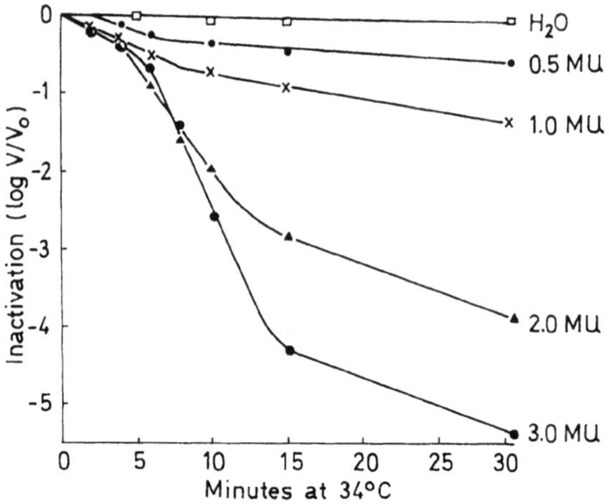

Fig. 29 A–B. The inactivation kinetics of poliovirus in urea and its stabilization by Mg^{++}

Fig. 29 A. Inactivation kinetics of poliovirus at 34° in water and urea
□——□ water; ●——● 0.5 M urea; x——x 1.0 M urea; ▲——▲ 2.0 M urea; ●——● 3.0 M urea. Inactivation is expressed as log reduction in virus titer (PFU/ml)

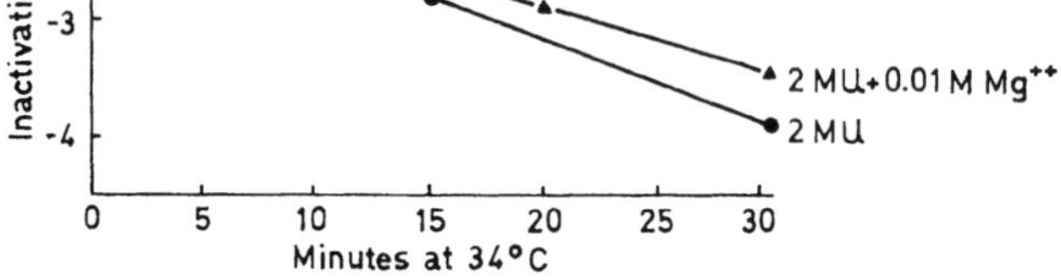

Fig. 29 B. Stabilization of poliovirus against urea inactivation by Mg^{++}
▲——▲ + 0.01 M $MgCl_2$; ×——× + 0.03 M $MgCl_2$; ·—— + 0.09 M $MgCl_2$; ■——■ + 0.5 M $MgCl_2$; ●——● 2.0 M urea, no $MgCl_2$. – Figures from Fujioka and Ackermann, 1975 a [Proc. Soc. Exp. Bio. Med. *148*, 1064 (1975)]

The final slow decrease of residual virus infectivity during urea inactivation indicates that a certain fraction of the virus population is resistant to urea inactivation. The persistent virus population is larger in the presence of Mg^{++} and decreases with an increase in the concentration of urea. It is unclear whether the persistent fraction is induced by the action of Mg^{++} and urea on an initially homogeneous virus population or whether it reflects the selection of a preexisting stable population of virus from an initially heterogeneous virus population (for example, in the microheterogeneity of capsid proteins or in the number of VP_0 molecules per virus particle).

Five urea resistant mutants were analyzed for alterations in the SDS electrophoretic mobility of their capsid polypeptides. Differences in VP_1, VP_2 or VP_3 were found in three mutants; two mutants revealed no altered capsid polypeptides by this method of analysis (Fennell and Phillips, 1974), *i.e.*, resistance to urea degradation cannot be attributed to a particular capsid protein. The result supports the notion that urea destabilizes a number of different bonds between the different capsid polypeptides (see Fig. 28B above).

F. Reassociation of Poliovirus from the Products of Urea Degradation

When the components obtained by treatment of the virus particle with 9 M urea—the individual capsid proteins and the intact RNA—are suspended in a solution of lower urea concentration (5 M), spontaneous and specific aggregation of the components is observed (Drzeniek and Bilello, 1972 b; Drzeniek, 1975; Yamaguchi-Koll *et al.*, 1975). A ribonucleo-polypeptide (RNPP) complex sedimenting at 45S, containing all of VP_1 and intact RNA, and 7–8S oligomers containing predominantly VP_3 and some VP_2, are obtained in addition to the monomers at 2S (the bulk of VP_2 and VP_4 and some VP_3). When the dissociation with 9 M urea was performed at higher ionic strength, subsequent reduction of the urea concentration to 5 M resulted in the formation of empty capsids instead of the RNPP. Further dilution (up to 32-fold) resulted in the reappearance of infectivity: a maximal 5,000-fold increase in virion infectivity above that of the dissociated sample was obtained. However, this represented a maximum efficiency in reconstituted infectivity of less than 0.05% (Drzeniek and Bilello, 1972 b).

G. Conclusions

The described degradation studies indicate that the capsid is held together by hydrogen bonds and by hydrophobic interactions between capsid proteins, and it appears to be stabilized by ionic interactions between the proteins and probably also by hydrogen and ionic bonds between the capsid proteins and the genomic RNA. The anatomy of the capsid, (*i.e.*, the tight packing of the capsid proteins, which constitute 1/3 of the particle diameter; the "penetration" of the 3 larger capsid proteins through the width of the capsid, interacting with the RNA in the inside and also forming part of the external surface), indicates that there exist a multitude of bonds and interactions, though each of different strength and significance. For example, the forces that are responsible for holding the small protein VP_4 internally in close association with the RNA are stabilized by Mg^{++}. These forces are disrupted easily and lead to the release of VP_4 from the capsid unless they are substituted by other stabilizing forces (see next section). In contrast, interactions with and between clusters of VP_1 and VP_3 are relatively strong and are sufficient to maintain the basic capsid structure.

The stepwise process of virion disruption indicates that the capsid proteins occupy discrete domains within the capsid. The consecutive release of separate VP2 aggregates and aggregates of VP1 and VP3 supports, but does not prove, the notion that VP2 is clustered in pentamers at the icosahedral 5-fold vertices, VP1–VP3 in hexamers at the icosahedral 3-fold symmetry centers.

VII. Conformational Forms of the Poliovirus Capsid

Under physiological conditions, the vast majority of poliovirus particles assumes a stable conformational state, which is defined as the "native" state of poliovirions. Native virions are characterized, as described, by their infectivity, ability to attach to susceptible host cells, D(N)-antigenicity, a sedimentation coefficient of 150S–160S in sucrose gradients, an isoelectric point of pH 7.0, a buoyant density of 1.34 g/cm^3 in CsCl gradients, resistance to proteases and RNases, impermeability to a number of dyes and low molecular weight substances, relative stability to extreme environmental conditions, and a composition of 35S–VPg-RNA plus sixty VP4-2-3-1 protomers (one or two of which may contain VP0 instead of VP4-2, Rueckert, 1971, 1976).

However, poliovirus does not always exhibit these structural characteristics of the "native" virion. Reversible conformational shifts occur already upon mild heat treatment or in acidic environments. Certain conformational shifts are characteristic–and probably even obligatory–aspects of the poliovirus life cycle: for

Table 18. *Conformational forms of intact poliovirus*

	Dense particles		Mature virions		Antibody bound
			A-form	B-form	
Sedimentation (S)	220	160	155	140	?
Antigenicity	(D)	(D)	D	D	D
Attachment-capacity	+	+	+	+	++, (±)
Composition					
– Protein (VP)	(4–2–3–1)	(4–2–3–1)	(4–2–3–1)	(4–2–3–1)	(4–2–3–1)
– RNA	35 S	35 S	35 S	35 S	35 S
Sensitivity to					
– RNase	+	–	–	–	–
– Protease	+	(+)	–	–	–
Stability in					
– SDS	–	(+)	+	+	+
– EDTA	?	?	+	+	+
Isoelectric point	?	?	7.2	4.5	4.5
Buoyant density in CsCl (g/cm^3)	1.44	1.34	1.34	1.32	1.34
Permeability to Cs^+	–	+	–	–	?
Electron microscopy core	+	–	–	–	?

example, in the passage through the acidic environment of the stomach to reach its natural replication habitat (the cells of the intestinal tract), in adaptation to different environmental conditions (survival in sewage, in waste water, in serum or at elevated temperatures), upon binding of the virion to its cell receptor (see Chapter 7), during assembly (see Chapter 10), and upon reaction with antibodies (see Section VIII below). Shifts in capsid conformation are manifested by alterations in antigenicity, in isoelectric point, in protein composition, in capacity to attach to cells, in pH stability, in susceptibility to proteolytic enzymes and to chemical modifications, in rate of ion permeation, and in appearance in the EM. The conformational forms of poliovirus are summarized in Tables 18–20.

A. Poliovirus Capsid Structures During Early Interactions with the Host Cell and upon Experimentally Induced Virion Disruption: Two Conformational States of the Shell

Attachment of virions to host cell may result in marked alterations of virion structure, leading to or accompanied by the complete loss of the small capsid protein VP4 (Table 19) (see also below, Joklik and Darnell, 1961; Fenwick and Cooper, 1962; Mandel, 1967; Habermehl *et al.*, 1974; Lonberg-Holm *et al.*, 1975; Kohn, 1979; Lonberg-Holm and Philipson, 1980). Some modified virions (sometimes called A-particles) may elute spontaneously from the host cell. They are then no longer capable of reattaching to their host cell. Unlike mature virions, A-particles sediment at 130–135S; the slower sedimentation coefficient may indicate a more expanded state of virion structure. In accordance with the concept of a more ex-

Table 19. *Conformational forms of poliovirus during degradation and uncoating*

	A particles	*In vitro* RNP	Artificial top component ATC	Skeleton capsid
Sedimentation (s)	135	80–90	80	60
Antigenicity	C	C	C	C
Attachment-capacity	–	–	–	–
Composition				
– Protein	2–3–1	2–3–1	2–3–1	3–1
– RNA	35 S	35 S	–	–
Sensitivity to				
– RNase	(–)	+	–	–
– Protease	+	?	+	?
Stability in				
– SDS	(+)	+	+	?
– EDTA	+	(+)?	+	?
Isoelectric point	4.5	4.5	4.5	4.5?
Buoyant density in CsCl (g/cm^3)	unstable	unstable	1.3	?
Permeability to Cs^+	+	+	+	+
Electron microscopy core	+	+	+	+

panded, "leaky" state of the capsid, are the observations of an increased sensitivity of A-particles to proteases and of an increased permeability to stains used for electron microscopy, such as PTA (phosphotungstic acid). As a consequence, A-particles appear "coreless" in the electron microscope, as do empty shells (see below). However, they do contain intact virion RNA, *i.e.*, the "coreless" appearance does not correlate with absence of RNA. The RNA is still resistant to digestion by RNase indicating that the holes responsible for the "leakiness" are not very large.

The structural alterations of the capsid are reflected also in distinct changes in antigenicity and in the isoelectric point. A-particles exhibit the C(H) antigenicity which is typical of empty shells. They also show a greatly increased tendency to aggregate in aqueous environments and an increased stickiness to membrane filters. This presumably reflects an exposure of hydrophobic groups on the capsid surface in A-particles, that would exert mutual attractive forces in aqueous solutions.

A variety of experimental procedures are known by which native virions (with D antigenicity) can be converted irreversibly to particles carrying C antigenicity, a conversion that appears to be analogous to that occurring after attachment and elution of virions from their host cells. Such treatments of virions include gentle heating (50° C), irradiation with ultraviolet light, high pH, mercurials, phenol, desiccation (Le Bouvier, 1955, 1959 a+b; Roizman *et al.*, 1959; Hummeler *et al.*,

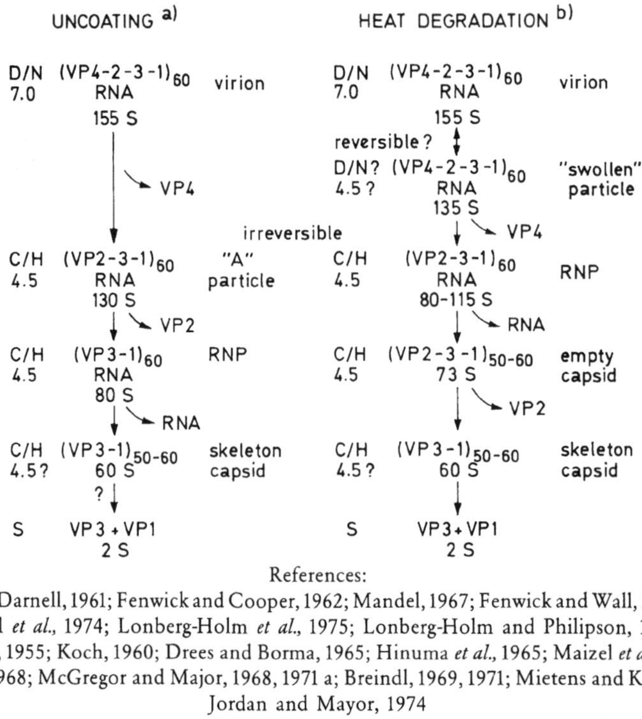

References:

[a] Joklik and Darnell, 1961; Fenwick and Cooper, 1962; Mandel, 1967; Fenwick and Wall, 1973; Habermehl *et al.*, 1974; Lonberg-Holm *et al.*, 1975; Lonberg-Holm and Philipson, 1980

[b] Le Bouvier, 1955; Koch, 1960; Drees and Borma, 1965; Hinuma *et al.*, 1965; Maizel *et al.*, 1967; Van Elsen *et al.*, 1968; McGregor and Major, 1968, 1971 a; Breindl, 1969, 1971; Mietens and Koschel, 1971; Jordan and Mayor, 1974

Fig. 30. Comparison of *in-vivo* uncoating and heat-induced dissociation of poliovirus

1962; Katagiri *et al.*, 1967) (see also Section VI above). Usually these treatments eventually lead to the loss of RNA as well, yielding an empty 80S shell lacking VP4 and RNA, although—as studied in detail in the case of heat dissociation—sequential steps can sometimes be separated in time. For example, 135S and 90S RNA-containing shells can be obtained after gentle heating. Since the viral RNA remains intact, such ribonucleoprotein particles are as infectious as isolated RNA (one hundred to one thousand fold less infectious than native virions) (Koch, 1973). Figure 30 provides a schematic comparison of the steps in in vivo uncoating and heat-induced dissociation of poliovirus.

The described antigenic and other structural alterations accompanying the loss of VP4 are independent of the loss of RNA (Taylor and Graham, 1959; Katigiri *et al.*, 1968, 1971; Breindl, 1971 a, b). The artificially formed empty shells also carry C antigenicity. They are permeable to PTA and have a low isoelectric point at pH 4.5. These shells are also called artificial top component (ATC) in analogy to the naturally occurring top component (NTC) as they also band above virions in CsCl gradients at a buoyant density of 1.29 g/cm^3, a value typical of pure protein complexes (Rueckert, 1971).

B. Viral Structures During Maturation

A number of virus capsid structures can be observed in the host cell during the replication cycle of poliovirus (Scharff and Levintow, 1963; Ghendon and Yakobson, 1971; Rueckert, 1976). Late in infection, immature capsids, mature virions and some intermediate ribonucleoprotein particles can be isolated from the infected cell. The properties of these structures are described in detail in Chapter 10; here we will present only a brief description of the characteristics of the capsid structures (Table 20).

The immature capsids sediment at 80S, they contain no RNA, and usually are made up entirely of VP1, VP3 and VP0 (precursor to VP2 and VP4). It was long taken to be certain that immature capsids contain the full set of 60 protomers (12 pentameric units), although recent experiments with mengovirus suggest that the 80S particles may consist of merely 10 pentameric units, having two holes through which the RNA might be "threaded" to complete the virion (Scraba, 1979). The immature capsids are permeable to phosphotungstate (PTA), which is used as a negative stain in electron microscopy. Hence, the immature capsids appear "empty" in electron micrographs. They band above virions in cesium chloride gradients at a densitiy of 1.30, and have therefore also been named naturally occurring top component (NTC) (in contrast to the artificial top component ATC, the empty shells that are formed during virion degradation and that lack VP4).

A recent study on 80S particles, carefully isolated from infected cells, demonstrated that there are actually at least two different forms of 80S particles (Marongiu *et al.*, 1981b) (see Table 67, Chapter 10). A more labile form appears to be the state actually present in infected cells. It can be disassembled into 14S pentameric subunits under relatively mild conditions, but is even more readily

Table 20. *Conformational forms of poliovirus during assembly*

	NTC (natural top component)		RNP	Provirion
Sedimentation (S)	80	65–75	80	$^{125}/_{150}$
Antigenicity	C	D	?	?
Attachment-capacity	–	+	?	±
Composition				
– Protein (VP)	0–3–1	0–3–1	0–3–1	$\frac{4-2}{0}$–3–1
– RNA	–	–	35S	35S
Sensitivity to				
– RNase	–	–	?	+
– Protease	+	+	?	?
Stability in				
– SDS	+	+	±	+
– EDTA	+	?	(+)	–
Isoelectric point	5.0	6.8	?	?
Buoyant density in CsCl (g/cm^3)	1.3	1.3	unstable	unstable
Permeability to Cs^+	+	?	?	?
Electron microscopy core	+	?	?	?

converted to stable 80S shells. It is not clear whether the characteristics of NTC described in the literature and summarized below are those of the instable or stable form (or both). The conditions routinely employed during their isolation are expected to convert instable into stable shells, therefore, the data of the older literature probably correspond only to the stable form. Isolated poliovirus NTC were reported to have a shell conformation which differed from that of native virions but seemed to be the same as that assumed by empty shells derived from virions. The characteristic features of the empty shell conformation are listed in Table 20. Isolated poliovirus NTC have a different set of predominant surface antigenic determinants–the C(H)-antigenic sites. They cannot attach to host cells, and they probably also exhibit a lower isoelectric point than native virions (see below). NTC of other picornaviruses, for example of rhinovirus, may assume one of two conformational states, one resembling that of native virion (D-antigenicity, attachment capacity, pI 6.3), the other resembling that of ATC (C-antigenicity, no attachment capacity, pI 4.5, Korant *et al.,* 1975).

There has been some other indirect evidence also for two different conformational states of poliovirus 80S immature shells. 80S shells self assembled in vitro from isolated 14S subunits were reported to exhibit a low isoelectric point of pH 5.0, shells assembled in vitro in the presence of rough endoplasmic membranes from infected cell extracts or shells isolated from infected cells supposedly exhibit a high isoelectric point at pH 6.8 (Putnak and Phillips, 1981). To date, this data has not been related to antigenicity, attachment capacity, or stability. A monoclonal antipoliovirus neutralizing antibody was recently obtained that also precipitated NTC but did not bind to ATC (Rueckert *et al.,* 1981). It seems that the NTC used in this study was in a D rather than in a C-antigenic state. Unfortunately, these results have also not yet been related to isoelectric points or attachment capacity. In analogy to the rhinovirus system, it is likely that poliovirus NTC may also acquire two conformational states, an empty shell like state (pI 5.0, C-antigenicity, no attachment capacity, stable) and a native virion-like state (pI 6.8, D-antigenicity, attachment capacity, "unstable" ?). Indeed, a recent study indicates that empty shells isolated from infected cells under careful conditions are in the D/N antigenic state, sedimenting at 65S. These particles are converted irreversibly to C/H antigenic, 80S capsids by incubation at 37° C and at alkaline pH (Rombaut *et al.*, 1982).

Ribonucleoprotein particles have been less well characterized in terms of capsid features. Distinct particles sedimenting at 80S, 125S and 150S have been observed (Agol *et al.*, 1970; Fernandez-Tomas and Baltimore, 1973; Guttmann and Baltimore, 1977a; Marongiu *et al.*, 1981a) (see Table 68, Chapter 10). The 80S and 125S particles probably represent complete (12 pentamer units) or near-complete (10 pentamer-units) immature shells ($VP_{0,3,1}$) in some association with 35S viral RNA. The 150S particles are immature virions with varying proportions of uncleaved VP_0 to VP_2 –VP_4. The lower sedimentation coefficients of 80S and 125S indicate a more expanded or awkwardly shaped form of the particles. Just as the 80S empty shells of NTC resemble those of ATC, the in vivo 80S and 125S ribonucleoprotein particles may resemble the 80S heat-induced RNPs and the 135S A-particles. In all cases, the fundamental difference is that VP_4 is lacking in the dissociation-induced particles, whereas it is present–still in covalent linkage to VP_2 as part of VP_0–in the naturally occurring particles during infection. Nevertheless, characteristic features of shell structures appear to be very similar in all cases.

The complete progeny virions isolated from infected cells usually are in the native state, although under certain conditions a significant proportion of progeny virions may occur in still another form,the dense particles (see below).

C. Two Distinct, Reversibly Interconvertible Conformational States of Intact Poliovirions

From the discussion above, it is evident that there are at least two antigenically distinct conformational states of the poliovirus capsid. NTC particles of several picornaviruses can occur in either state; poliovirus NTC particles also seem to assume the D conformation in vivo but are readily and irreversibly converted to the C-antigenic conformation by conditions traditionally employed for their isolation. Upon maturation (RNA encapsidation and cleavage of VP_0 to VP_2 and VP_4) and under physiological conditions, the virions assume the D-antigenic, pI 7.0 conformation, thereby acquiring capacity to attach to and infect their host cells. Either by natural (attachment to and elution from host cells) or experimental modification (heating, high pH, etc.), the virion can be converted–irreversibly–to the C-antigenic, pI 4.5, conformational state. This conversion is accompanied by loss of VP_4 and attachment capacity, and may or may not be accompanied by the loss of RNA.

It is very probable that the antigenic changes reflect a concerted reorientation of the viral protomers, *i.e.*, changes in overall virion conformation. It is unlikely that VP4 alone is responsible for D antigenicity and interaction with the receptor and thus accounts for the loss of attachment capacity and conversion to C antigenicity, since NTC contains the same VP4 sequence (merely still covalently connected to VP2 in VP0) as virions, but NTC also occurs in the C reactive state. Furthermore the all-or-none nature of the D- and C-antigens, *i.e.*, absence of mosaic particles with both D and C antigenicity or of particles with intermediate isoelectric points, is characteristic of concerted transitions.The C-antigenic pI 4.5 conformation appears to be a more expanded form of capsid, as evidenced by increased permeability for ions and a decrease in the sedimentation coefficient.

There is a good deal of evidence that the capsid of intact poliovirions may also assume a conformational state analogous, albeit with some important differences, to the expanded C-antigenic state described for NTC, ATC, and A-particles. During isoelectric focusing poliovirus distributes into two infective and interconvertible forms with isoelectric points of 7.0 (pI 7.0 form = A-form; not the same as A-particle) and 4.5 (pI 4.5 = B-form) (Mandel, 1971). The A-form predominates above pH 7, the B-form below pH 4.5, and both A and B forms exist in the pH region in between (Mandel, 1971). This implies that the reversible interconversion can be slow, that the individual particles exist in one form or the other, and that they do not behave as hybrids.

The B-form of intact virions thus has an isoelectric point that is very similar if not identical to that of ATC and A-particles. As pointed out above, the isoelectric point of a virion-like particle is determined by its proteinacious surface structure, the RNA (the pI of isolated poliovirus RNA would be expected to lie at about pH 2.0) provides little if any contribution. The surface charge is neutral at the isoelectric pH, it is negative at pH values above the pI, and positive at pH values lower than the pI. The marked pI differences between the A and B forms imply that the B-form has more negative (or less positive) residues exposed on the surface. The sedimentation coefficient of the poliovirion pI 4.5 form has not yet been reported. The acid sensitive picornavirus relative, rhinovirus, may also assume two different interconvertible low and high pI forms (Korant *et al.,* 1975). The low pI 4.5 form has a decreased sedimentation coefficient of 140S (compared to 150S of the pI 7.0 form), which is consistent with a more expanded state of the capsid. The low pI form of rhinoviruses also has a reduced attachment capacity. The pI 4.5 forms of intact virions as well as of A particles and empty shells all show a high tendency to aggregate and to stick to membrane filters and many other surfaces indicating hydrophobicity.

There are important features unique to the poliovirion pI 4.5 form (see Table 21). The C-antigenic A-particles and ATC have lost VP4 and are thus irreversibly locked in the pI 4.5 form, whereas the virion pI 4.5 form still contains VP4 and can revert back to the pI 7.0 form. The attachment capacity of the virion pI 4.5 form is reduced though not entirely absent as in ATC and A-particles. The attachment capacity of the pI 4.5 form of native poliovirions may even be higher than that of the pI 7.0 form, for example after incubation of poliovirions with dilute neutralizing antisera (Mandel, 1967a, 1971; Emini *et al.,* 1983a). The pI 4.5 form of

Table 21. *Characteristics of the pI 4.5 forms of poliovirion and related particles*

	NTC	Virion	A particles	ATC
Composition	$VP_{0,3,1}$	$VP_{4,2,3,1}$ + RNA	$VP_{2,3,1}$ + RNA	$VP_{2,3,1}$
Sedimentation coefficient	80	140(?)	135	80
Permeability to PTA	+	–	+	+
Sensitivity to protease	+	–	+	+
Antigenicity	C	D	C	C
Convertible to pH 7.0 form	–	+	–	–
Attachment-capacity	–	+	–	–

the native virion is much more resistant to attack by proteolytic enzymes than the C-antigenic A-particles or ATC. The pI 4.5 form of intact viurs is still in the D-antigenic conformation, since binding of some neutralizing antibody (see below), which converts the virus particle from the A to the B form, does not prevent the binding of additional anti-D-antibodies (Mandel, 1971; Icenogle *et al.*, 1983; Emini *et al.*, 1983a, b). As discussed in the following section the intrinsic tryptophan fluorescence of A and B forms of intact virions do not differ significantly, whereas conversion of the B-form to the C-antigenic pI 4.5 A-particles is associated by a marked transition in the intrinsic fluorescence of tryptophan. These observations indicate that the conversion from the A to B form of intact virions involves only relatively subtle changes in conformation and that the marked rearrangement of the capsid conformation only occurs when VP4 is lost.

A number of conditions have been described, whereby intact virions can be stabilized in the low pI 4.5 form, without liberating VP4, including acid pH, binding of neutralizing antibodies, treatment with arildone, and complexing with factors of the host cell membrane (Mandel, 1971, 1979; Caliguiri *et al.*, 1980; DeSena and Torian, 1980; Emini *et al.*, 1983b). Indeed, the described treatments confer an increased stability of the B-form virions to heat or alkaline-induced loss of VP4 from the capsid, and to the VP4 release induced by the host cell as the first step of uncoating.

When native virions are treated with D-specific neutralizing antibodies, they are converted to and locked in the B-form (Mandel, 1971). It is not clear, how many antibody molecules must bind to susceptible sites in order to induce this conformational transition. Neutralization kinetics are first order, indicating that it is possible for a single antibody molecule to neutralize the virion (Dulbecco *et al.*, 1956). This implies that a single antibody molecule may suffice to induce the transition from the pI 7.0 to the pI 4.5 form. The number of "critical sites" on the virion surface that must interact with antibody molecules to assure complete virion neutralization is on the order of 8 to 12 (Dulbecco *et al.*, 1956; Rueckert *et al.*, 1981). Recent analysis of the kinetics of neutralization by a D-specific monoclonal antibody revealed that binding of 4 antibody molecules suffices to assure 99% neutralization (Icenogle *et al.*, 1983). It is thus also possible that more than one antibody molecule must bind to native pI 7.0 virion, and that they must act together to induce the conformational transition to pI 4.5 virions. The VP4 component becomes exposed by antibody binding (Carthew, 1976), but is retained associated to the capsid surface in the antibody-stabilized pI 4.5 form. VP4 might contribute to a site that remains D-reactive in the pI 4.5 form of intact virion, but that is destroyed when VP4 is lost in a conversion to the pI 4.5 form (as in A-particles or ATC) prior to antibody binding (Breindl, 1971 b).

As mentioned above, attachment capacity is retained for the pI 4.5 B form of virions in contrast to the pI 4.5 A-particles (which lack VP4). Neutralized virions may even have an increased ability to attach to cells. Only when a virion is completely saturated with antibody, is it incapable of attaching to cells, evidently because all potential receptor recognition sites on the virus surface are either covered or "blocked" by steric-hindrance by antibody molecules (Holland and Hoyer, 1962). Antibody-bound virions may even enter the cell, but they cannot be uncoated or induced to release VP4 and RNA. Neutralized virus particles can be freed from neutralizing antibody and regain infectivity by repeated freezing and thawing or heating provided that divalent cations are present and that the pH is below 7.0 (Keller, 1965; Mandel, 1973; Wallis *et al.*, 1973). Mg^{++} ions are most effective, certain anions, such as phosphate, acetate, and sulfate, are strong inhibitors of reactivation, monovalent cations and anions have no effect. Acid pH alone can also dissociate neutralizing antibodies from virus-antibody-complexes and thereby restore infectivity (Mandel, 1973). The antibody-induced conformational shift can also be reversed and infectivity restored by digestion of the antibody with papain (Keller, 1968; Emini *et al.*, 1983b).

Arildone, an antiviral drug that selectively inhibits the replication of some RNA and DNA viruses (Diana *et al.*, 1977 a, b) has a comparable effect on the conformational state of intact poliovirions as the neutralizing antibodies. It stabilizes the virion against heat and alkaline treatment, it prevents the cell-mediated uncoating step that dissociates VP4 from the capsid (McSharry *et al.*, 1979; Caliguiri *et al.*, 1980), and it induces the conformational shift to the low pI 4.5 form (Eggers pers. communication). Arildone-treated virions also retain attachment capacity and enter into the cell. It will certainly be of interest to determine the antigenic reactivity of arildone-treated virions. The arildone-induced effect is also reversible upon removal of the drug. Arildone presumably interacts via hydrophobic interactions with the viral capsid, since the hydrophobic phenylring of arildone is necessary for its antiviral activity.

Intact virions with similar features of increased stability have been recovered from the host cell membrane shortly after adsorption (Lonberg-Holm *et al.*, 1975; DeSena and Torian, 1980). Such virions probably were first physically inserted into the lipophilic membrane bilayer from where they were released with membrane disrupting detergents. Some membrane components remain attached to these virions, and these are responsible for increased stability (see Chapter 7). These more lipophilic particles have not yet been characterized in terms of pI, antigenicity, sedimentation coefficient, or attachment capacity, yet it seems probable that they are in the isoelectric B-form. Their physical insertion into the lipophilic membrane implies hydrophobic surface characteristics which are also typical of the B-form.

The results discussed so far seem to indicate that the conversion of native virions from the pI 7.0 A form to the pI 4.5 form are not necessarily associated with a marked rearrangement of the capsid structure, moreover, that the dramatic structural alteration of capsid conformation which exposes a new set of antigenic determinants is related to the loss of VP4 and not to the conversion to the pI 4.5 form. In other words, the conversion to the pI 4.5 form is possible without dramatic alterations in the structure of the capsid. This conversion, however, might be a prerequisite for the following rearrangement of the capsid resulting in the loss of VP4. Based on the available evidence, it is tempting to speculate that the conversion to the pI 4.5 form is the result of a release of Mg^{++} ions from the capsid. This could explain the increase in negative charge on the virion surface. Under certain conditions, the viral capsid may be stabilized in this Mg^{++} free or depleted state by antibody, by arildone and at low pH. If VP4 is not released at this point, Mg^{++} may bind again to the capsid, restoring the pI 7.0 A form of the capsid. This evidently occurs when the pH is raised again from 4.5 to 7.0 and during reactivation of neutralized virions by freezing and thawing in the presence of divalent cations. The 110 Mg^{++} ions bound to a native virus (Mapoles, 1980) are suitable candidates for this interaction. The bond which is stabilized by Mg^{++} seems to be accessible to chelators of divalent cations only in the provirion stage (see Chapter 10), and not in the native pI 7.0 A form of the virus.

D. Conformational Alterations Monitored by Changes in the Intrinsic and Induced Fluorescence of Poliovirus Components

Structural alteration in the capsid of poliovirions can be analyzed by monitoring changes in the intrinsic fluorescence of poliovirions or changes in the fluorescence of added compounds (Grimmel *et al.*, 1983).

Proteins excited by UV light at 285–295nm emit fluorescence at 330–350 nm. The intensity of the emission is dependent on the local environment of the aromatic amino acid residues, notably tryptophan (Udenfriend, 1969). Sudden transitions in the fluorescence are observed when proteins snap into alternate tertiary structures. Structural alterations in virus particles can be followed in this way (Grimmel *et al.*, 1983). The permeability of the viral coat to small molecules can be assayed with ethidium bromide (EB). EB is a dye which intercalates between double stranded regions of RNA; its fluorescence increases thereby. As the RNA is converted from a partially double stranded to a more extended state during its release from the virions, EB is freed and the fluorescence decreases.

The application of these methods to polioviruses yielded the data presented in Figure 31. The intrinsic fluorescence of poliovirions is almost identical at pH between 4.1 and 8.4 indicating that the environments of the aromatic acids in the A and B forms of native virions do not differ significantly. This supports the

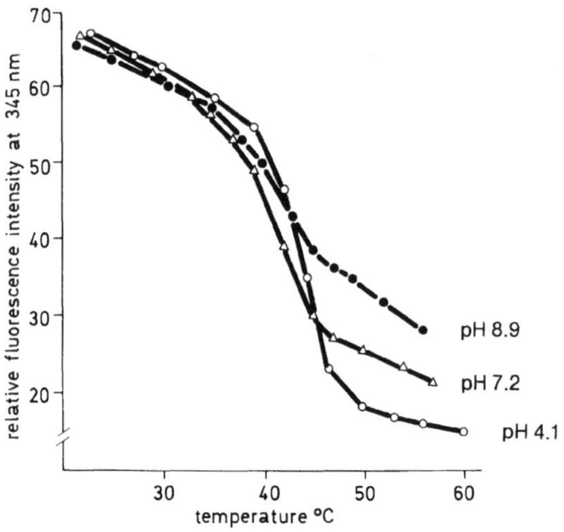

Fig. 31 A–B. Fluorescence spectrophotometric studies on structural alterations of the poliovirus capsid during heat-induced dissociation

Fig. 31 A. Transition in the intrinsic fluorescence of poliovirus during temperature elevation at different pHs
The figure shows the steep transitions in the intrinsic fluorescence of poliovirions accompanying the release of VP_4 during continuous temperature elevation (2°C/min). The initial decrease in fluorescence intensity between 22° and 35°C corresponds to the decrease of free tryptophan and of tryptophan in serum albumin in this temperature range. Quantitative evaluation of the data reveals a T_m of 43.3, 41.8 and 41.0°C for the transitions at pH 4.1, 7.2, and 8.9, respectively

notion that the conversion of poliovirions from the pI 7.0 to the pI 4.5 form involves only subtle changes in conformation as long as VP4 is not released. These changes may be accompanied by the loss of some positively charged components such as Mg^{++} ions. As soon as VP4 is released upon elevation of temperature, however, a sudden steep transition in the tryptophan fluorescence is seen with viruses both in the pI 4.5 and the pI 7.0 form. The extent of the transition indicates a major concerted conformational alteration in the capsid. The transition occurs at a temperature where the virus changes its antigenicity and looses its attachment capacity. Notice that the pI 4.5 form is more stable to elevated temperatures than the pI 7.0 form at neutral or alkaline pH, and that the transition in fluorescence at acidic pH occurs in a very narrow tmperature range whereas the transitions at neutral and alkaline pH are biphasic. A biphasic transition of the B form is consistent with a consecutive release of Mg^{++} and VP4. The increased stability and single phase transition at acidic pH is consistent with the release of VP4 from a particle that has already released most of its Mg^{++} and is stabilized by acidic pH. The transition only slowly proceeds to completion at neutral or alkaline pH in accord with a proposed heterogeneity of virions with regard to Mg^{++} ions. Poliovirions are converted to the pI 4.5 form by arildone even at neutral pH and are stabilized. The transition of the tryptophan fluorescence is shifted to higher temperatures in the presence of arildone (Grimmel *et al.*, 1983).

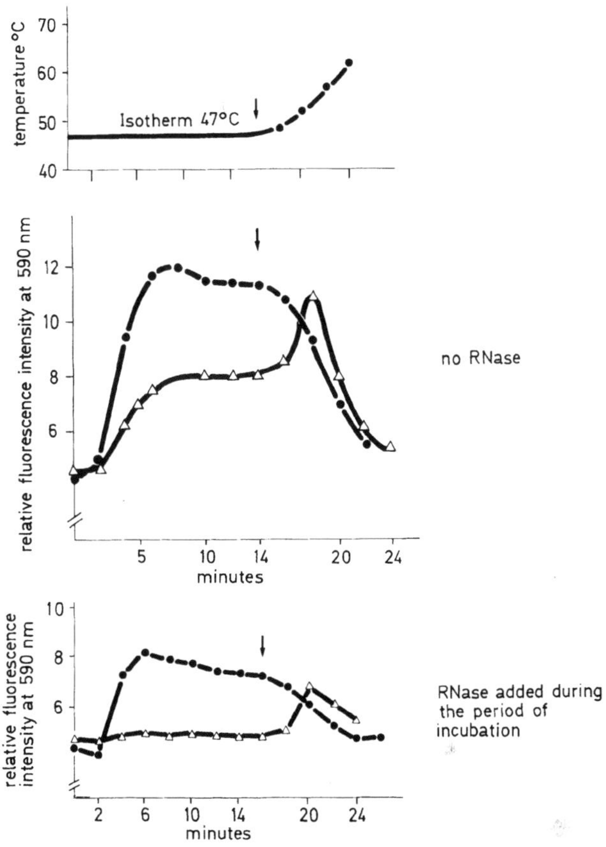

Fig. 31 B. The interaction of poliovirus RNA with ethidium bromide during temperature-induced uncoating

The figure illustrates the temperature induced permeability of the poliovirus capsid to ethidium bromide, and the associated increase in the fluorescence of ethidium bromide as it binds to the double stranded regions of RNA in the particles. At 47 °C, the VP_4 component is released from untreated virions (●——●), permitting ethidium bromide to interact with the viral RNA. Further elevation of temperature results in release of RNA from the particle, which causes the fluorescence to decrease again. In the presence of RNase (bottom figure), a large portion of the RNA is no longer accessible to labeling by ethidium bromide. Arildone (Δ——Δ) stabilizes virions against part of the permeability increase to ethidium bromide (top figure), but does not protect the RNA which is accessible to ethidium bromide binding from degradation by RNase (bottom figure). This indicates either heterogeneity of the virus population or two distinct compartments for RNA in all virions

VP_4 contains no tryptophan, VP_2 harbors eight, VP_1 and 3 three tryptophan residues each. The steep transition in the intrinsic tryptophan fluorescence accompanying the release of VP_4 reflects major structural rearrangements of VP_2 and probably also of VP_1 and 3. The rearrangement of VP_2 may be related to the disappearance of the D antigenicity, and that of VP_1 and 3 may result in the exposure of the C antigenic determinants and the loss of attachment capacity.

Figure 31B illustrates that the loss of VP4 from virions at 47°C renders the capsid permeable to small compounds such as EB. EB permeates into virus particles and intercalates with double stranded regions in the RNA. The association with the RNA is stable at 47° C. Further elevation of the temperature, however, causes a release of the RNA and thereby also a dissociation of EB from the RNA, detectable in a decrease in the EB fluorescence.

In the presence of arildone, where most of the VP4 component is retained in the virion, the permeability of virions incubated at 47° C is markedly reduced. Only half as much RNA as in untreated particles is accessible to ethidium bromide, the rest becomes accessible only after further elevation of temperature. This indicates either that the virus population is heterogeneous with respect to stabilization by arildone, or that the RNA is distributed in two distinct compartments within each virion, only one of which becomes permeable to ethidium bromide after arildone stabilization.

When the same experiments are repeated in the presence of RNase (Fig. 31B), half of the RNA regions bound by ethidium bromide in untreated virions at 47° C are resistant to RNase digestion and half are destroyed by the RNase. All of the RNA regions that were accessible to ethidium bromide in arildone treated particles at 47° C are destroyed by the RNase at 47° C. Further temperature elevation of arildone treated particles renders the remaining fraction of RNA accessible, first to ethidium bromide, and then also to RNase. These results can also be interpreted as reflecting particle heterogeneity with respect to RNase sensitivity after loss of VP4, or as reflecting two distinct compartments for RNA in all virion particles. It is not possible at present to decide between these two alternatives. The two compartment interpretation is in accord with observations from the studies on heat and alkaline induced dissociation of poliovirus discussed above. The sedimentation property of 90S RNA containing particles after heat dissociation of VP4 from the virions was shown to depend on the ionic strength of the sedimentation medium, indicating surface exposure of the RNA (Breindl, 1971 a). Under conditions where VP4 is released from the capsid by alkali, only part of the RNA became accessible to RNase digestion (Vanden Berghe and Boeye, 1973c). The RNA was cleaved into 10–12 fragments but was not degraded further by alkali. The initial 10% decrease in ethidium bromide fluorescence after loss of VP4 at 47° C (Fig. 31B) may reflect the melting of portions of RNA.

In any case, the techniques described should be useful in further analyses of compounds which induce conformational alterations in virions and which stabilize and destabilize virions.

E. Dense Particles

A third form of complete poliovirus particles has been discovered, that seems to differ from the interconvertible isoelectric A and B forms. These particles are slightly smaller and denser than native virions, they exhibit surface features similar to that of native virions, but they appear to have disturbed or partially disrupted interactions between the capsid proteins and the viral RNA (see Table 22).

Table 22. *Comparison of the properties of native virion and dense particles (DP)*

Properties	Virion	DP
Density (g/cm^3)		
in CsCl	1.34	1.44
in Cs_2SO_4	1.32	1.38
Sedimentation (s)	155	220 and 160
RNA	35 S	35 S
Protein content	4–2–3–1	4–2–3–1
Diameter (nm)	30.1 ± 0.74	28.2 ± 0.76
Stability	+++	highly labile under isotonic conditions converted via 90 S ($VP_{1,2,3}$ RNA) to 80 S $VP_{1,2,3}$ + 35 S RNA
Antigenicity	D, N	D, N less efficient in eliciting formation of neutralizing antibodies
RNA-protein interaction upon urea dissociation	RNA-VP_1 complex	no RNA-VP_1 complex
Specific infectivity	1.0	0.25

In addition to the major infective component which bands at the density of 1.34 g/cm^3 in cesium chloride, a distinct minor component, the so-called "dense" particle with an unusually high buoyant density of 1.44 g/cm^3 in cesium chloride has been found in preparations of poliovirus and other enteroviruses (Yamaguchi-Koll *et al.*, 1975; Rowlands *et al.*, 1975b; Cova and Aymard, 1979, 1980). A host cell dependence for the appearance of dense particles has been reported for Echovirus 11 (an enterovirus): dense particles are not produced during passage of Echovirus in primary monkey kidney cells, whereas they are produced upon passage in HeLa cells (Cova and Aymard, 1980). Dense particles contain the normal structural proteins $VP_{1\text{-}4}$ (small amounts of VP_0) and intact 35S RNA. The proportion of protein and RNA is identical to that of standard particles. The morphology of dense particles is similar to that of standard particles, but the diameter (28 nm) is slightly smaller (standard particle 30 nm) (Rowlands *et al.*, 1975b).

Dense particles sediment at 220S in sucrose gradients in the presence of cesium ions (Wiegers *et al.*, 1977, 1978). In the absence of Cs^+ ions, a portion of dense particles behaves physically like the standard particles, sedimenting at 160S and banding at 1.33 g/cm^3 (*e.g.*, in urografin), and the rest sediments at 220S and bands at 1.38 g/cm^3. These changes in the sedimentation coefficients and buoyant densities of dense particles are reversible. When dense particles were centrifuged in cesium sulfate gradients, they banded as one peak at 1.38 g/cm^3. However, no interconversion between dense and standard particles is observed on recycling in fresh cesium chloride gradients.

Dense particles have a somewhat (approximately four-fold) lower specific infectivity, that may be due in part to broken RNA (Dernick, 1981). Dense particles are less efficient in eliciting the formation of neutralizing antibodies due to their inherent instability; nonetheless, their infectivity is neutralized by standard type-specific poliovirus antisera. Capsid polypeptides in dense and standard virions are labeled by iodination in a similar pattern and to a similar extent, indicating no major differences in surface structure. In contrast to standard particles, dense particles are highly labile: They are rapidly degraded in isotonic phosphate-buffered saline, the degradation proceeds via an RNAcontaining 90S particle lacking VP4, to RNA and 80S empty capsids. No RNA-VP1 complex can be obtained upon urea-mediated degradation of dense particles in contrast to degradation of standard poliovirus. Native virions can be converted to dense particles by treatment with formaldehyde (Agol *et al.*, 1970).

The described properties of dense poliovirus particles are indicative of some conformational differences to standard poliovirus particles. The nature of these differences, however, is still unclear. The changes are evidently minor (similar iodination pattern retained, but reduced infectivity and antigenicity) if compared, for example, to the changes accompanying loss of VP4. Yet the alterations are also significant as evidenced by an increased permeability to Cs^+ and binding capacity for these ions and by the increased lability in isotonic phosphate buffer.

A buoyant density of 1.44 in cesium chloride gradients corresponds to a binding of approximately 7,500 cesium ions (Mapoles *et al.*, 1978; Mapoles, 1980) to the RNA; and this implies a cesium-mediated neutralization of all phosphate groups of the RNA. Standard poliovirus particles, which are impermeable to cesium ions have a maximal binding capacity of 5,800 cesium ions as determined by propagation of poliovirus in cesium-rich medium (see above). It was proposed that the remaining phosphate groups are neutralized in standard poliovirus particles by basic amino acid residues of the capsid proteins, and that the increased buoyant density of dense particles reflects the disruption of the usual RNA-protein interaction, which permits the binding of a greater amount of Cs^+ ions. The lack of a VP1-RNA complex after dense particle degradation by urea is consistent with this interpretation. In sum, the properties of dense particles may be explained by a "loosening" of the capsid structure caused or accompanied by a disruption of the RNA-protein interactions.

It would be of interest to determine the isoelectric point(s) of dense particles to find out if there might be a correspondence to the conformational A and B states of the native virion (see above), and to investigate the reactivity of the dense particles with anti-C antibodies in order to check if the conformational alterations might be accompanied by an exposure of the C-antigenic determinants.

F. Possible Functions of the Alternate Conformational States of the Poliovirus Capsid

In sum, the capsid of poliovirus is not as rigid a structure as is often supposed, instead it exhibits plasticity. The viral proteins in the capsid possess the remarkable inherent capacity to undergo concerted conformational transitions between distinct structural states. Three different states seem to be involved. Under certain conditions native virion may revert between two alternative states. These states exhibit markedly different surface charge characteristics, differences in overall architecture seem to be subtle. Dramatic structural rearrangements seem to occur only during loss of VP4, and this conversion is then irreversible. A reversible dissociation of divalent cations, presumably magnesium ions, may account for the marked increase in negative surface charge in the conversion to the pI 4.5 form of the virion. Figure 32 shows a hypothetical scheme of the structural basis for the conformational forms of poliovirus.

Under physiological conditions the virion capsid assumes the native, tightly packaged, state which is characterized by its relatively low permeability even to small molecules, a net surface charge of zero, a high affinity for the host cell receptor, resistance to uncoating by the host cell, and a characteristic set of antigenic surface determinants, the D(N) reactive sites. A multitude of natural and experimental conditions are known which destabilize this native conformation. The diverse treatments which include heating, alteration of pH or salt concentrations in the suspension buffer, interactions with either antibodies or the host cell receptor complex, probably all at first affect one or more – and perhaps different ones – of the many binding forces within and between the individual capsid proteins. For example, a change in pH may ionize or neutralize some residues which might disrupt an intracapsid linkage or it may destabilize nearby linkages. Binding of an antibody molecule or a host cell receptor complex to certain complementary sites on the virion surface may lead to an allosteric type of destabilization of other intracapsid linkages. Heat and denaturing agents may disturb binding forces by enhancing flexibility of proteins. As a result, the capsid proteins acquire a certain degree of transitional and rotational freedom, the capsid as a whole begins to "breathe" and expand. Also the capsid may become "leaky" allowing water or other small substances to channel through the shell, and some hitherto hidden antigenic sites may become exposed on the surface. When a critical number of the intracapsid bonds have been disrupted, all of the building blocks of the capsid simultaneously undergo a stereotype conformational transition. As a result, the surface charge of the particle turns negative, the pI of the new conformation is 4.5. It was proposed that this step involves the release of divalent cations, presumably Mg^{++}, from ionic bonds of great importance to virus stability. The release of Mg^{++} ions labilizes the bonds holding the small capsid protein VP4 within the particle, and – unless the virion is stabilized by acid pH, arildone, or antibody binding–VP4 is released in the same step.

When VP4 (and RNA) are retained as part of the virus during the conversion to the pI 4.5 form, the virus retains the capacity to revert back to the pI 7.0 form in a "reverse" conformational shift. This does not seem to hold also for the low pI form of the empty shells formed in vivo (NTC), which retain the VP4 component as part of the precursor protein VP_0. The virion can be locked reversibly in the low pI form by binding to neutralizing antibody, by treatment with arildone, and possibly also by certain lipophilic fractions of the host cell membrane. The low pI form of intact virion apparently retains capacity to attach to host cells and to bind

D-specific antibodies and still shows the same intrinsic fluorescence as the pI 7.0 form.

When the virus is not stabilized in the pI 4.5 form or when the disrupting conditions are increased, a second more dramatic structural alteration of the capsid ensues leading to a loss of VP4 and of the native antigenic determinants of the virion. This second transition presumably is driven by an influx of water into the interior of the shell, that would tend to expand the shell, a phenomenon that should be augmented by hypotonic and reduced by hypertonic solutions. The transition is probably controlled and directed by the sequential disruption of some bonds, the formation of new bonds, and by other prior bonds that remain

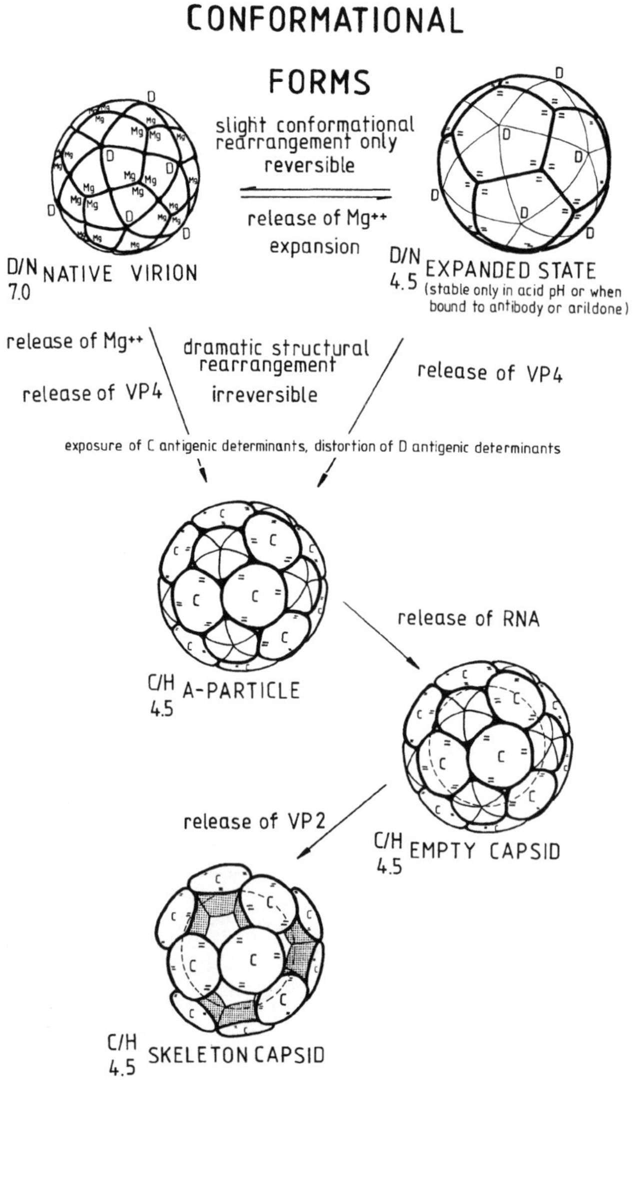

stable. At a certain point, the shell will approach a state that allows for a number of new, stable bonds to form. This potential state is quite stable and the capacity to acquire this state is an intrinsic, genetically determined feature of the capsid proteins. Depending on the severity of the disturbing influences the small capsid protein VP4 is released alone or together with the 35S viral RNA from the shell as it snaps into its alternative state.

The characteristic common features of the alternative capsid state are a different set of antigenic determinants, its low isoelectric point at pI 4.5, an increased surface hydrophobicity, a slightly more expanded state, and an increased stability (see Table 21 above). The high stability, *i.e.*, low energetic and thus thermodynamically favorable state of this form of the capsid is evidenced by the finding that the shell may snap into this state no matter whether it contains VP4 still as part of the VP0 precursor (as in NTC), or whether it has lost VP4 alone (A-particles), or both VP4 and RNA (ATC), and that the conversion is irreversibel when VP4 is released.

When VP4 is lost during the conversion to the pI 4.5 form, the capsid becomes permeable to PTA (and consequentially appears "coreless" in the EM), becomes sensitive to protease, looses its attachment capacity to host cells as well as its binding site for D-specific antibodies, and acquires the C-reactive surface determinants.

It appears as though the conformational shift to the expanded pI 4.5 form opens or widens "holes" in the shells through which the small capsid protein VP4 may be liberated from its internal position in the native state. In the pI 4.5 form of intact virions, the VP4 component apparently still covers the corresponding holes, since pI 4.5 virions are not permeable to PTA. The locations of the proposed holes for VP4 are not known. Assuming that pentamers of VP2 in

---

Fig. 32. Hypothetical structural basis of the conformational forms of poliovirus

This figure illustrates schematically the different conformational states of poliovirus

Top: The native virus particle may exist in one of two alternate states characterized by isoelectric points of 7.0 and 4.5. The interconversion between these two states is reversible. The pI 4.5 form is slightly larger, but retains the D/N antigenicity. Indirect evidence is consistent with a release of small cations (possibly Mg^{++}) in the conversion of the pI 7.0 to the pI 4.5 form. The pI 4.5 form is stable only in acid pH or when bound to antibody or arildone. Under physiological conditions or upon dissociation of antibody in the presence of Mg^{++}, the particle reverts to the pI 7.0 state

Bottom: A wide variety of treatments induce an irreversible rearrangement of capsid structure concomitant with the release of the smallest capsid protein VP_4. This modified state–the so called "A-particle"–is characterized by a low isoelectric point of 4.5 and a different set of antigenic surface determinants (C/H) typical of empty capsids, but still contains RNA. Subsequent release of RNA and the VP_2 capsid proteins occurs without further alterations of the antigenic properties or the isolelectric point

Explanation of symbols: Native virion is symbolized by a spherical 60 subunit (protomer) icosahedral lattice with D-antigenicity (D). The hypothetic release of Mg^{++} is symbolized by the appearance of negative charge on the capsid surface (–); the exposure of C antigenic determinants by an alternative clustering pattern of the capsid proteins and the letters C; the release of RNA by an "empty core" (dashed line); the release of VP_2 by removal of the corresponding cluster of capsid proteins about the 5-fold vertices–note that the latter steps do not alter the C-antigenic determinants (which evidently reside in the VP_1–VP_3 clusters) virions and VP_0 in NTC occupy the position at the twelve icosahedral 5-fold vertices, and further that VP_4 still lies in vicinity to VP_2 in native virion, two good possibilities for the location of the VP_4-holes come to mind: either directly at the icosahedral vertices (there would then be twelve holes) or in the center of a protomer between VP_2, VP_3 and VP_1 (*i.e.*, also near the "grooves" between the VP_2 pentameric clusters and the VP_3–VP_1 hexameric clusters; in this case, there would be 60 holes).

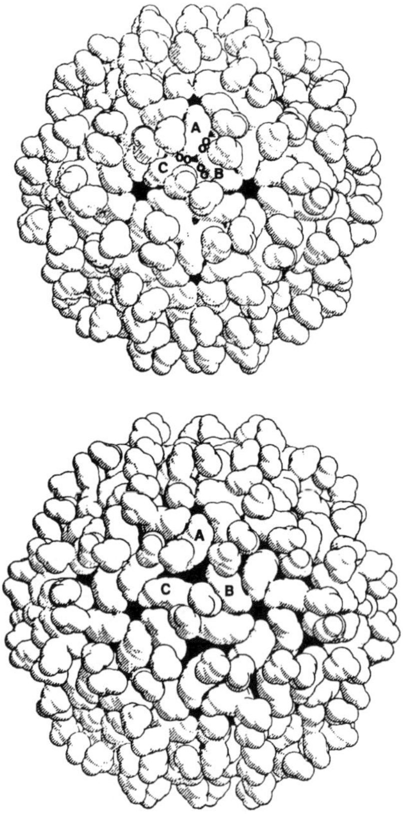

Fig. 33. Models of the native and expanded states of TBSV
This figure shows a pictorial view of the expansion of tomato bushy stunt virus
The compact particle of TBSV (top figure) is the native state of the virus particle
This state can be converted to an expanded state by a reversible, cooperate expansion at pH values above neutrality and in the absence of divalent cations. The expansion mechanism is triggered by deprotonation of asparate residues of the calcium binding sites of TBSV, provided the Ca^{++} salt bridges were removed by chelation (open circles)
As a result of expansion, a branched opening 80 Å long and large enough for a 20 Å sphere to pass through is formed in the center of each of the 60 promoters. Each promoter contains the capsid proteins A, B and C which are actually identical proteins in different conformational states. The protruding domains (P-domains, see Fig. 18 A) are rotated by relatively large angles (30° in the AB case and 103° for the CC dimer). Note the appearance of the inter-subunit opening and the rotation of the projecting domains in the expanded virus. – Figure courtesy of S. Harrison, Boston

It is quite interesting to note the similarities between the poliovirus conformational states and those of the plant picornaviruses; TBSV also can occur in a native state and in an expanded state (Fig. 33). X-ray diffraction studies on crystals of the expanded form have allowed an elucidation of its structure at a resolution of 8 A° (Robinson and Harrison, 1983). It can be seen that in passing from the native to the expanded state the individual structural subunits all have undergone a stereotype rotation and spread apart. Some of the interactions between subunits in the native state that are disrupted in the transition, and some of the new interactions formed in the expanded state, have been characterized on the molecular level. As a result of the expansion, a small hole appears to open in the center of each of the 60 icosahedral structure units (composed of three polypeptides each, see above). The N-terminal arm of one of the capsid proteins (presumably of the A- or B-protein), which occupies an internal position in the native state may move to the surface of the virion, presumably through the hole in the center of the structure unit. The arm thereby becomes sensitive to cleavage from its "parental" protein at a specific point, liberating a 12,000 dalton protein. The analogy with VP4, which can be considered as the N-terminal arm of VP_0, immediately comes to mind.

From the wide spread occurrence of the expanded and empty-shell-like conformational states and their conservation throughout evolution (shells of other picornaviruses also exhibit the capacity to assume an expanded state), it is evident that the expanded state must play an important role in picornavirus "life". It is probable that the poliovirus shell assumes this conformational state transiently during virion maturation —presumably to permit the cleavage of VP4 from VP_0 in the provirion. Complete virion can be expected to assume the pI 4.5 form again transiently during its passage through the acidic environment of the stomach, perhaps thereby providing some unknown protection or stability to the virion against dissociation. A similar statement can perhaps be made for certain conditions in the environment, where the virus is known to survive in sewage, waste water, etc. During the earliest interactions with the host cell, the surface hydrophobicity of the pI 4.5 form of virions may be a requirement for insertion of the virus particle into the lipophilic cell membrane. In order to pass through the lipid bilayer and to be released again on the cytoplasmic side of cell membrane, a reversion to the more hydrophilic pI 7.0 form may be required. VP4-lacking A particles cannot revert back to the more hydrophilic pI 7.0 form, whereas stabilized VP4-containing particles can revert back, providing a mechanism whereby the VP4 components of the virus particle can get to the cytoplasmic side of the cell as has been observed (see Chapter 7). The capacity of capsid components to assume alternative more hydrophobic or more hydrophilic conformational states may also play an important role in the coordination of virion assembly late in infection, since many of the involved steps occur in association with membranes. Finally, a coordinated capsid expansion mechanism may play an important role in the initiation of uncoating where holes are created through which the VP4 "uncoating plugs" and eventually also the genomic RNA may be liberated.

VIII. Antigenicity

To this date, some picornaviruses are still world wide health hazards to humans and livestock. Two prominent examples are the foot and mouth disease viruses and the human enteroviruses (including polio, hepatitis A, haemorrhagic conjunctivitis and coxsackie viruses). Many investigations of the antigenic properties of polioviruses and the other picornaviruses have been performed over the years and – correspondingly – a vast amount of literature has accumulated. This is not surprising, since the results of such studies are expected to be of help to solve the practical problems concerned with the control of picornavirus-induced diseases. Such problems include:

1. Production of efficient live or inactivated vaccines;
2. Antigenic "drifts" of wild type or vaccine viruses in response to antibody pressure;
3. Development of useful procedures for comparative diagnosis of closely related etiological viral agents (*e.g.*, different types of human enteroviruses, or the foot-and-mouth disease viruses of livestock).

Before one discusses the antigenicity data, it is of value to recall a few important points about capsid structure and the immune system:

1. Heterogeneity of antibodies is an intrinsic feature of the immune response in animals: An animal antiserum produced in response to immunization by a proteinacious antigen always contains a heterogeneous set of antibodies. It may contain antibodies of different fine-specificities, *i.e.*, antibodies directed against different antigenic sites on the same protein. However, the antiserum of a given animal will usually exhibit one or a few dominant specificities. The immune systems of different animals (even of the same species) may show preferences for different sites on the same antigen. Usually, certain antigenic sites–the main antigenic sites –however, will be statistically much more efficient than others in eliciting antibody production, even in different animals. Heterologous antisera can be purified and made monospecific by precipitation or adsorption with related structures. Typically the crude antisera are passed through columns containing covalently bound viral structures, such as intact virions, empty shells or isolated viral proteins. Contaminating antibodies, or antibodies directed against common sites, remain bound to the column and are thus filtered out from monospecific antibodies that are found in the flow-through.

– Even animal antibodies of a given specificity are not a single molecular species, rather they are a heterogeneous set of antibody molecules with a wide range of binding affinities for the same antigenic determinant. Such antibodies can be separated on the basis of different binding strengths to affinity columns containing the covalently bound antigen.

– Antibodies produced by a single cell–or by a clone from such a cell–are homogeneous. With modern hybridoma technology, antibody-producing lymphocytes from immunized animals can be immortalized in culture by fusion with tumor cells. Clones can be obtained from such hybrid cells which produce identical antibodies of a single specificity, so-called monoclonal antibodies

(Köhler and Milstein, 1975). With proper screening methods, clones synthesizing antibodies to any particular antigenic determinant can thus be selected.

2. Antibodies that inactivate virus infectivity as a consequence of their interaction with virus are termed neutralizing antibodies. Other, non-neutralizing antibodies may bind to a complementary antigenic site on the virion without abolishing infectivity. There are in all likelihood a number of different neutralizing antibodies directed against a given type of poliovirus, corresponding to a number of different neutralization—sensitive antigenic sites. Different neutralizing mechanisms may be involved in the virus—antibody interaction. Most simply, antibodies could directly block the host cell receptor recognition site on the virus particle or alter this site in an allosteric manner. Alternatively, antibody binding could lock the virion capsid in a structural state that prevents the dissociation of the capsid which is required for the liberation of the genome inside the host cell. Antibodies could induce neutralization by crosslinking and precipitating virions or by saturating and sterically blocking functional sites on the capsid surface. Or neutralization could involve the alteration of structural proteins, thereby inactivating a hypothetical essential function that these proteins might have in some step in the initiation of viral replication. The main neutralizing mechanism of poliovirus exhibited by antisera obtained after immunization with native virion apparently is of the second type, *i.e.*, inhibition of uncoating. Recently, a great number of monoclonal antibodies obtained after immunization with intact virions or subviral particles have been characterized; some of these apparently neutralize infectivity by one of the other mechanisms listed.

3. Polioviruses are relatively large protein antigens; the surface of the virion is covered with a number of different potential antigenic sites. The viral capsid is constructed from 60 repeating building blocks. Each building block has a molecular weight of close to 100,000 and is composed of some 880 amino acid residues; each building block can be expected to have on the order of 100 surface exposed amino acid residues. Antigenic determinants are usually constructed from only a few (6—10) amino acid residues. The size of the antigen binding site on antibodies is on the order of 2.5 nm (Eisen, 1974), the diameter of the surface exposed area of a poliovirus building unit is approximately 7 nm. An antigenic site is determined by the sequence of the component amino acid residues and the particular three dimensional conformation of the site.

4. The icosahedral geometry of the poliovirus capsid defines the number of possible antigenic sites. An antigenic site may be composed of polypeptide stretches from one of the capsid proteins, or it may be combined from the stretches of adjacent capsid proteins. Table 23 lists some theoretical possibilities for the constitution of antigenic sites and respective possible numbers on the poliovirus capsid. The sites are listed in terms of the involved capsid proteins and the geometrical features that follow from the structure of the proposed poliovirus model.

5. The surface of a virion is ruffled rather than smooth. Certain polypeptide stretches protrude from the capsid backbone; the polypeptide stretches of the capsid backbone in contrast are relatively more depressed or less surface-exposed; still other polypeptide stretches of the capsid surface may lie in grooves, particularly those at the borders of adjacent proteins or of adjacent morphological clusters.

Table 23. *Theoretical antibody binding sites on poliovirus*

Viral peptides	Number of peptides	Location on the virus particle	Surface area nm^2	Number of sites/virion
VP_1	1	part of 3-fold apices	15	60
VP_2	1	part of 5-fold apices	15	60
VP_3	1	part of 3-fold apices	10	60
$VP_1 + VP_3$; $VP_1 + VP_2$ $VP_2 + VP_3$	2	within protomers or across grooves between adjacent protomers	25	60
$(VP_1 + VP_2 + VP_3)$	3	across grooves between adjacent pentamer (VP_2) and hexamer (VP_3, VP_1) clusters	40	60
$(VP_3 + VP_1)_2$	4	groove between 2 hexameric clusters (2-fold apices)	50	30
$(VP_3 + VP_1)_3$	6	3-fold apices	75	20
$(VP_1)_3$ or $(VP_3)_3$	3	facet center	45	20
$(VP_2)_5$	2–5	5-fold apices	30–75	12

Upon conformational rearrangements of the capsid surface or upon breakdown of the capsid, such sites may become exposed and acquire immunogenicity.

6. Not all antigenic sites on native virion are also immunogenic; there may be sites exposed on the capsid surface of native virion which—in principle—can interact with complementary antibody (the sites are antigenic), but which cannot—as part of the native virion elicit the responsible B lymphocytes to produce the corresponding antibody (the sites are not immunogenic). Antigenic sites vary in the extent of their immunogenicity, that is, in their capacity to elicit susceptible B cells to begin production of the complementary antibodies. Although the virus surface may harbor many potential antigenic sites, only one or a few sites—the main antigenic sites—will usually dominate in the interactions with the immune system of the host. Potent immunogenic antigenic sites are likely to be constituted by special surface features such as protrusions which provide greatest accessibility. Nonimmunogenic, but antigenic, sites on native virion are probably less accessible sites hidden in grooves or valleys on the capsid surface.

An overall conformational rearrangement of the capsid surface—for example by an expansion of the entire capsid structure or an inside-out flopping of certain capsid components—may expose hitherto burried antigenic sites or it may create new antigenic sites. Similarily, complete dissociation of the virion capsid into the component capsid proteins may expose antigenic sites that are burried or hidden in native virion (for example sites involved in intra-capsid bonding or bonding to RNA). In addition, an isolated capsid protein may assume a substantially different conformation in the isolated state than as a component of a large ribonucleoprotein complex such as the virion. Although the major antigenic determinants of native virions, capsid shells, capsid subcomponents, and isolated capsid proteins may be very different, it is probable that all of these structures still contain common—potentially antigenic—structures or surface regions even if such regions exhibit only poor immunogenicity. Consequently, it may be merely a matter of fortune to obtain antibodies capable of interacting with native virion by immunization with isolated capsid proteins.

7. Antigenic sites on the virus capsid may overlap or they can be entirely distinct. Similar antigenic sites (slight differences in conformation or amino acid composition) may elicit the production of antibodies that can interact with either site, though usually with different affinities. Such antibodies are said to be cross-reactive.

Antibodies produced against "overlapping" antigenic sites will reveal similar specificities though they may differ in their interaction characteristics with the whole antigen. For example, stretches of two neighboring polypeptides may constitute overlapping antigenic determinants. An antibody directed against a combination of the two polypeptides may exhibit the same specificity as an antibody directed against the stretches from only one of the two polypeptides.

8. There are type-specific and group-specific antibodies. Type-specific antibodies are highly specific and interact exclusively with a particular type of virus (for example with poliovirus type 1 but not with types 2 or 3). Group-specific antibodies interact with different virus types of a related group (polioviruses types 1, 2, and 3, but not other enteroviruses). In other words, group-specific antibodies react

with similar sites on related viruses, sites which evidently either have been well conserved or covaried in evolution (capsid backbone, receptor recognition site ?). Type-specific antibodies distinguish between closely related viruses, *i.e.*, they interact with sites that have changed significantly in evolution (surface protrusions ?).

Anti-poliovirus neutralizing antibodies are type specific. Neutralizing antibodies against a type 1 poliovirus, for example, usually do not interact with or neutralize type 2 or 3 polioviruses. The poliovirus receptor on HeLa cells, on the other hand, has the characteristics of a group-specific antibody: It binds all three types of poliovirus.

9. The antibody response of an animal to the primary challenge (infection) by an antigen usually reveals some marked qualitative differences to the response upon subsequent exposure to the same or related antigens. The primary antibody response is usually highly specific. Once challenged by a particular antigen, the immune-response to subsequent infections acquires more generalized features, as though the immune system had learned or remembered something from the primary infection, antibodies produced in response to reinfection with the same antigen may reveal greater crossreacting capacities with related antigens. The antibody response to a subsequent infection by a related antigen may be much more vivid than the isolated primary response to such an antigen.

A. The Main Poliovirus Antigenic Sites: The D- and C-Antigenic States

Purified or crude suspensions of each of the three poliovirus serotypes may contain several particular forms that differ in size, density, chemical composition, infectivity, and antigenic nature (see also Section VII above). The antiserum produced by animals immunized with such virus preparations, or with poliovirus-infected tissue material, contain mainly two heterologous sets of antipoliovirus antibodies of different specificities, directed against the two distinct predominant antigenic determinants in poliovirus preparations (see Fig. 34).

These two main antigenic determinants were termed C- and D-antigens, respectively, based on their association with two of the four main fractions (A, B, C, D) obtained upon separation of poliovirus concentrates by fractionation in sucrose density gradients (Fig. 34A) (Le Bouvier *et al.*, 1957, Mayer *et al.*, 1957).

Fig. 34 B. The C and D antigen are immunologically unrelated
The center well of this agar plate contained antiserum directed against D and C antigen. D antigen was a suspension of 1 mg/ml purified virus, banded by equilibrium centrifugation in CsCl, C antigen was a similar suspension heated for 15 min at 50 °C. The well at the top (TC) contained empty capsids separated from virions by centrifugation in a CsCl gradient. The bottom well (PPV) contained the same concentration of partially purified virus preparation prior to centrifugation in CsCl
The crossing of the respective arcs of the C and D antigen indicate that these antigens are immunologically unrelated. On the other hand, heated virus particles ("C") is immunological identical to empty capsids obtained from viral material in a CsCl gradient (TC = top component). – Figure from Scharff *et al.*, 1964 [Proc. Nat. Acad. Sci. *51*, 330 (1964)]

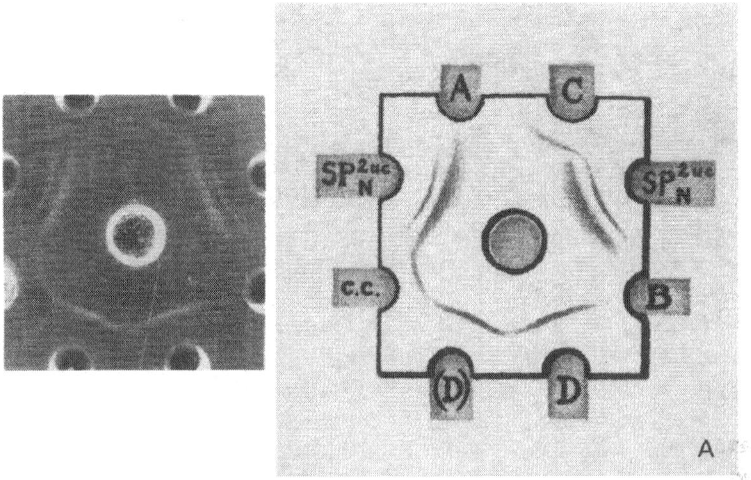

Fig. 34 A–B. Characterization of the antigens on poliovirus particles

Fig. 34 A. The original characterization of poliovirus antigens "C" and "D"

Poliovirus concentrates were obtained from the supernatant of infected monkey kidney cultures, partially purified, and separated into four components—A, B, C, and D—by fractionation in sucrose density gradient centrifugation. These four fractions were placed in the peripheral cups as indicated and allowed to precipitate with type specific antisera obtained from monkeys immunized with infected monkey cord (poliovirus type 2, MEF-1 strain). No precipitates were formed by heterotypic virus-serum combinations (not shown). SP_N^{2uc} = partially purified poliovirus concentrate prior to sucrose density gradient centrifugation, cc = 100 fold, crude virus concentrate. Fraction D consisted of virus particles and contained most of the infectivity. Particles in fraction C were less uniform in shape and flatter than those in fraction D when examined by electron microscope

One of the SP_N^{2uc} bands fused with the single band formed by fraction D, which appeared to be identical also to the "major" band produced by crude concentrate of virus. The second SP_N^{2uc} band fused with the single band formed by fraction C, and with a similar one, where present, formed by fraction B. – Figure from Le Bouvier *et al.*, 1957 [Virology *4*, 591 (1957)]

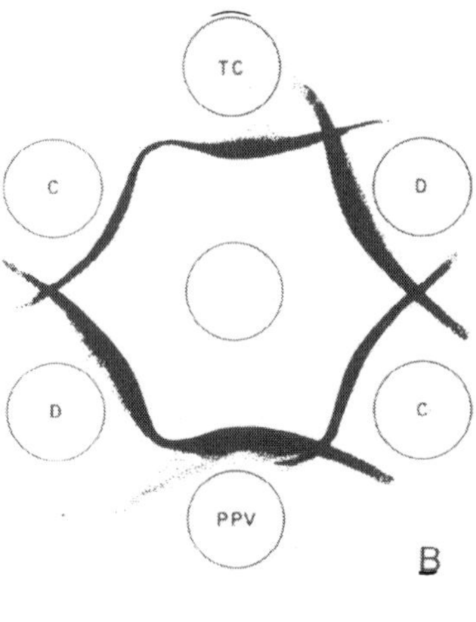

B

The D-antigen is associated with infective RNA-containing virus particles (native virus). The C-antigen is associated with noninfectious virus particles that partially or totally lack RNA. C-antigenic particles are less uniform in shape (flatter, less sharply delineated) than the D-antigenic virions. The D-antigenicity of virion is unstable at 56°C and at alkaline pH. It can be irreversibly converted to C-antigenicity by a number of degrading procedures that all cause the loss of the small capsid protein VP4 from the capsid, no matter whether the RNA is also lost or retained within the capsid under these conditions (Le Bouvier 1955, 1957, 1959; McBride, 1959; Roizman *et al.*, 1958, 1959; Taylor and Graham, 1959; Hummeler *et al.*, 1962; Breindl, 1971a).

The C-antigen is also formed in the course of poliovirus infection of tissue culture cells. Crude preparations of poliovirus infected tissue culture fluids contain D-antigen and varying amounts of C-antigen (Hummeler and Hamparian, 1958). The relative proportion of C-antigen production is related to the condition of the host cells and conditions of virus growth (Le Bouvier, 1959a+b). At least part of the C-antigen in tissue culture cells can be chased into D-antigen under certain conditions (Scharff *et al.*, 1964), reflecting the precursor-product relationship of poliovirus procapsids ($VP_{0,3,1}$) and virions ($VP_{4,2,3,1}$) (Jacobson and Baltimore, 1968a) (see Fig. 106, Chapter 10).

The D-antigenicity of virions is responsible for the induction of the specific neutralizing antibody response. This property disappears with a change in the native conformation and partial or complete degradation of virion or with its conversion to C-antigenicity (Le Bouvier, 1955; Hummeler and Hamparian, 1958). The neutralizing D-antibodies are type specific antibodies. In contrast, antibodies produced upon immunization with disrupted C-antigenic viral particles were poliovirus group-reactive antibodies (Hummeler and Hamparian, 1958; Svehag and Mandel, 1964; Hinuma *et al.*, 1970). There is only very little crossreactivity between the C-and D-antigenic states of poliovirus. Native virions exhibit only trace amounts of C-antigenicity. Conversely, the stable procapsids formed during isolation from infected cells exhibit only trace amounts of D-antigenicity (Roizman *et al.*, 1958). The empty shells (lacking VP4 and RNA) produced by heating of virions do not exhibit any capacity to bind complement-fixing D-antibodies. On the other hand, immunization of animals with artificially produced empty shells may produce very low titers of neutralizing antibodies (Hinuma *et al.*, 1970; Urasawa *et al.*, 1979).

As discussed, the antibodies produced by an animal upon immunization with D-antigenic native virions are usually a heterogeneous set of antibodies, some of which may interact with sites (minor antigenic determinants see below) common to native virions and empty shells or capsid proteins. A heterogeneous set of antibodies against native virion can be made monospecific for D-antigens by precipitation with C-antigen or by passage through sepharose columns to which D-antigenic particles are attached (Borriss and Koch, 1975). The so purified C- and D-antibodies do not react with any of the isolated capsid proteins from dissociated virions, indicating that the two major antigenic determinants of the poliovirus capsid are defined by special conformational states of the capsid proteins that are present only in native virions or C-antigenic shells, respectively.

In contrast, crude antibody preparations obtained after immunization with native D-virions or antigenic shells, and some monoclonal antibodies may be able to precipitate capsid proteins from disrupted virions or assembly intermediates (Scharff *et al.*, 1964; Beatrice *et al.*, 1980; Minor *et al.*, 1980; Rueckert *et al.*, 1981; Thorpe *et al.*, 1982).

In the subsequent discussion we will use the terms D-antigenic and C-antigenic determinants only in the purified sense, *i.e.*, to represent only those antigenic determinants which are unique to the native conformational state of virions, or to the empty shells, respectively. Based on the predominant characteristics of crude antisera induced by virons and empty shells, we assume that the so defined D-antigenic determinants and C-antigenic determinants represent the major in vivo antigenic determinants of virions and empty capsids. The additional antigenic determinants common to intact virions, empty shells and/or isolated capsid proteins we shall call minor antigenic determinants (see Section VIIIB below).

1. *Attempts to Identify the C- and D-Antigenic Determinants in Terms of the Constituent Capsid Proteins with Polyclonal Antisera*

The C-antigenic determinants appear to be specified predominantly by the VP1/VP3 backbone of the capsid. The stepwise removal of VP4, VP2 and RNA from the poliovirion by increased alkaline pH finally leaves a skeleton capsid backbone that sediments at 60S and is composed exclusively of VP1 and VP3 (see Section VI above). This skeleton capsid still exhibits the C-antigenicity (Katagiri *et al.*, 1971). Similar results have been obtained upon the mild stepwise dissociation of other enteroviruses. In the case of coxsackie B-3 virus, even 20S ringlike capsid-substructures composed exclusively of VP1 and VP3 still exhibited C-antigenicity (Beatrice *et al.*, 1980). Antibodies produced upon immunization with these 20S VP1-VP3 rings were reactive with the major group antigen (C-antigenicity) shared by the group B-coxsackie viruses. They could also bind to virion (as measured by the enzyme-linked immuno absorbent assay; Katze and Crowell, 1980) but could not neutralize the virions (as measured by the plaque reduction method).

The identification of the D-reactive neutralizable antigenic determinant(s) of the native poliovirion has been more difficult, since the D-reactive determinant is very sensitive to even mild degradation conditions or conformational rearrangements of the viral capsid. The D-reactive neutralizable determinant appears to be highly dependent on the correct conformation of the constituent polypeptide stretches, a conformation which is destroyed even by relatively mild degradation conditions, and whenever the capsid is dissociated into its individual polypeptides. A multitude of conditions that remove the small capsid protein VP4 from the capsid greatly reduce, perhaps completely abolish, the D-antigenicity.

Based on the good correlation between loss of VP4 and loss of antigenicity, it was tempting to assume that VP4 is the D-reactive component. On the basis of its tiny size (one twelfth of the capsid weight) and recent evidence that most of VP4 occupies internal positions close to the viral RNA (Wetz and Habermehl, 1979),

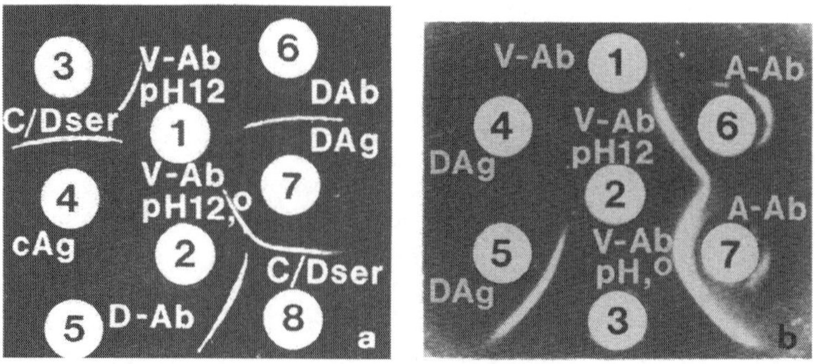

Fig. 35 A–B. Alkaline disruption of precipitates of poliovirus linked by D-antibody: implication of VP_4 as part of the D-reactive antigenic site on poliovirus

Fig. 35 A. Immunodiffusion tests with poliovirus/D-antibody precipitates and various antisera
The figure shows immunodiffusion tests with poliovirus/D-antibody precipitates at neutral pH, at pH 12 prior to heating, and at pH 12 after 1 min at 56 °C and various antisera: monoreactive D-antiserum (D-Ab), hyperimmune monkey serum after poliovirus innoculation (C/Dser), and anti-monkey rabbit serum (A-Ab)

Contents of holes

a

1 poliovirus/D-antibody precipitates at pH 12
2 virus-D Ab precipitates at pH 12, 1 min 56°C
3 poliovirus hyperimmune monkey serum
4 C-Antigen
5 monoreactive D-antiserum
6 monoreactive D-antiserum
7 native virus
8 poliovirus hyperimmune monkey serum

b

1 poliovirus/D-antibody precipitates
2 virus-D Ab precipitates at pH 12
3 virus-D Ab precipitates at pH 12, 1 min 56°C
4 native virus
5 native virus
6 anti-monkey rabbit serum
7 anti-monkey rabbit serum

Figure from Breindl, 1971 b [Virology *46*, 963 (1971)]

Fig. 35 B. Schematic interpretation of the data of the immunodiffusion tests
This figure illustrates a plausible interpretation of the data in a) and b) of Fig. 35 A. The contents of the central wells in a) and b) are illustrated pictorially: top = well b1); center = wells b2) and a1); bottom = wells b3) and a2). The precipitation bands between wells are represented by the corresponding numbers, i.e. a 1–3 = precipitate between wells 1 and 3 in figure a). ∅ = no precipitation-band, + = precipitation-band. Symbols for capsid structure units are as in Fig. 32 (p. 110). 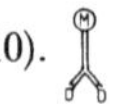 = monoreactive monkey D-antiserum;  = anti-monkey rabbit serum

After one week of incubation of native virion with monoreactive D-antiserum, large virus-antibody precipitates are formed, no free D-antibody is present (top). After incubation of the virus-antibody precipitates, in pH 12 at 0 °C, the precipitate is dissociated into empty C-reactive skeleton capsids, large D-antigen-D-antibody precipitates which cannot move into the agar to react with anti-Ig rabbit serum, and some segregated D-antibody bound to VP_4 which move into the agar and react with anti-Ig rabbit serum; no free D-antibody or D-antigen are present under these conditions (center). After heating of the mixture for 1 min at 56 °C and pH 12, the components are dissociated further into C-reactive aggregates (composed of VP_1–VP_3), and dissociated non-reactive D-antigen; D-antibody is liberated under these conditions and can react with D-antigen and with anti-Ig rabbit serum (bottom)

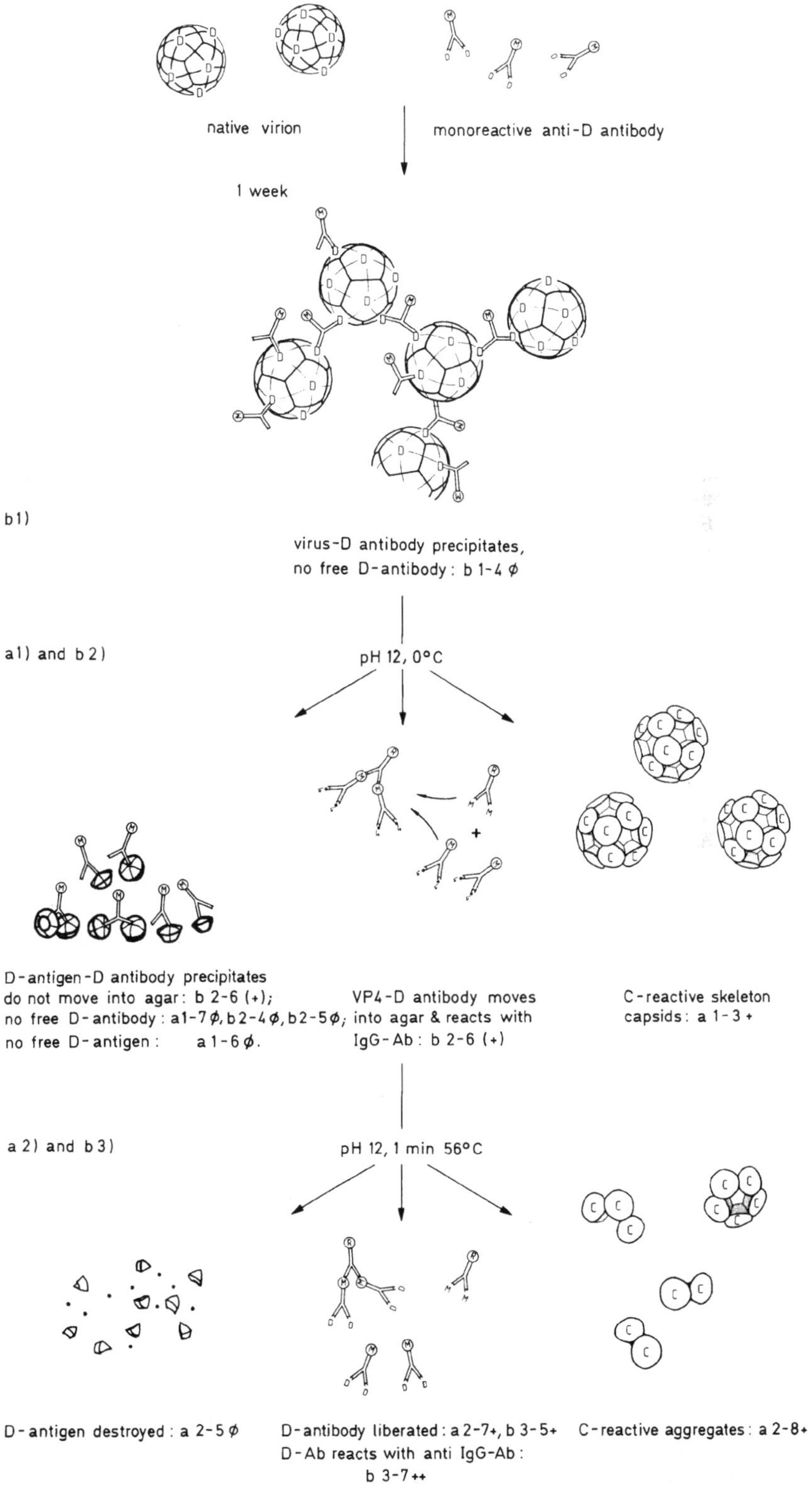
native virion
monoreactive anti-D antibody
1 week
b1)
virus-D antibody precipitates,
no free D-antibody: b 1-4 ϕ
a1) and b 2)
pH 12, 0°C
+
D-antigen-D antibody precipitates
do not move into agar: b 2-6 (+);
no free D-antibody: a1-7 ϕ, b 2-4 ϕ, b 2-5 ϕ;
no free D-antigen: a 1-6 ϕ.
VP4-D antibody moves
into agar & reacts with
IgG-Ab: b 2-6 (+)
C-reactive skeleton
capsids: a 1-3 +
a 2) and b 3)
pH 12, 1 min 56°C
D-antigen destroyed: a 2-5 ϕ
D-antibody liberated: a 2-7+, b 3-5+
D-Ab reacts with anti IgG-Ab:
b 3-7 ++
C-reactive aggregates: a 2-8+

however, it is highly unlikely that the small capsid protein VP4 alone is responsible for constituting the D-reactive determinant. Isolated VP4 by itself neither can induce the production of neutralizing antibodies, nor does it bind to D-reactive antibody. Heat inactivated poliovirus preparations contain empty capsids and free VP4 (Maizel *et al.*, 1967; Breindl, 1971b), yet they bind only C-reactive antibodies significantly (Hummeler and Hamparian, 1958). These observations, however, could not exclude the possibility, that short, surface-exposed portions of VP4 in the conformational state of native virion may contribute to the formation of the D-antigenic determinant.

Attempts were made to dissociate virions bound to purified D-reactive antibodies into capsid subcomponents without dissociating the specific bonds between the D-antibody and the D-reactive antigenic determinant (Breindl, 1971b). The virion/D-antibody bond indeed is more stable than some intracapsid bonds to certain degradation conditions, such as alkaline pH. Since the original experiment gave rise to an extended controversy, we will briefly describe and discuss it here (see Fig. 35).

Monoreactive D-antiserum was produced by precipitation of anti-poliovirus hyperimmune monkey serum with purified empty poliovirus capsids. Poliovirus-antibody precipitates were treated with increasing concentrations of alkali (0.1 M NaOH) at 0°C. The suspension cleared at about pH 12, and now contained C-antigen but only traces of free D-reactive antibodies (Fig. 35A, a1). Evidently the alkaline treatment had brought about a dissociation of the antibody-virion precipitate into free C-antigen and D-reactive antibody still associated with the D-antigenic component of the capsid. D-reactive antibody could be liberated from the complex by heating at 56°C for 1 minute at either high or neutral pH, a procedure which, however, also destroyed the D-antigenic component (Fig. 35A,a2).

These results indicate that the C- and D-reactive sites on poliovirus are located on two separable positions on the poliovirus capsid, *i.e.*, that they are constituted by different antigenic sites and are not constituted simply by two different conformational states of the same site(s). In the latter case the alkali-induced shift from D to C conformation should have released free C-antigen as well as free D-antibody.

Unfortunately, no direct attempts were made to separate and characterize the composition of the dissociated C-antigenic component and the complex of the D-antibody with the D-reactive capsid component. Instead it was attempted to isolate the complex of D-antigen and D-antibody by precipitation with anti-monkey rabbit serum in Ouchterlony plates (Fig. 35A,b2/3). Faint precipitate-bands were formed indeed between the well containing pH 12 dissociated virion and the well containing the anti-monkey rabbit serum. When these bands were cut out, dissociated, and analyzed by PAGE, they were found to contain significant amounts only of VP4. It was concluded that VP4 is the D-reactive part of poliovirus (Breindl, 1971b).

A more plausible interpretation of the results follows if one assumes that the complex of D-antigen and D-reactive antibody was present at 0° and pH 12 in larger aggregates (separate from empty shells) and thus not capable of diffusing into the agar (Fig. 35B). Its identity could have been a stabilized multifunctional aggregate (such as would be derived from dissociation of a VP4/VP2 complex from the skeleton capsid of VP1/VP3 (Katagiri *et al.*, 1971). The multifunctional aggregate could, of course, also have been composed of other capsid components (*e.g.* VP1), stabilized in the D conformation by the bound antibody. Heating of such an antibody-antigen complex would have caused its dissociation and the release of the D-antibody, explaining the observed reappearance of D-specific antibody and the concurrent disappearance of D-reactive antigen (Fig. 35A,b3).

The precipitated VP4-antibody complex that was mistaken for the D-antibody D-antigen complex can be explained in a number of ways. The original hyperimmune poliovirus antiserum used in these studies may have contained some VP4-specific antibodies, that would not have been removed by adsorption of the antiserum with empty VP4-lacking capsids. Alternatively, a small fraction of D-antibody may have been released from virions bound to VP4 upon alkali induced distortion of the D-antigen with consequential dissociation of a VP4-D-antibody complex from the D-antigen. As discussed, a set of antibodies of given specificity, such as anti-D, usually is heterogeneous and may contain antibodies with overlapping antigenic determinants. Some of the anti-D antibodies might recognize more of the VP4 component portion than of the rest of the D-antigenic determinant. Binding of such hypothetical antibodies to virions would not stabilize the antigenic site as efficiently as the dominant anti-D antibodies. The former antibodies would have a greater affinity to VP4 than to other portions of the D-antigen and would be liberated more readily in association with VP4 from virions by alkaline treatment. Of interest in this context is the finding that monospecific anti-D antibodies neutralize poliovirus by preventing its uncoating (Mandel, 1976). The neutralizing antibodies block the uncoating step in which the small coat protein VP4 is liberated from the capsid. This step can be envisioned as a conformational destabilization and rearrangement providing a hole or channel at the vertices or in the center of the protomers through which VP4 "slips" out from its internal position through the capsid. The D-specific antibodies may directly block this step by binding VP4 as part of the antigenic site, or sterically by standing in the way of VP4 at the "exit" of the channel or by binding to the D-antigen in a way that prevents the conformational shift that creates the hole for VP4. Neutralizing of the poliovirus family relative bovine enterovirus, indeed leads to the exposure of VP4 portions that are hidden in native virions (as measured by sensitivity to labeling with ^{125}I) (Carthew, 1976).

Recently, a similar study was carried out to characterize the D-reactive part of coxsackie B-3 (Beatrice *et al.*, 1980). We will briefly discuss these results with the precaution that immunological events with different viruses are likely to reflect differences of viral surface details even of closely related viruses. Coxsackie virus B-3 is an enterovirus, *i.e.* it is more closely related to poliovirus than the other picornaviruses, and its properties are more likely to pertain to poliovirus than those of more distantly related picornaviruses such as the foot and mouth disease viruses.

Mild disruption of coxsackie virus (3 M urea, pH 9.0, 20 mM EDTA, 20 mM DTT, 15 min at 37°) yielded two components: a 20S ringlike VP1—VP3 structure, and a 5S aggregate of VP_2 (possibly pentamers) with varying amounts of associated VP_4. Immunization with the dialyzed VP_2-aggregate (VP_4 removed) induced levels of type-specific neutralizing antiserum comparable to those raised against native virions; immunization with the 20S VP_1–VP_3 structure induced a group-specific antiserum that could bind to but not neutralize virions. Unfortunately, a more detailed comparison of the anti(VP_2)-aggregate neutralizing antibody with D-reactive antibody raised against native virion was not made in these studies. No attempts were made to obtain monospecific D-antiserum by adsorption of the anti-virion antiserum with empty shells or isolated capsid proteins. The antiserum against native virion was thus rather heterologous, precipitating also the capsid proteins from disrupted virions. The anti(VP_2)-aggregate antiserum in contrast; precipitated only VP_2 or VP_0 from disrupted virions or procapsids, respectively.

It would certainly be of interest to determine whether the anti(VP_2)-aggregate antiserum contains any monospecific neutralizing anti-D antibody that reacts only with the particular conformational state of the D-antigenic determinant in native virion but that does not react with isolated VP_2, analogous to the specificity of purified anti-D antibody against native virion. This could easily be tested by measuring the neutralizing capacity of anti(VP_2)-aggregate antibody after adsorption with isolated VP_2 or VP_0.

2. *Studies with Monoclonal Antibodies*

As this book neared completion, a wealth of data from studies with monoclonal antibodies against polioviruses began to appear in the literature. A number of laboratories reported the isolation of monoclonal antibodies, some of which exhibited potent neutralizing capacities (*e.g.* Icenogle *et al.*, 1981; Osterhaus *et al.*, 1981a, b; Crainic *et al.*, 1981; Ferguson *et al.*, 1981; Blondel *et al.*, 1982; Brionen *et al.*, 1982; Humphrey *et al.*, 1982; Minor *et al.*, 1982). The properties of one monoclonal neutralizing antibody are illustrated in Figure 36 (Icenogle *et al.*, 1981). These properties resemble but do not perfectly match those of polyclonal neutralizing D-specific antisera. The monoclonal antibody fails to react with ATC and isolated capsid proteins; yet it precipitated NTC as efficiently as virions and 14S structures to 20%, in contrast to D-specific antisera which do not interact with these subunits. These discrepancies could originate in differences in the conformational states of different NTC or 14S subunit preparations rather than in differences of antibody reactivities.

Recently a number of monoclonal antibodies obtained after immunization with intact type 1 virions were characterized in detail (Blondel *et al.*, 1983; Icenogle *et al.*, 1983; Emini *et al.*, 1983a, 1984; Wychowski *et al.*, 1983; Vrijsen *et al.*, 1984). Most of these antibodies reveal characteristics similar to those of polyclonal D-specific antisera (neutralizing, D-specific, type-specific, non-reactive with isolated viral proteins and empty shells, capable of inducing the shift in isoelectric point). The Fab-fragments of two monoclonal antibodies with these characteristics were crosslinked chemically to neutralized virion and were shown to bind to VP_1 (Emini *et al.*, 1982b, 1983a). Similarly, D-specific monoclonal antibodies against poliovirus type 3 were obtained, which did not interact with any of the iso-

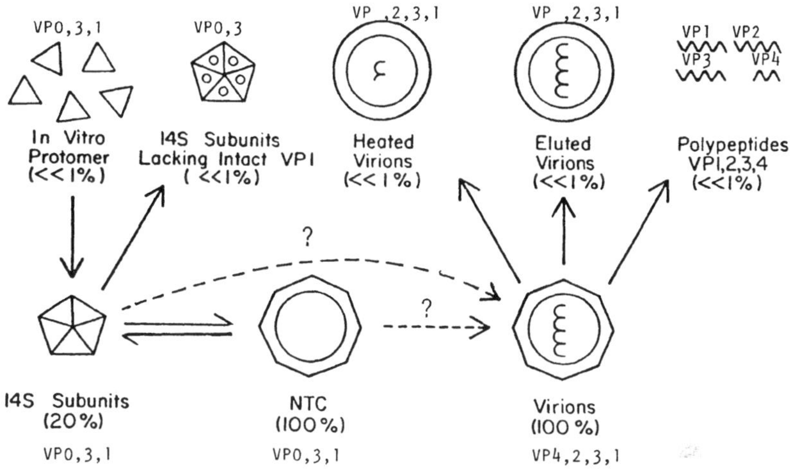

Fig. 36. The interaction of a monoclonal antibody with poliovirus structures
The figure illustrates the reactivity of monoclonal antibody "F 7.12" with native virions and subviral particles of poliovirus. The antibody reacted not only with native virions, but also with natural top component, and with 14 S assembly subunits. It did not react with A particles or heated virions. The reactivity of this antibody with natural capsids, but not with heat-induced empty capsids, indicates that the natural top components characterized in this study were still in their *in vivo* D-antigenic conformation, i.e. had not been transformed into stable empty capsids, as usually occurs upon their isolation from the infected cell. The antibody had neutralizing capacity; the number of critical sites was determined to be 8–10, the number of maximal binding sites on the virion was determined to be 30. The reactivity of the antibody with 14 S subunits may indicate that the antibody does not have the same specificity as the dominant D-antibodies from sera, which usually do not react with 14 S subunits. Alternatively, the 14 S subunits isolated in this study may have been in a different conformation from those usually obtained during isolation from infected cells, analogous to the different antigenic states of natural top component. – Figure from Icenogle *et al.*, 1981 [Virology *115*, 211–215 (1981)]

lated capsid proteins, but by detailed genetic analyses were shown to recognize an antigenic site determined by VP_1 (Evans *et al.*, 1983; Minor *et al.*, 1983; Schild *et al.*, 1984). The immunodominant region in VP_1 was identified as the segment between amino acid residues 93–100. One monoclonal antibody raised against C-particles was also shown to interact with the same region in VP_1, indicating that this region is also exposed on C-particles, albeit in a different immunogenic conformation (Ferguson *et al.*, 1984).

Many other D-reactive monoclonal antibodies have been characterized which evidently recognize other sites on the capsid surface (minor antigenic determinants, see below). Some monoclonal antibodies have D + C or D + capsid protein reactivities, or are D-specific but non-neutralizing; some even distinguish between strains of the same subtype, others show a braod range of interactions with viruses also from other subtypes (see for example Osterhaus *et al.*, 1981b; Minor *et al.*, 1982b; Blondel *et al.*, 1983; Schild *et al.*, 1984).

Most monoclonal antibodies obtained after immunization with C particles also exhibit C-specificity, are non-neutralizing, and do not react with capsid proteins. Others have been shown to immunoprecipitate VP_1, VP_3, or VP_2; another recognized an antigenic determinant present on C particles, isolated VP_1, and D-virions and was capable of immunoprecipitating and neutralizing infectious D particles (Blondel *et al.*, 1983; Ferguson *et al.*, 1984).

In sum, these results confirm the conclusion derived from elaborate studies on serodifferentiation of poliovirus strains, namely that the poliovirus capsid carries many functionally distinct types of antigenic determinants (Urasawa *et al.*, 1979). The results also illustrate the complex nature of the D-specific antigenic determinants of native virions. The segment of amino acids 93–100 of VP_1 appears to constitute an immunodominant site on the viral surface, capable of eliciting the production of type-specific neutralizing antibodies with the properties of monospecific polyclonal D-antisera. Additional sites with similar properties exist (*e.g.* amino acid segment 132–143 of VP_2, see below). Further characterization of monoclonal antibodies in conjunction with the elucidation of the three-dimensional structure of poliovirus in terms of surface protrusions, exposed and hidden segments of the capsid proteins, should permit the definitive characterization of immunogenic sites on the virion surface.

The very existence of distinct serotypes of poliovirus shows that the poliovirion has evolved structures on its capsid surface that can be varied rapidly in response to environmental pressures—sites which probably are not essential to virion structure and function. These regions—though immunodominant—are not necessarily those best suited for the production of synthetic vaccines since they are so variable. For the preparation of vaccines, rather those particular segments of the capsid should prove most useful which are exposed on the surface, but at the same time are essential for virion function (*i.e.* portions required for capsid stability or the receptor recognition site).

B. Other, Minor Antigenic Sites on Poliovirus

Aside from the major D- and C-antgenic determinants, there evidently are a number of other minor antigenic sites on poliovirions and poliovirus shells. Some of these sites may be common sites on isolated capsid proteins and viral particles. Binding of antibody to such sites on virions often does not result in neutralization. There may, however, also be neutralization sensitive sites that differ from the main D-antigenic site, for which the mechanism of neutralization may also differ. For example, an antibody against the receptor recognition site on poliovirus would also be expected to neutralize virions. Indeed, serological studies of different poliovirus mutants with refined techniques have indicated the existence of at least five distinct neutralization sensitive sites on the surface of the Mahoney Strain of poliovirus type 1 (Urasawa *et al.*, 1976, 1979; Sawinskaya *et al.*, 1979).

Various attempts have been made to obtain anti-poliovirus antibodies upon immunization of animals with isolated poliovirus capsid proteins derived by a variety of disruption and isolation procedures (Table 24). Since the capsid is a very tightly bonded structure, severe disruption conditions (*e.g.*, boiling in SDS, 6 M guanidine) have usually been employed to completely dissociate virions into their component proteins. Such procedures are expected to cause a more or less severe denaturation of proteins. The conformation shown by the capsid proteins upon resuspension in aqueous buffers is different from that present in the viral capsid. The capsid proteins probably never occur as individual proteins in infected cells (Korant, 1973). During or after its synthesis the large common capsid precursor protein NCVP1a becomes associated with intracellular membranes (see Chapter 10). The lipophilic environment within membranes exerts different constraints on the conformation of the four capsid proteins than those present within the context of the virion capsid. Conformation of the capsid proteins is influenced by hydrophobic interactions with intracellular membranes, by the continuous mutual interactions with, and binding to, the other capsid proteins. Furthermore, the conformation is altered by specific proteolytic cleavages during the process of capsid assembly. Therefore, the conformation of the capsid proteins within the cell and in the viral capsid are expected to be quite different from the conformation that the isolated capsid proteins will spontaneously acquire upon transfer to an aqueous environment.

In addition, some of the capsid proteins may show a tendency to aggregate in aqueous solutions, depending on the method employed for virion disruption (Scharff *et al.*, 1964, Yamaguchi-Koll *et al.*, 1977). For example, after virion disrup-

Table 24. *Reactivities of some antibodies against poliovirus structure units*

	Neutralize virion	Bind to virion	Bind to empty capsid		Bind to capsid proteins	Bind to 14 S	Bind to 5 S
			NTC	ATC			
Anti-D[a]	+ (blocks uncoating)	+	–	–	–	–	–
Anti-C[a]	–	–	+	+	–	–	
Anti-VP_2[b]	±	(+)	+	+	+	n.d.	+
Anti-VP_3[b]	±	(+)	+	+	+	n.d.	+
Anti-VP_1[b, c]	±	(±)	+	+	+	n.d.	+
"S" (dissociated[d] anti capsid protein)	–	–	(±)	(±)	+	+	+
Monoclonal[e] Osterhaus	+	+	?	?	+(VP_1)	?	?
Monoclonal[e] Icenogle	+ (neutralizing mechanisms unknown)	+	+	–	–	(+)	–
Anti 5 S[f]	–	–	–	n.d.	n.d.	–	+
Anti 14 S[f]	–	–	–	n.d.	n.d.	+	–

[a] Le Bouvier *et al.*, 1957; Roizman *et al.*, 1958; Hummeler *et al.*, 1962.
[b] Meloen *et al.*, 1979; Dernick *et al.*, 1983.
[c] Chow and Baltimore, 1982.
[d] Scharff *et al.*, 1964.
[e] Icenogle *et al.*, 1981; Osterhaus *et al.*, 1981a.
[f] Ghendon and Yakobson, 1971.

tion with concentrated urea and dilution of the suspending medium, the separated capsid proteins show a particularly high tendency to associate nonspecifically into larger aggregates—they have even been reported to aggregate, albeit with very low efficiency, into complete infectious virions (Drzeniek and Bilello, 1976b).

Antisera obtained after immunization with a mixture of capsid proteins (sedimentation coefficient 2S) obtained from disruption of virions by 6.5 M guanidine did not neutralize or precipitate virions. They did however precipitate 5S protomers and to a lesser extent also 14S subunits and 80S particles obtained from extracts of infected cells (Scharff *et al.*, 1964).

In a number of studies, antibodies obtained from immunization of animals with different preparations of isolated poliovirus capsid proteins did not show any significant neutralizing capacity (as measured by the plaque reduction method) (Meloen *et al.*, 1979; Blondel and Crainic, 1980; Wiegers and Dernick, 1981; Schärli and Koch, 1984); although a few of these antibodies could bind to native virion. In one case, rabbit anti-VP3 antibodies (Wiegers and Dernick, 1981), and in another guinea pig anti-VP2 antibodies, (Meloen *et al.* 1979) were found to react with virions (detected by adsorption to Staphylococcus A). The isolated capsid proteins had been obtained either by urea degradation of virions and subsequent isoelectric focusing in urea containing gels (VP3) or by SDS-PAGE separation of heat-disrupted particles (VP2) (Meloen *et al.*, 1979). After mild heating of virions at 37° for 1 hour at pH 8 and concomitant ATP mediated phosphorylation of capsid proteins, virions became more sensitive to precipitation by anti VP2, VP3 and VP1 antibodies under these conditions irrespective of the method of preparation of antigen and of the animal source of antibodies (Schärli and Koch, 1984).

Next to these common antigenic, yet not neutralization sensitive sites on intact virions and capsid proteins, other sites present on isolated capsid proteins have been found that constitute neutralization sensitive sites also on the intact virion. In one case, polyclonal, type specific neutralizing antisera to poliovirus type 3 bound also to VP1 and VP2 of poliovirus type 3 (Thorpe *et al.*, 1982); the antibody recognizing VP2 exhibited type-specificity, the antibody recognizing VP1 also bound to VP1 of poliovirus types 1 and 2. In another study, immunization of rats with isolated capsid proteins—though not eliciting a neutralizing response—did lead to a booster like reaction upon subsequent infection with small almounts of intact virions with respect to the production of neutralizing antibodies (Van Wezel *et al.*, 1981). In more recent studies, each of the large isolated capsid proteins VP1, VP2, and VP3 have been found to elicit a neutralizing antibody repsonse, albeit inconsistently or very weakly (Blondel *et al.*, 1982; Chow and Baltimore, 1982; Dernick *et al.*, 1983; Emini *et al.*, 1983d; Van Wezel *et al.*, 1983). Isolated VP1 was the first capsid protein found to elicit a neutralizing immune response (Blondel *et al.*, 1982; Chow and Baltimore, 1982). Subsequently it was shown that all three large capsid proteins could elicit a weak neutralizing response, provided the capsid proteins were isolated from formalin-inactivated (and stabilized) virus (Van Wezel *et al.*, 1983), or by high performance liquid chromatography of formic-acid disrupted virions (Dernick *et al.*, 1983). Immunization of rabbbits with VP4 (+ contaminating fragments of VP3) led to the formation of neutralizing antibodies directed against VP3 (Emini *et al.*, 1983a). The latter antibody is interesting in that it precipitates and neutralizes virions, but fails to induce the transition of the isoelectric point (Emini *et al.*, 1983b).

Finally, synthetic peptides of hydrophilic segments in VP_2 (amino acids 132–142) and in VP_1 (amino acids 93–100) have been shown to elicit neutralizing responses (Emini *et al.*, 1983c, 1984; Schild *et al.*, 1984). In addition, other synthetic peptides (amino acids 70–80 of VP_1 and 71–82 of VP_3) have been found

Table 25. In vitro *genetic marker tests for characterization of poliovirus strains*

Genetic marker test	Function of the test	Reference for the original or the early usage of the test
1. Antigenic	characterization of viral antigen (intratypic serodifferentiation of strain)	Dubes *et al.*, 1959 Wenner *et al.*, 1959 McBride, 1959 Gard, 1960 Wecker, 1960 Hahnemann *et al.*, 1963 Diwan *et al.*, 1963 Smit and Wilterdink, 1966 Chumakov *et al.*, 1974
2. RCT or "T"	reproductive capacity of a strain at supraoptimal temperatures	Lwoff and Lwoff, 1958 Benyesh-Melnick and Melnick, 1959 Sabin, 1961 Yoshioka *et al.*, 1959
3. "d"	delayed growth of a strain in medium containing low bicarbonate	Vogt *et al.*, 1957 Hsiung and Melnick, 1958
4. "MS"	plaquing capability of a strain in "MS" cells (a stable line of monkey kidney cells)	Kanda and Melnick, 1959
5. "E"	determines whether a strain is absorbed or eluted through a chromatograph column	Hodes *et al.*, 1960 Hollinshead, 1960
6. "A"	stability of a strain when heated at 50 °C in the presence of aluminum ions	Wallis *et al.*, 1962
7. "m"	a) plaquing capability of a strain under agar (minute plaque) b) agar containing sulfated polysaccharide c) agar containing dextran sulfate	a) Nomura and Takemori, 1960 b) Agol and Chumakova, 1963 c) Takemoto and Kirschstein, 1964
8. "PG"	plaquing capability of a strain under agar containing polyethylene glycol (MW 40,000)	Lokteva *et al.*, 1973
9. "Aa"	adsorption capacity of a strain to aluminum salts	Wallis *et al.*, 1963

Table from Nakano *et al.*, 1978 (Progr. Med. Virol. *24*, 178–206).

to bind to neutralizing and non-neutralizing monoclonal antibodies raised against polioviruses types 1 and 3 (Emini *et al.*, 1984).

These results reflect a simple feature of the poliovirus capsid, namely that the three major capsid proteins VP2, VP3, VP1, all exhibit some kind of surface exposure. Immunization of animals with isolated capsid proteins may yield a variety of antibodies with different specificities, depending on the conditions during the immunization process. The reaction of such anticapsid protein-antibodies with intact virion seems to depend on which particular portion(s) of the isolated capsid proteins were originally responsible for the induction of the immune-response. Some antibodies recognize common sites on capsid proteins and virions, these are usually non-neutralizing. Under carefully controlled conditions, however, the individual capsid proteins or defined peptides thereof may induce the production of antibodies that can recognize both, neutralization sensitive sites on the virion and antigenic determinants on the isolated capsid proteins.

C. Serodifferentiation of Polioviruses

In the laboratory and in the field, poliovirus genomes show quite a rapid rate of mutation, in particular in the capsid protein coding regions. These mutations sometimes result in corresponding substitutions in the amino acid sequence of the capsid proteins which in turn may affect the antigenic (and other structural) characteristics of the virion and its empty capsid. Such changes can be detected by a variety of serological tests that measure, for example, changes in the sensitivity to neutralization by, or interaction with, antibodies directed against a reference strain (Vonka *et al.*, 1962; Nakano *et al.*, 1978; van Wezel and Hazendonk, 1979; Brown, 1980; Blondel *et al.*, 1982).

With refined serological methods, at least five distinct independent antigenic sites have been differentiated on the poliovirus capsid (Urasawa *et al.* 1979). Table 25 lists some of the features that are often used to characterize and distinguish poliovirus isolates (Nakano *et al.*, 1978). These features vary not only between the three different poliovirus serotypes but also within strains of one particular serotype. It is important to keep this variability in mind when discussing antigenic properties of poliovirus in general terms as done in this chapter.

IX. Summary

A. Composition

Poliovirus is a small spherically symmetric RNA-containing virus. In the electron microscope, poliovirions appear as compact particles of uniform size, approximately 30 nm in diameter. When arranged within crystals, the virus particles have a hexagonal contour. Individual particles have a sphere-like appearance; a fine structure is difficult to detect. Surface details, which may be observed occasionally, are indicative of a morphological clustering of the capsid components into 32 apex-like clusters. The native virion has a molecular weight of 11.5 x 10^6. Approximately 25% by weight is RNA (2.6 x 10^6 daltons), 50% is protein (5.9 x 10^6) and 25% is water. Native virion also contains on the order of 5,000–9,000 ions, predominantly mono and divalent cations that are required for charge neutralization. The virion, by chance, may trap other small molecules within its core during virion-assembly in the host cell. No lipid or sugar components could be detected in poliovirus. The RNA component is a single strand of RNA, approximately 7,400 nucleotides long. More than half of the negatively charged phosphate groups are neutralized by cations, predominantly K^+, the rest is neutralized by interactions with the protein component of the virion. The entire protein component (with the exception of a single copy of the small (2,000 MW) protein VPg, which is covalently bound to the RNA) is involved in the construction of the compact viral capsid which surrounds and protects the compact globular virion RNA.

B. Structure

X-ray crystallography of virus crystals reveals an icosahedral construction principle of the virus particle with 5:3:2-fold symmetry and indicates that the capsid is composed of 60 equivalent structural units. The icosahedral lattice is one of three possible lattice types suitable to construct an isometric, spherical shell from a larger number of distinct, identical structure units (tetrahedral and octahedral lattices provide the other possibilities). The icosahedral lattice has the advantage of permitting the greatest number of identical equivalently situated structure units to be arranged in an isometric shell, namely 60 (as opposed to 24 or 12 for octahedral and tetrahedral lattices, respectively). Any icosahedral lattice has 12 centers of 5-fold symmetry, 20 centers of 3-fold symmetry and 30 centers of 2-fold symmetry. For the regular icosahedron these symmetry centers correspond to 12 pentagonal apices, 20 triangular facets, and 30 edges. A number of regular polyhedra are known that can be derived from the regular icosahedron and thus bear icosahedral symmetry. Some of these may be better suited than the regular icosahedron as models for poliovirus, since they illustrate the 60-structure unit pattern better and show that there usually are more than just the 12 apices implied by the regular icosahedron. Examples are provided by the pentakis dodecahedron or hexakis triacontrahedron (each with 60 facets and 32 apices).

C. The Capsid Proteins

1. Number of Peptides and Type of Association on the Structure Units

Separation of the capsid proteins by biochemical procedures and stoichiometrical calculations reveal that the capsid is indeed constructed from 60 equivalent asymmetric structural units, the protomers. Each unit consists of a single copy of each of the four viral capsid proteins. The proteins are named in order of decreasing molecular weight: VP1, 35,000; VP2, 30,000; VP3, 24,000; VP4, 5,500. Three of the capsid proteins are of roughly equivalent size VP1, VP2, VP3; the fourth peptide is very small (less than one-tenth the size of the structure unit). The four proteins arise by cleavage from one common precursor protein (NCVP1 a, MW 95,000) during viral morphogenesis, but they remain tightly associated to one another throughout the entire process of assembly. It is not well understood why the structure unit should be cleaved into four subcomponents. Two of the cleavages – yielding VP0, VP1, and VP3—occur prior to the association of structure units; the third cleavage of VP0 into VP2 and VP4 takes place only after the viral RNA has been encapsidated by a complete 60-structure unit shell. Perhaps the cleavage of the structure units into 3 similarly sized major subcomponents allows the subcomponents of adjacent structure units to form 32 specific morphological clusters: 12 pentametric clusters about the icosahedral 5-fold symmetry centers and 20 hexameric clusters about the icosahedral 3-fold symmetry centers. This type of clustering pattern is observed also in plant picornaviruses, whose 60 icosahedral structure units are each composed of three identical polypeptide chains. The advantage here is that it allows the arrangement of 180 identical structure units (genome economy) into quasi-equivalent environments (*i.e.*, pentameric and hexameric clusters) that employ similar types of built-in intersubunit bonding domains. What structural or functional advantages this type of clustering pattern might confer to poliovirus whose structure units are composed of three different major polypeptide chains is not known. The cleavage of VP0 to yield the small capsid protein VP4 may provide the capsid with VP4-uncoating plugs that might be required for initiation of the uncoating processs to liberate the viral genome in the cytoplasm of the host cell. One or two of the units may contain the viral protein VP0 (MW 41,000) instead of VP2 and VP4. The presence of this VP0—which must confer some conformational incongruity to the overall virion structure—is observed in many picornavirus classes, *i.e.*, it appears to be well conserved throughout evolution. This indicates that it may have an important function, for example in binding to the host cell, uncoating, or RNA-binding.

2. *Relative Localizations*

The accumulated evidence from a variety of experimental approaches has provided some insight into the relative positions and orientations of the individual capsid proteins within the capsid: In native virions the three major capsid proteins VP1, VP2, and VP3 all occupy discrete domains in space and they all penetrate the width of the capsid backbone. Each can be labeled in intact virions by one or more of several chemical and biological labelling techniques. Certain stretches of VP1 seem to be particularly exposed on the external surface. Each of the capsid proteins can also be crosslinked to viral RNA by UV irradiation, VP3 seems to be least involved in interaction with RNA. VP4 in contrast is located internally in close proximity to the RNA; VP4 cannot be labeled in native virions, but it can be crosslinked extensively to the RNA by UV irradiation.

Poliovirions can be degraded by a variety of procedures in a sequential stepwise manner. Typically virions first swell with an influx of H_2O; then the entire component of the small capsid protein VP4 is lost through holes in the shell, the entire VP2 component may also be released, leaving a skeleton capsid backbone composed of only VP3 and VP1. The RNA usually is released along with or after VP4. The results indicate that VP1 and VP3 supply the backbone structure of the virion, and may cluster in the form of (VP1—VP3) trimers at the 20 icosahedral facets. Since VP4 and also—somewhat less readily—VP2 may be released from the capsid without destroying the typical capsid structure, VP2 and VP4 probably lie in clusters in the form of (VP2—VP4) pentamers about the 12 icosahedral apices. The confirmation of these interpretations must await the results of X-ray diffraction studies at sufficient atomic resolution.

Optical rotatory dispersion studies indicate that there are less than 5% α-helical stretches in the capsid proteins. Inspection of the amino acid sequence (deduced from the nucleotide sequence), indicates the potential for extensive β-pleated sheet formation. Some possible membrane insertion sites can also be detected, which may be important during entry into the host cell, and during virion formation which occurs in close association with intracellular membranes.

3. *Heterogeneity*

A given population of virus particles is usually not homogeneous: Fractions can be found that deviate from the behavior of standard particles; some show a resistance to degradation conditions, or neutralization; some acquire a higher density in the presence of Cs^+ ions; the content of VP0 may vary from as low as 0 to several copies per virion. The nature of these apparent heterogeneities is not known. There have been a number of reports suggesting the occurrence of slightly different copies of the major capsid proteins, presumed to originate in ambiguous cleavage or other posttranslational modification, such as phosphorylation. The nucleotide sequence, however, does not indicate any ambiguous cleavage sites. The general validity and possible biological significance of this so-called microheterogeneity is still uncertain. It may simply reflect the presence of mutants in a heterogeneous population.

D. Two Conformational States of the Capsid

Despite its tight and compact structure, the capsid nonetheless reveals a good deal of flexibility which may culminate in a concerted conformational shift of the entire capsid protein component. Some flexibility is evident from the different appearances of the particles in the electron microscope, depending on whether they occur as individual particles (spherical appearance) or within a crystal array of particles (hexagonal contours).

Intact poliovirus may exist in one of two alternative conformational states. The native state is characterized by its D-antigenicity, isoelectric point at pH7.0, efficient attachment capacity, and a sedimentation coefficient of 155S. Under certain conditions (during assembly and uncoating, upon mild degradation conditions, at low pH), the virion may snap into a more expanded, more hydrophobic state, with reduced D-antigenicity and attachment capacity, isoelectric point at pH 4.5, sedimentation rate of 140S, and VP4 exposed on the surface. Neutralizing antibody and host cell membrane components can lock the virion in the expanded state. Release of Mg^{++} ions may be responsible for the lowering of the isoelectric point. The conformational alterations in the capsid accompanying the conversion between the two states appears to be rather subtle. A dramatic and irreversible structural rearrangement occurs when VP4 is lost during the conversion to the more expanded state. Attachment capacity and D-antigenicity are then lost, a new set of antigenic determinants is exposed on the surface, and the particles become sensitive to protease. When RNA is also lost, C-reactive empty shells remain.

E. Antigenicity

Poliovirus is a large protein antigen with numerous antigenic determinants. Native virions and empty capsids contain unique determinants which are not expressed in the isolated capsid proteins or in assembly intermediates, as well as determinants present also on the isolated structure units. At least five different antigenic determinants of native poliovirus have been characterized with refined serological techniques, and many more seem to be revealed with the aid of monoclonal antibodies.

The sum of antigenic determinants unique to native virion are termed D-antigen, those unique to empty capsids, C-antigen. Anti-D antibodies are type specific and have neutralizing capacity, anti-C antibodies are group specific and are non-neutralizing as they do not interact with native virion. Interaction of virions with anti-C antibodies resembles interaction with the host receptor which also exhibits group specificity. The C-antigen is determined by the VP1—VP3 component of the capsid: skeleton capsids composed only of VP1 and VP3 are C-reactive. Anti-D antibodies appear to interact with the VP2—VP4 components of the capsid or with an immunodominant site on VP1, thereby inducing the shift to the pI 4.5 form of intact virions. The neutralizing mechanism acts by blocking uncoating, adsorption to the specific host cell receptor is not blocked by neutralizing antibody. The vaccine strain of poliovirus type 1 carries several amino acid substitutions in a small part of VP1, which is reflected also in certain alterations of the capacity to interact with the host cell receptor. The D-antigenic determinant, however, is not altered significantly by these substitutions; antibodies induced by the vaccine strain also neutralize the parental strain viruses.

Injection of poliovirus suspensions into animals usually results in the induction of a heterogeneous set of antibodies with different specificities. Some of the monoclonal antibodies obtained from such immunizations resemble monospecific D-antisera in type-specificity, neutralizing capacity, induction of the shift in isoelectric point, and failure to react with empty shells or isolated capsid proteins. Other monoclonal antibodies are D-specific, but non-neutralizing; others recognize empty shells or isolated capsid proteins in addition to intact viral particles; some are heterotypic; some can distinguish between strains of the same subtype of poliovirus. Antibodies induced after injection of isolated capsid proteins usually do not exhibit any neutralizing capacity, although some of these antibodies may interact with or even precipitate native virions. Under carefully controlled conditions, isolated capsid proteins and defined peptide fragments of VP_2 or VP_1 have been shown to elicit a neutralizing response.

The main antigenic determinants are probably specified by those portions of the capsid surface which protrude from the surface. Such structures appear to be nonessential to virion structure or function since they can vary rapidly in response to environmental pressures. They may represent a simple viral defence mechanism to evade the host's immune system and to „hide“ essential portions of the capsid surface such as the receptor recognition site. Correspondingly, certain determinants may be exposed on the capsid which are inefficient in inducing antibody formation as part of intact virions. Upon virion dissociation, such sites may acquire immunogenicity and induce the formation of antibodies which can then react with intact virions. Such antigenic, but not immunogenic sites are likely to be located in grooves or valleys of the capsid surface. For the construction of synthetic vaccines, conserved essential regions of the capsid surface—such as the receptor recognition site—may proove more useful than the highly variable immunodominant sites.

4

Structure and Function of the Genome

I. Introduction

The genome of poliovirus is a single-stranded molecule of RNA of messenger RNA polarity. Since significant amounts of viral RNAs can be conveniently obtained in high purities, the RNA of poliovirus and other picornaviruses has in the past been used as a model for studying the translation of eukaryotic mRNA (Smith, 1975; Fellner, 1979). With more detailed structural investigations of the viral RNAs, however, it has become clear that they possess some special features, which in part have not been encountered in other eukaryotic mRNAs.

Within the past two years, the entire genome sequences of the wildtype virulent poliovirus type I (Mahoney strain) and the corresponding attenuated poliovirus type 1 live vaccine strain (LSc2ab = Sabin 1 strain) have been determined (Kitamura *et al.*, 1981; Racaniello and Baltimore, 1981a; Nomoto *et al.*, 1982). cDNA of the poliovirus type 1 genome has been cloned in plasmids of E. coli (van der Werf *et al.*, 1981), and a plasmid containing a complete complementary DNA copy of the poliovirus RNA genome has been shown to be infectious in mammalian cells (Racaniello and Baltimore, 1981b). These experiments have paved the way for many new approaches to studies of the structure and function of the poliovirus genome and its expression in host cells. Already, the analyses of the genomic nucleotide sequence, and of the amino acid sequence determined by it, have revealed many interesting features of poliovirus biology.

The ultimate function of any genome is the storage of information for its survival and its identical reproduction in a suitable environment. Genetic information is expressed by the translation of the genomic nucleotide code into amino acid sequences in the synthesis of proteins. The poliovirus genome has a rather limited coding capacity which allows the production of only 5—10 average-size proteins. This information nevertheless sufficies to completely reorganize the metabolism of the infected host cell for the massive replication of the infecting genome.

Polioviruses can be classified into three major sereologically distinct groups, which differ in as much as 30 % of their genomic nucleotide sequences. The advancement of experimental techniques during the past two decades permits the rapid typing of isolated poliovirus and makes possible a tracing of these viruses to their genetic origin. It has become clear that polioviruses adapt rapidly to changing environments with profound modifications of their genetic information. A deep understanding of the genetics and evolutionary progress of polioviruses should prove useful for the development of strategies to control poliomyelitis and diseases by related viruses, as well as for the construction of models of viral evolution in general.

In the following chapter we first summarize the characteristic features of the poliovirus genome, discuss implications of its nucleotide sequence, then provide an overview of its gene order and of potential functions of its genome products, and finally point out some of the vast amount of data on poliovirus genetics and its application to practical aspects of poliovirus replication.

II. Genome Structure

A. Characteristic Features

Some of the characteristics of poliovirus genomes are summarized in Table 26. The genome of poliovirus is a continuous, linear, single-stranded molecule of RNA. The isolated viral RNA has a sedimentation coefficient of 35 S and a molecular weight of about 2.4–2.7 x 10^6 (Granboulan *et al.*, 1969, Tannock *et al.*, 1970). The genome RNA thus comprises approximately 32 % of the dry mass of the

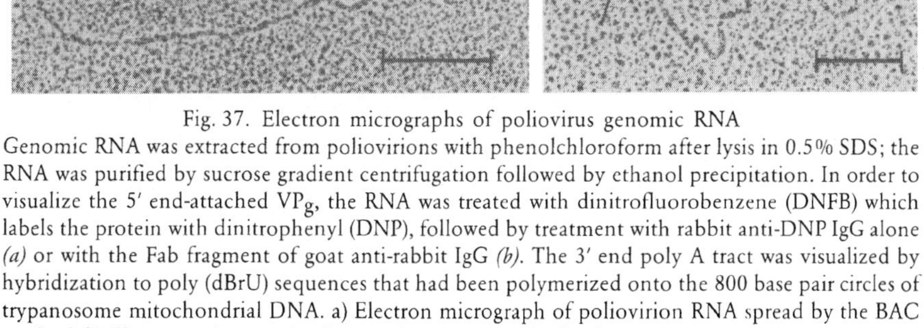

Fig. 37. Electron micrographs of poliovirus genomic RNA
Genomic RNA was extracted from poliovirions with phenolchloroform after lysis in 0.5% SDS; the RNA was purified by sucrose gradient centrifugation followed by ethanol precipitation. In order to visualize the 5′ end-attached VP_g, the RNA was treated with dinitrofluorobenzene (DNFB) which labels the protein with dinitrophenyl (DNP), followed by treatment with rabbit anti-DNP IgG alone *(a)* or with the Fab fragment of goat anti-rabbit IgG *(b)*. The 3′ end poly A tract was visualized by hybridization to poly (dBrU) sequences that had been polymerized onto the 800 base pair circles of trypanosome mitochondrial DNA. a) Electron micrograph of poliovirion RNA spread by the BAC method. b) Electron micrograph of a cytochrome c spread of poliovirion RNA. The bar represents 1 kilobase. The arrows point to 5′ end terminal protein dots. – Figure from Wu *et al.*, 1978 [Nucl. Acid Res. *5*, 4711 (1978)]

Table 26. *Characteristics of the poliovirus genome*

Composition:	one linear, single stranded molecule of RNA and a small covalently linked protein VPg
Genome polarity	+
Sedimentation coefficient	35 S
Molecular weight	2.6×10^6
Percentage of dry weight of virus particle	32 %
Base composition of RNA (%)	C G A U
vRNA (chemical)	24.1 22.4 30.3 23.3
vRNA (nucleotide sequence)	23.7 23.0 29.7 23.6
cRNA (chemical)	23.3 23.3 23.2 30.0
dsRNA	23.2 23.1 26.8 26.9
5′ end	pU covalently linked to 22 amino acid long VPg
3′ end	50–150 nucleotide long tail of poly-A
Length of RNA (no. of nucleotides)	7,500
of noncoding region	800
at 5′ end	740
at 3′ end	70
of coding region	6,700
Initiation site(s) for translation	the ninth AUG at position 743 from the 5′ end (other initiation sites might function in the initial phase of virus replication)
Coding domains for genome products: P-1, P-2, P-3	
5′, P-1	capsid proteins
middle, P-2	regulatory proteins
3′, P-3	protease, RNA polymerase
Infectivity of viral RNA	+
% homology of nucleotide sequence in different poliovirus strains	70 % total, 95 % in the first 20 nucleotides at the 5′ end

virus particle. Each virion particle contains only one RNA molecule. Figure 37 shows an electron micrograph of the unraveled virion RNA (Wu *et al.*, 1978).

The genomic RNA is (+)-stranded, *i.e.*, it has the same 5'–3' polarity as the mRNA synthesized after virus infection, and it must itself be utilized as mRNA, at least initially upon infection. Isolated viral RNA is infectious under suitable conditions (Koch, 1973), indicating that the viral genome encodes all the information necessary to support a productive infection. It is generally thought that the RNA contains only one major initiation site for translation near the 5' end (Baltimore, 1970). However, circumstantial evidence for one or two additional initiation sites which may be in different structural and functional states ("on" or "off") at different times of the infection cycle has been obtained, and investigations on this matter continue (Ehrenfeld, 1979; Koch *et al.*, 1980b; Stewart *et al.*, 1980).

B. Structural Organization

Figure 38 illustrates the overall structural organization of the poliovirus genome. The genome contains approximately 7,500 nucleotides (Lee *et al.*, 1979;

Fellner, 1979) with similar molar amounts of adenylate (30 %), cytidylate (24 %), guanylate (22.5 %), and uridylate (23.5 %) (Schaffer *et al.*, 1960; Roy and Bishop, 1970; Newman *et al.*, 1973), which, however, are not evenly distributed throughout the genome. The 5' end is linked to a small protein, VPg, the 3' end to a stretch of poly A, 40–100 nucleotides long (see below).

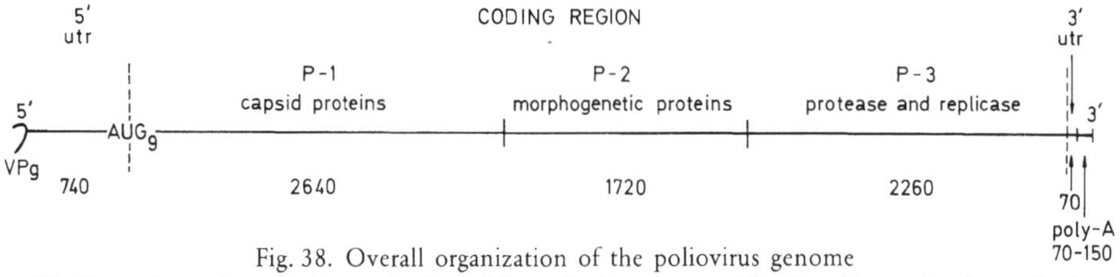

Fig. 38. Overall organization of the poliovirus genome

The figure shows the overall organization of the poliovirus genome. The 5′ end is covalently attached to the 22 amino acid long genomic protein VP_g, the 3′ end carries a tail of poly-A. An unusually long stretch of untranslated nucleotides is located at the 5′ end, a relatively short stretch of untranslated nucleotides is located at the 3′ end. Translation of the poliovirus genome is initiated on the ninth AUG codon at position 743. The coding region falls into three functionally distinct domains P-1, P-2, and P-3. The stretches of the poliovirus genome are drawn to scale, the numbers below the figure indicate the approximate lengths (in numbers of nucleotides) of the different regions as determined from the consensus nucleotide sequence of poliovirus type 1, Mahoney strain (see Fig. 44)

There are a total of more than 800 non-coding nucleotides located at the two ends of polio RNA: 740 untranslated bases at the 5' end and 70 untranslated bases preceeding the poly (A) tract at the 3' end. The untranslated regions presumably play regulatory roles in the initiation of protein synthesis (binding of ribosomes and initiation factors) and RNA synthesis (polymerase recognition and binding sites), and perhaps also for the interaction of viral RNAs with elements of the cytoskeleton and intracellular membranes, and in the stability of the RNA (poly A, secondary structure), in its interaction with the viral capsid during packaging of the RNA, and for the binding of presumptive regulatory molecules that control the consecutive translation and transcription of the RNA.

Approximately 6,700 nucleotides are left as a coding capacity for 2,200 amino acids or a protein of about 250,000 daltons. There are no indications for an increased coding capacity by a double reading via a frameshift for polio RNA as has been reported for the bacteriophages Qß (ss RNA virus) and ØX174 (ss DNA virus) (Barrell *et al.*, 1976). A little more than one third of the coding capacity is used to code for the capsid proteins VP4, VP2, VP3, and VP1. The rest is used to code for viral proteins and enzymes –NCVPs 1–10 (*N*on *C*apsid *V*iral *P*roteins)– which carry out important functions in the reorganization of cellular activity, in processing of viral proteins, in viral RNA replication, and in virion assembly.

Coding for all poliovirus proteins (genome products) falls into three major domains of the poliovirus genome. These domains evidently are structurally and functionally distinct regions of the genome, reflected in the three primary gene products NCVP1a (P-1 domain), NCVP3a (P-2 domain), and NCVP1b (P-3 domain). The P-1 domain codes for the structural proteins, the P-3 domain for replicative proteins, and the P-2 domain for proteins of unexplored function(s) (protease, inductor for membrane synthesis, compartmentalization, regulator of RNA synthesis) (see Section IV below).

C. Secondary Structure

Substantial secondary structure has been detected by biochemical means in infectious poliovirus RNA isolated from mature purified poliovirus (Koza, 1975). A computer search of the total genome sequence has identified many long potential stem structures, but their significance at this time is uncertain (Racaniello and Baltimore, 1981a). In particular, two stable secondary hairpin structures are found in the 5' untranslated region (see below). In contrast, preliminary computer analysis suggests that no larger stable secondary structures can be formed within an 840 base region at the 3' end (Kitamura *et al.*, 1980a). A single-stranded structure at the 3' terminus might be important in providing an easy access to the viral replicase during initiation of replication (Porter *et al.*, 1978). Stable secondary structures at the 5' end may be important for stopping ribosome binding on

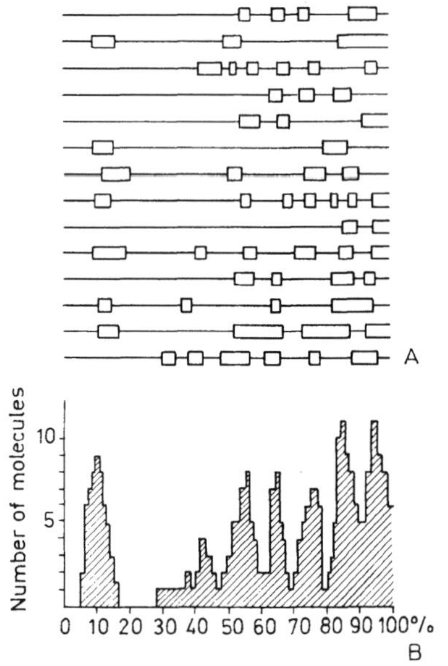

Fig. 39. The denaturation map of poliovirus type 1 replicative form RNA
RF-RNA molecules of poliovirus type 1 RNA (double helices of complementary RNAs) were partially denatured in 60% dimethyl sulfoxide and 8% formaldehyde at 27 °C and then prepared for examination by electron microscopy. Under these conditions most molecules contained two to four denatured regions
A Oriented and normalized schemes of individual RF-RNA molecules at 27 °C. The boxes represent denatured regions
B Statistical distribution of denatured regions along the genome in a. The length of the genome is expressed in % total length. – Figure from Čumakov *et al.*, 1979 [Virology *92*, 259 (1979)]

the parental RNA temporarily in order to allow its transcription by the newly synthesized replicase in the 3'—5' direction (see Chapter 9).

Further characterization of the secondary structure of poliovirus RNA has been obtained by denaturation mapping of double stranded RNA (Čumakov *et al.*, 1979; Agol *et al.*, 1980). When poliovirus double stranded RNA is subjected to mild denaturing conditions (69 % DMSO, 8 % formaldehyde), those regions containing a clustering of A-U base pairs (bound by two hydrogen bonds, in contrast to C-G base pairs, which are bound by three hydrogen bonds) are preferentially melted. Such regions are distinguishable from unmelted regions by electron microscopy, and a denaturation map may be established which is a distinguishing characteristic of the RNA. The denaturation map for poliovirus type 1 obtained by this method is presented in Figure 39.

D. The Genome-Linked Protein VPg

At its 5' end the genomic RNA of poliovirus is covalently linked to a small viral-coded protein, VPg (Lee *et al.*, 1976, 1977). All other picornaviruses also seem to possess a similar small protein linked to the 5' terminus of their genomic RNAs (Hruby and Roberts, 1978; Vartapetjan *et al.*, 1979). Calciviruses and the small plant picornaviruses also have been found to contain VPgs (Burroughs and Brown, 1978; Zabel *et al.*, 1982).

The nature of the covalent VPg-RNA linkage is a O^4-phosphodiester bond between the OH of the side chain of the tyrosine residue at position 3 of VPg and of the O^5 of the 5' terminal uridylic acid of the RNA (Fig. 40) (Ambros and Balti-

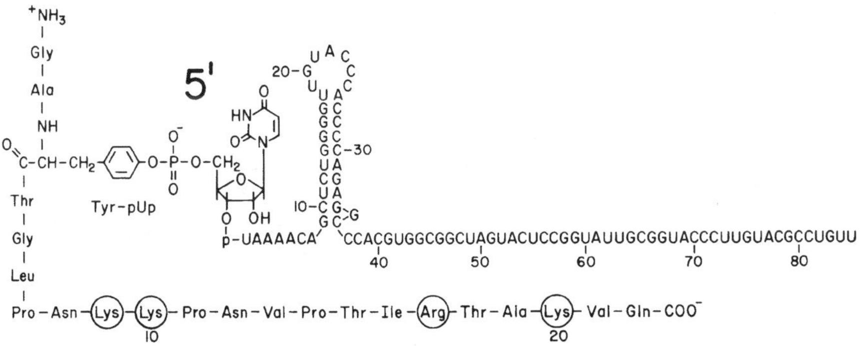

Fig. 40. Linkage of VPg to the 5' terminus of poliovirus RNA

The small genomic viral protein VPg is linked to the viral RNA via an O^4-(5'-uridylyl)tyrosine (Tyr-pUp) bond between its third amino acid residue and the 5' terminal nucleotide of the RNA. The phosphodiester bond in Tyr-pUp can be considered energy rich by analogy to O^4-(5'-adenylyl)tyrosine. The stem and loop structure near the 5' end of the RNA has a G° of −21 kcal/mol. The basic amino acid residues of VPg are encircled. The nature of the interaction between other portions of the VPg and RNA is not known. VPg is thought to play an important role in the initiation of RNA synthesis and possibly also in the recognition of capsid proteins for RNA encapsidation. — Figure from Kitamura *et al.*, 1981 [Nature *291*, 547—553 (1981)] more, 1978; Rothberg *et al.*, 1978). The linkage is energy rich (Holzer and Wohlhueter, 1972), but the mechanism by which it is formed is still not known.

From a combination of protein and nucleic acid sequence data, the length of VPg has been determined to be 22 amino acids, corresponding to a molecular weight of 2,400 (Kitamura *et al.*, 1980b, 1981; Semler *et al.*, 1981b). The protein is positively charged at neutral pH; its unusual properties include acid solubility (insolubility in water) and an exceptional ability to adhere to different surfaces (Lee *et al.*, 1976, 1977; Flanegan *et al.*, 1977; Nomoto *et al.*, 1977a).

VPg is encoded in the viral genome, as is evidenced by differences in the molecular masses and amino acid compositions of VPgs from different picornaviruses, and the independence of the molecular mass of a particular virus VPg on the kind of cell used for viral propagation (Nomoto *et al.*, 1976; Golini *et al.*, 1978; Sangar, 1979). The coding region for VPg on the viral genome lies within the replicase region (Pallansch *et al.*, 1980; Kitamura *et al.*, 1980b, 1981). Two different types of VPg—a more acidic and a more basic form—have been detected in linkage to poliovirus RNAs from infected cells (Richards *et al.* 1981), as well as in other picornaviruses. Whereas the three different types of VPg of FMDV have corresponding tandem coding sequences on the FMDV genome (Fors and Schaller, 1982), the poliovirus genome codes for only one type of VPg, indicating that the two poliovirus VPg types are modified versions of a single peptide.

VPg is attached to the 5' end of all newly synthesized plus and minus viral RNA strands and to all nascent strands of the replicative intermediate (Nomoto *et al.*, 1977a; Flanegan *et al.*, 1977; Pettersson *et al.*, 1978), the structure in which viral RNA is synthesized (Baltimore and Girard, 1966). VPg is also attached to the double stranded replicative-form RNA (Wu *et al.*, 1978). These observations suggest that VPg might function as primer for the initiation of viral RNA synthesis (Nomoto *et al.*, 1977a; Kitamura *et al.*, 1980b).

No significant change in infectivity of the RNAs is observed when the RNA is not linked to VPg, (Flanegan *et al.*, 1977; Nomoto *et al.*, 1977b). Thus VPg does not seem to be required during the early stages following infection.

In contrast to the virion RNA, the viral RNA associated with polyribosomes in infected cells, *i.e.* the mRNA involved in translation, is not linked to VPg (Nomoto *et al.*, 1976). In the cell cytoplasm, an enzymatic activity has been found that can cleave the VPg-RNA linkage (Ambros *et al.*, 1978; Ambros and Baltimore, 1980). Viral RNA lacking the VPg linkage apparently is not encapsidated during

---

Fig. 41. Oligonucleotide maps from RNase digests of poliovirus type 1 RNA (^{32}P) labeled RNA was digested simultaneously with RNase T1 and calf intestine alkaline phosphatase *(A)* or with RNase A *(B)*. After precipitation with ethanol, the RNase resistant fragments were separated by two dimensional polyacrylamide gel electrophoresis. Electrophoresis in the first dimension was from left to right (in 8% polyacrylamide, 6 M urea, pH 3.3), and in the second dimension from bottom to top (in 22% polyacrylamide, 50 mM Tris-borate, pH 8.2). Large, structurally unique, well-resolved oligonucleotides occupy the lower half of each map. In B, the base composition of the smaller oligonucleotides is indicated, as determined by secondary cleavage of the oligonucleotides with KOH followed by separation with paper electrophoresis. — Figure from Lee *et al.*, 1980 [J. gen. Virol. *44*, 311–322 (1980)]

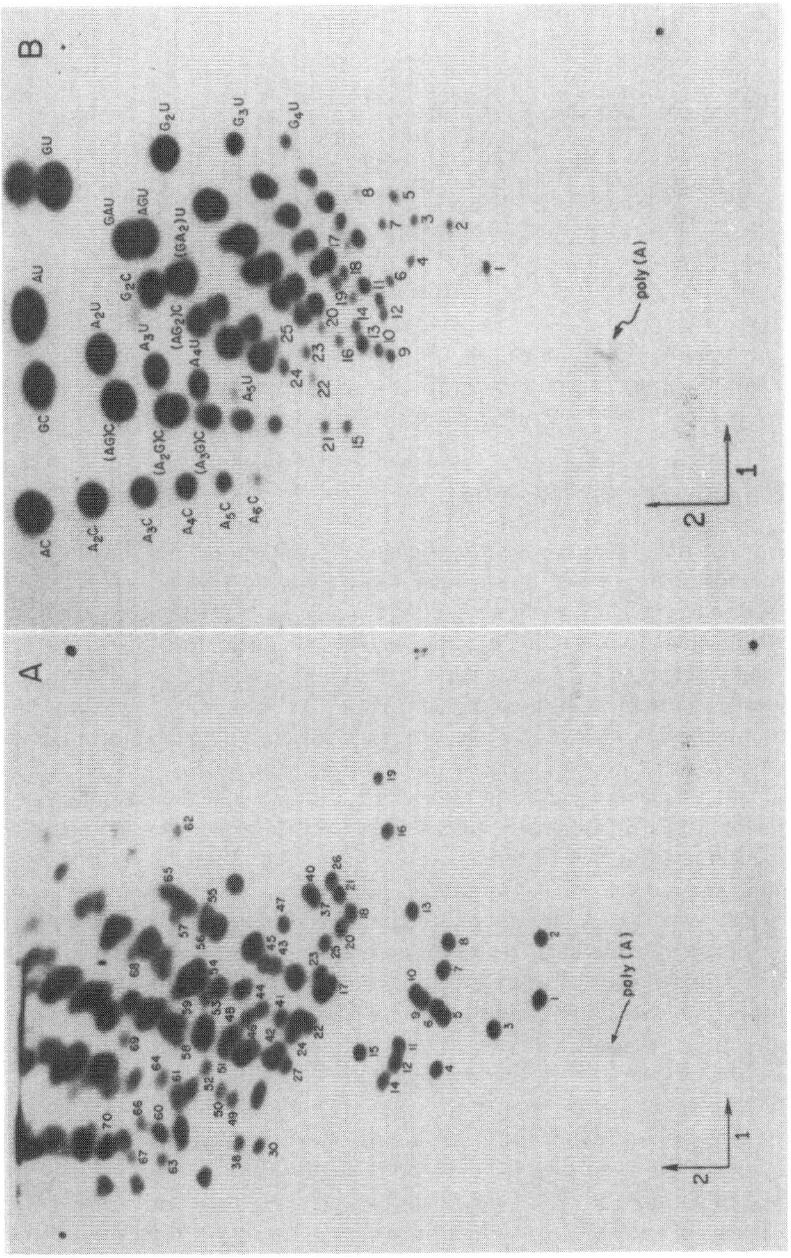
A
B
poly (A)
poly (A)
AC
GC
AU
GU
A_2C
(AG)C
A_2U
GAU
AGU
G_2U
A_3C
$(A_2G)C$
A_3U
G_2C
$(GA_2)U$
$(AG_2)C$
A_4C
$(A_3G)C$
A_4U
A_5C
A_5U
A_6C
G_3U
G_4U

assembly (Fernandez-Munoz and Lavi, 1977), suggesting that VPg may also be important for the interaction of viral RNA with capsid proteins in the assembly of virions.

III. Nucleotide Sequences

Early sequencing studies of poliovirus made use of the specific markers at the termini of the viral RNA (5' VPg and 3' poly A). This allowed the selective isolation of terminal oligonucleotides, whose sequences were either determined directly or after transcription into cDNA (Porter *et al.*, 1978, 1979). Unique, large oligonucleotides obtained from extensive digestion of poliovirus RNA with ribonucleases (RNase T1, RNase A, ribonuclease III) were sequenced and mapped onto the poliovirus genome (Stewart *et al.*, 1980; Nomoto *et al.*, 1981a). The elucidation of the entire nucleotide sequence of poliovirus type 1 RNA owed its success to modern genetic engineering technology.

A. Oligonucleotide Mapping

An elegant method for characterizing RNA genomes is the oligonucleotide mapping of RNAs by two dimensional electrophoresis (Frisby *et al.*, 1976b; Lee and Wimmer, 1976). Similar to the tryptic peptide fingerprinting of proteins described below, RNA can be fingerprinted after digestion with specific nucleases into oligonucleotides. Upon nuclease digestion (for example by RNase T1) the products are separated by two dimensional gel electrophoresis resulting in the typical fingerprint of the respective RNA. The corresponding fingerprint of poliovirus type 1 is shown in Figure 41 (Lee *et al.*, 1979).

The oligonucleotide spots in fingerprints can be mapped within the genome. The virion RNA is fragmented either by weak alkaline conditions (Nomoto *et al.*, 1979a) or by treatment with ribonuclease III, an endoribonuclease from Escherichia coli (Harris *et al.*, 1978; Stewart *et al.*, 1980). Poly (A)-containing fragments (3' end) are then isolated by affinity chromatography on appropriate columns, and may be separated according to size on SDS-agarose-acrylamide gels. Each fragment is isolated, digested with T1 ribonuclease and the product is fingerprinted on two dimensional gels as above. The fingerprint of each successively larger fragment contains all the oligonucleotide spots of its preceeding smaller fragment plus one (or a few) new spot(s). Thus, the mapping position of the spots can be accurately determined in succession from the 3' (poly (A)) end to the 5' (VPg-containing) end. 76 oligonucleotides of the fingerprint of poliovirus type 1 (Mahoney, Maizel laboratory strain) were thus mapped (Stewart *et al.*, 1980). Figure 42 shows the corresponding physical map. Such maps are expected to be useful in identifying segments of the polio RNA captured in biologically active complexes with ribosomes (McClain *et al.*, 1981), polymerase molecules, or other proteins that may be involved in the regulation of polio-RNA translation or replication, as well as in defining segments of the genomic RNA that have been deleted (as in DI particles) or otherwise rearranged.

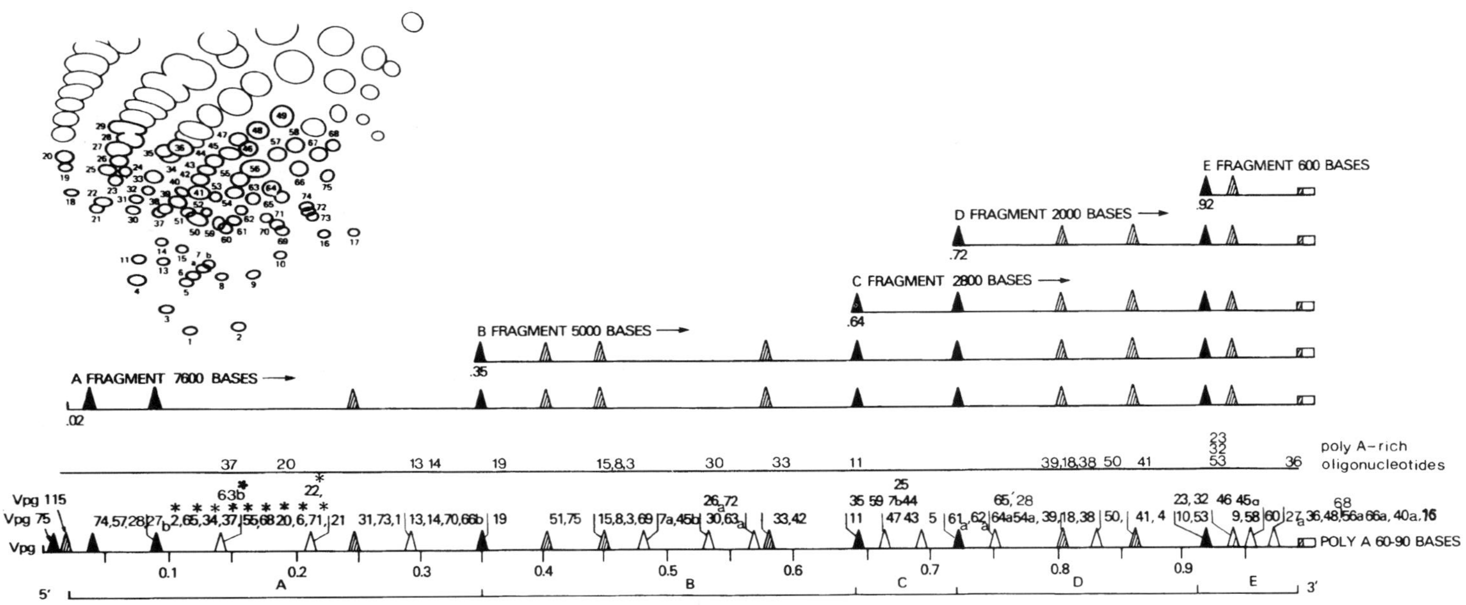

Fig. 42. Physical map of the RNase T1 resistant oligonucleotides of poliovirus type 1 RNA

A diagrammatic representation of a typical RNase T1 oligonucleotide map is presented in A). Note that the labelling of spots is slightly different from that in Figure 41. For physical mapping of the oligonucleotide spots on the viral genome, virion RNA was first digested with RNase III producing specific fragments of different sizes. Poly-A containing fragments (3′ terminal fragments) were isolated by chromatography on poly-U sepharose columns and separated according to size by polyacrylamide gel electrophoresis. Subsequent digestion of the different fragments by RNase T1 allowed the mapping of 71 RNase T1 resistant oligonucleotides (B). The upper portion of B) shows the major fragments produced by RNase III cleavage of poliovirus RNA. ▲ denote preferred RNase III cleavage sites producing major fragments (fragments A–E). ▲ denote additional RNase III cleavage sites producing minor fragments. △ denote the position of the 5′ ends of fragments cut from gel in the absence of detectable bands for use in mapping of the intermediate fragments. The distribution of poly A rich nucleotides is indicated in the middle of B). The Asterisks denote spots reduced or absent in defective interfering particle DI (A) (see below). – Figure from Stewart *et al.*, 1980 [Virology *104*, 375–397 (1980)]

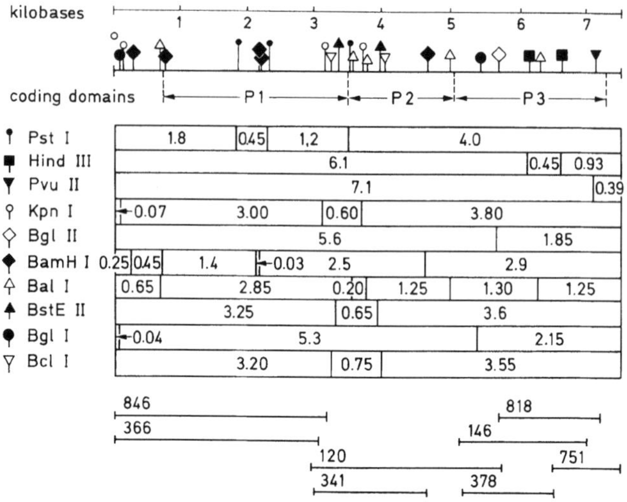

Fig. 43 A–B. Molecular cloning of the genome of poliovirus type 1

Fig. 43 A. The restriction endonuclease map of poliovirus 1 cDNA and the positions of some representative cloned cDNA inserts

Poliovirus cDNA-RNA hybrids were prepared from the Mahoney strain of poliovirus type 1 by using reverse transcriptase. The hybrids were tailed with poly(dC) and were inserted into the poly(dG)-tailed Pst I site of Escherichia coli plasmid pBR322. Bacterial clones containing recombinant DNA were selected on the basis of their resistance to tetracycline and sensitivity to ampicillin. Clones carrying recombinant DNA were screened for the presences of virus specific sequences by in situ hybridization with various RNase T1 resistant oligonucleotides (see Fig. 41 and 42). Cloned virus specific cDNA inserts were analyzed by electron microscopy and were mapped by double digestion with restriction endonucleases

The figure shows the restriction endonuclease map of poliovirus type 1 cDNA and the mapping of some of the clones of poliovirus cDNA inserts. The viral RNA is shown with the 5′ end to the left. The mapping positions of the three coding domains are indicated in the second row. The mapping positions of the cleavage sites by different restriction endonucleases is indicated by the corresponding symbols. The individual restriction fragments generated by cleavage with the different enzymes are shown in the center, the sizes of the fragments are given in kilobases. The lines in the bottom of the figure represent the cloned cDNA inserts. – Figure modified from Van der Werf *et al.*, 1981 [Proc. Natl. Acad. Sci. U.S.A. *78*, 5986 (1981)]

Fig. 43 B. The construction of a full-length cDNA clone of poliovirus type 1

This figure shows the construction of a full-length poliovirus cDNA clone from the two cDNA clones pVR101 and pVR103, and of an additional plasmid, pVR105, containing the 200 5′ terminal nucleotide sequence. Single lines: DNA of E. coli plasmid pBR322; double lines: poliovirus specific cDNA [light areas in the poliovirus-specific region represents oligo (A) · oligo (T), the oligo (G) · oligo (C) tails on both ends of the inserts are not shown]. Numbers refer to nucleotide position on the viral RNA.

The plasmid DNAs were grown in E. coli, extracted, purified by CsCl gradients, cleaved with restriction endonucleases; the resulting fragments were separated by electrophoresis on agarose gels and recovered from the gels by the glass powder technique. In order to join plasmids pVR101 and pVR103, and then pVR104 and pVR105, the plasmids were cleaved by Bgl II, Eco RI, and Bam HI as indicated in the figure, corresponding fragments were isolated and ligased by T4 DNA ligase. The resulting plasmid pVR106 contained a full length cDNA copy of the poliovirus genome at the Pst I site of E. coli plasmid pBR322. Cultured mammalian cells transfected with this hybrid plasmid produced infectious poliovirus. Cells transfected with plasmid pVR104, in contrast, which lacks the first 115 bases of the poliovirus genome did not produce virus. – Figure from Racaniello and Baltimore, 1981 [Science *214*, 917 (1981 b)]

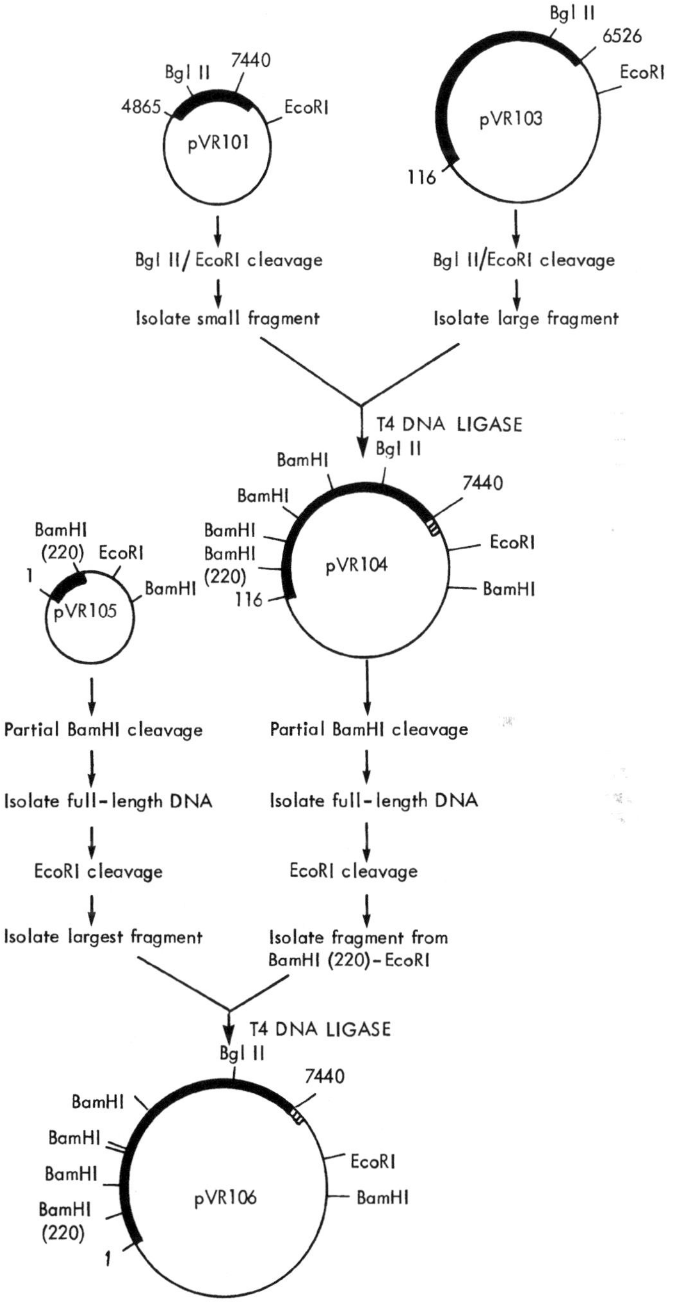

pVR101
Bgl II
7440
4865
EcoRI
pVR103
Bgl II
6526
EcoRI
116
Bgl II/EcoRI cleavage
Isolate small fragment
Bgl II/EcoRI cleavage
Isolate large fragment
T4 DNA LIGASE
pVR104
Bgl II
BamHI
BamHI
7440
BamHI
BamHI
(220)
116
EcoRI
BamHI
pVR105
BamHI
(220)
EcoRI
1
BamHI
Partial BamHI cleavage
Isolate full-length DNA
EcoRI cleavage
Isolate largest fragment
Partial BamHI cleavage
Isolate full-length DNA
EcoRI cleavage
Isolate fragment from
BamHI (220)-EcoRI
T4 DNA LIGASE
pVR106
Bgl II
7440
BamHI
BamHI
BamHI
BamHI
(220)
1
EcoRI
BamHI

B. Cloning of Poliovirus cDNA

The poly (A) tract at the 3' end of poliovirus RNA permits its reverse transcription into cDNA using oligo (dT) primers and reverse transcriptase (Kacian and Myers, 1976a; Porter *et al.*, 1979; Kitamura and Wimmer, 1980). Within the restrictions of national guidelines for cloning of recombinant DNA, various fragments of double stranded poliovirus cDNA were cloned in plasmids of Escherichia coli (van der Werf *et al.*, 1981; Girard *et al.*, 1981; Kopecka *et al.*, 1981; Racaniello and Baltimore, 1981a; Nomoto *et al.*, 1982b). The cDNA fragments within bacterial plasmids could be mapped onto the poliovirus genome with the aid of the large oligonucleotide markers and with electron microscopic analysis of cDNA containing plasmid-poliovirus RNA hybrids (Grunstein and Hogness, 1975; Lee *et al.*, 1979; Girard et al., 1981; van der Werf *et al.*, 1981). The restriction maps of the cDNAs of poliovirus types 1 and 2—the mapping positions of the cleavage sites on poliovirus cDNA for dsDNA specific restriction endonucleases—have been reported (Fig. 43A) (Kopecka *et al.*, 1981; Nomoto *et al.*, 1982a).

Plasmids containing defined fragments of poliovirus cDNA have been obtained. These could be classified into three major groups, carrying cDNA sequences corresponding to the three major gene domains (5' region, middle, 3' region) of the poliovirus genome (Girard *et al.*, 1981; van der Werf *et al.*, 1981). Poliovirus type 1 cDNA clones, that contain a complete complementary DNA copy of the RNA genome, have recently been obtained (Fig. 43B). Cultured mammalian cells transfected with the complete poliovirus DNA containing plasmid produced infectious poliovirus, whereas cells transfected with a plasmid which lacked the first 115 bases of the poliovirus genome did not produce virus (Racaniello and Baltimore, 1981b).

The bacterial clones carrying plasmids that contain poliovirus cDNA are expected to become a very powerful tool for studying the relationship between the structure of the poliovirus genome and its biological significance. It will now be possible to specifically mutagenize the infectious cloned poliovirus DNA, generating mutants of poliovirus with defined defects in any part of the genome. The expression of virus specific proteins in E. coli could open new routes to the production of new types of vaccine (Enger-Valk *et al.*, 1984). Poliovirus cDNA should also turn out to be useful for the detection of low quantities of poliovirus genomic sequences in RNA isolated from tissue (Tracy and Smith, 1981).

C. The Consensus Sequence of Poliovirus Type I

Different sequencing techniques have been followed to sequence either poliovirus cDNA directly (Kitamura *et al.*, 1981) or defined fragments of cloned cDNA (Racaniello and Baltimore, 1981a; Nomoto *et al.*, 1982). The particular techniques employed include modifications of Sanger's chain termination method (Sanger *et*

Fig. 44. The complete consensus nucleotide sequence and deduced amino acid sequence of poliovirus type 1, Mahoney strain. Figure courtesy of E. Wimmer, Stony Brook

```
TRANSLATION OF POLIOCON                         IN PHASE  2

                        16                  31                  46
VPg-pU UAA AAC AGC UCU GGG GUU GUA CCC ACC CCA GAG GCC CAC GUG GCG
       ### ASN SER SER GLY VAL VAL PRO THR PRO GLU ALA HIS VAL ALA

                        61                  76                  91
       GCU AGU ACU CCG GUA UUG CGG UAC CCU UGU ACG CCU GUU UUA UAC
       ALA SER THR PRO VAL LEU ARG TYR PRO CYS THR PRO VAL LEU TYR

                       106                 121                 136
       UCC CUU CCC GUA ACU UAG ACG CAC AAA ACC AAG UUC AAU AGA AGG
       SER LEU PRO VAL THR *** THR HIS LYS THR LYS PHE ASN ARG ARG

                       151                 166                 181
       GGG UAC AAA CCA GUA CCA CCA CGA ACA AGC ACU UCU GUU UCC CCG
       GLY TYR LYS PRO VAL PRO PRO ARG THR SER THR SER VAL SER PRO

                       196                 211                 226
       GUG AUG UCG UAU AGA CUG CUU GCG UGG UUG AAA GCG ACG GAU CCG
       VAL MET SER TYR ARG LEU LEU ALA TRP LEU LYS ALA THR ASP PRO

                       241                 256                 271
       UUA UCC GCU UAU GUA CUU CGA GAA GCC CAG UAC CAC CUC GGA AUC
       LEU SER ALA TYR VAL LEU ARG GLU ALA GLN TYR HIS LEU GLY ILE

                       286                 301                 316
       UUC GAU GCG UUG CGC UCA GCA CUC AAC CCC AGA GUG UAG CUU AGG
       PHE ASP ALA LEU ARG SER ALA LEU ASN PRO ARG VAL *** LEU ARG

                       331                 346                 361
       CUG AUG AGU CUG GAC AUC CCU CAC CGG UGA CGG UGG UCC AGG CUG
       LEU MET SER LEU ASP ILE PRO HIS ARG @@@ ARG TRP SER ARG LEU

                       376                 391                 406
       CGU UGG CGG CCU ACC UAU GGC UAA CGC CAU GGG ACG CUA GUU GUG
       ARG TRP ARG PRO THR TYR GLY ### ARG HIS GLY THR LEU VAL VAL

                       421                 436                 451
       AAC AAG GUG UGA AGA GCC UAU UGA GCU ACA UAA GAA UCC UCC GGC
       ASN LYS VAL @@@ ARG ALA TYR @@@ ALA THR ### GLU SER SER GLY

                       466                 481                 496
       CCC UGA AUG CGG CUA AUC CCA ACC UCG GAG CAG GUG GUC ACA AAC
       PRO @@@ MET ARG LEU ILE PRO THR SER GLU GLN VAL VAL THR ASN

                       511                 526                 541
       CAG UGA UUG GCC UGU CGU AAC GCG CAA GUC CGU GGC GGA ACC GAC
       GLN @@@ LEU ALA CYS ARG ASN ALA GLN VAL ARG GLY GLY THR ASP

                       556                 571                 586
       UAC UUU GGG UGU CCG UGU UUC CUU UUA UUU UAU UGU GGC UGC UUA
       TYR PHE GLY CYS PRO CYS PHE LEU LEU PHE TYR CYS GLY CYS LEU

                       601                 616                 631
       UGG UGA CAA UCA CAG AUU GUU AUC AUA AAG CGA AUU GGA UUG GCC
       TRP @@@ GLN SER GLN ILE VAL ILE ILE LYS ARG ILE GLY LEU ALA

                       646                 661                 676
       AUC CGG UGA AAG UGA GAC UCA UUA UCU AUC UGU UUG CUG GAU CCG
       ILE ARG @@@ LYS @@@ ASP SER LEU SER ILE CYS LEU LEU ASP PRO

                       691                 706                 721
       CUC CAU UGA GUG UGU UUA CUC UAA GUA CAA UUU CAA CAG UUA UUU
       LEU HIS @@@ VAL CYS LEU LEU ### VAL GLN PHE GLN GLN LEU PHE
                            START ┌► VP4
                       736                 751                 766
       CAA UCA GAC AAU UGU AUC AUA|AUG GGU GCU CAG GUU UCA UCA CAG
       GLN SER ASP ASN CYS ILE ILE MET GLY ALA GLN VAL SER SER GLN
```

```
                    781                 796                 811
AAA GUG GGC GCA CAU GAA AAC UCA AAU AGA GCG UAU GGU GGU UCU
LYS VAL GLY ALA HIS GLU ASN SER ASN ARG ALA TYR GLY GLY SER

                    826                 841                 856
ACC AUU AAU UAC ACC ACC AUU AAU UAU UAU AGA GAU UCA GCU AGU
THR ILE ASN TYR THR THR ILE ASN TYR TYR ARG ASP SER ALA SER

                    871                 886                 901
AAC GCG GCU UCG AAA CAG GAC UUC UCU CAA GAC CCU UCC AAG UUC
ASN ALA ALA SER LYS GLN ASP PHE SER GLN ASP PRO SER LYS PHE

                    916                 931                 946
ACC GAG CCC AUC AAG GAU GUC CUG AUA AAA ACA GCC CCA AUG CUA
THR GLU PRO ILE LYS ASP VAL LEU ILE LYS THR ALA PRO MET LEU

                    961                 976                 991
AAC UCG CCA AAC AUA GAG GCU UGC GGG UAU AGC GAU AGA GUA CUG
ASN SER PRO ASN ILE GLU ALA CYS GLY TYR SER ASP ARG VAL LEU

                   1006                1021                1036
CAA UUA ACA CUG GGA AAC UCC ACU AUA ACC ACA CAG GAG GCG GCU
GLN LEU THR LEU GLY ASN SER THR ILE THR THR GLN GLU ALA ALA

   ┌─►VP2          1051                1066                1081
AAU|UCA GUA GUC GCU UAU GGG CGU UGG CCU GAA UAU CUG AGG GAC
ASN SER VAL VAL ALA TYR GLY ARG TRP PRO GLU TYR LEU ARG ASP

                   1096                1111                1126
AGC GAA GCC AAU CCA GUG GAC CAG CCG ACA GAA CCA GAC GUC GCU
SER GLU ALA ASN PRO VAL ASP GLN PRO THR GLU PRO ASP VAL ALA

                   1141                1156                1171
GCA UGC AGG UUU UAU ACG CUA GAC ACC GUG UCU UGG ACG AAA GAG
ALA CYS ARG PHE TYR THR LEU ASP THR VAL SER TRP THR LYS GLU

                   1186                1201                1216
UCG CGA GGG UGG UGG UGG AAG UUG CCU GAU GCA CUG AGG GAC AUG
SER ARG GLY TRP TRP TRP LYS LEU PRO ASP ALA LEU ARG ASP MET

                   1231                1246                1261
GGA CUC UUU GGG CAA AAU AUG UAC UAC CAC UAC CUA GGU AGG UCC
GLY LEU PHE GLY GLN ASN MET TYR TYR HIS TYR LEU GLY ARG SER

                   1276                1291                1306
GGG UAC ACC GUG CAU GUA CAG UGU AAC GCC UCC AAA UUC CAC CAG
GLY TYR THR VAL HIS VAL GLN CYS ASN ALA SER LYS PHE HIS GLN

                   1321                1336                1351
GGG GCA CUA GGG GUA UUC GCC GUA CCA GAG AUG UGU CUG GCC GGG
GLY ALA LEU GLY VAL PHE ALA VAL PRO GLU MET CYS LEU ALA GLY

                   1366                1381                1396
GAU AGC AAC ACC ACU ACC AUG CAC ACC AGC UAU CAA AAU GCC AAU
ASP SER ASN THR THR THR MET HIS THR SER TYR GLN ASN ALA ASN

                   1411                1426                1441
CCU GGC GAG AAA GGA GGC ACU UUC ACG GGU ACG UUC ACU CCU GAC
PRO GLY GLU LYS GLY GLY THR PHE THR GLY THR PHE THR PRO ASP

                   1456                1471                1486
AAC AAC CAG ACA UCA CCU GCC CGC AGG UUC UGC CCG GUG GAU UAC
ASN ASN GLN THR SER PRO ALA ARG ARG PHE CYS PRO VAL ASP TYR

                   1501                1516                1531
CUC CUU GGA AAU GGC ACG UUG UUG GGG AAU GCC UUU GUG UUC CCG
LEU LEU GLY ASN GLY THR LEU LEU GLY ASN ALA PHE VAL PHE PRO
```

```
                 1546                1561                1576
CAC CAG AUA AUA AAC CUA CGG ACC AAC AAC UGU GCU ACA CUG GUA
HIS GLN ILE ILE ASN LEU ARG THR ASN ASN CYS ALA THR LEU VAL

                 1591                1606                1621
CUC CCU UAC GUG AAC UCC CUC UCG AUA GAU AGU AUG GUA AAG CAC
LEU PRO TYR VAL ASN SER LEU SER ILE ASP SER MET VAL LYS HIS

                 1636                1651                1666
AAU AAU UGG GGA AUU GCA AUA UUA CCA UUG GCC CCA UUA AAU UUU
ASN ASN TRP GLY ILE ALA ILE LEU PRO LEU ALA PRO LEU ASN PHE

                 1681                1696                1711
GCU AGU GAG UCC UCC CCA GAG AUU CCA AUC ACC UUG ACC AUA GCC
ALA SER GLU SER SER PRO GLU ILE PRO ILE THR LEU THR ILE ALA

                 1726                1741                1756
CCU AUG UGC UGU GAG UUC AAU GGA UUA AGA AAC AUC ACC CUG CCA
PRO MET CYS CYS GLU PHE ASN GLY LEU ARG ASN ILE THR LEU PRO

             →VP3
                 1771                1786                1801
CGC UUA CAG|GGC CUG CCG GUC AUG AAC ACC CCU GGU AGC AAU CAA
ARG LEU GLN GLY LEU PRO VAL MET ASN THR PRO GLY SER ASN GLN

                 1816                1831                1846
UAU CUU ACU GCA GAC AAC UUC CAG UCA CCG UGU GCG CUG CCU GAA
TYR LEU THR ALA ASP ASN PHE GLN SER PRO CYS ALA LEU PRO GLU

                 1861                1876                1891
UUU GAU GUG ACC CCA CCU AUU GAC AUA CCC GGU GAA GUA AAG AAC
PHE ASP VAL THR PRO PRO ILE ASP ILE PRO GLY GLU VAL LYS ASN

                 1906                1921                1936
AUG AUG GAA UUG GCA GAA AUC GAC ACC AUG AUU CCC UUU GAC UUA
MET MET GLU LEU ALA GLU ILE ASP THR MET ILE PRO PHE ASP LEU

                 1951                1966                1981
AGU GCC ACA AAA AAG AAC ACC AUG GAA AUG UAU AGG GUU CGG UUA
SER ALA THR LYS LYS ASN THR MET GLU MET TYR ARG VAL ARG LEU

                 1996                2011                2026
AGU GAC AAA CCA CAU ACA GAC GAU CCC AUA CUC UGC CUG UCA CUC
SER ASP LYS PRO HIS THR ASP ASP PRO ILE LEU CYS LEU SER LEU

                 2041                2056                2071
UCU CCA GCU UCA GAU CCU AGG UUG UCA CAU ACU AUG CUU GGA GAA
SER PRO ALA SER ASP PRO ARG LEU SER HIS THR MET LEU GLY GLU

                 2086                2101                2116
AUC CUA AAU UAC UAC ACA CAC UGG GCA GGA UCC CUG AAG UUC ACG
ILE LEU ASN TYR TYR THR HIS TRP ALA GLY SER LEU LYS PHE THR

                 2131                2146                2161
UUU CUG UUC UGU GGA UCC AUG AUG GCA ACU GGC AAA CUG UUG GUG
PHE LEU PHE CYS GLY SER MET MET ALA THR GLY LYS LEU LEU VAL

                 2176                2191                2206
UCA UAC GCG CCU CCU GGA GCC GAC CCA CCA AAG AAG CGU AAG GAG
SER TYR ALA PRO PRO GLY ALA ASP PRO PRO LYS LYS ARG LYS GLU

                 2221                2236                2251
GCG AUG UUG GGA ACA CAU GUG AUC UGG GAC AUA GGA CUG CAG UCC
ALA MET LEU GLY THR HIS VAL ILE TRP ASP ILE GLY LEU GLN SER

                 2266                2281                2296
UCA UGU ACU AUG GUA GUG CCA UGG AUU AGC AAC ACC ACG UAU CGG
SER CYS THR MET VAL VAL PRO TRP ILE SER ASN THR THR TYR ARG
```

```
                    2311                2326                2341
CAA ACC AUA GAU GAU AGU UUC ACC GAA GGC GGA UAC AUC AGC GUC
GLN THR ILE ASP ASP SER PHE THR GLU GLY GLY TYR ILE SER VAL

                    2356                2371                2386
UUC UAC CAA ACU AGA AUA GUC GUC CCU CUU UCG ACA CCC AGA GAG
PHE TYR GLN THR ARG ILE VAL VAL PRO LEU SER THR PRO ARG GLU

                    2401                2416                2431
AUG GAC AUC CUU GGU UUU GUG UCA GCG UGU AAU GAC UUC AGC GUG
MET ASP ILE LEU GLY PHE VAL SER ALA CYS ASN ASP PHE SER VAL

                    2446                2461                2476
CGC UUG UUG CGA GAU ACC ACA CAU AUA GAG CAA AAA GCG CUA GCA
ARG LEU LEU ARG ASP THR THR HIS ILE GLU GLN LYS ALA LEU ALA

   → VP1
                    2491                2506                2521
CAG|GGG UUA GGU CAG AUG CUU GAA AGC AUG AUU GAC AAC ACA GUC
GLN GLY LEU GLY GLN MET LEU GLU SER MET ILE ASP ASN THR VAL

                    2536                2551                2566
CGU GAA ACG GUG GGG GCG GCA ACA UCU AGA GAC GCU CUC CCA AAC
ARG GLU THR VAL GLY ALA ALA THR SER ARG ASP ALA LEU PRO ASN

                    2581                2596                2611
ACU GAA GCC AGU GGA CCA ACA CAC UCC AAG GAA AUU CCG GCA CUC
THR GLU ALA SER GLY PRO THR HIS SER LYS GLU ILE PRO ALA LEU

                    2626                2641                2656
ACC GCA GUG GAA ACU GGG GCC ACA AAU CCA CUA GUC CCU UCU GAU
THR ALA VAL GLU THR GLY ALA THR ASN PRO LEU VAL PRO SER ASP

                    2671                2686                2701
ACA GUG CAA ACC AGA CAU GUU GUA CAA CAU AGG UCA AGG UCA GAG
THR VAL GLN THR ARG HIS VAL VAL GLN HIS ARG SER ARG SER GLU

                    2716                2731                2746
UCU AGC AUA GAG UCU UUC UUC GCG CGG GGU GCA UGC GUG ACC AUU
SER SER ILE GLU SER PHE PHE ALA ARG GLY ALA CYS VAL THR ILE

                    2761                2776                2791
AUG ACC GUG GAU AAC CCA GCU UCC ACC ACG AAU AAG GAU AAG CUA
MET THR VAL ASP ASN PRO ALA SER THR THR ASN LYS ASP LYS LEU

                    2806                2821                2836
UUU GCA GUG UGG AAG AUC ACU UAU AAA GAU ACU GUC CAG UUA CGG
PHE ALA VAL TRP LYS ILE THR TYR LYS ASP THR VAL GLN LEU ARG

                    2851                2866                2881
AGG AAA UUG GAG UUC UUC ACC UAU UCU AGA UUU GAU AUG GAA CUU
ARG LYS LEU GLU PHE PHE THR TYR SER ARG PHE ASP MET GLU LEU

                    2896                2911                2926
ACC UUU GUG GUU ACU GCA AAU UUC ACU GAG ACU AAC AAU GGG CAU
THR PHE VAL VAL THR ALA ASN PHE THR GLU THR ASN ASN GLY HIS

                    2941                2956                2971
GCC UUA AAU CAA GUG UAC CAA AUU AUG UAC GUA CCA CCA GGC GCU
ALA LEU ASN GLN VAL TYR GLN ILE MET TYR VAL PRO PRO GLY ALA

                    2986                3001                3016
CCA GUG CCC GAG AAA UGG GAC GAC UAC ACA UGG CAA ACC UCA UCA
PRO VAL PRO GLU LYS TRP ASP ASP TYR THR TRP GLN THR SER SER

                    3031                3046                3061
AAU CCA UCA AUC UUU UAC ACC UAC GGA ACA GCU CCA GCC CGG AUC
ASN PRO SER ILE PHE TYR THR TYR GLY THR ALA PRO ALA ARG ILE
```

```
                    3076                3091                3106
UCG GUA CCG UAU GUU GGU AUU UCG AAC GCC UAU UCA CAC UUU UAC
SER VAL PRO TYR VAL GLY ILE SER ASN ALA TYR SER HIS PHE TYR

                    3121                3136                3151
GAC GGU UUU UCC AAA GUA CCA CUG AAG GAC CAG UCG GCA GCA CUA
ASP GLY PHE SER LYS VAL PRO LEU LYS ASP GLN SER ALA ALA LEU

                    3166                3181                3196
GGU GAC UCC CUU UAU GGU GCA GCA UCU CUA AAU GAC UUC GGU AUU
GLY ASP SER LEU TYR GLY ALA ALA SER LEU ASN ASP PHE GLY ILE

                    3211                3226                3241
UUG GCU GUU AGA GUA GUC AAU GAU CAC AAC CCG ACC AAG GUC ACC
LEU ALA VAL ARG VAL VAL ASN ASP HIS ASN PRO THR LYS VAL THR

                    3256                3271                3286
UCC AAA AUC AGA GUG UAU CUA AAA CCC AAA CAC AUC AGA GUC UGG
SER LYS ILE ARG VAL TYR LEU LYS PRO LYS HIS ILE ARG VAL TRP

                    3301                3316                3331
UGC CCG CGU CCA CCG AGG GCA GUG GCG UAC UAC GGC CCU GGA GUG
CYS PRO ARG PRO PRO ARG ALA VAL ALA TYR TYR GLY PRO GLY VAL

                    3346                3361                3376
GAU UAC AAG GAU GGU ACG CUU ACA CCC CUC UCC ACC AAG GAU CUG
ASP TYR LYS ASP GLY THR LEU THR PRO LEU SER THR LYS ASP LEU
               -> 3b,8
                    3391                3406                3421
ACC ACA UAU|GGA UUC GGA CAC CAA AAC AAA GCG GUG UAC ACU GCA
THR THR TYR GLY PHE GLY HIS GLN ASN LYS ALA VAL TYR THR ALA

                    3436                3451                3466
GGU UAC AAA AUU UGC AAC UAC CAC UUG GCC ACU CAG GAU GAU UUG
GLY TYR LYS ILE CYS ASN TYR HIS LEU ALA THR GLN ASP ASP LEU

                    3481                3496                3511
CAA AAC GCA GUG AAC GUC AUG UGG AGU AGA GAC CUC UUA GUC ACA
GLN ASN ALA VAL ASN VAL MET TRP SER ARG ASP LEU LEU VAL THR

                    3526                3541                3556
GAA UCA AGA GCC CAG GGC ACC GAU UCA AUC GCA AGG UGC AAU UGC
GLU SER ARG ALA GLN GLY THR ASP SER ILE ALA ARG CYS ASN CYS

                    3571                3586                3601
AAC GCA GGG GUG UAC UAC UGC GAG UCU AGA AGG AAA UAC UAC CCA
ASN ALA GLY VAL TYR TYR CYS GLU SER ARG ARG LYS TYR TYR PRO

                    3616                3631                3646
GUA UCC UUC GUU GGC CCA ACG UUC CAG UAC AUG GAG GCU AAU AAC
VAL SER PHE VAL GLY PRO THR PHE GLN TYR MET GLU ALA ASN ASN

                    3661                3676                3691
UAU UAC CCA GCU AGG UAC CAG UCC CAU AUG CUC AUU GGC CAU GGA
TYR TYR PRO ALA ARG TYR GLN SER HIS MET LEU ILE GLY HIS GLY

                    3706                3721                3736
UUC GCA UCU CCA GGG GAU UGU GGU GGC AUA CUC AGA UGU CAC CAC
PHE ALA SER PRO GLY ASP CYS GLY GLY ILE LEU ARG CYS HIS HIS

                    3751                3766                3781
GGG GUG AUA GGG AUC AUU ACU GCU GGU GGA GAA GGG UUG GUU GCA
GLY VAL ILE GLY ILE ILE THR ALA GLY GLY GLU GLY LEU VAL ALA

                    3796                3811                3826
UUU UCA GAC AUU AGA GAC UUG UAU GCC UAC GAA GAA GAA GCC AUG
PHE SER ASP ILE ARG ASP LEU TYR ALA TYR GLU GLU GLU ALA MET
```

```
      ┌─► 5b,10
            3841                    3856                    3871
GAA CAA|GGC AUC ACC AAU UAC AUA GAG UCA CUU GGG GCC GCA UUU
GLU GLN GLY ILE THR ASN TYR ILE GLU SER LEU GLY ALA ALA PHE

            3886                    3901                    3916
GGA AGU GGA UUU ACU CAG CAG AUU AGC GAC AAA AUA ACA GAG UUG
GLY SER GLY PHE THR GLN GLN ILE SER ASP LYS ILE THR GLU LEU

            3931                    3946                    3961
ACC AAU AUG GUG ACC AGU ACC AUC ACU GAA AAG CUA CUU AAG AAC
THR ASN MET VAL THR SER THR ILE THR GLU LYS LEU LEU LYS ASN

            3976                    3991                    4006
UUG AUC AAG AUC AUA UCC UCA CUA GUU AUU AUA ACU AGG AAC UAU
LEU ILE LYS ILE ILE SER SER LEU VAL ILE ILE THR ARG ASN TYR

            4021                    4036                    4051
GAA GAC ACC ACA ACA GUG CUC GCU ACC CUG GCC CUU CUU GGG UGU
GLU ASP THR THR THR VAL LEU ALA THR LEU ALA LEU LEU GLY CYS

            4066                    4081                    4096
GAU GCU UCA CCA UGG CAG UGG CUU AGA AAG AAA GCA UGC GAU GUU
ASP ALA SER PRO TRP GLN TRP LEU ARG LYS LYS ALA CYS ASP VAL
                                    ┌─► X
            4111                    4126                    4141
CUG GAG AUA CCU UAU GUC AUC AAG CAA|GGU GAC AGU UGG UUG AAG
LEU GLU ILE PRO TYR VAL ILE LYS GLN GLY ASP SER TRP LEU LYS

            4156                    4171                    4186
AAG UUU ACU GAA GCA UGC AAC GCA GCU AAG GGC CUG GAG UGG GUG
LYS PHE THR GLU ALA CYS ASN ALA ALA LYS GLY LEU GLU TRP VAL

            4201                    4216                    4231
UCA AAC AAA AUC UCA AAA UUC AUU GAU UGG CUC AAG GAG AAA AUU
SER ASN LYS ILE SER LYS PHE ILE ASP TRP LEU LYS GLU LYS ILE

            4246                    4261                    4276
AUC CCA CAA GCU AGA GAU AAG UUG GAA UUU GUA ACA AAA CUU AGA
ILE PRO GLN ALA ARG ASP LYS LEU GLU PHE VAL THR LYS LEU ARG

            4291                    4306                    4321
CAA CUA GAA AUG CUG GAA AAC CAA AUC UCA ACU AUA CAC CAA UCA
GLN LEU GLU MET LEU GLU ASN GLN ILE SER THR ILE HIS GLN SER

            4336                    4351                    4366
UGC CCU AGU CAG GAA CAC CAG GAA AUU CUA UUC AAU AAU GUC AGA
CYS PRO SER GLN GLU HIS GLN GLU ILE LEU PHE ASN ASN VAL ARG

            4381                    4396                    4411
UGG UUA UCC AUC CAG UCU AAG AGG UUU GCC CCU CUU UAC GCA GUG
TRP LEU SER ILE GLN SER LYS ARG PHE ALA PRO LEU TYR ALA VAL

            4426                    4441                    4456
GAA GCC AAA AGA AUA CAG AAA CUA GAG CAU ACU AUU AAC AAC UAC
GLU ALA LYS ARG ILE GLN LYS LEU GLU HIS THR ILE ASN ASN TYR

            4471                    4486                    4501
AUA CAG UUC AAG AGC AAA CAC CGU AUU GAA CCA GUA UGU UUG CUA
ILE GLN PHE LYS SER LYS HIS ARG ILE GLU PRO VAL CYS LEU LEU

            4516                    4531                    4546
GUA CAU GGC AGC CCC GGA ACA GGU AAA UCU GUA GCA ACC AAC CUG
VAL HIS GLY SER PRO GLY THR GLY LYS SER VAL ALA THR ASN LEU

            4561                    4576                    4591
AUU GCU AGA GCC AUA GCU GAA AGA GAA AAC ACG UCC ACG UAC UCG
ILE ALA ARG ALA ILE ALA GLU ARG GLU ASN THR SER THR TYR SER
```

```
              4606           4621           4636
CUA CCC CCG GAU CCA UCA CAC UUC GAC GGA UAC AAA CAA CAG GGA
LEU PRO PRO ASP PRO SER HIS PHE ASP GLY TYR LYS GLN GLN GLY

              4651           4666           4681
GUG GUG AUU AUG GAC GAC CUG AAU CAA AAC CCA GAU GGU GCG GAC
VAL VAL ILE MET ASP ASP LEU ASN GLN ASN PRO ASP GLY ALA ASP

              4696           4711           4726
AUG AAG CUG UUC UGU CAG AUG GUA UCA ACA GUG GAG UUU AUA CCA
MET LYS LEU PHE CYS GLN MET VAL SER THR VAL GLU PHE ILE PRO

              4741           4756           4771
CCC AUG GCA UCC CUG GAG GAG AAA GGA AUC CUG UUU ACU UCA AAU
PRO MET ALA SER LEU GLU GLU LYS GLY ILE LEU PHE THR SER ASN

              4786           4801           4816
UAC GUU CUA GCA UCC ACA AAC UCA AGC AGA AUU UCC CCC CCC ACU
TYR VAL LEU ALA SER THR ASN SER SER ARG ILE SER PRO PRO THR

              4831           4846           4861
GUG GCA CAC AGU GAU GCA UUA GCC AGG CGC UUU GCG UUC GAC AUG
VAL ALA HIS SER ASP ALA LEU ALA ARG ARG PHE ALA PHE ASP MET

              4876           4891           4906
GAC AUU CAG GUC AUG AAU GAG UAU UCU AGA GAU GGG AAA UUG AAC
ASP ILE GLN VAL MET ASN GLU TYR SER ARG ASP GLY LYS LEU ASN

              4921           4936           4951
AUG GCC AUG GCU ACU GAA AUG UGU AAG AAC UGU CAC CAA CCA GCA
MET ALA MET ALA THR GLU MET CYS LYS ASN CYS HIS GLN PRO ALA

              4966           4981           4996
AAC UUU AAG AGA UGC UGU CCU UUA GUG UGU GGU AAG GCA AUU CAA
ASN PHE LYS ARG CYS CYS PRO LEU VAL CYS GLY LYS ALA ILE GLN

              5011           5026           5041
UUA AUG GAC AAA UCU UCC AGA GUU AGA UAC AGU AUU GAC CAG AUC
LEU MET ASP LYS SER SER ARG VAL ARG TYR SER ILE ASP GLN ILE

              5056           5071           5086
ACU ACA AUG AUU AUC AAU GAG AGA AAC AGA AGA UCC AAC AUU GGC
THR THR MET ILE ILE ASN GLU ARG ASN ARG ARG SER ASN ILE GLY
                                → 1b,9a
              5101           5116           5131
AAU UGU AUG GAG GCU UUG UUU CAA|GGA CCA CUC CAG UAU AAA GAC
ASN CYS MET GLU ALA LEU PHE GLN GLY PRO LEU GLN TYR LYS ASP

              5146           5161           5176
UUG AAA AUU GAC AUC AAG ACG AGU CCC CCU CCU GAA UGU AUC AAU
LEU LYS ILE ASP ILE LYS THR SER PRO PRO PRO GLU CYS ILE ASN

              5191           5206           5221
GAC UUG CUC CAA GCA GUU GAC UCC CAG GAG GUG AGA GAU UAC UGU
ASP LEU LEU GLN ALA VAL ASP SER GLN GLU VAL ARG ASP TYR CYS

              5236           5251           5266
GAG AAG AAG GGU UGG AUA GUC AAC AUC ACC AGC CAG GUU CAA ACA
GLU LYS LYS GLY TRP ILE VAL ASN ILE THR SER GLN VAL GLN THR

              5281           5296           5311
GAA AGG AAC AUC AAC AGG GCA AUG ACA AUU CUA CAA GCG GUG ACA
GLU ARG ASN ILE ASN ARG ALA MET THR ILE LEU GLN ALA VAL THR

              5326           5341           5356
ACC UUC GCC GCA GUG GCU GGA GUU GUC UAU GUC AUG UAU AAA CUG
THR PHE ALA ALA VAL ALA GLY VAL VAL TYR VAL MET TYR LYS LEU
```

```
                     ┌─► VPg
                 5371│                  5386                    5401
UUU GCU GGA CAC CAG|GGA GCA UAC ACU GGU UUA CCA AAC AAA AAA
PHE ALA GLY HIS GLN GLY ALA TYR THR GLY LEU PRO ASN LYS LYS
                                                   ┌─► 2,7c,6a
                 5416                   5431       │        5446
CCC AAC GUG CCC ACC AUU CGG ACA GCA AAG GUA CAA|GGA CCA GGG
PRO ASN VAL PRO THR ILE ARG THR ALA LYS VAL GLN GLY PRO GLY

                 5461                   5476                    5491
UUC GAU UAC GCA GUG GCU AUG GCU AAA AGA AAC AUU GUU ACA GCA
PHE ASP TYR ALA VAL ALA MET ALA LYS ARG ASN ILE VAL THR ALA

                 5506                   5521                    5536
ACU ACU AGC AAG GGA GAG UUC ACU AUG UUA GGA GUC CAC GAC AAC
THR THR SER LYS GLY GLU PHE THR MET LEU GLY VAL HIS ASP ASN

                 5551                   5566                    5581
GUG GCU AUU UUA CCA ACC CAC GCU UCA CCU GGU GAA AGC AUU GUG
VAL ALA ILE LEU PRO THR HIS ALA SER PRO GLY GLU SER ILE VAL

                 5596                   5611                    5626
AUC GAU GGC AAA GAA GUG GAG AUC UUG GAU GCC AAA GCG CUC GAA
ILE ASP GLY LYS GLU VAL GLU ILE LEU ASP ALA LYS ALA LEU GLU

                 5641                   5656                    5671
GAU CAA GCA GGA ACC AAU CUU GAA AUC ACU AUA AUC ACU CUA AAG
ASP GLN ALA GLY THR ASN LEU GLU ILE THR ILE ILE THR LEU LYS

                 5686                   5701                    5716
AGA AAU GAA AAG UUC AGA GAC AUU AGA CCA CAU AUA CCU ACU CAA
ARG ASN GLU LYS PHE ARG ASP ILE ARG PRO HIS ILE PRO THR GLN

                 5731                   5746                    5761
AUC ACU GAG ACA AAU GAU GGA GUC UUG AUC GUG AAC ACU AGC AAG
ILE THR GLU THR ASN ASP GLY VAL LEU ILE VAL ASN THR SER LYS
                                                           ┌─► 4a
                 5776                   5791               │5806
UAC CCC AAU AUG UAU GUU CCU GUC GGU GCU GUG ACU GAA CAG|GGA
TYR PRO ASN MET TYR VAL PRO VAL GLY ALA VAL THR GLU GLN GLY

                 5821                   5836                    5851
UAU CUA AAU CUC GGU GGG CGC CAA ACU GCU CGU ACU CUA AUG UAC
TYR LEU ASN LEU GLY GLY ARG GLN THR ALA ARG THR LEU MET TYR

                 5866                   5881                    5896
AAC UUU CCA ACC AGA GCA GGA CAG UGU GGU GGA GUC AUC ACA UGU
ASN PHE PRO THR ARG ALA GLY GLN CYS GLY GLY VAL ILE THR CYS

                 5911                   5926                    5941
ACU GGG AAA GUC AUC GGG AUG CAU GUU GGU GGG AAC GGU UCA CAC
THR GLY LYS VAL ILE GLY MET HIS VAL GLY GLY ASN GLY SER HIS

                 5956                   5971                    5986
GGG UUU GCA GCG GCC CUG AAG CGA UCA UAC UUC ACU CAG AGU CAA
GLY PHE ALA ALA ALA LEU LYS ARG SER TYR PHE THR GLN SER GLN
┌─► 4b
│                6001                   6016                    6031
|GGU GAA AUC CAG UGG AUG AGA CCU UCG AAG GAA GUG GGA UAU CCA
GLY GLU ILE GLN TRP MET ARG PRO SER LYS GLU VAL GLY TYR PRO

                 6046                   6061                    6076
AUC AUA AAU GCC CCG UCC AAA ACC AAG CUU GAA CCC AGU GCU UUC
ILE ILE ASN ALA PRO SER LYS THR LYS LEU GLU PRO SER ALA PHE

                 6091                   6106                    6121
CAC UAU GUG UUU GAA GGG GUG AAG GAA CCA GCA GUC CUC ACU AAA
HIS TYR VAL PHE GLU GLY VAL LYS GLU PRO ALA VAL LEU THR LYS
```

```
                    6136                     6151                     6166
AAC GAU CCC AGG CUU AAG ACA GAC UUU GAG GAG GCA AUU UUC UCC
ASN ASP PRO ARG LEU LYS THR ASP PHE GLU GLU ALA ILE PHE SER

                    6181                     6196                     6211
AAG UAC GUG GGU AAC AAA AUU ACU GAA GUG GAU GAG UAC AUG AAA
LYS TYR VAL GLY ASN LYS ILE THR GLU VAL ASP GLU TYR MET LYS

                    6226                     6241                     6256
GAG GCA GUA GAC CAC UAU GCU GGC CAG CUC AUG UCA CUA GAC AUC
GLU ALA VAL ASP HIS TYR ALA GLY GLN LEU MET SER LEU ASP ILE

                    6271                     6286                     6301
AAC ACA GAA CAA AUG UGC UUG GAG GAU GCC AUG UAU GGC ACU GAU
ASN THR GLU GLN MET CYS LEU GLU ASP ALA MET TYR GLY THR ASP

                    6316                     6331                     6346
GGU CUA GAA GCA CUU GAU UUG UCC ACC AGU GCU GGC UAC CCU UAU
GLY LEU GLU ALA LEU ASP LEU SER THR SER ALA GLY TYR PRO TYR

                    6361                     6376                     6391
GUA GCA AUG GGA AAG AAG AAG AGA GAC AUC UUG AAC AAA CAA ACC
VAL ALA MET GLY LYS LYS LYS ARG ASP ILE LEU ASN LYS GLN THR
                                                      → 6b
                    6406                     6421                     6436
AGA GAC ACU AAG GAA AUG CAA AAA CUG CUC GAC ACA UAU|GGA AUC
ARG ASP THR LYS GLU MET GLN LYS LEU LEU ASP THR TYR GLY ILE

                    6451                     6466                     6481
AAC CUC CCA CUG GUG ACU UAU GUA AAG GAU GAA CUU AGA UCC AAA
ASN LEU PRO LEU VAL THR TYR VAL LYS ASP GLU LEU ARG SER LYS

                    6496                     6511                     6526
ACA AAG GUU GAG CAG GGG AAA UCC AGA UUA AUU GAA GCU UCU AGU
THR LYS VAL GLU GLN GLY LYS SER ARG LEU ILE GLU ALA SER SER

                    6541                     6556                     6571
UUG AAU GAC UCA GUG GCA AUG AGA AUG GCU UUU GGG AAC CUA UAU
LEU ASN ASP SER VAL ALA MET ARG MET ALA PHE GLY ASN LEU TYR

                    6586                     6601                     6616
GCU GCU UUU CAC AAA AAC CCA GGA GUG AUA ACA GGU UCA GCA GUG
ALA ALA PHE HIS LYS ASN PRO GLY VAL ILE THR GLY SER ALA VAL

                    6631                     6646                     6661
GGG UGC GAU CCA GAU UUG UUU UGG AGC AAA AUU CCG GUA UUG AUG
GLY CYS ASP PRO ASP LEU PHE TRP SER LYS ILE PRO VAL LEU MET

                    6676                     6691                     6706
GAA GAG AAG CUG UUU GCU UUU GAC UAC ACA GGG UAU GAU GCA UCU
GLU GLU LYS LEU PHE ALA PHE ASP TYR THR GLY TYR ASP ALA SER

                    6721                     6736                     6751
CUC AGC CCU GCU UGG UUC GAG GCA CUA AAG AUG GUG CUU GAG AAA
LEU SER PRO ALA TRP PHE GLU ALA LEU LYS MET VAL LEU GLU LYS

                    6766                     6781                     6796
AUC GGA UUC GGA GAC AGA GUU GAC UAC AUC GAC UAC CUA AAC CAC
ILE GLY PHE GLY ASP ARG VAL ASP TYR ILE ASP TYR LEU ASN HIS

                    6811                     6826                     6841
UCA CAC CAC CUG UAC AAG AAU AAA ACA UAC UGU GUC AAG GGC GGU
SER HIS HIS LEU TYR LYS ASN LYS THR TYR CYS VAL LYS GLY GLY

                    6856                     6871                     6886
AUG CCA UCU GGC UGC UCA GGC ACU UCA AUU UUU AAC UCA AUG AUU
MET PRO SER GLY CYS SER GLY THR SER ILE PHE ASN SER MET ILE
```

```
                  6901                6916                6931
AAC AAC UUG AUU AUC AGG ACA CUC UUA CUG AAA ACC UAC AAG GGC
ASN ASN LEU ILE ILE ARG THR LEU LEU LEU LYS THR TYR LYS GLY

                  6946                6961                6976
AUA GAU UUA GAC CAC CUA AAA AUG AUU GCC UAU GGU GAU GAU GUA
ILE ASP LEU ASP HIS LEU LYS MET ILE ALA TYR GLY ASP ASP VAL

                  6991                7006                7021
AUU GCU UCC UAC CCC CAU GAA GUU GAC GCU AGU CUC CUA GCC CAA
ILE ALA SER TYR PRO HIS GLU VAL ASP ALA SER LEU LEU ALA GLN

                  7036                7051                7066
UCA GGA AAA GAC UAU GGA CUA ACU AUG ACU CCA GCU GAC AAA UCA
SER GLY LYS ASP TYR GLY LEU THR MET THR PRO ALA ASP LYS SER

                  7081                7096                7111
GCU ACA UUU GAA ACA GUC ACA UGG GAG AAU GUA ACA UUC UUG AAG
ALA THR PHE GLU THR VAL THR TRP GLU ASN VAL THR PHE LEU LYS

                  7126                7141                7156
AGA UUC UUC AGG GCA GAC GAG AAA UAC CCA UUU CUU AUU CAU CCA
ARG PHE PHE ARG ALA ASP GLU LYS TYR PRO PHE LEU ILE HIS PRO

                  7171                7186                7201
GUA AUG CCA AUG AAG GAA AUU CAU GAA UCA AUU AGA UGG ACU AAA
VAL MET PRO MET LYS GLU ILE HIS GLU SER ILE ARG TRP THR LYS

                  7216                7231                7246
GAU CCU AGG AAC ACU CAG GAU CAC GUU CGC UCU CUG UGC CUU UUA
ASP PRO ARG ASN THR GLN ASP HIS VAL ARG SER LEU CYS LEU LEU

                  7261                7276                7291
GCU UGG CAC AAU GGC GAA GAA GAA UAU AAC AAA UUC CUA GCU AAA
ALA TRP HIS ASN GLY GLU GLU GLU TYR ASN LYS PHE LEU ALA LYS

                  7306                7321                7336
AUC AGG AGU GUG CCA AUU GGA AGA GCU UUA UUG CUC CCA GAG UAC
ILE ARG SER VAL PRO ILE GLY ARG ALA LEU LEU LEU PRO GLU TYR

                                          --STOP
                  7351                7366   |            7381
UCA ACA UUG UAC CGC CGU UGG CUU GAC UCA UUU|UAG UAA CCC UAC
SER THR LEU TYR ARG ARG TRP LEU ASP SER PHE *** ### PRO TYR

                  7396                7411                7426
CUC AGU CGA AUU GGA UUG GGU CAU ACU GUU GUA GGG GUA AAU UUU
LEU SER ARG ILE GLY LEU GLY HIS THR VAL VAL GLY VAL ASN PHE

                  7441
UCU UUA AUU CGG AGG
SER LEU ILE ARG ARG
```

al., 1977) and chemical sequencing of extension products (Maxam and Gilbert, 1980).

To date, three complete sequences of poliovirus type 1 genomes have been published from independent investigations, two from the virulent Mahoney strain (Kitamura *et al.*, 1981; Racaniello and Baltimore, 1981a) of type 1 poliovirus. This virus has been propagated in different laboratories under conditions of low antigenic pressure and essentially non-selective conditions for many decades. The third elucidated genome sequence is that of the Sabin vaccine strain of poliovirus type 1 (Nomoto *et al.*, 1982). The complete consensus sequence of the virulent Mahoney strain is presented in Figure 44. The consensus sequence includes the corrections for minor sequencing errors that had been made in earlier publications. As this book was going into print, the complete nucleotide sequences of poliovirus types 2 and 3 (see Appendix IV) were also published (Stanway *et al.*, 1983; Toyoda *et al.*, 1984). Implications for genetic relationship, gene function, and antigenic determinants are discussed in these papers and the reader is advised to consult them as useful supplements to the present text.

D. Terminal Sequences

The terminal regions of viral RNA are of utmost physiological significance, as they are involved in critical steps of viral reproduction, namely the initiation of protein synthesis and the initiation of RNA synthesis.

1. 5' End Sequences

742 nucleotides at the 5' end of poliovirus type 1 RNA are not translated. The initiating AUG codon beginning at position 743 is the ninth AUG codon from the 5' terminus. The position of the initiator codon at 743 has recently been confirmed by sequence analysis of viral capsid proteins and of mRNA (Dorner *et al.*, 1982).

Poliovirus thus contains an unusually long 5' terminal untranslated stretch, much longer than it is generally found in other eukaryotic and viral mRNAs. For reovirus mRNA, for example, the corresponding length is only 36–60 nucleotides (Kozak, 1977).

Unlike the viral genomes of other picornavirus genera (EMC-virus, mengovirus, and FMDV), the poliovirus genome does not contain a large poly C tract (the length of which varies from 100–250 nucleotides) near the 5' end of the genome. (The distance from 5' end to poly C tract is 140 nucleotides for EMC, 400 nucleotides for FMDV). The poly C tract in these viruses is not translated and its function remains obscure (Brown *et al.*, 1974; Fellner, 1979; Sangar, 1979; Agol, 1980).

A long, untranslated 5' stretch seems to be common in picornaviruses. The 5' untranslated region in EMC virus RNA may be 1,000 nucleotides in length (Fellner, 1979). Studies on the in vitro translation of the mRNA of EMDV, and defined fragments of its mRNA (Sangar *et al.*, 1980), locate the initiation site for protein synthesis a good distance from the 5' end (more than 400 nucleotides) of the RNA.

The 5' untranslated region of poliovirus also contains 29 termination codons distributed among the three reading frames. All but two of the eight internal AUG codons in this region are quickly followed by stop codons within their reading frame. Figure 45 shows the positions of the AUG codons and termination codons within the 5' 1,000 terminal nucleotides of the poliovirus genome and the respective polypeptides that could by synthesized if any of these AUG codons were functioning in addition to the major AUG initiator codon at 743. It is uncertain whether any of these smaller peptides are ever produced in the infected cell. Translation of poliovirus mRNA in cell free extracts, however, does result in the translation of an additional small peptide under conditions of high ionic strength,

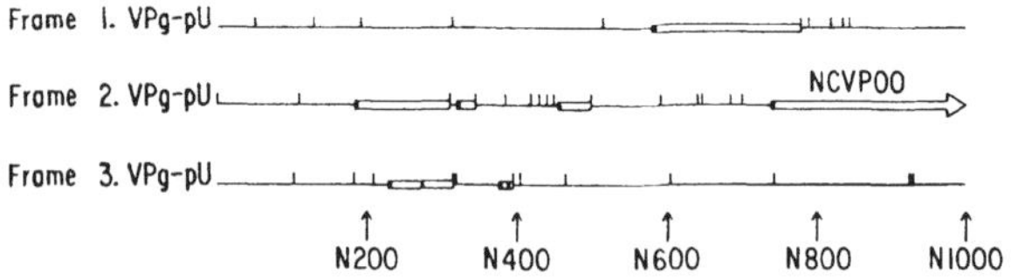

Fig. 45. Potential translation regions within the 5′-terminal 1,000 nucleotides of the poliovirus type 1 genome

The three lines represent the three reading frames beginning with the first, second, or third nucleotide of the RNA. The vertical lines represent termination codons. The solid boxes represent AUG codons, and the open boxes indicate open translation regions following an AUG codon. The long open reading frame (nucleotides 743 through 7,369) present in the second reading frame is labeled NCVP00. The numbers below the vertical arrows indicate distances from the 5′ end in nucleotides. The nucleotide assignments are from the consensus sequence of poliovirus type 1 (see Fig. 44). — Figure from Dorner *et al.*, 1982 [J. Virol. *42*, 1017–1028 (1982)]

in the presence of 4 mM $MgCl_2$ (Ehrenfeld, 1979). Whether this peptide corresponds to any of the potential peptides indicated in Figure 45 remains to be determined (see also Chapter 8). Translation initiation at the 5'-proximal AUG codon at nucleotide 185 could produce a protein of 4,720 daltons; the eighth AUG codon at nucleotide 586 could produce a protein of 7,180 daltons, overlapping the beginning of the long open reading frame. The functions of the additional AUG codons preceeding the initiator AUG and of the potential open reading frames are still not known. It has been proposed that the additional AUGs may serve to tie up ribosomes during the switch from translation to replication early in infection (Dorner *et al.*, 1982). Only three of the AUG codons in the 5' non-coding region (at positions 322, 464, 593) have been conserved among the serotype genomes, the lengths of the short open reading frames following the AUGs, however, are not conserved among the three serotypes (Toyoda *et al.*, 1984).

A number of very stable secondary stem and loop structures have been identified in the 5' untranslated region by computer search. An example of such a structure at the very 5' terminus is shown in Figure 40 (Kitamura *et al.*, 1981), another includes the 6th AUG codon at position 390 (Nomoto *et al.*, 1982b). The stable hairpin structure close to the 5' end of the RNA of poliovirus type 1 (Fig. 40) exists also in the other two serotype genomes, albeit with slightly different free energies (Toyoda *et al.*, 1984). The secondary structure of 5' terminal sequences in poliovirus RNA may play an important role in the recognition of the correct initiation codon by ribosomes. A 60 nucleotides long adenine-uridine rich segment is found in the nucleotide region just preceeding the AUG initiator codon at 743. This correpsonds to a distinct melting region observed upon denaturation mapping of poliovirus type 1 replicative form RNA (Čumakov *et al.*, 1979) (see section II-C above). It is possible that this region does not participate in the formation of stable secondary structures in the single stranded mRNA, thus leaving the initiator codon relatively exposed.

It has been confirmed by amino acid sequence analysis that the protein coded for by the region directly following the initiator codon 743 of poliovirus is the capsid protein VP_4 (Dorner *et al.*, 1982). A splicing mechanism that could produce an mRNA capable of directing the synthesis of a leader peptide preceeding VP_4

was ruled out by an S1 nuclease analysis of poliovirus mRNA isolated from polysomes of infected cells (Dorner *et al.*, 1982). Absence of a leader peptide is one more feature that distinguishes poliovirus from its picornavirus relatives FMDV and EMC, for which the synthesis and genomic coding of a small 20,000 dalton leader peptide preceeding the capsid proteins has been demonstrated (Sangar *et al.*, 1980; Boothroyd *et al.*, 1981; Kazachkov *et al.*, 1982).

The terminal nucleotide of the virion RNA of all picornaviruses investigated so far (poliovirus, FMDV, EMC- and mengo-viruses, and rhinoviruses) (Agol, 1980) is uridylic acid. Beyond this, the sequences at the 5' ends are highly conserved in different picornaviruses (see Fig. 46). At least the first 20 nucleotides at the 5' end of poliovirus types 1 and 2 are identical (except for nucleotide No. 19) (Babich *et al.*, 1980; Hewlett and Florkiewicz, 1980), the poliovirus relative from the enterovirus genus, coxsackie B4, has 16 of the first 20 nucleotides at the 5' end identical to those of poliovirus type 1, and the sequence homology to the virion RNA of a different picornavirus genus FMDV is 70% in this region (Harris, 1979).

```
                 1                 10                   20                  30
PV 1       VPg-pU-U-A-A-A-A-C-A-G-C-U-C-U-G-G-G-G-U-U-G-U-A-C-C-C-A-C-C-C-C-A-

PV 2       VPg-pU-U-A-A-A-A-C-A-G-C-U-C-U-G-G-G-G-U-  G-
                                                   C-

COX B1     VPg-pU-U-A-A-A-A-C-A-G-C-         G-G-G-U-U-G-
                                    C-U-G-U-

FMDV A10   VPg-pU-U-  A-A-A-       G-C-  C-U-  G-G-G-U-U-     A-C-C-C-         C-A-
                    G-      G-G-G-    G-      A-           U-C-        C-U-A-G-
```

Fig. 46. 5′ end sequences of some picornavirus RNAs

The figure shows the high conservation of sequence at the 5′ end of different picornaviruses. Sequence homology between the two poliovirus types 1 and 2 is 95% within the 20 5′ nucleotides. Another enterovirus, coxsackievirus B1, also has a very similar 5′ sequence, differing from those in the poliovirus RNAs in a single 4 nucleotide long stretch. The 5′ end sequence of one of the Foot and Mouth Disease Viruses A 10, a more distant relative to the polioviruses, exhibits a higher sequence divergence, sequence homology to poliovirus type 1 RNA is still higher than 60% within the first 30 nucleotides. – Data from Hewlett and Florkiewicz, 1980 [Proc. Nat. Acad. Sci. *77*, 303–307 (1980)]; Harris, 1979 [Nucl. Acid. Res. *7*, 176–178 (1979)] and consensus nucleotide sequence of poliovirus type 1 (see Fig. 44)

The high sequence homology (of more than 90%) in the 5' non-coding region in the three poliovirus serotype genomes extends through the first 650 nucleotides. This marked conservation suggests that the 5' end terminal region serves important function(s) in ribosome binding and initiation of protein synthesis, and/or also in the binding of the replicase at the complementary 3' end of the minus strand. Interestingly, the nucleotide sequences in the 5' untranslated region immediately preceeding the VP4 coding region (nucleotides 650–760) are an exception in that they differ strikingly from each other in the three poliovirus serotypes (Toyoda *et al.*, 1984).

The importance of the 5' terminal region is further illustrated by the observation that cells transfected with plasmids, containing all but the first 115 bases of the poliovirus genome, did not produce virus in contrast to plasmids containing a complete copy of the poliovirus genome (Racaniello and Baltimore, 1981b). However, variation seems to occur more often in the adjacent regions, as is evidenced by greater differences of oligonucleotide fingerprints of the 100base sequence at the 5' end of two different strains of poliovirus (Harris *et al.*, 1978; Stewart *et al.*, 1980).

The sequences of the ribosomal binding sites of different eukaryotic and viral mRNAs have been determined by binding ribosomes to the mRNA, digesting away the "free" ends of the RNA with nucleases, and then isolating and sequencing the ribosome bound RNA fragment (Kozak, 1978; Kozak and Shatkin, 1979; Darzynkiewicz and Shatkin, 1980). Recently, ribosomal binding sites have been

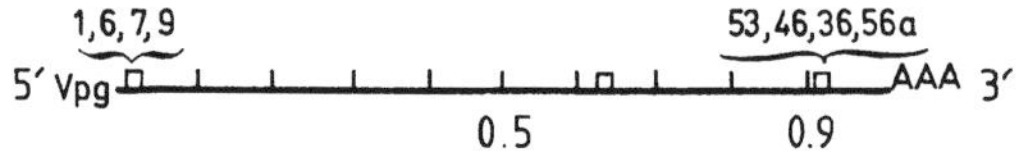

Fig. 47 A–B. Ribosomal binding sites on poliovirus type 1 RNA

Fig. 47 A. The boxes indicate the localizations of the RNA fragments protected against RNase digestion by ribosome binding. 80 S ribosomes were bound to poliovirus RNA under conditions specific for initiation of translation and were prevented from translocation with the antibiotic sparsomycin. The ribosome bound RNA was digested with RNase T1. Oligonucleotides protected by the ribosomes against RNase digestion were extracted from the ribosomes with phenol, separated by gel electrophoresis and then subjected to secondary RNase T1 treatment. The ribosome protected oligonucleotides could then be mapped by comparing these secondary digests to the RNase T1 oligonucleotide map as determined in Figure 42. Three ribosome protected fragments, RP1–3 from left to right, were thus detected. RP1 and RP2 were determined to be approximately 33 bases long and RP3 approximately 30 bases. The numbers above the boxes refer to the mapped oligonucleotides as in Figure 42. Spot 53 of RP3 spans a 120 long nucleotide stretch between 650 and 780 bases from the 3' end; spots 46, 36, and 56 a are located to the 3' terminal end of 53. The fingerprint pattern of RP2 was similar to those of RNase III fragments mapping to the 3' side of midgenome; the RP2 oligonucleotides, however, are not in unique regions and could thus not be mapped precisely

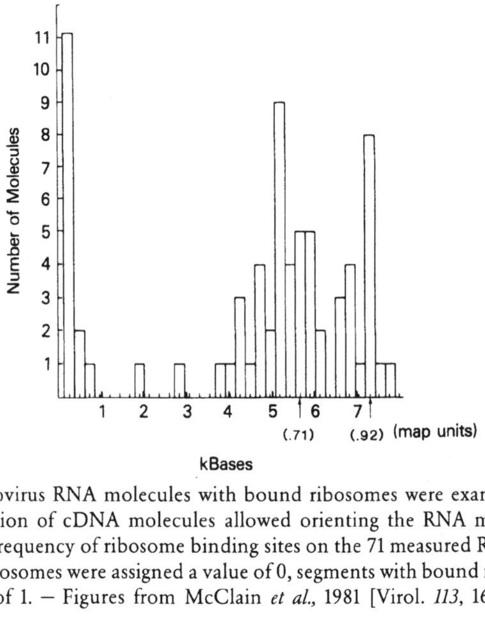

Fig. 47 B. 71 poliovirus RNA molecules with bound ribosomes were examined by electron microscopy. Hybridization of cDNA molecules allowed orienting the RNA molecules. The figure is a histogram of the frequency of ribosome binding sites on the 71 measured RNA molecules. Segments with no bound ribosomes were assigned a value of 0, segments with bound ribosomes were assigned a value of 1. – Figures from McClain *et al.*, 1981 [Virol. *113*, 162, 165 (1981)]

determined for poliovirus type 1 mRNA by analysis of the nucleotides protected from T1 RNase by ribosome binding. 80S ribosomes were allowed to bind to poliovirus RNA under conditions specific for initiation of translation, but were prevented from translocation with the antibiotic sparsomycin (McClain *et al.*, 1981). Three ribosome binding sites were detected and mapped by comparing the protected fragments to T1 digests of RNase III fragments of known position, and by electron microscopical examination of ribosome bound RNA. Figure 47 shows the location of the binding sites. Only one ribosome binding site was detected in the 5' untranslated region, approximately 115 bases from the 5' end. A second site was located approximately 5,000 nucleotides from the 5' end, close to the beginning of the replicase coding region, the third site at approximately 6,800 nucleotides from the 5' end, near the center of the replicase region. The functional significance of these ribosome binding sites has not yet been clarified. With similar techniques, four ribosomal binding sites were detected for mengovirus RNA, three of these also bind eIF-2 with high specifity (Pérez-Bercoff and Kaempfer, 1982).

2. 3' End Sequences

40–90 nucleotide residues at the very 3'end of the virion RNA are adenylic acids (the poly (A) tract). The termination codon of poliovirus polyprotein synthesis is located at nucleotide 7361, 73 nucleotides preceeding the poly (A) tract (Emini *et al.*, 1982). The poliovirus untranslated 3' end region thus is unusually short compared to those in eukaryotic mRNAs (Hamlyn *et al.*, 1978; Catterall *et al.*, 1978). (An earlier report (Kitamura and Wimmer, 1980) locating the termination codon at position 562 preceeding the poly (A) tract was erroneous due to two minor sequencing errors).

The nucleotide sequence at the 3' end of poliovirus RNA does not contain an A-A-U-A-A-A hexanucleotide that characteristically occurs within 20 bases of poly (A) in all untranslated regions of mammalian cytoplasmic RNAs known to date (Tucker *et al.*, 1979). This hexanucleotide is also not found in FMDV-RNA, but it is found at position −15 to −10 in the cardioviruses EMC and mengo (Porter *et al.*, 1978; Fellner, 1979). This hexanucleotide has been implicated as a signal for polyadenylation of transcripts in the cell nucleus (Proudfoot and Brownlee, 1976). The absence of this sequence in poliovirus, however, is not in disagreement with this hypothesis, since poly (A) in poliovirus RNA is synthesized by transcription from poly (U) and not by posttranscriptional polyadenylation (Dorsch-Häsler *et al.*, 1975).

Corresponding 3' terminal sequences (40–150 nucleotides long) preceeding the poly (A) of several other picornaviruses have been determined (Porter *et al.*, 1978; Drake *et al.*, 1982; Marquardt, 1982). Within the same picornavirus genus, the 3' terminal sequences are at least 60 % homologous (polio and swine-vesicular disease virus) and up to 90 % homologous (EMC and mengo virus). Viral genomes from different genera show less similarity, although comparison of the 3'terminal sequence of rhinovirus RNA to that of poliovirus RNA still revealed a strong homology of 66% in the noncoding region, which did not extend into the coding region (Colonno and Cordova, 1981).

Comparison of the genomes of the two serologically unrelated viruses of the enterovirus genus, poliovirus and swine vesicular disease virus showed some other interesting features. There is a highly conserved region in these viruses in the vicinity of the poly (A) sequence, a sequence of 30 nucleotides is identical. Part of this region and the section directly next to this region is also found in EMC virus RNA. Interestingly, this homologous region contains a palindrome sequence. Palindromes have been found previously in the 3' untranslated regions of host mRNAs from several different sources (Seeburg *et al.*, 1977; Shine *et al.*, 1977; Wilson *et al.*, 1977; Proudfoot, 1977), and it has been suggested that they may provide binding sites for oligomeric proteins in the corresponding DNA sequence.

It is possible that this sequence and others in the untranslated portion have similar functions, *i.e.* to bind the virus polymerase complex during initiation of replication. It should be noted that there are no 3' sequences complementary to the 5' sequences described above. Thus, identical conserved sites are not present in the 3' sequences of the plus and minus viral RNA strands, so that the mechanism of replicase recognition must differ for the plus and minus viral RNA strands.

3. *The Poly (A) Tract*

Similar to many eukaryotic mRNAs (Kates, 1970), poliovirus RNA has at its 3' end a terminal stretch of poly (A). The length of the poly (A) tract in a given viral RNA population is usually heterogenous, the poly (A) tract is approximately 60—90 (Yogo and Wimmer, 1972) or 40—80 (Ahlquist *et al.*, 1979) bases long. The poly (A) tracts in the RNA of other picornavirus genera are either shorter (EMC virus) (Giron *et al.*, 1975/1976; Burness *et al.*, 1977), or of similar size (EMC virus) (Chatterjee *et al.*, 1976), or even longer (rhinovirus) (Nair *et al.*, 1976). The length of the poly (A) tract seems to be independent from either the host cell system (Spector and Baltimore, 1975d) or conditions (Hruby and Roberts, 1976) used for the propagation of virus.

The poly (A) tracts in cellular mRNAs are added posttranscriptionally (Weinberg, 1973), while that of polio RNA is transcribed. The minus strand RNA of poliovirus, prepared from either the replicative form or the replicative intermediate, contains a tract of poly (U) at the 5' terminus (Yogo and Wimmer, 1973, 1975; Yogo *et al.*, 1974; Spector and Baltimore, 1975c). This poly (U) tract contains about 60—100 nucleotides, *i.e.* it is approximately the same size as the poly (A) tract at the 3' end of the (+)strand RNA. These observations together with studies of polyadenylation of poliovirus RNA *in vitro* (Dorsch-Häsler *et al.*, 1975) strongly suggest that the poly (A) and poly (U) tracts are transcribed from each other at the successive steps in the replicative process, *i.e.* that the poly (A) is genetically coded, rather than added posttranscriptionally. However, some posttranscriptional addition apparently can also occur: plaques arising from cells infected by RNA with artificially shortened poly (A) tracts yielded progeny virions with poly (A) tracts of normal length (Spector and Baltimore, 1975b).

The biological role of the poly (A) tract is not clear. It is conceivable that the 3' poly (A) plays some role comparable to that of the poly (A) found in many eukaryotic mRNAs. Such poly (A) tracts have been suggested to be involved in a) processing of precursors to mRNA, b) transport of mRNA from the nucleus to the cytoplasm, c) efficient translation of mRNA, d) binding to the trabeculae system of the cell cytoskeleton (Penman *et al.*, 1982), and e) protection of mRNAs against hydrolysis by exonucleases (Johnston and Bose, 1972; Fellner, 1979). It has been suggested that the poly (A) tract is important for RNA infectivity since RNA molecules with artificially shortened poly (A) tracts (poliovirus) (Spector and Baltimore, 1974), or naturally short RNA tracts (EMC) (Goldstein *et al.*, 1976; Hruby and Roberts, 1976) have a lower specific infectivity than those with longer poly (A) tracts. On the other hand, virion RNAs of the family relative FMDV with very short poly (A) sequences (less than 10 nucleotides) do not show a reduction in infectivity (Baxt *et al.*, 1979). For poliovirus, no appreciable differences in the translational capacity of poly (A)-poor and poly (A)-rich mRNA species were noticed (Spector *et al.*, 1975). Yet, in contrast again, for another family relative, EMC virus, a subpopulation of mRNAs containing a relatively short poly (A) stretch is translated *in vitro* about 2 to 3 times less efficiently than a subpopulation of mRNAs with a longer poly (A) tract (Hruby and Roberts, 1977), suggesting that the poly (A) may play a role in the functional stability of the mRNA (Hruby, 1978).

E. Features of the Coding Region

The open reading frame consists of 6,600 nucleotides corresponding to a coding potential of 2,200 amino acids (see Fig. 44 above). 71% of the nucleotides are common in the genomes of the three poliovirus serotypes; the homology in the predicted amino acid sequences is 88% (Stanway *et al.*, 1983; Toyoda *et al.*, 1984). Reflecting the evolutionary pressure to conserve the primary structure of the viral proteins, and the greater variability of the third codon position in accord with the wobble-hypothesis, 93% and 86% of the nucleotides are conserved at the first and second residue positions, respectively, whereas only 32% of the nucleotides at the third residue position are conserved.

1. Codon Usage

The codon usage in the poliovirus open reading frame is presented in Table 27 (Racaniello and Baltimore, 1981a). There is a strong bias against CpG sequences in the poliovirus genome similar to the situation in vertebrate cellular DNA and eukaryotic mRNA (Grantham *et al.*, 1980; Russel *et al.*, 1976), which may suggest a DNA origin of poliovirus (Rothberg and Wimmer, 1981). The codon usage of the poliovirus serotypes has been conserved (Toyoda *et al.*, 1984).

2. Cleavage Signals for Proteolytic Processing

The amino acid sequence that is determined by the consensus sequence (Fig. 44) implicates three different types of cleavage sites for the processing of the primary translation product of the poliovirus genome: predominantly recognized

A) gln/gly cleavage sites recognized by the poliovirus protease

VP2 ← 340 → VP3
LEU ARG ASN ILE THR LEU PRO ARG LEU GLN | GLY LEU PRO VAL SER ASN THR PRO GLY SER

VP3 ← 578 → VP1
THR HIS ILE GLU GLN LYS ALA LEU ALA GLN | GLY LEU GLY GLN MET LEU GLU SER MET ILE

NCVP8 ← 1029 → NCVP5b, 10
TYR ALA TYR GLU GLU GLU ALA PET GLU GLN | GLY ILE THR ASN TYR ILE GLU SER LEU GLY

NCVP10 ← 1160 → NCVPX
VAL LEU GLU ILE PRO TYR VAL ILE LYS GNL | GLY ASP SER TRP LEU LYS LYS PHE THR GLU

NCVPX ← 1455 → NCVP 1b, 9
ILE GLY ASN CYS MET GLU ALA LEU PHE GLN | GLY PRO LEU GLN TYR LYS ASP LEU LYS ILE

NCVP 9b ← 1542 → VPg
VAL MET TYR LYS LEU PHE ALA GLY HIS GLN | GLY ALA TYR THR GLY LEU PRO ASN LYS LYS

NCVP 9a ← 1564 → NCVP 2, 7c, 6a
VAL PRO THR ILE ARG THR ALA LYS VAL GLN | GLY PRO GLY PHE ASP TYR ALA VAL ALA MET

NCVP 7c ← 1739 → NCVP 4b
LEU LYS ARG SER LEU PHE THR GLN SER GLN | GLY GLU ILE PRO TRP MET ARG PRO SER LYS

NCVP ? ← 1686 → NCVP 4a
TYR VAL PRO VAL ARG ALA VAL THR GLU GLN | GLY TYR LEU ASN LEU GLY GLY ARG GLN THR

B) Additional potential gln/gly cleavage sites

188 (VP2)
VAL GLN CYS ASN ALA SER LYS PHE HIS GLN | GLY ALA LEU GLY VAL PHE ALA VAL PRO GLU

927 (NCVP8)
ASP LEU LEU VAL THR GLU SER ARG ALA GLN | GLY THR ASP SER ILE ALA ARG CYS ASN CYS

1356 (NCVPX)
PRO SER HIS PHE ASP GLY TYR LYS GLN GLN | GLY VAL VAL ILE MET ASP ASP LEU ASN GLN

2006 (NCVP 4b)
GLU LEU ARG SER LYS THR LYS VAL GLU GLN | GLY LYS SER ARG LEU ILE GLU ALA SER SER

C) tyr/gly cleavage sites recognized by protease

VP2 ← 880 → VP3
PRO LEU SER THR LYS ASP LEU THR THR TYR | GLY PHE GLY HIS GLN ASN LYS ALA VAL TYR

NCVP 6a ← 1894 → NCVP 6b
LYS GLU MET GLN LYS LEU LEU ASP THR TYR | GLY ILE ASN LEU PRO LEU VAL THR TYR VAL

D) Additional potential tyr/gly cleavage sites

20 (VP4)
GLY ALA HIS GLU ASN SER ASN ARG ALA TYR | GLY GLY SER THR ILE ASN TYR THR THR ILE

807 (VP1)
GLN SER ALA ALA LEU GLY ASP SER LEU TYR | GLY ALA ALA SER LEU ASN ASP PHE GLY ILE

858 (VP1)
PRO ARG PRO PRO ARG GLN LEU ALA TYR TYR | GLY PRO GLY VAL ASP TYR LYS ASP GLY THR

1838 (NCVP 4b)
GLU GLN MET CYS LEU GLU ASP ALA MET TYR | GLY THR ASP GLY LEU GLU ALA LEU ASP LEU

2072 (NCVP 4b)
ASP LEU ASP HIS LEU LYS MET ILE ALA TYR | GLY ASP ASP VAL ILE ALA SER TYR PRO HIS

2096 (NCVP 4b)
SER LEU LEU ALA GLN SER GLY LYS ASP TYR | GLY LEU THR MET THR PRO ALA ASP LYS SER

E) asn/ser cleavage sites in procapsid protein VPO

VP4 ← 69 → VP2
VAL LEU ILE LYS THR ALA PRO MET LEU ASN | SER PRO ASN ILE GLU ALA CYS GLY TYR SER

Fig. 48. Cleavage sites determined by the genome of poliovirus type 1
The figure shows the potential cleavage sites in the poliovirus polyprotein as inferred from the consensus nucleotide sequence (Figure 44) and limited amino acid sequence analysis of poliovirus proteins (underscored amino acids). Additional potential asn/ser sites are not shown. The localization of the cleavage sites is indicated by the corresponding amino acid position in the polyprotein (the numbering beginning at the N-terminus of the polyprotein). – Data courtesy of E. Wimmer, Stony Brook

Table 27. *Codon usage in the poliovirus open reading frame*

		U		C		A		G	
U	Phe	38	Ser	20	Tyr	40	Cys	22	U
		42		32		58		19	C
	Leu	23		48	Term	0	Term	0	A
		38		10		0	Trp	28	G
C	Leu	25	Pro	30	His	18	Arg	7	U
		26		20		33		7	C
		33		56	Gln	38		3	A
		32		13		43		7	G
A	Ile	52	Thr	53	Asn	47	Ser	21	U
		48		53		70		20	C
		33		50	Lys	63	Arg	48	A
	Met	67		13		60		23	G
G	Val	22	Ala	50	Asp	54	Gly	37	U
		29		35		63		26	C
		28		58	Glu	63		46	A
		59		18		48		33	G

Codon first position is at left, second position at top, third position at right. The total occurrence of each codon through the open reading frame of the poliovirus genome is shown. Term, chain termination.
Table from Racaniello and Baltimore, 1981.

sites are at gln/gly residues, others at tyr/gly, and a third special type of cleavage occurs between an asn/ser pair. The latter cleavage occurs only in the assembled virion yielding VP_2 and VP_4 from VP_0. The main cleavage sites have been confirmed by amino acid sequencing studies (Emini *et al.*, 1982; Larsen *et al.*, 1982). The eleven major poliovirus specific stable cleavage products are accounted for by 9 gln/gly cleavages, two tyr/gly and one asn/ser cleavage. The consensus sequence reveals another four potential gln/gly, seven tyr/gly, and five asn/ser cleavage sites (see Fig. 48). At least one of these additional gln/gly sites and one of the tyr/gly sites seem to be utilized in alternative cleavage pathways, yielding NCVP4a and NCVP6a plus 6b, respectively. It is conceivable that some of the other sites are also utilized to yield other short-lived products of unknown functional significance during poliovirus replication. Figure 49 shows the sites which are known to be cleaved, as well as those which apparently are not utilized. All cleavage sites are conserved among the three poliovirus serotypes, the amino acids neighboring the cleavage sites, however, show a high degree of variation (Toyoda *et al.*, 1984).

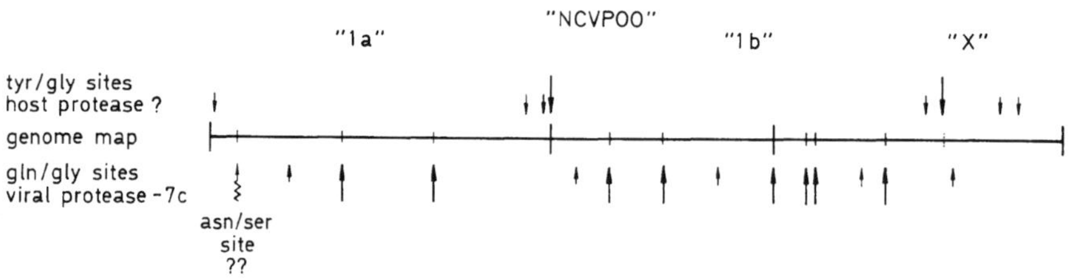

Fig. 49. Potential and actual cleavage sites on the poliovirus polyprotein
The figure shows the potential and actually employed cleavage sites on the poliovirus polyprotein
The arrows in the top row point to the potential tyr/gly sites. The long arrows indicate the sites acutally cleaved, short arrows point to sites which apparently are not used
The center line shows the relative mapping of the poliovirus proteins as deduced from the nucleotide sequence. Long lines indicate the borders of the primary cleavage products NCVP1 a, NCVP3 b, and NCVP1 b corresponding to the P-1, P-2, and P-3 domains, respectively
The arrows in the bottom row point to the potential gln/gly sites, long arrows = actual cleavage sites, short arrows = sites apparently not used
The asn/ser cleavage site is in a susceptible conformation only in (almost) completely assembled virions (ragged arrow). Other potential asn/ser sites are not indicated
The tyr/gly sites are presumably cleaved by a cellular protease. Cleavages at the gln/gly sites can all be prevented by monoclonal antibodies against NCVP-7 c
The top of the figure illustrates alternative primary cleavage products after inhibition of viral protease by protease-inhibitors. The designations of these proteins have been marked by quotation marks in order to show possible erroneous assignments to these products. This is still purely speculative, but may explain the mysterious results of tryptic peptide analyses of Abraham and Cooper, 1975, and Cooper, 1977 (see discussion on p. 176)

An interesting observation has come from the analysis of RNA fragments generated by treatment of polio RNA with E. coli ribonuclease III. This enzyme recognizes specific structures in many RNA molecules of viral and cellular origin (*e.g.* Dunn, 1976). The preferred ribonuclease III cleavage sites on the polio genome correlate with the cleavage sites on the polyprotein leading to the primary and stable translation products (Stewart *et al.*, 1980, see Fig. 42 above).

The translatability of the ribonuclease III fragments of poliovirus RNA by reticulocyte lysates has recently been studied (McClain *et al.*, 1981). Interestingly, the synthesis of a 56,000-dalton protein is enhanced. This protein is apparently encoded in the D fragment, which codes for NCVP4, the viral replicase activity. These observations have revived the idea of a possible role for specific secondary structures in exerting negative translational control, and as sites for processing of the poliovirus RNA to generate specific subgenomic RNA fragments in infected cells (Stewart *et al.*, 1980), as has been suggested or demonstrated for other RNA containing viruses (tobacco mosaic virus and T7 early mRNA) (Dunn and Studier, 1975; Hunter *et al.*, 1976). Enzymes having ribonuclease III-like activity, which would be required for such cleavages, have been found also in eukaryotic (chicken embryo) cells (Hall and Crouch, 1977).

IV. Genome Products and Gene Order

During a typical poliovirus growth cycle, up to 35 viral coded proteins and many more smaller peptide fragments can be detected. Classically, these proteins are distinguished according to their electrophoretic mobility in polyacrylamide gels.

Table 28. *Features of poliovirus proteins*

Protein	Position in consensus nucleotide sequence	Number of amino acids	Molecular weight ($\times 10^{-3}$)	Lys + arg + ½ his	Glu + asp	Net charge	Cys
NCVPOO	743–7369	2209	246.7	245.5	227	+ 18.5	42
P1-1a	746–3385	880	97.3	82.5	82	+ 0.5	15
VP_0	746–1765	340	37.5	27.5	28	− 0.5	8
VP_4	746– 949	68	7.4	7.5	6	+ 1.5	0
VP_2	950–1765	272	30.1	20	22	− 2	8
VP_3	1766–2479	238	26.4	21.5	26	− 4.5	5
VP_1	2480–3385	302	33.5	33.5	28	+ 5.5	2
P 2-3b	3386–5110	575	65.0	68	59	+ 9	18
NCVP 8	3386–3832	149	16.7	14	15	− 1	6
NCVP 5b	3833–5110	426	48.3	54	44	+ 10	12
NCVP 10	3833–4123	97	10.7	9	9	0	1
NCVP X	4124–5110	329	37.6	45	35	+ 10	11
P 3-1b	5111–7369	753	84.2	95	86	+ 9	9
NCVP 9b	5111–5371	87	10.0	9.5	9	+ 0.5	2
VPg	5372–5437	22	2.4	4	0	+ 4	0
NCVP 2	5438–7369	644	72.1	81.5	77	+ 4.5	7
NCVP 7c	5438–5986	183	19.8	19.5	16	− 3.5	2
NCVP 4b	5987–7369	461	52.4	62	61	+ 1	5
NCVP 6a	5438–6430	331	36.6	39.5	41	− 1.5	3
NCVP 6b	6431–7369	313	35.6	42	36	+ 6	4
NCVP?	5438–5803	122	13.5	12.5	16	− 3.5	0
NCVP 4a	5804–7369	522	57.7	69	61	+ 8	7

Table modified from Kitamura *et al.*, 1981.

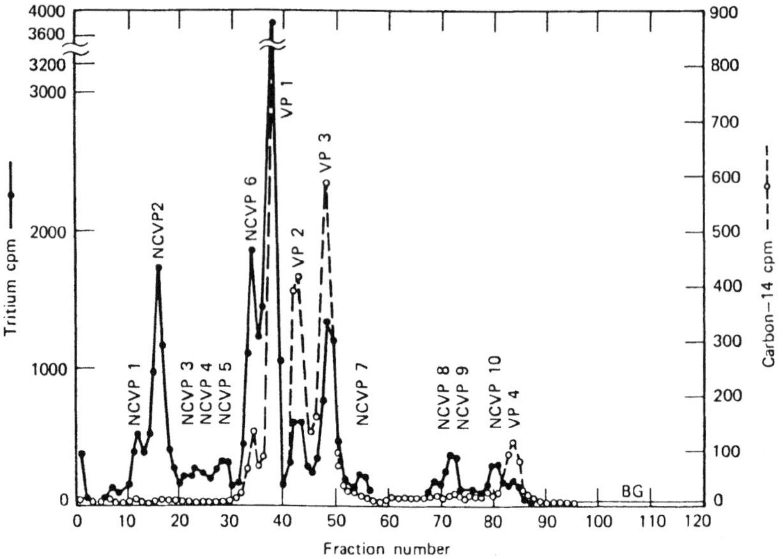

Fig. 50 A–B. The poliovirus-specific polypeptides formed in infected HeLa cells

Fig. 50 A. The original nomenclature of the poliovirus specific proteins
This figure shows the original polyacrylamide gel electrophoretic separation of the poliovirus protein products in the infected cell, and the original nomenclature of the proteins. Extracts from infected cells labeled with ^{3}H amino acids during the peak of viral protein synthesis were mixed with ^{14}C-labeled polio virions. Solid line = ^{3}H label, dashed line ^{14}C label. – Figure from Summers *et al.*, 1965 [Proc. Nat. Acad. Sci. U.S.A. *54*, 505–513 (1965)]

Figure 50 shows the typical profile of PAGE of proteins synthesized in poliovirus infected HeLa cells. The proteins are named according to the generally accepted nomenclature and are assigned with their apparent molecular weights. Upon translation of the poliovirion RNA in cell free extracts, a similar profile of proteins is obtained (Ehrenfeld, 1979; Agol, 1980). Table 28 provides an overview of the features of the poliovirus proteins. Table 29 illustrates the relative distribution of charged amino acids and cysteine in the poliovirus proteins (see also Fig. 91, Chapter 8).

As discussed in Chapter 3, proteins migrate in polyacrylamide gels in a good correlation to their molecular weights. The molecular weights of the polioviral proteins can be estimated in comparison with the migration of selected standard proteins whose molecular weight is known. It has recently been determined, however, that the substitution of even one single hydrophobic amino acid, such as leucine by a related amino acid, may result in an altered mobility, corresponding to a change in the molecular weight of the protein by up to 3 % (DeJong *et al.*, 1978). Furthermore, different gel conditions may result in slightly different migration patterns (Kew *et al.*, 1980; Vande Woude and Bachrach, 1971; Abraham and Cooper, 1976). Thus, the molecular weights assigned to poliovirus proteins on the basis of PAGE-migration should be regarded as provisional estimates. The exact

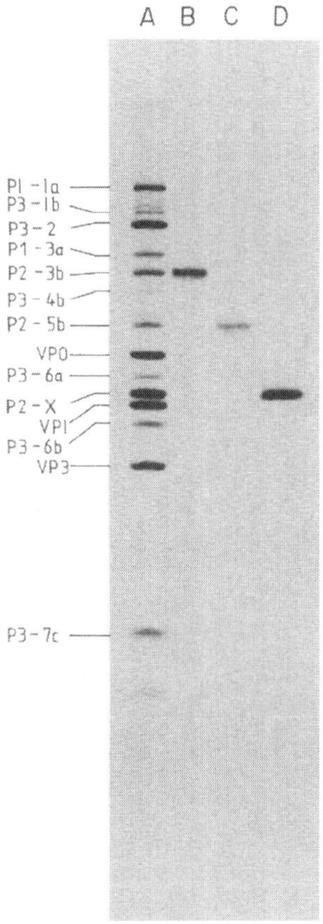

Fig. 50 B. Example of recent polyacrylamide gel electrophoretic separation of the intracellular poliovirus-specific proteins

Poliovirus-infected HeLa cells were pulse-labeled with ^3H alanine from 3 to 3.25 h.p.i. The proteins were separated by electrophoresis on a 10–18% linear gradient gel of polyacrylamide. The original nomenclature of poliovirus proteins shown in figure 50 A) has been adapted by addition of small letter subscripts (a, b, . . .) to signify additional bands that have been separated by improved gel techniques from the original bands in figure A). Furthermore, the prescript NCVP (for nonviral-capsid protein) has been replaced by the terms P1, P2, and P3 in order to indicate the relation of the protein to the three main coding domains and the respective primary cleavage products of the polyprotein. The nomenclature of the poliovirus capsid proteins has been retained. Notice that the precursor to VP_4 and VP_2 was originally named NCVP6. It has since been renamed VPO because it is regularly found in one or a few copies in the uncleaved form in most virions

Purified P-2 proteins –3 b, – 5b, and –X were coelectrophoresed in this gel in separate columns. The smaller viral proteins NCVP-8, -9, -10, and VP_4 have been run off the bottom of the gel under these electrophoresis conditions. The virus employed in this study was polivirus type 1, Mahoney strain, which is known to utilize the alternate cleavage pathway of the replicase precursor to NCVP6 a and -6 b to a significant extent (see also Fig. 51). – Figure from Semler *et al.*, 1981 [Virol. *114*, 590 (1981)]

Table 29. *Relative distribution of charged amino acids and cysteine in the poliovirus proteins*

	Basic amino acids	Acidic amino acids	Cysteine
NCVPOO	0 (−11 %)*	0 (−3 %)	0 (−50 %)
NCVP 1a	−2**	−2	0
VP_4	0	−3	∅
VP_2	−5	−4	+3
VP_3	−3	+1	0
VP_1	0	−2	−3
NCVP 3b	+1	0	+3
NCVP 8	−2	0	+5
NCVP 10	−4	−3	−4
NCVP X	+3	+1	+3
NCVP 1b	+2	+2	−2
NCVP 9b	0	0	+1
VPg	+5	∅	∅
NCVP 7c	−1	−3	−2
NCVP 4b	+3	+5	−2

* Indicates the average distribution in poliovirus polyprotein, the numbers in paranthesis indicates the percent deviation from random distribution.

** The numbers indicate the % difference from the average distribution in the poliovirus polyprotein NCVP00; the values are arbitrarily assigned in the following manner:

Value		Basic amino acids	Acidic amino acids	Cysteine
		% deviation from average distribution		
0	=	± 3.5	± 2.5	± 12.5
1	=	± 10.5	± 7.5	± 27.5
2	=	± 17.5	± 12.5	± 52.5
3	=	± 24.5	± 17.5	± 77.5
4	=	± 31.5	± 22.5	± 102.5
5	=	> 31.5	> 22.5	> 102.5
∅	=	no such amino acid in this protein		

molecular weigths can be calculated from the amino acid sequences as deduced from the nucleotide sequence of the poliovirus genome (Table 28).

The capsid proteins of polioviruses were named according to their migration in PAGE VP1–VP4; VP1 is the slowest and VP4 the fastest migrating viral protein. It was observed later that empty virus coats (lacking viral RNA), called natural top components, had no VP2 and VP4 but a larger protein (a precursor for VP2 and VP4) which was named VP0 (NCVP6 in earlier studies). To differentiate the other virus coded proteins from the capsid proteins, they were first classified as *n*on *c*apsid *v*iral *p*roteins: NCVP. Again they were numerically ordered according to their electrophoretic mobility in PAGE from NCVP1–NCVP10. Due to improvements in the separation technique and by further analysis, more and more virus specific proteins were detected. Therefore, new subscripts were introduced, as NCVP1a and NCVP1b, and larger precursor proteins were named NCVPO, and then even NCVPOO, which denotes the full translate of the poliovirus genome, the poliovirus specified polyprotein. A viral protein which comigrated in earlier PAGE systems with VP1 was later identified as a unique protein by Korant (1973) and named NCVPX. It became apparent that the total translate of the poliovirus genome under "normal" experimental conditions was found in infected cell extracts in the form of three primary cleavage proteins. They were named P1–P3 (in genomic order) and correspond to NCVP1a, NCVP3b and NCVP1b, respectively. As the details of the complex precursor–product relationships are elucidated, the intermediate and final cleavage products can be correlated to the three primary translation products P1, P2 and P3, and are named accordingly (see Table 28). With genomic and biochemical mapping procedures, the individual viral proteins could be assigned to specific regions of the viral genome (Fig. 51 and Table 28).

A. Mapping of the Genome Products on the Poliovirus Genome

A variety of different approaches were followed in the 1960s and 1970s in order to determine the gene order of poliovirus and other picornaviruses. Such approaches included tryptic peptide mapping of viral proteins (Jacobson and Baltimore, 1968b; Cooper and Bennet, 1973; Abraham and Cooper, 1975; Rueckert *et al.*, 1979; Kew *et al.*, 1980; Pallansch *et al.*, 1980), pactamycin and hypertonic initiation block mapping (Summers and Maizel, 1971; Taber *et al.*, 1971; Saborio *et al.*, 1974), and genetic recombination experiments (reviewed by Cooper, 1977). With the aid of these studies, the basic pattern of the genomic map of poliovirus was elucidated (reviewed by Rekosh, 1977). This pattern was finally confirmed and extended in fine detail when the entire nucleotide sequence of poliovirus type 1 was determined and correlated to partial amino acid analysis of the virus specific proteins. Figure 51 summarizes the present day information on the genomic map of poliovirus.

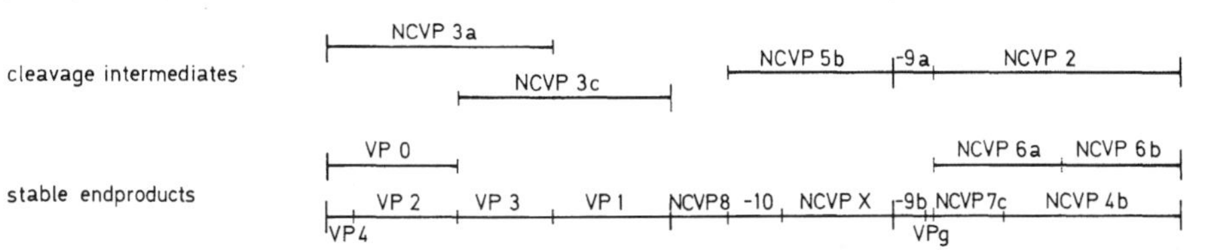

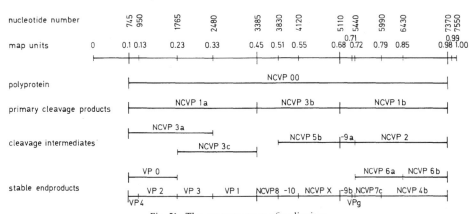

Fig. 51. The genome map of poliovirus

The figure illustrates the mapping of the poliovirus proteins on its genome and the processing pattern of the polyprotein. The localizations of the cleavage sites are marked by nucleotide number and relative map units. The figure is compiled from the data of the consensus sequence of poliovirus type 1 (Fig. 44); the figure is drawn to scale

1. Relatedness of Viral Proteins: Tryptic Peptide Analysis

Much useful information concerning the relatedness of proteins was obtained by an analysis of their tryptic peptide digests. Since proteases in general recognize and thus cleave only at specific amino acid sequences, each protein is digested by such a protease into a characteristic set of small peptides. When these peptides are separated by one- or two-dimensional gel electrophoresis, a characteristic pattern or "fingerprint" of the protein is obtained. Such fingerprints are very useful in establishing or ruling out precursor-product relationship between larger and smaller proteins, *i.e.* all peptides of a product (except for the one or two peptides which contain the cleavage sites) should also be contained within their corresponding precursor. In this manner, neighborhood relationship on the genetic map of the corresponding coding regions of the analyzed proteins can also be established.

The first poliovirus proteins to be analyzed in this manner were the capsid proteins (Jacobson and Baltimore, 1968b; Cooper and Bennett, 1973). It was determined that NCVP1a is the precursor to the four viral capsid proteins, and that their genome order is from 5' to 3' VP4, VP2, VP3, VP1. In the only study published so far where this procedure was also applied to large poliovirus precursor polyproteins (Abraham and Cooper, 1975), a large precursor protein (NCVPOO) was analyzed and found to contain the peptides of NCVP1a (capsid precursor), and of NCVP1b (the replicase precursor), but not the peptides of NCVPX. This suggested that NCVPX does not lie between NCVP1a and NCVP1b. Correlation of the elucidated nucleotide sequence with results from partial amino acid sequence analyses of viral proteins, however, has established with certainty that the coding region for NCVPX lies between those for NCVP1a and NCVP1b (Semler *et al.*, 1981a, b). The results from the earlier tryptic peptide studies still stand unexplained. One hypothetical way to explain the results is by postulating that the large precursor protein designated NCVPOO actually corresponded to the consecutive proteins NCVP1a (P1), NCVP3b (P2), and NCV-P6a (the N-terminal half of NCVP1b = P3), and that the protein designated NCVPX actually corresponded to NCVP6b, the C-terminal end of NCVP1b and of the poliovirus polyprotein. The cleavage site between 6a and 6b is try/gly, which presumably is recognized by a host specific protease. NCVP6b is of a similar size as NCVPX (35 vs 38 kd) and may have been purified instead of NCVPX (see Fig. 49, p. 170).

The tryptic peptide mapping was also employed to locate the genome position of the small genome protein VPg (Pallansch *et al.*, 1980). It was found for EMC virus that the two tryptic peptides of VPg were contained within the replicase precursor protein C (corresponding to polio NCVP1b). The corresponding experiments for poliovirus VPg led to similar results, although the evidence here was inconclusive. The conclusions derived from these studies were that the coding region for VPg probably lies directly next to, or close to, the 5' end of the coding region for NCVP2 and within the 5' coding region of NCVP1b. Although some controversy about this location came from early RNA sequencing studies, which appeared to locate VPg at more internal positions of the NCVP1b coding region (Kitamura and Wimmer, 1980; Kitamura *et al.*, 1980b), correlation of the nucleotide sequence with amino acid analysis finally confirmed the positioning of VPg derived from the tryptic peptide studies next to the 5' end of NCVP2 (Semler *et al.*, 1981a+b).

In a more recent study, tryptic peptide analyses of the proteins of different poliovirus strains were presented (Kew *et al.*, 1980; Wiegers and Dernick, 1981), which establish definite precursor-product relationship amongst the three primary poliovirus proteins and their products. These studies also demonstrated that tryptic peptide analyses can be useful for determining the relatedness of different strains of poliovirus or the localization of mutations that are expressed in proteins of vaccine strains (see Fig. 53 below). Tryptic peptide analyses were also employed in the first biochemical demonstration of recombination events between two strains of poliovirus (Tolskaya *et al.*, 1983).

2. *Biochemical Mapping*

The biochemical evidence for a genetic map of the polio genome came mainly from experiments employing irreversible or reversible inhibitors of initiation of protein synthesis (pactamycin, hypertonic initiation block = HIB) during late stages of poliovirus replication near the peak of viral protein synthesis (reviewed by Rekosh, 1972, 1977; Cooper, 1977; Rueckert *et al.*, 1979). The validity of the genetic maps determined in this way is based on the assumption that only one single initiation site near the 5' end of the polio mRNA is operating in the course of these experiments (Jacobson *et al.*, 1970), and that, once a ribosome has initiated the synthesis of polypeptides at this point, it wanders down the entire length of the mRNA in the 3' direction, and continuously "reads-off" the consecutive coding regions on the mRNA.

In the classical study (Summers and Maizel, 1971; Taber *et al.*, 1971), the antibiotic pactamycin was used to elucidate the gene order of poliovirus RNA. Pactamycin inhibits the initiation of polypeptide synthesis irreversibly, but does not significantly effect the elongation and termination of already initiated polypeptide chains. When short pulses of radioactively labeled amino acids are applied to the infected cells shortly after the pactamycin induced block of polypeptide chain initiation, the labeled amino acid will of course be incorporated only into those polypeptides whose synthesis was just in progress at the time of addition of label. By increasing the time interval between application of pactamycin and labeled amino acids, those proteins whose coding region lies near to the end of the mRNA (*i.e.* the 3' end of the genome) will be preferentially labeled. The coding map of the mRNA can thus easily be read backwards from the 3' end to the 5' end. This map has come to be known as the pactamycin map.

The "pactamycin map" was confirmed in similar experiments using elevated tonicity of the growth medium to block reversibly the initiation of protein synthesis (HIB) (Saborio *et al.*, 1974). This method has the additional advantage of providing a completely reversible block of initiation. Elongation and termination of already initiated polypeptide chains are not significantly affected. The ribosomes can thus be "run off" from their mRNAs. Upon subsequent reestablishment of isotonicity of the growth medium, the synthesis of polypeptide chains is synchronously reinitiated. Here, the addition of short pulses of radiolabeled amino acids at increasing time intervals after reinitiation, leads to the consecutive labeling of poliovirus proteins according to their distance from the initiation site (5' end). The coding map of the mRNA can thus easily be read forwards from the 5' end to the 3' end. This "HIB map" is in good agreement with the pactamycin map.

3. The Genome Map as Deduced from Genetic Studies

Essential characteristics of the pactamycin and HIB maps are confirmed by evidence from studies on poliovirus genetics (reviewed by Cooper, 1977). A large number of poliovirus mutants are described in the literature. The mutants could be characterized as falling into three main functional domains: One, determing the virion coat, and two others, determing two types of replicase functions, one of which is responsible for the synthesis of dsRNA (replicase II activity to produce the complementary (–)strand), and the other for the synthesis of ssRNA (replicase I activity to produce new (+)strand virion RNA or mRNA). By analyzing many recombinants between different mutants of poliovirus, the relative localizations and distances of the genetic markers were determined and a genetic map of poliovirus was established. Figure 52 shows a correlation of this „genetic map" to the map determined from nucleotide sequencing studies. A more detailed description of poliovirus mutants and their genetics is presented in section VI below.

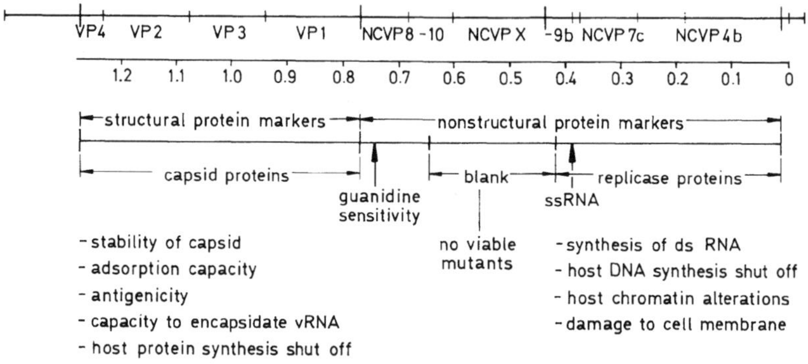

Fig. 52. The genetic map of poliovirus

The figure shows a schematic map of the poliovirus genome. The functions directed by the different segments of the poliovirus genome are inferred from genetic studies, their relative localizations determined by genetic recombination experiments, covariation, and complementation. The genetic map is tentatively correlated to the map deduced from the nucleotide sequence of poliovirus type 1 (top line, see also Fig. 44). The map units (second line) are arbitrary units assigned to the genome for mapping of recombinant distances (Cooper, 1975). A more detailed version of the recombination map is presented in Fig. 57 below

B. Functions of the Viral Proteins

Certain and probable functions and some properties of viral proteins are listed in Table 30. The functions of some viral proteins are still obscure, some proteins seem to have dual functions. It is also conceivable that alternative cleavage

Table 30. *Functions of poliovirus proteins*

a	b	c	
P-1	*capsid proteins*		
– 1a			capsid protein precursor, assembly factor
	– 3a		? (VP-0 + VP-3)
	– 3c		? (VP-3 + VP-1)
	VP-0		procapsid protein (VP-4 + VP-2)
		VP-4	uncoating plug, RNA binding, protein synthesis shut-off
		VP-2	D-antigen
		VP-3	
		VP-1	capsid backbone, C-antigen, receptor recognition site
P-2	*regulatory proteins*		
– 3b			?
	– 5b		? formation of virus-induced vesicles
		– 8	guanidine sensitivity, cysteine dependence
		–10	? membrane formation
		– X	? VP-0 protease, RNA encapsidation
P-3	*replicase proteins*		
– 1b			replicase precursor
	– 9a		membrane insertion site
	– 2		soluble replicase
		– 9b	?
		– VPg	genomic protein, RNA initiation, capsid recognition
		– 7c	gln/gly protease
		– 4b	RNA polymerase, inhibition of host RNA synthesis
		– 6a	alternative protease
		– 6b	alternative replicase
Non-assigned proteins *			– 4a – 5a – 7a – 7b – 7d
Additional potential functions			
Capsid proteins (P-1)			proteinkinase (phosphorylation of cap binding protein and eIF 2), S-7-, ox- and Hy-sensitivity
X-proteins (P 2)			alterations of cell cytoskeleton, induction of specific membrane permeability for monovalent cations, inhibition of Na^+ K^+ ATPase
Replicase proteins (P-3)			inhibition of host DNA synthesis, chromatin alteration, induction of membrane leakiness (possibly induced by dsRNA)

[a] Primary cleavage products.
[b] Cleavage intermediates.
[c] Stable end products.
* 4a = 60 kd from C-terminus of P-3; 5 a = 49 dk from N-terminus of P-3; 7 a = P-2–8 + P-2–10; 7 d = 17 kd from C-terminus of P-3–6 a; in addition trace amounts of polypeptides X/9 (= P-2–X + P-3–9 a) and 3 b/9 (P-2 + P-3–9 a) have been detected in infected cells (Pallansch *et al.*, 1984, J. Virol. 49, 873–880).

pathways may generate additional proteins with different functions. A number of small peptides are cleaved from the amino terminal sides of the precursors of the P-2 and P-3 domains which are quite stable (NCVP8, NCVP9a + b, NCVP10). Some of these peptides may present leader peptides for the insertion of the primary polyproteins NCVP3a and NCVP1b into intracellular membranes or for the translocation of these proteins through intracellular membranes into specific compartments, or they may serve as control mechanism for the site or time of activation of the proteins from which they are cleaved, analogous to prohormones. Whether these and other small, stable cleavage products have any specific functions by themselves is not yet known.

Conversely, a number of viral induced effects on the host cell could not be assigned as yet to any specific viral protein(s). Although further investigation may close this gap in our knowledge, there is of course no *a priori* necessity for the association of each of the virus induced effects on host cells with a specific viral protein(s). Some of the virus induced effects on biochemical pathways of the host cell may well represent certain "latent" negative control mechanisms that are inherent to the host cell, which merely are activated by the presence of the virus (see *e.g.* Chapter 8), or they may be secondary effects due to inhibition of macromolecular synthesis, membrane damage, or effects of dsRNA.

1. The P-1 Proteins: Coat-Proteins, Proteinkinase, Shut-Off of Protein Synthesis

P-1 Proteins: 1a, 3a, 3c, VP_0, VP_4, VP_2, VP_3, VP_1

The final proteins derived from P-1 are the structural components of the viral capsid: VP_0, VP_4, VP_2, VP_3, VP_1. These proteins are of course involved in the various capsid specific functions described in detail in chapter 3 (protection of viral RNA, interaction with plasma membranes and antibodies, etc.). A variety of experimental evidence indicates that all or some of the individual capsid proteins may have additional functions: shut-off of cellular protein synthesis, enhancement of assembly, proteinkinase activity, ionophore, and guanidine sensitivity. In light of more recent evidence, however, it seems probable that guanidine sensitivity is determined by gene products of the P-2 domain (Koch *et al.*, 1980b; Romanova *et al.*, 1980; Tolskaya *et al.*, 1983).

2. The P-2 Proteins: Guanidine Sensitivity, VPO Protease

P-2 Proteins: 3b, 5b, X, 8, 10

P-2—3b is very rapidly cleaved via 5b (and NCVP8) to X (and NCVP10) so that infected cells labeled in pulses of 5 min or less contain little of the proteins 3b and 5b. In infected cell lysates, X is predominantly found attached to smooth membranes in association with the replication complex. Korant (1979) has proposed that X harbors protease activity (or copurifies with a protease). Monoclonal antibodies directed against NCVPX did not, however, block any of the major cleavage steps in poliovirus protein processing in vitro, and it seems certain that the major protease is NCVP7c from the P-3 coding region. The protease for cleavage of VP_0 has not been identified, and the possibility remains that this function is carried out by one of the P-2 proteins. Other functions, hypothetically assigned to the P-2 proteins include guanidine sensitivity, induction of membrane formation, and linkage of procapsid to replicase. It is interesting to note that the NCVP-X coding region is the most highly conserved region of the poliovirus genome. The Sabin vaccine strain, which contains multiple mutations, does not harbor any amino acid substitution in NCVP-X. No viable mutants mapping in the X coding region have been reported to date. These observations indicate that X plays a central role in poliovirus replication.

3. *The P-3 Proteins: gln/gly Protease, VPg, Replicase, Cytopathic Effect*

P-3 Proteins: 1b, VPg, 2, 4a, 4b, 7c, 6a, 6b, 9

The proteins of P-3 are important factors for the replication of the viral RNA (see Chapter 9). 1b, 2, and 4b in particular show RNA-copying capacity in vitro. 7c has been implicated as the main virus-specific protease which is responsible for the cleavage of gln-gly pairs (Hanecak *et al.*, 1982).

It is likely that the alternate cleavage pathway of NCVP2 yielding 6a and 6b instead of 4b and 7c, which is preferred by certain viral strains (Kew *et al.*, 1981) similarly separates the protease and RNA-polymerase activities into two distinct proteins 6a and 6b, respectively.

VPg might serve as a primer in RNA replication, possibly also in virion assembly. VPg is not required for the initiation of the virus growth cycle. Unbound VPg has not been found in uninfected cells (Nomoto *et al.*, 1977a; Kitamura *et al.*, 1980c; Baron and Baltimore, 1982). After addition to cellfree extracts, it is rapidly degraded (Dorner *et al.*, 1981; Sangar *et al.*, 1981). More recently, free VPg as well as a discrete molecule with the structure VPg-pUpU were demonstrated in infected cells by antibody directed against chemically synthesized VPg (Crawford and Baltimore, 1983); and VPg-pUpU was shown to be synthesized *in vitro* by a membrane fraction from poliovirus infected cells (Takegami *et al.*, 1983b).

Immunoprecipitation experiments with antibodies directed against small synthetic peptide sequences of polio VPg (Semler *et al.*, 1982; Baron and Baltimore, 1982) have also identified larger precursor polypeptides containing the VPg sequence. The predominant VPg-containing precursor was identified as NCVP9a, the 12,000 molecular weight cleavage product produced upon cleavage of NCVP1b to NCVP2. In contrast to VPg, NCVP9 appears to be stable, both in the infected cell and in cell free extracts. The highly polar VPg sequences are located at the carboxy terminal end of NCVP9a, corresponding to residues 88 through 120; a highly hydrophobic region is located somewhat upstream in NCVP9 at residues 59 through 80. NCVP9 was shown to be a membrane associated protein (Semler *et al.*, 1982; Takegami *et al.*, 1983), and it was proposed that the hydrophobic region might serve as a membrane anchor, leaving the VPg moiety available to RNA replication enzymes for utilization as a primer for the initiation of RNA synthesis. Since many more polyprotein molecules are synthesized in the infected cell than RNA molecules, the amount of VPg utilized in RNA synthesis is small, explaining why NCVP9a is relatively stable in infected cells.

Two other, larger viral precursor polypeptides were precipitated specifically with anti VPg antibodies. These polypeptides were mapped as NCVP3b/9 (77kd) and NCVPX/9 (50kd) (Takegami *et al.*, 1983a; Pallansch *et al.*, 1984). They appear to be rapidly cleaved in infected cells, since detectable quantities are not routinely observed, unless protein processing is inhibited.

V. Genomic Variation of Polioviruses

Historically, polioviruses are classified into three serologically defined serotypes. The "Brunhilde" strain, poliovirus type 1, was found to be responsible for approximately 85 % of all poliomyelitis illnesses; the "Lansing" strain, type 2, and "Leon" strain, type 3, together were found to cause only approximately 15 % of the cases. The amino acid sequences of the three poliovirus serotypes are 88% homologous–the amino acid homology in the replicating enzymes is even 95% (Toyoda *et al.*, 1984). These observations confirm the notion that the three poliovirus serotypes are all derived from a common prototype poliovirus ancestor by evolutionary divergence.

Within each serotype of poliovirus there exist many naturally occuring variants as well as mutants induced and selected in the laboratory. Of particular importance are the vaccine strains that were liberated from laboratory confinement in the early sixties in mass vaccination programs and which have since spread over large regions of the world, often displacing the wildtype viruses. The vaccine strains exhibit considerable genetic variation during their field propagation in response to environmental pressures. Even long-term multiplication in a single host–of a Sabin vaccine derived strain in the intestinal tract of a small boy with agammaglobulinemia–was accompanied by substantial antigenic variation in the progeny isolates, giving rise to a non-vaccine like virus (Hara *et al.*, 1981). Genetic recombination between genomes of simultaneously infecting subtypes occurs at a frequency similar to recombination events in eukaryotic DNA genomes, thereby providing a basis for augmented exchange and spread of genetic information among the polioviruses (Cooper, 1971, 1977; Tolskaya *et al.*, 1983).

The subtypes of poliovirus and their variants can be distinguished and characterized by a variety of procedures (see Table 25, Chapter 3). Sereologic methods include complement fixation (Wenner *et al.*, 1959), measurement of neutralization kinetics (McBride, 1959), modified Wecker analysis (Nakano and Gelfand, 1962), and neutralization with cross-absorbed antisera (van Wezel and Hazendonk, 1979). Subtypes are also identifiable by other phenotypic characteristics, like stability against changes in pH or temperature, sensitivity to detergents, and to a number of protein reactive agents. All of these phenotypic properties are caused by an alteration in the viral protein coat, which reflects only about one third of the coding capacity of the poliovirus RNA. Analysis of poliovirus variability by the above means has indicated that several of the phenotypic markers are unstable during human intestinal passage (Melnick, 1961), and that the genetic information in the viral RNA is highly unstable (Cooper, 1969). This is in contrast to what occurs during virus propagation in cell culture, where the same phenotypic markers are conserved to a remarkable extent (Ghendon, 1963).

Many of the experimental methods described in this chapter also allow the detection of genomic variations in the regions coding for non-capsid proteins. Denaturation mapping, oligonucleotide mapping, nucleotide sequencing of the viral RNA, polyacrylamide gel electrophoresis and tryptic peptide mapping of the viral proteins have been employed increasingly in the past decade to analyze the relatedness of polioviruses isolated from patients, and to follow the genetic

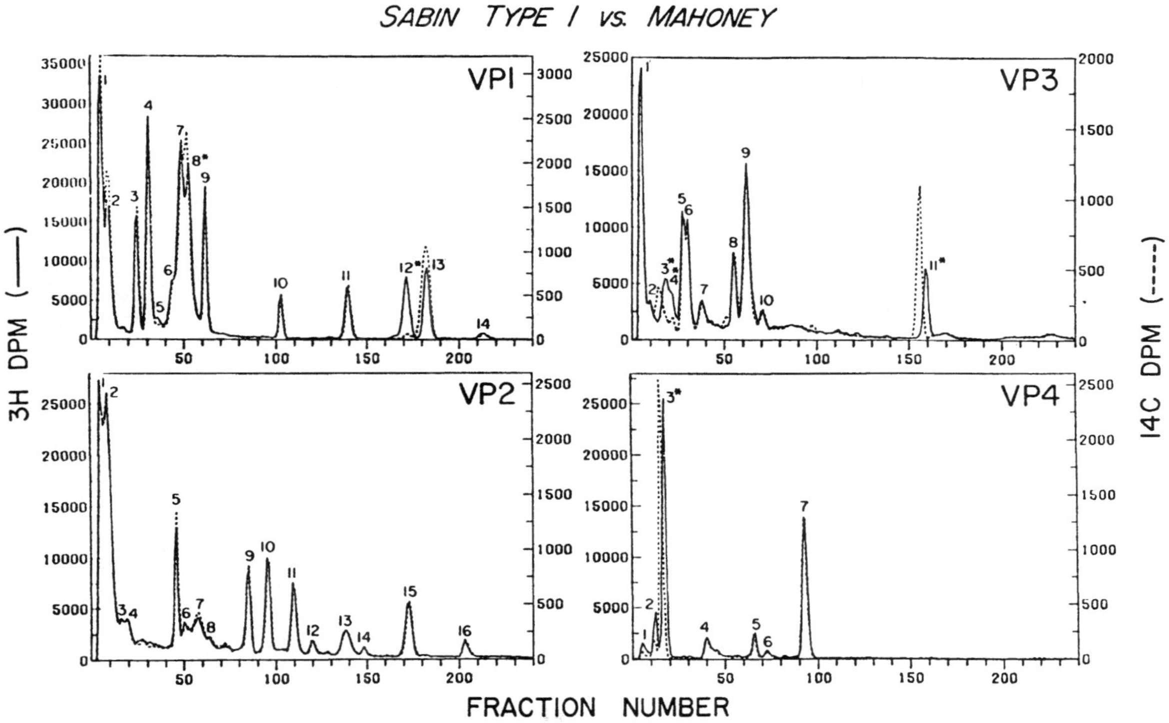

Fig. 53. The tryptic peptide maps of the capsid proteins of poliovirus type 1

Tryptic comparisons of the capsid proteins of wild type poliovirus type 1: Mahoney and its vaccine strain, LSc 2 ab. Homologous capsid proteins, purified on polyacrylamide gels were mixed and digested with trypsin, and the peptides were resolved by cation exchange-chromatography. LSc 2 ab, ^{3}H label (solid line); Mahoney, ^{14}C label (dashed line). The numbers refer to peaks in the vaccine strain; asterisks indicate peptides that differ between the two strains. Recoveries of label (^{3}H, ^{14}C) from prefiltered tryptic digests were: VP_1, 82%, 81%; VP_2, 78%, 79%; VP_3, 82%, 78%; VP_4, 81%, 90%. – Figure from Kew *et al.*, 1980 [J. Virol. *33*, 260 (1980)] adaptation of polioviruses to changing environments. Figure 53 shows a comparison of the tryptic peptide maps of the capsid proteins of the parental and vaccine strains of poliovirus type 1 (Kew *et al.*, 1980).

The regions, where the denaturation maps of the two poliovirus genomes differ from each other, can be correlated to sequences of considerable non-homology between these RNAs. Differing regions are determined by electron microscopic examination of heteroduplex RNA molecules constructed from complementary RNA chains of different poliovirus types. Figure 54 shows a correlation of the denaturation maps of poliovirus types 1 and 3 with the corresponding heteroduplex map of these viruses. Tentatively, dissimilar sequences in the genomes of the two serotypes seem to be largely concentrated in a region coding for capsid proteins, a relatively small non-homologous region being present also in a non-capsid part of the genome. There is no evidence for large deletions or insertions of the genetic material accompanying the evolutionary divergence of these poliovirus serotypes (Agol, 1980). These observations have been confirmed by nucleotide sequence analyses (Toyoda *et al.*, 1984) (see Appendix IV).

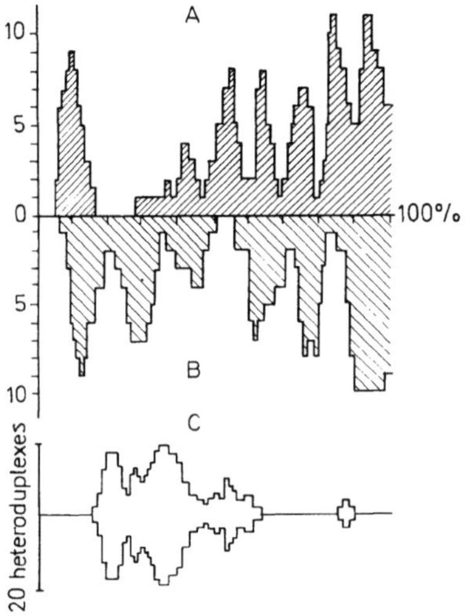

Fig. 54. Comparison of the denaturation and heteroduplex maps of poliovirus types 1 and 3
The figure shows a comparison of the denaturation and heteroduplex maps of two strains of poliovirus. The denaturation maps of poliovirus type 1 *(A)* and poliovirus type 3 *(B)* were obtained as described in Fig. 39
C The polio 1/polio 3 heteroduplex RNA map, illustrating the statistical distribution of unpaired regions along the heteroduplex RNA. The central line represents double-stranded regions; the separated lines represent single stranded loops. The distance between separated lines reflects the proportion of molecules displaying strand separation in the given region of heteroduplexes. – Figure from Čumakov *et al.*, 1979 [Virol. *92*, 268 (1979)]

Oligonucleotide maps of the three poliovirus strains are shown in Figure 55A (RNase T 1 digests) (Lee *et al.*, 1979). All three fingerprints differ strikingly in the separation pattern of the large oligonucleotides which represent approximately 5 to 8 % of the virus genome. Smaller oligonucleotides shared by the different virus

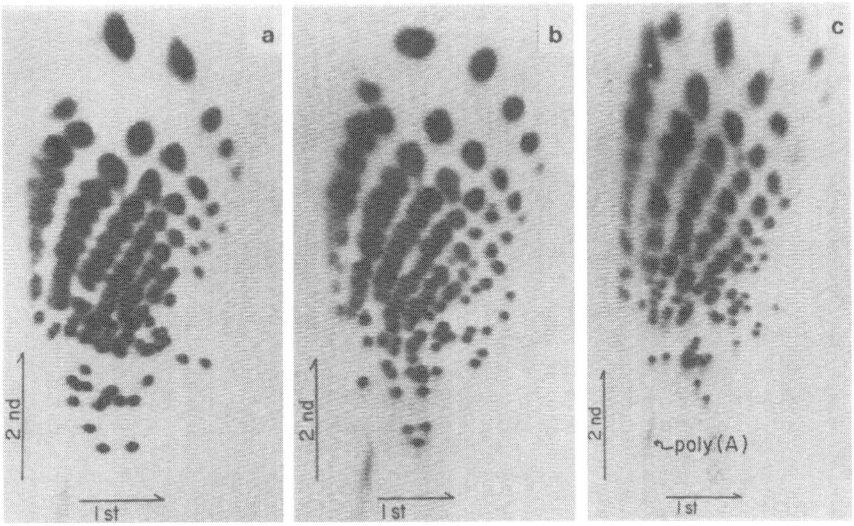

Fig. 55 A–B. Differentiation of poliovirus strains by oligonucleotide mapping of their RNAs

Fig. 55 A. The RNase T1 oligonucleotide maps of poliovirus type 1 *(a)*, type 2 *(b)*, and type 3 *(c)* The figure shows the separation of RNase T1 digests of virion ^{32}P-RNA of poliovirus types 1, 2, and 3 by two dimensional gel electrophoresis. The fingerprints are very similar in the area of short oligonucleotides (less than 10 bases in length) which migrate furthest in the second dimension (top of figure). In contrast, little if any similarity can be observed in the lower portion of the gel which contains larger oligonucleotides. The heterogeneously sized poly (A) tract is present as a smear in the lower left part of the fingerprint. – Figure from Lee *et al.*, 1979 [J. gen. Virol. *44*, 314 (1979)]

RNAs may be found. The fingerprints of the vaccine strain LSc2ab and its parent poliovirus type 1 show a much closer resemblance (see Fig. 55B). It can be seen that two large oligonucleotides are missing, while four others move at different positions during gel separation (Nomoto *et al.*, 1979a; Nottay *et al.*, 1981).

Oligonucleotide mapping has been applied successfully to analyze the relatedness to the vaccine or to "wild" strains of polioviruses isolated from patients (Fig. 55B) (Nottay *et al.*, 1981; Kew *et al.*, 1981; Minor, 1982b; Yoneyama *et al.*, 1982; Kew and Nottay, 1984). The fingerprinting system separates oligonucleotides according to their base composition and size (DeWachter and Fiers, 1972; Lee *et al.*, 1979). Therefore, even a single base change can result in a change in the positioning of one oligonucleotide. The most similar oligonucleotide maps were those of isolates from members of the same family. In one such case, one spot difference was detected. The oligonucleotide maps of successive isolates provide a measure of the rate of changes in the viral genome. One world-wide transmission of an epidemic strain over a period of thirteen months was analyzed by oligonucleotide mapping of eighteen independent isolates. The number of base changes between the first and the last isolate were estimated to be on the order of 100, or about 1–2 % of the total bases of the poliovirus RNA. Despite these changes, the oligonucleotide patterns still indicated the close relationship between these isolates. Oligonucleotide spots in maps from successive isolates often disappeared

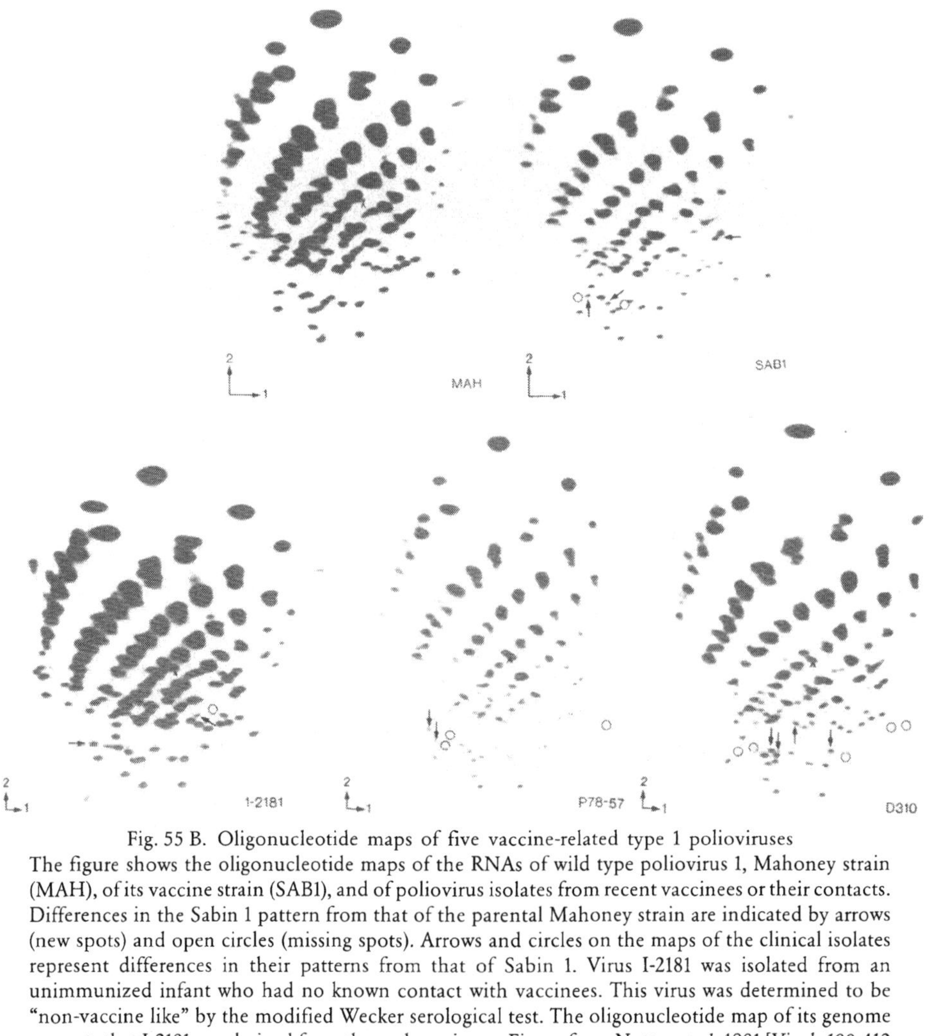

Fig. 55 B. Oligonucleotide maps of five vaccine-related type 1 polioviruses
The figure shows the oligonucleotide maps of the RNAs of wild type poliovirus 1, Mahoney strain (MAH), of its vaccine strain (SAB1), and of poliovirus isolates from recent vaccinees or their contacts. Differences in the Sabin 1 pattern from that of the parental Mahoney strain are indicated by arrows (new spots) and open circles (missing spots). Arrows and circles on the maps of the clinical isolates represent differences in their patterns from that of Sabin 1. Virus I-2181 was isolated from an unimmunized infant who had no known contact with vaccinees. This virus was determined to be "non-vaccine like" by the modified Wecker serological test. The oligonucleotide map of its genome suggests that I-2181 was derived from the oral vaccine. — Figure from Nottay *et al.*, 1981 [Virol. *108*, 412 (1981)]

and then reappeared in later isolates, suggesting the existence of highly mutable region(s) of the genome, but many of the large, unique oligonucleotides did not change their position during the epidemic, indicating conserved regions.

The elucidation of the nucleotide sequences of the genomes of virulent poliovirus type 1 and its vaccine derivative have revealed several clusters of mutation sites. Figure 57 (see p. 190, below) shows the positions of the mutations in the Sabin vaccine strain as compared to the parental virulent strain (Nomoto *et al.*, 1982b). Most striking is a cluster of mutations in the genome region coding for VP1, which has been implicated also as the capsid protein predominantly responsible for the antigenic drift of polioviruses. Comparable mutation clusters in VP1 are

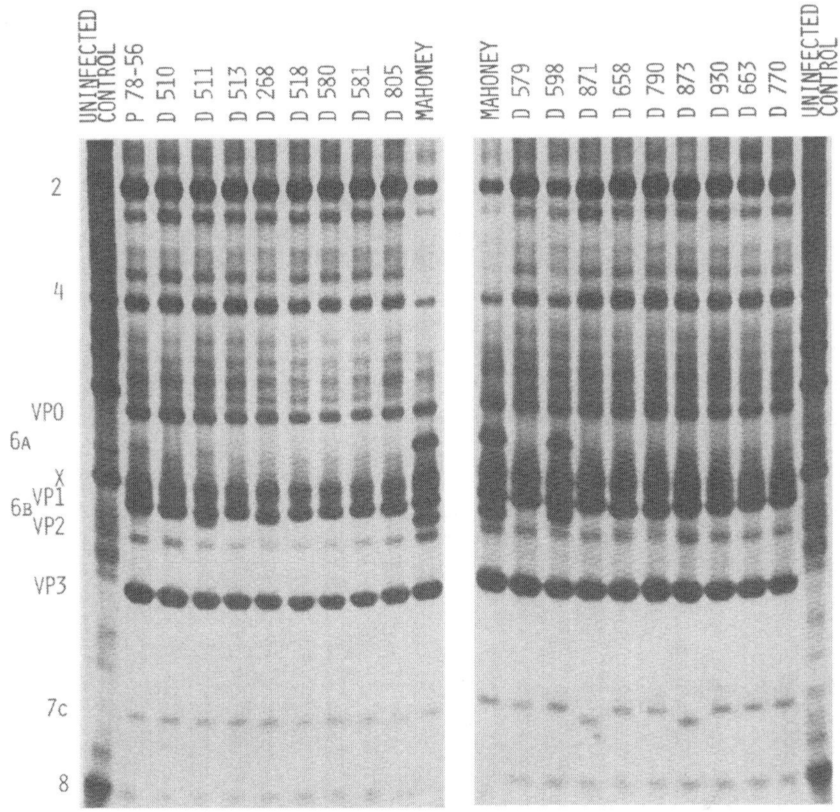

Fig. 56. Comparison of viral proteins from isolates of the 1978 Netherlands-Canada-U.S.A. epidemic The figure shows a comparative analysis of the viral proteins from epidemic isolates by polyacrylamide gel electrophoresis. The mobilities and relative amounts of most proteins were the same for all isolates. However, small but reproducible differences in some proteins were detected among the patterns. Notice, in particular, the shifts in the migration pattern of VP_1, the major virion surface polypeptide. In two of the U.S.A. isolates, D871 and D873, the pair of replicase related proteins, NCVP2 and its cleavage product NCVP7 c (the viral protease) showed marked mobility shifts. Notice also, the unusually high amounts of NCVP6 a and NCVP6 b, the alternate cleavage products of the replicase precursor, in D598. The corresponding oligonucleotide maps from these isolates (not shown) revealed discrete differences in consecutive isolates from the epidemic, demonstrating the molecular evolution of poliovirus during its transmission in an epidemic

The epidemic was caused by a poliovirus type 1 virus. The first paralytic case occurred in April 1978 among members of a Dutch religious community who rejected immunization. The disease was apparently transmitted by healthy carriers to Canada, and from here to the U.S.A., where it spread in Amish communities in several states. The epidemic was finally controlled by an oral-vaccine immunization program, the last case occurred in June 1979. A total of 101 poliomyelitis cases were reported, including 101 paralytic diseases and 32 cases of aseptic meningitis (Morbidity and Mortality Weekly Report, 27 and 28, 1979, Center for Disease Control, Atlanta, U.S.A., and Furesz *et al.*, 1978). – Figure from Nottay *et al.*, 1981 [Virol. *108*, 417 (1981)] also observed in distant picornavirus relatives of poliovirus, such as FMDV (Boothroyd *et al.*, 1982). In contrast, the regions coding for NCVPX and NCVP9a (the membrane bound precursor to VPg) are highly conserved and do not reveal any base substitutions that are also expressed in amino acid substitutions.

Interestingly, those genome regions shown to contain the highest rates of mutation from the wild type parent to the vaccine strain of poliovirus type 1, correspond roughly to the regions exhibiting the least homology between the genomes of different poliovirus types as determined by denaturation mapping (see above). This supports the notion, that the three polioviurs types are derived from a common ancestor in a series of single mutations clustered in regions of the genome which are least essential to virus function.

Mutations which are expressed by amino acid substitutions in the viral proteins can often also be detected by alterations in the migration behavior of the proteins during polyacrylamide gel electrophoresis and by their tryptic fingerprints (Kew *et al.*, 1981; Nottay *et al.*, 1981). Figure 56 provides one example of the application of these techniques to poliovirus isolates from the epidemic in Pennsylvania in 1978. Such studies also demonstrated that different variants of polioviruses prefer different alternative cleavage pathways in the processing of their precursor proteins, especially of the P-3 domain.

VI. Genetics

Early work on the genetics of polioviruses was stimulated by the search for attenuated strains which were not neurovirulent and thus suitable for the use as live vaccines (Sabin *et al.*, 1954). Many mutant strains were isolated and their mutant traits were characterized. Mutagenic agents were employed to increase the number and variety of mutants (Dulbecco and Vogt, 1958). Later, conditionally lethal mutants were exploited for extensive genetic analysis (Lwoff, 1958; Cooper, 1969, 1977), which finally lead to the establishment of a genetic map for poliovirus (Cooper, 1977) (see Fig. 52, p. 178). The recent application of recombinant DNA technology to poliovirus genomes has opened an entirely new field of research on poliovirus genetics which is expected to complement and extend much of the earlier studies.

A. Mutations

Mutations may occur either spontaneously (naturally occuring variants), or they can be induced by mutagenic agents. Amongst the mutagenic agents which have been used to obtain poliovirus mutants suitable for genetic analysis are growth of virus in the presence of 5-fluorouracil (5-FU) (Cooper, 1964), or nitrosoguanidine (Cooper *et al.*, 1975), or treatment of virus or free viral RNA with nitrous acid (Boeye, 1959; Cooper *et al.*, 1971, 1975), or proflavine (Dulbecco and Vogt, 1958). Mutants may contain single mutations, more commonly however, mutants contain multiple mutations. Ideally, mutants which are exploited for genetic studies should contain only one single mutation. Isolation of mutants after a single passage increases the chance of obtaining single mutation mutants. On the other hand, special mutagenic treatments may increase the number of base substitutions. It can thus be expected that mutants—in particular those derived after mutagenic treatments—often contain multiple mutations. Only some of the mutations which arise from base substitutions in the nucleic acid of the viral genome also result in an amino acid substitution in the corresponding gene product. Some, but again not all, amino acid substitutions may lead to minor or more severe alterations in the conformation of the respective protein structure. A significant change in the structure or surface charge of a protein may alter—or even completely abolish—its function.

Alterations of protein structure and function are usually evidenced as a change in the phenotype of the virus. Characteristic changes in phenotype can be detected with experimental procedures and have been termed mutant markers. Many such markers have been studied and classified for poliovirus (Table 31).

Mutant phenotypes often revert back to the parental phenotype. In principle, this could reflect simply the replacement of the substituted amino acid by one resembling the original parental amino acid. Statistically it is more likely however,

Table 31. *Genetic markers of poliovirus*

Universal markers	
rct, ts	sensitivity of replication to elevated temperatures
Structural markers P-1 proteins	
ΔH (T)	thermal lability, heat sensitivity
IC	changes in antigenicity
psr	repression of host protein synthesis
bo	sensitivity to inhibitors in bovine serum
S-7	sensitivity to inhibition by ethyl-2-methylthio-4-methyl-5-pyrimidine carboxylate
ox	sensitivity to inhibition by 2-(3-chloro-p-tolyl)-5-ethyl-1, 3, 4-oxadiazole
Hy	sensitivity to inhibition by 5-methyl-5-3, 4-dichlorophenylhydantoin
dex	sensitivity to dextran sulfate
m	sensitivity to inhibitors in agar overlay
d	sensitivity to acid overlay (growth capacity at low bicarbonate concentrations)
ho	sensitivity to inhibitors in horse serum
MS	growth capacity in a stable monkey kidney cell line
n	neurovirulence
EC_{50}	adsorption capacity
Non structural markers P-2 proteins	
cy	cysteine dependence
g	sensitivity to inhibition by guanidine
Non structural markers P-3 proteins	
ssRNA	capacity to synthesize ssRNA
dsRNA	capacity to synthesize dsRNA
pti	repression of host DNA synthesis
chr	damage to cell chromatin
tb	damage to cell membrane (permeability to trypane blue)

that reversion to parental phenotype is caused by a second amino acid substitution—usually at a nearby locus. The additional substitution induces a configurational change which "corrects for" the change induced by the first amino acid substitution. In this way, many different mutants with properties intermediate between those of parent and mutant may be formed. In general single mutants show a high rate of reversion to the parental phenotype, whereas mutants containing multiple mutations exhibit relatively more stable mutant phenotypes.

The Sabin vaccine strain of poliovirus type 1 LSc2ab was derived from its parent by spontaneous mutations in serial passage and contains 57 base substitutions dispersed over the genome, 21 of which result in amino acid substitutions in the corresponding proteins (Nomoto *et al.*, 1982b) (see Fig. 57). Only some of these lead to any detectable alterations of the phenotype in the mutant strain. The mutation map of the attenuated strain of poliovirus type 1 is shown in Figure 57. The nucleotide sequence of the type-3 Sabin vaccine strain (P3/Leon/12a$_1$b) differs from that of its parental neurovirulent progenitor (P3/Leon/37) at only 10 positions. Only three of these result in an amino acid change (Stanway *et al.*, 1983; Schild *et al.*, 1984), namely those at positions 2034 (in VP3), 3333 (VP1), and 3464 (NCVP8). None of the mutations observed in the type 1 strain have an identical counterpart in the type-3 vaccine strain, indicating that the mutational events responsible for attenuation are different in these two serotypes.

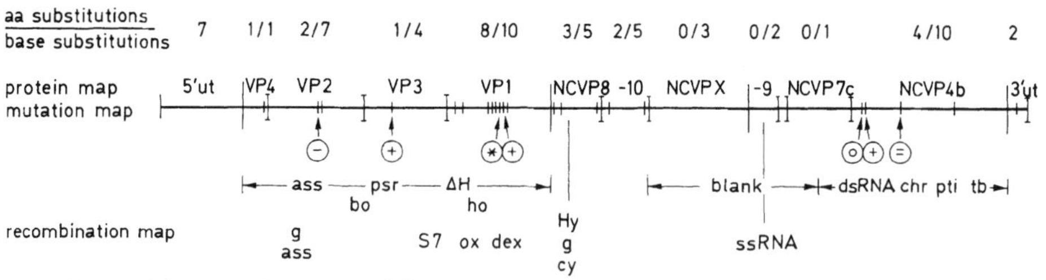

Fig. 57. The "mutation-map" of the attenuated poliovirus Sabin 1 vaccine strain LSc 2 ab
− = substitution of neutral amino acid by a negatively charged amino acid
+ = substitution of neutral amino acid by a positively charged amino acid
O = substitution of negatively charged amino acid by a neutral amino acid
= = substitution of positively charged amino acid by a negatively charged amino acid
* = substitution of proline amino acid residue
The figure illustrates the distribution of amino acid substitutions (short dashes) in the proteins of the Sabin vaccine strain of poliovirus type 1 relative to the parental wild type poliovirus type 1 Mahoney strain resulting from numerous mutations in the genome. Notice that there are no amino acid substitutions in proteins NCVPX, NCVP9, and NCVP7 c. This corresponds to the "blank" region in the recombination map of poliovirus type 1 (see Fig. 57 and Fig. 52), for which no viable mutants have been detected to date. The genetic markers of the vaccine strain are altered in ts, ΔH, and dex, they are wild type in g and ho. The dextrane sensitivity of the vaccine strain may well be related to the positively charged amino acid substitution in VP$_1$. An attempt has been made to correlate the mutation map of the vaccine strain to the recombination map. Symbols are used as defined in Table 31. — Data from Nomoto *et al.*, 1982, and Cooper, 1977

Mutants selected for a given phenotypic variation are often found to be changed in other properties as well. This may either reflect multiple mutations or the involvement of a single viral protein with several properties of the virus (pleiotropism). Mutation of a single viral protein is likely to have occurred if two or more properties also revert together (covariant reversion).

Pleiotropism has been exploited in the selection of live poliovirus vaccine strains as well as in the characterization of poliovirus isolates from patients. Some of the phenotypes which are associated with attenuated non virulent strains are indicated in Table 31 (p. 189). Using this correlation, it is far easier to select such markers in vitro than to test for neurovirulence in animals. Attenuated viruses, indeed, are usually selected first for avirulence-related markers, reserving the animal testing for final characterization.

B. Mutant Types

1. Temperature Sensitive Mutants

Often, the altered function of a protein is not expressed at physiological temperatures, but becomes evident only at elevated temperatures. Such mutants are said to be conditionally lethal, temperature sensitive (ts) mutants. Ts mutants are the only class of conditionally lethal mutants occuring in animal viruses. Ts mutants have been extremely valuable since they occur in most—possible all—genes, *i.e.* they are "universal markers".

In general, optimal growth of picornaviruses takes place within a relatively narrow range of temperature which lies between $37 \pm 2^{\circ}$ for wild type strains. Temperature sensitive mutants replicate efficiently only at a physiological temperature and replication is interrupted with a shift-up in temperature. Elevated temperatures exert general effects on cellular functions which may also complicate viral replication events. Examples are the interference with polypeptide chain initiation (disaggregation of polysomes), and enhanced activation of lysosomes with release of nucleases (Fiszman *et al.*, 1970; Adler *et al.*, 1973). In HeLa cells, for example, single stranded viral RNA not enclosed in virions is rapidly degraded late in infection, a process which is enhanced by an increase in temperature.

In addition to these general detrimental effects of elevated temperatures, high temperatures may also exert specific effects on individual mutant proteins. Correspondingly ts mutants usually are said to bear one or more specific temperature-sensitive lesions. These specific lesions probably result from a temperature induced alteration of the conformation of the mutant protein. One or more amino acid substitutions evidently can render a protein more sensitive to temperature mediated alterations or modifications of conformation. The Sabin vaccine strain LSc2ab is an illustrative example of a temperature sensitive strain. At 36° C this virus replicates normally, but the formation of both infectious virus and viral RNA is considerably reduced at 40° C (Tershak, 1969). At 38.5° C the formation of mature, infectious virus is inhibited by 90 % with only moderate reduction in the formation of viral RNA (Fiszman *et al.*, 1972). The inhibition at the restricted temperature is the result of a block in virion morphogenesis and a restricted processing of viral precursor proteins (Garfinkle and Tershak, 1971).

Many of the temperature-sensitive mutants obtained through the use of mutagenic agents revert with high frequency, others are reasonably stable. The latter mutants may show defects of various sorts at non-permissive temperatures (McCahon and Cooper, 1970; Cooper *et al.*, 1970; Shea and Plagemann, 1971). Among the properties affected in the mutant are the resistance to or dependence upon inhibitory substances during replication, the sensitivity or resistance to inactivation by a variety of agents, its surface properties reflected in altered physical or antigenic behavior, its pathogenicity—neurovirulence, and its capacity to interfere with host cell metabolism and to direct the synthesis of viral RNAs in the infected cell.

Table 31 (p. 189) summarizes the more widely studied mutant markers of poliovirus temperature sensitive mutants. Many of the markers are related to alterations in the conformation or charge of capsid proteins and thereby often affect the physical and antigenic properties of the entire virus particle. Other markers reflect the properties of poliovirus proteins required for the intracellular replicative activities of the virus, such as the RNA polymerase, maturation factors, and factors which interfere or alter the host cell metabolism. Of particular interest are the markers for the sensitivity or resistance to specific inhibitors of virus replication. It will be one of the great challenges of modern gene-technology to correlate these markers with specific sites on the poliovirus genome.

2. *Structural Markers*

Structural markers are those directly concerned with the function of the viral capsid proteins in the complete virus particle. Mutations that cause conformational alterations in capsid proteins may cause alterations of capsid stability, or in the capacity of the virus to interact with factors which recognize specific surface properties of the virus such as antibodies and the host cell receptor. Alterations in capsid stability may be detectable by an increased sensitivity of virus particles to disruption by high temperatures, urea, or other agents.

Alterations in capsid surface charge or capsid configurations required for interaction with the host cell receptor or antibodies, are reflected by different capacities to grow in certain host cells, different sensitivities of mutant viral strains to inhibition by antibody-like factors in the sera of domestic animals, or in the sensitivity to inhibition by factors present in agar, such as dextran sulfate, or special conditions such as acid pH of the culture medium. Genetic variation of poliovirus influencing the antigenic properties of the capsid surface have been discussed in detail already in Chapter 3. The correlation of structural markers expressed in poliovirus mutants to their antigenic reactivity to defined monoclonal antibodies should be an important theme for future investigations.

Several markers that are concerned with the virion surface, also appear to interfere with the adsorption of virions to the host cell receptor. Inhibitors of poliovirus growth occur in certain normal sera of domestic animals in particular in equine (ho inhibitors) and bovine sera (bo inhibitors) (Takemori *et al.*, 1957). When added in high concentrations (10—15 %) to the medium overlaying infected cultures, these inhibitory sera often reduce the size and number of plaques formed. Poliovirus strains can then be classified as inhibitor-sensitive or -resistant. The ho and bo inhibitors are rather specific, and are usually active against only one or a limited number of poliovirus strains (Pagano *et al.*, 1965). Both inhibitor classes consist of a number of different factors and presumably are derived from some form of immunoglobulins. Both, bo and ho inhibitors, bind to the virion and thereby cause a loss of infectivity. In some aspects, the actions of bo and ho inhibitors resemble those of type-specific antipoliovirus antibodies. Sensitivities to bo and ho inhibitors nevertheless are distinctive markers which usually do not covary, i.e. the sensitivity to these inhibitors seems to reside in different capsid proteins. Strains displaying resistance to bo inhibitors retain the sensitivity to ho inhibitors (Takemoto and Habel, 1959; Hirst, 1962; Pagano, 1965).

The bo marker does not display any covariation with other genetic markers of poliovirus (Ledinko, 1963; Carp, 1964; Pagano and Böttiger, 1964; Kanamitsu *et al.*, 1967). In sharp contrast, the ho marker usually covaries with the markers for lability to heat inactivation, sensitivity to inhibition by dextran sulfate (see below), and ability to multiply in MS cells. Some bo inhibitors show neutralizing activity which is distinct from that of the normal poliovirus neutralizing antibodies, but resembles it in some aspects (Takemori *et al.*, 1958). Neutralization kinetics of bo inhibitors are first order, and virus-bo inhibitor complexes dissociate at acid pH, similar to complexes between virus and type specific antibodies (Pagano *et al.*, 1965). D and C type precipitins were detected in bovine sera. In some bovine sera, a 7S inhibitor was detected which displayed virus precipitating as well as virus neutralizing activity, while a 19S (IgM) inhibitor had only neutralizing activity (Urasawa *et al.*, 1968a and b).

Ho inhibitors also bind directly to virions. The association can be dissociated by acid pH with a recovery of infectivity (Pagano, 1965; Thomssen *et al.*, 1966). Inhibitor-resistant mutants have been obtained which no longer bind to the ho inhibitors, but which remain antigenically indistinguishable from the parent virus and retain the same binding capacity for homologous 7S rabbit antibodies as the parental virus (Takemoto and Habel, 1959; Thomssen *et al.*, 1966). This indicates that the site of ho binding is distinct from that binding neutralizing type specific antibody.

It is tempting to speculate on the basis of the described characteristics of the ho and bo markers and genetic mapping experiments for these markers (see below), that the bo marker reflects alterations mainly in capsid protein VP_2 and VP_4 in analogy to one of the presumptive D-determinants (see Section VIII in Chapter 3), whereas the ho marker reflects alterations mainly in VP_1 and VP_3 which may also be involved in the recognition of the host cell receptor.

A number of factors present in agar, influence the growth capacity of certain mutant strains of poliovirus (Takemoto, 1966), presumably by interfering with adsorption of the mutants to the host cell. Spontaneously occuring mutant populations of poliovirus were observed which were distinguished by forming minute plaques (the m character). Such mutants often gave rise to revertants forming large plaques when they were propagated in the presence of agar extracts (Nomura and Takemori, 1960). Subsequently it was shown that agar contains several factors—mostly polyanionic substances—which interfere with the growth of these mutant strains of poliovirus.

Of all agar inhibitors, the mode of action of dextran sulfate has been studied most extensively. Dextran sulfate is a sulfated polysaccharide, *i.e.* a highly negatively charged polyelectrolyte. As a rule, the inhibitory effect of dextran sulfate is proportional to the content of sulfate groups and to its molecular weight, the most active compound having a molecular weight of 2 x 10^6, *i.e.* about 1/3 of the virion molecular weight (Bengtsson, 1965; Takemoto and Spicer, 1965). Wild type polioviruses grow even in the presence of dextran sulfate. Mutants which are sensitive to inhibition by dextran sulfate have been observed, as well as mutants whose growth is enhanced by dextran sulfate (Takemoto and Liebhaber, 1962; Voss, 1964).

Dextran sulfate interacts with the capsid protein of sensitive mutants thereby preventing the reversible virus adsorption and subsequent attachment and penetration (Voss, 1964; Bengtsson, 1965). The effect of dextran sulfate is much lower at low pH (Takemoto and Kirschstein, 1964). Polycations, such as DEAE dextran or protamine, counteract the inhibitory effect of dextran sulfate and other sulfated agar polysaccharides (Wallis and Melnick, 1968a, b). Sensitive poliovirus strains are more strongly adsorbed to dextran sulfate below the isoelectric point of the virus, *i.e.* when the virus is positively charged (Bengtsson *et al.*, 1964). These results indicate that the "dex character" is correlated to changes in the surface charge of the mutant virions. Indeed, poliovirus strains differing in their dex character, were shown to exhibit corresponding differences in their adsorption and elution pattern in various types of chromatographic columns (Bengtsson *et al.*, 1964).

Dextran sulfate sensitivity is a distinguishing characteristic of the LSc2ab type 1 Sabin vaccine strain, and the dex marker was therefore believed to be potentially useful as a marker for tracing attenuated vaccine strains in vaccinated areas (Pagano and Sedwick, 1966), however, spontaneous reversion of the dex-sensitive character to the original wild type dex resistant character seems to be a common event in LSc2ab populations (Bengtsson, 1965; Sergiescu *et al.*, 1967). Passage through the human intestine even increases the spontaneous reversion frequency by a factor of 10. When a dextran sulfate sensitive strain such as LSc2ab is plated under an agar overlay containing dextran sulfate, normal size plaques were detected among the minute plaques formed by the dextran sensitive particles. From the larger plaques, purified stocks of dextran resistant strains could be obtained which resembled the wild type parental dextran insensitive strain (Takemoto and Liebhaber, 1962). In addition some even larger plaques were observed (Voss, 1964). From the kinetics of induction of dextran resistant revertants from sensitive mutants, it could be inferred that the transition can be brought about by a point mutation (Klein *et al.*, 1966). The observation that dextran resistant strains can be obtained after a single exposure of sensitive mutants to dextran sulfate also suggests a single step pattern for the mutation.

Several other virion surface related markers have been shown to covary frequently with the dextran marker. Among these, the m marker for minute plaque variants reflecting the sensitivity to inhibition by different agar inhibitors, was already mentioned (Rouhandeh *et al.*, 1965). The d marker–sensitivity for inhibition by acid pH of the agar overlay–(wild type virus is resistant to acid pH) (Vogt *et al.*, 1957; Hsiung and Melnick, 1958), also seems to lie close to the dex marker (Agol and Chumakova, 1962; Takemoto and Kirschstein, 1964). However, of many attenuated virus strains of all three poliovirus types, all of which have the d marker and are acid sensitive, only LSc2ab strain of poliovirus type 1 is also dex sensitive. All others are dextran resistant, and no poliovirus type 3 strains sensitive to dextran sulfate have ever been found (Bengtsson, 1965), indicating that the d and dex markers indeed are different (Wallis and Melnick, 1968a,b).

The dex marker was also shown to vary independently of other structural markers, such as heat sensitivity or capacity to grow in MS cells (Sergiescu *et al.*, 1967), bo and ho inhibitors (Pagano, 1965; Bengtsson, 1968), and to other inhibitor characters such as sensitivities to guanidine, HBB (Sergiescu *et al.*, 1969), and gliotoxin (Sergiescu and Aubert-Combiescu, 1969). In studies on the recombination of the dex marker with ho and guanidine markers, it was shown that there probably exist more than one site for dextran sensitivity, and that these seem to lie close to the site for the ho inhibitor (Bengtsson, 1968). On the basis of observations described above, it is tempting to speculate that the dex character or LSc2ab corresponds to its altered VP_1 capsid polypeptide, and in particular to the two charge alterations by the amino acid substitutions near the center of VP_1 (see Fig. 57 above).

3. *Non Structural Markers*

Poliovirus strains may differ markedly in their capacity to alter certain host cell functions, such as the inhibition of host cell protein, RNA and DNA synthesis, and the modification of intracellular membranes, or in the capacity to efficiently replicate their own RNA. Poliovirus strains may also differ markedly in their sensitivity to inhibitors, which act at stages of virus replication subsequent to attachment and penetration, during the peak of viral replicative events.

The capacity to inhibit host cell protein synthesis—the psr marker—correlates with structural markers, *i.e.* it often covaries and recombines with structural markers. In contrast, the capacity to inhibit host cell DNA synthesis, the pti marker, and the capacity to cause membrane leakiness late in infection, the tb marker, always covary with markers of the RNA polymerase, in particular the capacity to synthesize dsRNA, the dsRNA marker. Mutants defective in RNA synthesis usually are still effective in inducing shut-off of host cell protein synthesis (Cooper, 1977; Hewlett *et al.*, 1982). Inhibition of host cell DNA synthesis and alteration of host cell chromatin structure, expressed in the chr marker, are events which occur at relatively early stages in virus replication and may directly reflect the activity of a functioning replicase protein. Induction of membrane leakiness, in contrast, is a relatively late event, which occurs after virus maturation, and may thus be an indirect effect of the replicase activity, being caused for example by the accumulation of dsRNA (also a late event), which depends on a functioning polymerase molecule.

C. Genetic Recombination

The ultimate goal of genetic studies is twofold: The total mapping of a genome and the assignment of specific functions to the individual regions of the genome. The classical methods of genetic analysis are complementation and recombination upon mixed infection with two temperature sensitive mutants which are defective in different functions. During mixed infection at the elevated temperature –*i.e.* under conditions where neither virus can replicate by itself–the functioning product of one virus should be able to substitute for the defective product of the other (complementation).

Generally, complementation permits the delineation and mapping of genes or cistrons without involving an actual change in the genetic structure of the participating viruses. Ideally, mutants with defects in the same gene will not complement, and thus can be classified into groups with corresponding defects in the same gene or cistron. On the basis of complementation, mutant types should be classifiable into different functional groups. Complementation in picornaviruses, however, is very inefficient and has not been of much value for the study of poliovirus genetics (Cooper, 1965, 1977; Ghendon, 1966). The reason for the general inefficiency of complementation among picornaviruses is not clear. Utilization by one of a pair of infecting viruses of a product produced by the other, however, can occur under certain circumstances. Use of heterologous capsid proteins takes place efficiently during mixed infection with two different poliovirus serotypes. This leads to the production of phenotypically mixed progeny particles exhibiting antigenic properties of both parental viruses, *i.e.* particles that are doubly neutralizable (Ledinko and Hirst, 1961). Guanidine resistant and guanidine sensitive strains are also able to provide products which may be utilized by the defective virus upon mixed infection in the absence or presence of guanidine (Agol and Shirman, 1965). In each case, the "rescued" virus is phenotypically masked by the heterologous viral coat (Wecker and Lederhilger, 1964a; Cords and Holland, 1964; Holland and Cords, 1964; Agol and Shirman, 1965). Rescue takes place, however, only under conditions which permit the replication of one of the strains (Ikegami *et al.*, 1964).

All of the above are examples of nongenetic interactions between mixedly infecting viruses. Truly genetic interactions between temperature sensitive mutants of poliovirus–by crossing over or gene conversion–may also lead to the formation of progeny virions with characters from both parents (recombination). Recombination during mixed infection by viruses with mutations in different loci may produce novel progeny virions with characteristics of both parents. In general, novel combinations of genetic material may arise either by reassortment of larger linkage groups, such as between different chromosomes or different fragments of RNA in the segmented RNA viruses, or by reassortment within a single linkage group, such as in single chromosomes or single stranded RNA molecules as in poliovirus. In the latter case of molecular recombination, nucleic acids are covalently broken and rejoined by enzymes.

Genetic recombination between two mutants can be inferred from the observation of progeny double mutants after mixed infection in significant excess over the value of spontaneous double mutations. Recombination between poliovirus strains was first demonstrated with the use of the three noncovariant markers ho, bo, and g (Hirst, 1962; Ledinko, 1963). An increase in double mutants (to 0.4 %) which exceeded by 20 times the value of spontaneous double mutants was observed. It was shown that the three markers studied are equally spaced in the sequence bo-ho-g by appropriate crosses. Since no recombinants displaying intermediate degrees of guanidine resistance were observed, it seems that the loci involved in guanidine resistance are very closely linked. Subsequently, the markers for thermolability (Δ H) and resistance to inhibition by dextran sulfate were shown to lie closer to ho than to g. Variable recombination frequencies obtained were taken as an indication for occurrence of several different sites of the dex marker.

Isolation of temperature sensitive mutants and extremely careful control of the experimental procedures finally permitted quantitative mapping of many ts mutants (Cooper, 1968, 1969, 1977). The properties of the so derived genetic map gave confidence to the notion of true recombination. Recently, direct biochemical evidence for recombination between two types of poliovirus was presented (Romanova *et al.*, 1980; Tolskaya *et al.*, 1983).

The recombination frequencies per codon of poliovirus markers are not very different from those of conventional DNA systems. Due to the smallness in size of the poliovirus genome, however, the absolute values of recombination frequencies are very small. In poliovirus, the total recombination frequency (assuming reciprocal crossovers) between the most distant markers averages 2.2 %, corresponding to 1 % per 1250 nucleotide pairs of the double stranded form (Cooper, 1975).

A summary of the genetic map deduced from genetic recombination studies is presented in Figure 58. The genetic map in Figure 58 is correlated to the known coding sequence for poliovirus proteins as determined from the consensus nucleotide sequence of poliovirus type 1 (Fig. 44), and amino acid sequence analyses of its proteins. Since it is not exactly known what portion of the poliovirus genome is actually encompassed between the most distant mutants of the genetic map, the maps are not precisely colinear. Additional uncertainty can arise from the fact that some of the mapped mutants may contain more than one altered phenotype. Nevertheless, an interesting pattern emerges.

An important observation is the localization of the guanidine locus. In contrast to many reports in the literature (Cooper *et al.*, 1970), the guanidine locus does not seem to lie within the coding region for the capsid proteins, but near the N-terminal region of the P-2 domain, most probably in the region coding for NCVP8. On the basis of a tryptic peptide analysis of a large poliovirus precursor protein which apparently did not contain the NCVP-X peptides, but which contained peptides of the capsid and replicase proteins (Abraham and Cooper, 1973), Cooper placed the capsid protein coding region directly next to that coding for the replicase proteins in the interpretation of his genetic map (Cooper, 1977). With more recent evidence, however, it has become clear, that the central portion of the genome is occupied by the P-2 proteins NCVP8, - 10, and -X. The alingment of Cooper's genetic map with the processing map from the consensus sequence places the guanidine locus outside the region of the capsid proteins, near the N-terminal region of the P-2 domain, most probably in the region coding for NCVP8. PAGE studies of the poliovirus proteins synthesized in the presence of guanidine by guanidine sensitive strains of poliovirus revealed altered migration

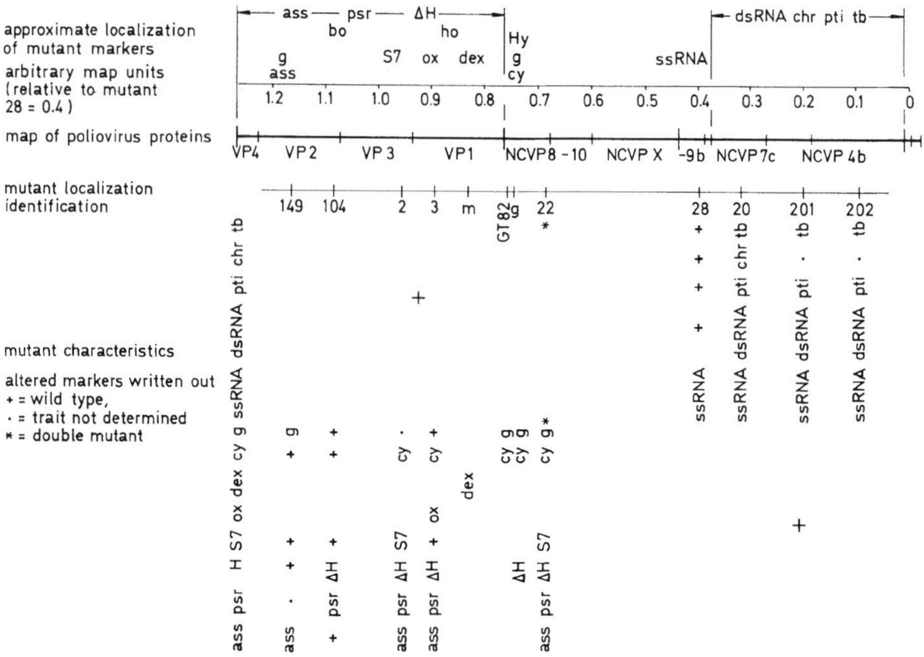

Fig. 58. The genetic recombination map of poliovirus

The figure shows a tentative correlation of the genetic recombination map of poliovirus to the map deduced from the consensus nucleotide sequence (Fig. 44). The symbols for the mutant markers are as in Table 31. Data compiled from Cooper *et al.*, 1975, and Cooper, 1977. Note that the orientation of the recombination map has been turned around to permit a correlation to the maps written in the more conventional 5′—3′ direction

behavior of NCVP-X and its precursors (Koch et al., 1980). Localization of the guanidine locus outside the region coding for capsid proteins has been confirmed in biochemical characterization of the recombination from a mixed infection of polioviruses type 1 and type 3: The recombinant inherited the entire capsid protein coding region from its type 3 parent, whereas the nonstructural protein coding region, including the locus of guanidine sensitivity, was inherited from the type 1 parent (Romanova *et al.*, 1980; Tolskaya *et al.*, 1983). Similar results have been obtained for aphtoviruses (King *et al.*, 1982). Electrofocusing studies of the polypeptides induced by guanidine resistant mutants of FMDV revealed alterations only in the P2 protein p34 (corresponding to poliovirus NCVP-X) and its precursor p52 and the unmapped p34 related protein p20c (Sanders and King, 1982).

The cysteine marker (cy), which usually covaries with guanidine sensitivity also appears to map within the P-2 coding region. Certain strains of poliovirus are dependent on the presence of cysteine for efficient replication in the infected cell. The highest concentration of cysteine residues occurs in the N-terminal protein of the P-2 domain, NCVP8, and in NCVPX (see Table 29, p. 174). It is possible that one of these cysteine residues is responsible for the cy marker. The importance of these P-2 proteins for poliovirus replication is evidenced also in the high conservation of this sequence during attenuation of wild type viruses. No amino acid substitutions have occurred in this region during the attenuation of poliovirus type 1 (see Fig. 57, p. 190), and no mutants have yet been described for NCVPX in poliovirus.

The loci of the marker for dextran sensitivity seems to lie in the coding region for VP_1. VP_1 indeed reveals substantial surface exposure in intact virion (see Chapter 3), and part of VP_1 may be important for interaction with the host cell receptor (see Chapter 7).

Another interesting observation comes from mutant 28 in the coding region of NCVP9, the VPg containing cleavage product of NCVP1b. This mutant does not synthesize any single stranded RNA at the restricted temperature, whereas it does synthesize double stranded RNA. Synthesis of double stranded RNA by this mutant was originally interpreted as evidence for two separate replicase activities, one responsible for the synthesis of double stranded RNA (replicase II), the other for the synthesis of single stranded RNA (replicase I) (Cooper *et al.*, 1970). On the basis of additional information available today, it is more likely that the mutant property of ts 28 rather reflects the lack of a requirement for NCVP9 and associated factors for the initiation of cRNA synthesis on single stranded vRNA, and efficient elongation of RNA synthesis by the intact NCVP4 replicase, yielding dsRNA. Initiation of plus strand RNA synthesis, yielding single stranded RNA of virion RNA polarity, on the other hand seems to be a more complicated event, requiring additional host cell factors and a functioning membrane associated NCVP9 (see also Chapter 9).

D. Mechanism of Recombination

Little is known about the mechanisms of recombination between the progeny RNAs of two mixedly infecting polioviruses. Additivity of recombination frequencies requires the equal and reciprocal involvement of both parental strands. Recombination between polioviruses appears to be an early event: 68 % of the recombinational events have taken place by the time in which only 25 % of the total RNA has been synthesized (Cooper, 1968). Some increase in the proportion of recombinants still takes place during the rest of the replication cycle (Ledinko, 1963). Interestingly, the presence of guanidine early in the replication cycle—which increases cRNA content and the proportion of double and multistranded forms of RNA—approximately doubles the recombination frequency (Cooper, 1977).

Recombination thus takes place at a time during the phase of exponential RNA synthesis, *i.e.* at a time when RNA templates occur in a soluble—not membrane associated—state in which they are relatively accessible to newly formed replicase molecules and other relevant enzymes.

With the switch to linear kinetics of RNA synthesis and increasing membrane association of RNA synthesis, the frequency of recombinational events decreases.

This may simply reflect a decrease in the accessibility of the RNA template to enzymes required for recombination, or reflect the more stringently organized mechanisms of plus strand RNA synthesis.

Two different types of mechanisms could in principle account for the appearance of true recombinant progeny viruses. A precisely ordered molecular exchange or crossing over similar to that of DNA systems (Hotchkiss, 1971) is suggested from the self-consistent nature of the genetic map determined from recombination frequencies. Suitable host enzymes have in part been identified. For example, the presence of an RNA ligase has been described in poliovirus infected HeLa cells (Yin, 1977b). On the other hand, reassortment of nascent chains plus replicase between two mutant templates, followed by precise realingment and continued replicase action on the new template could also lead to the formation of recombinants, without the requirement for specific enzyme systems.

With the application of techniques of modern biotechnology, such as gene surgery and defined genetic modification of the poliovirus genome, and monoclonal antibody studies, it should be possible within the next decade to provide answers to the still open questions regarding the precise functions of the poliovirus proteins, the colinearity of biochemical and genetic maps, the molecular basis of the mutant traits, and mechanisms of recombination.

VII. Summary

The genome of poliovirus is a single stranded molecule of RNA of mRNA polarity, some 7.500 nucleotides in length with a molecular weight of 2.6×10^6. Covalently attached to the 5' end of the virion RNA is a small, 22 amino acid long, basic protein, VPg. The 3' end is polyadenylated. 740 nucleotides at the 5'end and 73 nucleotides at the 3'end are not translated. These untranslated regions are highly conserved in evolution and probably play important roles in the initiation of protein synthesis and viral RNA replication. Approximately 10 stable gene products are encoded by the remaining 6,700 nucleotides. The entire coding sequence is translated into a 2,200 amino acid long polyprotein with a molecular weight of 250,000, which under normal conditions is co- and posttranslationally processed into the stable products. The initiating AUG at position 743 is the ninth AUG from the 5' end. Coding for all poliovirus proteins falls into three major domains on the poliovirus genome which are distinguished on structural and functional bases.

The P-1 domain codes for the capsid proteins VP_4, VP_2, VP_3, and VP_1. In addition to their role as protecting agents for the RNA, the capsid proteins have been implicated in the shut off of host protein synthesis, with a protein-kinase activity associated with intact virions, and as regulators of orderly sequential translation and replication of the parental virion RNA.

The P-2 domain codes for three proteins of ill-defined function: NCVP8, 10, and NCVPX. The loci for guanidine sensivity and cysteine dependence seem to reside in the P-2 domain. The P-2 proteins may be involved in the induction of the synthesis of virus specific membranes required for replication and encapsidation of RNA. The P-2 proteins may directly participate in and coordinate the synthesis and encapsidation of single stranded progeny plus strand RNA. The importance of the P-2 proteins is underlined by the striking conservation of sequence in their coding regions, as reflected also in the identity of the amino acid sequence of NCVP-X in the attenuated Sabin vaccine strain and in the corresponding wildtype strain, and in the very low incidence of viable mutants containing alterations in the P-2 domain.

The P-3 domain encodes the VPg containing membrane protein NCVP9a, the major virus specified gln/gly protease NCVP7c and the viral RNA polymerase NCVP4. An alternate cleavage pathway preferred by some strains of poliovirus yields (protease?) NCVP6a and (replicase?) NCVP6b.

The basic pattern of the genomic map of poliovirus proteins was elucidated with the help of biochemical and genetic methods. The pattern was confirmed and extended with the deciphering of the entire nucleotide sequence of the poliovirus genome.

Polioviruses adapt rapidly to environmental conditions with modifications of their genetic information. The naturally occurring variants of wild type polioviruses have been displaced to a large extent by vaccine viruses, which have been introduced artificially into the environment in large vaccination programs since the early sixties. The variants of poliovirus can be characterized, and their genetic divergence can be traced by a variety of immunological procedures, as well as by oligonucleotide mapping of their RNAs, polyacrylamide gel electrophoresis and tryptic peptide mapping of their proteins, and lastly by sequence analyses of their genomes.

Many mutants of poliovirus have been induced and selected in laboratories throughout the world during the past thirty years. Many of these mutants have been of use in defining functions of the poliovirus proteins, in the development of attenuated vaccine strains, and in the elucidation of the replicative strategy of polioviruses in the infected host cell. With the application of modern recombinant DNA technologies to the poliovirus genome, these early classical studies will be complemented in the coming years. The greatest challenges of modern genetics will be to correlate the traditional genetic markers with specific sites on the poliovirus genome, to define the functions of the viral proteins with the aid of constructed mutants, and to elucidate the molecular basis of neurovirulence and attenuation.

Part II

The Replication of Poliovirus

5
Introduction

I. The "Life-Cycle" of Poliovirus

Outside of and apart from their host cells poliovirions behave like stable inert ribonucleoprotein particles. They appear as if there is no metabolism, none of the dynamic activities that are associated with and are often taken to define living matter. Poliovirus particles can be crystallized and may remain in this stable state almost indefinitely. Only when a virus particle encounters a susceptible host cell, does it "come to life". With a successful virus-host cell encounter, a dramatic reorganization of the host cell activities begins. Within minutes the virus "takes over command" of the vital host cell metabolic pathways.

Within a few hours thousands of viral particles are synthesized and assembled. The release of the viral progeny—200.000 particles per infected cell—back into the environment is usually associated with or followed by the disruption of the host cell membrane and cell death. The freshly formed virus particles again behave as dead, inert matter, and will continue to do so until they encounter a susceptible host cell, or until they are inactivated or destroyed by some other environmental influences (antibodies, temperature, radiation, heat, etc.). The "life-cycle" of a poliovirus thus involves first a phase of inertness and "inateness" of variable time, the exact extent of which is determined by the timepoint of the chance encounter with a susceptible host cell. Here the second "living" phase may begin resulting in the "disappearance" of the infecting particle itself for the sake of the production of a great number of progeny viruses, and at the cost, often, of host cell life. When one considers the smallness and the simplicity of the virus particle in contrast to the complexity and sophistication of the host cell—the polio genome at a molecular mass of 3×10^6 is a mere one millionth of the size of the host cell genome at 3×10^{12}—, it is fascinating that a single virus particle can interfere with and reorganize such a gigantic machinery to serve its own interests.

The aim of the infecting viruses is to use the metabolic machinery of the cell for its own replication. The strategy of the poliovirus must be an ingenious one, and must involve a number of intriguing but very efficient regulatory mecha-

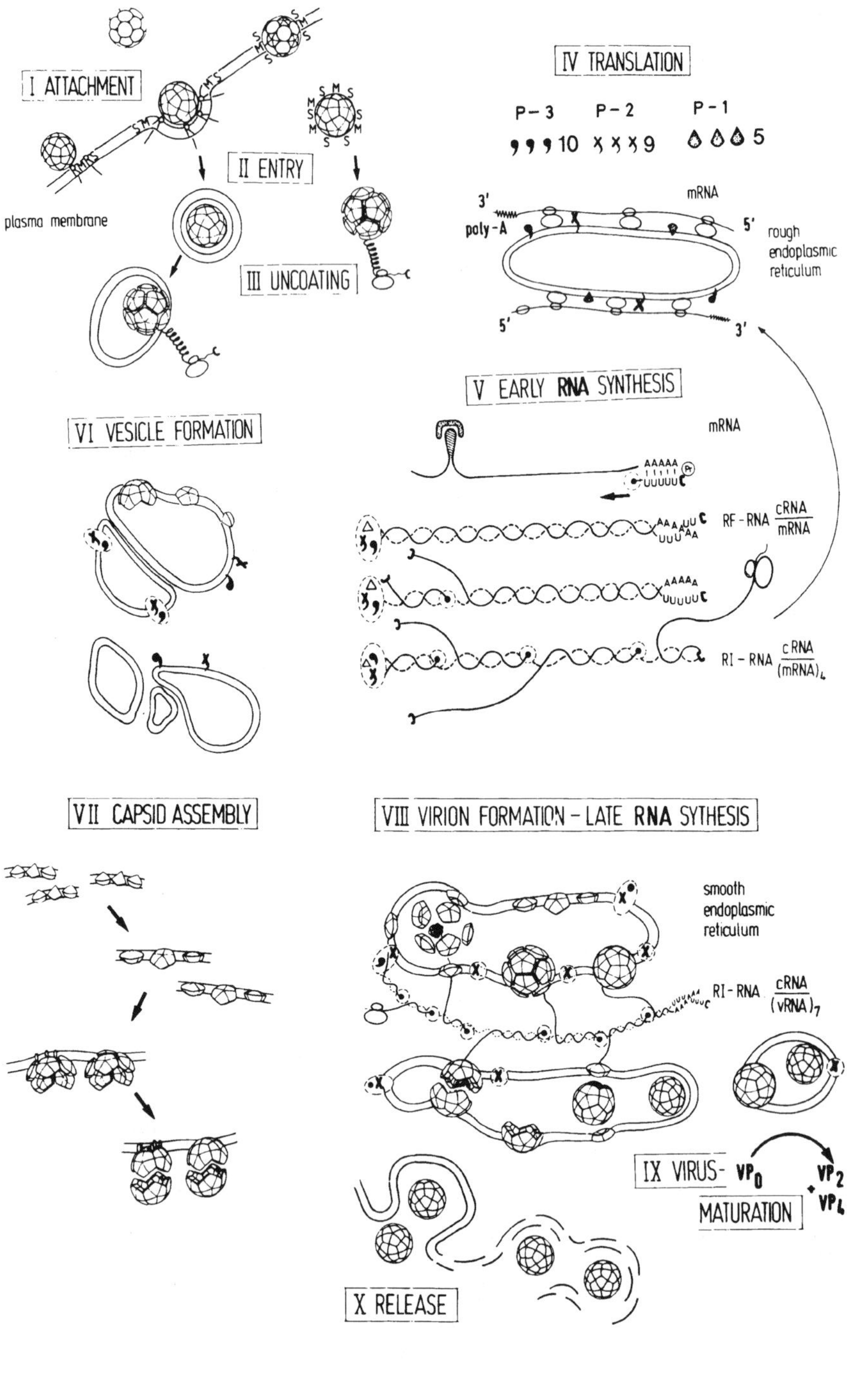
I ATTACHMENT
plasma membrane
II ENTRY
III UNCOATING
IV TRANSLATION
P-3
P-2
P-1
10
9
5
3'
poly-A
mRNA
5'
rough endoplasmic reticulum
5'
3'
V EARLY RNA SYNTHESIS
mRNA
AAAAA
UUUUU
RF-RNA cRNA/mRNA
RI-RNA cRNA/(mRNA)4
VI VESICLE FORMATION
VII CAPSID ASSEMBLY
VIII VIRION FORMATION - LATE RNA SYTHESIS
smooth endoplasmic reticulum
RI-RNA cRNA/(vRNA)7
IX VIRUS- MATURATION
VP0
VP2 + VP4
X RELEASE

nisms. Some of these have been unraveled in the course of many examinations and experiments on poliovirus infected host cell tissue culture systems over the past 30 years, while other aspects of the viral genome strategy still remain a mystery.

In the following sections we will present an overview of the virus induced reorganization of host cell structure and function and briefly describe a typical poliovirus host cell in tissue culture—the HeLa cell—with respect to the structural and functional components that are of relevance for poliovirus replication. In the following chapters, then, we shall attempt to summarize the insights achieved, and the questions remaining concerning the individual steps of the replication cycle of poliovirus.

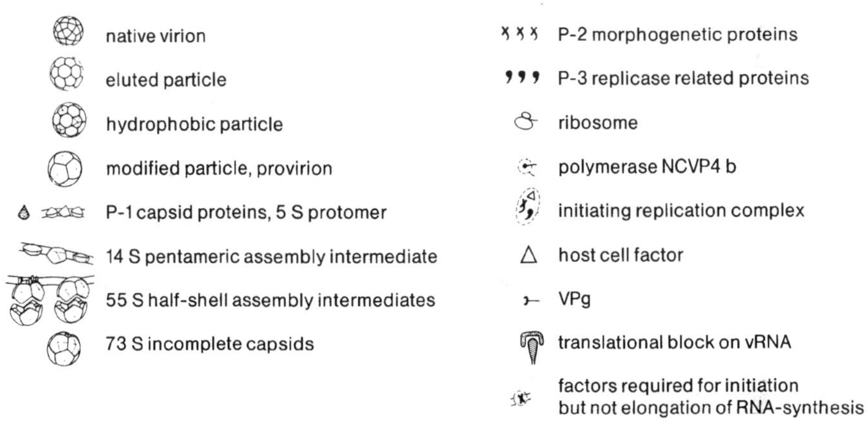

Fig. 59. Schematic overview of poliovirus replication

The interaction of the virion with membrane receptors—attachment (I)—is followed by entry (II) and uncoating (III) of the virion. Attachment and penetration of virus induce changes in conformation and antigenicity of the virus particle. Polioviral RNA is infectious without its coat protein because the RNA can be translated by host ribosomes to synthesize polymerase needed for RNA replication. The free virion RNA directs the synthesis of three distinct units of viral proteins which are formed by proteolytic cleavage of a larger precursor (IV): P-1 = the capsid proteins, P-2 = membrane interacting proteins, and P-3 = protease and polymerase

The parental +RNA also serves as template for the synthesis of −RNA (RF–RNA) that, in turn, will direct the synthesis of more +RNA (RI–RNA) (V). This newly synthesized +RNA will be used in the infected cell in three different ways: early in infection 1) as template to direct the synthesis of −RNA (V), and 2) as mRNA to direct the synthesis of viral proteins (IV), and late in infection 3) it will become encapsidated as vRNA (attached to the genomic protein VPg) into new virions (VIII)

Cellular membranes play a central role in poliovirus replication. Viral protein synthesis is associated with membrane bound polysomes on the rough endoplasmic reticulum (IV). Some of the virus proteins may become inserted cotranslationally into the membranes of the rough endoplasmic reticulum. Capsid assembly and vRNA synthesis are intimately associated with the smooth endoplasmic reticulum of newly formed virus specific vesicles (VI–VIII)

Cleavage of the capsid protein VP_O induces a final change in virion conformation and antigenicity and completes maturation of the progeny virions (IX), which are then released from the cell by exocytosis or upon lysis of the host cell (X)

II. Timecourse of Poliovirus Replication

The important stages and events during poliovirus replication are summarized in Figure 59 and Table 32. Poliovirus can interact only with susceptible host cells, *i.e.* with cells which carry specific receptor sites complimentary to certain structural sites of the viral surface. Virus-host cell receptor binding and interaction constitutes the first step in virus replication. This process, termed *adsorption*, triggers a series of events in both reaction partners: it brings about dramatic conformational shifts of the virus capsid as well as certain alterations of host cell membrane functions, including capping of virus receptor complexes, decrease in membrane fluidity, increase in permeability for small molecules and activation of a negative pleiotropic response. The next steps involve the *entry* of the viral genome and at least part of its protein component into the host cell, which is accompanied or followed by the release of the viral RNA from the virus structural proteins, the *uncoating* process. These early steps in the replication cycle are difficult to study experimentally because infection often involves only one or a few viral particles on a "background" of more than a hundredfold excess of virus particles which do not participate in the replication process. It is difficult if not impossible to follow by present day biochemical means the few viruses which are responsible for viral replication.

The early phase of the replication cycle *i.e.* the time between penetration or the apparent "disappearance" of the infecting virus and the first detectable appearance of progeny virions has been termed the *eclipse*. In this period, the metabolic machinery of the host cell is reorganized and redirected almost entirely to the massive production of progeny virions. The degree of reorganization varies from cell to cell and is influenced by the nutritional state of the cell. The biochemically detectable alterations are accompanied by marked morphological alterations of the host cell. Table 32 correlates the poliovirus induced biochemical and morphological alterations of host cell functions with respect to the time—course of poliovirus replication. Within the first two hours after the onset of poliovirus replication, the synthesis of host cell protein, RNA, and DNA are drastically reduced, a phenomenon called the *shut-off*. The only cellular activity that is stimulated dramatically in the course of poliovirus replication is the synthesis of intracellular membranes. Many of the virus specific metabolic activities and maturation processes occur in close association to virus induced membranes. In addition, the cytoskeleton is rearranged and changes in the intracellular ionic environment are induced. It seems that the construction of specialized compartments and membrane systems are prerequisites for the synthesis of virus components and for the assembly of virions.

The incoming poliovirus particle itself contains no replicase activity, and the host nucleic acid polymerase systems are incapable of copying the viral genome. Therefore, translation of the viral genome (or at least translation of the viral replicase coding region) must take place to provide the viral replicase. The incoming virus genome is then transcribed into a complementary RNA which in turn serves as template for the synthesis of viral messages, and later also for new viral genomes. These early events must involve delicate regulatory mechanisms, which

Table 32. *Correlation of morphological and biochemical alterations in poliovirus infected cells*

Time in hours post infection	Microscopic observations	Biochemical events
0–1	intact virus particles in the cytoplasm (parental virions) directly below the plasma membrane and in micropinocytotic vesicles	transient increase in amino acid uptake, elevated Na^+/K^+ pump activity, decrease in membrane fluidity, parental capsid proteins and RNA in polysomes and lysosomes
1–2	decrease in cell size, distortion of nucleus, wrinkling of nuclear membrane, chromosome condensation	inhibition of host protein and RNA synthesis, release of host mRNA from cytoskeleton, beginning of viral protein synthesis, peak activity of Na^+/K^+ pump, exponential phase of RNA synthesis (mainly cRNA and mRNA production)
2–2.5	membrane proliferation, appearance of large clusters of membrane bound polysomes in cytoplasmic periphery, structural rearrangement of the cell cytoskeleton	increase in choline incorporation, host cell mRNA inactive, viral protein synthesis and all newly formed viral proteins membrane associated, inactivation of NA^+/K^+ pump, increase in intracellular Na^+, decrease in intracellular K^+
2.5–3	several foci of vesicle formation	switch to linear phase of viral RNA synthesis: predominantly membrane associated vRNA synthesis, peak of RNA synthesis
3–4	perinuclear conglomeration of membrane enclosed vesicles into one large mass, nuclear extrusions, appearance of progeny virions	peak of virion assembly, increased membrane permeability for monovalent cations, declining rates of protein- and RNA synthesis
4–6	autophagic vacuoles, redistribution of lysosomal enzymes over mass of membrane bound vesicles, release of progeny virions	accumulation of ds RNA, depletion of metabolic precursors, release of lysosomal enzymes
6–8	viral crystals in cell periphery lysis of host cell	

again are difficult to study experimentally, for instance, the regulation of the head-on traffic along the parental viral RNA: During translation the ribosomes travel continuously down the viral genome in the 5' to 3' direction; during the synthesis of complementary RNA, on the other hand, the replication complexes travel along the viral genome in the 3' to 5' direction.

The viral replication phase thus commences with the binding of the viral RNA to host ribosomes. Viral proteins are synthezised by the repeated formation of polypeptide initiation complexes on probably only one initiation site near the 5' end of the RNA. Most viral protein synthesis occurs on membrane bound polysomes (rough endoplasmic reticulum). Figure 60A and B show the rates of overall protein and RNA synthesis in poliovirus infected cells. In Figure 60C, the formation of virus specific protein, RNA and infectious virions is plotted, as percent of maximal yield. This allows a better comparison of the relative kinetics of synthesis and assembly of the virion components. *Synthesis of viral proteins* becomes detectable at about 1.5 hours post infection, from then on the rate of viral protein synthesis proceeds at an ever increasing rate up to 2.5 hours, at which time it levels off and continues at a constant rate for another 1/2–1 hours (Fig. 60A). Until the end of the replication cycle and host death, protein synthesis again steadily declines.

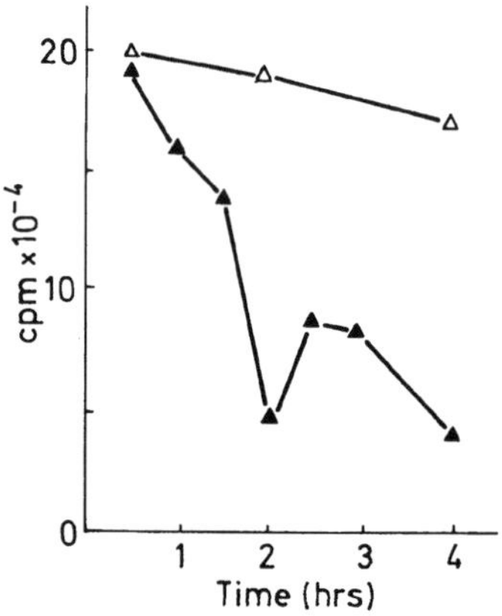

Fig. 60 A–C. The formation of poliovirus-specific protein and RNA in infected HeLa cells

Fig. 60 A. Uninfected and infected HeLa cells were pulse labeled 15 min at 37 °C with (^{35}S) methionine at the times indicated. (▲) Poliovirus infected cells, (△) mock infected cells

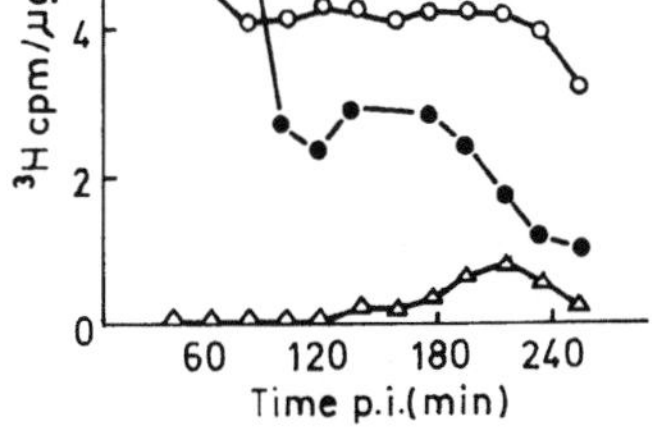

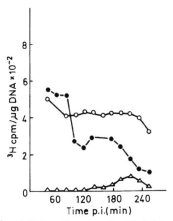

Fig. 60 B. Uninfected and infected HeLa cells were pulse labeled 5 min at 37 °C with (^{3}H) uridine. (●) Poliovirus infected cells, (△) poliovirus infected cells in the presence of actinomycin D to block host cell specific RNA synthesis, (O) mock-infected cells. – A) from Koch *et al.*, 1982 b, in: Protein Biosynthesis in Eukaryotes, p. 341, 1982; B) from Flores-Otero *et al.*, 1982 [Virology *116*, 621 (1982)]

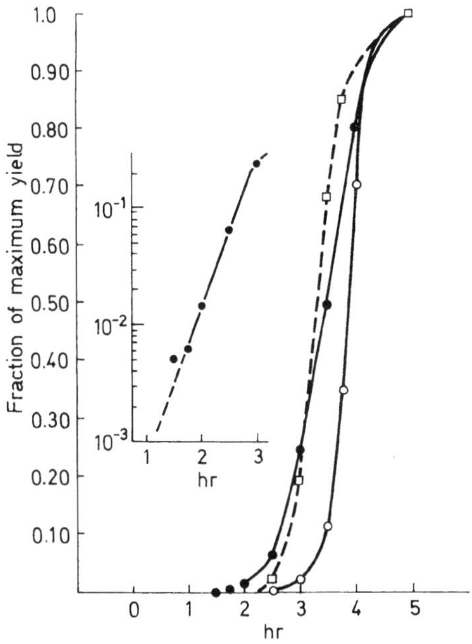

Fig. 60 C. This figure shows the time course of appearance in poliovirus-infected HeLa cells of infectious virus (○); viral protein, measured as the sum of detectable soluble and particulate antigens (□); and viral RNA, measured as uridine incorporated into RNA in the presence of actinomycin D (●). The initial portion of the RNA curve is replotted on semilogarithmic coordinates in the insert. — From Levintow, 1974, in: Comprehensive Virology, Vol. II, 109—169, 1974; compiled from Data from Levintow *et al.*, 1963, Scharff *et al.*, 1963, 1964; and Baltimore *et al.*, 1966

Virus specific proteins are required for the synthesis of RNA, complementary to the viral RNA, and of new viral progeny RNA. Viral RNA synthesis takes place on smooth membranes. *Viral RNA synthesis* is first detected with biochemical methods between 1 and 2 hours post infection, shortly before the onset of detectable viral protein synthesis and reaches a maximum approximately at 3 hours post infection, usually about 1/2 hour after the peak of viral protein synthesis. Early RNA synthesis proceeds with exponential kinetics, later it continues linearly until about 4 hours post infection. Thereafter a considerable decline in the rate is observed (Fig. 60B). This reduction may reflect the reduced synthesis of virus-specific proteins, energy depletion, or the release of nucleases from disrupted lysosomes. The bulk of viral RNA synthesized early during infection is used as mRNA; later on most of the synthesized RNA is genomic RNA that becomes encapsidated into progeny virions.

The production of poliovirus proteins involves a number of specific proteolytic *processing* steps from larger precursor molecules to final functional products. Processing involves distinct enzymes (enzyme systems) of host and viral origin. Processing begins already during synthesis on nascent chains, and continues posttranslationally when the primary translation products are cleaved further. Cleavages can be characterized as formative cleavages (processing to functional peptides) or morphogenetic cleavages (cleavage of coat proteins during assembly).

Newly synthesized viral RNA and newly synthesized viral proteins specifically interact in the last step of virus replication, virus maturation. Virus maturation can be detected by the apearance of virus specific antigens or of infectious units (Fig. 60C). The first new virus particles that can be detected intracellularly with the electron microscope appear at 3 hours post infection. At high multiplicities of infection the first infectious progeny virions are completed already 2 hours p. i. The orderly, sequential *assembly* of the viral capsid components usually leads to the encapsidation of the genomic RNA and culminates in a final rearrangement of the viral capsid into its stable native configuration. The process of maturation occurs on smooth membranes in close association to viral RNA synthesis. Assembly may occur in a manner comparable to that of several phages whose nucleic acids enter a preformed protein shell, or the coat proteins may aggregate around a condensed viral RNA. Assembly is rendered irreversible by the cleavage of the viral precursor protein VP_0 into two viral coat proteins (VP_4 and VP_2).

In the last stages of the replication cycle many more completed viral particles accumulate in the cytoplasma of the host cell until finally—usually at 6 to 8 hours post infection—up to several thousand infectious progeny viruses are *released* from the infected cell. The cellular membrane becomes leaky or disrupts and the cell dies. Intracellular crystals of poliovirus develop occasionally, usually as very late events after considerable disintegration of host cell constituents.

The kinetics of poliovirus replication may vary considerably within a population of infected cells. The overall kinetics depend on the multiplicity of infection, the type of host cell, the stage in the cell cycle, and the nutritional state of the host cell. The relative rates of synthesis and assembly of virion components as shown in Figure 60C, however, are quite consistent findings in different poliovirus—host cell systems. Note that the metabolically most active phase of poliovirus replication lasts only two hours, whereas the entire replication cycle—from attachment of the parental virion to the release of progeny virions—may last four times as long.

The effects of a number of *drugs and environmental influences* on host cell and viral functions during poliovirus infection have been studied. Some of these have shed light on host cell regulatory mechanism as well as on the strategy of the viral genome and will therefore be discussed in the respective chapters. Interference with virus replication by drugs is also of interest with respect to a possible medical application in the prevention and treatment of polio or other viral infections.

III. The Host Cell

Virus replication depends on certain basic functions of the host cell. It is, therefore, not possible to describe and understand virus specific events without an insight into these functions of the host cell. We will focus our attention mainly on features of the cell which directly participate in virus replication. Figure 61 shows an electronmicrograph of an uninfected HeLa cell.

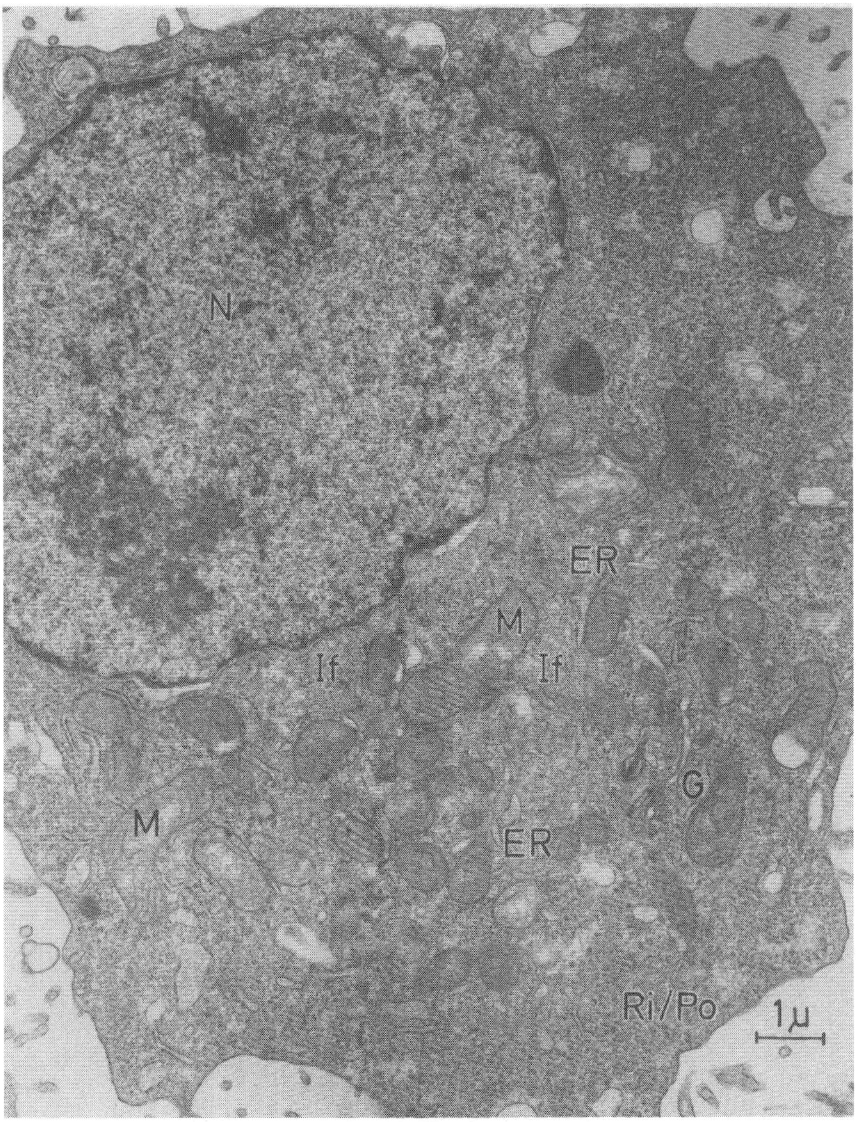

Fig. 61. Electron micrograph of a typical HeLa suspension-culture cell
N nucleus; *M* mitochondria; *If* intermediate filaments; *ER* endoplasmic reticulum; *G* Golgi apparatus; *Ri/Po* ribosomes in polysomes. – Figure from Lenk and Penman, 1979 [Cell *16*, 290 (1979)]

A. Constituents of the Cell

Table 33 lists the main constituents of a HeLa cell. Two thirds of the dry weight of the cell is protein, 10% lipid and 7% RNA. One cell contains approximately 10^{10} proteins (asssuming an average molecular weight of 35 K) and up to 10^7 ribosomes (see Table 34, p. 222). The genetic complexity of the DNA of the host cell is almost 10^6 times larger than that of the polio-RNA. 72% of the wet weight of the cell is water. Although most of the constituents of the cell are water soluble, not all are in solution. Instead they may be associated in a gel like state or they may be bound tightly to other componentes. Therefore, the contribution of the soluble constituents, in particular the ions, to the intracellular osmolarity may be less than is indicated by their intracellular concentrations. The HeLa cell cycle usually lasts about 20 hours. All constituents of the cell are at least duplicated once during one growth cycle; components which show a high turnover rate like mRNA or hnRNA are synthesized several times during the time required for one cell division. Table 35 (see p. 223, below) compares the maximal rates of macromolecular synthesis in uninfected and infected cells.

Table 33. *Features of the HeLa host cell*

A. Constituents of an uninfected HeLa cell

	Total mass per cell	Percent wet weight	Percent dry weight
H_2O	1500 picograms[a]	72	
Protein	400 picograms[b]	19	66
RNA	40 picograms[c]	1.9	6.6
DNA	10 picograms[d]	0.5	1.6
Amino acids	8 picograms[e]	0.4	1.3
Kations	20 picograms[f]	0.9	3.3
Lipid	66 picograms*	3.1	10.1
Carbohydrates and other components	66 picograms*	3.1	10.1
	2110	100 %	100 %

B. Comparative measurements

Total volume of a HeLa cell	2.1×10^{12} nm^3
Surface area	8×10^9 nm^2**
Diameter	1.6×10^4 nm
Volume of nucleus	1.1×10^{11} nm^3
Surface area	1×10^8 nm^2
Diameter	6×10^3 nm
Volume of poliovirus	11,500 nm^3
Surface area	2,500 nm^2
Diameter	28 nm
Volume of hemoglobin	87 nm^3

Calculations based on:

[a] Distribution of ^3H labeled H_2O and ^{14}C inulin.

[b] Lowry assay.

[c] Phenol extraction and optical density.

[d] HeLa cells are heteroploid, maintaining 50–70 chromosomes; Littlefield and Gould (1960); Puck and Marcus (1955).

[e] Methanol extraction and HPLC.

[f] Flame photometry.

* Estimates.

** Taking into account surface protrusions, sufrace area is > 10 × this value.

B. The Nucleus

The nucleus is the dominating compartment of the cell. It harbours in the DNA the genetic information of the cell. Only a small part of this information is transcribed into RNA at any given time, and only part of the primary RNA transcripts reach the cytoplasm in the form of processed mRNA for the synthesis of cellular proteins (Miller, 1981; Perry, 1981). An extensive network of regulatory processes operating in the nucleus governs the metabolic machinery of the cell.

The synthesis of ribosomal RNA proceeds in a subcompartment of the nucleus, the nucleolus. In addition, the nucleolus provides the site for association of ribosomal proteins with ribosomal RNA. Ribosomal proteins move rapidly into the nucleolus from the site of their synthesis in the cytoplasm and return to the cytoplasm in the form of ribosomes. This unequal distribution of proteins between nucleus and cytoplasm is shared by other proteins which can be found both in the nucleus and in the cytoplasm. Steroid binding proteins are found in a free state in the cytoplasm and complexed with hormones in the nucleus. Most nuclear proteins, histones, non-histone chromosomal proteins, enzymes for the synthesis of nucleic acids and the proteins of the nuclear skeleton are present in the cytoplasm only during the time of their synthesis (Bonner, 1978).

Not only proteins are specifically compartmentalized between cytoplasm and nucleus. The distribution of nucleic acid precursors between nucleus and cytoplasm is also regulated. Most nucleosidetriphosphates are present in higher concentrations in the nucleus.

Although all known steps of poliovirus replication occur in the cytoplasm, poliovirus infection interferes markedly with nuclear functions, including RNA and DNA synthesis, compartmentalization of nucleic acid precursors, and the distribution of certain nuclear proteins.

C. The Plasma Membrane

Most of the lipids of the cell are constituents of the plasma membrane and the various intracellular membranes. Due to intensive research during recent years, many basic features of the plasma membrane have been elucidated (Bretscher and Raff, 1975). A continuous bilayer of lipids, mainly phospholipids and sphingoglycolipids, oriented with non polar groups inwards constitute the matrix for the imbedment of proteins and glycoproteins. The disposition of the glycoproteins and proteins have been studied by controlled proteolysis or chemical and isotope labeling (Wallach, 1972; Juliano, 1978). The proteins are anchored in the hydrophobic lipid bilayer by their hydrophobic regions whereas their hydrophilic parts project from the bilayer into the aqueous environment on the internal and/or external sides of the membrane. The orientation outward–inward of plasma membrane proteins is fixed: the oligosaccharides of glycoproteins are always found on the external side of the plasma membrane (Rothmann and Lenard, 1977).

The lateral movement of membrane proteins is generally rather free but dependent on the microviscosity or the fluidity of the membrane (Singer and

Nicolson, 1972; Quinn and Chapman, 1980). The components of the plasma membrane are normally in a dynamic state. The fluidity is determined by the lipid and protein composition of the membrane and by temperature. Incubation of cells at temperatures below 23°C results in a decrease of membrane fluidity, protein-movement is nearly abolished at 4°C. Paradoxically, the permeability for cations is drastically enhanced at low temperatures.

The cytoskeleton is a complex meshwork of protein filaments and tubules which passes through the cytoplasmic matrix and may interact with the glycocalix cortex underlying the cytoplasmic side of the cell membrane (Branton *et al.*, 1981; Oliver and Berlin, 1982). The cytoskeleton may control the movement of certain membrane proteins that are attached to or "anchored" to filamentous proteins of the cytoskeleton. The interaction of membrane and cytoskeleton becomes apparent in changes in cell shape. The shape of a cell in suspension culture is usually round, its surface rather smooth. The cell, however, is capable of dramatic changes in structure: it may send out surface protrusions of various sorts: from microvilli, fingerlike protrusions, to membrane-lamellae, wide sheet-like extensions of the plasma membrane with little separating space. During these reformations, the volume of the plasma membrane may change substantially. It is not known where the extra membrane components come from.

The composition and physical state of the membrane may vary greatly during the cell cycle (Folkman and Moscana, 1978). Tissue culture cells change their morphology during the cell cycle, especially when they grow attached to surfaces. Different proteins may become exposed on the cell surface at different times of the cell cycle: When HeLa cells are grown in suspension, for example, certain proteins are more exposed to attack by proteases during S phase than during other phases of their cell cycle (Kalvelage and Koch, 1982). Exposure of the poliovirus receptor may also vary with the cell cycle.

The interaction of various membrane proteins with each other plays a major role in the cellular control mechanisms for proliferation and differentiation of cells. The plasma membrane also plays an important mediatory role in the response of a cell to changes in the external environment. As the limiting barrier of the cell to the outside, the plasma membrane—or specific receptors that are anchored within it—registers signals in the form of hormones or other factors in the environment (Kahn, 1976). For the task of transmiting such signals to the cellular interior, the plasma membrane and its receptors have a variety of mechanisms at their disposal. One of the best studied examples is the adenylate cyclase system (Johnson *et al.*, 1980) which responds to the specific adsorption of peptide hormones with the conversion of ATP to cAMP, which, in turn activates intracellular proteinkinases with a variety of regulatory functions. Elements of the cell cytoskeleton which interact extensively with membrane proteins may also aid in the transmission of external stimuli to the cellular matrix (see below).

D. The Ionic Environment

In recent years it has become evident that the movements of inorganic ions (H^+, Na^+, K^+, Ca^{2+}, Cl^- etc.) across the plasma membrane by means of a variety of specific transport systems are not only subject to a high degree of regulation, but themselves exert control over intracellular processes, ranging from initiation of lymphocyte proliferation to differentiation of neurons (Gomperts, 1976; Wilson, 1978; Kaplan and Pasternak, 1983). The plasma membrane of eukaryotic cells is considerably more permeable for water than for ions. For most ions, specific ion gates exist for passive flow according to concentration gradients and electrical potential across the membrane. Figure 62 illustrates the relative intracellular and extracellular concentrations of some important ions and the electrical potential across the cell membrane for a prototype tissue culture cell.

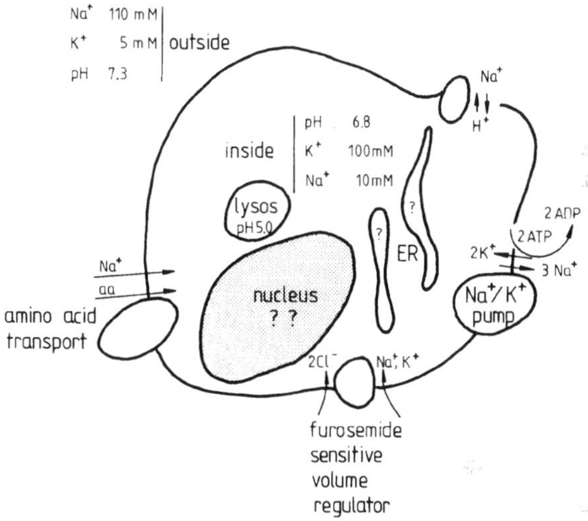

Fig. 62. Schematic representation of the ionic environment of a suspension culture HeLa cell This figure illustrates schematically the ionic conditions within and surrounding a typical suspension culture HeLa cell. The ion concentrations within intracellular compartments are not known, the conditions within the endoplasmic reticulum probably resemble those of the extracellular environment. Four transport systems for sodium and other small monovalent cations are illustrated: the well characterized Na^+/K^+ ATPase, the amino acid co-transport, the ionically neutral Na^+, $K^+/2\,Cl^-$ co-transport, and the Na^+/H^+ exchange system

Movement of monovalent cations plays a critical regulatory role in normal cells and possibly also in many virus infected cells (Carrasco, 1978; Sweadner and Goldin, 1980; Kyte, 1981; Nair, 1981). In the resting cell, there is a continous influx of Na^+ ions via the Na^+ coupled transport systems, notably for amino acids, and other ways (see below). One of the main tasks of the Na^+ K^+ ATPase (Na^+K^+ pump) is to correct for this influx. Indeed, most of the ATP in a resting state cell is consumed by the Na^+ pump. The pump extrudes three Na^+ ions with a concomitant uptake of two K^+ ions for each ATP hydrolyzed. The Na^+ K^+ pump is a principle regulatory target in cell proliferation and differentiation of cultured cells, an essential early event in the proliferative response of lymphocytes and in poliovirus infected cells.

Another ion transport system is essential for the regulation of the cell volume. The Na^+ K^+2Cl^- cotransport system (operationally defined by its sensitivity to diuretics like furosemide) is present in many cells (Geck *et al.*, 1980). This system organizes the coordinate entry of one Na^+, one K^+, and two Cl^- ions into cells, thereby increasing the intracellular tonicity. Water that follows these ions increases the cell volume.

A Na^+–H^+ exchange via the plasma membrane participates in the regulation of the intracellular pH (Moolenar *et al.*, 1981; Pouysségur *et al.*, 1980). pH gradients across membranes have been implicated in the control of other cellular processes (Skulacher, 1977; Fillingame, 1980; Khan and Macnab, 1980). A proton pump regulates the ADP/ATP exchange across the mitochondrial membrane and the uptake of catecholamines into intracellular membrane compartments (in the adrenal medulla). The proton gradient across the membrane of E. coli is dissipated upon infection with the bacteriophage T1, and the proton gradient across the membrane of lysosomes modulates the penetration and uncoating of certain enveloped viruses.

A specific transport system—the Ca^{++} ATPase—regulates the transport of the divalent cation Ca^{++}, an essential component in the activation of contractile elements in muscle and other cells (De Meis and Vianna, 1979). In certain cells, Ca^{++} may even be sequestered in specialized vesicles within the cell cytoplasm, especially during mitosis.

Changes in intracellular pH and ionic environment may also profoundly influence the integrity of the cell cytoskeleton. Unfortunately, the ionic conditions within intracellular compartments are largely unknown (see below).

It should be noted that some "soluble" components of the cytoplasm, *i.e.* the cytoplasmic matrix, may be in a gel like state rather than fluid. Ions are often complexed to cytoplasmic proteins and do not necessarily occur only as osmotically active, "free ions".

E. The Cytoskeleton

Recent studies have elucidated the role of the subcellular framework—the cytoskeleton—in cell architecture and in metabolic activity of cells. In eukaryotic cells, cell shape and spatial distribution of organelles is underlaid by a cytoarchitecture of great complexity (Fulton, 1981; Penman *et al.*, 1982). Microtubules, microfilaments, intermediate filaments, contractile proteins and microtrabeculae are cytoskeletal components of many cells (Brinkley *et al.*, 1975; Osborn and Weber, 1976, 1977, 1982; Jorgensen *et al.*, 1976; Porter, 1976; Lazarides, 1976, 1980; Heggeness *et al.*, 1977; Cohen, 1979). They interconnect with each other and with centrioles, ruffles, ribosomes, and other cellular organelles. The internal cellular architecture can be seen as a quasi rigid framework or "cytoskeleton" (Porter, 1976) at a given movement. However, in time, the proteins rearrange in the cytoskeleton, but only during continuing protein synthesis (Fulton, 1981).

The cytoskeleton determines organelle localization and cell shape. Its contractile parts are involved in cell motility, chromosome movement, membrane transport and phagocytosis. RNA processing, transport and localisation of mRNA occur in association with the subcellular framework (Cervera *et al.*, 1981). The rate and extent of polymerisation of the protein building blocks of the cytoskeleton are dependent on protein phosphorylation and on the ionic environment of the cytoplasm (Lazarides, 1981). Some of the cytoskeletal components are markedly altered by infecting viruses, the function of others may be essential to virus replication (Lenk and Penman, 1979; Hunter, 1980).

The cytoskeleton is involved in extensive interaction with the cell membrane. The interaction of membranes and the cytoskeleton becomes apparent in changes of the cell shape. The cytoskeleton may control the movement of certain membrane proteins that are attached or "anchored" to filament proteins of the cytoskeleton.

The cytoskeletal components can be characterized into structurally and functionally distinct subgroups. The microtubules are hollow cylinders constructed from 13 long filaments of α and β tubulin repeats, to which are attached a variety of maps—*m*icrotubule *a*ssociated *p*rotein*s*. Microtubules are polymerized on a specialized paranuclear microtubule forming center near the cell center. The centrioles and mitotic spindle apparatus are constructed from microtubules. Directed transport of cellular substrates often occurs along microtubules (Kirschner, 1978; Weber and Osborn, 1979; Amos and Baker, 1979; Bergen and Borisy, 1980).

The microfilaments are long cables of polymerized actin—one of the contractile components of muscle fibers. Classes of actin binding, capping, and severing proteins have been characterized. The microfilaments are the motile constituents of cell surface portrusions, membrane lamellae, and microvilli. Interaction of myosin with actin filaments may provide the basis of motility of cell surface projections and of entire cells. α-actinin intercalates actin filaments into ladder-like structures, whereas vinculin bundles actin filaments together. The rigid focal contacts of culture cells with the surface of culture dishes is formed by such vinculin—actin—α actinin interactions. Actin filaments reach into the cell and arcade to and about the cell nucleus, sheaths of actin filaments may lie below the cell surface, and as such may be important for transmembrane events and ion transport (Korn, 1982; Weeds, 1982; Pollard and Craig, 1982; Lazarides and Weber, 1974).

The intermediate filaments predominate in the paranuclear region of the cell center. They are 10 nm wide, with 21 nm repeats and are constructed from rod-shaped subunits. Most cells express only one type of intermediate filament protein. These are specified according to the origin of the cell as vimentin, cytokeratin, desmin, and neurofilament from mesenchymal, epithelial, contractile, and neuronal cells, respectively. Most cells in culture contain vimentin in addition to the specific intermediate filament protein. HeLa cells, for example, which are derived from an epithelial cervix cancer, contain vimentin and prekeratin. The function of intermediate filaments is still uncertain (Frank *et al.*, 1978; Lazarides, 1980, 1981; Anderton, 1981).

The microtrabeculae are heterogeneous in shape and composition. They are connected to essential components of the protein synthesizing machinery (Lenk and Penman, 1979; Wolosewick and Porter, 1979; Schliwa and van Blerkom, 1981).

Lenk *et al.* (1977) described a technique for the preparation of the cytoskeleton framework of suspension-grown HeLa cells by extraction of cells with Triton X-100 in hypertonic buffer. The isolated framework retains many of the morphological features of the intact cells. The nucleus, nearly all the cellular polyribosomes, initiation factors for protein synthesis, notably the cap binding protein (see Chapter 8) and the mRNA are attached to it. Polyribosomes are linked to the network of trabeculae by their mRNA and are excluded from the zone of intermediate filaments.

F. Intracellular Membranes

Animal cells contain numerous membrane enclosed organelles and compartments. Membranes of organelles of the smooth and rough endoplasmic reticulum and of the Golgi apparatus show characteristic lipid and protein compositions and harbor specific enzymes. The latter have frequently been used to identify specific membrane fractions (Novikoff, 1976).

Morphological observations suggest that most intracellular membrane systems are discontinous (Jamieson and Palade, 1967; Palade, 1975). Transfer of substances between compartments occurs in vesicles involving a process of fusion and fission of membranes (Rothman and Fine, 1980). Characteristic changes in membrane composition and structure may result from the transfer of proteins and lipids by selective flow in the plane of the membrane and from interstitial addition of components from the cytoplasm. Overlapping enzyme and protein contents in various intracellular membranes suggest limitations in selective flow and selective insertion of proteins into membranes.

Secreted and integral membrane proteins are usually cotranslationally transported across or inserted into the membrane of the rough endoplasmic reticulum by a specific protein translocation system (Sabatini *et al.*, 1982; Walter and Blobel, 1981). The portions of the proteins that protrude into the cysternae of the endoplasmic reticulum are often processed further by specific modifying enzymes. Secretory proteins and membrane proteins are modified further and packaged in the GERL and Golgi complexes (Farquhar and Palade, 1981; Rothman, 1981). Transport of proteins occurrs not only within the channels and membranes of endoplasmic reticulum, but also along elements of the cytoskeleton, in particular the microtubules.

Not very much is known about the ionic environments of the intracellular vesicles and compartments. The pH of lysosomes appears to be more acidic than that of the cytoplasmic matrix, certain vesicles of some cells are capable of sequestering Ca^{++} ions. It is conceivable that the ionic conditions within the cysternae of the rough endoplasmic reticulum, Golgi complex, and other vesicles are also quite different from that of the cytoplasmic matrix.

Intracellular membranes are altered dramatically after infection with viruses, some of these membranes are essential for virus replication. Picornaviruses even induce the formation of specific vesicles for their replication.

G. The Cell Cycle

The HeLa cell cycle usually lasts between 16 and 25 hours (Terasima and Tolmach, 1963; Rao and Engelberg, 1965; Griffin and Ber, 1969; Erlandson and De

Harven, 1971; Pardee *et al.*, 1978). Figure 63 illustrates the biological clock of HeLa cells. Protein and RNA syntheses are interrupted during cell division and resume with entry of the daughter cells into G1 which lasts from 8 to 10 hours. Most of the anabolic activities of the cell occur during G1 and S. DNA and centriole replication, in contrast, occur only during the S phase, which lasts 7 to 9 hours. In the shorter phase between chromosome replication and beginning of cell division, the G2 phase, the cell is relatively inactive metabolically. The actual process of mitosis with its dramatic reorganization of cellular organelles and cytoarchitecture culminating in the division of the cell, is quite rapid, lasting only 1/2 to 1 hour (M in Fig. 63).

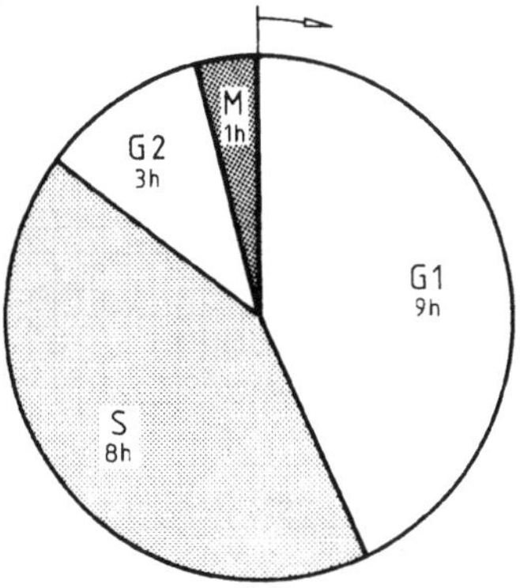

Fig. 63. The HeLa cell cycle

H. Employment of the Metabolic Machinery of the Host Cell by the Virus

The virus specific, metabolically active phase of a poliovirus replication cycle lasts approximately 2/10 of the time of a cell cycle. In order to compare the amounts of viral RNA and viral proteins synthesized to the optimal synthetic activities of an uninfected cell, we can take into account data which help us to derive estimates (see Tables 33 and 34). In infected cells, only one third to half as many ribosomes as in uninfected cells are found in polysomes, but elongation of protein synthesis proceeds with equal rates in infected and uninfected cells (Saborio *et al.*, 1974). Isotope tracer studies also indicate that infected cells incorporate amino acids into proteins at maximally 50% of the rate of uninfected cells (see Fig. 60A, p. 210). The amino acid pools in infected and uninfected cells do not differ markedly. Taking the values of Tables 33 and 34 as a basis, one obtains an estimate of approximately 20pg of viral protein made per infected cell.

Another independent way to estimate the amount of viral proteins synthesized is to determine the number of progeny particles per infected cell. Poliovirus infected HeLa cells yield 100–2 000 PFU/cell or 2×10^5 progeny particles (Oppermann and Koch, 1973). For each virus particle formed, the viral mRNA has to be translated at least 60 times (60 copies of each coat protein/virus particle). This means that for the maturation of 2.5×10^5 progeny virions, at least 1.5×10^7 viral

Table 34. *Comparison of constituents in uninfected and infected HeLa cells*

	Number per cell	
	Uninfected	Infected
Genome complexity	DNA: 15×10^9 base pairs	RNA: 7,300 bases
Ribosomes	8×10^{6a}	
Ribosomes in polysomes	4×10^{6b}	3×10^6
tRNA	6×10^7	
Host mRNA	6.5×10^{5c}	
Polio mRNA		$1 \times 10^{5*}$
Total viral RNA		5×10^5 (2.0 pg)[d]
RNA in progeny virions		2×10^{5e}
Host proteins	6.8×10^{9f}	
Total viral proteins		$1.7–2.2 \times 10^{8*}$ (15 pg)[g]
Total number of copies of each viral protein		$2.4–3.6 \times 10^{7g}$
Progeny virion		2×10^{5e}
Kations	1.4×10^{11h}	
Kations in all progeny virions		1.2×10^{9i}
Mitochondria	100*	100
Smooth vesicles	10^{3*}	$2 \times 10^{5*}$
Golgi	1–10	–

Basis for calculations:
[a] Calculated from RNA content of 40 pg (Tab. 33), assuming that 80 % of the RNA is ribosomal vRNA.
[b] Sucrose gradients: 50 % of total ribosomes in polysomes.
[c] Assuming 1 mRNAs per 6 ribosomes, all mRNAs engaged in translation.
[d] 5 % of total RNA in infected cells (Hewlett *et al.*, 1977).
[e] (Oppermann and Koch, 1973).
[f] Lowry assay, assuming an average molecular weight of 35 Kd.
[g] Assuming that 30–50 % of synthesized capsid proteins are incorporated into progeny virions.
[h] Flame photometry.
[i] 6000 kations per virion (Mapoles, 1980).
* Estimates.

polyproteins corresponding to 6.3 pg of viral proteins must be made. These calculations suggest that approximately one third of the capsid proteins synthesized are incorporated into progeny virions.

A comparable calculation can be made for the synthesis of virus specific RNA. 2×10^5 progeny RNA molecules are equal to 0.8 pg. The infected cells contain 2–3 times as much virus specific RNA (RF-RI and ss RNA) as present in progeny virions indicating synthesis of approximately 2.0 pg of total viral RNA or a maximal rate of 1.4 pg per hour (Table 35).

IV. Some Speculations on Abortive Infections of Poliovirus

Not all virus host cell encounters lead to a productive infection, *i.e.* one in which poliovirus replication leads to the formation of many progeny virions. Indeed

Table 35. *Maximal rates of macromolecular synthesis in uninfected and infected HeLa cells*

	Uninfected (pg/h)	Infected (pg/h)	% rate in uninfected cells
Protein	30	10–15*	20– 50
RNA	3	1.4**	30– 50
Lipid	5	8	160

* Only virus specific protein synthesis (total amount of viral proteins in the infected cell: 15–20 pg).

** Only virus specific RNA synthesis (total amount of viral RNA in the infected cell: 2–3 pg).

more than 90% of all poliovirus-host cell encounters are "abortive" infections in which no detectable progeny virions are found even though the virion RNA may have entered the cell. Thus, the usual course of infection is abortive; productive infections are relatively rare events, each one, however, leads to the production of hundreds of thousands of progeny virions.

Not very much is known about abortive infections. A possible explanation of abortive infections is that they reflect either some structural genetic defect in the invading virus, or an efficient protection mechanism by the host cell against foreign genetic elements, for example by partial degradation of viral RNA. Another possibility is that non-productive infections reflect some kind of symbiotic course of host cell and virus development. Some close relatives of poliovirus, the enteroviruses, hepatitis A virus and Theiler's murine encephalomyelitis virus have recently been implicated to establish persistent infections in their natural hosts and in tissue culture (Lipton, 1980; Siegl *et al.*, 1981; Vallbracht *et al.*, 1984).

It is unlikely that the genetic information of poliovirus can become integrated into the host cell genome as in the case of the lysogenic bacteriophages or the RNA tumor viruses. The enzymes that would be required for this process, in particular an RNA-dependent DNA polymerase for transcription of the poliovirus RNA into DNA, have never been found to date in eukaryotic cells (except in those harboring the RNA tumor viruses). To our knowledge, there has been only one unconfirmed report of poliovirus-like DNA sequences in human cells, namely in the gut cells of patients suffering from amyotrophic lateral sclerosis (Prassad, personal communication). Otherwise, attempts to locate poliovirus related DNA sequences in human tissue have not yielded any positive findings (Miller *et al.*, 1980).

Considering the high rate of abortive infections, and the high specificity and evolutionary conservation of the host cell receptors for poliovirus (see Chapter 7), however, the question arises whether the normal abortive course of infection might have any advantageous function for the host cell, or a given host cell population such as the epithelial cells of the gut, or for a host organism such as the human. The usual course of enterovirus infections of the human intestine is indeed reminiscent of a kind of symbiotic relationship between the infecting viruses and the host cell population rather than of an antagonistic relationship. Normally, only a limited number of gut cells are infected productively; the majority of the cells harbor abortive infections and they survive the encounter with the virus and then appear to have acquired a life-long resistance against further infections. This local resistance of epithelial cells to repeated poliovirus infections is acquired only after an encounter with a wild type virus and possibly also with the live vaccine virus, but not after vaccine injections with formaldehyde inactivated virus. Acquired resistance to infection could be an immunological phenomenon—due to local secretion of IgA for example. On a cellular level, it could, in principle, reflect the disappearance of receptors or genetic transformation of the epithelial cells themselves.

In general, when a virus-like genetic element enters a cell, a "companionship" of varying duration between the viral and cellular genetic elements ensues. Both, cellular and viral genetic elements may be modified during the companionship, or the viral genetic element may replicate and leave the host cell again in the form of many progeny virions. Or the viral genome becomes integrated for some time or permanently into the cellular genome. The concept of possible noncytocidal courses of viral infections, that might even be of some kind of advantage to the host cell is closely related to the question of the origin of viruses (see for example Luria *et al.*, 1978).

One theory on the origin of viruses holds that they have evolved from one or several cellular genetic elements to become independent transmittable elements of genetic information. This could have occurred either by regressive evolution or by the acquisition of a viral transfer mechanism. A virus is then essentially regarded as part of a cell that has become independent enough to pass from one cell to another, in contrast to other components that are more tightly tied up with the whole system. At least during early stages of evolution, such virus-like genetic elements may have played an important role in the exchange of genetic information between cells, as is still seen today, for example, in the transduction of genetic information between bacteria by some bacteriophages.

It is conceivable that virus mediated transfer and merging of genetic information, or the temporary companionship of genetic information may still be playing an important role in evolution or development today. Alternatively, these phenomena may represent the evolutionary "left-overs" of such an advantageous relationship from a long time ago, *i.e.* an abnormal evolutionary side-tract leading to the formation of complexes of low evolutionary value—diseased cells. A virus may act as a regressed parasite or as an advantageous independent genetic element depending on the character of its interaction with the host cell or on the particular phase of its evolutionary history (Luria *et al.,* 1978).

In tissue culture cell systems, polioviruses display mainly parasitic characteristics. For the in vivo interaction of poliovirus with its human host, the question of whether this interaction represents solely the battle between a parasite and a potential host, or a kind of symbiosis or companionship offering advantage to both partners is unanswered at present. Certainly, the virus and its host have coexisted for a very long time. Up until the improved sanitary conditions in Western civilization at the turn of this century, nearly all human beings were infected with poliovirus in the earliest stages of their life. Yet a detrimental outcome for the human being of this encounter with the poliovirus was and still is an extremely rare event. The chances of a detrimental outcome increase with the age of the human host if there had been no previous infection by the poliovirus.

Such observations are compatible with the notion of a possible advantageous role of poliovirus infection early in life. The high specificity of the poliovirus receptors on human cells and the high conservation of receptors during evolution—contrasting with the low rate of productive infection—, and the acquired life long resistance to infection by epithelial gut cells are also intriguing phenomena which indicate that it may be of some kind of advantage for the host cell to be infected by poliovirus.

A closer study of abortive infections with more refined techniques or a successful tracing of the behaviour of poliovirus in the infected organism and a more detailed study of the ensuing reactions in the host organism may shed further light on these questions. The comparison of the elucidated genome sequences of different picornaviruses and of these and cellular genetic elements may provide further insights into the origin and evolution of picornaviruses. The high ratio of CG content in poliovirus RNA, for example, has been taken to suggest a DNA origin of the virus (Kitamura *et al.*, 1981).

In the following chapter we will discuss the molecular biology of poliovirus infection mainly from the point of view of the virus, *i.e.* the genomic strategy which the virus genome follows in a productive infection in order to reproduce itself as efficiently as possible within the given environment of the infected host cell. In tissue culture cells the virus persues this strategy with relatively high efficiency. When a given population of susceptible tissue culture cells is infected with a smaller quantity of virus, eventually all cells become infected. Several rounds of replication cycles in even a few cells will yield sufficient progeny virus to infect and eventually kill all cells. Nearly all experimental evidence available on poliovirus infection to date is concerned with productive infection for a number of reasons: productive infection is much easier to study with present day experimental techniques than abortive infection, the resulting data provide some very basic and important insights into the fundamentals of cellular life and of virus replication, and finally the results are relevant for approaches to combat the clinical courses of poliovirus and other picornavirus infections.

6

Morphological Alterations of the Host Cell as an Essential Basis for Poliovirus Replication

I. Introduction

The infection of cells by viruses is often followed by specific morphological alterations (Schrom and Bablanian, 1981). These alterations have been studied both with the light and the electron microscope (EM). Early investigations described them as cellular injury (Ackermann *et al.*, 1954), cellular lesion (Barski *et al.*, 1955), or cytopathic change (Dunnebacke, 1956).

Such alterations include a marked increase in intracellular membranes with the formation of numerous masses of membrane enclosed vesicles, "vacuolization" in the perinuclear cytoplasm, a rearrangement of polyribosomes, shrinkage in cell size, deformation and displacement of the nucleus, proliferation of the external nuclear membrane and the formation of nuclear extrusions, changes in the cell-cytoskeleton, aggregation of cellular material, and the conglomeration of densely staining amorphous material. Poliovirus infection eventually leads to death and disruption of the host cell. The observed morphological alterations were thought to reflect increasing cell "sickness", hence they were collectively named "cytopathic effect" (CPE), a term which has persisted in the literature to the present day. Correspondingly, viruses causing changes in morphology of the host cell eventually leading to cell death have been named "cytocidal" viruses. The early demonstration that poliovirus infection causes a drastic inhibition of host cell protein, RNA and DNA synthesis (Darnell and Levintow, 1960; Fenwick, 1963) reinforced the concept of poliovirus cytotoxicity and even led to the suggestion that virus induced cell pathology is the direct result of the inhibition of host directed RNA- and protein synthesis (Holland, 1963, 1964).

Today we consider the "cytopathic" alterations accompanying viral replication not primarily as viral-induced damage to the host cell, but rather as a process in which cellular structures are modified and altered from their normal role in cell metabolism in order to serve specific functions in viral replication (Blinzinger *et al.*, 1969; Lenk and Penman, 1979). The term cytophatic effect should be reserved for the final morphology of the dying host cell: pyknosis of the cell nucleus, rounding of cells, loss of surface contact, degeneration of mitochondria, and rupture of the cell membrane. These "late" alterations can be detected even in the light microscope and have been of great help in the analysis and semiquantitations of virus infections.

More refined analysis with the EM and biochemical studies have revealed that the early alterations are not simply a consequence or cause of the virus induced cell damage. Many of the early changes reflect the "strategy" of poliovirus in the reorganization of the cellular metabolic machinery to serve the one purpose of producing large amounts of progeny virus. This pertains in particular to the dramatic reorganization of intracellular membranes and the increase in synthesis of new membrane components. Essential virus specific processes, including viral protein synthesis, RNA synthesis and virion-maturation proceed efficiently only in association with membranes.

Membrane proliferation goes along with a readily visible, extensive compartmentalization of the cytoplasmic matrix. A corresponding spatial segregation of viral functions: local separation of the sites of synthesis of complementary viral RNA, viral mRNA, and genomic RNA, as well as separation of sites of synthesis and assembly of viral capsid proteins, might be prerequisites for productive replication. Synthesis of new cellular membranes is a specific virus controlled process which depends on the accumulation of virus coded proteins.

Early virus induced cytological changes include not only accumulation of membranes, but also alterations in the cytoskeleton and in the intracellular ionic enviroment. A reduction in cell size may indicate a loss of water due to a change in the potential across the membrane and a loss of osmotically active intracellular salt. Recent observations (Schaefer *et al.*, 1983) indicate, however, that the intracellular concentrations of both Na^+ and K^+ increase up to one hour p.i. and then decrease again. An internal sequestering of osmotically active cations (by binding to RNA or to elements of the cytoskeleton) could account for the increase in intracellular concentration of cations and concommitant shrinkage of the cell. As discussed below, we believe that an intracellular compartmentalization of ions is a prerequisite for a successful replication of poliovirus.

Morphological alterations have to be discussed in correlation with known accompanying biochemical events. Table 32 of chapter 5 provides an overview of the time course of poliovirus replication and correlates morphological and biochemical alterations in the infected cell (p. 209).

II. Microscopic Observations

The process of poliovirus replication has been investigated in detail by light- and electron-microscopic examinations (Ackermann *et al.*, 1954; Reissig et al., 1956; Horne and Nagington, 1959; Fogh and Stuart, 1960; Dales *et al.*, 1965; Mattern and Daniel, 1965; Bienz *et al.*, 1973, 1980, 1983). In the course of infection, a number of distinct alterations in the structure and organization of the cell cytoplasm can be observed.

In early studies on poliovirus infected tissue culture cells, morphological descriptions were limited to the late stages of infected cells, after the peak of progeny virion formation and release (Robbins *et al.*, 1950; Ackermann *et al.*, 1954). In these final stages of virus replication, the changes are so prominent that they can be detected in unstained preparations. Cell rounding and shrinkage, pyknosis of the nucleus, and karyorhexis (fragmentation of nuclear chromatin) are typical of dying cells. These characteristics do not differ from those in cells in the last stages of toxic or senile cell degeneration.

A. Light Microscopy

The basic morphological alterations during early and intermediate stages of poliovirus replication (exponential phase) were first described and classified in a detailed lightmicroscopical investigation of poliovirus infected monolayer monkey kidney cells (Reissig *et al.*, 1956). Infected cells were classified into seven different stages. Figure 64 summarizes their properties and correlates the timecourse of appearance of the different morphological changes to steps in virus replication.

The entire sequence of changes ending with the death of the cell was usually completed in about 7 hours after virus inocculation, although there were some variations in the responses of individual cells. Infectious virions were formed, and most viruses released *before* cell rounding and the typical nuclear pyknosis of degenerating cells were observed. A period of 3 to 4 hours—corresponding well to the viral eclipse period in these cells—was noted between the onset of virus adsorption and the first alterations in cell morphology detectable by light microscopy.

Already Reissig *et al.* (1956) emphasized that the morphological alterations of the infected host cell cytoplasm and nucleus during the intermediate stages of poliovirus replication were very different from those of classical autolytic degeneration. The nucleus becomes distorted and appears to be pushed aside by an eosinophilic mass developing near the center of the cell. The nucleoli remain prominent throughout the virus replication cycle, and the chromatin condenses on the nuclear membrane. One of the earliest changes in these cells is the appearance of small acidophilic inclusion bodies in the nucleoplasm. Similar inclusions had been described in the motoneurons of monkeys following poliovirus infection (Sabin and Ward, 1941; Bodian, 1948). The peripheral cytoplasm appeared unaltered over the early and intermediate stages of virus replication. Near the end of the replication cycle, as the cells began to round off, the basophilia of the peripheral cytoplasm was observed to increase somewhat. Basophilic granules, presumably representing aggregation of host material, were seen occasionally in the final stages in some cells.

Early cytoplasmic alterations had been observed also in the nerve cells of infected monkeys and in cultured nerve cells, including chromatolysis of the cytoplasmic Nissl bodies (Bodian, 1948), presumably corresponding to the breakdown of host cell polysomes, and a retraction of cytoplasmic processes (Hogue *et al.*, 1955).

Under the phase microscope, the distorted nucleus and paranuclear mass were often masked by a cluster of fat droplets and mitochondria. The intranuclear

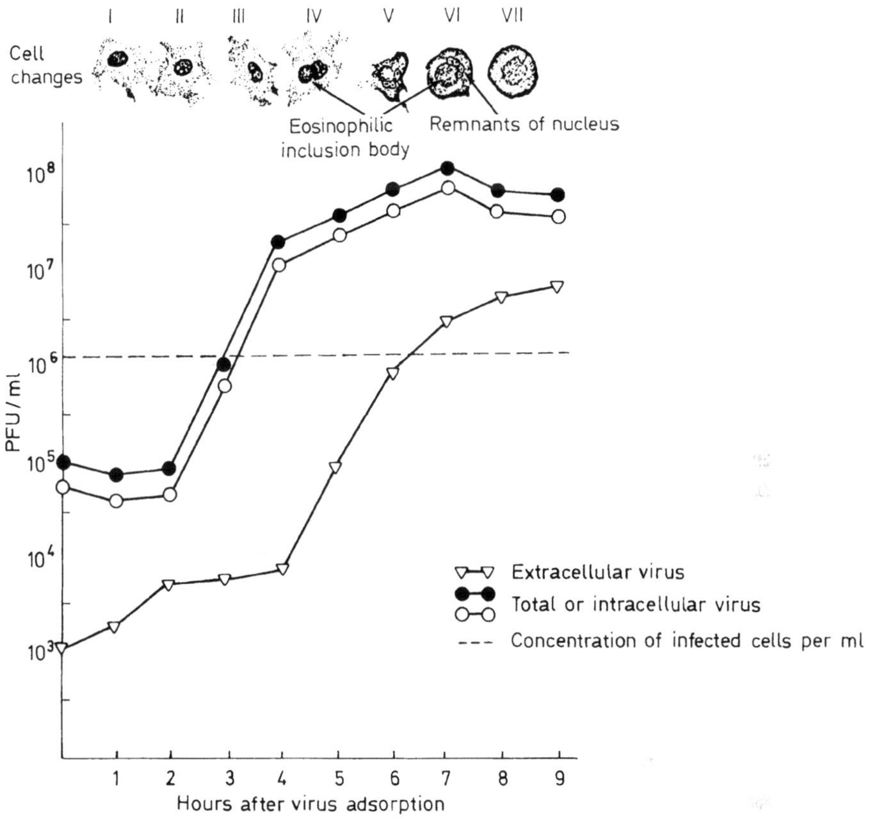

Cell type	Time of appearance (h.p.i.)	Morphological appearance of the cytoplasm	nucleus
I	0– 4	pale, containing fat droplets and mitochondria	round, fine chromatin network, 1 or 2 nucleoli
II	3– 5	normal	beginning distortion, small acidophilic inclusion bodies, condensation of chromatin on the nuclear membrane
III & IV	4– 8	appearance of an eosinophilic paranuclear mass	stronger distortion and wrinkling of the nucleus
V	5–10	cells pull away from surface, increasing peripheral basophilia	nucleus is pushed aside by the paranuclear mass
VI	6.5–10	rounding of cells, short-lived large vacuoles and basophilic granula in some cells	eccentric position of nucleus, beginning pyknosis
VII	7–15	acidophilic cytoplasm, lysis of cells	strong pyknosis, karyorhexis

Fig. 64. Growth curve and light microscopy of poliovirus replication in monolayer cultures of monkey kindey cells. Data from Reissig *et al.*, 1956 [J. Exp. Med. *104*, 289–304 (1956)]

inclusions could not be seen with the phase microscope. The phase microscope, however, was particularly suited to detect and follow the appearance of characteristic short lived cytoplasmic vacuoles in the last stages of virus replication (Barski *et al.*, 1955; Lwoff *et al.*, 1955; Reissig *et al.*, 1956). These vacuoles originated in hyaline perinuclear areas and underwent constant changes in shape and location until they finally developed into clearly defined vacuoles in the peripheral zone of the cytoplasm. These vacuoles disappeared from the cell within an hour or so of their formation and were no longer present in the rounded cells of the last stage.

When the release of progeny virions is monitored from single cells, a major burst is observed releasing most of the virions within 30 minutes (Lwoff *et al.*, 1955) just prior to cell pyknosis. Virus release from populations of infected cells occurs over a longer time period of approximately 3 hours, but seems to preceed cell death (Reissig *et al.*, 1956; Kallmann *et al.*, 1958; Howes, 1959).

B. Electron Microscopy

The findings of the early light microscopic studies were soon confirmed and extended by electron microscopy in the early 1960ies (Kallmann *et al.*, 1958; Horne and Naginton, 1959; Stuart and Fogh, 1959; Fogh and Stuart, 1960a+b; 1961; Mattern and Danniel, 1965; Dales *et al.*, 1965). Figures 65–68 demonstrate the characteristic alterations in the course of poliovirus replication in HEp2 cells.

1. The Nucleus

The nucleus is the first of cellular organelles that is visibly modified. As soon as 2 hours post infection, the nucleus becomes distorted and its membrane is wrinkled. Autoradiography and electron microscopy of cells labeled with radioactive nucleosides revealed an early inhibition of cellular hnRNA synthesis, whereas rRNA synthesis in the nucleoli was not affected significantly (Bienz *et al.*, 1982). The cellular chromatin condenses and becomes associated in clumps to the nuclear membrane. It is not surprising, that the virus should interfere with nuclear functions; the nucleus, after all, is the central control organ of the cell. However, we are still far from understanding how the virion interferes with nuclear activities and what relevance this interference bears on viral replication. All virus specific synthetic activities as well as virion morphogenesis are cytoplasmic events. Viral replication is possible–though not very efficient–in enucleated cells, yet viral proteins are regularly detected in the nuclei of infected cells (Bienz *et al.*, 1982; Fernandez-Tomas, 1982). It is still not possible to say whether the observed alterations of nuclear functions merely reflect a virus induced block of the central control organ, or whether they reflect the dependence of the virus on some nuclear factors–for example, enzymes and factors required for nucleic acid synthesis. We have not been able to find an electron microscopic description in the literature of alterations corresponding to the nuclear "inclusion bodies" described in early light microscopic investigations of poliovirus infected tissue.

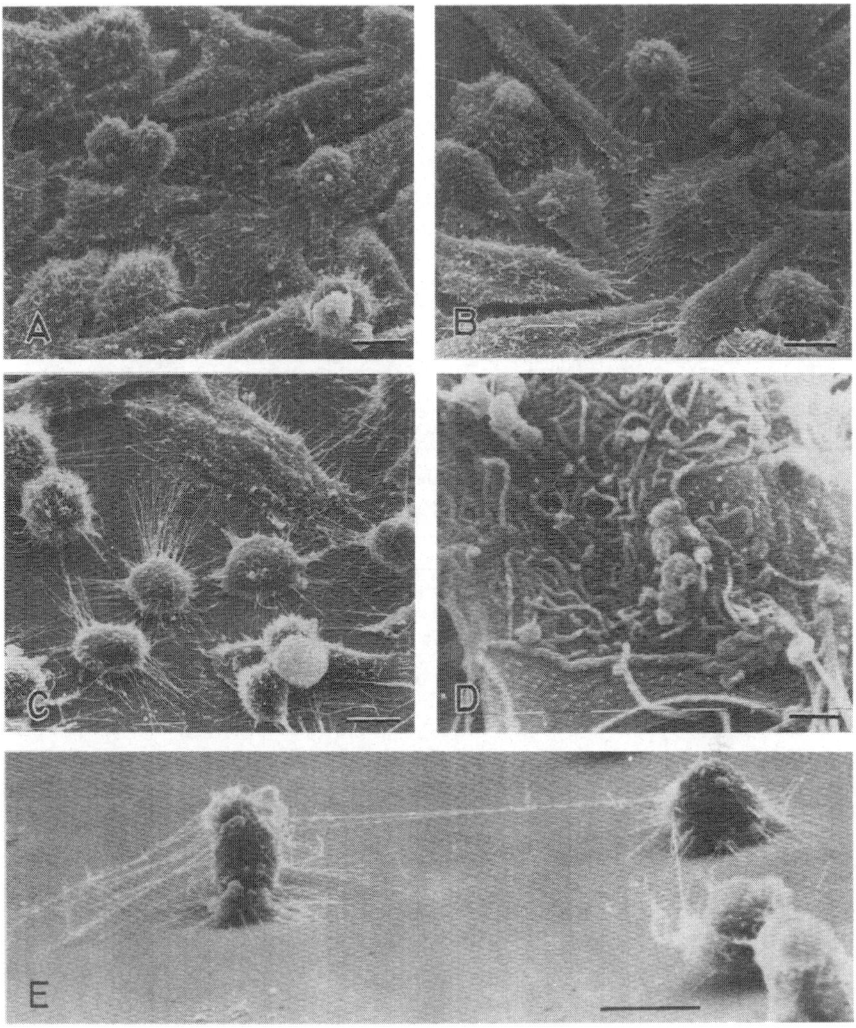

Fig. 65. Scanning electron-microscopy of surface alterations of poliovirus-infected HEp-2 cells In comparison to uninfected HEp-2 cells (A), cells infected with poliovirus type 1 (Mahoney strain) for 3 hours show first signs of rounding up and pyknosis, and formation of filopodia (B). 8 h.p.i. most cells are strongly rounded up and pyknotic and the filopodia are extremely elongated (up to 70 μm) (C). At this time, the microvilli are collapsed and condensed at the cell surface (D). Later in infection (10 h.p.i.) the filopodia of neighboring cells are merging
bar in A–C, E = 10 μm; bar in D = 1 μm
Figures A, C–E from Zeichhardt *et al.*, 1982 (J. Gen. Virol. 58, 417–428, 1982), Fig. B courtesy of H. Zeichhardt, Berlin

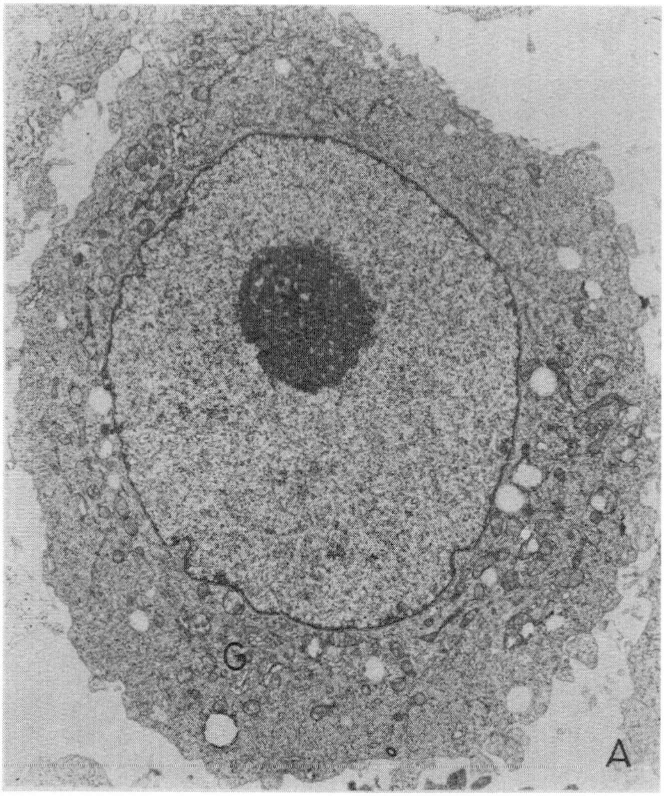

Fig. 66 A–C. Electron micrographs of uninfected and poliovirus infected HEp-2 cells, 3 h.p.i. *Nu* Nucleus; *Nl* Nucleolus; *C* Chromatin; *M* mitochondria; *MV* virus induced membraneous vesicles; *rER* rough endoplasmic reticulum; *Go* remants of Golgi complex; *Vp* Viroplasm. Figs. A and B courtesy of K. Bienz, Basel, Fig. C from Bienz *et al.*, 1980 [Virology *100*, 390 (1980)]

Fig. 66 A. Conventional EM of an uninfected HEp-2 cell. *G* Golgi complex

---

Fig. 66 B. Conventional EM of a HEp-2 cell, 3 h.p.i.
Beginning of the distortion of the nucleus and condensation of the chromatin on the nuclear membrane. Small clusters of virus induced vesicles appear at separate locations throughout the cytoplasm, often in proximity to rough endoplasmic reticulum and remnants of Golgi complexes

Fig. 66 C. EM autoradiograph of same type of cell as in B), after (^{3}H) uridine labelling in the presence of actinomycin D
The silver grains demonstrate viral RNA synthesis in connection with the clusters of newly formed vesicles

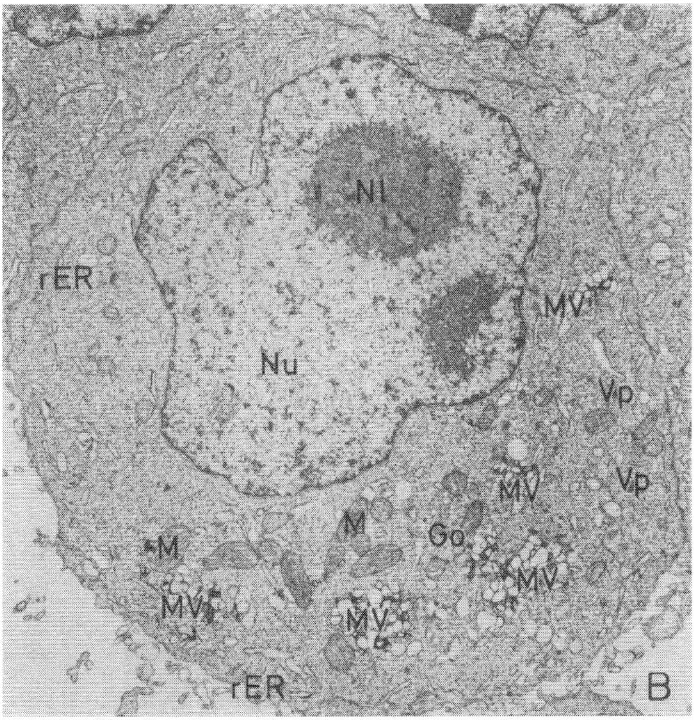
Nl
rER
Nu
MV
Vp
Vp
MV
M
M
Go
MV
MV
MV
rER
B

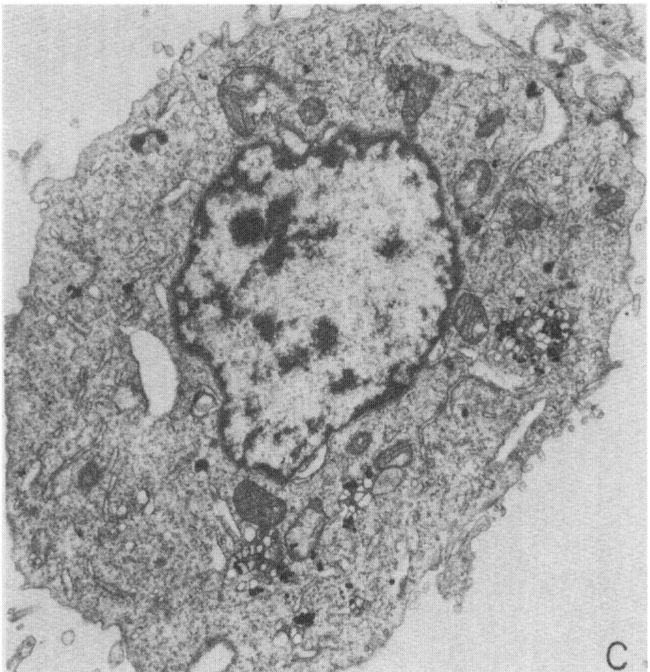
C

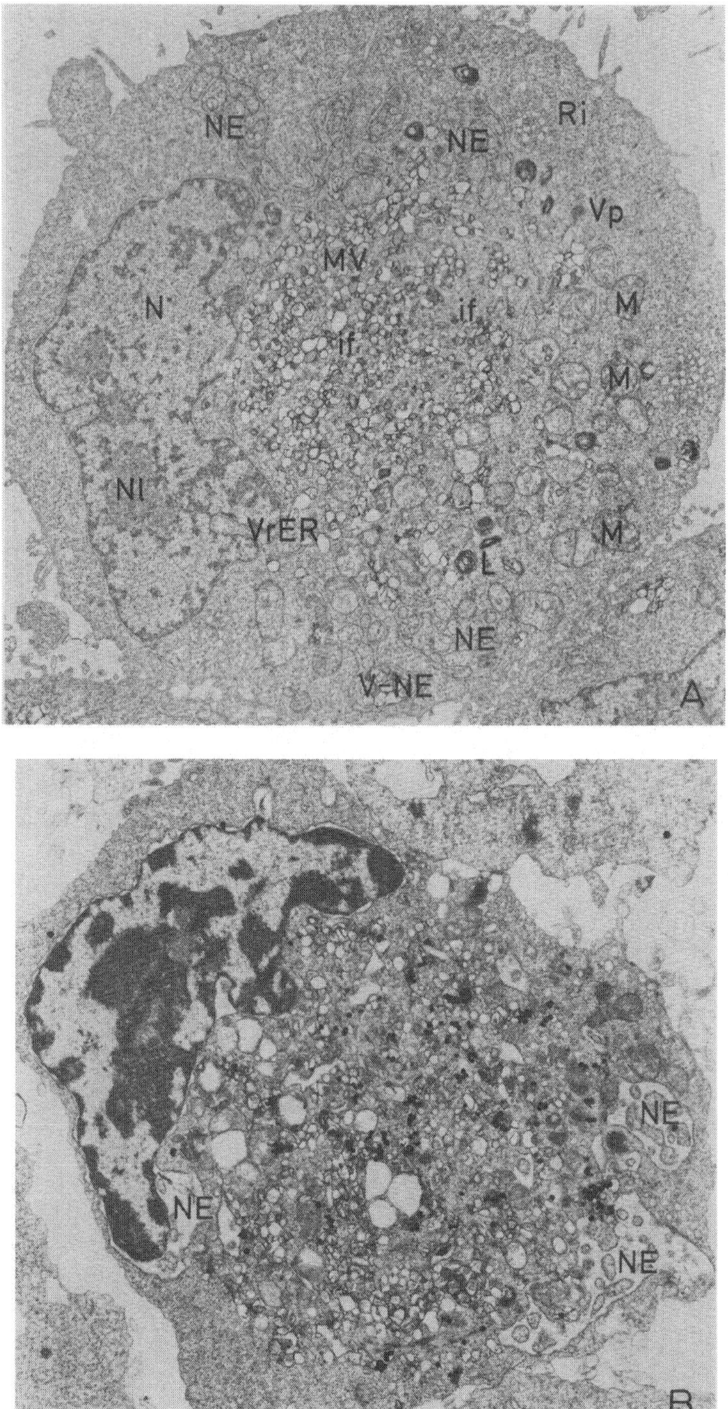
NE
NE
Ri
Vp
MV
N
if
if
M
M
Nl
VrER
M
L
NE
V-NE
A
NE
NE
NE
B

2. *Ribosomes*

Ribosomes were found in infected cells in large aggregates (large polysomes) often attached to the endoplasmic reticulum, whereas in uninfected cells ribosomes were present in the cytoplasm in smaller aggregates (Figure 69) (Dales *et al.*, 1965). These large aggregates of ribosomes are observed in the cytoplasmic matrix within 3 hours post infection and represent the large polyribosomes active in the synthesis of virus specific proteins (Penman *et al.*, 1963; Scharff *et al.*, 1963) (see Chapter 8). Polyribosomes appear to undergo a redistribution within the cytoplasm: whereas polysomes are scattered rather uniformly throughout the cytoplasm in uninfected cells, poliovirus specific polyribosomes are concentrated near the cytoplasmic periphery (Lenk and Penman, 1979). In some cell lines, such as HEp-2 cells, the relative proportion of membrane bound polyribosomes does not increase as dramatically as in other cell lines upon poliovirus infection. Large polysomes disaggregate again in infected cells shortly before the appearance of larger amounts of progeny virions, corresponding to the decline of viral protein synthesis in the late stages of the replication cycle (see also Fig. 60A, Chapter 5).

3. *"Viroplasm"*

Aggregates of dense material (so-called "foci of viroplasm") are found throughout the entire replication cycle, beginning at 2—3 h. p. i. (Fig. 66, 67). At first, the aggregates are small, averaging 0.2 μm across. By 5 hours p. i. larger aggregates measuring over 1 μm across are common. The frequency of occurence of these viroplasmic foci decreases again late in the virus infection cycle. At first, the dense material consists of tightly packed filamentous and granular elements. At 5—7 h. p. i. there are associated with these foci dense particles, 17—25 nm in diameter. These particles were referred to as "viroplasm" and thought to be poliovirions in intermediate states of condensation (Dales *et al.*, 1965).

However, "viroplasm" appears even when virus replication is blocked by guanidine and therefore probably consists of cellular material altered to an aggregated form (Lenk and Penman, 1979). It is likely that viroplasm is the result of agglutination of degenerating ribosomes or of the released host cell mRNA, rather than structures actively involved in virus production. Chromatolysis of Nissl substance in infected nerve cells presumably reflects an analogous process (Bodian,

Fig. 67. Electron micrographs of poliovirus-infected HEp-2 cells, 4 h.p.i.
Symbols as in Fig. 66; *NE* nuclear extrusions; *if* intermediate filaments; *VrER* vesicles associated to rough endoplasmic reticulum; *L* lysosome; *NE-V* nuclear extrusion engulfing cluster of vesicles. — Fig. A courtesy of K. Bienz, Basel; Fig. B from Bienz *et al.*, 1980 [Virology *100*, 390 (1980)]
A and B same as in Fig. 66 B and C respectively, but 4 h.p.i. The different centers of RNA synthesis have fused into a large area of vacuoles, viral RNA synthesis is still found associated with the vacuolated region. The nucleus assumes the typical crescent shape and is pushed aside by the mass of vacuoles. Rearrangements of intermediate filaments are observed. Nuclear "extrusions" appear and tend to surround the central vacuolated region. Notice the different types of vesicles: some are electron-translucent, others very dense; portions of the membranes of some vesicles are thick and fuzzy; many vesicles appear to stick together and the contours of many vesicles are awkward; it is often difficult to distinguish the borders of adjacent vesicles. Mitochondria and ribosomes are conspicuously absent in the central region

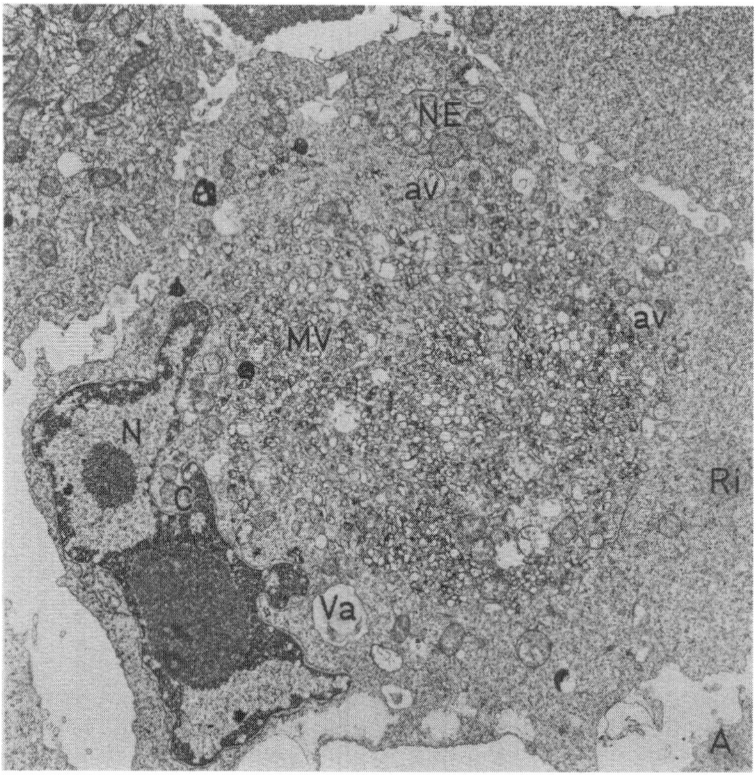

Fig. 68 A–C. Electron micrographs of poliovirus-infected HEp-2 (A & B) and HeLa (C) cells, 7 h.p.i. Symbols as in Figs. 66 and 67; *av* autophagic vacuole; *Va* large vacuole; *VC* virus crystal; *iv* individual virus particles; *B* membraneous bodies; *C* chromatin. Figures A and B courtesy of K. Bienz, Basel, Figure C from Dales *et al.*, 1965 [Virology *86*, 389 (1965)]

Fig. 68 A. Same as in Figs. 66 and 67, but 7 h.p.i. These cells no longer exhibit any metabolic activities. The nucleus is very pyknotic. Lysosomal enzymes are redistributed in these cells over the conglomeration of vesicles in the cell center. Small and very large autolytic vacuoles appear. Ribosomes are no longer organized into polysomes

1948; Price and Porter, 1972). The cellular localization of "viroplasm" as seen in the EM is similar to that of the basophilic granules observed late in infection with the light microscope.

Indirect immunoferritin labeling of infected cells showed some ferritin coating of viroplasm at 3 hours post infection, but little or none in the 4–6 hour period of the peak of viral replication (Levinthal *et al.*, 1969), indicating that association of viral antigen with these structures is temporary. Similar results were obtained in autoradiographic studies that followed the fate of newly synthesized viral proteins (Bienz *et al.*, 1980). These observations suggest that viroplasm represents a virus capsid protein-induced aggregate of cellular material.

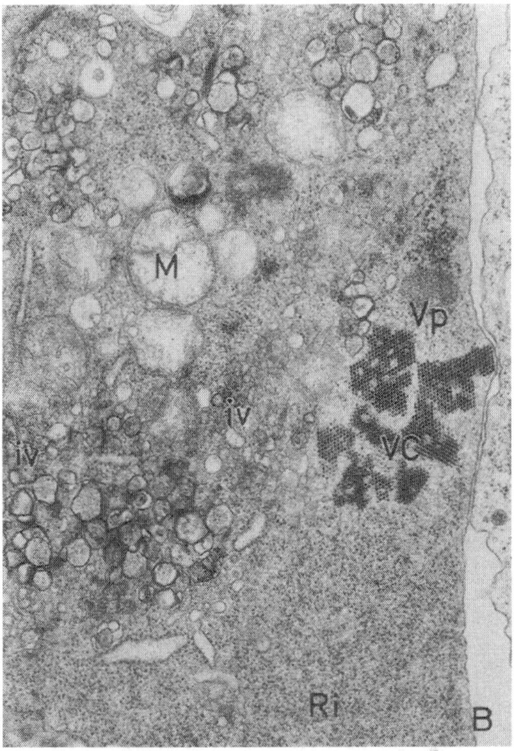

Fig. 68 B. Higher magnification of part of a late stage HEp-2 cell. A crystalline aggregate of virus particles lies directly beneath the plasma membrane. Individual virus particles are scattered between remnants of virus induced membranes. Advanced autolysis

4. *Alterations of Intracellular Membranes*

Two types of membrane rearrangements proceed in the cell cytoplasm, beginning in the third or fourth hour of infection, leading to the formation of an intricate system of intracellular channels and vesicles. One type appears to originate from the nuclear membrane, the other from independent cytoplasmic foci of vesicle formation.

a) Nuclear "Extrusions"

Separation of the two nuclear membranes and the formation of so called nuclear extrusions is often observed in poliovirus-infected cells (Fig. 67A, B). In some locations the outer nuclear membrane appears to be pushed apart from the inner nuclear membrane and it begins to surround and engulf portions of the cytoplasm (Mattern and Daniel, 1965). As a result, round and elongated membrane bound vesicles containing cytoplasm-like material become embedded in a matrix which resembles the ordinarily latent space between the two nuclear membranes. These large vesicles are surrounded by a membrane that apparently is derived from the outer nuclear membrane.

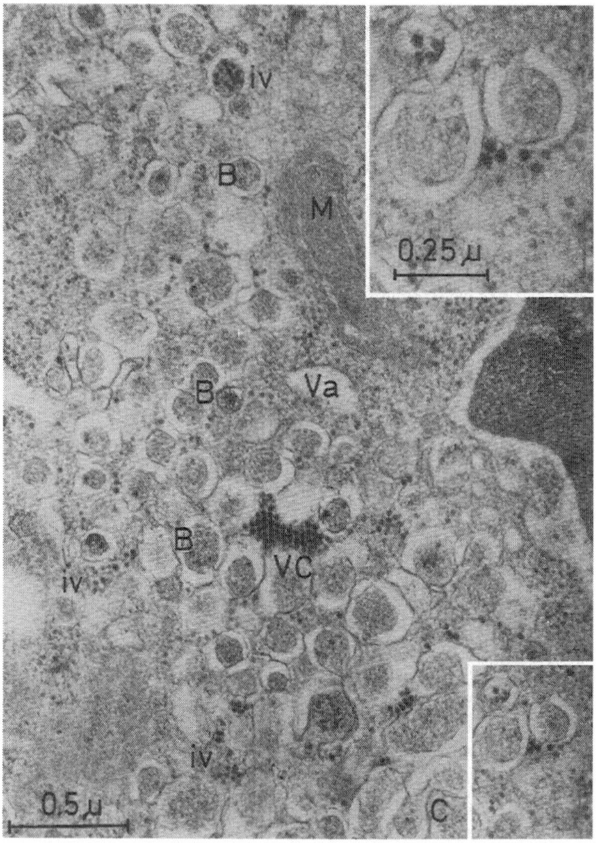

Fig. 68 C. Portion of a late stage HeLa cell. A small crystalline aggregate of virus particles and individual particles between and within virus induced membranous bodies

Electron dense particles, indistinguishable from ribosomes, are alligned on the outer side of this membrane as well as on the inner side of smaller elongated vesicles contained within the larger vesicles. This effect may begin very early after virus inoculation (Mattern and Daniel, 1965); usually, however, it is most pronounced by the fourth hour of infection (see Fig. 67). The number and size of nuclear extrusions may increase substantially through the sixth hour. In cell cross-sections the extrusions often appear entirely surrounded by cytoplasm. Occasionally, they are connected still by stalks to the outer nuclear membrane. The extrusions appear to successively surround the system of aggregated vesicles that is accumulating at this time in the cell center (see below). A schematic diagram of the development of the extrusions is presented in Figure 70.

An analogous phenomenon has been observed in epithelial cells of the subcomissural organ in rats that had been kept under the stress of immobilization (Krstic, 1979). It is conceivable that the nuclear extrusions which are often observed in poliovirus infected cells also reflect a cellular stress reaction. Autoradiography of infected cells labeled with radioactive amino acids or nucleosides have not

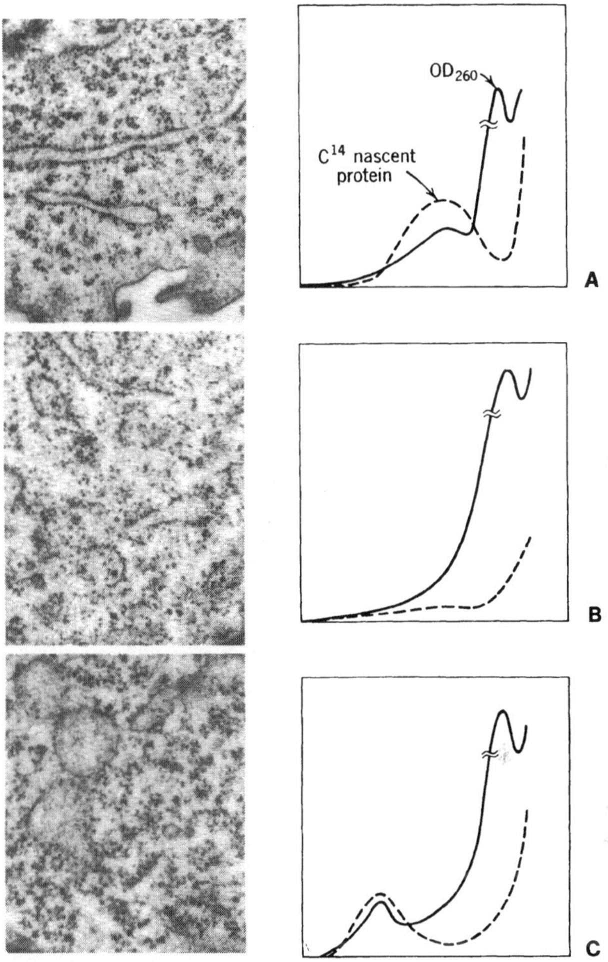

Fig. 69. Electron micrographs showing the distribution of ribosomes in poliovirus-infected HeLa cells

A Uninfected HeLa cell in which ribosomes occur in small clusters and occasionally on rough endoplasmic reticulum

B Virus infected cell about 2 hours after infection, showing loss of normal polyribosomal clusters—a reflection of the virus induced shut-off of host protein synthesis

C Infected cell about 3 hours after infection, in which large clusters of polyribosomes appear, often in proximity to membraneous structures

The graphs on the right show the polyribosome patterns in sucrose gradients of extracts from cells pulse labeled for 5 minutes with (^{14}C) amino acids. Solid lines: OD_{260}, corresponding to ribosomal material; broken lines: radioactivity in nascent proteins

Figures from Luria *et al.*, 1978 [General Virology, 3rd. ed., p. 318 (1978)]. Photographs courtesy of T. Borun; graphs redrawn from Penman *et al.*, 1963, and Summers *et al.*, 1965

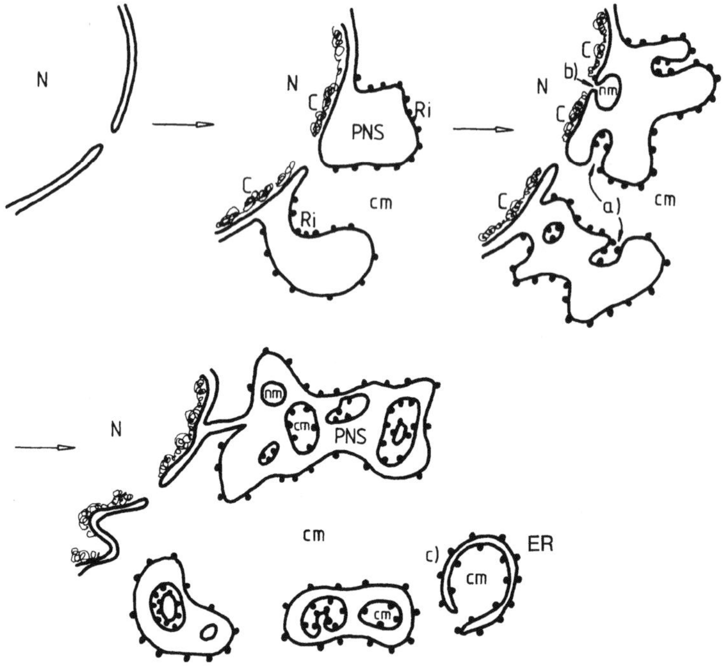

Fig. 70. Schematic representation of the formation of nuclear extrusions or protrusions in poliovirus-infected cells
N nucleus; *C* condensed chromatin; *PNS* space between the two nuclear membranes; *Ri* ribosomes; *cm* cytoplasmic matrix; *nm* nuclear matrix; *ER* endoplasmic reticulum
a formation by engulfing portions of cytoplasmic matrix
b formation by expulsion of portions of nuclear matrix
c during late stages in the infected cell, nuclear extrusions may fuse with portions of the rough endoplasmic reticulum

revealed any association of virus-specific protein- or RNA synthesis with the nuclear extrusions (Bienz *et al.*, 1980). It could not be ruled out, however, that transport of viral protein away from the site of synthesis had occured during the pulse labeling time of 10 minutes in these experiments. Also in accord with a stress/defense interpretation of the nuclear extrusions is the apparent disintegration of the contents of their vesicles. In addition, some of the nuclear extrusions seem to be expelled from the cell during later stages of infection as so-called "cytoplasmic blebs" (Yilma *et al.*, 1978).

b) Membraneous Cisternae

The second type of vesicle formation seems to originate between the 2nd and 3rd hour of infection at independent points throughout the cytoplasm in the form of groups of small round vesicles (Mattern and Daniel, 1965; Dales *et al.*, 1965). The groups of vesicles, the so-called "membraneous cisternae" (Amako and Dales, 1967), are at first of a size similar to that of mitochondria (Fig. 66). By 5 hours p. i. the individual groups of membraneous cisternae have coalesced into a large paranuclear mass (Figs. 67 and 68), which evidently corresponds to the large eosinophilic paranuclear mass described by light microscopy (Fig. 64).

The small vesicles are quite heterogeneous, their diameters range from 70–200 nm. Most are limited by a single membrane and contain material of similar or lesser density and texture than the cytoplasmic matrix. Some of the vesicles appear fuller than others, some appear empty. The vesicles seem to arise from the trans-side of the Golgi complex (Bienz *et al.*, 1983, see Fig. 102B, p. 423) and develop into an extensive network of sacs or vacuoles bound by smooth-endoplasmic membranes. Kinetic analyses of the incorporation of radioactive lipid precursors into subcellular fractions revealed that they are first incorporated into fractions of rough endoplasmic reticulum and are then rapidly transported into the fractions of smooth endoplasmic reticulum that become active in the synthesis of viral RNA (see Fig. 74, p. 247, below).

Polyribosomes are conspiciously absent from the membrane-bound bodies. Radiolabeled viral proteins are seen to be transported from the rough endoplasmic reticulum to the central system of vesicles (Caliguiri and Tamm, 1970b; Bienz *et al.*, 1983). Autoradiography of cells after incorporation of radiolabeled nucleosides reveal that poliovirus specific RNA synthesis is associated with these vesicles (Bienz *et al.*, 1980) at least during the linear phase of RNA replication (Fig. 66C + 67B). In addition, the earliest specific labeling of infected cells with fluorescent poliovirus antibody is observed in the perinuclear region near the vesicle system (Buckley, 1956, 1957; Levy, 1961; Mayor, 1961; Mayor and Jordan, 1962).

It is conceivable that the newly formed membranes are directed to their sites of function by some viral proteins that become inserted into the membrane of the rough endoplasmatic reticulum during or after their synthesis. Association of viral proteins with newly formed membranes, in turn, may provide a transport mechanism for these proteins to their site of function. The proteins of the RNA replication complex, as well as the capsid proteins which appear to be intimately associated to this complex, for example, need to be transported to the sites of RNA synthesis and virion RNA encapsidation. The following viral proteins have been found to copurify with the virus-induced vesicles: NCVP5b (a P-2 precursor), NCVP4b, NCVP6a, NCVP7c, and NCVP9 (all of these P-3 products) (Bienz *et al.*, 1983). It has been proposed that the formation of transport vesicles for secretory proteins is induced by an accumulation of the corresponding leader peptide in the membranes of the endoplasmic reticulum. The P-2 as well as the P-3 proteins contain such potential leader peptides with hydrophobic stretches suitable for membrane insertion (see Fig. 91, p. 360, Chapter 8). The observation that guanidine prevents the migration of freshly made vesicles from the rough to the smooth membrane fraction (Mosser *et al.*, 1971) implicates the guanidine-sensitive P-2 proteins (Tolskaya *et al.*, 1983) as mediators of vesicle formation. Recently, the formation of the poliovirus-induced vesicles has been observed to correlate best with the appearance of the P-2 precursor NCVP5b (Bienz *et al.*, 1983).

The poliovirus induced membraneous vesicles have a number of unusual features (Fig. 67 + 71A, B). In electron microscopic images, they often appear to stick together in clusters of 10–20 small, but heterogeneously sized, vesicles. Although the vesicles seem to stick together, they retain their integrity, so that the region of

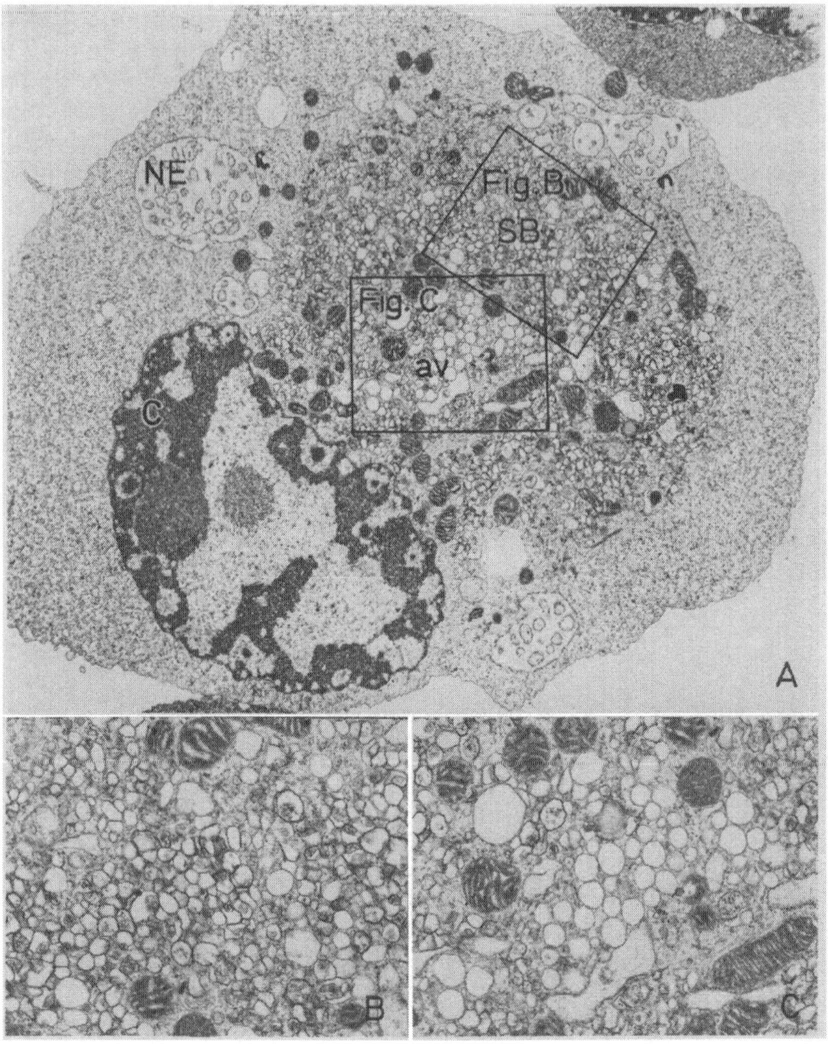

Fig. 71. Electron micrographs of a poliovirus infected HEp-2 cell, 6 h.p.i.

Fig. 71 A. This section of a poliovirus-infected HEp-2 cell at 8 h.p.i. illustrates the two types of vesicles induced by poliovirus infection. The small-body type of vesicles are formed at discrete localities during the peak of viral replication (see Fig. 66 B) and accumulate near the cell center (see Fig. 67 A). The autolytic vacuole type of vesicles are formed near the end of the replication cycle. The latter probably correspond to the rapidly expelled vacuoles seen by phase microscopy

Figs. 71 B and C. Higher magnifications of the virus induced "small body" type vacuoles *(B)* with which replication of viral RNA and virion assembly are intimiatly associated, and of late stage autolytic vacuoles *(C)*. Notice the marked difference in appearance: The autolytic vacuoles *(C)* are homogeneous in appearance, round, they have smooth surface contours, and do not stick together. The poliovirus induced membraneous vesicles *(B)*, in contrast, are very heterogeneous in appearance, exhibit awkward shapes, have irregular contours, and appear to stick together. Their membranes often appear thicker than other intracellular membranes and often carry a fuzzy coating
av autolytic vesicles; *SB* poliovirus induced membraneous vesicles. — Photographs courtesy of K. Bienz, Basel

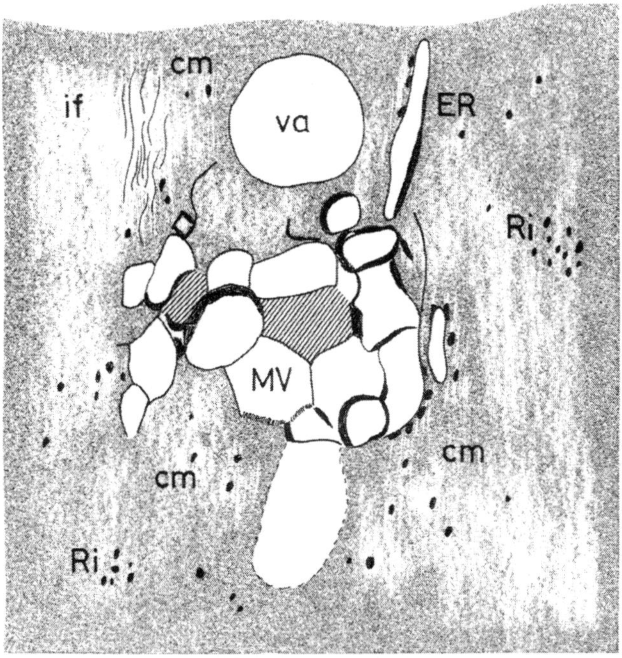

Fig. 72. Schematic representation of a typical poliovirus-induced cluster of small membraneous vesicles in infected HEp-2 cells, 3 h.p.i.
Poliovirus specific RNA replication proceeds in association with these clusters of membraneous vesicles. In contrast to the membranes of smooth vacuoles and most endoplasmic reticulum, the membranes of the poliovirus specific vesicles are heterogeneous in appearance. The membranes are coated in some places with densely staining material, portions are thick, other portions very thin, some are fuzzy in outline. The apparent stickiness of the vesicles is striking: the membranes of other intracellular vesicles and organelles never stick together in a similar fashion. Distinct corners, straight edges, and angular appearance are other unusual features of the vesicles. Portions of rough or smooth endoplasmic reticulum are often seen in the vicinity of the clusters. Segments of this endoplasmic reticulum also appear coated in a fashion similar to some of the membraneous vesicles
va normal vacuole; *MV* membraneous vesicles; *ER* endoplasmic reticulum; *if* intermediate filaments; *cm* cytoplasmic matrix; *Ri* ribosomes

contact between two vesicles appears as a single membrane. Single vesicles usually have round or rather smooth elongated contours, whereas the "sticky" vesicles within clusters often have awkward, irregular contours. Distinct segments of the vesicle membranes are very densely stained and thicker or fuzzier than other typical intracellular membranes. Figure 72 shows a schematic representation of a typical poliovirus induced vesicle cluster.

Intracellular membranes in uninfected cells—membranes of the endoplasmatic reticulum, mitochondria, vacuoles, transport vesicles, lysosomes etc.—are very rarely found to stick together. Different membranes may lie close together, but usually there is a clearly distinguishable space between them. Whenever the membranes of these vesicles come in contact, they usually coalesce, thus forming a single vesicle, as in the fusion of primary lysosomes and autophagic vesicles, or in the fusion of coated vesicles with the Golgi apparatus. The irregular contour and varying surface thickness is also not found in intracellular membranes of uninfected cells. It is, of course, possible that these features of the poliovirus induced vesicles are artifacts of the preparation procedure for electron microscopy: A swelling of closely opposed vesicles could lead to a distortion of the surrounding membranes and could provide a simple explanation for the increased translucence of the vesicles. Another more intruiging interpretation is that there is something within the infected cell that causes stickiness of vesicle membranes: RI or RF forms of poliovirus RNA bound to the surface of the vesicle membrane, for example, could be responsible for the unusual features.

In colcemid arrested mitotic cells, poliovirus infection also induces the formation of clusters of virus specific vesicles; the vesicle clusters, however, are prevented from coalescing into the single large paranuclear mass typical for interphase cells (Bienz *et al.*, 1973). Kinetics and extent of virus replication are similar in the

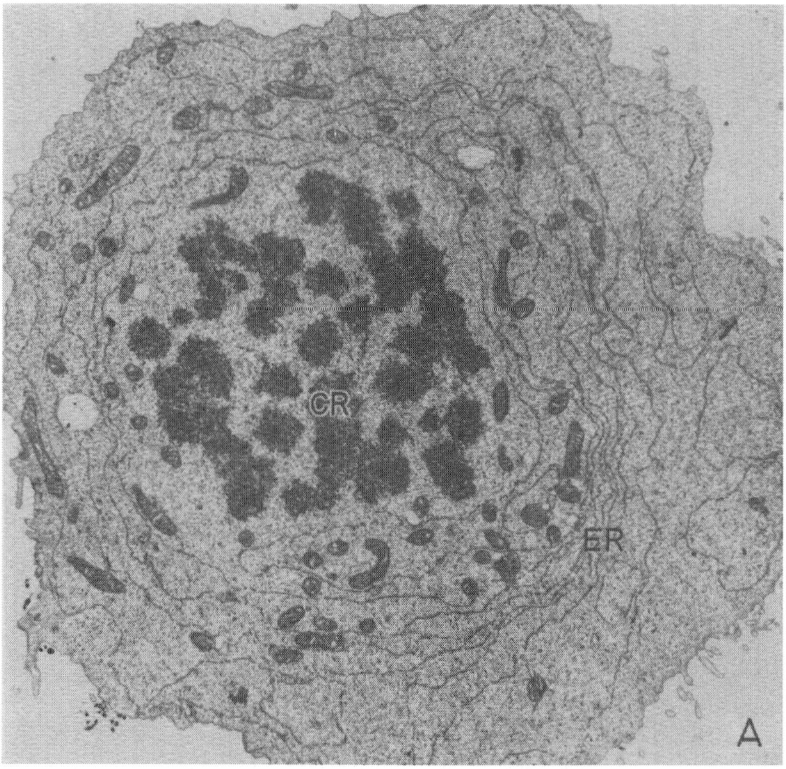

Fig. 73. Electron micrographs of uninfected and poliovirus-infected HEp-2 cells during mitosis

Fig. 73 A. The uninfected colcemid-arrested mitotic cell does not contain any nuclear membrane. The endoplasmic reticulum is arranged about the chromosomes in concentric sheaths. There are no clusters of vesicles

CR chromatin; *ER* endoplasmic reticulum; *SB* poliovirus-induced membraneous vesicles. – Photographs courtesy of K. Bienz colcemid arrested cells and untreated interphase cells. Figure 73 shows electron-micrographs of uninfected and poliovirus infected mitotic cells. The unique features of the virus specific vesicles are readily discernable in these figures.

c) Biochemical Aspects of Membrane Formation

The microscopic observations on poliovirus induced changes in intracellular membranes were soon confirmed and extended by biochemical studies. Poliovirus RNA polymerase activity was detected in the microsomal fraction of cytoplasmic extracts (Baltimore *et al.*, 1966). Virus specific polyribosomes were also found attached to membranes of vesicles, termed "virus synthesizing bodies" (Penman *et al.*, 1964). After infection with polioviruses, cells incorporate increasing amounts of ^{32}P and ^{3}H-choline into cellular phospholipids (Miroff et al., 1957; Cornatzer *et al.*, 1961; Penman, 1965). Stimulation of incorporation of ^{3}H-choline begins between 2.5 and 3 hours p. i. and is dependent on prior virus-directed protein syn-

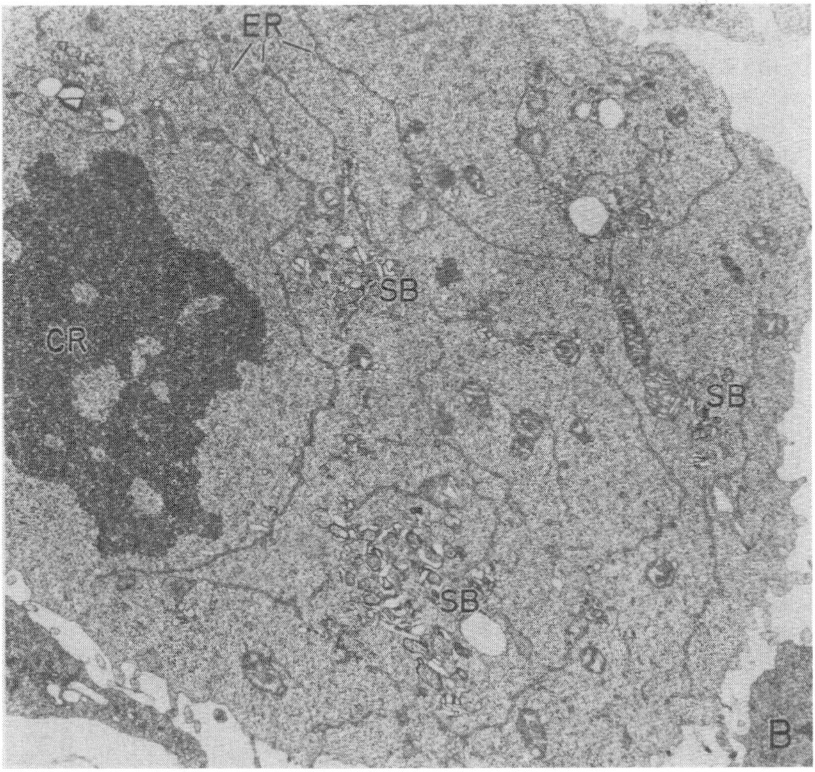

Fig. 73 B. The poliovirus infected colcemid-arrested mitotic cell also does not contain any nuclear membrane. The endoplasmic reticulum is prominent, but no longer arranged concentrically. Several clusters of poliovirus specific membraneous vesicles with typical morphology are distributed throughout the cytoplasm. The mitochondria show signs of disintegration. Lysosomal enzymes are not redistributed in such cells. The final yield and kinetics of formation and release of progeny virions are similar to those in normal interphase cells. — Photographs courtesy of K. Bienz

thesis (Penman, 1965). Membrane synthesis continues at a high rate even late in infection when viral RNA synthesis ceases (Mosser *et al.*, 1972a) indicating that a continued rapid rate of viral RNA synthesis is not required for the accumulation of these virus specific membranes. An increase in the rate of phosphatidylcholine biosynthesis was obtained in the postmitochondrial supernatants of extracts from uninfected cells by addition of CTP (100 μM). The CTP content in the cytoplasm of poliovirus infected cells is higher. It is possible that the concentration of CTP in the cytoplasm determines the rate of phosphatidyl biosynthesis, and that the higher content of CTP at least late in infection (5 h) contributes to the enhanced rate of accumulation of membrane lipids (Vance *et al.*, 1980; Choy *et al.*, 1980).

The synthesis and fate of intracellular membranes was examined with the help of radioactive precursors and fractionation of the isolated intracellular membranes by sucrose gradient centrifugation (Caliguiri and Tamm, 1970 a+b). The data are summarized in Table 36 and Figure 74.

Synthesis of membranes as determined by ^{32}P, ^{3}H-choline or ^{3}H glycerol incorporation occurs predominantly in the rough endoplasmic reticulum fraction (fractions 5 and 6, Fig. 74) both in uninfected and infected cells. In uninfected cells, all label incorporated within a 3 min pulse remains in membranes of identical density during a 30 min chase, whereas in infected cells about 1/3 of the incorporated label moves rapidly into membranes with lighter densities. These results indicate that in poliovirus infected cells there is preferential synthesis in the rough endoplasmic reticulum of lipid which is destined to become part of the smooth membranes in fraction 2 (Fig. 74) (Mosser *et al.*, 1972b).

The membrane fractions differ considerably in their content of protein and lipids (Table 36). Fractions 6 and 7 consist mainly of proteins with attached membrane lipid components. The relatively high density of ribosome containing fractions (4 and 5) is due to the presence of ribosomal RNA. Mosser *et al.* (1972a) analyzed the lipid and protein composition of fraction 2 at hourly intervals during the course of poliovirus replication in HeLa cells. The protein content increased 8 fold and the lipid content 13 fold between 2.5 and 6 hours. Membranes with the

Table 36. *Features of membrane fractions from poliovirus infected cells*

Fraction in Sucrose Gradient	Density g/cm^3	mg Lipid / mg Protein: Infected	Uninfected	Probable composition
2	1.12	0.62	0.37	smooth membrane vesicles (s.m.v.)
3	1.18	0.38	0.32	smooth membrane vesicles + scattered ribosomes
4	1.21	0.28	0.30	mixture of 3 + 5
5	1.25	0.40	0.33	vesicles with attached clusters of ribosomes
6	1.27	0.15	0.05	rough membrane vesicles (r.m.v.)
7*	1.31	0.07	0.11	proteins attached to membrane components

* Contains mature virions.

Data from Caliguiri and Tamm, 1970 a.

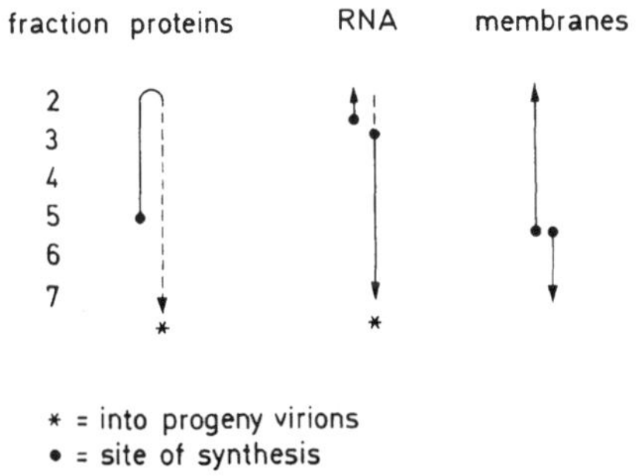

Fig. 74. The flow of virus specific macromolecules with respect to intracellular membranes
This figure illustrates schematically the synthesis and flow of virus-specific macromolecules with respect to intracellular membranes in the infected cell as determined by the incorporation of radioactively labeled precursors into protein, RNA, and lipid and the subsequent pattern of distribution of label. Membrane fractions were obtained from infected HeLa cells by sucrose gradient centrifugation, the fractions are numbered as in Table 36. Viral proteins are synthesized in the rough membrane fraction 5, become transiently associated to the virus-induced smooth membrane fraction 2, and are incorporated into progeny virions which appear in fraction 7. Membrane lipids are synthesized in the rough membrane fraction 5 and are redistributed to the other fractions. Viral RNA synthesis is tightly coupled to the smooth membrane fractions 2 and 3, where virion assembly presumably is initiated. — Data from Caliguiri and Tamm, 1970 a and b; Mosser *et al.*, 1972 a and b

low density of 1.1 in sucrose (fraction 2, Table 36) are virtually absent in uninfected cells (after removal of plasma membrane) and are virus specific in several aspects: a) They show a high content of lipids and a specific phospholipid composition with a very low (2.8 %) content of sphingomyelin. b) They contain viral RNA-replicase activity, RI and RF-RNA. c) They contain newly synthesized viral capsid proteins. d) They are not present in infected cells treated with guanidine.

Guanidine does not prevent or severely inhibit membrane synthesis, but does not allow the movement of synthesized lipids from the rough to the smooth endoplasmic membranes (Mosser *et al.*, 1971). The production of virus specific smooth membranes may depend on the continued synthesis and the accumulation of large amounts of certain viral proteins. These proteins are not produced when the required viral mRNA is not formed due to the presence of guanidine. Guanidine interferes with poliovirus replication in four ways (see Table 64, p. 407, Chapter 9): a) It rapidly inhibits initiation of RNA synthesis, b) it prevents—in vivo—release of 35 S RNA from the replication complex and virus maturation, and c) it inhibits the movement of newly synthesized membranes from their place of formation (rough endoplasmic reticulum, fraction 5) to the smooth virus specific membranes (fraction 2), and d) it prevents the association of procapsids and NCVPX to the membrane bound replication complex. Genetic analysis correlates guanidine sensitivity or resistance with the P-2 proteins.

The newly synthesized smooth membranes in HEp-2 cells induced by poliovirus infection in the presence of guanidine can be exploited by superinfecting mouse Eberfeld (ME) virus (Zeichhardt *et al.*, 1982). The latent period of ME virus replication is then shortened by up to 3 hours. The replication of the ME virus takes place on the poliovirus induced smooth membranes after insertion of ME virus specific proteins, especially P-3 proteins (replicase). Thereby the sedimentation rate of the membrane fraction increases continuously with time after infection by ME virus from 470 S (polio induced) to 570 S (2 hours p. i.) to 620 S (4 hours p. i.) to 700 S (6 hours p. i.).

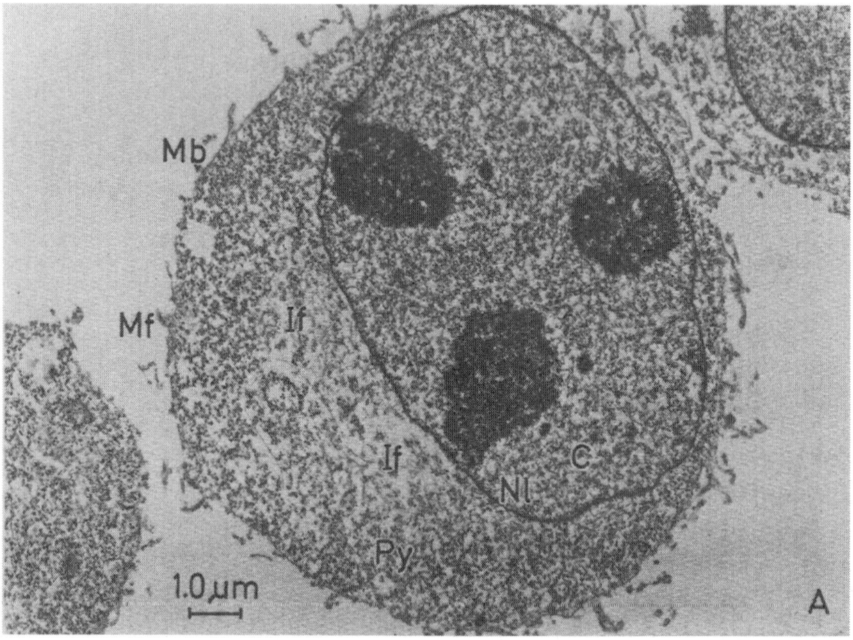

Fig. 75. Electron micrographs of cytoskeletal preparations from uninfected and poliovirus-infected suspension culture HeLa cells

NL nuclear lamina; *C* chromatin; *If* intermediate filaments; *Mf* microfilaments; *Mb* remnant plasma membrane sheet; *Py* polyribosomes associated with trabeculae; *r* ribosomes bound to remnant structures; *V* "viroplasm", aggregates of host cell material; *cf* coated filaments; *ib* irregularly shaped objects. – Figures from Lenk and Penman, 1979 [Cell *16*, 289–301 (1979)]

Fig. 75 A. The cytoskeleton of an uninfected HeLa cell was prepared by gentle extraction with Triton X-100. The nucleus is bordered by the densely staining nuclear lamina. The outer periphery of the cytoskeleton is formed by a sheet of protein apparently derived from the plasma membrane. Microtubules are not preserved under the lysis conditions. Microfilaments are bundled beneath the cell surface in some places. Intermediate filaments are concentrated in small clusters in a region adjacent to the nucleus. Ribosomes are attached to the network of amorphous trabeculae, distributed throughout the cytoplasm–but are excluded from the region of intermediate filaments. Ribosomes are attached to the trabeculae by their mRNA

Fig. 75 C. The cytoskeleton of a poliovirus infected HeLa cell, 3 h.p.i. which had been treated with low concentrations of guanidine from 1 h.p.i. Viral RNA replication is severely inhibited under these conditions, translation proceeds at a reduced rate. Alterations of the cytoskeleton are profound, though not as dramatic as during uninhibited viral replication. The chromatin collapses. Intermediate filaments seem clustered about several foci. Ribosomes are released almost completely from the cytoskeleton. Remaining ribosomes are studded on densely staining sheets–possibly endoplasmic reticulum Viroplasm, presumably aggregated host material, is retained in part in association to the cytoskeleton

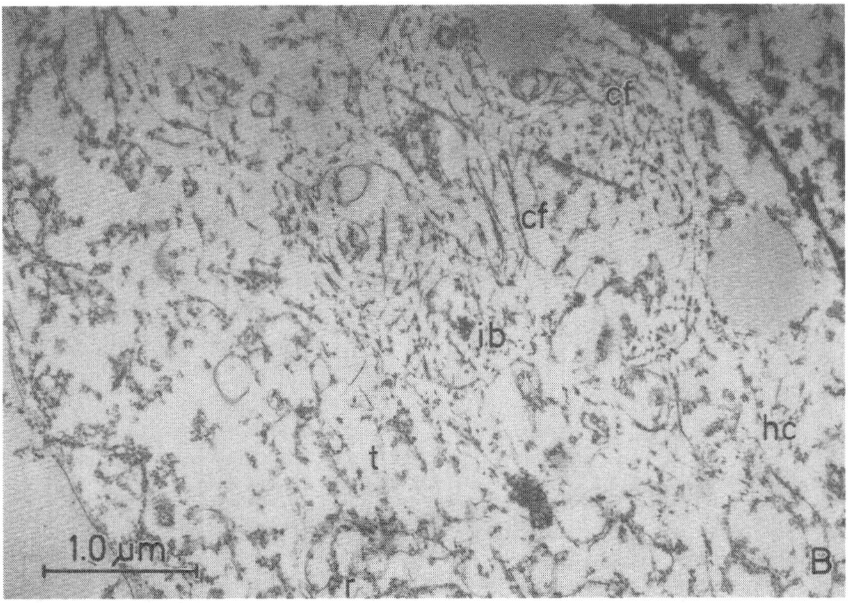

Fig. 75 B. The cytoskeleton of a poliovirus infected HeLa cell, 3 h.p.i.—near the peak of viral replication—prepared as the cell in A. The structures in the central region are no longer recognizable as intermediate filaments. The objects now are much thicker and fuzzy in outline, in several places they branch into thinner fibers. Fewer ribosomes remain attached to the cytoskeleton, they are now distributed nearer the cell periphery and occur in larger clusters. The poliovirus specific membraneous vesicles are dissolved by the extraction procedure. Poliovirus specific RNA and replication complex are associated to the cytoskeleton, whereas progeny virions do not show any affinity to the cytoskeleton, they are released with the extraction procedure

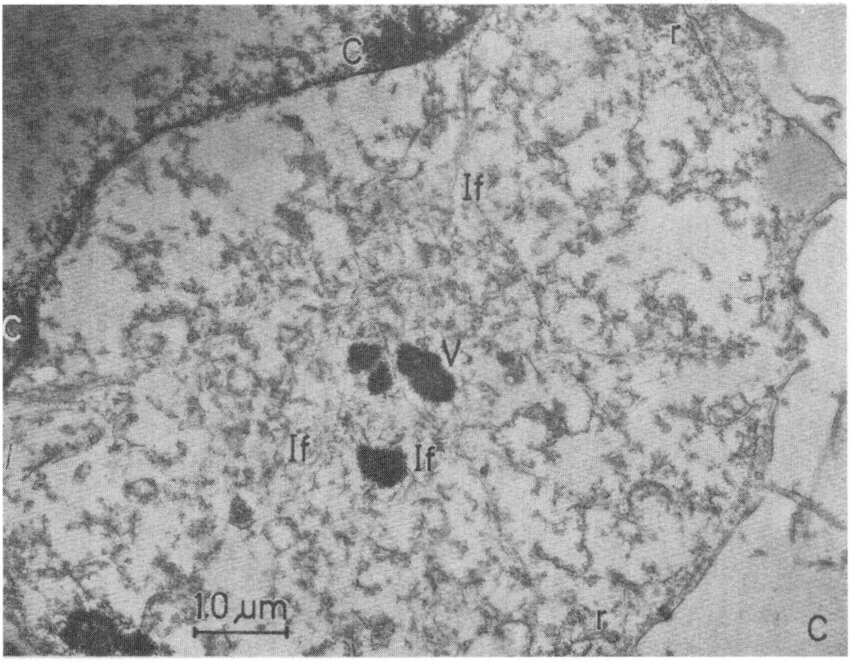

5. Changes in the Cytoskeletal Framework

Lenk and Penman (1979) analyzed the effect of poliovirus infection on the isolated cytoskeletal framework (Fig. 75). The preparation of the cytoskeleton with Triton dissolves all intracellular membranes including those of the vesicles in the central region. This region displays a dramatically altered pattern of intermediate filaments in skeleton preparations from infected cells. Instead of intermediate filaments, the central region shows structures which are straighter, thicker and fuzzy in outline and which branch in several places into thinner fibers, suggesting that the thick portions represent filaments that are aggregated. They are coated with darkly staining material, possibly RNA, of viral origin. In addition, the central region harbors a large number of small particles about 20 nm in diameter, which may represent precursors of mature virions. In contrast, mature virions are found only in the soluble fraction from infected cells. Most (75—80 %) of the viral RNA synthesized 2.5 hrs post infection in the presence of ^{3}H uridine is found in 15 min in association with the cytoskeleton and sediments with 200—400 S in sucrose gradients, indicating that the replication complex is attached to the cytoskeleton. The remaining 20—25 % newly labeled RNA are not associated to elements of the cytoskeleton and sediment with 100S.

Two—three hours p. i. ribosomes are found in large clusters (large polyribosomes), all attached to the cytoskeleton and principally near the cell periphery. Poliovirus specific polyribosomes can be released from the cytoskeleton by exposure of infected cells to cytochalasin B prior to lysis. Exposure of tissue culture cells to cytochalasin D causes disturbances in the disposition of microfilaments because cytochalasin D caps actin filaments and prevents further polymerization. Addition of cytochalasin to cells prior or shortly after infection with low multiplicities of polioviruses (or with isolated poliovirus RNA) increases the infective center yield by a factor of two to four (Deitch *et al.*, 1973; Koch and Oppermann, 1975; Koch and Koch, 1978). Cytochalasin has no effect on adsorption or penetration of the virions (although it causes alterations in the structure and function of the plasma membrane). It was suggested that cytochalasin somehow facilitates the initiation of viral RNA translation. Cytochalasin also enhances the development of cytopathic changes in the host cell.

Even when extensive synthesis of viral protein is prevented by addition of low concentrations of guanidine, the initial stages in the rearrangement of components of the cytoskeleton still take place (Fig. 75C). The infected cell cytoskeleton of guanidine treated cells is nearly completely devoid of ribosomes. Apparently all host cell mRNA is released from the cytoskeleton unchanged in size, stable, and with normal translation properties (S. Farmer cited in Lenk and Penman, 1979). These results are in accord with previous studies from other laboratories which showed that shut-off of host protein synthesis may still proceed in the presence of low concentrations of guanidine (see Chapter 8).

6. Assembly and Release of Progeny Virions

The first particles of progeny virus appear at 2.5—4 hour p. i. The identification of single viral particles or of particles in the process of assembly is complicated by the fact that they are similar in size to ribosomes. Progeny virions are readily recognized when aggregated into intracytoplasmic crystals (see Fig. 68) (Stuart and Fogh, 1959; Fogh and Stuart, 1960). Such crystals, however, are regularly seen only very late in infection, usually long after the biochemically detectable peak in the appearance of infectious progeny virions (Scharff and Levintow, 1963), and well after the peak of virion release from cells. Special preparatory procedures for electron microscopy sometimes enhance the number and size of intracellular crystals, presumably by causing aggregation of virions, that had been diffusely distributed throughout the cytoplasm, into viral crystals.

Some individual progeny virions, nonetheless, are recognized during intermediate time points of viral infection. The complete virus particles are 26–28 nm in diameter. They are predominantly localized in the central region of the cell, where they appear either trapped within the membrane-bound bodies or are dispersed in the intervening cytoplasm (Dales *et al.*, 1965). Empty capsids of the same size as complete virus particles are also observed in the same location. Other virions are observed as free particles in the cytoplasm, within a tubular system which might open through the plasma membrane, and aligned along fibrils in the cytoplasm (Rifkind *et al.*, 1961; Dales *et al.*, 1965; Blinzinger *et al.*, 1969; Dunnebacke *et al.*, 1969; Levinthal *et al.*, 1969; Friedmann and Lipton, 1980).

Further insight into the site of virion formation was obtained with the indirect immunoferritin technique (reaction of fixed infected cells with rabbit antipoliovirus serum followed by ferritin-labeled goat antirabbit globulin) in a variety of different poliovirus infected tissue culture cells. This technique allows the identification and localization of viral antigens both as assembled capsid and within intermediate stages of assembly (Dunnebacke *et al.*, 1969; Levinthal *et al.*, 1969). At no time after infection was there retention of ferritin antiglobulin in the nucleus, upon mitochondria, Golgi apparatus, intact plasma membrane, or membrane bound ribosomes. At 3 hours post infection ferritin is evenly dispersed over the cytoplasm with some concentration about vesicles, fibrils, and viroplasm. In the 4–6 hour period ferritin continuously becomes more localized, in particular to the clusters of vesicles that accumulate during this period. The ferritin coats single dense or emtpy particles among these vesicles. These particles are presumed to be complete or incomplete progeny virion. Heavy ferritin coating was also observed on the sheaves of fibrils found near the clusters of vesicles. An increased association of ferritin with other cytoplasmic structures implies that free viral proteins increase in amount and are widely distributed during advanced stages of virus replication. Indirect immunofluorescence of poliovirus infected human embryonic kidney cells with antipolio-rabbit sera revealed small peripheral accumulations of cytoplasmic fluorescence at 4 hours post infection. In the 3–6 hour period after inocculation the intensity of the fluorescence increased with time and in a few cells was concentrated in the diffuse paranuclear mass.

Full and empty ferritin ringed capsids were so regularly found among the clustered vesicles in the 4–6 hour period that some function in viral assembly was proposed for the vesicle surface (Levinthal *et al.*, 1969). Progeny virions were also observed in a tubulo-vacuolar system appearing first at 3 hours post infection and enclosing only complete virus particles. These particles were not ferritin coated except where the tubules opened through the plasma membrane. Virions within the tubules were not ferritin coated, presumably because ferritin was not able to penetrate the enclosing membranes. The diameter of the tubules was found to be fairly constant except near the plasma membrane, where the tubules occasionally appeared to coalesce into vacuolar forms. These observations were interpreted as a tubulo-vacuolar excretion system for poliovirus (see also Fig. 102, Chapter 10) (Levinthal *et al.*, 1969).

A similar suggestion for the involvement of endoplasmic reticulum derived tubules and cysts in the formation of poliovirus was made upon electron microscopical examinations of mononuclear cells in the spinal cord of infected monkeys (Blinzinger *et al.*, 1969). Crystalline arrays of poliovirus were observed within membrane bound cysts, which often were confluent with tubules of the rough endoplasmic reticulum. It was proposed that intracytoplasmatic poliovirus crystals result upon membrane disintegration of the cysts.

In this respect, it is tempting to speculate that the phase-microscopically observed vacuoles that arise in the perinuclear region of the infected cell and migrate to the cell surface within an hour of their formation (see p. 230) may correspond to the presumptive tubulo-vacuolar excretion system (Reissig *et al.*, 1956). Such an excretion system operating over a limited period of time prior to cell lysis would also explain the reported burst of progeny virus release from individual cells prior to cell death (Lwoff *et al.*, 1955).

7. Lysosomes and Autophagic Vesicles

Autophagic vacuoles appear to form as a result of the engulfment of portions of cytoplasm by crescent shaped channels of endoplasmic reticulum at the end of viral replication. The contents of these double-layered vesicles seem to disintegrate more rapidly than the cytoplasmic matrix late in infection (Figs. 68

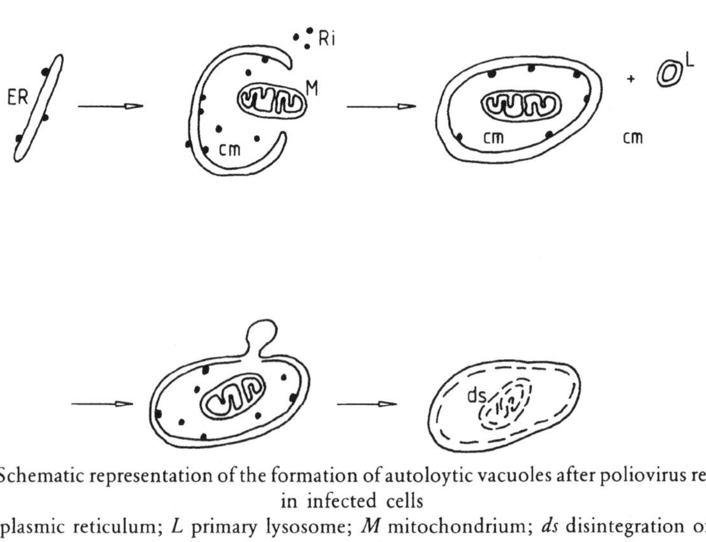

Fig. 76. Schematic representation of the formation of autoloytic vacuoles after poliovirus replication in infected cells
ER endoplasmic reticulum; *L* primary lysosome; *M* mitochondrium; *ds* disintegration of vacuole contents; *cm* cytoplasmic matrix

and 71). The pattern of formation of these vacuoles corresponds to autophagocytosis in the early phases of physiological cell dying (Hirisimäki *et al.*, 1975; Krstic, 1979). Figure 76 illustrates schematically the formation of autophagic vacuoles.

In the late stages of the poliovirus replication cycle, there appears to be a controlled release or redistribution of lysosomal enzymes (Macieira-Coelho *et al.*, 1965; Flanegan, 1966; Dusing and Wolff, 1969; Guskey *et al.*, 1970; Heding and Wolff, 1973; Bienz *et al.*, 1973, Koschel *et al.*, 1974; Rice and Wolff, 1975). The redistribution occurs gradually over a period of several hours, beginning at approximately 4–5 hours post infection, *after* the peak in appearance of infectious progeny virions.

Specific chemical labeling of acid phosphatase showed localized acid phosphatase in lysosomes in the perinuclear region of uninfected cells and poliovirus infected HEp-2 cells up to 2 hours post infection (Heding and Wolff, 1973). In most virus infected cells, a redistribution of the enzyme developed between 4–5 hours post infection, following and accompanying the proliferation of the network of vacuolar structures in the juxtanuclear region of the cytoplasm. The enzyme appears to become diffusely associated with the aerea of newly formed membraneous cisternae. It is absent from the peripheral and other regions of cytoplasm lacking the membraneous vesicles. Here, however, intact lysosomes containing large amounts of enzyme, were occasionally observed.

From these observations it has been proposed that released lysosomal enzymes may indirectly mediate the proliferation of the membranes in picornavirus infected cells (Amako and Dales, 1967; Dales, 1969). Products from the hydrolase-mediated breakdown of lipids could stimulate the enzymes responsible for phospholipid synthesis and the formation of new membranes.

In light of the kinetics of membrane formation—which begins relatively early—and lysosomal enzyme redistribution—which sets in later—it appears unlikely that there is a direct causal relationship in the sense that released lysosomal enzymes induce membrane proliferation. However, release or redistribution of lysosomal enzymes may certainly influence the cytoplasmic morphology during the final stages of cell dying.

Experiments with cells infected during mitosis support the interpretation that poliovirus specific proteins rather than host cell lysosomal enzymes are the mediators responsible for the induction of the metabolically active clusters of vesicles. Metaphase arrested HEp-2 cells produced and released the same amount of virus as randomly growing interphase cells, but there were only a few clusters of vesicles, no lytic cytopathology, and no detectable redistribution of lysosomal enzymes (Bienz *et al.*, 1973).

8. *Inhibitors of Morphological Alterations*

A variety of metabolic inhibitors have been employed in order to investigate the possible mediators of the early morphological alterations and the late cytolopathic changes in the infected cell, and the time point of their formation or activation. Early studies already indicated that the shut-off of host macromolecular synthesis, virus replication and the development of late cytopathic changes are independent processes. A mere inhibition of protein and RNA synthesis by puromycin or actinomycin D, respectively, in uninfected cells never led to morphological changes reminiscent of those in poliovirus infected cells, indicating that these changes are not merely a cause of the shut-off events in infected cells.

Addition of specific inhibitors of viral replication, such as guanidine (Rightsel et al., 1961; Loddo *et al.,* 1962) or HBB, to infected cells during the first two hours of infection (eclipse-phase) typically blocked or markedly delayed the development of cytopathic changes and completely blocked the formation of progeny virion, whereas the shut-off of host protein and RNA synthesis was not prevented (Bablanian *et al.*, 1965a+b; Summers *et al.*, 1965; Skinner *et al.,* 1968; Koschel, 1971; Mosser *et al.,* 1971; Bienz *et al.*, 1980). If the time point of treatment with guanidine was delayed until 3–4 hours post infection, after the onset of the exponential phase of viral replication, the total yield of progeny virion could still be reduced significantly. The typical cytological changes, however, were no longer prevented.

Addition of specific inhibitors of protein synthesis–such as puromycin–to infected cells, also blocked development of cytopathic changes if added during the first two hours of infection. Unlike guanidine, however, puromycin was capable of preventing the major part of the virus induced cytopathic changes even when added 3.5 hours after infection, i.e. after the first detectable onset of viral protein synthesis between 2–3 hours p. i. (Bablanian *et al.,* 1965b). This was taken to indicate that a viral product–most likely a virus coded protein–made soon after the onset of the logarithmic phase of virion replication was responsible for the induction of the major morphological alterations in infected cells and that continued synthesis and accumulation of this viral product were required for the induction. The viral RNA formed during the exponential phase prior to the blockage of virus replication by late addition of guanidine at 3.5 hours post infection would no longer be replicated, but would continue to make protein even in the presence of the compound (Halperen *et al.*, 1964), explaining why development of cytopathic changes is no longer prevented after late addition of guanidine. Similar studies of the effects of inhibitors in Mengovirus infected L cells also indicated that membrane proliferation and cytopathic changes are mediated by a viral protein product, and that these alterations could even be induced by a precursor protein (Collins and Roberts, 1972).

Addition of the amino acid analogue fluorophenylalanin (FPA) early during infection, which interferes with protein processing in infected cells, completely prevents virion formation, but does not affect development of the typical cytopathology (Ackermann *et al.,* 1954; Jacobson *et al.,* 1970). The inhibitor was much less effective when added 3 hours p. i., when added 4 to 5 hours p. i. it produced no inhibition of virion formation. Evidently the viral function responsible for the induction of the morphological alterations is active in a precursor from or it is not as sensitive to distortion by an amino acid analogue as are the essential functions of RNA replication and virion assembly. Inhibition of virion formation by FPA could be completely reversed by phenylalanine if the amino acid was added within 6 hours but not later after the induction of virostasis. By this time the infected cell had become so altered or extensively damaged that it could no longer support poliovirus replication. Unfortunately, the effects of FPA have never been analyzed more closely by electron microscopy. Recently, the development of membraneous vesicles in poliovirus infected cells has been found to correlate with the appearance of the P-2 region precursor protein NCVP5b (NCVP10 + NCVPX) when protein processing is inhibited by $ZnCl_2$ (Bienz *et al.*, 1983).

During the entire period up to 10–12 hours of guanidine block, infected human embryonic lung or HEp-2 cells retain a normal ultrastructure without signs of the typical morphological alteration when examined by conventional light or electron microscopy (Bablanian *et al.*, 1965a+b; Bienz *et al.*, 1980). Upon release from the guanidine block virus replication commences with usual kinetics.

Extensive rearrangements of the host cell cytoskeleton, however, have been observed to occur in poliovirus infected cells within 3 hours p. i. in the presence of low concentrations of guanidine added at 1 hour p. i. (Fig. 75C) (Lenk and Penman, 1979). The central region in which the smooth cytoplasmic vesicles accumulate in infected cells in the absence of guanidine, is altered also in infected cells in the presence of guanidine. The intermediate filaments are rearranged and seem tightly clustered about several foci where large numbers of filaments converge, while other filaments are arranged in parallel bundles, and radiate from the nucleus to the periphery of the cell, arrangements which have never been found in uninfected cells. Alterations in the infected cell in the absence of guanidine, however, are much more extensive: intermediate filaments are no longer recognizable as such by 3 hours p. i. Instead, straigthened objects much thicker and fuzzyer in outline than the intermediate filaments of uninfected and guanidine treated cells are observed (see p. 249). The rearrangements seen in the presence of guanidine, nevertheless, indicate that some modifications can be brought about already by small amounts of viral proteins accumulating under these conditions.

Temperature sensitive mutants of poliovirus, defective in the induction of late cytopathic changes at the nonpermissive temperature have been isolated and characterized (Garwes *et al.*, 1975). The defect of these mutants has been located in the genome region coding for the virus replicase functions responsible for synthesis of double stranded RNA. On the basis of these observations it was suggested that dsRNA is the mediator of late cytopathic changes in poliovirus infected cells.

Isolated dsRNA of poliovirus was found to be cytotoxic when added to uninfected cells at concentrations comparable to those observed in infected cells at late stages during infection (Cordell-Stewart and Taylor, 1971, 1973; Celma and Ehrenfeld, 1974). Late in infection, near the end of virion formation, the virus specific RNA, still present in the cytoplasm, seems to be converted progressively into dsRNA. As initiation of RNA synthesis comes to a halt, the last round of RNA copying seems to convert the replicative intermediate into dsRNA (see Chapter 9). It is conceivable that this dsRNA mediates late cytopathic changes in the infected cells, perhaps by inducing some inherent cellular defense or "cleaning" system. The kinetics of appearance of dsRNA, however, seem to rule out a role for dsRNA in the early induction of membrane proliferation.

III. Speculations on the Function of Compartmentalization in Virion RNA Synthesis and Assembly

In order to speculate on possible roles of the virus induced vesicles and the mechanisms by which they are utilized for virion production, a number of points have to be considered.

A. Concerning the Mode of Vesicle Formation

Vesicle formation in the uninfected cell may occur by budding of vesicles from channels of rough or smooth endoplasmic reticulum (ER) either directly or via the Golgi -ER –luminal complexes, thus conserving the luminal/cytoplasmic polarity of the ER (Krstic, 1979). A second type of vesicle formation by in-budding of endoplasmic reticulum membranes has occasionally been observed or proposed. This mechanism leads to the formation of double layered vesicles (with two surrounding membranes) and with a reversed polarity: inside of the vesicle = cyto-

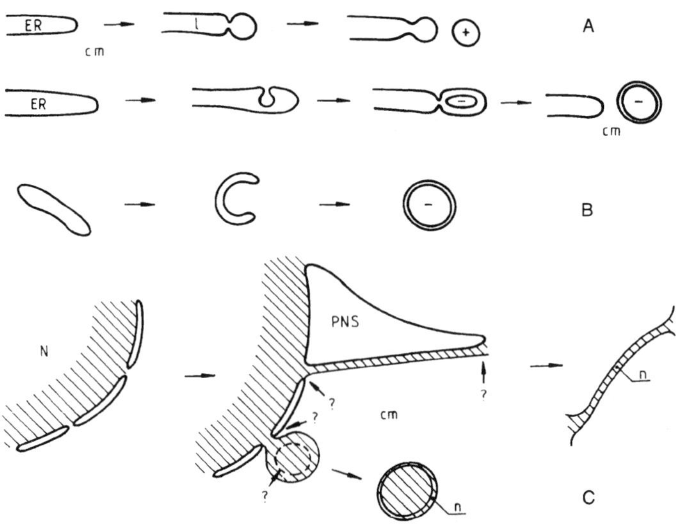

Fig. 77. Possible types of vesicle formation

A Simple budding from elements of endoplasmic reticulum (rER, sER, Golgi apparatus, GERL complex): polarity conserved, vesicles single-membraned

B Inbudding of endoplasmic reticulum or engulfment by endoplasmic reticulum: polarity reversed, vesicles double-membraned

C Outgrowth of nuclear membrane and de novo membrane formation: polarity uncertain, vesicles double-membraned

ER endoplasmic reticulum; *cm* cytoplasmic matrix; *l* lumen of endoplasmic reticulum; *N* nucleus; *PNS* perinuclear space; + same polarity as endoplasmic reticulum; – polarity of vesicle opposite to that of endoplasmic reticulum; *n* nuclear polarity; ? questionable mode of de novo membrane formation. – Figure C redrawn from Bienz *et al.*, 1970 [Arch. ges. Virusforsch. *31*, 262 (1970)] plasm, the original luminal portion of the ER being restricted to the small space between the two surrounding membranes of the vesicle. This type of vesicle has been described as a specific kind of virus induced vesicle in flavivirus infected cells and has been proposed to be a specialized assembly factory (Leary and Blair, 1980). Although vesicles with double layers of membranes have been described in poliovirus infected cells (Dales *et al.*, 1965), these types of vesicles do not seem to be formed in the early phases of vesicle formation in poliovirus infected cells and thus probably represent autophagic vacuoles rather than specific structurs involved in virus replication. Figure 77 shows schematic representations of the theoretically possible modes of vesicle formation.

A third type of vesicle or channel formation has been proposed from electron-microscopic observations in coxsackie virus (a picornavirus) infected muscle cells. Here, a canicular system of membranes has been proposed to arise from an outgrowth of the nuclear membrane originating at the nuclear pores (Fig. 77C), creating channels bound by a single layer of membrane (Bienz *et al.*, 1970). The lumen of these channels, however, was envisioned as being continuous with the intranuclear matrix and in turn with the cytoplasm rather than with the space between the two nuclear membranes. In other words, these channels would be of a polarity opposite to that of channels of the endoplasmic reticulum, with their contents corresponding to the cytoplasmic matrix. We think that this type of vesicle formation is unlikely to occur.

It has been proposed that vesicle formation of secretory proteins in uninfected cells is induced by an accumulation of cleaved leader peptides in the membranes of the endoplasmatic reticulum. An analogous mechanisms can be considered for the poliovirus induced formation of specific vesicles. Synthesis of a protein on membrane bound polyribrosomes rather than on soluble ribosomes can be taken as an indication for its destiny: to become a membrane associated or secreted protein. A major part of poliovirus protein synthesis does indeed proceed on membrane bound polysomes, and it can be expected that some of the poliovirus proteins are integral membrane proteins or are translocated through the endoplasmic reticulum, although it is not known as yet which of the poliovirus proteins actually participates in the membrane incorporation system of the host cell. There are several extended stretches of hydrophobicity within the poliovirus polyprotein (see Fig. 91, p. 360), and thus several potential membrane-anchoring sites.

The accumulation of integral poliovirus membrane proteins may provide the signal for the formation of poliovirus specific vesicles. Possible candidates are the "leader peptides" NCVP8, NCVP10 = both of these are N terminal fragments of P-2 proteins; or NCVP9, the N-terminal P-3 protein. Since guanidine sensitivity resides in the P-2 proteins, and since guanidine prevents the formation of virus specific vesicles, the P-2 proteins may well play a decisive mediatory role in the modification of poliovirus induced membranes. The vesicles are probably formed by budding from the rough endoplasmatic reticulum—either directly or via the Golgi or GERL complex (see Fig. 102B, p. 423). The Golgi complexes appear to disintegrate in the infected cell when the formation of vesicles begins. This could lead to the formation of several centers of small clusters of vesicles scattered throughout the cytoplasm.

B. Concerning the Relative Localization of RNA Synthesis and Virion Assembly with Respect to the Vesicle Membrane

Poliovirus RNA replication proceeds in two separable phases: During an early exponential phase of replication, most of the complementary RNA component in the infected cells is synthesized. cRNA is copied from mRNA (incoming parental RNA and RNA made early in infection from freshly formed cRNA) and, in turn, mRNA from cRNA in an exponential manner. The second phase begins near the midpoint of the replication cycle near the peak of protein synthesis, when a switch in the rate of RNA synthesis occurs from exponential to linear. This happens at a time when most of the cRNA component–in the form of a replicative intermediate–has been completed. 2,000–3,000 replicative intermediate molecules are present in the infected cell, and only little additional cRNA is made. Most of the RNA made from then on is of plus strand polarity, and most of this is virion RNA which is rapidly encapsidated into mature particles (see Chapter 9).

Very little can be said about the early phase of RNA synthesis with respect to host cell morphology, since the host cell retains the appearance of an uninfected cell during most of this phase, and RNA quantities are so low that they are difficult to detect. The parental viral RNA must be located on the cytoplasmic side of the endoplasmic reticulum since it is being translated by ribosomes. At some point after sufficient quantities of viral replicase proteins have been made, the parental RNA is temporarily removed from the translational machinery and copied into cRNA. cRNA in turn is copied into RNA of mRNA polarity, this very RNA back into cRNA and so on. A major portion of the RNA of plus strand polarity formed during this phase is removed from the replication process and returned to the ribosomes for translation. It is probable that this early RNA replication also takes place on the cytoplasmic side of the endoplasmic reticulum (we find it difficult to envision a specific transport mechanism by which cRNA is transfered to the luminal side of the ER and mRNA back to the cytoplasmic side). Since no virus specific vesicles can be detected during this early phase, it is also likely that RNA replication in this phase occurs in the soluble phase of the cytoplasm, *i.e.* not membrane associated. It is conceivable that the cytoskeleton provides binding sites for the replicative intermediate and mRNA at this time.

With the first appearance of virus specific vesicles in the form of small clusters, RNA synthesis seems to become intimately associated with some of these vesicles. The moment of the switch from the exponential phase of RNA replication to the linear phase seems to correspond to the first appearance of virus specific vesicles. A vast amount of biochemical evidence has provided support for the notion that RNA synthesis and encapsidation of virion RNA by capsid precursors are coupled processes that occur in tight association with these membranes during the linear phase of RNA replication.

Assuming that virion RNA synthesis and assembly into intact virus particles are coupled processes occuring on the membranes of sacs, channels or vacuoles, four theoretical possibilities for the relative localization of RNA synthesis (RNA template and replicase enzymes), and of virion assembly (capsid proteins engulfing the freshly synthesized RNA) exist with respect to the vesicle membrane.

Template copying and encapsidation of progeny RNA could occur on the same side of the membrane, the membrane merely providing a surface for three-dimensional organization and catalysis of the different processes. Both processes could be located either on the luminal or on the cytoplasmic side of the membrane compartments. Alternatively, template copying and encapsidation could occur on opposite sides of the membrane. In this case the membrane would serve a special secretory function. A mechanism whereby the freshly synthesized RNA is transferred from its site of synthesis to the site of its encapsidation would then have to be postulated. Figure 78 shows schematic representations of the four different possibilities.

A number of arguments support mechanisms 2 and 4 as most plausible for poliovirus replication. The models certainly call for experimental testing in the future. A strong argument against mechanisms 1 and 3 is the difficulty in envisioning how the RI could get to the luminal side of the membrane. mRNA certainly is cytoplasmic, cRNA made from mRNA is first found in a ds form, and in all likelyhood is also cytoplasmic. With the onset of copying of the cRNA on this ds template, the typical replicative intermediate is formed (a complete single strand of cRNA with 4 to 5 strands of incomplete +RNA). A second argument is that RNA synthesis is a process requiring energy. Each nucleotide substrate is a triphosphate, which cannot permeate across membrane barriers. It is likely that luminal RNA synthesis would quickly use up the substrate available within the restricted lumen of a vesicle.

Mechanism 2 provides the least conceptual difficulties. RNA synthesis and encapsidation would occur both on the cytoplasmic side of the vesicles, the vesicle-membrane may serve to organize and catalyze the individual steps. The consistent observation of progeny particles within membrane bound cisternae and channels (see section IIB6 above and Chapter 10), however, is not explained by this model. One would need to postulate engulfment of portions of cytoplasm by endoplasmic reticulum or de novo formation of membranes around virus particles (see Figs. 76, 77). Mechanism 4, in contrast, would result in the enclosure of progeny virions within vesicles. Cytoplasm-located particles in this case, could originate from lysis of vesicles late in infection.

For mechanism 4 to operate, a number of requirements must be met, most notably the transfer of a significant portion of capsid proteins across the endoplasmic membrane. In contrast to a presumptive transport mechanism of RNAs across intracytoplasmic membranes, which to our knowledge have not been described to date, transport of proteins across the membrane of the endoplasmatic reticulum is a well established and well studied phenomenon. Many examples for the insertion into or translocation across the ER membrane of secreted and transmembrane proteins are known (for review see Sabatini *et al.*, 1982). The final characteristic orientation of membrane proteins with respect to the membrane is thought to be determined by the sequence of transfer and half-transfer signals within the polypeptide chain of the protein. Insertion of proteins into the endoplasmic membrane or translocation into the lumen of the endoplasmic reticulum usually occurs during ongoing synthesis of these proteins.

As far as the capsid proteins are concerned, mechanism 4 would require that they are either translocated into the lumen of the ER or are integrally incorporated into the ER membrane with a significant portion exposed on the internal side of the membrane. The replicase proteins, at least those required for the initiation of vRNA synthesis may also be integral membrane proteins, the enzymatically active portions being exposed on the cytoplasmic side of the ER. The N-terminal fragment of the P-3 proteins, NCVP9, has been shown to be an integral membrane protein (Semler *et al.*, 1982; Takegami *et al.*, 1983a). It contains the genomic protein VPg and may provide a binding site for the RI template to the vesicle membrane and the associated P-3 proteins. VPg is cleaved from NCVP9 soon after initiation of RNA synthesis. The actual replicase activity responsible for elongation of RNA synthesis (NCVP4) is probably a soluble protein, possibly liberated from a more tightly membrane associated complex by proteolytic cleavage from NCVP2 after initiation of RNA synthesis. The template itself, the replicative intermediate, is probably also bound somehow to the cytoskeleton on the cytoplasmic side of the ER (Lenk and Penman, 1979).

A presumptive component of the transmembrane capsid protein should have a specific recognition site for the 5' VPg containing terminal of freshly initiated vRNA. Transmembrane localization of the capsid protein precursor (in the form of NCVP1a or as a complex of VP-$_{0,\ 3,\ 1}$) would provide an environment that enhances the polymerization of five such precursors into the 14S pentamers. There are several examples for the formation of such protein multimers within the plane of cellular membranes (Klingenberg, 1981). The presumptive poliovirus transmembrane pentamer would have a center of symmetry at the future five fold apices of the virion capsid (see Chapter 3). The 5' end of freshly initiated virion RNA could be recognized by portions of the transmembrane capsid pentamers, which at the same time may provide a channel for translocating the vRNA molecule through the vesicle membrane. Since progeny virions are found in the soluble phase of cell extracts, the virions must be released from membrane-association after completion of assembly—either on the cytoplasmic (mechanism 2) or on the

Fig. 78. Possible localization of linear phase poliovirus RNA synthesis and virion assembly with respect to the membranes of intracellular vesicles

A Mode 1: same side, luminal

B Mode 2: same side, cytoplasmic

C Mode 3: opposite sides: RNA synthesis luminal virion morphogenesis cytoplasmic

D Mode 4: opposite sides: RNA synthesis cytoplasmic virion morphogenesis luminal

provirion

native virion

initiating replication complex

viral RNA polymerase NCVP4 b

dsRNA unwinding protein (host factor?)

RI = replicative intermediate, template for ss + RNA synthesis

factors required for initiation, but not elongation of RNA synthesis (NCVPX, NCVP9, NCVP7c)

presumptive membrane channel for RNA

mRNA

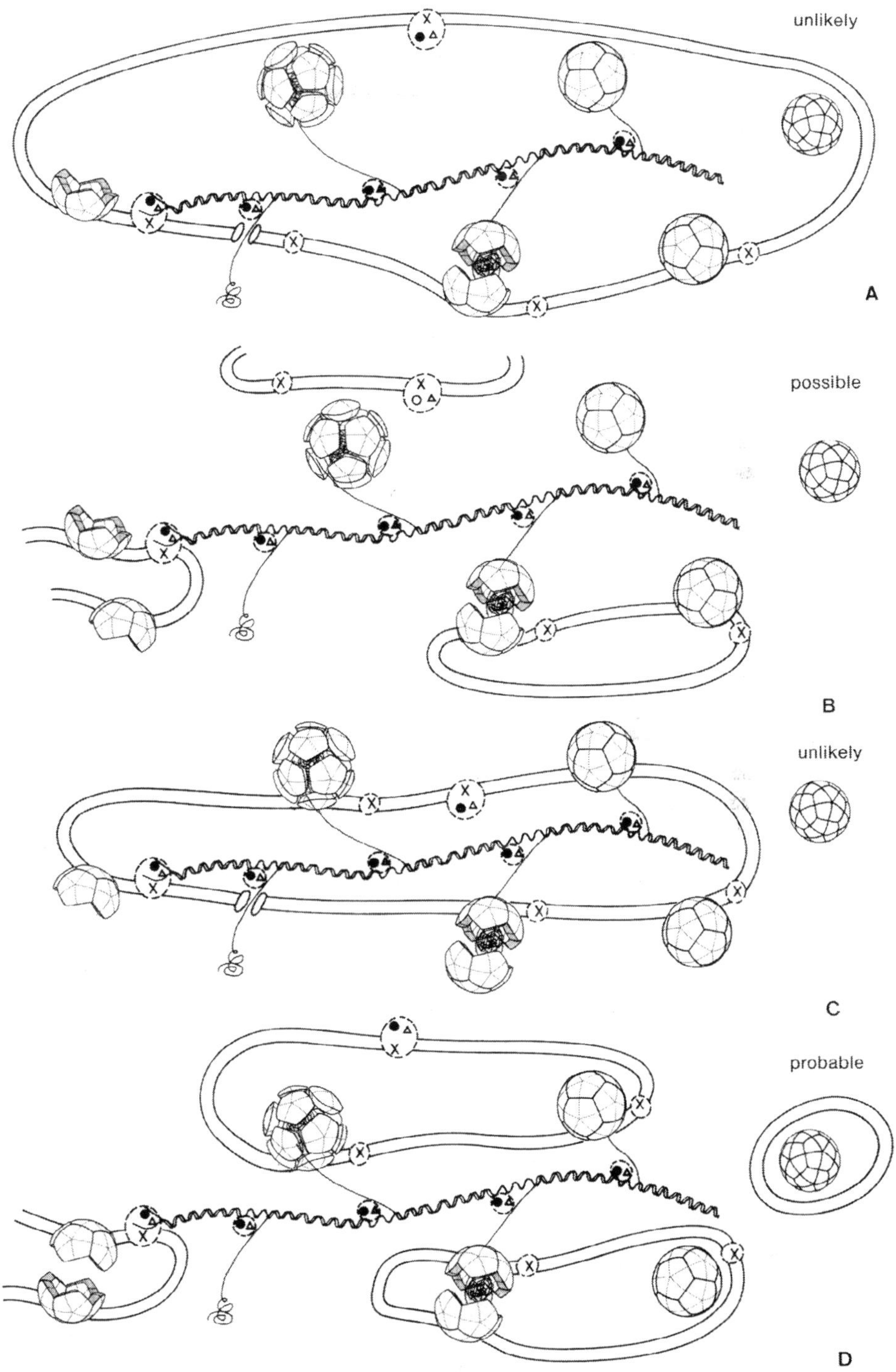
unlikely
A
possible
B
unlikely
C
probable
D

cisternal (mechanism 4) side of the vesicles. A number of specific predictions follow from the models, which should be open to experimental testing with present day technology.

IV. Poliovirus Induced Alterations in Functions of the Plasma Membrane, in the Intracellular Ionic Environment, and in Cell Size

Alterations in the structure and function of the cell membrane appear to be essential for several steps involved in preparing the cell for virus replication. Table 37 lists distinct phases of characteristic membrane alterations in the poliovirus-HeLa cell system. Figure 65 (see p. 231, above) shows electron micrographs of membrane alterations during poliovirus replication in HEp2 cells (Zeichhart *et al.*, 1980).

Table 37. *Phases of plasma membrane alterations in poliovirus infected HeLa cells*

Phases	Time post infection
1. *During or directly following adsorption and penetration:* DPH-increased amino acid transport, transient membrane leakiness	10 minutes
2. *At the onset of virus replication:* cell volume decrease, activation of Na^+/K^+ ATPase, increase in membrane potential, increase in size of amino acid pool, inhibition of furosemide sensitive $Na^+/K^+/Cl^-$-uptake	0.5–1.5 hours
3. *During the peak of virus replication:* Na^+/K^+ATPase inhibited, membrane potential decreases, Na^+ increases, K^+ decreases, cell volume increases, membrane permeability increases	2.5–4 hours
4. *Cessation of virus replication, death of host cell* further increase in membrane permeability, membrane leakiness, loss of life-sustaining components (ATP)	5–7 hours

A. Membrane Changes Accompanying Adsorption and Penetration of Poliovirus

The poliovirions first interact with the plasma membrane of susceptible cells (see Chapter 7). Both reacting partners are altered, the viral coat is labilized and the virion enters the cell with active participation of the plasma membrane and the cytoskeleton. In some virus host cell systems virus adsorption and penetration is accompanied by an immediate increase in membrane permeability notably for K^+, resulting in the collapse of the K^+ gradient (Pasternak and Micklem, 1981). So

Table 38. *Plasma membrane alterations during the first two hours after poliovirus infection*

	Maximal change %	Time of maximal change post infection
DPH fluorescence polarization	− 13	5 min
Cell volume	− 18	1 hour
Na^+ and K^+ contents	no change	—
Na^+/K^+ ATPase activity	+ 34	1 hour
Membrane potential	+ 10	1 hour
Amino acid pool	+ 30	1 hour
Protein synthesis	− 80	2 hours

Data from Zibirre and Koch, 1983; Schaefer *et al.*, 1983 a, b.

far, comparable results have not been reported with a member of the picornavirus group. However, very early changes in membrane functions after poliovirus infection are indicated by the following observation (Table 38): Within five minutes following poliovirus infection of HeLa cells a transient increase in the uptake of amino acids (AIB) into cells and a decrease in the fluorescence polarization of DPH become apparent (Zibirre and Koch, 1983). These very early events might be caused by adsorption and penetration of the virus. The rapid decline in DPH fluorescence polarization was shown to occur in several virus host cell systems immediately after virus adsorption and was attributed to an increase in membrane fluidity (Kohn, 1979; Zibirre and Koch, 1983). Poliovirus particles bind DPH. It cannot be excluded that the decrease in DPH fluorescence polarization is simply due to trapping of DPH by virions (Zibirre, personal communication).

Even though the intracellular contents of Na^+ and K^+ ions do not change in the first two hours following infection, the activity of the Na^+K^+ ATPase appears elevated within one hour following infection when determined either directly in isolated plasma membrane fractions or by following the influx of Rb^+ after preloading of the cells with Na^+ (by incubation at low temperature). An increased influx of Rb^+, however, is not observed without preloading of the cells with Na^+ (Schaefer *et al.*, 1983). The activation of the Na^+K^+ ATPase may be a direct result of adsorption or penetration of poliovirus to or into HeLa cells. The relatively small increase in membrane potential as well as the marked increase in the size of the amino acid pool at 60 min. p.i. may be due to the increase in the activity of the Na^+K^+ ATPase. The increased activity of the pump may contribute to the reduction in cell volume, which is lowered by 20% within one hour p.i. A comparable activation of the pump is observed after addition of serum or of growth factors to HeLa cells, suggesting that this may be a defensive response of the cell against virus. Indeed, inhibition of the pump by ouabain increases virus replication.

Another function of the cell membrane becomes impaired at 60 min p. i., that is the furosemide or piretanide sensitive $Na^+K^+2Cl^-$ cotransport (Schaefer *et al.*, 1983). This cotransport is involved in the regulation of the cell volume in uninfected cells (see Figure 62, p. 217).

The increase in the activation of the Na^+K^+ pump, the decrease in the cell size, and the increase in the size of the amino acid pool are greatest at one hour p. i. and are reversed then. From this point on the virus is in full command. The cell size increases again, membrane potential and amino acid transport decline concomitantly with an ever increasing inhibition of host protein synthesis.

B. Membrane Alterations Accompanying Virus Replication at the Maximal Rate

At two hours p. i., a sudden drastic reduction in the Na^+ and K^+ gradients occurs, the intracellular Na^+ concentration rises, the K^+ concentration declines leading to near collapses of both gradients across the plasma membrane. This is accompanied by an increase in the size of the cells. Soon thereafter viral replication (both viral RNA and protein synthesis) proceed at a maximal rate. Viral RNA and protein synthesis—unlike cellular synthesis—do not require high concentrations of K^+ and are not dependent on low intracellular concentrations of Na^+.

There are large differences in the values found for intracellular Na^+ and K^+ concentrations depending on the method used for the estimation of the cell volume. The Coulter Counter indicates a large increase due to a drastic reduction in cell size at 3 hours p.i. (Nair, 1981). The partitioning of ^{14}C-inulin and tritiated water, however, indicates a swelling of cells beginning at 2–3 hours after poliovirus infection (Schaefer *et al.*, 1983). It is conceivable that a change in certain physical or biochemical properties of infected HeLa cells can distort Coulter Counter measurements. An increase in the conductivity of the membrane due to an increased permeability of the membrane for ions would result in a decreased signal when cells pass through the channel in the Coulter Counter (Geck, personal communication).

At late times in poliovirus replication the permeability of the cell membrane increases further which leads to release of ATP and the cessation of virus replication.

Virus induced alterations of the plasma membrane have to be discussed under two aspects: the take over of the metabolic machinery of the host cell by the virus, and the defense of the host cell against this take over. The early activation of the Na^+ K^+ ATPase, the small increase in the membrane potential and in the size of the amino acid pool mimics events following the addition of growth factors or serum and suggests that the cell attempts to counteract the take over by the virus. Two observations support this view. Addition of serum to HeLa cells following infection with poliovirions delays or inhibits virus replication (Schärli and Koch, unpublished), whereas addition of ouabain or other cardioactive steroids increases the yield of virus in monkey kidney cells (Koch and Fehèr, 1973).

The early decrease in cell volume and the accompanying rise in the intracellular concentration of cations might be involved in the triggering of the shut-off of host macromolecular synthesis. The increase in membrane permeability to ions and other low molecular weight substances, and the collapse of the Na^+ and K^+ gradients across the plasma membrane cause a cessation of host protein synthesis. The virus induced inhibition of host protein synthesis can be amplified by an increase in medium tonicity (Nuss *et al.*, 1975) and can be counteracted by a decrease in the tonicity of the medium (Tolskaya *et al.*, 1966; Alonso and Carrasco, 1981).

V. Summary

The infection of cells by poliovirus leads to a dramatic reorganization of cellular structures. Most striking are the distortion and displacement of the nucleus, an extensive proliferation of membraneous sacs and vesicles in the perinuclear region, and profound rearrangements of the cytoskeleton. Two partly opposed mechanisms seem to underlie the observed alterations: 1. a specific virus induced compartmentalization and functional reorganization of cellular membranes for the synthesis and assembly of virion components, and 2. a cellular defense mechanism that recognizes the foreign viral material and attempts to degrade and expel it from the cytoplasm. In the late stages of virus replication, after the bulk of virion formation, the effects of cellular autophagic processes and virus related cytotoxic products predominate and eventually kill and disintegrate the host cell.

During the early and linear phases of virus replication, modifications of cellular structure and metabolism to suit viral interests seem to predominate. Presumably under the direction of a freshly synthesized virus protein(s), proliferation of membranes sets in. Freshly synthesized membranes are of specific chemical composition. The site of membrane synthesis is the rough endoplasmic reticulum. The newly synthesized membranes are then transported—presumably via the Golgi/GERL complex and/or in the form of small vesicles—to smaller clusters of vesicles and to the perinuclear area. It is likely that the formation of membranes and their transformation into a transportable form are induced by a virus specific protein or leader peptide, analogous to the formation of coated and secretory vesicles during the synthesis of integral membrane and secreted proteins. Potential candidates for such functions in the poliovirus infected cell are the P-2 proteins NCVP3b, -5b, -X and the corresponding "leader peptides" NCVP8 and -10. The locus for guanidine sensitivity lies within the coding region for these proteins, and addition of guanidine indeed prevents the migration of membraneous vesicles. Some of the freshly formed vesicles are intimately involved in poliovirus RNA synthesis and virion assembly. The replication complex, containing replicative intermediate RNA (a complete single strand of complementary RNA and 4—6 tails of plus strand polarity in the process of synthesis), the viral replicase NCVP4 and other P-2 proteins, as well as host proteins, are very tightly associated with the membranes and with capsid proteins. This association can only be disrupted by strong detergents, which may conserve the elongation capacity of the replicase, but destroy the capacity to initiate RNA replication as well as the capacity to encapsidate the freshly synthesized RNA.

The major steps of virion assembly evidently occur in conjunction with specialized vesicles, presumably already during RNA synthesis. The relative localization of replicase, RNA template (RI), capsid proteins and freshly synthesized virion RNA with respect to the vesicle membrane (outside/inside) are still matters of speculation. A useful experimental approach to answer such questions has yet to be found. Some arguments, however, can be put forward for a concept that envisions RNA template and replicase to be located on the one side (the outside) of the vesicle membrane, virion capsid proteins on the opposite side (corresponding to the lumen of the endoplasmic reticulum). Virion RNA synthesis then is visualized as a transmembrane phenomenon, the RNA being guided through the membrane shortly after initiation of RNA synthesis, and becoming encapsidated sequentially on the other side of the membrane.

The number of vesicles that accumulate in the interphase infected cell by far surpasses the number that would theoretically be required for RNA replication on 2,000 - 4,000 replicative intermediate templates and for the production of 10,000 - 200,000 progeny virions. (Assuming that 5—20 virion particles are formed from each RI). Evidently a large number of vesicles formed in the infected cell do not directly participate in RNA replication or virion formation.

At least some of these vesicles seem to participate in a normal host cell defense mechanism. As virion-replication proceeds, a large number of "foreign" molecules accumulate in the host cell, in particular the double and multistranded forms of RNA (RF RNA, RI RNA), as well as large quantities of non-cellular proteins and virus particles. The cell certainly has inherent mechanisms for the recognition and destruction of foreign materials, and some of these seem to be activated by poliovirus infection. The vesicles that accumulate in poliovirus infected cells are very heterogenous, and it is as yet rather difficult to distinguish their functional significance on the basis of morphological criterea alone. Individual vacuoles can of course often be identified as typical autophagic vacuoles, primary vacuoles, etc. Nevertheless, the significance of the different types of vesicles formed remains to be elucidated.

Cellular defense mechanisms seem to predominate late in infection. When the bulk of virion formation has been completed, a redistribution of lysosomal enzymes over the entire paranuclear mass of vesicles can be followed by cytochemical procedures. Dissolution of fragments of cytoplasm and of cellular organelles (mitochondria) can be observed in the late stages. Energy depletion, substrate depletion, shut-off of host macromolecular synthesis, and autodigestive processes may be involved in these late stage phenomena that ultimately lead to death and lysis of the cell. Genetic evidence indicates that dsRNAs that accumulate near the end of virion replication directly mediate some of these cytotoxic processes.

Morphological alterations resembling typical stress reactions can also be seen in infected cells: separation of the two nuclear membranes and the formation of large nuclear extrusions. Neither protein nor RNA synthesis appears to be associated to the nuclear extrusions, and it is doubtful that they play any significant functional role during poliovirus replication.

Not much is known about the mechanism of release of progeny virion from the host cell. A large proportion of progeny virion seems to be released from the host cell prior to cell lysis, so that simple cellular breakdown does not appear to be the predominant cause of virion release. Specific canicular extrusion systems have been observed in some cells, as well as seemingly less specific expulsions of large fragments of cytoplasm (cytoplasmic blebs) that may contain virions.

Intracellular virus crystals are usually observed only very late in infection and only in few cells, when significant degradation of host cell cytoplasm has already occured. This may simply reflect an enhanced diffusion potential for virions within the cytoplasm, or the release of virions from the confinement of cellular compartments.

7

Early Interactions of Virus and Host Cell

I. Introduction

A. Overview and Definition of Terms

Molecular biologists have concentrated their efforts in studies of the composition and structure of polioviruses and on the intracellular steps in viral replication. Compared to these topics only a moderate number of publications in the last 25 years have appeared dealing with the early events of virus host cell interaction. Discussing these events we will have to refer repeatedly to observations made with other picornaviruses. It is important to keep in mind that there is probably no other phase in viral replication in which members of the picornavirus groups differ so markedly. Terms pertaining to the early interactions of poliovirus and its host cell are explained in Table 39.

The first encounter of a virus with its host cell during natural infection takes place on the plasma membrane. The binding of poliovirus to the host cell membrane is a highly specific process that involves a series of steps: the recognition and binding of poliovirus by the host cell receptor, the formation of a receptor complex, and insertion of the virion into the lipid bilayer of the cell membrane. *Adsorption* denotes the initial complementary interaction between a virus and specific sites on receptor(s) of the host cell, resulting in loose binding of the virion to the host cell. The term *attachment* refers to the sum of the reactions leading to a firm binding of virions. Attachment may involve or trigger *insertion* of the virion into the cell membrane. At physiological temperatures the binding process is so rapid that the individual steps cannot be distinguished in time. Earlier studies in particular did not attempt to put forward a distinction of these steps, and the terms adsorption and attachment were often used synonomously.

As a consequence of the first interaction between virus and host, rapid and dramatic dynamic events ensue in the course of which both reaction partners are markedly modified. The diverse effects on the virion include:

a) changes in virion conformation from native to "sticky", with increased surface hydrophobicity, loosening of the capsid, and swelling of virions

Table 39. *Definition of terms pertaining to the early events in poliovirus infection*

Term/Phase	Definition
Adsorption	specific interaction between complementary sites on the virus capsid (receptor recognition site) and on the receptor (virus recognition site) leading to relatively loose binding
Attachment	
a) tight binding	binding by multiple components of receptor complexe(s)
b) Insertion	physical insertion into the lipid bilayer of the cell membrane by hydrophobic interactions
Penetration, entry	movement of virus from the outside to the inside of the cell
a) endocytosis or phagocytosis	virus is located within membrane enclosed vesicles (phagocytotic vesicles)
b) fusion – viropexis	virus inserts into and passes through the membrane barrier
Uncoating	stepwise destabilization and consecutive loss of virion capsid proteins from encapsidated RNA
Start of replication	initiation of translation of incoming genome on host cell ribosome
Virus	
Receptor recognition sites	specific surface structures complementary to the virus recognition site on the receptor
a) Virus like particles	
Native virion	intact virus with unaltered coat composition and structure; isoelectric 7.0 = A form, 155S
"Sticky" virion	like native but altered structure; isoelectric 4.5 = B form, 155S
Membrane complexed or stabilized virion	physically complete virus, stabilized by non virus components against degradation
b) Altered particles	change in composition of coat
A or M-particles (modified)	VP_4 lost
C-particles (core particles)	VP_2 lost
Stabilized A-particles	stabilized by components of the receptor complex
Host cell	
a) Receptor	Integral membrane glycoprotein that harbors the virus recognition site
virus recognition site(s)	specific portion of the receptor complementary to the receptor recognition site on virus
mono- or multifunctional	one or more virus recognition sites per receptor
b) Receptor-complex	multifunctional system of different—possibly associated—membrane components involved in virus attachment and viropexis
Constituents:	
receptor	protein with virus recognition site
virus stabilizing activity	responsible for stabilizing intact virions and altered particles
modifying activities	responsible for destabilization and alteration of viral particles

Term/Phase	Definition
inserting activity	promotes insertion of virus into membranes
viropexis mediator	mediates passage of viral particle through the membrane
membrane anchor(s)	hydrophobic portions of receptor complex proteins
cytoskeleton attachment site(s)	site(s) on receptor complex protein(s) for interaction with components of the cytoskeleton
c) Attachment site	region on the membrane where virion is tightly bound by several receptors or receptor complex(es)
d) Attachment patches	aggregation of two or more attachment sites with associated virions
e) Attachment caps	aggregation of many attachment patches
Infection:	introduction of genomic RNA into host cell
a) productive	resulting in production of progeny virions
b) abortive	interaction of virus and host cell without formation of progeny virions

b) changes in coat composition, loss of VP4 or

c) restabilization of the virion by membrane components.

Alterations in the cell membrane involve lateral movements of proteins in the membrane resulting:

a) in the formation of receptor complex(es) and attachment sites and

b) in (local) changes in membrane permeability and fluidity.

Entry or *penetration* refer to the process whereby the viral genome—usually still in some encapsidated form—overcomes the membrane barrier and reaches the cytoplasmic matrix. In the first step of penetration, the virion is moved from the extracellular space across the cell membrane to a location within the host cell cytoplasm. Two theoretical possibilities exist for this step:

a) the virion passes through the hydrophobic lipid bilayer of the plasma membrane after attachment and insertion, either intact or in modified form—this process is termed *fusion-viropexis*

b) the virus is engulfed by the cell membrane into a membrane enclosed vesicle (phagocytic vesicle) that is "pinched off" from the plasma membrane and comes to lie in the cytoplasmic matrix. This process is termed *endocytosis.* Early intracellular appearance of parental virus has often been observed within such vesicles. Yet in this state, the virus still faces the same problem: The viral genome has to overcome a membrane barrier—now in form of the vesicle membrane—in order to reach the ribosomes.

The term *uncoating* is used to denote the sum of the structural alterations of the viral capsid that lead to the release of infectious genomic RNA into the cell cytoplasm.

Principally, entrance into the cell cytoplasm of the genomic RNA alone suffices for initiating infection. Yet in contrast, for example, to the T-even phages of bacteria, the genome of poliovirus appears to enter the cell still in some association with capsid proteins. The parental capsid proteins of poliovirus may play an important role during early steps of infection. Indeed, the infectivity of poliovirus RNA encapsidated in its native coat is a thousandfold greater than that of isolated RNA, even though comparable amounts (50%) of isolated viral RNA and virion enter the cell. The hydrophobic capsid proteins may serve as carrier for the hydrophilic RNA through the hydrophobic membrane barrier (either of the cell membrane or of the phagocytic vesicle). It is conceivable that the capsid proteins protect the RNA against RNAse not only during penetration of the membrane barrier but also in the cytoplasm. The ultimate fate of the parental viral proteins is still uncertain. Some viral proteins of parental origin are found in polysome preparations. They may remain associated with the parental viral RNA during translation. It is conceivable that the capsid proteins facilitate transport of the parental genome to ribosomes. In addition, they might exert a local modifying effect on the host's translational machinery and promote—perhaps locally—the shut-off of host protein synthesis, thereby facilitating initiation of parental viral RNA translation. Another intriguing possibility is that the parental viral proteins mediate the switch from translation to replication of the parental RNA. Whatever the answers to these questions about the function of parental viral proteins may be, most of the viral capsid proteins must eventually be removed from the genomic RNA so that translation of the viral RNA can commence and initiate the synthesis of the viral proteins that are required for productive infection.

Adsorption, attachment, insertion, penetration and uncoating are treated as distinct processes mainly for the sake of clarity. They are distinguished by different functional concepts. During the early interactions of poliovirus with its host cell, these processes are actually not so clearly separated in time and space. Each phase involves a series of steps, and some steps may overlap in time or may occur in the same locality of the host cell. For example, modification of the virus on the cell surface during attachment might also be considered as the initial step of uncoating.

Some of the events involved in the early interactions of poliovirus and the host cell membrane have been studied in detail and are well characterized. Still it is difficult to assess the biological significance of the accumulated data. The fundamental question is, do reported observations represent the first essential steps of a productive infection or the steps of "dead-end pathways" leading to abortive infection?

B. The Superposition of Abortive and Productive Pathways in Infection

Addition of 1,000 poliovirus particles to a monolayer of 1,000,000 cells results in adsorption of 90 % of the virus particles to the cells within 15 min at 37°C, but leads to only 10 productive infections (formation of plaques by infective centers) (Schwerdt and Fogh, 1957). Adsorption of a poliovirus particle to a host cell does not predetermine its fate, nor is adsorption or attachment always the initiating event of a productive infection. In fact, the fates of individual viruses differ markedly after attachment: Only about 50 out of a hundred adsorbed poliovirus particles succeed in penetrating into the cell and of these only one or two are able to induce a virus replication cycle. Hence 97–99% of initiated virus-host cell interactions are abortive, only 1–3% proceed to productive infection. Whether the step determining the successful initiation of infection by an adsorbed virus resides in the viral capsid structure or in activities of the viral genome or in properties of the host cell is not known.

Presently, no biochemical or other approach is available to follow specifically that one virus particle or viral genome which is responsible for infection, or to determine the special features and conditions that permit it to carry out a successful infection. There is at present no definite way that could help to distinguish between the phenomena that are related to abortive and those related to productive pathways.

Since molecular biologists are primarily interested in virus-specific events in macromolecular synthesis, they have mainly studied the late phases of productive infections. To obtain quantifiable results on events in the productive pathway in spite of the high abortive rate, investigations on poliovirus in general, and in particular those on early events, are carried out at high multiplicities of infection. Infective ratios of at least 10 plaque forming units (PFU) per cell, corresponding to 1,000 virus particles per cell, are used in order to infect nearly all cells. Thus for each 10 productively infective virus particles there are also 300–600 virus particles that interact with the cell and enter it, but are mere participants in abortive pathways.

An unavoidable consequence is the superposition of concomitant abortive and productive pathways. It is very important that the reader keeps in mind this restriction in the discussion of all observations dealing with attachment and the ensuing early events in virus infection.

II. Adsorption and Attachment

Although adsorption and attachment of picornaviruses have been studied for three decades, our understanding of these processes is still rather limited (for detailed reviews see Kohn, 1979; Lonberg-Holm and Philipson, 1974, 1980; Crowell and Landau, 1978, 1979, 1983). A precise understanding of virus adsorption and attachment requires biochemical and structural analysis of both reaction partners in terms of

a) the specific sites on the virus involved in receptor recognition

b) the cellular receptor in the context of the cell membrane (receptor complex)

c) a kinetic analysis of their specific interaction, and

d) the modifications of both reaction partners resulting from the interaction.

Adsorption of polioviruses to their host cells which is due to the interaction of one receptor recognition site on the viral capsid with a specific site on the plasma membrane is reversible. Adsorption is followed by or "triggers" further interaction between virus and host cell which can lead to a firm binding = attachment of the virion to the host cell. One virus particle might bind to several receptor sites and to other components of the plasma membrane (receptor complex).

A. The Reaction Partners

1. The Virus Particle

The identity and specific features of the receptor recognition sites on the poliovirus capsid are still largely unknown. Present ideas are based only on indirect experimental evidence and on general considerations pertaining to capsid structure. The receptor recognition sites are likely to be located on (specific) protrusions from the capsid backbone. These are the most exposed and sterically most accessible regions of the virus particles, features that may be important for ready interactions with a complementary receptor. Of course, other surface exposed regions including "grooves" or "valleys" between adjacent capsid proteins or adjacent surface projections can not be excluded with certainty as being potential receptor recognition sites. Many icosahedral viruses are known to have special attachment devices protruding from their 12 five-fold vertices. Well-known examples are the twelve antennas of adenoviruses and bacteriophage ØX174 (Philipson *et al.*, 1968; Dales, 1973).

Rather than being specified by a particular single protein out of several viral capsid proteins, a receptor recognition site may very well be constituted from the projections of two or more adjacent capsid proteins. The icosahedral capsid of poliovirus contains 60 copies of each of the four viral proteins and, therefore, theoretically up to 60 potential receptor recognition sites. The geometrical arrangement of the capsid proteins, *e.g.* the clustering of individual capsid proteins about the twelve 5-fold apices or the twenty 3-fold facets of an icosahedral lattice (see Chapter 3), provides the possibility that various capsid peptides may act in a concerted manner to establish receptor recognition sites for adsorption. Various theoretical possibilities for the constitution of poliovirus receptor recognition sites that follow from the poliovirus structure model are presented in Table 40. They are similar to those listed in the Chapter 3 for the antigenic sites (see Table 23, p. 116). In most picornaviruses, one or two of the 60 VP4 - VP2 precursor proteins VP0 are not cleaved during virus maturation. It is possible, though unlikely, that these specify significant or high affinity receptor recognition sites.

The receptor recognition site of poliovirus is in an active or accessible state only when poliovirus is in its native antigenic D state. The conversion of the virions form the A state (isoelectric point 7.0) to the B state (isoelectric point 4.5) does not alter the accessibility of the receptor recognition site. Various conditions that convert polioviruses from the D to the C conformation (characteristic of empty capsids) and cause a selective loss of VP4 from polioviruses, also abolish the capability of the virus to attach to the receptor. Since VP4 is only a small capsid protein (10% by weight of the capsid protein), which occupies an internal position close to the RNA, it is quite unlikely that VP4 by itself constitutes the receptor recognition site that is lost by these treatments.

The relationship between the receptor recognition site and the site responsible for binding of neutralizing antibodies on the surface of poliovirus is still unclear. Any portion of the viral capsid, that can interact specifically with a complementary proteinacious cell receptor, should also constitute an antigenic determinant. As discussed in Chapter 3, the capsid sites that may induce plasma

Table 40. *Theoretical receptor recognition sites on picornaviruses*

Viral peptides	Number of peptides	Location on the virus particle	Surface area nm^2	Number of sites/virion	Remarks
VP_1	1	part of 3-fold apex	15	60	possible
VP_2	1	part of 5-fold apex	15	60	possible
VP_3	1	part of 3-fold apex	10	60	possible
$VP_3 + VP_1$	1	part of 3-fold apex	25	60	possible
(VP_1, VP_2, VP_3)	3	groove between clusters	40	60	improbable
$(VP_3 + VP_1)_2$ VP_3, VP_1 / VP_1, VP_3	4	groove between 3-fold apices	50	30	unlikely
$(VP_3 + VP_1)_3$	6	3-fold apex	75	20	most likely
$(VP_3 + VP_1)_6$	12	$2 \times$ 3-fold apices	150	10	possible
$(VP_2)_{2-5}$	2–5	5-fold apex	30–75	12	improbable
$(VP_1)_3$	3	facet center	45	20	likely

cells to initiate production of antibodies are in all probability constituted by specific surface projections (see p. 114 ff.). Potential antigenic determinants, that are hidden in grooves or valleys on the native virion capsid, may be able to interact with complementary antibodies, but may be unable to induce plasma cells to initiate production of these antibodies until the virion capsid is altered or dissociated. Based on the evidence presented in Chapter 3, it was proposed that type-specific neutralizing antibodies against poliovirus are specified by surface projections involving VP_2 or VP_1 near the 12 icosahedral 5-fold vertices, whereas group-specific, non-neutralizing antibodies were directed towards less exposed evolutionary-conserved sites on the capsid backbone, presumably involving VP_1 and VP_3. The poliovirus receptor does indeed resemble the group-specific antibody in that it does not distinguish between the three serotypes (but does differentiate polioviruses from other enteroviruses). However, the receptor recognition site is certainly not identical to the C-reactive or antigenic site of the capsid since C-reactive A particles and ATC do not have attachment capacities.

Binding of one or a few neutralizing anti-D antibodies to a virus particle—which "locks" the virion capsid into its hydrophobic pI 4.5 state—interferes with the normal sequence in infection (in particular with the uncoating process) (Mandel, 1962b, 1967a,b, 1976, 1979). The capability to adsorb to HeLa cells of poliovirus after interaction with neutralizing antibodies was found to vary according to several conditions. Soon after virus-antibody interaction or after incubation with dilute antisera virus antibody complexes adsorb as well as unexposed virions. When neutralized virus was incubated at $5^{o}C$, its capability to adsorb even improved up to three-fold over the control. This time-dependent improvement was observed only with 7S but not with 19S antibodies. These results indicate that the conformational shift of the viral capsid induced by antibody binding does not distort or hide the receptor recognition site and that receptor and antibody bind to different areas (peptides) of the viral capsid. Larger virus-antibody aggregates are less efficiently bound to cells (Mandel, 1967b), probably due to steric interference. Even after prolonged incubation with antisera, polioviruses possess residual infectivity of the order of 0.01—1.0%.

In the case of foot and mouth disease virus, proteolytic cleavage of VP_1 on intact virion inactivates the attachment capacity of the virus (Cavanagh *et al.*, 1977). Poliovirus mutants have been described which cannot attach to host cells in the presence of the polyanion dextran sulfate (Bengtsson, 1965). The dextran marker apparently maps in the coding region for VP_1, further suggesting that portions on this protein may be important for attachment (see p. 198—199, Chapter 4). Recent immunological studies have implicated VP_1 as an immunodominant peptide of the viral surface (see p. 130—131, Chapter 3). It is possible that some of the neutralizing antibodies obtained after inoculation of rats or rabbits with isolated VP_1 (Blondel *et al.*, 1982; Chow and Baltimore, 1982) neutralize infectivity by blocking the receptor recognition site.

As discussed in Chapter 3, the basic arrangement of capsid proteins within the capsid backbone is highly conserved in evolution, in contrast to particular surface protrusions which may be very different even in closely related viruses. Such variations in surface protrusions presumably correspond to changes in antigenicity as a consequence of antibody pressures in the host organism. The peptide regions of the receptor recognition site on the viral capsid should be relatively conserved among the three poliovirus serotypes, since they recognize the same receptor on the host cell. In comparison to other picornaviruses, in contrast, the receptor recognition site should differ markedly, reflecting the adaptation of these viruses to other hosts and host cell receptors. Comparison of the deduced amino acid sequences of the three poliovirus serotypes (Toyoda *et al.*, 1984; Chow *et al.*, 1984) and of other picornaviruses may provide indirect clues for the peptide regions constituting the receptor recognition site. Crosslinking studies with polioviruses attached to isolated receptors fixed on polysterene (Krah and Crowell, 1982) should provide additional information on the viral peptides or peptide regions involved in adsorption. On the basis of the elucidated three-dimensional structure of poliovirus (Hogle, 1984) and with the aid of monoclonal antibodies and site directed mutagenesis, the precise definition of the receptor recognition site should prove possible.

2. The Virus Receptor Complex on the Host Cell

The plasma membrane harbors the virus receptors. The entire plasma membrane system responds rapidly to virus attachment with selective changes in permeability, microviscosity (Kohn, 1979), patch-formation and capping (Gschwender and Traub, 1979) and in reduced transport of amino acids (Koch *et al.*, 1980a; Schaefer *et al.*, 1983).

The presence or absence of specific receptors on the plasma membrane of a given cell determines the susceptibility of this cell to infection by a given virus. Resistance to virus infection in general is due to a lack of receptors rather than to an inability to support productive infection. This conclusion is based on the observation that cells of non-primate origin which lack poliovirus receptors can be infected by isolated poliovirus RNA (Holland *et al.*, 1959a and b, 1961). Consequently receptors are important for virus tropism and, therefore, also in pathogenesis. In analogy with other virus-host cell systems, poliovirus susceptible primate cells are thought to possess components on the surface of their plasma membrane—the receptor site—which in a specific structural conformation show a high affinity for polioviruses. A specific site of a single protein or portions of several molecules together might provide the required receptor specifity. Receptor monomers might adsorb viruses only weakly unless coupled into polymeric structures (see below).

A virus receptor might not only be responsible for efficient binding of viruses to the surface but might also function alone or together with several other membrane components to induce a conformational shift in the viral capsid and/or to direct the virus to the intracellular location of virus replication (Fenwick and Wall, 1973). The sum of the membrane components that are responsible for the binding and modification of poliovirus are termed the receptor complex (Crowell and Landau, 1979). To date it is not yet possible to ascertain whether all, or which one, of these components actually are associated with the attached virus in a

Table 41. *Properties of the poliovirus receptor complex*

Treatment	Inactivation	Conclusion
Protease	yes	protein
High ionic strength	no	receptor not loosely bound to membrane
Organic solvents: ether, chloroform	yes	integral membrane component
Partial membrane solubilization with 1 % Triton and short ultrasonic treatment, isopycnic centrifugation	no	receptor activity associated with lipids
Complete membrane solubilization	yes	active only as membrane component
SH reactive substances	no	SH groups not essential
Periodate	no	carbohydrates not essential
Con A	yes	prevention of lateral movement of membrane components and formation of receptor complex
Heat 56 °C	yes	protein (multi-component)
Competition binding:		
– to different poliovirus subtypes	yes	group-specificity for polioviruses
– to other picornaviruses	no	distinct from other picornavirus receptors

larger complex, or whether they act relatively independently in a consecutive fashion.

a) Properties of the Poliovirus Receptor

Table 41 summarizes the properties of the poliovirus receptor. Treatment of intact cells with proteolytic enzymes destroys the receptor activity for polioviruses as well as for coxsackie virus B3 (Zajac and Crowell, 1965; Levitt and Crowell, 1967), indicating that a protein carries the receptor site.

When protease treated cells are washed, lysed and assayed for their capacity to bind polioviruses, no receptor activity is detected. It was concluded that cells contain poliovirus receptors only at the outer plasma membrane (Zajac and Crowell, 1965; McLaren *et al.*, 1968). The complete removal of poliovirus receptor activity by proteases also suggests that there are no cryptic receptor sites in the cell membrane. The regeneration of poliovirus receptor activity occurs relatively fast, within 1–2 hours (Levitt and Crowell, 1967), much faster than the regeneration of receptors for other viruses. Sensitive cells efficiently adsorb polioviruses during all phases of the cell cycle (Eremenko et al., 1972).

Poliovirus receptors are integral membrane proteins. They are not removable from cells by hypertonic treatment, indicating that they are not merely bound to the cell surface by ionic interactions. Treatments that disrupt cells into plasma membrane fractions but conserve membrane integrity do not destroy receptor activity. "Receptors" for polioviruses can be detected in broken cells by their ability to inactivate virus infectivity (McLaren *et al.*, 1959; Holland and McLaren, 1961; Quersin-Thiry, 1961; Baron *et al.*, 1963) or to alter the electrophoretic mobility of the virions (Thorne, 1963). Attempts were made to solubilize and purify virus receptors by lysis of cells with the non-ionic detergent Triton X 100 (1%) and ultrasonic treatment (McLaren *et al.*, 1968). The partially solubilized membrane components contained receptor activity towards different picornaviruses. Virus receptor activity was only detected when susceptible cells were used as starting material. Polioviruses adsorbed to plasma membrane fractions also after partial membrane-disruption by homogenization or short ultrasonic treatment of cells. CsCl gradient analysis of poliovirus receptor activity indicated association with a complex of lipids and glycoproteins.

Complete solubilization of membranes, however, destroys the poliovirus receptor activity, for example, during attempts to isolate receptors from cells by detergents. Receptor activity is no longer detectable after removal of detergent (De Sena and Mandel, 1976; Crowell and Siak, 1978). The activity is also lost by exposure of isolated receptor preparations to organic solvents like ether, chloroform and butanol (Holland and McLaren, 1959; McLaren *et al.*, 1968). Integral membrane proteins commonly lose biological activity upon extraction from their lipophilic environment, since membrane proteins undergo conformational shifts upon isolation and transfer into hydrophilic environment. In aqueous solution, hydrophobic regions of proteins tend to cluster in the center of the proteins as far away as possible from the hydrophilic surface. Proteins with extended hydrophobic regions on the outside have a greater tendency to aggregate. The view that poliovirus receptors are integral membrane proteins which lose their activity outside of the membrane is supported by the observation that firm binding of solubilized viral receptors to hydrophobic polystyrene microtiter plates restores the virus binding activity (Krah and Crowell, 1982).

Poliovirus receptors in disrupted HeLa cells are inactivated at 56°C (Holland and McLaren, 1959) and do not withstand prolonged sonic treatment (Holland and McLaren, 1959; Quersin-Thiry and Nihoul, 1961; Philipson and Bengtsson, 1962). These results indicate that membrane fragments or a multicomponent complex is required for receptor activity.

Exposure to SH reactive substances does not destroy receptor activity suggesting that sulfhydryl groups of receptors are not required for binding of polioviruses (Baron *et al.*, 1963).

Con A inhibits adsorption and attachment of poliovirus and of a human rhinovirus to HeLa cells (Lonberg-Holm, 1975). Con A does not bind to poliovirus. It specifically reacts with D-mannosyl-like residues and it was therefore suggested that picornavirus receptors contain mannose and are glykoproteins. However, poliovirus receptor activity is not destroyed by periodate exposure. This suggests that carbohydrates are not required for poliovirus attachment. In contrast, the receptors for other picornaviruses, bovine enterovirus and equine rhinovirus, are glykoproteins since their activity is lost by exposure to neuraminidase (Stoner *et al.*, 1973; Lonberg-Holm and Philipson, 1980). After preexposure of cells to Con A, more polioviruses elute from cells, indicating that Con A might prevent virus interaction by steric interference. In addition, Con A might interfere with movement of receptors in the membrane, and thereby prevent interaction of virions with several receptors as a prerequisite for firm binding.

The possibility of inserting integral membrane proteins into artificial membranes (liposomes) (Engelhard *et al.*, 1978) may open a new approach to the study of virus-receptor interactions. The purification of receptors may be facilitated by the use of affinity chromatography of membrane fractions with intact viruses coupled to sepharose (Borriss and Koch, 1975) and with anticellular receptor-blocking antibodies (Axel and Crowell, 1968).

b) Other Functions and Components of the Receptor Complex

As mentioned, viral receptors in the plasma membrane are in close contact and/or association with other components of the host cell membrane which might specifically or unspecifically interact with the attached virus particle. Together they form the hypothetic receptor complex (see Table 40). At least two additional membrane factors are believed to be distinct from the receptor site responsible for virus adsorption; one has properties modifying the capsid, the other properties stabilizing the capsid. These factors have been discovered in studies with disrupted membranes, they are discussed in section B.3 below.

c) Number of Attachment Sites/Cell

The number of attachment sites are experimentally determined by the quantitation of virions bound per cell and by the number of virions modified by one cell. Since more than one receptor may interact with one virus particle at an attachment site, the plasma membrane probably contains more virion receptors than attachment sites. The first encounter of a virus particle on the surface of the cell may attract other nearby receptor molecules to the virion. It is conceivable that a considerable area of the virus particle is thereby covered with membrane components. The lateral movement of virus receptors is inhibited by low temperatures or by previous binding of Con A to the cell surface.

Poliovirions attach to HeLa cells with low efficiency when more than 3×10^3 virus particles are bound per cell (Lonberg-Holm and Philipson, 1974). The total surface area of a HeLa cell is at least 5×10^8 nm^2, the total surface of a virus particle measures 2.4×10^3 nm^2. The HeLa cell surface is, therefore, at least 200,000 times larger than that of a virus particle. The virus particle has a diameter of 30 nm and therefore covers an area of only 100 nm^2, provided there is no severe encarvation of the plasma membrane after virus adsorption; maximally 5×10^6 particles would cover or plaster the surface of the HeLa cell completely. The area of the plasma membrane covered by 10^4 picornaviruses is only about 0.2 %, therefore, saturation is not caused by spatial limitation. In comparison, partial saturation of receptors by other picornaviruses are slightly higher: 1–2 x 10^4 for rhinovirus and 1 x 10^5 for coxsackie B3 (Crowell, 1966, 1967). Fibroblasts have been reported to contain 10–100 times more attachment sites for picornaviruses than HeLa cells (Medrano and Green, 1973).

Guttman and Baltimore (1977) determined that each intact HeLa cell can alter more than 5 x 10^3 poliovirions. These particles correspond to about 50 % of bound virions and elute from the cell lacking VP4. They are modified but do not register as cell bound viruses. Since only 1–3% of bound polioviruses register as PFU in the infective center assay, only 30 to 100 PFU of poliovirus can be efficiently bound by one host cell.

d) Specificity and Genetics of Virus Receptors

Interaction between viruses and their host cells are known to be highly specific. In principle, the nature of this reaction could be one of two types: a) Viruses interact with specific virus receptors on the host cell. The host cell codes for and synthesizes specific virus receptors which have no other functions. b) Viruses interact with membrane components that normally carry out different functions for the host. Viruses would then adapt surface configurations that mimic the normal reaction partner of the host cell membrane component: The virus fools the host cell into "believing" it is interacting with one of its natural reaction partners. This is the case in certain phage-bacterial systems: Phage receptors function in the transport of iron (Braun *et al.*, 1976; Schweiger and Wagner, 1979) or vitamin B 12 (Bradbeer *et al.*, 1976).

Which type applies for picornavirus receptors is not known. Growth of tissue culture cells is not inhibited when picornavirus receptor activity is inactivated by antibodies (Axler and Crowell, 1968; Much and Zajac, 1974), which indicates that picornavirus receptors do not serve essential functions in tissue culture cells. The sensitivity to poliovirus infection may depend on the state of differentiation: Human leukocytes usually are resistant to poliovirus infection because they lack receptors. However, they gain infectability —through synthesis or membrane exposure of receptors— when proliferation is stimulated by PHA (van Loon, 1977). Similarly, in myoblasts, receptors to coxackie viruses are expressed concomitantly to differentiation (Schultz and Crowell, 1980). It is possible that these virus receptors might have some functional significance in differentiation. Polioviruses usually only adsorb to cells of primate origin, and poliovirus receptors are readily detectable in tissue samples derived from human and monkey CNS and intestine, but are generally absent in non-primate tissue and samples from human lung, heart, skeletal muscle and skin (Kaplan, 1955; Holland, 1961; Kunin and Jordan, 1961; Kunin, 1962, 1964; Baron *et al.*, 1963; Harter and Choppin, 1965). It has been known for a long time that all three types of polioviruses can also be adapted to grow in different species of laboratory animals, including cotton rats, white mice, hamsters, and chicken embryos (Armstrong, 1939; Li and Habel, 1951; Moyer *et al.*, 1952; Roca-Garcia and Jervis, 1955; Koroleva *et al.*, 1973). Multiplication of certain poliovirus strains has been observed after intracerebral injection in suckling cotton rats and new born mice (Koroleva *et al.*, 1974).

Heterokaryons between virus permissive and non permissive cells are often permissive, indicating that receptor activity is dominant. Virus susceptibility of human-mouse hybrid cell lines depends upon the presence of human chromosomes. Loss of permissiveness correlates well with loss of a human chromosome and poliovirus receptor activity. In each case, cells are resistant to all three poliovirus serotypes (Kusano *et al.*, 1970), but in addition also to echovirus 11 and coxsackie B3 (Couillin *et al.*, 1976a,b), indicating that coding regions for the receptors for different picornaviruses are located on one chromosome. Presence and loss of poliovirus receptor activity in man-mouse hybrid cell lines correlates well with phosphoglucose isomerase (GPI); this enzyme's coding region is known to lie on the human chromosome F 19 (Miller *et al.*, 1974; Couillin *et al.*, 1975).

In order to gain insight into the genetics of receptors, a search for poliovirus resistant cell lines was conducted. Loss of infectability by polioviruses of mutant cells might be due to either missing receptors or to failure in cell mediated eclipse or uncoating (Vogt and Dulbecco, 1958; Darnell and Sawyer, 1960; Holland, 1962a). Only one poliovirus resistant cell line has been reported to show a specific inability to attach poliovirus (Soloview *et al.*, 1968). In addition, a SV 40 transformed human amnion cell was found to be resistant to infection by intact poliovirus but not to poliovirus RNA (Hahn and Fogh, 1970). Attachment of virus was normal; it was concluded that the process of uncoating was defective.

The classification of picornaviruses into Genera and Subgroups (see Chapter 2) correlates well with their receptor specificity. Evidence that all three poliovirus serotypes use the same receptor site on HeLa cells, was first obtained by exposing HeLa cell fragments to formalized poliovirus type I (Quersin-Thiry, 1961). The attachment of infectious virus of all three serotypes was reduced to the same extent by prior saturation of receptors with one type of poliovirus, while the attachment of another picornavirus (coxsackie B) was not affected, indicating that polioviruses and other picornaviruses use different sites on one cell (Lonberg-Holm and Philipson, 1980). This conclusion was confirmed in studies with intact cells by measuring the binding of purified radioactive viruses (Lonberg-Holm and Philipson, 1974) or by following the loss of virus infectivity (Crowell, 1966). Although the three poliovirus serotypes use the same receptor site on HeLa cells, they are adsorbed with different rates (Crowell and Landau, 1983). This may simply reflect slight differences in the capsid structures and receptor recognition sites of the three serotypes, and therefore also in the binding affinity to the host cell receptors.

B. The Interaction of Poliovirions with the Host Cell Membrane

The entire process of virus interaction with the host cell membrane has been variously referred to as adsorption, attachment, binding, or fusion, and there still exists some unclarity in the terminology today. In the following, we will use the term adsorption to refer to the specific, but rather loose initial binding of poliovirus to the host cell at 4° C; the term attachment to refer to the tighter binding of virion to the host cell membrane which only occurs at or near physiological temperatures; the term insertion to refer to the penetration of the virion into the lipid bilayer of the host cell membrane which may occur on the outer cell membrane or only after the virion has been ingested by the host cell into a phagocytic vesicle; and the term binding to refer to the sum of the steps, or whenever a clear distinction between adsorption and attachment is not possible (see Table 39). It should be kept in mind, that the major portion of virions which interact with the host cell membrane do not become firmly bound to the host cell or inserted into the membrane: Up to 90% of bound virions may be released again from the cell membrane in an altered non-infectious state. Another fraction of virions may be taken up by endocytosis and digested in lysosomes without ever inserting into the membrane bilayer. Table 42 summarizes the steps involved in the early interaction of poliovirus with the host cell membrane and lists their characteristic features.

Table 42. *Proposed early steps in poliovirus infection*

a) *Adsorption by interaction of one recognition site on the viral capsid with a specific site on the receptor on the plasma membrane:* hydrogen bonds dependent on low, and stabilized by high, concentrations of salt; ionic bonds reversed by low pH and high salt concentrations; hydrophobic bonds reversed by urea and nonionic detergents. Virus is elutable intact. Intact virus might elute spontaneously (dilution of cell virus mixture).

b) *Attachment or tight binding by interactions of several recognition sites on the virion with multiple receptors and other components of the membrane:* dependent on lateral movement of membrane components, inhibited by temperatures below 20 °C and by the presence of Con A. Virus is elutable with ionic detergents.

c) *Stabilization of intact virus by "stabilizing factors"* (membrane derived glykolipoprotein complex): leads to a reduction in virion-density and sedimentation rate; the proportion of stabilized virions is enhanced by pretreatment of virus with SH reactive compounds or the lipophilic compound arildone which stabilize the virions in the D conformation and which inhibit the next step. Virus is elutable with ionic detergents.
with multiple receptors in attachment site): results in complete loss of VP_4. Modified virions have a more lipophilic surface and become sensitive to protease and to detergents but not to RNAse. Virus elutable in a modified (VP_4 minus) form. Modified virus might elute spontaneously.

e) *Loss of VP_2 and gain of RNAse sensitivity.*

f) *Release of RNA.*

Many studies have been performed to elucidate the underlying molecular mechanisms of poliovirus host cell membrane interactions (for reviews see Kohn, 1979; Crowell and Landau, 1983). The most common approach is to examine the effects of different experimental conditions on the interaction, and to determine how the interaction can be inhibited or reversed. The effects of temperature, pH, ionic conditions, for example have been studied in detail. When interpreting the results, one should keep in mind that both reaction partners may be modified by most of these conditions. For example, varying ionic culture conditions affect the cell size and thereby also the geometric arrangement of the host cell receptors. They may also disturb ionic interactions between virion and cell receptor. Low pH can affect hydrogen bonding between virion and host cell receptor, but may

Table 43. *Types of bonds involved in the interaction of virus with the host cell membrane*

Interaction Condition	Result			Conclusion
	Adsorption	Attachment	Uncoating	
4 °C	+	–	–	specific receptor
25 °C – energy	+	+	(+)	lateral movement of membrane components required for attachment
25 °C + energy	+	+	+	energy required for uncoating but not for attachment
Detergents	–	–	–	hydrophobic interactions important
Sucrose: alone	(+)	–	–	ionic interactions
Sucrose + NaCl	+	+	+	required but alone
Sucrose + Mg	+	+	+	not sufficient;
High salt	–	–	–	greater
Low salt	++	+	+	accessibility of receptors in swollen cells
Exposure of virions				
– to SH reactive substances, DEP	+	+	–	SH groups involved in uncoating
– to dextran sulfate	–	–	(+)	positive charge required for adsorption
– to SIP and Bolter reagent	+	+	+	E amino of lysine not essential
– to arildone	+	+	–	poliovirus stabilized against uncoating by lipophilic compound

+ = reaction can occur under said condition.
– = reaction is blocked under said condition.

also cause a conformational rearrangement of the proteins in the capsid of the virion. Table 43 provides an overview of the factors and conditions that influence virus-receptor interaction and lists the probable sites of action.

1. *Adsorption*

Adsorption of viruses to host cells is generally determined by measuring the disappearance of either radiolabeled virus or infectious units from the supernatant of virus cell mixtures. With polioviruses, both methods yield comparable results, however, only when virus cell mixtures are incubated at temperatures below 25°C or for relatively short times at 37°C. The variable results obtained during longer periods of incubation at temperatures above 25°C are a consequence of the dynamic events which follow adsorption. In the course of these events the virus is either tightly bound (attachment) or modified and released from the cell membrane. These processes do not occur at lower temperatures, probably because of the decreased membrane fluidity. Presence of serum in the culture medium may also interfere with measurements of adsorption. Serum proteins and other compounds might inactivate polioviruses without inhibiting adsorption. Anticellular serum can block receptors and thereby also abolish adsorption (Axler and Crowell, 1968).

Adsorption of poliovirus at temperatures below 20°C leads to a relatively loose binding: Polioviruses which have interacted with cells at temperatues below 25°C can be detached from their receptor sites in an intact form by dilution of the virus-cell mixture. Lonberg-Holm and Whitely (1976) determined the kinetics of adsorption of poliovirus type 2 as well as of a human rhinovirus at temperatures from 0°C to 37°C. Adsorption is relatively slow at temperatures below 20°C but is efficient even at 4°C. The energy necessary for the activation of adsorption of poliovirions at 4°C was calculated to be 13 Kcal/Mol (Lonberg-Holm and Whitely, 1976).

The rate of adsorption of poliovirus is dependent on and directly proportional to cell concentration (number of receptors) but not influenced by virus concentration (within a wide range from 0.01 to 50 PFU). Of course, approaching saturation of receptors by viruses will also result in a decreased rate of adsorption. The three types of poliovirus attach to HeLa cells in suspension culture with velocity constants, K, of 1.54 (type 3), 1.94 (type 1), and 2.81 (type 2) x 10^{-9} ml/min x cell (Holland and McLaren, 1959; Lonberg-Holm and Whiteley, 1976; Crowell, 1976). These rates are comparable to those observed for other picornaviruses (Crowell, 1976).

Early virus-cell interaction is most likely due to formation of ionic and hydrogen bonds. Adsorption of poliovirus is poor in salt free sucrose, but is enhanced to normal levels as soon as NaCl is added (McLaren *et al.,* 1959) and further increased by Ca^{++} and Mg^{++} (Bachthold *et al.*, 1957). Some picornaviruses (rhinoviruses, coxsackieviruses A 9 and A 13, and FMDV) but not poliovirus even require divalent cations for adsorption (Noble-Harvey and Lonberg-Holm, 1974). Divalent cations might form salt bridges between negatively charged groups on the virus and on the receptor. These picornaviruses can be detached by EDTA, a chelator of divalent cations (Lonberg-Holm *et al.,* 1975; Baxt and Bachrach, 1980).

Addition of sucrose in order to increase the tonicity of the culture medium slows the rate of virus adsorption. Hypotonic salt solutions, in contrast, have an opposite effect. The increase in the cell volume by influx of water under hypotonic conditions might lead to greater accessibility of receptors. The cell surface contains numerous microvilli, which–in close contact–can hide receptors or prevent adsorption of virions by steric hindrance. Such contact of microvilli might be enhanced in hypertonic and diminished in hypotonic environments.

In the presence of detergents, poliovirus and host receptors do not associate. Bound viruses can be freed from their receptors by detergents (Fenwick and Cooper, 1962). This suggests that hydrophobic interactions are important for the binding of virus to the host cell receptor. They are certainly relevant to the insertion of virus into the membrane bilayer.

The optimal pH range for the adsorption of various enteroviruses to HeLa cells varies (Crowell, 1976). Polioviruses show a rather broad permissive pH range for adsorption (4.5–8.5), whereas echovirus adsorb efficiently only between pH 6.5 and 8.5. Coxsackie B4 binds better at pH 3.0 than at pH 7.0 (50 fold). The cell surface is relatively more positively charged at low pH, more negatively charged at high and neutral pH. The pI 7.0 form of poliovirus carries a net positive charge at lower pH, the pI 4.5 form carries a net negative charge at pH above 4.5. It is not known in which conformational state the virion binds to the receptor. It is possible that the receptor itself induces a conformational shift in the virus particle analogous to the pI 7.0 to pI 4.5 conversion. The virus particle is more hydrophobic in the pI 4.5 form, which may be important for the physical insertion of the particle into the cell membrane (see section III, p. 291 ff.). The efficient attachment of polioviruses above pH 7.4 where both host cell and virus carry a net negative charge suggests that the primary attractive force between virus and host receptor is not solely due to electrostatic interactions at this pH. Ionic interactions, however, between positively charged groups on polioviruses and negatively charged groups on receptors are not excluded by these observations. Indeed, there are numerous observations which indicate participation of positively charged groups of the virions in adsorption: Treatment of certain mutant strains of poliovirus with sulfated polysaccharides (negatively charged) inhibits the infectivity of these mutants by interfering with their adsorption to host cells (Takemoto and Spicer, 1965). Other mutants are prevented from binding to cells by acidic components in agar medium (see Chapter 4). All of these mutants seem to map in the region of the genome coding for VP_1.

Exposure of poliovirus to crosslinking agents like dimethyl suberimidate (DMS) or dimethyl adipimidate (DMA) results in a loss of infectivity. Analysis of disrupted DMS modified virus preparations revealed almost exclusive crosslinkage between one VP_1 and one VP_3. When one out of four VP_1 peptides was crosslinked to one VP_3 the infectivity was reduced 45 fold. It should be of great interest to see whether loss of infectivity of DMS treated particles can be correlated with altered capacity of adsorption.

Surface exposed lysine residues in polioviruses are apparently not essential for attachment since extensive modification of the amino groups of lysine by monofunctional reagents (N-succinimidyl-2,3-^{3}H-proprionate (SIP) or the Bolter and Hunter (1973) reagent (N-succinimidyl 3-(4-hydroxy, 5-^{125}I-iodophenyl proprionate) does not decrease the infectivity of polioviruses (Wetz and Habermehl, 1979). However, exposure of poliovirus to diethylpyrocarbonate (DEP), which leads to substitution of amino and SH groups, results in a diminished binding capability (Oberg, 1970; Breindl and Koch, 1972). DEP treatment of polioviruses does not result in any detectable structural alterations of the particles. Moderatly DEP exposed viruses sediment like untreated viruses and have the same buoyant density in CsCl and the same protein composition. The RNA in the virus particle is fully protected against RNase, and after isolation it shows the same infectivity as the RNA extracted from untreated particles. The latter results suggest that amino groups or SH groups participate in binding.

The proposal that sulfhydryl groups of picornaviruses participate in binding (Philipson and Choppin, 1962) has been a matter of controversy for a long time. SH reactive chemicals failed to react with intact polioviruses (Wetz and Habermehl, 1979), but glutathione, cysteine and thiopyrimidines decrease virus infectivity and stabilize polioviruses against heat inactivation (Fenwick and Cooper, 1962; Yamazi *et al.*, 1966, 1970; La Colla *et al.*, 1972; Lonberg-Holm *et al.*, 1975). This type of stabilization suggests the presence of S-S bonds; such bonds, however, could not be detected in polioviruses (Dernick, 1981). More recent evidence suggests that SH groups of polioviruses do not participate directly in attachment, but play a role in the host membrane induced alteration of virus particles (see below).

Poliovirus binds the lipophilic antiviral compound arildone. It is thereby converted to the pI 4.5 form and stabilized against heat, pH degradation and the membrane induced modification of the viral capsid (see below) (Eggers, personal communication). Arildone, however, does not interfere with adsorption and attachment.

2. *Attachment*

The reversible adsorption is followed by further interactions of the virus with the cell membrane, resulting in changes in both of the reacting partners (see Tables 42 and 43). These interactions occur with measurable rates only at temperatures above 20° C. At these temperatures membrane proteins begin to diffuse freely in the lipid bilayer of the plasma membrane and more than one receptor or other membrane components may bind to one virus particle.

The virus particle may become inactivated (liberation of VP4) and released again from the membrane (Joklik and Darnell, 1961), or it may become bound to the membrane so tightly that it can only be detached again by very acidic pH, high concentrations of urea (8M), LiCl (6M) or SDS (McLaren *et al.*, 1959; Fenwick and Cooper, 1962; Holland, 1962a; Mandel, 1962). Roughly half of the adsorbed virus particles become firmly bound to the membrane, and half are released in a modified form. Neutralized poliovirus particles which are locked in the pI 4.5 form have a five to seven fold decreased probability of being eluted from HeLa cells compared to non-neutralized virions (Mandel, 1967b, 1976). This indicates either that the hydrophobic pI 4.5 form has a higher affinity for the cell membrane or that it is more stable against the membrane activity responsible for dissociating VP4.

The progression of cell associated poliovirus to a complex which requires 0.2% SDS for its dissociation occurs within a few minutes at 37°C. The conversion from a loosely associated to a tightly bound state can be retarded or prevented by cooling or by the presence of 100 μg/ml of Con A (Lonberg-Holm, 1975). These observations were taken to indicate that tight bindng requires selective lateral movement of additional receptors which then all bind to one virion. Other components of the receptor complex may then interact with the tightly bound virions. In the course of these events the virus particle becomes sorrounded and then inserted into or engulfed by the plasma membrane (see p. 293, below) (Lonberg-Holm and Philipson, 1980).

A small fraction of viruses, which are tightly bound to the cell at temperatures above 20°C, can still be recovered in an intact manner (containing VP4) after lysis of the cell by non-ionic detergents (Fenwick and Wall, 1973). These viruses are then still bound to membrane components (and sediment with 130S). Viruses with bound membrane components are more stable to heat than free viruses. Intact, native virions can be recovered from these complexes by exposure to ionic detergents (0.2% SDS). Tightly bound poliovirus - unlike other enteroviruses (McLaren *et al.*, 1959)—can also be intactly recovered from cells by incubation at acid pH between 1.5 to 2.0, by treatment of cells with deoxycholate (Fenwick and Cooper, 1962), by 6 M LiCl, or 8 M urea (Zajac and Crowell, 1969), indicating that alteration of the virus particle is not a prerequisite for tight binding.

Another fraction of the attached virions can be eluted from cells in a modified form by detergents. Like spontaneously eluted particles, these bound and modified virions are characterized by the loss of the smallest capsid protein VP4, a corresponding reduction in the sedimentation coefficient, altered antigenicity, and the loss of capability of attachment. It is possible that these modifications correspond to the first step of the uncoating process during productive infection. Alternatively, they may be the products of an inherent cellular defense mechanism against viral infection (see also section IV).

3. *Response of the Plasma Membrane*

a) Changes in Membrane Fluidity and Capping of Viruses

Attachment of poliovirions to the cell membrane is reflected in alterations of membrane fluidity (Levanon and Kohn, 1978; Kohn, 1979). Generally, virus interaction with the cell membrane could alter the fluidity of the membrane by a number of different mechanism. Binding of multifunctional, but not of monofunctional lectins induces changes in membrane fluidity. Polioviruses might induce a change in membrane fluidity by binding simultaneously to several receptors, thereby acting like multifunctional lectins. Alternatively, modified or unaltered poliovirions might directly insert into the membrane and thereby effect fluidity. In addition, attached or inserted virus might lead to a reorganization in membrane structure by activation of phospholipase or by an effect on the cytoskeleton. Lastly, virus induced phagocytosis may cause a decrease in microviscosity (Berlin and Ferra, 1977).

The degree of fluidity of lipid bilayers can be determined by a number of different methods such as fluorescence polarization, electron spin resonance, nuclear magnetic resonance and freeze fracture electron microsocopy. In studies with artificial membranes it was shown that membrane rigidity increases with the insertion of viral proteins (Stoffel *et al.*, 1976). Protease exposure of these viral protein containing membranes can result in an increase in fluidity (Sefton and Gaffney, 1974). Based on these and other observations it was suggested that only changes in the integral proteins of the membrane alter the fluidity of membranes (Levanon *et al.*, 1977). Kohn and coworkers (Levanon *et al.*, 1977, 1979) studied the effect of poliovirus adsorption on HeLa cells, using the fluorescent probe DPH (1,6 diphenyl 1,3,5 hexatriene). A reduction in microviscosity was observed within a few minutes following addition of 1-2 PFU of poliovirus per cell at 37°C. The maximum change in microviscosity was obtained after incubation of cells with 50-100 PFU/cell for 5-6 minutes at 37°C. There is a good correlation between number of virus particles adsorbed and degree of change in membrane fluidity. This correlation even allows a rapid estimation of the titer of a virus stock. Type 1 and 2 polioviruses are equally effective in reducing membrane viscosity. Preincubation of polioviruses with homologous but not with heterologous antisera abolishes about 50% of the virus induced change in membrane viscosity. Neutralized polioviruses still adsorb to host cells, but at a reduced rate. Therefore the response of the host cell membrane to virus antibody complexes in form of smaller changes in fluidity might simply reflect their slower rate of virus adsorption. Polioviruses inactivated by UV light still trigger the alteration in membrane fluidity. It should be noted that binding of DPH by polioviruses might contribute to the decrease in DPH fluorescence, since DPH is inserted into virus particles (Zibirre, personal communication).

In addition to decreasing membrane viscositiy, adsorption of picornaviruses to the cell membrane may initiate distinct movements of proteins and other components of the cell membrane. Morphologically, tight binding of virions may correspond to the formation of patches on the membrane surface of several receptor-virion complexes. Studies on the distribution of poliovirions on the cell surface after adsorption have not yet been performed, however, the fate of mengoviruses after adsorption to cells was followed with the aid of fluorescent labelled antibodies (Gschwender and Traub, 1979). Viruses rapidly appear in patches. Patch formation is independent of the metabolic state of the cell. Virus patches begin to migrate to one pole of the cell comparable to the movement of antigen-antibody complexes (capping). Capping only occurs when cellular activity is not disturbed.

The plasma membrane region containing virus aggregates is subsequently engulfed by the cell in a process resembling phagocytosis (Gschwender and Traub, 1979). These observations were taken to indicate that mengoviruses enter cells by phagocytosis. However, the capping of mengovirions might have been induced or influenced by the antibodies themselves. The interaction of bivalent antibodies with immunoglobulin molecules on the surface of lymphocytes and with surface bound plant lectins can itself result in patch formation (Pernis *et al.,* 1970; De Petris and Raft, 1973). Similarly, antibodies can simultaneously interact with two virus particles which might enhance patch formation and capping. Virus antibodies can also trigger conformational changes in virions which might influence the mode of interaction with the cell membrane. Nevertheless, the use of fluorescent labelled antibodies indicates that receptor bound virions can be moved within the membrane. The receptor itself or other components of the receptor complex seem to be connected to the cytoskeleton. The signal for movement may be due to an allosteric structural change in the receptor or critical components of the receptor complex. Inactivation of membrane proteins (for example binding of ouabain to the Na^+K^+ pump) is often followed by internalization of these proteins; binding of virus to a receptor may have a similar effect.

b) Changes in Membrane Permeability and Membrane Potential

Already in 1948 Doermann reported that bacteria become leaky after infection with bacteriophages. Colicines and phage ghosts (Phillips and Cramer, 1973) also increase the permeability of the membrane of their target cells with resulting loss of intracellular K^+, a concomitant depolarization of the mebrane, and death of the cell. Comparable effects of virus adsorption on animal cells were first reported by Klemperer (1960). He found that hemolysis of erythrocytes by newcastle disease virus was preceded by loss of intracellular K^+ which occured within 5 minutes at 37°C after virus addition. So far comparable early alterations in picornavirus infected cells have not been reported. In contrast, the intracellular concentration of K^+ increases early in infection of cells by picornaviruses (Egberts et al., 1977). Only later in infection with picornaviruses do cells become leaky for ions (see Chapter 6).

The membrane potential in animal cells can be determined with the use of the ^3H labelled lipophilic cation TPP^+ (tetraphenylphosphonium) (Lichtsthein *et al.*, 1979). Significant changes in membrane potential where not observed with this method during the first hour after poliovirus infection of HeLa cells (Hiller *et al.*, 1984), inspite of the fact that poliovirus infection results in reduced uptake of several amino acids which are transported in a Na^+ and membrane potential dependent manner (Koch *et al.*, 1980a).

Tissue culture cells release low molecular weight ether soluble substances immediately after infection with several picornaviruses (EMC-, echo- and polioviruses). These substances can be detected by gas chromatography (Levanon *et al.*, 1977a). The observations were taken as an indication for a selective increase in membrane permeability. However, the results could also be interpreted as phospholipase activation and liberation of membrane components.

c) Interactions with Modifying and Stabilizing Membrane Components

The interaction of a virus with several receptors can in itself be considered as a modification of the virus particle. Studies with membrane fractions of dissociated cells have revealed that there are additional membrane components that can affect virus stability and structure. Such interactions can increase virion stability but can also result in structural alterations of the viral capsid ultimately culminating in liberation of the viral RNA.

In order to identify virus interacting membrane components, plasma membranes from infected cells have been isolated by a number of procedures (Chan and Black, 1970; Guttmann and Baltimore, 1977b; De Sena and Torian, 1980).

These include: a) dounce homogenization after swelling of cells in hypotonic medium and subsequent differential centrifugation (Bosmann *et al.*, 1968); b) dounce homogenization of cells in hypotonic buffer in the presence of 10^{-3} M $ZnCl_2$ to stabilize membranes and subsequent interphase separation (Brunette and Till, 1971), and c) lysis of cells by low concentrations of non ionic detergents, followed by differential centrifugation (Helenius and Simons, 1975). The thus isolated membrane fractions were purified further by density gradient centrifugation (Roesing *et al.*, 1975; Eggers *et al.*, 1979; Rosenwirth and Eggers, 1979). Different components of the membrane were extracted by prolonged incubation at 5°C or by exposure to salt or non ionic detergents (McGeady and Crowell, 1979; De Sena and Torian, 1980).

Characterization of these membrane fractions revealed that the plasma membrane contains a "modifying activity" which attacks and alters virus particles. This modifying activity is sensitive to pronase, lipase, ether, deoxycholate, wheatgerm agglutinin, and exposure to Con A, indicating that both lipid and glycoprotein components are required for activity (De Sena and Torian, 1980). The modifying activity seems to be distinct from receptor activity. A number of antiviral drugs (arildone, thiopyrimidines (S7), gluthatione, and rhodanine) interfere specifically with the modifying activity without disturbing adsorption of virus to the receptor. The modifying activity appears to be a multicomponent complex, the activity of which depends on the native conformation of the complex. Degradation, dissociation or blockage of one component of the complex destroys its activity (De Sena and Torian, 1980). The membrane fractions with modifying activity also contain protease as well as RNase activities.

The virus derived particles obtained after incubation of virus with modifying activity in isolated plasma membranes from HeLa cells (for 45 min at 37°C) were characterized by centrifugation through sucrose density gradients and analysis of their peptide composition (Guttmann and Baltimore, 1977). Most of the polioviruses were modified to particles with the following properties: 130S, loss of VP_4, stability in deoxycholate, NP_{40}, and Brij, each at concentrations of up to 1% (see Table 44). These particles were named M particles and evidently correspond to the A particles eluted from cells after adsorption. The viral RNA in these particles was still protected against degradation by high concentrations of ribonuclease (1.25 mg/ml, 45 min at room temperature). The modifying reaction was inhibited either by NaCl or $MgCl_2$ at concentrations higher than 50 mM, and it did not occur at 22°C. Similar experiments (incubation for 40 min at 36°C) were performed, though with a different procedure for the preparation of plasma membranes, and comparable results were obtained (De Sena and Mandel, 1976, 1977; De Sena and Torian, 1980). In these experiments, the modifying activity was inhibited by Triton X 100. The modified particles became sensitive to protease. Further incubation of M particles at 37° with isolated membrane fractions or with chymotrypsin resulted in further alterations of the particles: the RNA became sensitve to RNase, and the viral protein VP_2 was liberated completely from the altered particles. The chymotrypsin derived particles were referred to as C particles (see Table 44). An orderly conversion of M to stable C particles, however, occurred only when the nonsolubilizable membrane fraction was also present. In

Table 44. *Properties of virus related particles during uncoating*

	M or A particles (modified)	D particles (native virion)	C particles (core-particle)
VP_2	+	+	−
VP_4	−	+	−
RNA	+	+	±
D reactive	−	+	−
C reactive	+	−	+
Receptor binding	−	+	−
RNase resistance	+	+	−
Sedimentation rate (S)	135	150	60
Isoelectric point	4.5	7.2	?
Hydrophobicity	+	±	?
Sensitivity to 1% DOC, NP_{40} or Brig	−	−	+
Protease sensitivity	+	−	?
Permeability to CsCl	+	−	+
Density in CsCl	unstable	1.34	unstable

the absence of the latter, the soluble complex degraded M particles more extensively and most of the RNA was converted to oligonucleotides.

In addition to the modifying activity, the plasma membrane contains a virus stabilizing factor which counteracts the factor with modifying activity. Both factors might convert the virus particle into a more lipophilic particle which more readily inserts into the membrane. Coxsackie B3 virus particles were eluted from intact cells and from two different plasma membrane fractions which had been isolated from infected cells (McGeady and Crowell, 1979). Virus particles recovered from intact cells and from the heavier membranes are considerably more stable against degradation induced by incubation at 37° C and higher temperatures than particles recovered from the lighter membrane fraction. This has been interpreted to indicate that the viruses are eluted with different amounts of membrane components. Only the heavier membrane fraction offered components which efficiently prevented the uncoating of the virus particle. These stabilizing components were solubilized successfully simply by incubation of the heavier membrane fractions with saline. The origin of the two membrane fractions (differing in density) has not yet been investigated. They might not be derived from the same cell but rather from cells at different stages of their growth cycle. The amount of surface exposed protease sensitive glykoproteins, for example, varies greatly during the growth cycle of tissue culture cells (Kalvelage and Koch, 1982). Comparable experiments have been performed with echoviruses (Rosenwirth and Eggers, 1979) and almost identical results were obtained. It was concluded that one membrane fraction contained a virus stabilizing component and/or was devoid of an uncoating activity.

It is difficult to interpret the results from these in vitro studies since it is not known which activities represent cellular defense mechanisms against the viral particles and which represent essential steps on the virion's route to productive infection. The modifying activity of the cell membrane may reflect a cellular defense mechanism designed to disrupt the particle and render it noninfectious (eluted then as A particles), or to sensitize the particle to degradation by proteases and RNases in the cell membrane, cytoplasm, or lysosomes. Similarly, the stabilizing activity could present a defense mechanism designed to stabilize virions so as to inhibit their uncoating, analogous to the action of neutralizing antibody. Alternatively, the stabilization of the virion might be a prerequisite for protection of the viral RNA on its way from the cell membrane to the site of onset of viral replication. Replication may be initiated more efficiently when the release of the viral RNA occurs only after virus particles have reached the site of synthesis of viral proteins, the rough endoplasmic reticulum. The stabilization of virus particles by cellular membrane components may prevent premature uncoating and release of viral RNA which otherwise would be degraded by RNase before reaching ribosomes. Lastly, it could be just the orderly, sequential activation of both, stabilizing and modifying activities, that leads to productive infection.The defense mechanism hypothesis gains support from recent studies on the infectivity of polioviruses enclosed in liposomes (Wilson *et al.*, 1979). Poliovirus containing liposomes fuse with the cell membrane and release the virus particles intact into the cell cytoplasm. When the membrane barrier is circumvented by this procedure 20% of the virus particles are infectious, compared to 0.1-1% for native polioviruses. This observation suggests that the cell membrane normally contributes considerably to reducing the ratio of infectious units to physical particles of polioviruses.

The different binding, modifying, and stabilizing activities of the cell membrane which the virion encounters in the first few minutes of interaction with its prospective host cell may not even be the greatest of obstacles in its way. If taken up by pinocytosis, aggressive enzymes may be waiting in lysosomes; the viral RNA then still has to be transferred across the membrane, and translation must be initiated on one of the many ribosomes which at the moment are still active in translating host mRNAs.

III. Penetration of Virus Particles into the Cell: Insertion and Phagocytosis

The mechanism of penetration is still a matter of controversy. It has been proposed that poliovirus particles enter cells either by insertion into and penetration through the plasma membrane or by active pinocytotic uptake by cells (Dunnebacke *et al.*, 1969; Dales, 1973). Some authors have also suggested phagocytosis— after capping of viruses (Gschwender and Traub, 1979). These alternative modes of entry are illustrated schematically in Figure 79. Lysates obtained from infected cells 10–60 minutes after adsorption contain RNase sensitive viral RNA, modified particles, and also intact infectious unaltered virus particles, indicating that both intact and modified virus particles are taken up by cells (Fenwick and Wall, 1973; Habermehl *et al.*, 1974; Lonberg-Holm *et al.*, 1975). Evidence for the suggested modes of entry for picornaviruses are listed in Table 45.

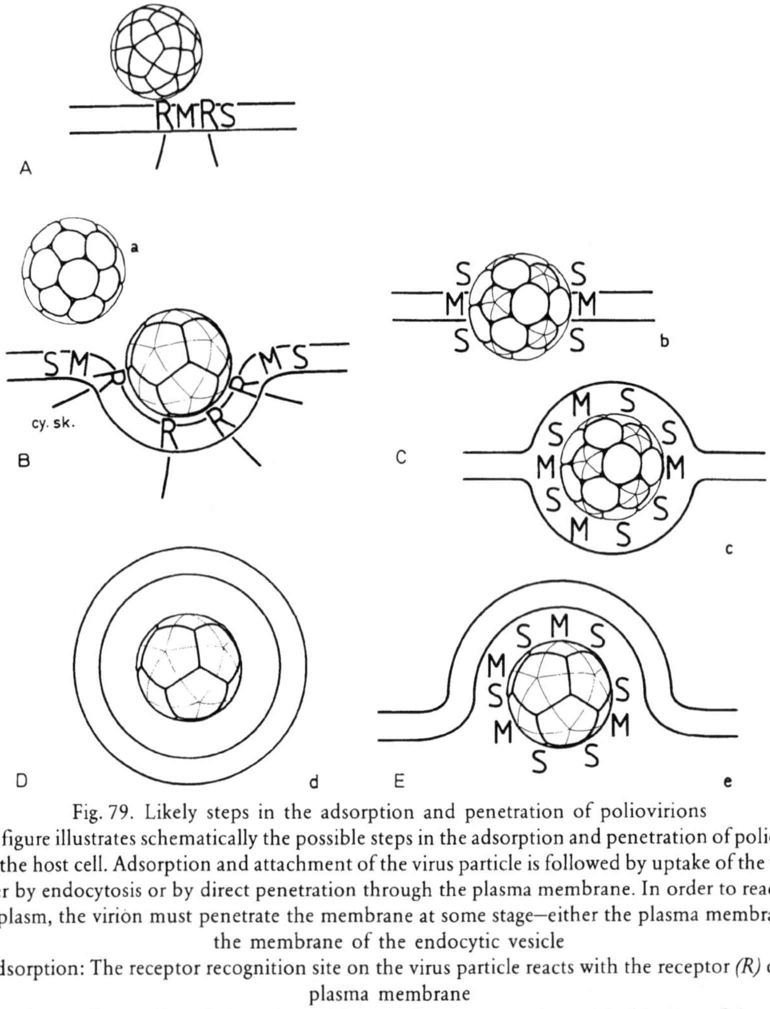

Fig. 79. Likely steps in the adsorption and penetration of poliovirions

This figure illustrates schematically the possible steps in the adsorption and penetration of poliovirus into the host cell. Adsorption and attachment of the virus particle is followed by uptake of the virion either by endocytosis or by direct penetration through the plasma membrane. In order to reach the cytoplasm, the virion must penetrate the membrane at some stage—either the plasma membrane or the membrane of the endocytic vesicle

A Adsorption: The receptor recognition site on the virus particle reacts with the receptor *(R)* on the plasma membrane

B Attachment: Interaction of the virion with several receptors leads to tight binding of the virion. Modifying *(M)* and stabilizing *(S)* activities may interact with the attached particle. *a* eluted noninfectious A—particle; *cy.sk.* presumptive anchorage of receptor to cytoskeleton

C–E Penetration of the virion into the cytoplasm by fusion-viropexis: *b* virus as integral membrane protein; *c* virus inserted between the membrane bilayer; *e* virus modified and stabilized free in cytoplasm

B–D Penetration of the virion into the cell by phagocytosis: *d* virus in micropinocytotic vesicle

Table 45. *Evidence in support of insertion-penetration or receptor mediated endocytosis as modes of entry of poliovirus into the host cell*

1. Neutralizability of infectivity by virus-specific antibodies is lost within minutes after adsorption, indicating rapid entry of virion into the cell (Mandel, 1967).
2. Tagging of adsorbed virions (mengovirus) with fluorescent-labeled antibodies leads to capping and engulfment of virions, suggesting that these virions are taken up by phagocytosis (Gschwender and Traub, 1979).
3. Electron microscopy of infected cells shortly after adsorption reveals either distorted particles beneath the cell membrane, suggesting direct penetration through the membrane (Dunnebacke *et al.*, 1969), or virions within pinocytic vesicles, indicating uptake by endocytosis (Dales, 1973; Bienz, Zeichhardt, personal communication).
4. Fractionation of cytoplasmic extracts from infected cells shows that radiolabeled virions are taken up by lysosomes, where viral protein and RNA is degraded (Habermehl *et al.*, 1974). Intact and modified virions can be isolated from cell extracts (Lonberg-Holm *et al.*, 1975), and radiolabeled parental virions can be recovered in association with polysomes (Fenwick and Wall, 1973; Habermehl *et al.*, 1974), indicating that intact particles can penetrate the cell membrane or that of endocytic vesicles.
5. Infection of HeLa cells by poliovirus is insensitive to inhibition by the ionophore monensin under conditions which completely block infection by enveloped viruses (Nobis *et al.*, 1983). Other authors have reported that entry of poliovirus is sensitive to inhibition by monensin and other ionophores, protonophores, and amines (albeit less so than entry of enveloped viruses and certain protein toxins) (Zeichardt *et al.*, 1983; Madshus *et al.*, 1984). These results suggest that acidification of endocytic vesicles may be important for membrane penetration by poliovirus under certain conditions.
6. Exposure of surface-bound poliovirus to low pH enhances the efficiency of virus entry, indicating that an acid mediated conformational shift in the viral capsid (as may be induced by fusion of an endocytic vesicle with a lysosome) can enhance membrane penetration (Madshus *et al.*, 1984).
7. Metabolic inhibitors which deplete cells of ATP inhibit entry of poliovirus into the host cell, indicating that membrane penetration is an energy-requiring process (Lonberg-Holm and Whitely, 1975; Madshus *et al.*, 1984).
8. The ratio of physical: infectious particles is decreased from 1,000 : 1 to 5 : 1 when the membrane barrier is circumvented by releasing virions directly into the cytosol after fusion of virus-containing liposomes with the cell membrane (Wilson *et al.*, 1979), indicating that the vast majority of infecting virus particles are degraded during the normal route of entry.

It should be kept in mind that these observations reflect the superposition of abortive and productive pathways of infection. Many virus particles evidently enter the cell by endocytosis and are subsequently degraded in lysosomes. Whether the productively infecting particles enter by direct penetration of the cell membrane or by receptor mediated endocytosis remains a matter of controversy.

Electron microscopic evidence has been presented in support of either concept (see Fig. 80). In one case, particles the size of intact virions were detected free (not enclosed in vesicles) in the cytoplasm as soon as 3 min after infection at 37°C. Since mock infected cells did not contain comparable particles, it was concluded that virions had entered cells directly (Dunnebacke *et al.*, 1969). The virion like particles appeared slightly skewed and misshaped in these electron micrographs, the cell membrane directly above the particles was always intact (Fig. 80A). Such pictures support the concept of fusion-viropexis. The awkward shape of the particles may reflect structural rearrangements of the protein coat required for penetration through the membrane or may indicate modified virions. Ultrastructural analysis of echovirus infected cells also revealed intracellular particles resembling virions soon after infection (Eggerstedt,1963). Other authors have not been able to confirm these observations (Dales, 1973; Bienz, Zeichhardt, personal communication). In their observations, virus particles were found regularly only in micropinocytic vesicles (Fig. 80B). Such observations support the concept of entry by receptor mediated endocytosis.

Entry of poliovirions in phagocytic vesicles alone, however, is not sufficient to enable the viral RNA to initiate viral replication. As pointed out by Lonberg-Holm and Philipson (1974), the viral RNA then has still to pass the membrane of the phagocytic vesicle, transfering the membrane obstacle only from the outside to the inside. On the other hand, components of radioactively labelled virions have been detected in lysosomes, and it was even proposed that uncoating in phagocytic vesicles after fusion with lysosomes may be a prerequisite for the initiation of virus replication (Habermehl *et al.,* 1974). Enveloped virions (VSV, Semiliki Forest Virus) enter cells in endocytic vesicles. Acidification which can occur by fusion with lysosomes is a prerequisite for their release and uncoating (White and Helenius, 1980; White *et al.*, 1981; Matlin *et al.,* 1982; Marsh, *et al.*, 1983). Addition of monensin (a Na^{+}-H^{+} exchanger)–or other drugs which prevent acidification of endocytic vesicles–prior to or at the time of infection abolishes the infectivity of these viruses. Infection of HeLa cells by poliovirus is not inhibited by monensin under conditions which completely block infection by VSV, indicating that poliovirus can penetrate into the cytoplasm without the aid of acidified endocytic vesicles (Nobis *et al.,* 1983). Other authors have recently reported that compounds which dissipate proton gradients across membranes, like monensin, protonophores, and amines do inhibit the entry of poliovirus into HeLa cell monolayer cultures, but not the binding of poliovirus of these cells (Zeichhardt *et al.,* 1983; Madshus *et al.,* 1984). When cells with surface bound virus were exposed to low (5.5) pH, the virus entered efficiently even in the presence of these drugs (Madshus *et al.,* 1984). The authors conclude that low pH is required for poliovirus entry–conditions encountered in intracellular vesicles–although the entry of poliovirus requires less acidification than the entry of Semliki Forest virus or protein toxins (Helenius *et al.,* 1980; Olsnes and Sandvig, 1983), and, further, that under normal conditions, poliovirus does not inject its RNA at the cell surface but that this takes place in acidic intracellular vesicles.

In either case, fusion of the viral capsid with the membrane seems to be required to get the viral RNA to the cytoplasmic side of the membrane–be it plasma, pinocytic vesicle, lysosome or endoplasmic reticulum membrane. It is certainly easier to envision the insertion of capsid proteins into the membrane than the passage of free viral RNA through the hydrophobic lipid bilayer (although under certain carefully defined conditions, even isolated viral RNA can permeate through the membrane, see section V below).

Virus insertion into the membrane may require changes in the hydrophobicity of virions. The surface of native poliovirions is in part hydrophobic (Farrah *et al.*, 1981). An increase in hydrophobicity may occur on the virion surface after interaction with the plasma membrane in two ways: a) by binding of lipophilic membrane components (Fenwick and Wall, 1973; Lonberg-Holm *et al.*, 1975), b) by turning hydrophobic regions of coat peptides from the inside to the outside,

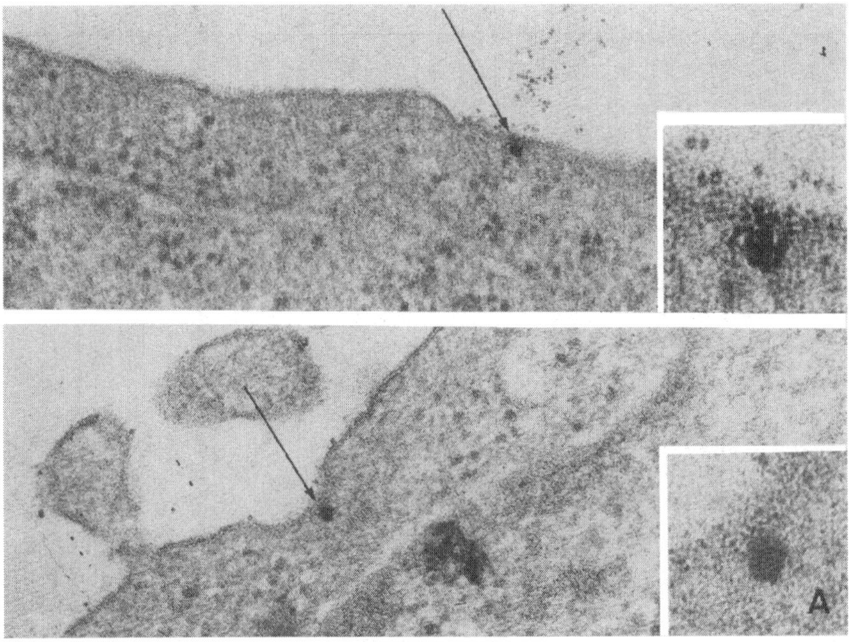

Fig. 80. Electron micrographs of the entry of poliovirions into the host cell

A Direct penetration by fusion viropexis. KB cells at 2 and 5 min post-innoculation. – Figure from Dunnebacke *et al.*, 1969 [J. Virol. *4*, 508 (1969)]

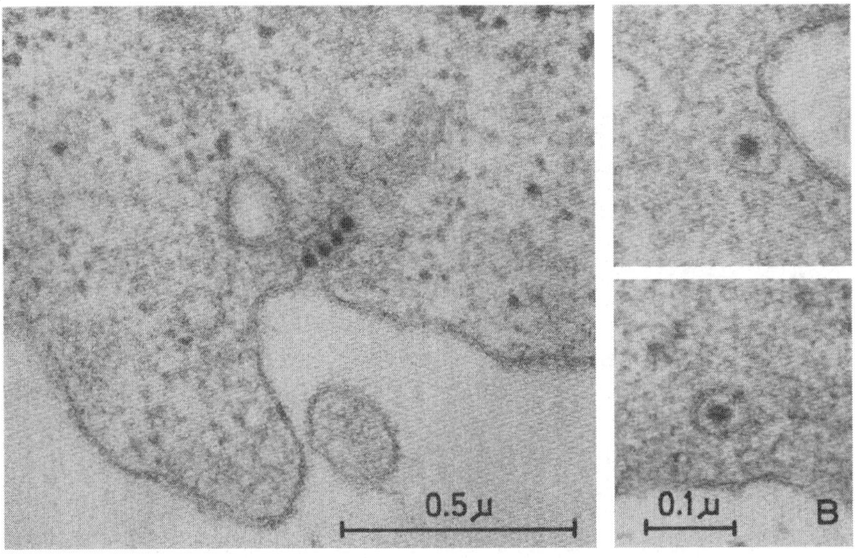

B Indirect penetration by uptake into micropinocytic vesicles. HeLa cells 10 min post-innoculation. – Figure from Dales, 1973 [Bacteriological Reviews *37*, 118 (1973)]

perhaps with a concomitant conversion to the pI 4.5 form and release of cations from virions. An increased hydrophobicity of the virion surface would facilitate insertion of the virion into the plasma membrane as well as the interaction of virions with modifying and stabilizing components of the receptor complex. In fact, it has been shown that modified virions insert more readily into artifical membranes (Lonberg-Holm *et al.,* 1976).

Insertion of a virus particle which is four times as thick as the membrane will result in an enormous local deformation of the membrane. We are not aware of any EM-picture of a virus inserted (in the real sense, Fig. 79C, p. 292) into the membrane. Recent work on the structure of artificial membrane bilayers has revealed that inversed micelles can indeed be "inserted" into the membrane (Sen *et al.*, 1981), and this might be a model also for the insertion of a large component like poliovirus. After insertion into the membrane, penetration could be triggered through activity of the cytoskeleton, and might be driven by the electric potential or proton gradient across the membrane. Membrane penetration seems to be an energy requiring process, since metabolic inhibitors which deplete cells of ATP inhibit virion entry into the cell (Lonberg-Holm and Whiteley, 1976; Madshus *et al.,* 1984).

Certain conditions may shuffle the virus from the abortive to the productive route of infection or vice versa. Incubation of cells at pH 5.5 during the first hour after adsorption enhances the proportion of productively infecting particles, and treatment of cells with low concentrations of weak amines has a similar effect (Madshus *et al.,* 1984). In contrast, interaction of poliovirus with antibodies seems to direct the virion via endocytic vesicles to lysosomes and thereby increase the rate of the abortive pathway. In accord with this view is the observation that the RNA from antibody neutralized virus complexes is degraded to a greater extent than the viral RNA from unneutralized virions after infection of HeLa cells (Mandel, 1967a).

The relative proportion of intact virus particles in extracts of infected cells shortly after adsorption can be greatly increased by prior exposure of virions to SH reactive compounds such as glutathione or to arildone (Fenwick & Cooper, 1962; McSharry et al., 1979). The effect of these substances in many ways resembles that of neutralizing antibody. Virions are stabilized by these substances against in vitro degradation by heat or alkali; adsorption and attachment are not inhibited and may even be enhanced, and the stabilized particles enter the cell but are not uncoated.Comparable effects on poliviruses are shown by the SH reactive antiviral compound S7 (ethyl-2-methylthio-4-methyl-5-pyrimidine carboxylate) (Lonberg-Holm *et al.*, 1975) and by arildone (Caliguiri *et al.*, 1980). In the presence of S7 poliovirions are adsorbed to cells with unaltered kinetics. Intact poliovirions can be recovered from cells after adsorption in the presence of S7 by lysis of the cell with an ultrasonic generator in the presence of 0.5% Nonidet P 40 (NP 40), a nonionic detergent (Lonberg-Holm *et al.,* 1975). The recovered virions remain bound to some cellular constituents and sediment in sucrose gradients with 130S, somewhat slower than intact virions. The attached cellular material can be removed from the virions by exposure to 0.2% SDS, and intact infectious virions are obtained. Spontaneous elution of virus particles from cells and the loss of their VP4 does not occur in the presence of S7. It is interesting to note here that several thiols, but also disulphides, including both reduced and oxidized glutathione shorten the latency phase and the rate of replication of several viruses, including polioviruses, in tissue culture cells (Marcialis *et al.*, 1977). These results indicate that the structural alterations associated with release of VP4 from virions are not essential for insertion and penetration. Moreover, stabilization of the viral capsid seems to enhance the chance for penetration of the cell membrane.

In sum, it remains a matter of controversy at what point and how the virus particle penetrates the membrane, since it is not possible to distinguish clearly between abortive and productive pathways of infection. Appearance of virus particles in micropinocytic vesicles and in lysosomes may reflect the abortive pathway. Particles which succeed in penetrating the plasma membrane directly may be in an advantage in that they evade the harsh conditions of lysosomes. On the other hand, acidification of endocytic vesicles may serve as a trigger of virus penetration through the membrane. Finally, both phenomena may occur concomitantly during normal poliovirus infection. Adsorption, modification, insertion, membrane penetration, and endocytosis can be envisioned as continuous processes—the precise moment of membrane penetration (plasma or vesicle membrane) depending on the relative rates of the individual steps. Certain membrane components, transmembrane potential, or proton gradients may enhance membrane penetration. In any case, the virus is very inefficient in surmounting the membrane barrier, and less than 5% of the infecting RNA survives the early interactions of virus with membrane components and intracellular vesicles (Joklik and Darnell, 1961).

IV. Uncoating: A Multistep Process

A. Possible Steps and Sites of Uncoating

One aspect of uncoating is absolutely certain: It must occur. The viral RNA has to be released from the confinement of the protein coat in order to initiate viral replication on a host cell ribosome. Almost all other aspects are uncertain. To what extent must the viral capsid be dissociated? Where does uncoating start, when is it complete? What happens to the viral capsid proteins? The uncertainties arise from the superposition of abortive and productive pathways of infection. It is generally supposed that the consecutive dissociation of capsid components from virus derived particles occurs as discussed in section II above; capsid destabilization, release of VP4, release of VP2, release of RNA, correspond also to the uncoating steps required for productive infection (see Table 44, p. 290). It is questionable whether all of these steps really are required for the release of RNA. One model of virion morphogenesis holds that nascent viral RNA is injected into an (almost) complete procapsid (see Chapter 10). The insertion of RNA into procapsid is thought to occur in close association with intracellular membranes and may be driven by ongoing RNA synthesis. It might be sufficient for uncoating, if only a small segment of the viral RNA—its 5' end—penetrated through a tiny hole in the viral capsid to interact with ribosomes, translation of the RNA then providing the energy to "suck" out the RNA. Infecting particles do differ in some important aspects from the procapsids in progeny assembly: Many penetrated particles contain VP_4 and VP_2 instead of the uncleaved precursor VP_0 in procapsids, other penetrated particles have already lost their VP_4 component. The release of VP_4 at least seems to be necessary for RNA uncoating; it may be important, however, that this release occurs only in the vicinity of ribosomes. Substances which stabilize virions against VP_4 dissociation in vitro (antibodies, arildone, reducing substances, see section III above) also inhibit the initiation of viral replication in vivo presumably by blocking the release of RNA from the capsid.

Uncoating in vivo is usually inferred from indirect observations (Table 46). Release of RNA can be analyzed by measuring the sensitivity of radiolabeled intracellular viral RNA to digestion by RNase. Capsid alterations are indicated by loss of neutralizability of the infectivity by specific antibody and by the induction of capsid leakiness to dyes. Joklik and Darnell (1961) analyzed the RNase sensitivity of radiolabeled intracellular viral RNA at different times after infection of HeLa cells with labeled polioviruses. Release of viral RNA commences as soon as 10 minutes p.i. and is almost complete after 30 minutes of incubation at 37°C. By one hour after infection, most of the viral RNA found after lysis of cells is already degraded and in soluble form even without addition of RNase.

A detailed investigation on the onset of viral capsid leakiness during the uncoating of polioviruses after adsorption to cells was carried out by Mandel (1967b) (Table 46). Polioviruses grown in the presence of photosensitizing dyes such as acridine orange, proflavine or neutral red incorporate these dyes into virus particles during maturation (Crowther and Melnick, 1961; Schaffer and Hackett, 1963; Wallis and Melnick, 1963; Wilson and Cooper, 1963) and thereby become sensitive to inactivation by exposure to light. This light sensitivity is lost after infection as a result of the release of the dyes from the virions during one of the first steps in uncoating, implying leakiness of the viral coat. Mandel (1967a, b) showed that light resistance is gained only after the infectivity of polioviruses is no longer

Table 46. *Evidence for uncoating of poliovirus in the host cell*

1. RNase sensitivity of prelabeled virion RNA begins within 10 minutes after adsorption and is almost complete at 30 minutes p.i. Less than 5 % of the infecting RNA can be recovered in an acid insoluble form (Joklik and Darnell, 1961).
2. Light sensitivity of virions containing photosensitizing dyes is lost within minutes after adsorption, indicating that the capsid has become permeable and that the dyes can diffuse away from the RNA (Mandel, 1967).
3. The adsorption capacity of virus particles isolated from infected cells is lost within minutes post infection (Joklik and Darnell, 1961; Lonberg-Holm *et al.*, 1975).
4. Modified particles can be obtained upon fractionation of freshly infected cells (Fenwick and Wall, 1973; Lonberg-Holm *et al.*, 1975) or after *in vitro* incubation of virus with isolated membrane fractions (Chan and Black, 1970; DeSena and Mandel, 1976; Guttmann and Baltimore, 1977; DeSena and Torian, 1980).
5. With the detection of virus-specific protein synthesis and RNA synthesis entry and uncoating of the infecting virus particle can be inferred.

neutralizible by antibodies. He concluded that poliovirus particles penetrate the cell as intact particles and are uncoated within the infected cell with complex kinetics in at least two stages. The initial phase (loss of neutralizibility by antibodies) lasted about 20 min at 37° C or 40 min at 32° C. The second stage—gain of light resistance—occurs more rapidly and probably is accompanied or followed by loss of VP4. Both stages of uncoating require high energies of activation (about 70 Kcal).

These uncoating steps can be mimicked in vitro. The coat of native poliovirions is so tight that cations do not exchange. The viral capsid becomes leaky in vitro when polioviruses are exposed to high pH or temperature (see Chapter 3). The virions take up water, swell, and cations may exchange. VP4 is usually released in this step. A more subtle destabilization and structural alteration of poliovirus particles occurs in vitro when polioviruses are incubated in the presence of ATP at pH 8.0 and 37° C for 90 min (Schärli and Koch, 1984). A virus bound-kinase is activated which phosphorylates all coat proteins. This results in further structural alterations in the viral coat which are detectable with a number of antisera prepared against isolated viral coat peptides. Most of these antisera do not react with native virions but all of them react with intact virions after phosphorylation (Schärli and Koch, unpublished).

The loss of VP4 in vivo and the conversion from a light sensitive to a light resistant state were inhibited by compounds which stabilize polioviruses. The stabilizing compounds belong to three chemical classes and may act in different ways: a) by hydrophobic interactions with virions (arildone) b) by stabilization of SH groups and c) by binding of neutralizing antibodies. Since arildone, SH protecting compounds as well as DEP prevent the phosphorylation of viral proteins in vitro (Schärli and Koch, 1984) we propose a still hypothetical scheme for the individual early steps in uncoating of polioviruses (see Table 47). It is uncertain where in the cell these steps occur. By binding to multiple receptors or

Table 47. *Proposed steps in virus uncoating*

1. Interaction with receptor complex.
2. Expansion and loosening of the capsid.
   a) exposure of SH groups
   b) activation of virus associated proteinkinase
   c) formation of S-S bonds within VP_2 or VP_0 or between VP_2
   d) phosphorylation of viral proteins notably VP_0 resulting in further destabilization of the capsid.
3. Shift from the pI 7.0 to the pI 4.5 form, release of cations, uptake of water and swelling of virions. Loss of VP_4, change to C antigenicity, loss of attachment capacity, spontaneous elution of altered virions.
4. Newly exposed hydrophobic regions on the surface of virions result in insertion into the membrane and binding to lipophilic (virus stabilizing) membrane components.
5. Loss of VP_2—loss of hydrophobic interaction.
6. Dissociation of VP_1–VP_3 and RNA after penetration into the cytoplasm.
7. Attachment of viral RNA to ribosomes.

intracellular membrane components, SH groups in VP_0 or VP_2 become exposed to the surface of virions and form S-S bridges which keep the viral coat now in a more expanded and more open conformation. Phosphorylation of the viral coat proteins by activation of a capsid bound or cellular proteinkinase leads to an opening of the coat structure. This in turn, might allow water to enter, cations to leave the coat and results in a swelling of the virus particles. Sudden, yet orderly release of VP_4 allows the exit of photosensitizing dyes from virus particles and can be detected in that way. In vitro studies with isolated membrane fragments indicate that the next step in uncoating is the complete release of VP_2 giving rise to altered virions which still contain their full complement of RNA (see section II above). Such particles have not yet been obtained after interaction with intact cells. This could reflect the inherent instability of such particles. On the other hand it is well known that the RNA can be released even from VP_{2-3-1} containing shells. The in vitro observed release of VP_2 may represent an abortive pathway. In any case, it seems probable that the dissociation of RNA from viral capsid of productively infecting particles occurs near cellular ribosomes.

B. The Fate of the Parental Capsid Protein

By one hour after infection most of the viral RNA found after lysis of cells is already degraded and in an acid-soluble form even without addition of RNase (Joklik and Darnell, 1961). Both intact and modified parental virus particles are bound to the rough endoplasmic reticulum of infected cells and part of labeled parental viral proteins are bound to virus specific polysomes (Fenwick and Wall, 1973; Habermehl *et al.*, 1973). The rest is degraded in lysosomes.

It is quite possible that some of the capsid proteins are required for effective initiation of translation of the parental RNA. Poliovirus temperature sensitive mutants have been characterized which are defective in inhibiting host protein synthesis at elevated temperatures. The defect in these mutants maps in the coding region for the capsid proteins, to the 5' side, *i.e.* possibly within the coding region for VP_4 (or VP_2) (Steiner-Pryor *et al.*, 1973). These are just the proteins released during virion uncoating.

Furthermore, addition of VP_4 to HeLa cells sensitizes cells for infection by isolated viral RNA indicating that VP_4 is bound to cells or might even enter cells independently of virions. Intact virus particles at a ratio of one per 20 ribosomes and isolated VP_4 inhibit polypeptide chain initiation in vitro in rabbit reticulocyte extracts (Racevskis *et al.*, 1976). Inhibition of polypeptide chain initiation accelerates and amplifies viral replication (Koch *et al.*, 1980a; Ramabhadran and Thach, 1980). Invasion of the host cell by virions as intact particles may facilitate the transport to the rough endoplasmic reticulum and permit uncoating to occur at the site of initiation of viral protein synthesis. This may be important for two reasons. First, the release of the viral RNA at the site of action as mRNA reduces the chance of prior degradation of the RNA. Second, interference of cellular protein synthesis by poliovirus capsid proteins may increase the efficiency of polypeptide chain initiation by viral RNA (see Chapter 8).

V. Infection of Cells Lacking Receptors

A productive infection can be initiated by incubating cells lacking specific receptors with extremely high concentrations of virions. This procedure allows the adaptation of poliovirions to hosts other than monkey or man (Armstrong, 1939). In addition, many cells of non primate origin—including even bacterial cells—can be infected with isolated poliovirus specific RNAs (for review see Koch, 1973). Recently, the artificial entry of intact virions into cells lacking receptors was achieved by the enclosure of virions within liposomes (Wilson *et al.*, 1979).

In the following sections we will discuss some aspects of studies on the infection of cells by isolated poliovirus RNAs. These studies have shed some light on the mechanisms of uptake of RNA-species by cells and on factors which are important for the succesful initiation of replication by the infecting poliovirus RNA.

A. Introduction

The first successful infection of cells with isolated nucleic acid free of viral proteins was achieved with the RNA from tobacco-mosaic-virus (TMV) in 1956 (Gierer and Schramm, 1956; Fraenkel-Conrat *et al.*, 1957). The infectivity of isolated poliovirus RNA was detected in 1957 (Colter *et al.*, 1957; Alexander *et al.*, 1958a, b; Holland *et al.*, 1959). Shortly thereafter several laboratories reported the productive infection of a variety of cells with isolated viral nucleic acids, RNA as well as DNA, from different viruses (review by Wecker, 1962; Schaffer, 1962).

RNA isolated from intact polioviruses by different methods as well as the three virus specific RNAs isolated from poliovirus infected cells (see Chapter 9), namely single-stranded RNA, double stranded RNA (RF-RNA), and multistranded RNA (RI-RNA), are infections for tissue culture cells only under special experimental conditions (Table 48, p. 303). The first method applied to tissue culture cells was based on treatment of cells with hypertonic salt or sucrose solutions. Infection with RNA occured only in a hypertonic environment (Alexander *et al.*, 1958 a+b; Koch *et al.*, 1960). Later it was shown that the infectivity of poliovirus-specific RNAs, including doublestranded (RF-RNA) and multistranded (RI-RNA) molecules, can be determined with an agar cell suspension plaque assay, provided the host cells are sensitized by polycation treatment (for review see Pagano, 1970; Koch, 1973). Several other methods to stimulate cells or to determine RNA infectivity have been applied. These include exposure of cells to DMSO and to dehydration (Ludwig and Smull, 1963), and binding of RNA to insoluble carriers (Dubes *et al.*, 1964; Dubes, 1971).

All these conditions show two effects on the host cell, a) they lead to a more efficient adsorption and penetration of the viral RNA and b) they interfere with macromolecular synthesis of the cell, notably protein synthesis.

The infectivity of isolated viral nucleic acids is in general nevertheless several orders of magnitude lower than the infectivity of the viruses from which the nucleic acids were obtained, provided the infectivity of both the isolated RNA and the intact virus are assayed on the same host cell. Finally, it has been shown that poliovirus proteins interfere with cellular protein synthesis in vitro (Racevskis *et al.*, 1976). On the basis of these observations it was suggested that poliovirus proteins carry out three functions: a) efficient adsorption and uptake of the viral RNA b) protection against RNAse and c) interference with macromolecular synthesis of the host cell.

Intact viruses show a very narrow host range, that is they interact under physiological conditions only with cells carrying specific receptors for them on their surface. Isolated nulceic acids show a broad host range, they can induce one virus growth cycle (under specific experimental conditions) in a variety of cells. Polioviruses for example interact only with human or monkey cells, the isolated viral RNA infects cells from all mammals and from birds. Picornavirus RNA is even translated in plants. Injection of isolated mengovirus RNA into the nuclei of acetabularia, followed by implantation of these nuclei into or fusion with anucleated acetabularia results in the synthesis of virus specific proteins detectable with immunofluorescence 1–5 days after the injection of RNA (Cairns *et al.*, 1978). Translation of viral RNA was also observed in cellfree extracts from bacteria (Baltimore *et al.*, 1969), and poliovirus specific RNA is synthesized in E. coli (Koch and Vollertsen, 1972a, b; Koch, 1973).

B. Adsorption of Viral RNA to Cells

While in physiological salt solutions, up to 80 % of single-stranded viral RNA is absorbed to HeLa cells within 5–10 minutes at 37° C (Borris and Koch, 1964a), but only 20 % of RF-RNA or RI-RNA (Wentzky and Koch, 1971; Wiegers and Koch, 1972). RNAs of lower molecular weight (*i.e.* ribosomal RNA and transfer RNA) are adsorbed less efficiently than viral RNA. 28S ribosomal RNA is adsorbed to a higher extent than 18S ribosomal RNA. Thus, the adsorption of single-stranded RNA to HeLa cells appears to depend on the molecular weight. The amount of adsorbed RNA increases with the size of the RNA (Borriss and Koch, 1964a), whereas the adsorption of RF-RNA and RI-RNA seems to be limited by the rigid structure of the molecules rather than molecular weight.

The concentration of the RNA may vary over a wide range–from 0.1 to 125μg/ml for viral RNA and up to 20μg/ml for RI-RNA–without altering the rate or percentage of RNA adsorption; the adsorption of single-stranded viral RNA is, however, highly dependent on the cell concentration. At 6 x 10^6 cells/ml only 24 % of the viral RNA is adsorbed in 15 minutes at 37° C; at 4 x 10^7 cells/ml, however, 70–80 % is adsorbed (Borris and Koch, 1964a). In contrast, RF-RNA adsorption is little affected by variation in cell density, whereas RI-RNA adsorption at cell concentrations above 2 x 10^6 cells/ml follows more the pattern observed with single-stranded RNA: 5 x 10^6 cells bind 6.0 % of RI-RNA, but 20 % is adsorbed at 4 x 10^7 cells/ml (Wentzky and Koch, 1971). However, this is considerably less than single-stranded RNA (72 %) and also lower than RF-RNA (30 %) under identical conditions (Wiegers and Koch, 1972).

When 10μg DEAE-dextran/ml are added, viral RNA adsorption increases from 60 % to about 90 %. Adsorption of RF-RNA and RI-RNA is enhanced 5- to 9-

fold, respectively (Wentzky and Koch, 1971; Wiegers and Koch, 1972). Other polycations show comparable enhancing effects (Table 48).

Since viral RNA adsorption to HeLa cells is relatively high (up to 80 %) in the absence of polycations, the mechanism of the enhancement of RNA adsorption to cells was further examined only with RF-RNA and RI-RNA. The more cells are present in the incubation mixture, the more DEAE-dextran is needed to obtain 90% adsorption of RF-RNA.

Table 48. *Effect of various polycations on the infectivities of single- and double-stranded poliovirus RNAs*

Polycation(s) added	Polycation concentration	Infectious centers/μg RNA single-stranded	Infectious centers/μg RNA double-stranded
None	–	17	26
DEAE-dextran	50 μg/ml	1.7×10^5	1.0×10^6
DMSO	10 %	1.8×10^4	3.6×10^4
DEAE-dextran + DMSO	160 μg/ml + 10 %	2.0×10^6	2.0×10^6
Hydroxylamine virus	5 μg/ml	6.0×10^3	1.5×10^4
VP_4	0.001 μg/ml	2.0×10^4	8.0×10^4
VP_4	0.01 μg/ml	3.0×10^4	2.0×10^5
VP_4	0.05 μg/ml	4.0×10^4	5.0×10^5
VP_{1-3}	1.0 μg/ml	6.0×10^2	7.0×10^3

The biological activity of standard preparations of RNA was assayed with the agar cell suspension plaque assay. Polycations were added at the concentrations indicated one minute before the addition of RNA to the suspended cells. DEAE-dextran was added together with the RNA.

In an attempt to find out whether the stimulating effect of polycations on RI-RNA adsorption is exerted on the RNA or the cells, either the cells or RI-RNA were preincubated separately at 37° C for 5 min with 4 to 20μg/ml of polycations, and after an additional 15 min incubation the percentage of RNA adsorbed to cells was determined. The values for RNA adsorption were compared to those obtained when polycations were added together with the RNA. With three polycations (DEAE-dextran, poly-arginine, poly-L-lysine) tested at concentrations of 4μg/ml, the adsorption of RI-RNA was enhanced several fold after preincubation of the RNA with the polycations. With less than 20μg/ml of DEAE-dextran maximal RNA adsorption was observed when the RI-RNA was allowed to interact with the polycation prior to contact with the cells.

When the cells were preincubated with low concentrations of DEAE-dextran (4 or 5μg/ml) for different periods at 37° C before addition of RF-RNA or RI-RNA, an interesting picture emerged. The DEAE-dextran interacts with the cells but the enhancement of RF and RI-RNA binding is lost rapidly and almost completely with increasing time of incubation. This would indicate that DEAE-dextran penetrates into the cells or is attached to the cell membrane in a way that does not lead to improved binding of RNA.

C. Penetration of Isolated Viral RNA into Cells

There are three ways to force the entrance of nucleic acids into a cell: 1. passive influx via osmotic shock, aiding the penetration of RNA molecules already adsorbed to the cell membrane; 2. stimulation of the cell for active uptake of RNA by exposure to polycations (Ryser and Hancock, 1965; Koch *et al.*, 1966), or to insoluble facilitators (Dubes *et al.*, 1964), and 3. entrance of viral RNA via lipid vesicles (Wilson *et al.*, 1979).

1. RNA-Penetration by Passive Influx of RNA

The yield of infectious centers induced by a given amount of poliovirus RNA in cell monolayers is dependent on the salt concentration of the medium. Infection of the cells by RNA is extremely inefficient in an isotonic environment, but is strikingly greater when the cells are exposed or preexposed to hypertonic salt solutions. At 1M NaCl or higher, 1000 x more plaques are induced than in physiological saline (Koch *et al.*, 1960). In hypertonic salt solutions HeLa cells shrink.

The kinetics of RNA adsorption and penetration in hypertonic salt solutions have been studied by washing the cell monolayers at different times after seeding. The number of plaques increases up to 12 minutes after the addition of the RNA and is not influenced by the salt molarity of the washing solution. Therefore, at this time the viral RNA must be either inside the cell or firmly attached to it. To decide between the two alternatives, ribonuclease was added to the cells at different times after infection and in different salt solutions. All potentially infectious centers were abolished when the RNase was added in hypertonic salt, but 50 to 80% of the infectious centers survived treatment by RNase in isotonic solution.

The conversion of cell-bound RNA from an RNase-sensitive to an RNase-resistant state must occur immediately after transfer of the cells from the hypertonic to the isotonic milieu. The RNase is only able to destroy RNA infectivity as long as the RNA has not completely penetrated the cell. To gain support for this conclusion, cells were treated for 30 seconds with isotonic solution before the RNase was added in either isotonic or hypertonic solution. In neither case did RNase influence the number of plaques induced. Adsorbed RNA, therefore, enters the cell during the change from hypertonic to isotonic environment together with the influx of the solvent (Koch, 1963).

2. Stimulation of Active Uptake of Viral RNA

Ryser and Hancock (1965) reported that a number of polycations stimulate the active uptake of proteins by tissue culture cells. Based on their findings Koch and Bishop (1968) studied the interaction of polycations with suspended HeLa cells and their influence on the efficiency of viral RNA infection.

In an isotonic environment, viral RNA and RF-RNA infectivity is lost rapidly when exposed to RNase (Bishop and Koch, 1967; Mittelstaedt *et al.*, 1975). The penetration of infectious RNA can therefore be measured by following the effect of RNase on infectious center formation. Only those cells which have already engulfed one complete RNA molecule before RNase is added will register as infectious centers in the plaque assay. Judged by the effect of RNase on infectious center formation, 60 to 80% of viral RNA penetrates the cells within 15 to 20 minutes at 37°C. RF-RNA penetrates the cells within 5 to 10 minutes at 37° C (Koch & Bishop, 1968; Koch, 1971b). The infection of suspended HeLa cells by poliovirus RNA was stimulated as much as 10,000-fold by exposing the cells to poly-L-lysine, poly-L-ornithine, and DEAE-dextran. All polycations tested enhance the PFU titer of viral RNA and RF-RNA considerably. The highest number of PFU for a given RNA preparation is obtained with 20 to 50μg/ml of poly-L-ornithine or 100 to 200μg/ml DEAE-dextran. These results show that polycations enhance not only adsorption of poliovirus-specific RNAs, but also their penetration into cells (see Table 48, p. 303).

3. Entrance of Poliovirus RNA into Cells via Lipid Vesicles (Liposomes)

Poliovirus RNA can be encapsidated into large unilamellar vesicles (LUV) in an RNAse resistant form and thereby delivered efficiently to various tissue culture cells including cells which are not sensitive to infection by the intact virus. LUV-encapsidated poliovirus RNA shows infectivity comparable to that of isolated RNA after polycation exposure of cells (Wilson *et al.*, 1979).

At first sight, the results seem to indicate that an efficient delivery of the RNA into the cytoplasma of the cell is the only prerequisite for the initiation of a viral growth cycle. However, it is conceivable that liposomes trigger signals on the cell membrane which mediate local interference with macromolecular synthesis in cells which cannot be detected by presently available techniques. The specific infectivity of poliovirus RNA entrapped in liposomes is enhanced by exposure of the liposomes to either active or inactive ribonuclease. The polycation RNase may act on cells in a similar way as DEAE-dextran or other polycations (see above).

D. Cellular Competence for Infection by Viral RNA

In an isotonic environment, untreated tissue culture cells show a very low competence for infection by viral RNAs, in spite of the fact that all the cells can be infected by intact viruses (Borriss and Koch, 1964a+b; Koch and Bishop, 1968). Since—as described above—adsorption and penetration do not limit the infectivity of RNA, what are the factors which do control the cell's competence?

In the poliovirus-HeLa cell system isolated RNA is most infective under conditions where some of the cells have lost their viability due to exposure either to hypertonic saline or polycations. Drastic interference with the host cell's metabolism might, therefore, be a prerequisite to a successful infection by viral RNA (Koch *et al.*, 1960; Koch and Bishop, 1968).

During infection of cells with intact viruses this prerequisite for successful virus replication might be fulfilled by the action of viral capsid protein(s) (Breindl and Koch, 1972). Addition of poliovirus to reticulocyte lysates inhibits rapidly the initiation of protein synthesis (Racevskis *et al.*, 1976). Exposure of HeLa cells to the poliovirus capsid protein VP_4 enhances the competence for infection by isolated viral RNA (see Table 48 above). Of interest in this respect is also the observation that exposure of L 929 cells to the methyl ester of amphotericin B (AMBNE), a macrolide polyene antibiotic, at a concentration of 100μg/ml enhances the infectivity of isolated encephalomyocarditis (EMC) RNA by a factor of 10-100, even when the viral RNA is already cell associated. The site of action of AMBNE has not yet been elucidated. However, AMBNE was shown not to inhibit RNase. Exposure of cells to AMBNE does not alter the infectivity of intact EMC-virus. It would be of interest to study whether AMBNE enhances uptake of RNA, interferes with cellular protein synthesis or shows both of these effects in a way comparable to polycations. Exposure of cells to low concentration of polycations results in optimal adsorption and penetration of isolated viral RNA. However, higher concentrations of polycations are required for an optimal RNA infectivity.

These observations pose a number of question: a) Does the time and sequence of adding polycations and RNA affect the competence of the cells and the yield of infective centers? b) Is the concentration of RNA critical? c) How stable is the polycation-induced cell competence? d) Are all the cells of a given cell population competent for infection by viral RNA?

1. Optimal Conditions for the Use of Polyactions

The concentration of tissue culture cells can be varied from 2 x 10^5 to 9 x 10^7/ml without a significant influence on the titer of a standard RNA preparation. This is apparently due to the fact that the number of cells competent for infection is in excess to the number of RNA molecules capable to initiate infection. Within a certain limit, greater quantities of input RNA would give a higher number of infectious centers. The yield of infectious centers is determined by the polycation: cell ratio rather than the absolute concentration of the polybasic compound. The important practical implication of these data is that an appropriate polycation concentration must be present to obtain an optimal infection by viral RNA.

The sequence of adding RNA and polycation to the host cells, and the time interval separating these two events also effects the yield of infectious centers. Infectivity is stimulated by adding poly-L-ornithine or methylated albumine both before or after the RNA, but the exposure of the host cells to polycations at different times in relationship to the addition of RNA leads to variations in the final titer. In the case of methylated albumine, maximum titers are obtained only if the cells are exposed within 3 to 5 minutes before or after the addition of RNA. The timing of poly-L-ornithine addition is apparently much less critical. DEAE-dextran augments RNA titers only if used prior to, or simultaneously with, the addition of RNA.

The kinetics of sensitization of cells for infection with poliovirus RNA were analyzed in more detail with poly-L-lysine, poly-D-lysine and poly-L-ornithine of different chain length (Dubes and Wegrzyn, 1977), and with histones and protamine (Dubes and Wegrzyn, 1978). The sensitization rates were fastest for the polycation with the lowest molecular weights (poly-ornithine, 15,500, and poly-L-lysine 1700). The polycations with greater chain length (poly-L-lysine 160,000) sensitized at a slower rate. Desensitization effects were obtained with the two shorter polycations. Both histone and protamine rapidly sensitize cells for infection by isolated viral RNA.

2. Relationship Between RNA Concentration and Yield of Infectious Centers

Within a certain range of RNA concentrations, the yield of infectious centers is directly proportional to the amount of RNA present in the incubation mixture. This has the same implication as the similar relationship observed with intact virus: a single molecule of RNA is sufficient to initiate infection.

There is a striking similarity between the curve that relates poliovirus RNA concentration to the yield of infectious centers (Koch, 1973) and the curve that relates the concentration of transforming DNA to the yield of transformed bacteria (Ravin, 1958). In transformation studies no further increase in transformants occurs above a definite concentration of DNA. This plateau reflects the fact that only a limited percentage of the bacterial population is competent for transformation. Saturation of these cells with DNA generally occurs when 10 % of the bacteria are transformed. With single-stranded viral RNA it was observed that only 4 % of the cell population could be infected when the cells were exposed to polycations (Koch and Bishop, 1968). The combined use of DEAE-dextran and DMSO to stimulate the competence of HeLa cells leads to productive infection in 50 % of the cells (Oppermann and Koch, 1973). When poliovirus RNA is introduced into cells via liposomes, up to 90 % of the cells register as infectious centers (Wilson *et al.*, 1979). However, transforming DNA and infectious viral RNA differ in one aspect; above a certain concentration of RNA, the plateau in the yield of infectious centers is followed by a sharp decline. Concentrations of RNA higher than 10 μg/ml result in reduced yields of infectious centers. The relationship between the number of RNA molecules per host cell and the efficiency of the assay was analyzed to rule out the possibility that the reduction in plating efficiency was due to aggregation of RNA. These experiments were facilitated by the fact that the cell concentration in the assay mixture can be varied over a considerable range without affecting the infectious center yield from a given dilution of RNA. It is therefore possible to vary the RNA molecule: cell ratio (by changing both cell and RNA concentrations) while maintaining the absolute concentration of RNA below the inhibitory level of 1—5 μg/ml. Under these conditions, quantities of RNA ranging from 50 to 50,000 molecules per cell have no effect on the yield of infectious centers. However, if the number of RNA molecules exceeds 100,000 per cell, the efficiency of infection drops sharply, despite the fact that the concentration of RNA is well below the level which is inhibitory in the standard procedure. This contrasts with the observation that maximal adsorption of RNA continues until ratios of 50,000 molecules per cell (Borriss and Koch, 1964a). With the decrease in assay efficiency, there is a marked reduction in cell viability (as judged by cloning efficiency). The reduced yield of infectious centers at high cell: RNA input ratios parallels a reduction in host cell viability which occurs at high RNA: cell ratios. This toxic effect of RNA adsorbed to the plasma membrane can be partially reversed by addition of polycations (Oppermann and Koch, unpublished).

3. *The Combined Effect of Dimethylsulfoxide and DEAE-Dextran on the Competence of Cells for Infection by Viral RNA*

It has been shown that poliovirus RNA infects mammalian cells more efficiently in the presence of dimethylsulfoxide (DMSO) (Amstey, 1966; Tovell and Colter, 1967). However, it was not known whether DMSO affects primarily the cells or the RNA. DMSO alters the membrane permeability of mammalian cells (Jacob *et al.*, 1964), and melts double-stranded RNA at room temperature (Katz and Penman, 1966). Tovell and Colter (1967) compared the effect of DMSO and DEAE-dextran on RNA infectivity in the L cell-mengovirus system. When present during RNA-cell interaction, both augment the infectivity of the RNA to the same extent, but the addition of DMSO and DEAE-dextran together does not lead to further stimulation of infectivity. The results obtained in similar experiments carried out with the HeLa cell-poliovirus system are in general agreement with the observation of Tovell and Colter (1967), except that DEAE-dextran and DMSO show a synergistic effect in the poliovirus RNA-HeLa cell system. The PFU titer of poliovirus RNA is four to ten-fold higher when HeLa cells are sensitized to RNA infection by exposure to both DEAE-dextran and DMSO together. DMSO was preincubated with either viral RNA or cells in order to investigate the mechanism of action. Where cells were pretreated with 10–20 % DMSO in addition to DEAE-dextran, the infectivity titer increased tenfold compared to treatment with DEAE-dextran alone. In contrast, preincubation of viral RNA in various salt solutions containing 10–90 % DMSO alone did not alter the infectivity titer (Koch, 1971b). The experimental conditions were designed to prevent DMSO, present in the RNA diluent, from effecting the cells. The results reveal that the primary effect of DMSO is on the cells rather than on the RNA.

The viral RNA infectivity is increased when poly-L-ornithine and methylated albumin are added to HeLa cells either 30–5 minutes before, or 5–30 minutes after the addition of RNA. In contrast, DEAE-dextran stimulates cellular competence only if it is added shortly before the viral RNA. If the cells are kept at 37° C, they rapidly lose the competence for infection induced by DEAE-dextran and DMSO and this cannot be restored completely by adding DEAE-dextran once more together with the RNA. However, when cells are cooled to 0° C after the exposure to DEAE-dextran and DMSO (for one min at 37° C), they remain competent for infection by viral RNA for up to 48 hours (Koch, 1971b).

The loss of cell competence for infection by viral RNA and RF-RNA at 37° C is more pronounced when the cells are incubated in the absence of DMSO. Polycations are toxic for cells, therefore it was of interest to study the effect of DEAE-dextran in our system. The viability of the cells after different periods of incubation with DEAE-dextran was determined by counting the number of cells which were stained by trypanblue. HeLa cells (2.6×10^6/ml) were incubated with 160 μg DEAE-dextran for 30 minutes at 37° C in Eagle's medium without and with 10 % DMSO. In the absence of DMSO, 38 % of the cells lost their viability, but in the presence of 10 % DMSO only 12 % became trypanblue-positive. Thus, DMSO enables the cells to tolerate higher concentrations of DEAE-dextran.

4. Competence of HeLa Cells for Infection by Viral RNA at Different Stages in a Cell Growth Cycle

While, at an appropriate multiplicity of infection, native poliovirus can infect every cell of a given HeLa population, only a fraction of a polycation-sensitized HeLa cell population produces infectious centers after exposure to viral RNA (Oppermann and Koch, 1973). Since the competence of bacteria for transformation increases at the end of the log phase (Ravin, 1958), the competence for infection by viral RNA might likewise depend on a certain physiological stage of the cell population. Therefore, HeLa cells were synchronized by a double thymidine block as described by Tobia *et al.*, 1970, and exposed to viral RNA at hourly intervals after thymidine removal (Breindl and Koch, 1972). A maximum yield of infectious centers was obtained 8 to 12 hours and again at 23 hours after thymidine removal. Cells were 40-fold less sensitive during S-phase (phase of optimal thymidine incorporation). When the synchronized cell population was sensitized by exposure to DEAE-dextran and DMSO at different stages in the growth cycle, however, the yield of infectious centers remained constant throughout the cell cycle. The maximum percentage of cells (25 %) which could be infected with RNA also remained constant and comparable to the percentage of susceptible cells of the unsynchronized cell population.

The results show that the competence of unsensitized HeLa cells for infection by viral RNA depends on the stage of the cell cycle, the cells being most susceptible during phases with reduced protein synthesis (G2, M, beginning G1). This dependence can be abolished by treatment of cells with enhancers of infectivity such as DMSO and DEAE dextran, substances which themselves inhibit protein synthesis.

E. Conclusions

During the isolation of viral nucleic acids from purified virus preparations, the nucleic acid is freed from its protective coat and thereby rendered RNase sensitive. In addition, the RNA is deprived of a specific virus component important for attachment to and penetration into the host cell. It is, therefore, not surprising that infectivity titers of naked RNA preparations are found to be only 0.1 to 1% of those of the initial virus preparations, and that viral RNA infectivity is optimal when experimental conditions favor RNA adsorption to cells, and exclude RNA-degrading enzymes during the interaction of viral RNA with cells. These findings in turn led to the widespread acceptance of the following hypothesis: the main and perhaps only functions of viral proteins are to provide the viral nucleic acid with a protective coat and with a specific attachment site for interaction with sensitive cells. The observation that methods used to enhance the infectivity of viral RNA (including osmotic shock, exposure to polycations or DMSO) favor either RNA adsorption to cells or RNA penetration into cells, or both support this hypothesis. Indeed, it has been suggested that reversible cell damage, elicited by a variety of adverse conditions, augments penetration of the nucleic acid by virtue of pathological effects on the cell surface activity (pinocytosis, vacuolization) (Ryser, 1967).

However, a number of observations indicate that RNA adsorption and penetration are not the critical factors controlling the infectivity of isoloated viral RNA. First, adsorption and penetration of RNA appear to be quite efficient even under isotonic conditions. Yet, the level of productive infection is extremely low unless a polycation is applied to the RNA-cell complex. The highest RNA infectivities in different assay procedures were obtained when manipulations were used which were toxic for the cells (Koch *et al.*, 1960; Koch and Bishop, 1968). HeLa cells lose their cloning ability by exposure to polycations (Koch and Bishop, 1968) and their viability by prolonged exposure (30 min at 37°C) to DEAE-dextran (Koch, 1971b). Finally, the polycations show a high level of efficiency even if applied at a time when adsorption and penetration are known to be complete. These observations suggest that the effect of cell damage on the RNA-cell interaction has little to do with uptake of nucleic acid, but rather with a later stage of RNA-cell interaction.

An alternative explanation could be that interference with the metabolism of the host cell sensitizes cells for infection by viral RNA, and that this sensitization is a perequisite for obtaining an optimal infectious center yield with viral RNA. For example, the number of binding sites for ribosomes or initiation factors available for the incoming viral RNA might be extremely low because of a great surplus of cell messenger RNAs. It could be argued that agents such as polycations and osmotic shock alter this situation in favor of the viral RNA, perhaps by direct or indirect effects on the production of host cell mRNA or their ability to compete for initiation (see Chapter 8). The observation that cells are most sensitive to RNA infection in phases of the cell cycle with reduced protein synthesis (see p. 309, above) is also consistent with this explanation. Provided that interference with host cell translation is required for efficient initiation of a viral growth cycle, one could presume that viral proteins, in addition to their task of protecting the RNA and of promoting its adsorption, also have the function of interfering with host cell translation. VP4 is the only viral protein which sensitizes cells for infection by viral RNA. VP4 stimulates the competence of HeLa cells for infection by the viral RNA and RF-RNA up to 10,000 fold at concentrations from 10^{-3} to 10^{-1} μg protein/ml. None of the many polycations analyzed so far affected the competence of HeLa-cells for infection by RNA so much. We anticipate that further studies on the role of VP4 in viral infection might reveal additional support for the hypothesis that viral proteins interfere specifically with host cell functions, thereby enhancing the chance of viral RNA to initiate a virus growth cycle.

VI. Summary

Virus receptors and their complementary sites on virions are thought to determine virus tropism. Studies on the classical virus-host cell system, the T phages and E. coli, have shown that mutations in both virus and host can change the host range of viruses. Indirect evidence suggests that picornaviruses and their host cells have undergone comparable changes in evolution. Members of the different genera of picornaviruses adsorb only to their specific receptors on the same host cell. A comparison of the amino acid sequences of the capsid proteins of the different picornaviruses and of the poliovirus subtypes in terms of conserved and variable regions can be expected to provide valuable information on the constitution and evolution of the receptor recognition site on these viruses.

The poliovirus receptor has the specificity of a group specific antibody, binding all three poliovirus types but not other picornaviruses. The receptor is an integral membrane protein, which is active only when bound in its natural membrane or on artificial hydrophobic surfaces. Binding of poliovirus to the cellular receptor is rapid and efficient even at 0°C. It is not known what, if any, function the poliovirus receptor has for uninfected cells. Ionic as well as hydrophobic interactions during virus adsorption lead to a relatively loose binding of the virus particle to the host cell. The largest of the capsid proteins VP1 has been implicated by genetic studies as participating in the specific adsorption to the host cell. Binding of neutralizing antibodies to virions does not prevent adsorption and uptake of the neutralized particles, indicating that the receptor recognition site is located on regions of the capsid distinct from the main antigenic sites of the virus particle.

At physiological temperatures the early specific interactions between viruses and cells lead to alterations of both reaction partners. Approximately 50% of the adsorbed poliovirus particles are eluted again from the host cell in an altered non-infectious state. These so-called M or A particles which have lost the VP4 component, carry new, C antigenic determinants, and can no longer attach to host cells. The rest of the adsorbed virus particles become attached firmly to the membrane either in an intact form or also as modified particles. In this tightly bound state, the particles can only be removed from the cell membrane by strong detergents or denaturing agents. Whereas adsorption or loose binding might be due to interaction of one receptor recognition site on the virion with one receptor, subsequent binding of many receptor recognition sites on one virus particle by several receptors on the host cell may lead to the tight attachment.

Very early alterations in the plasma membrane following adsorption of picornaviruses include changes in membrane fluidity and permeability and patching and capping of attachment sites. Polioviruses do not enter cells below 20°C. A certain mobility of membrane proteins and a critical fluidity of the membrane seem to be essential for productive infection. Changes in membrane permeability and membrane functions might be signals or triggers for the virus induced shut-off of macromolecular synthesis of the host.

Polioviruses are taken up by cells either intact or in a modified form by fusion-viropexis or by endocytosis. Attachment itself seems to trigger a conformational change in the coat leading to a greater hydrophobicity of the coat which would favor insertion of virions into the plasma membrane of the host cell. Membrane components—other than receptors—function in inducing further alterations of poliovirus but also in stabilization of the virions. The stabilizing components might prevent premature uncoating and release of viral RNA which otherwise would be degraded by cellular RNAse.

Like bacteriophages, picornaviruses rapidly induce resistance of host cells to superinfection by the same virus, but not to viruses belonging to other genera of picornaviruses. This indicates either that the poliovirus receptors are inactivated by the binding interactions with virions or that the receptors are taken up by the cell together with the virus particles, analogous to the internalization of other allosterically modified membrane components (the $Na^{+}K^{+}$pump after inactivation by ouabain, for example).

It is impossible to say what route the productively infecting particle follows from its first interaction with the host cell receptor to the initiation of translation of its RNA on a host cell ribosome. Some of the uncoating activities characterized in isolated cell membrane fragments may represent a host defense mechanism against viral invasion, others may represent the first steps of virion uncoating. Most of the RNAs of firmly attached virus particles are degraded by the host cell, only 5% can be recovered in macromolecular form. Virus particles which had been taken up by phagocytosis may be degraded in lysosomes. Still some of the particles must succeed in penetrating through the lipid bilayer of a membrane—either that of the plasma membrane, or that of the endocytic vesicle. Part of the capsid proteins of infecting virus particles have been detected in association with the rough endoplasmic reticulum and with ribosomes. It is not known how much of the capsid must be disassembled in order for the RNA to gain access to binding to ribosomes. Local release of VP4 near the prospective initiating ribosomes may enhance the chances for initiation of translation by the viral RNA. Studies on the infectivity of isolated poliovirus RNA has supported the notion that the capsid proteins are important for early events following adsorption and penetration of parental virus particles or RNA. Substances which enhance the infectivity of isolated RNA may do so by substituting for some of these functions of the capsid proteins, notably by interfering with host-cell translation.

8

Translation of the Viral Genome

Viral protein synthesis is dependent on the functional protein synthesizing machinery of the host cell. Before dealing with the characteristics of viral protein synthesis, a summary of basic features and of the control of protein synthesis in uninfected cells is required. Particular aspects which are specific for the translation of poliovirus mRNA or those which are thought to be affected by poliovirus infection are pointed out in section I and are discussed in detail in section II.

I. Mammalian Protein Synthesis

A. The Protein Synthesizing Machinery

The basic features of protein biosynthesis were first elucidated in studies with bacterial systems. The eukaryotic protein synthesizing machinery follows the basic pattern of prokaryotes but exhibits greater complexity. The larger ribosomes consist of larger RNA and contain more protein components. There are a greater number and a higher complexity of initiation factors. The eukaryotic mRNAs show specific features (cap and polyA) and the mRNA is associated with specific proteins in mRNP particles (for review see Scherrer, 1979; Hershey, 1982b). Perhaps it is only natural to think that the basic machinery for protein biosynthesis is present in bacterial systems and that the additional components of the more complex eukaryotic system are in some way involved in specific control mechanisms. Indeed, mammalian protein synthesis is subject to extensive regulation at the translational level. Poliovirus infection markedly affects the translational machinery and its regulation. The virus even seems to exploit some of the host cell's regulatory mechanisms for its own benefit.

1. Ribosomes

a) Structure and Composition

Electron microscopic studies have revealed many similarities in the structure of prokaryotic and eukaryotic ribosomes. However, ribosomes of mammalian cells are larger and more complex than their prokaryotic counterparts (for review see Wool, 1982). They consist of 50 % protein and 50 % RNA. The small 40 S subunit is composed of one 18 S RNA molecule (MW 0.7 x 10^6) and 30 proteins. The larger 60 S subunit contains three RNA molecules: one of 28 S (MW 1.7 x 10^6), one of 5.8 S (MW 0.51 x 10^6), one of 5 S (MW 0.39 x 10^6), and 45–50 proteins. So far a reconstitution of eukaryotic ribosomes from their RNA and protein constituents has not been accomplished. Therefore, we know little about the function of the individual proteins of eukaryotic ribosomes.

b) Monosomes and Polysomes

A varying fraction of ribosomes does not participate in protein synthesis in eukaryotic cells. These ribosomes are in a nonfunctional state and sediment as mono-

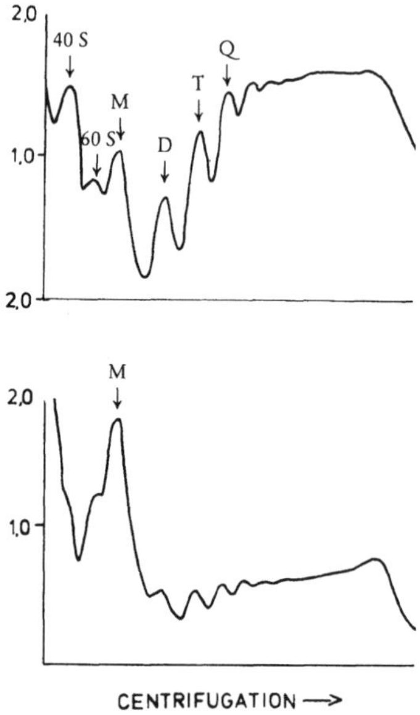

Fig. 81. Polyribosome profiles of uninfected and poliovirus-infected HeLa cells
Cytoplasmic extracts from uninfected HeLa cells (upper profile) and from cells 2.5 h.p.i. with 50 plaque forming units of poliovirus type I (lower profile) were analyzed by centrifugation in 15%–45% sucrose gradients in reticulocyte standard buffer (RSB). The UV-absorbancy at 260 nm was recorded continuously. 40 S subunits, 60 S subunits, monosomes *(M)*, disomes *(D)*, trisomes *(T)*, and quatrosomes *(Q)* can be distinguished. The poliovirus induced shut-off of host protein synthesis is reflected in the breakdown of polysomes and the accumulation of ribosomes in 80 S monosomes. The peak of very large polysomes in the infected cell contains the ribosomes engaged in the translation of the long viral mRNA (see also Fig. 69, p. 239)

somes with 80 S. In contrast, ribosomes engaged in protein synthesis are associated with mRNA in polymeric structures called polyribosomes. Polysomes can be separated from monosomes by sucrose gradient centrifugation (Fig. 81). Disomes, trisomes and so on can be identified. The number of ribosomes associated with one mRNA is determined by a) the size of the mRNA and b) the efficiency of polypeptide chain initiation (see below). An efficient mRNA binds one ribosome per 100 nucleotides under optimal conditions for protein synthesis. The ratio of ribosome distribution in monosomes and polysomes is affected by the nutritional state of the cell, the presence of growth factors and by virus infection (Koch *et al.*, 1980a). Shuttleing ribosomes between monosomes and polysomes is one important means by which the cell controls and regulates protein synthesis at the level of translation, both quantitatively and qualitatively (see below).

Whether or not the different functional states of ribosomes present in monosomes or polysomes have a structural basis is still a matter of controversy. The level of phosphorylation of ribosomal proteins in non-functional 80 S ribosomes and in functional ribosomes present in polysomes differs (Kruppa and Martini, 1978). Phosphorylation of ribosomal proteins might play an important role in the control of polypeptide chain initiation. The functional state of ribosomes can also be altered by the binding of low molecular weight substances activated or released by membrane mediated events (Koch *et al.*, 1976; Koch *et al.*, 1980a, see below).

c) Free and Membrane Bound Polysomes

Polysomes are present in the cytoplasm either attached to the membranes of the endoplasmic reticulum or in a free form. The function of the latter is the synthesis of soluble proteins. Polysomes engaged in the synthesis of plasma membrane proteins or secretory proteins are attached to membranes via their nascent peptide chains (Blobel and Dobberstein, 1975a, b). It is generally accepted that selection of polysomes for attachment to membranes is mediated by their nascent peptides. Once this selection has occured, anchorage on the membranes is stabilized by ribosomal protein-membrane interactions.

2. mRNA

Eukaryotic mRNAs show a number of specific features that are not found in the mRNAs of prokaryotes. These include a cap at the 5' end; larger non-translated regions at both ends; a 3' poly A sequence; a single initiation site; and binding of specific proteins producing mRNP particles. Figure 82 shows a model of eukaryotic mRNA.

a) The Cap

The 5' end of the mRNA is modified by specific enzymes in the nucleus during and following transcription and can be further modified in the cytoplasm (Perry 1982, Shatkin 1982). RNA transcription starts with a 5' triphosphate of a purine

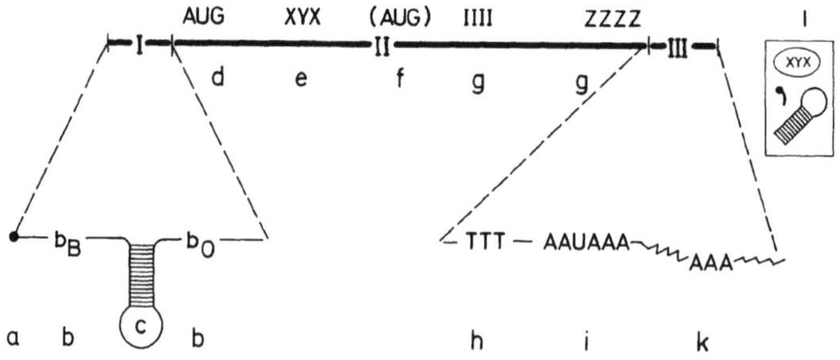

Fig. 82. The features of eukaryotic mRNAs

I 5′ end untranslated region

Length varies from 15 to more than 200 nucleotides, determines the translational efficiency of the mRNA

a cap at 5′ end, enhances efficiency of initiation

b contains ribosome-binding (b_B) and ribosome orientation (b_O) sites, and putative effector-binding sites

c the secondary structure is stabilized by high ionic strength and may affect the fidelity and efficiency of translation

II coding region

400 to 7500 nucleotides long; usually monocistonic, i.e. coding for one single polypeptide which may be cleaved co- or post-translationally

d AUG initiation codon, translation usually but not always begins at the AUG codon closest to the 5′ end

e putative splicing signals, weak termination sites, and ribosome binding sites that might be used in viral mRNAs for modulating the relative translation of encoded regions

f internal "cryptic" initiation sites are thought to be unrecognizable as such by eukaryotic ribosomes

g putative "slow down" or "halt" signals may cause a slow down in the elongation rate to permit the (co-translational) processing or membrane insertion of the nascent polypeptide chain

III 3′ end untranslated region

Usually 15 to 100 nucleotides long, conserved in evolution, function not well understood

h termination codon

i the AAUAAA hexapeptide functions as a signal for polyadenylation

k poly A-tail, determines mRNA stability and half life

l effector molecules: proteins, small oligonucleotides, or other substances may influence the translatability of mRNAs by binding to specific complementary sites

nucleoside. The terminal nucleoside-triphosphate interacts with guanosine-triphosphate to from a 5'5' ppp-linkage. Often both 5' purines are methylated. The 5' structure on the mRNA thus formed is called the cap. As methylation occurs to a different extent at the second and third nucleotide (on the base and/or the ribose) different cap structures are formed (Perry, 1982, Shatkin, 1982) . Evidently all mRNAs of eukaryotic cells are capped.

The function of the cap may be twofold: 1. The rate of initiation of mRNA translation appears to be influenced by the presence of the cap at the 5' end of most eukaryotic mRNAs. Caps enhance the efficiency of initiation manyfold (Both *et al.* 1975a, Mutukrishnan *et al.* 1975), but they are not absolutely required, nor do they affect the precision of initiation site selection (Kozak and Shatkin 1979, Shatkin 1982). Special cap binding proteins may moderate or enhance the binding of mRNA to ribosomes during initiation. Alternatively, the cap on the mRNA might enhance binding to the 40 S ribosomal subunit by also preventing a conformation of mRNA unfavourable for initiation (Kozak, 1982). 2. In addition, the cap – perhaps in conjunction with cap binding proteins - stabilizes mRNA against degradation by 5' exonucleases (Furuichi *et al.*, 1979), thereby prolonging the half life of the mRNA.

Thus it is of interest that poliovirus mRNA is not capped. The virus exploits this difference to host mRNAs to redirect the translational machinery for the preferential translation of its own mRNA (see below).

b) The 5' Terminal Untranslated Region

The length of the untranslated polynucleotide region between the cap and the AUG triplet where translation is initiated is surprisingly long in eukaryotic mRNAs (from 15 to over 100 nucleotides) and varies considerably between different mRNAs (reviewed in Kozak, 1982). Contained within this sequence are potential sites for binding of initiation factors, ribosomes and possibly specific regulatory factors. Since such binding occurs with different efficiencies and specificity depending on the nucleotide sequence and the structural conformation of the RNA, the untranslated 5' regions constitute potential controlling sites for the regulation of translation. The finding that the 5' untranslated regions are conserved in the evolution of related mRNAs, as in the case of ß globin mRNAs (Martin *et al.*, 1981), supports the notion that these regions carry important functions. The untranslated 5' ends of the RNAs of the three serotypes and their corresponding vaccine strains of poliovirus show greater sequence homology than the coding regions (Toyoda *et al.*, 1984).

The assignment of specific functions to the 5' untranslated regions, however, has not been an easy task. The corresponding sequences of almost 100 eukaryotic mRNAs have been determined to date, only to reveal a seemingly endless diversity in sequence and secondary structure (Kozak 1982). It has not been possible to find any apparent common structural features within these regions that might function as typical ribosome binding or ribosome orientation sites or by binding of regulatory molecules.

In prokaryotes, ribosome binding and thereby selection of mRNA for translation is dependent on basepairing between a purine rich tract at the 5' end of the mRNA and a pyrimidine rich region at the 3' end of the 16S RNA of the small ribosomal subunit (the so called SD sequence) (Shine and Dalgarno, 1974; Steitz and Jakes, 1975). To what extent such an interaction between mRNA and rRNA is involved in the selection of mRNA in eukaryotic cells is still uncertain. The 3' terminal nucleotide sequence of eukaryotic 18S rRNA is highly conserved and does exhibit a remarkable homology to the corresponding region of prokaryotic 16S rRNA, implying an important function for this region. The eukaryotic 18S rRNA, however, lacks the very sequence implicated in the binding of prokaryotic mRNAs to ribosomes; instead they contain a purine rich sequence which is complementary to a sequence appearing in many but certainly not all eukaryotic mRNAs (Hagenbüchle *et al.*, 1978).

It has been proposed that binding of ribosomes to eukaryotic mRNAs always occurs first via the cap structure at the 5' end, and is followed by movement of ribosomes along, and interaction with, the 5' untranslated region (Kozak, 1978, 1982). It is still unclear to what extent the nucleotide sequence and structural conformation of the adjacent untranslated regions contribute to this binding. Initiation and ribosome binding occur, albeit less efficiently, also when the cap is removed, indicating that some other structural features are also important in ribosome binding.

Attempts have been made to find common structural features of eukaryotic mRNAs involved in binding of initiation factors, of the small ribosomal subunit as part of the initiation complex, and finally of the entire 80S ribosome, by sequencing the RNA fragments that are protected by binding to these complexes against RNase digestion (Kozak, 1978, 1982). Again diversity prevails over common features: fragments protected by the 40S subunit vary greatly in size (from 18 to 52 nucleotides in length), most but not all contain the cap, some but not all the initiation AUG codon, and some both. Fragments protected by binding of initiation factor eIF-2 seem to be similar to those protected by the 40S ribosomal subunit (Kaempfer, 1982). Fragments protected by the 80S ribosome are usually smaller than those protected by the 40S subunit and seem to center around the initiation AUG codon. At present, it is difficult to interpret these findings. What can be said is that there is "something" in the 5' untranslated region that makes the mRNA recognizable for ribosome binding, and that this "something" evidently varies greatly between different mRNAs. Similarly, there is "something" that makes the initiating AUG codon recognizable as the correct site for the beginning of translation of the mRNA. In most, but again not all eukaryotic mRNAs, it is simply the AUG codon closest to the 5' end that is used.

One feature that can be explained by the diversity of the 5' untranslated regions is the variation in the intrinsic translational efficiencies of mRNAs (Tab. 49). The efficiency of initiation varies greatly between individual eukaryotic mRNAs and can be affected by changes in environmental conditions. Specific translational efficiencies and thereby competition between mRNA for initiation components play an important role in the regulation of translation (see below). Variations in nucleotide sequences or secondary structure of the 5' untranslated regions are likely to affect the initiation efficiency. Experiments with synthetic oligonucleotides have revealed that A-U rich sequences seem to be particularly efficient in binding to 40S subunits (Both *et al.*, 1976). So far, however, it has not been possible to correlate the translational efficiencies of mRNAs observed *in vivo* or *in vitro* with specific features of their 5' ends.

One last intriguing, albeit still speculative, function for the 5' untranslated region of mRNAs that could account also for the observed diversity in their sequences is the control of the translatability of individual mRNA species by binding specific regulatory molecules. The sequence of the 5' untranslated regions of mRNAs may harbour sites for the binding of specific regulatory molecules (RNA or proteins). This could provide a highly specific mechanism to control gene expression at the level of translation (Hershey 1982a). Although postulated for certain mRNAs—in particular for oocyte mRNAs and myosin mRNA during

Table 49. *Relative translational efficiencies (RTE) of some mRNAs*

Protein	Culture-condition	% of total protein synthesis	RTE of mRNA
Actin	optimal (o.)	5	
	inhibited (i.)*	0.5	0.1
Histones	o.	3	
	i.	12	4.0
Hemoglobin	o.	20	
	i.	80	4.0
Immunoglobulin			
L-chain	o.	6.9	
	i.	18.5	2.7
H-chain	o.	9.5	
	i.	12.7	1.3
Poliovirus	o.	12.5	
Proteins	i.	90	8.0
Friend virus	o.	0.3	
gp55	i.	0.03	0.1

The relative translational efficiency is determined from the ratio of the synthesis of a given protein with respect to total protein synthesis under optimal and restricted conditions of polypeptide chain initiation.

* Rate of initiation reduced to 10% of maximal rate.

differentiation—such specific regulatory molecules have never been clearly identified to date. The translation of mRNAs for the ribosomal proteins in prokaryotes has been shown to be sensitive to a feedback regulation by its end product: addition of isolated ribosomal proteins to in vitro protein synthesizing systems rapidly inhibits the further synthesis of these proteins (Nomura *et al.*, 1980). This selective inhibition correlates with the binding of proteins to their corresponding mRNAs, although it is not known whether these proteins bind to the 5' end of the mRNA. Such a control mechanism could be a working concept also for the regulation of the switch from translation to replication of the parental poliovirus mRNA.

In this context it is interesting to note that the poliovirus messenger RNA appears to contain one of the largest untranslated 5' regions known to date: 743 nucleotides in length (Dorner *et al.*, 1982), providing ample space for intrinsic regulatory mechanisms for initiation of protein synthesis as well as for the initiation of RNA synthesis on the 3' end of the complementary RNA strand in RI-RNA (see Chapter 9).

c) The 3' Terminal Untranslated Region and the Poly A Tract

Eukaryotic mRNAs except histone mRNAs carry on their 3' end poly A tracts of varying lengths (Browermann, 1977). These tracts are added post-transcriptionally in the nucleus by an enzyme which recognizes a specific nucleotide sequence (see Fig. 82, III). There is a good correlation between the length of the poly A tract and the halflife of an mRNA, indicating that the poly A protects the mRNA against exonucleases. This protection might be due to or enhanced by binding of specific proteins to the poly A tract (see p. 321). It has also been suggested that the poly A region has a nuclear function and mediates transport of RNA from the nucleus to the cytoplasm (Adesnik *et al.* 1972).

Between the poly A tract and the termination codon in eukaryotic mRNAs are untranslated regions of variable lengths (40–110 nucleotides long) (Wilson *et al.*, 1977). The 3' terminal regions of ß-globin mRNAs have been shown to be even more conserved in evolution than the 5' terminal untranslated regions (Martin *et al.*, 1981), indicating important functions for this region. What these function(s) could be is still unclear with one exception: the hexanucleotide AAUAAA has been found in all 3' untranslated regions of eukaryotic mRNAs examined to date. It is thought to serve as a signal for polyadenylation (Proudfoot and Brownlee, 1976; Fitzgerald and Shenk, 1981).

The length of the untranslated 3' region between the termination codon and the poly A tract of poliovirus mRNA is 78 nucleotides long. It does not contain the hexanucleotide signal for polyadenylation. This is not surprising, since the poly A of poliovirus RNA is genetically determined and does not seem to arise by post-transcriptional polyadenylation (Wimmer, 1979).

d) Monocistronic mRNAs and Potential Internal Initiation Sites

All eukaryotic mRNAs are considered to be monocistronic. They contain only one functioning initiation site and their primary translation product is one polypeptide. This polypeptide can serve as a precursor for several functional peptides which arise by cotranslational or post-translational protein processing (see below).

In contrast, prokaryotic mRNAs are polycistronic, containing several sites for attachment of small ribosomal subunits (30 S). The initiation of translation on the internal sites proceeds independently of initiation at the 5' end. The yield of translation for the individual cistrons on the polycistronic mRNA may differ.

In eukaryotic cells the protein synthesizing machinery does not usually recognize internal initiation sites. An exception to this rule may occur during the translation of foreign mRNAs. When poliovirus mRNA is translated in extracts of eukaryotic cells, two independent initiation sites are utilized. In addition, there are several experimental observations which indicate that ribosomes (and the initiation factor eIF2) are also firmly attached to internal sites on some viral mRNAs, notably on picornavirus RNA in vitro (McClain *et al.*, 1981; Pérez-Bercoff, 1982). Ribosomes are bound to 3 different sites on poliovirus RNA and to 4 sites on mengovirus RNA. There are "cryptic" internal initiation codons on several mRNAs of plant and animal viruses which are, at best, poorly functional until the mRNAs are processed into fragments exposing an AUG codon close to the newly formed 5' end. It is still a matter of controversy whether internal initiation sites on poliovirus mRNA function during poliovirus replication in intact cells (Pérez-Bercoff 1982) (see below, section IIC2).

e) mRNPs

In contrast to prokaryotes, where mRNAs exist solely in naked form, the mRNAs in all eukaryotic organisms are bound to a number of distinct proteins, from the time of synthesis in the nucleus to the end of translation. Cytoplasmic ribonucleo-protein particles (RNP) containing mRNA sequences were first described and called informosomes by Spirin *et al.* (1964). Shortly thereafter, such particles were also identified in cell nuclei (Samarino *et al.*, 1966). Research on mRNPs has focused on isolation and characterization of the constituents of mRNPs and on the elucidation of the functional role of the mRNP proteins. It is still difficult to evaluate whether some mRNP proteins are contaminants which become adventitiously bound during the isolation procedure or whether important mRNP constituents are lost during isolation. Nevertheless, evidence has been obtained that the mRNA associated proteins protect mRNA against degradation and influence the translatability of individual mRNAs (see also section IB6 below).

Poly A mRNPs can be isolated by affinity chromatography with oligo (dT) cellulose (Lindberg and Sundquist, 1974; Jain *et al.*, 1979). mRNPs have also been isolated successfully by density gradient centrifugation with metrizamide (Buckingham and Gros, 1975) sucrose gradient electrophoresis (Liautard and Kohler, 1976) and by density gradient centrifugation in Cs_2SO_4 (Liautard and Liautard 1977). The major difficulty in judging the effectivness of an isolation procedure is the lack of criteria for defining the in vivo form of RNPs. Nuclear and cytoplasmic RNPs are heterogeneous in size and sediment in sucrose gradients with 20 to 80S. Their buoyant densities of 1.4—1.42 g/cm^3 indicate a protein content of about 75% (reviewed in Hershey, 1982b).

The cytoplasmic RNPs are either free in the cytoplasm or polysome bound. The free mRNPs might represent RNPs in transit from the nucleus to the cytoplasm which have not yet participated in protein synthesis. Alternatively, they may represent storage forms of mRNAs whose entry into polysomes is blocked.

The polysome bound RNPs contain two major proteins with molecular weights of about 52 K and 78 K, as well as several minor proteins. The 78 K protein binds to the poly A of the mRNA. It has been suggested that this protein plays a role in transport of mRNPs from the nucleus to the cytoplasm (Schwartz and Darnell, 1976), and that it is the poly A polymerase (Rose *et al.*, 1979). The 78 K protein binds very tightly to poly A (Barrieux *et al.*, 1976). Association with other parts of the mRNA is indicated by its presence in histone polysomal mRNPs which lack poly A (Liautard and Jenteur, 1979), and by digestion of mRNPs with T_2 RNase. T_2 RNase digestion of mRNPs yields a pattern of fragments that are multiples of about 27 nucleotides (Baer and Kornberg, 1980). This pattern appears to be due to protection of the RNA by the 78 K protein.

An important function of the mRNP proteins is the protection of mRNA against degradation. In a recent study on the half life of increasing amounts of foreign mRNA after injection into oocytes, Richter and Smith (1981) concluded that mRNA association with protein(s) confers stability. Since mRNA degradation did not directly relate to concentration of injected mRNA, the authors suggested that association of the injected mRNAs with intracellular proteins is a relatively slow process and that the amount of "protective protein" present in a free state in oocytes is limiting. The "slow" association of foreign RNA with cellular proteins might be due to exchange of mRNP proteins between endogeneous mRNP and foreign naked mRNA. This should lead to a decrease in the half live of endogeneous mRNA.

The proteins in mRNPs might function in the enhancement or in the inhibition of translation and/or in anchoring mRNA to the cytoskeleton. Whether mRNP proteins are part of initiation factors or enhance interaction of mRNA with initiation factors has not yet been determined. All isolated polysomal mRNPs promote protein synthesis in cell free extracts as well as fully deproteinized mRNAs. Since cell extracts contain free proteins which associate with naked mRNA to form mRNP, it is not known whether the proteins which are already present in RNPs or those which are attached to mRNA only after addition of mRNA to the extracts, facilitate protein synthesis in any way.

In contrast to polysomal RNPs, a substantial part of free mRNPs is not active in protein synthesis in vivo in certain cells. It is expected that some of their constituent proteins are responsible for their non-translatability. This view is supported by the observation that native free mRNPs are inactive in protein synthesis *in vitro,* and that fully deproteinized mRNAs obtained from free mRNPs are fully competent for *in vitro* protein synthesis. Treatment of isolated free mRNPs with high salt solutions converts them from inactive to active messages (Geoghegan *et al.*, 1979; Liautard and Egly, 1980). Precisely which proteins are removed by high salt has not yet been determined.

It is interesting to note that translatable mRNPs can also be converted to untranslatable mRNPs by incubation of 40S preinitiation complexes *in vitro* under conditions of restricted polypeptide chain initiation (Buhl *et al.*, 1981). An increasing number of recent observations indicate that mRNA specific proteins are involved in this conversion. This would explain a) why the pools of untranslatable and translatable mRNPs do not exchange proteins readily, and b) why the protein compositions of free mRNPs show considerable variations.

3. Initiation Factors and the Process of Initiation

So far the identification of initiation factors has rested mainly on studies carried out with *in vitro* assays for initiation. A number of proteins were defined as initiation factors on the basis of their stimulation of *in vitro* protein synthesis mainly in rabbit reticulocyte extracts (Safer *et al.*, 1976; Schreier *et al.*, 1977; Trachsel *et al.*, 1977; Benne *et al.*, 1978; Hershey, 1982a). Another approach is presently under way using specific antibodies against invididual polypeptides and evaluating their inhibitory effect on *in vitro* protein synthesis (Trachsel *et al.*, 1981). The most active factor-dependent *in vitro* translation system used in these studies is a crude post-ribosomal supernatant fraction which shows a polymerization rate of only 3-5% of the rate of protein synthesis in intact reticulocytes. A rigorous identification of all required factors for initiation of protein synthesis is difficult to accomplish with the presently available experimental approaches. Nevertheless, twelve factors which participate in the formation of the initiation complex have been elucidated. Each of them has been purified to near homogeneity and has been characterized as indicated in Table 50.

The sequence of events in which these initiation factors are believed to promote initiation is shown in the flow diagram in Figure 83. The initiation

Table 50. *Initiation factors in enkaryotic cells*

Nomenclature	Peptides	Size of peptides in Kd	Function	Inactivation in poliovirus infected cells
eIF-1	1	15	stimulation of the binding of mRNA to 40 S	
eIF-2	3	38, β 52, 54	formation of ternary complex with met-tRNA and GTP	
eIF-3	9–11	24–210	promotes dissociation of 80 S	+[a]
Cap binding	5	24, 28, 50[1], 80[2], 200	binds to cap of mRNA, promotes binding of mRNA to 40 S	+[b]
eIF-4 A	1	49	binding of mRNA to 40 S, unravelling of secondary structure in mRNA	
eIF-4 B	1	80	binding of mRNA to 40 S, unravelling of secondary structure in mRNA	+[c]
eIF-4 C	1	17.5	not well defined, promotes dissociation of 80 S	
eIF-4 D	1	16.5	stimulates initiation in the abscence of polyamines	
eIF-5	1	168	joining 60 S to 40 S preinitiation complex, acts catalytically	
Co-eIF-2 A	1	19	stabilizes ternary complex	
eIF-2 A	1	65	promotes binding of met-tRNA to 40 S	
eSP	5	32, 40, 57, 65, 80	promotes activation of eIF-2 by binding to GTP, stimulates formation of ternary complex	

[1] May be equivalent to eIF-4 A.
[2] May be equivalent to eIF-4 B.
[a] Helentjaris *et al.*, 1979; Hansen *et al.*, 1982.
[b] Trachsel *et al.*, 1980; Tahara *et al.*, 1981; Etchison *et al.*, 1982; Lee and Sonenberg, 1982.
[c] Golini *et al.*, 1976; Baglioni *et al.*, 1978; Rose *et al.*, 1978.

process begins with the dissociation of 80S ribosomes into 40S and 60S subunits which is promoted by eIF-3, and to a lesser extent by eIF-4C. In a parallel reaction, a ternary complex is formed with eIF-2, met-tRNA and GTP. GTP is a potent inhibitor of this reaction, while Co-eIF-2A stimulates this reaction or stabilizes the ternary complex. This ternary complex binds to the 40S subunit to form a complex which contains eIF-2, eIF-3 and eIF-4C. Next, the mRNA – present as mRNP (with an associated cap binding protein) - attaches to this 40S complex to form the 40S preinitiation complex. This process is poorly understood. Three additional factors may be involved: eIF-1, eIF-4A, eIF-4B. The cap binding protein may aid in the unravelling of the secondary structure on the 5' region of the mRNA, thereby promoting ribosome binding and/or movement. This may be especially important during conditions of high ionic strength which stabilize the secondary structure of mRNAs. *In vitro,* the difference between the rate of trans-

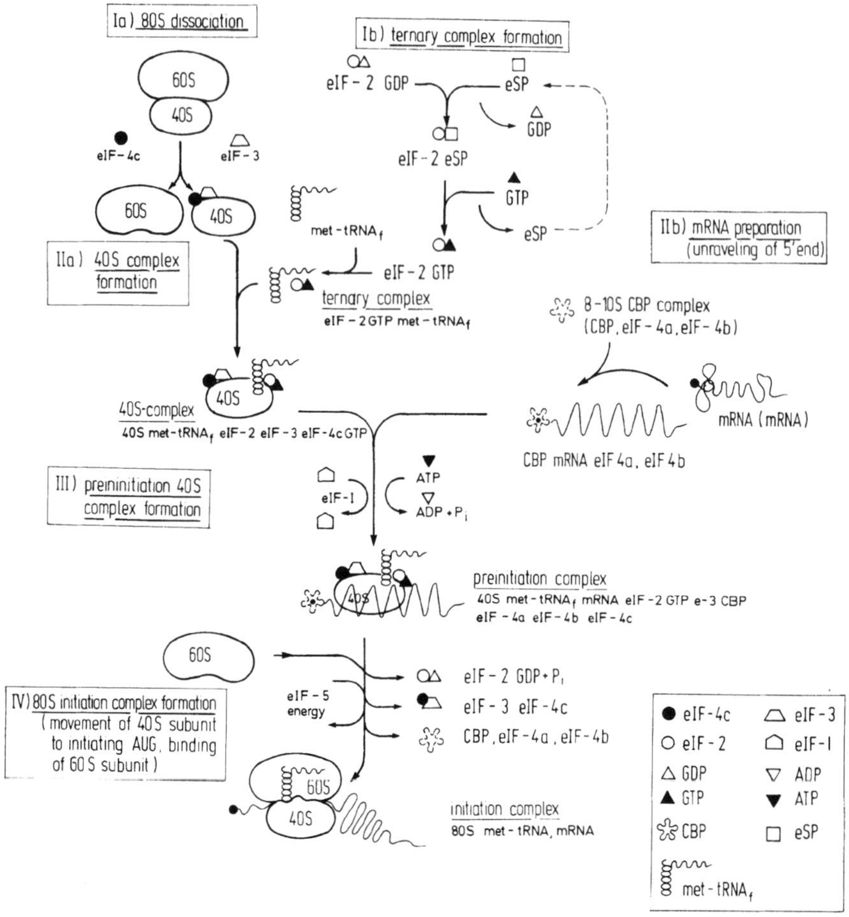

Fig. 83. Flowchart of polypeptide chain initiation in eukaryotic cells

lation of capped and uncapped mRNAs increases as the K^+ concentration rises (Weber *et al.*, 1977, 1978).

The 40S subunit is thought to bind always first to the cap. Polypeptide chain initiation then must be at least a two step event: (1) binding of 40S to the 5' cap and (2) movement of the 40S subunit from the cap to the initiation codon (Kozak, 1978, 1979, 1982). The concept of ribosome movement during initiation is supported by studies with edeine, an inhibitor of initiation which prevents the joining of the 60S subunit with the preinitiation complex (Hunt, 1974). In the presence of edeine several 40S subunits associate with a single mRNA molecule. The binding of all the 40S subunits however is sensitive to inhibition by cap analogues (Hickey *et al.*, 1976) indicating that each 40S subunit enters at the 5' end and then moves along the mRNA to allow a renewed binding of a further 40S subunit at the 5' end (Kozak and Shatkin, 1978).

It is still a matter of controversy whether or not the 5' entry at the cap is an absolute prerequisite for 40S subunit binding and therefore initiation of protein synthesis by the eukaryotic translation machinery. Studies with synthetic rings of RNA seem to support the notion. In contrast to their prokaryotic counterparts, eukaryotic ribosomes cannot initiate translation on synthetic circular oligonucleotides (Kozak, 1982). Cleavage of such circular RNAs provides a free 5' end, and indeed the eukaryotic ribosomes can then initiate translation. Yet eukaryotic ribosomes have also been shown capable of binding to internal initiation sites on prokaryotic or poliovirus mRNAs (Jense *et al.*, 1978; Atkins *et al.*, 1979; Pérez-Bercoff, 1982), thereby bypassing the requirement for "threading" in at the 5' end. It is still not known what signals cause the 40S subunit to stop at the correct initiation site. In most but not in all mRNAs, translation starts at the AUG codon closest to the 5' end. Initiation site selection for start of translation must include other, so far unknown features of the mRNA besides the AUG codon (Kozak, 1981). The lengths of untranslated regions preceeding the initiating AUG codon vary between mRNAs. Secondary structure of the mRNA could of course shorten the distance between the 5' end and the initiation codon for mRNAs with long untranslated 5' regions, or "hide" early AUG codons that are not destined for initiation (Kozak, 1980). This may be particularly important for the poliovirus mRNA, whose 5' region preceeding the initiating AUG is unusually long (743 nucleotides) and contains several AUG codons that apparently are not used for initiation.

As a last stage in the initiation process, finally, the 60S subunit joins the 40S preinitiation complex in a reaction catalyzed by eIF-5. GTP hydrolysis occurs and eIF-2 and eIF-3 and presumably other factors are released and the 80S mRNA initiation complex is formed. The 80S ribosome mRNA complex so formed at the initiating AUG then begins the translation of the mRNA. So far, only eIF-5 and eIF2-eSP have been shown to act catalytically.

4. Elongation and Termination of Translation

Elongation of protein synthesis proceeds in a cyclic manner. For each amino acid that is carried to the ribosome by a specific tRNA and attached to the nascent polypeptide chain with the help of two elongation factors (according to the instruction by the mRNA), energy in the form of two GTPs is consumed. Approximately 30–40 amino acids of the nascent chain are protected by the ribosome during elongation.

The rate of elongation is usually constant during translation of mRNA in eukaryotes and amounts to 200 peptide bonds formed per minute. An average mRNA coding for a protein with a molecular weight of 60 K is translated in three minutes (Saborio *et al.*, 1974).

The "normal" process of elongation is modified under a number of conditions, including either a slowdown or arrest of elongation in the synthesis of integral membrane proteins until the nascent peptides are anchored to membranes, interferon induced depriviation of specific charged tRNAs, or presence of certain antibiotics (puromycin). A slowdown in the rate of elongation has been observed in the *in vitro* translation of some picornaviral RNAs at specific regions of the mRNA.

Termination of translation occurs as soon as one of three possible stop codons is encountered. The nascent chain is cleaved from the last tRNA and released from the ribosome.

5. *Cotranslational Processing and Membrane Insertion of Nascent Polypeptide Chains*

The biosynthesis of several cellular and many viral proteins is accompanied by modification (e. g. phosphorylation, hydroxylation, glycosylation, ribosylation, acetylation) and proteolytic processing of the primary translation product. Both modification and processing can take place during polypeptide chain elongation, as well as long after the completion of protein synthesis (for reviews see Reich *et al.*, 1974; Korant, 1978, 1979; Koch and Richter, 1980; Koch *et al.*, 1982 a). By way of example, glycosylation (Behrens, 1974)—especially of membrane proteins—and hydroxylation of collagen (Lukens, 1965) occur on nascent polypeptide chains. Similarly, forms of specific proteolytic processing, e.g. removal of the leader sequence from membrane and secretory proteins (e.g. Milstein *et al.*, 1972; Blobel and Dobberstein, 1975 a, b; Jackson and Blobel, 1977; Wickner, 1979), and cleavage of the picornavirus polyproteins to three distinct precursor proteins occur during elongation (Rueckert *et al.*, 1979; Lucas-Lenard, 1979; Korant, 1979). Modification and processing of a polypeptide can occur concomitantly, and in some instances may be interdependent (Dickson *et al.*, 1976; Naso *et al.*, 1976; Shapiro *et al.*, 1976; Leavitt *et al.*, 1977; Garoff and Schwartz, 1978). Both processes often have important regulatory functions (activation, inactivation, transport, compartmentalization, secretion, virus assembly). Modification and processing enzymes may be associated to ribosomes or occur "free" in the cytoplasm or within intracellular compartments.

The synthesis of membrane proteins or of proteins destined for secretion usually proceeds on polysomes bound to the rough endoplasmic reticulum. Specialized hydrophobic peptide regions of the nascent polypeptide become inserted into intracellular membranes to provide anchors for future integral membrane proteins or routes for membrane passage of secreted proteins (reviewed in Sabatini, 1982). A halt in the elongation of translation of proteins may provide a mechanism to ensure correct membrane insertion or processing of nascent peptide chains. Nascent proteins are rapidly folded and acquire tertiary structures that are thermodynamically stable and suited for the surrounding environment: in the hydrophylic environment of the cell cytoplasm, hydrophobic peptide regions tend to cluster in the interior of the protein. If elongation of nascent chains proceeded unabated, hydrophobic regions required for membrane anchorage or stretches designed for recognition by proteases could become hidden during the folding of the nascent chain. Indeed, the in vitro synthesis of some eukaryotic integral membrane proteins comes to a halt unless membranes to which the nascent chain can bind are added to the system (Meyer *et al.*, 1982). Similarly, the *in vitro* translation of some picornaviral mRNAs has been observed to slow down at certain regions of the mRNA (see below).

The insertion of proteins into membranes during their synthesis involves some intriguing mechanisms, the details of which are not yet fully understood (for review see Sabatini, 1982). Several models have been proposed, however, it appears unlikely that there is only one universal pattern to which all proteins conform. The signal hypothesis as proposed by Blobel and Dobberstein (1975 a+b) suggests that the amino terminal signal peptide of a protein destined for export facilitates interaction between the ribosome from which it emerges and components in the membrane. Once this interaction is established, the signal sequence may be cleaved off to generate the mature amino terminus of the protein. The polypeptide is elongated to full length, concomitantly extruded through the membrane and finally released on the other side of the membrane. Special hydrophobic sequences in nascent chains provide start-transfer and stop-transfer signals for the insertion of integral membrane proteins. The final characteristic orientation of membrane proteins with respect to the membrane is determined by the sequence of the transfer and halt-transfer sequences.

The membrane trigger hypothesis, as proposed by Wickner (1979), states that the signal portion of the polypeptide originally serves as a means of keeping the polypeptide folded in a soluble conformation. After the polypeptide has been completed, or alternatively, when a substantial part of it has been synthesized, the polypeptide as a whole interacts with the membrane in an event that concomitantly translocates the polypeptide through the membrane and cleaves off the signal peptide. This cleavage would render the process irreversible.

Poliovirus mRNA appears to be translated primarily on membrane bound polysomes (Roumiantzeff *et al.*, 1971), and many of the poliovirus proteins become tightly associated to intracellular membranes and may in fact operate at least in part as integral membrane proteins. To date we do not know of any studies dealing with the question as to how membrane insertion of poliovirus proteins takes place.

B. The Regulation of Protein Synthesis

Whereas protein synthesis in prokaryotes is primarily regulated by control of transcription, mammalian protein synthesis is subject to extensive regulation at post-transcriptional levels (Revel and Groner, 1978; Scherrer, 1979). The mRNAs of prokaryotes have short half lives, so that the switch-on and switch-off of the synthesis of specific proteins can be accomplished rapidly by alterations in mRNA synthesis. Nevertheless, control at the level of translation has also clearly been demonstrated in bacterial systems (Schweiger *et al.*, 1978). The translation of the three lactose metabolizing proteins directed by the polycistonic lac- mRNA, for example, proceed with different efficiencies which can be regulated by changes in the environmental conditions (Queen and Rosenberg, 1981). In addition, virus infection can trigger or amplify translational control in prokaryotes by interference with the interaction of mRNA, initiation factors, tRNA and ribosomes.

The quantities of mRNA in eukaryotes, of course, are also primarily regulated at the level of transcription. Eukaryotic mRNAs, however, have half lifes of many hours or even days, so that alterations of transcription rates alone will not result in very significant rapid changes in the pattern of newly sythesized proteins. Mammalian cells seem to have acquired extensive control mechanisms of protein synthesis at the level of initiation of translation (Koch, 1976; Koch *et al.*, 1976; Lodish, 1976). We are just beginning to understand to what extent all the extra complexity of the eukaryotic translational system is involved in that control: translational control mechanisms are influenced by culture conditions and involve the inactivation or modification of initiation factors and ribosomes, the competition between mRNAs for limiting initiating components, and withdrawl of mRNAs from the translatable pool. Translational control may be mediated by the activity of protein-kinases, by intracellular ionic conditions (Carrasco and Smith 1976, Weber *et al.*, 1977, 1978) or by effector molecules, some of which may be released by signals from the cell membrane. Table 51 provides an overview of translational control mechanisms that could act at the site of initiation.

Table 51. *Possible regulation sites of polypeptide chain initiation*

I a)	Modification of eIF-3, eIF-4 c or of ribisomal proteins and binding of inhibitors may *prevent dissociation of 80 S complex*
I b)	Phosphorylation of eIF-2 *blocks formation of the ternary complex*
II a)	Modification of the 40 S ribosomal subunit or of the ternary complex *blocks formation of the 40 S complex*
II b)	Modification of CBP, eIF-4 a or eIF-4 b (by phosphorylation) and cap-analogs *inhibit preparation of mRNA for binding to 40 S subunit.* Binding of proteins to mRNA may yield *untranslatable mRNPs*
III)	Modification of eIF-1 may *block formation of the 40 S pre-initiation complex*
IV)	Energy depletion may *prevent movement of the 40 S subunit to the AUG initiation codon.* Modification of the 60 S ribosomal subunit or eIF-5 may *prevent formation of the 80 S initiation complex*
V)	Intracellular pH and ionic conditions influence efficiency of initiation

The stages are numbered as in Fig. 83, p. 324.

A different type of regulation might control the relative activities of the two pools of functioning ribosomes: the "free" polysomes in the cytoplasmic matrix on the one hand and on the other hand the endoplasmic reticulum associated polysomes involved in the synthesis of membrane and secretory proteins.

1. The Role of Culture Conditions

Animal cells respond rapidly to external signals with increased or decreased synthesis of macromolecules (Hassel and Engelhardt, 1976). Many external signals first interact with the cell membrane, which functions as both a receiver and a transmitter of regulatory signals. A number of studies have been performed to elucidate the role of the cell membrane in the induction of the proliferative response of cells to mitogens (Clarkson and Baserga, 1974). The binding of mitogens triggers a) changes in membrane permeability for ions, b) alterations in the level of cyclic nucleotides, and c) increased transport of ions, nucleosides, glucose, and amino acids. These very early events are soon followed by elevation in the rate of protein synthesis. This elevation takes place by activating preexisting pools of mRNs and ribosomes.

The corresponding negative response ("negative pleiotropic growth response" Hershko *et al.*, 1971; Koch *et al.*, 1979) is activated by a number of unfavorable growth conditions, such as amino acid starvation (Vaughan *et al.*, 1971; Austin and Clemens, 1981) or glucose depriviation (Van Venrooij *et al.*, 1979), high cell density (Levine *et al.*, 1965), ATP depletion (Giloh and Mager, 1975), exposure to reduced or elevated temperatures (McCormick and Penman, 1969; Craig, 1975), incubation of cells in medium with increased medium osmolarity (Wengler and Wengler, 1972; Saborio *et al.*, 1974), and during mitosis (Fan and Penman, 1970) (Table 52). In each case, the overall rate of protein synthesis is reduced by an interference with the initiation step of protein synthesis.

Table 52. *The pysiological regulation of protein synthesis*

A. Factors that stimulate or inhibit protein synthesis

Enhancement (proliferative response)	Inhibition (negative pleiotropic response)
Mitogens	Nutrient starvation
Serum	Amino acid depletion
Proteases	Isolated membrane glycopeptides
Low cell density	High cell density
	Medium hyperosmolarity
	Virus infection
	DMSO and other inducers of terminal differentiation in Friend cells
	Protease inhibitors (TPCK)

B. The effects of nutritional conditions on cell functions related to protein synthesis

Cell function	Step up in nutritional condition	Step down
Membrane potential	high	low
Amino acid transport	high	low
mRNA participation	almost all mRNA	differentially limited
Synthesis of individual proteins	determined by mRNA concentration	determined by relative translational efficiencies
Ribosomes	most in polysomes	most in monosomes
S 6 phosphorylation	high	low
Low molecular weight inhibitor	present in very small amounts	large amounts bound to ribosomes and also in free form

In analogy to its role in mediating the proliferative response, the cell membrane also seems to mediate the negative pleiotropic response (Koch *et al.*, 1976, 1980, 1982). A number of experimental conditions which interfere with the initiation of protein synthesis in intact cells show little or no effect on the synthesis of proteins in cell-free extracts, indicating membrane mediation (Koch *et al.*, 1976) (Table 53).

Table 53. *Inhibition of protein synthesis in vivo and in cell-free extracts*

Addition	Conc.	Exposure (min)	Amino acid incorporation % of control: in vivo	Amino acid incorporation % of control: in vitro	Reversible in vivo	References
Sucrose	0.2 M	5–10	5	100	+	1
DMSO	12 %	3	2	50	+	2
Polycations (DEAE-dextran)	160 μg	30	10	0	(+)	3
Ethanol	2 %	1	60	100	+	4
Ethanol	3 %	1	10	70	+	4
Cytochalasin B	20 μg	30	50	100	(+)	5
TPCK	30μg	5–10	2	80	–	6
Trypsin	500 μg	20	5	0	+	7
Pronase	100 μg	30	5	0	+	7
Glycopeptides	5–100 μg	5–10	0–80	0–80	+	8

Protein synthesis *in vivo* in suspended HeLa S_3 or L cells was studied by following the incorporation of [^{35}S] methionine or [^{3}H] alanine into acid soluble proteins. The reversibility of the inhibition of protein synthesis *in vivo* was determined by sedimenting the cells and resuspension in fresh growth medium. Recovery of protein synthesis of better than 50 % (+) or 80 % + is listed in the second to last column in the table.
The data on the inhibition of *in vivo* protein synthesis have been published in part previously (1. Oppermann *et al.*, 1973; 2. Saborio and Koch, 1973; 3. Saborio *et al.*, 1975; 4. Koch and Koch, 1974; 5. Koch and Oppermann, 1974; 6. Pong *et al.*, 1975; 7. Koch, 1974; 8. Koch *et al.*, 1974; Kalvelage and Koch, 1982 a + b).

We have obtained evidence for the release of small effector substance(s) which promote the inhibition of protein synthesis under these conditions. No detectable endogenous protein synthesis *in vitro* is present in crude extracts of cells 45 minutes after transfer to amino acid-deficient medium, after a 15 minute exposure to hypertonic culture conditions, and two to three hours after infection with poliovirus. Protein synthesis in these extracts can be partially restored upon Sephadex G-25 gel filtration of the extracts, indicating that neither ribosomes nor mRNAs are irreversibly inactivated by exposure of cells to these conditions. Late eluting fractions from the Sephadex G-25 column (containing amino acids) also contain low molecular weight substances which inhibit protein synthesis in cell-free extracts from untreated cells. Upon dialysis of the cell extract the inhibitor is found in the dialysate. Gel filtration over G-15 or G-10 sephadex reveals a molecular weight below 1,000. The inhibitors do not react with ninhydrin, indicating absence of amino or imino groups. During high voltage paper electrophoresis at pH 3.5, the inhibitors remain at the origin, indicating the absence of anionic groups, e. g. carboxyl or phosphate. The compounds are quite hydrophobic because they migrate on cellulose thin layer plates using butanol as solvent, and are bound to XAD2 columns at pH 2.0 in the presence of 2M $CaCl_2$. They also bind strongly to glass surfaces, which hampers their isolation and purification. The inhibitors reduce the rate of protein synthesis in cell-free extracts and in intact tissue culture cells. They bind to ribosomes from which they cannot be removed by high salt treatment. We propose that the inhibitors are released or activated by membrane mediated events and act at the level of initiation. Their activities are destroyed rapidly by counteractive membrane mediated stimuli (mitogens and growth factors).

2. Competition Between mRNAs

During the course of our work on the regulation of protein synthesis in tissue culture cells, it became apparent that the modulating effect upon protein synthesis was not unilateral, in that the synthesis of all proteins was not equally affected. Under the conditions of reduced rates of initiation, a shift in the pattern of proteins synthesized is observed. The translation of each mRNA proceeds with its own characteristic efficiency (Koch *et al.*,1976) and competition between RNAs for initiation components becomes an important regulatory mechanism.

Competition between mRNAs is related to different translational efficiencies. The relative translation rates of individual mRNAs under competing conditions are determined by (a) the amount of the mRNA present, (b) the relative translational efficiency of the mRNA, and (c) the extent to which the target of competition is limiting (for instance initiation factors, met-tRNA or ribosomal subunits).

a) The Role of mRNA Concentration

As long as constituents of the protein synthesizing machinery are in surplus, the amount of protein synthesized is strictly dependent on the concentration of the individual mRNA. However, at and above the level of saturation of the protein synthesizing system with mRNA, competition of mRNAs for the limiting initiation sites becomes apparent and amplified (Koch *et al.*, 1980a, 1982b; Kaempfer, 1982b). Upon addition of increasing amounts of a given mRNA to an mRNA dependent *in vitro* or *in vivo* protein synthesizing system, there is an equimolar synthesis only as long as mRNA is not saturating. Beyond that point, mRNAs are translated with different efficiencies, leading to preferential synthesis of some proteins, where the corresponding mRNA with high translational efficiency outcompetes the mRNA with lower efficiency (McKeehan, 1974; Kabat and Chapell, 1977; Di Segni *et al.*, 1979). This type of mRNA concentration related competition was first observed in experiments dealing with the synthesis of α and β globin chains in reticulocyte lysates (Lodish, 1971). Both globin chains were synthesized in equimolar amounts in spite of the fact that the extracts contain a 1.5-fold molar excess of α globin mRNA, indicating that ß globin mRNA is translated more frequently, i. e. with higher efficiency. A similar phenomenon can be observed upon injection of foreign mRNAs into oocytes.

b) Relative Translational Efficiencies of mRNAs

The translational efficiency of a given mRNA is determined by its capacity to form a functional initiation complex i. e. by the efficiency of its interaction with the components of the translational machinery involved in the formation of the initiation complex. Studies concerning the features of mRNAs which determine the translational efficiency have just been initiated. The role of nucleotide sequences, secondary structure, the cap, and of the distance between cap and initiator codon are being investigated (Kaempfer, 1982a; Kozak, 1982; Shatkin, 1982) (see section A 2 above). The relative translational efficiency (RTE) of an mRNA is probably determined by the affinity of the mRNA to initiation factors and to the small ribosomal subunit.

One experimental approach to measure the translational efficiencies of individual mRNAs is to analyze the effects of different experimental conditions restricting the rate of polypeptide chain initiation *in vivo* and *in vitro* (Nuss *et al.*, 1975; Koch *et al.*, 1976, 1980a; Lodish and Rose, 1977). Specific inhibition of initiation of translation affects the synthesis of every individual protein to a different extent (Nuss *et al.*, 1975). The ratio of the extent of isotope labelling of a given protein under unrestricted and restricted conditions of initiation gives a direct estimate of the RTE of its mRNA (Koch *et al.*, 1976, 1980a). Viral messages, including poliovirus mRNA, are generally very high efficient messages (see Table 49, p. 319).

Even in prokaryotes, the RTEs of mRNAs may play a role in the regulation of protein synthesis. The three genes of the gal operon are translated from one polycistronic message. Contrary to expectation, the gal genes are not always expressed coordinately. Under different metabolic conditions, the ratios of newly synthesized gal enzymes vary (Ullmann *et al.*, 1979). In the presence of high levels of intracellular cAMP, essentially equal amounts of the three enzymes are found; when cAMP levels are low, the protein encoded by the 5' terminal region is synthesized at a fourfold higher rate than the other two. This phenomenon, called discoordinate expression, is due to the synthesis of two mRNA transcripts which differ only in the extent of the 5' untranslated region. The longer mRNA translates the cistron closest to the 5' region of the mRNA with a 4-fold higher ratio than the shorter mRNA. This explains the discoordinate expression of the gal-proteins and reveals a form of cellular regulation based on differential translational efficiencies of mRNA which differ only in their 5' untranslatable region. This is the first clear evidence that the RTE of an mRNA is dependent on the nucleotide composition and length in this region of the RNA (Queen and Rosenberg, 1981).

c) The Role of Limiting Initiation Components

The correlate to increasing mRNA concentrations above saturation levels to initiate mRNA competition is the limitation of an essential initiation component(s). Any reduction in the overall rate of initiation results in a preferential inhibition of the translation of mRNA species with low translational efficiencies, *in vitro* (Lodish, 1976) as well as *in vivo* (Nuss *et al.*, 1975; Koch *et al.*, 1980a).

A wide variety of experimental conditions induce a reduction in the rate of initiation *in vivo*. The target initiation components that are affected may differ (see Tables 50 and 51 above). Sometimes, the block of initiation can be released by addition of the limiting factor, allowing a more precise determination of the affected initiation component.

3. Alteration or Inactivation of the Cap Binding Protein and Other Initiation Factors

The activity of some initiation factors evidently can be altered reversibly or inactivated completely. Well known examples are the cases of eIF-2 and the cap binding protein (see p. 352).

Most in vitro studies have been performed with the reticulocyte system. Initiation of protein synthesis in this system is inhibited reversibly under a variety of conditions, such as during hemin deprivation, in the presence of low concentrations of double-stranded RNAs at high temperature (42° C), and in the presence of oxidized glutathione (Jackson and Hunt, 1980, 1982). In all instances, the rate of protein synthesis declines rapidly to about 1-10 % of the control. The decline is preceeded by a reduction in 40S-met-tRNA complexes, although the levels of native 40S subunits, met-tRNA and GTP are unchanged. This suggests a depletion of the ternary complex or inactivation of eIF-2 and, indeed, the inhibition of protein synthesis can be overcome by the addition of eIF-2 in excess.

The primary function of eIF-2 is the formation of a ternary complex with met-$tRNA_f$ and GTP, and the direction and association of this complex to the 40S ribosomal subunit (see Fig. 83, p. 324). In addition, eIF-2 binds to mRNA at a specific site(s) which is often identical with the ribosome binding site (Kaempfer, 1982a). These properties of eIF-2 suggest an important role for eIF-2 in the recognition of mRNA and its binding to ribosomes and make this protein a suitable candidate for translational control. On this prediction, Kaempfer (1982b) has examined the effect of addition of isolated eIF-2 to reticulocyte lysates on the endogenous synthesis of α and ß globin chains. Increasing the concentration of eIF-2 in reticulocyte lysates results in a selective increase in the translation of α chain (up 1.7 fold) without an effect on overall protein synthesis. This specific influence of eIF-2 is completely abolished by addition of extra mRNA.

The competition between α and ß chain synthesis is enhanced by increasing the concentration of Cl^-. Again eIF-2 relieves the competition. Since elevated Cl^- ions inhibit protein synthesis primarily by interference with the binding of mRNA to the 40S initiation complexes without causing a significant reduction in the formation of 40S/met-$tRNA_f$ complex, the eIF-2 interaction with mRNA represents one of several critical steps in mRNA competition. This conclusion is supported by results from direct binding experiments of globin mRNA to eIF-2 (Di Segni *et al.*, 1979). There is a good correlation between efficiency of translation of mRNA *in vitro* and binding affinity to eIF-2.

Further experiments on the effect of increasing concentrations of KCl on mRNA competition and the relief from this competition by eIF-2 suggest that the mRNA does not interact with free eIF-2 but with eIF-2 present in the 40S met-tRNA complex. Viral and host mRNAs show also 30 to 35 fold differences in affinity for eIF-2 (Rosen *et al.*, 1981). Protein synthesis in rabbit reticulocytes is regulated by phosphorylation and dephosphorylation of the initiation factor eIF-2 through the action of several independently regulated proteinkinases and phosphatases (Jackson and Hunt, 1980, 1982). Other initiation factors have been shown to be impaired during poliovirus infection (see below). Addition of eIF-4B preferentially stimulates host mRNA over viral mRNA translation in cell extracts

(Golini *et al.*, 1976b). In direct binding experiments, this factor exhibits a preference for picornavirus RNA (Baglioni *et al.*, 1978).

4. Modification of Ribosomes

Another possible site for the control of initiation is the ribosome. The ratio of nonfunctioning 80S ribosomes to ribosomes in polysomes varies greatly with the nutritional state of the cells, with cell density, and with the presence or absence of growth factors or mitogens. In order to participate in protein synthesis, 80S monomeric ribosomes have to be dissociated into the 40S and 60S ribosomal subunit. This dissociation event might be controlled in several ways. It is promoted by the initiation factor eIF-3, which binds tightly to the 40S subunit and may also act as an anti-association factor. eIF-4C also promotes dissociation but to a lesser extent (Hershey, 1982a). Inactivation of eIF-3 or eIF-4C would result in inhibition of initiation. Alternatively, effector substances might prevent 80S dissociation by binding to ribosomes.

Indeed, some indirect evidence has been obtained in our laboratory for a low molecular weight effector substance that inhibits initiation by binding to ribosomes. The effector apparently can be activated by a variety of conditions, which all seem to involve a membrane trigger (see section B-1, p. 330). Ribosomes with bound inhibitors are less active or inactive in the formation of initiation complexes. Small molecular weight inhibitors of protein synthesis are also found to be released during the course of poliovirus infection (Koch *et al.*, 1982b).

The phosphorylation state of certain ribosomal proteins also seems to correlate with the activity of the ribosomes in protein synthesis. The phosphorylation is higher in polysomes than in monosomes. In MPC-11 cells, a positive correlation between the phosphorylation of the small subunit protein S6 and the amount of ribosomes engaged in protein synthesis was observed (Martini and Kruppa, 1979). Raising the tonicity of the growth medium by addition of 100 mM excess NaCl results in an increase of single ribosomes with a concomitant reduction in the phosphorylation of protein S6 in single ribosomes and in residual polysomes. Phosphoprotein S6 was almost completely dephosphorylated after glucose starvation of Ehrlich ascites cells (Kruppa, 1979). It is tempting to speculate that the massive dephosphorylation of a specific small subunit protein accompanying starvation and a tonicity shift especially affects the initiation process of translation. Several ribosomal proteins within the large subunit also incorporated ^{32}P, but the changes in the degree of phosphorylation were minimal upon raising the medium tonicity. SV 40 transformed African green monkey kidney cells contain a ribosome bound phosphoprotein which is absent in untransformed cells (Segawa *et al.*, 1977). Correlated with this phosphorylation is the ability of transformed cells to translate some late adenovirus mRNAs while untransformed cells lack this ability.

5. The Role of Uncharged tRNA

The rate of initiation of protein synthesis in a number of different tissue culture cells declines rapidly when the cells are incubated in media lacking amino acids.

Since starvation for any single amino acid can elicit this response, the regulatory switch must operate when not more than 5% of the total cellular complement of tRNA is uncharged, regardless of which species of tRNA is concerned. This result rules out a simple direct mechanism in which an uncharged tRNA acts as an inhibitor of one step in initiation. The situation resembles the events in prokaryotes where starvation for one essential amino acid results in the accumulation of uncharged tRNA which induces bacterial ribosomes to synthesize the effector ppGpp (for review see Koch & Richter, 1979). So far, numerous efforts to detect ppGpp in mammalian cells have failed (Silverman *et al.*, 1979). However, the search for other effector molecules has not yet been abandoned. The relatively small differences in tRNAs of different cells have recently been reviewed in detail (Ofengard, 1982).

Extracts from cells treated with interferon often show a reduced ability to translate endogenous mRNAs (Content *et al.*, 1975). Analysis of the in vitro translation products reveal the accumulation of incomplete polypeptide chains (Revel *et al.*, 1975, Sen et al., 1976). This defect can be corrected by addition of tRNA to the extracts (Zilberstein *et al.*, 1976). The type of tRNA required for restoration of translation depends on both, the mRNA used and the origin of the cell-free extracts. Prolonged preincubation (1-2 hrs) of the cell extracts–but only in the presence of ATP–can increase the interferon induced elongation block which is still reversible by tRNA (Falcoff *et al.*, 1978; Revel, 1979). Extracts of interferon treated cells contain inhibitor(s) of tRNA synthetases. Poliovirus replication is not impaired by pre-exposure of cells to interferon and it is unlikely that changes in the activity of tRNA synthetases accompany poliovirus replication although the proper experiments have not been performed.

6. *Transfer of mRNAs Between Untranslatable and Translatable Pools*

A more intriguing problem is the role of proteins in removing mRNA from the translatable pool and storing it in an untranslatable state. Well studied examples of untranslatable mRNAs are some maternal mRNAs in eggs (Denny and Tyler, 1964; Gross and Cousineau, 1964), host mRNAs in infected cells (Oppermann and Koch, 1976a), mRNAs of characteristic proteins in determined cells prior to terminal differentiation, e.g. myosin mRNA in prefusion myoblasts (Buckingham *et al.*, 1974), and globin mRNA in Friend cells at an early stage during induced differentiation (Bilello *et al.*, 1980). Also of relevance are mRNAs which are translated immediately after mitogenic stimulation of lymphocytes or other resting cells. Apparently untranslated mRNA can often rapidly be converted into a translatable form. It is still not known by which means mRNAs are converted from an untranslatable form to the translatable state and vice versa.

An interesting finding which focuses on the mechanism of conversion of untranslatable to translatable mRNPs is the induction of ferritin synthesis in tissue culture cells by addition of Fe^{++} (Zahringer *et al.*, 1976). Without Fe^{++}, only 50% of the cytoplasmic ferritin mRNA in liver is associated with polysomes, the rest occurs as free mRNPs. Administration of Fe^{++} results in the rapid accumulation of all ferritin mRNAs into polysomes. If the mRNPs are a specifically repressed form of mRNA, they should be untranslatable *in vitro* as mRNPs, whereas the corresponding deproteinized mRNA should be active. In accordance with this prediction are reports from Scherrer *et al.*, (1979) on free mRNPs of duck reticulocytes. These RNPs are not translatable *in vitro,* moreover, they inhibit the translation of deproteinized mRNA (Civelli *et al.*, 1976).

Selective conversion of mRNAs from untranslatable to translatable forms is particularly marked during the fertilization of oocytes. Fertilization can lead to a 2 to 4 fold increase in the rate of protein synthesis within ten minutes and is accompanied by a radical change in the pattern of newly synthesized proteins (Rosenthal *et al.*, 1980). The proteins synthesized in a reticulocyte mRNA-dependent lysate by crude RNP containing extracts from unfertilized and from fertilized eggs are characteristic for the source of mRNPs. However, free mRNA obtained from the same crude extracts of unfertilized eggs by phenol extraction, a process that dissociates proteins from RNA, is translated in the lysate into a mixture of proteins characteristic of pre- and post-fertilization states. These observations indicate that mRNA can be stored reversibly in an intact untranslatable form by binding of proteins.

In resting cells, or in cells in mitosis, inactive mRNAs can be withdrawn from the untranslatable pool by stimulation of protein synthesis, as after mitogenic stimulation. On the other hand, untranslatable mRNPs in differentiating cells and in certain virus infected cells can be activated by inhibition of polypeptide chain initiation. For instance, incubation of Friend erythroleukemic cells under conditions which inhibit total protein synthesis does not result in a decrease but in a concomitant quantitative increase in globin synthesis (Bilello *et al.*, 1980). A comparable result was obtained in vaccinia virus infected HeLa cells (Oppermann and Koch, 1976a) and in frog virus 3 infected cells (Willis et al., 1977) at late times in infection. Some early viral mRNA which does not participate in protein synthesis under "normal" conditions is withdrawn from the untranslatable pool as soon as polypeptide chain initiation is inhibited. In poliovirus infected cells, finally, cellular mRNA is withdrawn from the translatable pool, possibly by the formation of inactive mRNPs.

7. Control of Free and Membrane Bound Pools of Ribosomes

In general, mRNAs destined for translation on membrane bound polysomes do not compete efficiently with those translated on free polysomes after injection into Xenopus oocytes. This is at least in part due to a paucity of rough endoplasmic reticulum membranes in oocytes, since injection of dog pancreatic rough endoplasmic reticulum into oocytes stimulates the synthesis of membrane proteins (Richter and Smith, 1981). This observation also offers an explanation for the finding that addition of globin mRNA in a ratio of 1 : 10 to EMC-RNA-virus reduces translation of the latter by 50 % (Laskey *et al.*, 1977). Since EMC-RNA is translated on rough endoplasmic reticulum (Rekosh, 1977), only a small fraction of the EMC-RNA might be able to compete with globin mRNA.

II. Translation of the Poliovirus Genome

A. Overview and Introduction

Viral protein synthesis proceeds in general in the same manner as cellular protein synthesis albeit with interesting modifications. Within 30 to 60 minutes after infection of cells by poliovirus, host cell protein synthesis is so abruptly and dramatically reduced, that this phenomenon has been termed the shut-off (reviewed in Bablanian, 1975; Koch *et al.*, 1980a, 1982b; see Fig. 60A, p. 210). Host cell polysomes disintegrate, indicating a blockage of initiation (see Fig. 81, p. 314, and Fig. 69, p. 239) (Summers and Maizel, 1967a, b). The shut-off of host cell protein synthesis in poliovirus infected cells approaches completion by 3 hours post infection. The first virus specific proteins that can be detected by SDS-PAGE appear near the end of the second hour post infection, as the "background noise" of newly synthesized host cell proteins fades away (Fig. 84A, B). Viral protein synthesis reaches a peak at about 4 to 5 hours post infection. In the final stages of infection, overall protein synthesis decreases again dramatically, presumably reflecting a depletion of the cell's resources and signaling the imminent death of the cell (see Fig. 60A, p. 210).

The major aspects of the mode of translation of poliovirus mRNA and the formation of virus specific proteins were elucidated more than a decade ago by *in vivo* experiments, largely in the "middle phase" when the rate of viral protein synthesis is near its peak and host cell protein synthesis is very low. At this time, virus specific protein synthesis occurs on large newly assembled polyribosomes, which seem to be tightly bound to cytoplasmic membranes (Penman and Summers, 1965; Willems and Penman, 1966; Leibowitz and Penman, 1971; Caliguiri and Tamm, 1969, 1973). The virus-specific polyribosomes are much larger than the cellular ones in uninfected cells and appear to be far more homogeneous (Penman *et al.*, 1963), sedimenting in sucrose gradients as a relatively sharp band at 350S in contrast to the broad spectrum of host polysomes, which sediment with a broad peak at 200S (see Fig. 69, p. 293, and Fig. 81, p. 314). The large size of viral polysomes corresponds to 35 ribosomes bound per message and indicates that polypeptide chain initiation proceeds with high efficiency on the viral message. In the final stages of the viral replication cycle the size of virus-specific polysomes decreases again concomitant with the final fade out of protein synthesis.

The synthesis of poliovirus specific proteins on cellular membranes indicates that at least some of the viral proteins are destined for interaction with cellular membranes, although the nature of this interaction is not yet fully understood. Poliovirus proteins do not show the characteristic properties of secreted proteins (which are normally synthesized on membrane-bound ribosomes), *i.e.* they are probably not glycosylated and they are certainly not secreted. They are, however, transported from their site of synthesis to the smooth endoplasmic reticulum of the virus induced membraneous vesicles (see Chapter 6). Here poliovirus proteins exhibit specific functions related to RNA replication and virion assembly, which may in part be regulated by their interaction with membranes. Some viral (notably P3) proteins are transported into the nucleus.

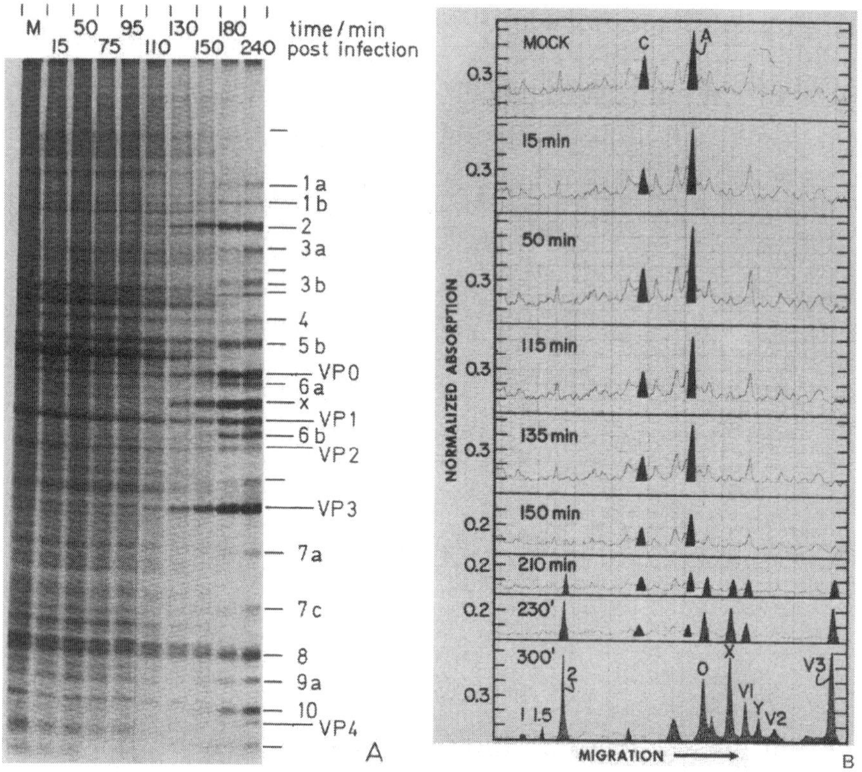

Fig. 84. Alterations of protein synthesis in poliovirus-infected HeLa cells
This figure illustrates the shut-off of cellular protein synthesis and the appearance of viral proteins above the decreasing background of cellular proteins in poliovirus infected HeLa cells in suspension culture. HeLa cells were pulse-labeled with (^{35}S) methionine for 15 min at different times after infection. Cell-extracts were prepared at the times indicated and freshly synthesized proteins were analyzed by polyacrylamide gel electrophoresis. The incorporated label was determined by autoradiography of the gels *(A)* and quantified by densitometric tracings of the autoradiographs *(B)*. Shaded areas in B illustrate the disappearance of cellular proteins (*A* actin, *C* an unidentified protein) and the appearance of virus specific proteins at 3.5 h.p.i. — Figures from Koch *et al.*, 1982 b [in: Protein Synthesis in Eukaryotic Cells, pp. 341–349 (1982)]

In a pioneering study, which introduced the now widely used SDS-PAGE technique, Summers *et al.* (1965) identified 14 different virus specific polypeptides in poliovirus infected HeLa cells (see Fig. 50, p. 172). The combined molecular weight of the 14 virus specific proteins exceeded 500 kd, while the coding capacity of poliovirus RNA is only sufficient to code for approximately 250 kd of protein. In subsequent experiments it was shown that certain virus specific peptides are preferentially labelled during a short pulse of radioactive amino acids. During subsequent incubation of pulse labelled cells with excess unlabelled amino acids (chase), radioactivity, previously associated with certain larger polypeptides, was found to be present in several lower molecular weight proteins. These observations were taken as an indication that the smaller peptides arise by cleavage of larger primary products (Maizel and Summers, 1968; Summers and Maizel, 1968; Holland and Kiehn, 1968; Jacobson and Baltimore, 1968b). Figure 85 depicts an example of the electrophoretic analysis of the viral peptides of poliovirus infected cells labelled by a brief pulse with ^{35}S-methionine, followed by a "chase" with a 10,000-fold excess of cold methionine. The protein pattern illustrates the formation of smaller polypeptides by secondary cleavage of large primary products.

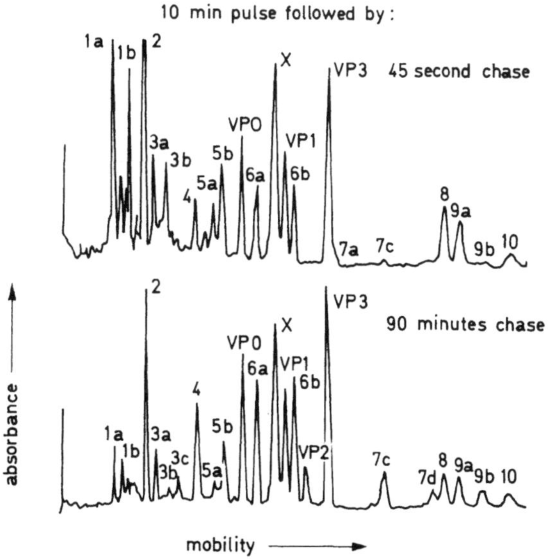

Fig. 85. Processing of poliovirus proteins in infected HeLa cells
This figure illustrates the processing of viral precursor proteins to smaller products in poliovirus infected cells by tracing the movement of incorporated labeled amino acids from large precursors into smaller products during a chase with an excess of unlabeled amino acids. At 3.5 h.p.i., cells were pulse labeled for 10 min with (^{35}S) methionine, and the label chased by adding an excess of unlabeled methionine. The cells were lysed at the indicated chase periods and electrophoresed on an SDS-polyacrylamide slab gel. – Figure from Rueckert *et al.*, 1979 [in: The Molecular Biology of Picornaviruses, pp. 113–125 (1979)]

In short pulse labelling experiments (up to 10 minutes), three major, larger primary virus specific peptides were typically observed: NCVP1a, NCVP3b and NCVP1b, implicating these proteins as precursor proteins. A large polypeptide with a molecular weight over 200 kd (NCVPOO) representing approximatly the entire coding capacity of the viral RNA (and corresponding to the combined weights of the three precursor proteins) was subsequently identified after incubation of infected HeLa cells in the presence of several amino acid analogues (Jacobson *et al.*, 1970; Fig. 86) or protease inhibitors. Evidently, the primary cleavages are normally so fast that special inhibitory treatments are required to demonstrate the larger primary translation products (Baltimore, 1969).

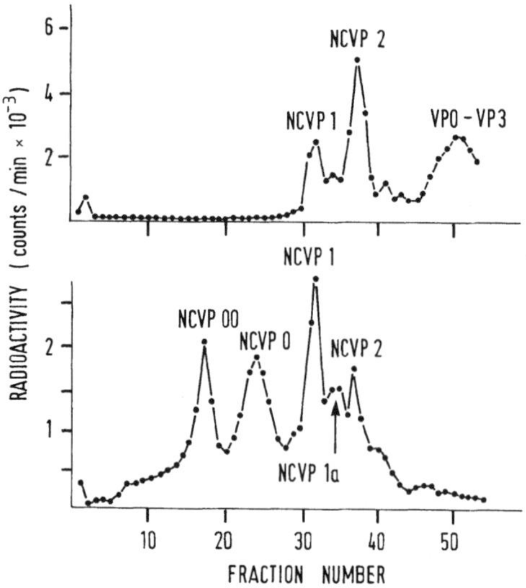

Fig. 86. Inhibition of poliovirus protein processing by amino acid analogues
This figure illustrates the accumulation of very large viral precursor proteins in infected cells when proteolytic processing is blocked by amino acid analogues. At 3 h.p.i., cells were treated with the amino acid analogues p-fluorophenylalanine, canavanine, azetidine-2-carboxylic acid, and ethionine. 8 min later, cells were pulse-labeled with (^{3}H) leucine for 30 min and chased for 5 min with an excess of unlabed leucine. Cytoplasmic extracts were made and analyzed by polyacrylamide gel electrophoresis. The upper profile shows polypeptides made in infected cells without amino acid analogues, the lower profile shows polypeptides made in the presence of amino acid analogues. — Figure from Jacobson *et al.*, 1970 [J. Mol. Biol. *49*, 660 (1970)]

From these studies, it was also concluded that poliovirus RNA is translated from a single initiation site close to the 5' end, the giant polyprotein being cleaved twice during its synthesis into three primary products which in turn are cleaved further post-translationally into mature virus specific proteins.

Processing maps were obtained in studies with inhibitors of polypeptide chain initiation (Rekosh, 1972; Saborio *et al.*, 1974). Such maps were later refined by analyzing precursor-product relationships by tryptic fingerprinting of proteins, and more recently also by evaluation of the primary nucleotide sequence of the genome (see Chapter 3). Figure 87 illustrates the present day version of the processing map.

The following sequence then is generally thought to take place during a typical translation of the poliovirus mRNA. The transit time of the ribosome through the capsid protein gene region takes about 3 minutes (Saborio *et al.*, 1974). As the ribosomes leave the region of the mRNA coding for the structural protein, the first cleavage of the nascent chain occurs liberating NCVP1a, a protein which is subsequently cleaved to the viral capsid proteins. Nevertheless, the ribosomes proceed further along the RNA normally without termination and without reinitiation and translate the next region on the mRNA. The second primary translation product, NCVP3b, is rapidly processed to NCVP5b and then to NCVPX. A second nascent cleavage occurs, and again the ribosomes proceed to translate the last gene, which codes for the replication proteins and a protease, yielding the primary translation product NCVP1b. This protein is again rapidly converted to NCVP2 and then more slowly processed further (see below).

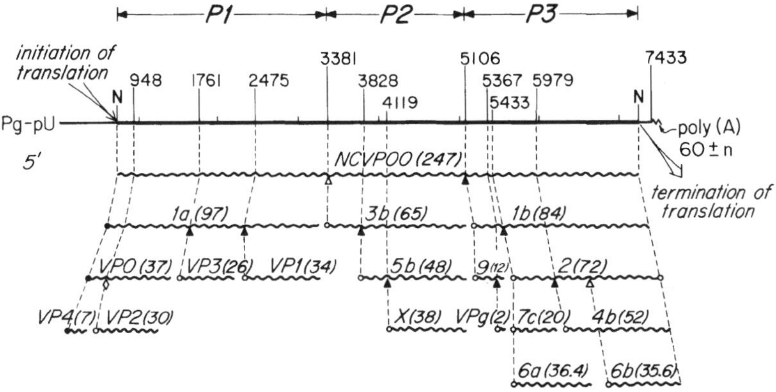

Fig. 87. The processing map of the poliovirus proteins
▲ gln/gly, △ tyr/gly, ◇ asn/ser. – Figure from Kitamura *et al.*, 1981 [Nature *291*, 552 (1981)]

An extension of the single initiation site proposal for poliovirus mRNA translation was the hypothesis that all eukaryotic cells–unlike prokaryotic cells–are unable to read "internal" initiation sites. This hypothesis has been supported by a number of experimental findings and has helped to understand why viruses, such as RNA tumor viruses, elaborate special mRNAs for the translation of envelope, internal and transforming proteins (Koch *et al.*, 1982a). It has also been assumed that there is no temporal control over the synthesis of individual proteins in poliovirus infected cells (Rekosh, 1977); translation of the poliovirus mRNA beginning at a single initiation site was pictured as resulting in equimolar production of different virus-specific proteins throughout the course of infection.

Recent evidence from *in vitro* and *in vivo* experiments, however, has revived the old question of whether internal initiation sites of poliovirus RNA might be functioning as such, in particular at early times during infection (Celma and Ehrenfeld, 1975a and b; Koch *et al.*, 1980a and b, 1982b; McClain *et al.*, 1981) (see below).

In the following sections we will first review recent data from investigations on the translation of poliovirus RNA in cell free systems as they have furnished valuable information on some of the mechanisms involved in the realization of the overall translation process of the poliovirus genome. Then, we shall discuss some of the additional complexity of the in vivo process, in particular: 1. the factors mediating the shut-off phenomenon, 2. more recent data on viral protein synthesis during the early stage of infection, 3. the role of association of viral proteins to the intracellular membranes, and 4. the regulation of viral protein processing.

B. Translation of Poliovirus RNA in Cell Free Extracts

1. General Comments

Many attempts have been made to translate *in vitro* genomic poliovirus RNA or poliovirus mRNA isolated from infected cells. Cell-free extracts were obtained from uninfected and infected cells, including systems from E. coli, HeLa cells, wheat germ, and rabbit reticulocytes (Rekosh *et al.*, 1969, 1970; Villa-Komaroff *et al.*, 1974, 1975; Shih *et al.*, 1978; Ehrenfeld, 1979; Agol, 1980). In nonpreincubated cell-free systems from infected HeLa cells, in which ribosomes initiate and elongate on endogeneous viral mRNAs, the newly synthesized polypeptides show excellent correlation with those made in infected cells (Helentjaris and Ehrenfeld, 1978). Neither premature terminations nor cleavage deficiencies are apparent.

Some of the difficulties encountered in earlier studies of poliovirus mRNA translation in uninfected cell extracts, such as poor efficiency and incompleteness of translation and processing, have been surmounted in recent years with a refinement of techniques and of the employed cell-free systems. The first definitive identification of poliovirus proteins synthesized *in vitro* in cell free extracts from uninfected HeLa cells was reported by Villa-Komaroff *et al.* (1975). They showed that several of the synthesized proteins had electrophoretic characteristics identical to those of poliovirus proteins synthesized in infected cells. In addition, they observed some uncleaved precursor proteins which are found in vivo only under conditions inhibiting processing. The final viral proteins VP_0, VP_1, VP_2 and VP_3 were, however, not detectable.

In cell free extracts from wheat germ, which are quite efficient at translating other eukaryotic mRNAs (Roberts and Peterson, 1973), poliovirus RNA was found to be translated poorly (Ehrenfeld and Lund, 1977) although poliovirus RNA is a highly efficient messenger *in vivo.* It has been suggested that the wheat germ system exhibits a high requirement for a capped 5' terminus on mRNAs for translation (Lodish and Rose, 1977), and that the lack of the usual cap structure in poliovirus mRNA which carries the VPg on its 5' end, and the inability of the extract to provide the cap for poliovirus RNA may be responsible for the poor translation in this system. The results of recent studies support this notion (see below).

More successful were the attempts to translate poliovirus RNA in extracts from uninfected and poliovirus infected HeLa cells (Celma and Ehrenfeld, 1975a+b; Villa-Komaroff *et al.*, 1975; Rose et al., 1978) and in extracts from rabbit reticulocytes (Shih *et al.*, 1978; Ehrenfeld, 1979). In all these studies RNAs isolated from purified virions were used as mRNAs. These RNAs carry the covalently linked VPg at their 5' termini in contrast to viral mRNAs isolated from polysomes of infected cells which terminate in 5'pUp.

Complete translation of poliovirus mRNA, and processing of the primary translation products to yield most of the poliovirus specific proteins found also during *in vivo* translation, was finally obtained in rabbit reticulocyte lysates (Shih *et al.*, 1978, 1979) in which the translation of endogenous mRNA had been eliminated by treatment with micrococcal nuclease (Pelham and Jackson, 1976).

2. *Initiation of in vitro Translation*

Since the N-terminal amino acid of poliovirus is methionine (Chaterjee *et al.*, 1973) as in all eukaryotic mRNAs, initiation of polypeptide synthesis from picornaviral RNAs can be analyzed by following the incorporation of ^{35}S-methionine into peptides donated by ^{35}S-formyl-methionyl-tRNA$_f^{met}$. This approach was first used with extracts from mouse ascites cells and EMC virus RNA (Öberg and Shatkin, 1972). One major tryptic peptide (90%) and, in addition, a small proportion (10%) of a second ^{35}S-met containing peptide were found, indicating that two different initiation sites were being utilized. Two different initiation sites are also utilized when wild type poliovirus RNA is translated in extracts from uninfected HeLa cells, and also in extracts from poliovirus infected cells which contained only endogenous viral template (Celma and Ehrenfeld, 1975a+b). The two amino terminal tryptic peptides labelled with ^{35}S-formyl-methionyl-tRNA$_f^{met}$ were readily resolved by high voltage electrophoresis or by paper chromatography. The penultimate amino acids adjacent to the N-terminal f-met are different in the two peptides. These experiments were repeated with RNA isolated from the LSc strain of poliovirus type 1, a multistep temperature sensitive mutant (Jense *et al.*, 1978). Translation of RNA isolated from LSc virus *in vitro* also occurs at two different initiation sites and yields two tryptic peptides which are identical to those produced by translation of RNA isolated from wild type virions (Jense *et al.*, 1978). When the f-met labelled polypeptides synthesized *in vitro* were analyzed directly by SDS-PAGE, two major bands were resolved with apparent molecular weights of 115 Kd and 6-9 Kd. The 115 Kd polypeptide was identified as the authentic precursor for the coat protein NCVP1a. The identity of the small molecular weight f-met labelled polypeptide has not yet been accomplished (Villa-Komaroff *et al.*, 1975; Jense *et al.*, 1978; Knauert and Ehrenfeld, 1979). The location of the second initiation site on the viral genome is also an unsolved question. The relative rates of initiation of the two sites varied markedly as a function of the Mg^{++} concentration. Thus the relative rates of initiation at each of the two sites can be experimentally manipulated. At lower Mg^{++} concentrations, the high molecular weight f-met labelled polypeptide predominates, whereas at higher Mg^{++} concentrations the small f-met labelled polypeptide accumulates almost exclusively in the cell free extracts (Celma and Ehrenfeld, 1975a and b).

Evidence for additional potential internal initiation sites on poliovirus mRNA has also been presented by recent studies on the binding of 40S ribosomal subunits to poliovirus RNA under conditions that block elongation. Three high affinity binding sites were discovered, one near to the 5' end, one to the 3' side of the midgenome, and one close to the 3' end (McClain *et al.*, 1981) (see Fig. 47, p. 164). More recently, the utilization of an internal initiation site on the 3' side of the midgenome of poliovirus RNA in reticulocyte extracts was demonstrated by immunoprecipitation, time course, and ^{35}S-formyl-methionyl labelling studies of the products, namely proteins from the P-3 region of the polyprotein (Dorner *et al.*, 1984).

The biological significance of the existence of two or more initiation sites for translation of poliovirus RNA in vitro depends upon whether or not these sites are utilized for translation in vivo. As mentioned above, a number of experimental data indicate that *in vivo* all polioviral translation products are made in equal molar amounts, an observation which speaks for the utilization of only one single initiation site. These data, however, were obtained in studies on infected cells at a time when viral protein synthesis proceeds at a maximal rate (*i.e.* at later times of infection). Different results are obtained when viral protein synthesis is analyzed at early times after infection (see p. 356, below).

3. *In vitro Elongation and Termination*

Shih *et al.* (1979) have observed that the rate of ribosome movement on EMC RNA in reticulocyte lysates decreases considerably after the ribosomes traverse the region coding for the capsid precursor. The presence of weak internal termination sites on picornavirus RNA has been suggested (Paucha *et al.*, 1974; Paucha and Colter, 1975). Adding exogenous tRNA partially overcomes this block in translation. It is conceivable that tRNAs influence the secondary structure of mRNA and thereby facilitate unwinding of RNA. Part of the blockage may be due to a paucity of membranes which may be required for efficient elongation of a long mRNA that in vivo ist translated on membrane bound polysomes.

C. The Additional Complexity of *in vivo* Translation During Infection

Upon invasion of a host cell which is a million times its size, the parental virion(s) faces the vast task of getting to and redirecting the host's metabolic machinery for the production of virion components. The situation in the host cell evidently is far more complicated than that during the translation of large pools of isolated virion RNAs which are simply added to a functional *in vitro* translation system. Notably, the invading virion activates a variety of regulatory mechanisms, some of which are quite complex.

Modification of the host's translational machinery attains a central role during poliovirus infection. The "tools" (viral proteins) required for a successful infection must first be produced, a task which is accomplished by translation of the genome RNA of the invading virus. Practically no information, only speculation is available on the events leading to the first few rounds of translation of the parental viral RNA. Later, after a more solid basis of viral "tools" has been established, and when the parental RNA has been copied a sufficient number of times to produce a fair-sized pool of fresh viral mRNAs, the machinery of the host cell is altered to meet the viral demand for the production of massive amounts of viral protein. These alterations are so dramatic that they can be readily studied with available biochemical techniques.

In the following sections we shall first discuss the mechanisms which bring about the dramatic alterations of the translational machinery, "the shut-off" which becomes apparent by the first hour or so post infection, and which then proceeds at an exponentially increasing rate during the next two to three hours of infection. Finally, we will summarize some of the ideas which have been put forward to explain the early steps controlling translation of the parental virion RNA(s).

1. *The Shut-Off Phenomenon*

As mentioned above, one of the first and most dramatic virus induced alterations of the host cell's translational apparatus is the shut-off phenomenon, which renders host cell mRNAs untranslatable and leads to the preferential translation of poliovirus mRNAs (see Fig. 60A, p. 210, Fig. 81, p. 314, and Fig. 84, p. 338).

The shut-off phenomenon has been under investigation for over 20 years, and although many details have been observed and described (reviews: Bablanian, 1975; Lucas-Lenard, 1979; Koch *et al.*, 1982b), the exact mechanism(s) by which it is induced is still unclear. The site affected appears to be the translational apparatus itself and not the host cell mRNAs. Host cell mRNAs are neither destroyed nor irreparably damaged after infection: both the 3'(poly A) and 5'(cap) ends are normal, the cellular mRNAs are not detectably modified, and they retain the ability to prime protein synthesis in cell-free extracts from uninfected cells (Colby *et al.*, 1974; Koschel, 1974; Fernandez Munoz and Darnell, 1976).

A priori there is no reason to expect only one unique mechanism for the virus induced shut-off effects. (Indeed it is known that bacteriophages kill their host cell by several independent mechanisms (Schweiger *et al.*, 1978)). Table 54 lists a number of hypotheses have been developed to explain the shut-off phenomenon in cells infected by poliovirus and other viruses. For the sake of clarity, we shall discuss some of the proposed mechanisms separately. However, it is important to note that these mechanisms are not mutually exclusive (as has occasionally been assumed); moreover, that several or all of them may act in a concerted manner. The mechanisms which we shall discuss below, include:

a) One model put forward by our group (Koch *et al.*, 1976; 1980a, 1982b), which proposes that the virus somehow triggers a normal host cell regulatory mechanism for the inhibition of the synthesis of cellular macromolecules.

b) Competition between viral and host cell mRNAs, as a result of increased mRNA concentrations or limitations in the supply of active initiation component(s) (Lawrence and Thach, 1974; Nuss *et al.*, 1975; Kaempfer, 1982b).

c) Another model which postulates ionic disturbances, in particular an increase in the intracellular Na^+ concentration induced by altered membrane

Table 54. *Comparison of shut-off mechanisms employed by poliovirus and other viruses*

	Activation of negative pleio-tropic response	Competition of mRNA	Membrane leakiness	Inactivation of initiation components	Interaction of viral proteins with cellular components
Polio (ssRNA +)	+	+	+	+	+
VSV (ssRNA –)	+	+	?	?	?
Reo (dsRNA)	–	+	?	?	?
SV 40 (DNA)	–	+	?	?	+
Vaccinia (DNA)	–	+	?	?	+

permeability (membrane leakiness) as a cause for the inhibition of host cell protein synthesis (Carrasco and Smith, 1976; Carrasco, 1977).

d) Alterations of the protein synthesizing machinery, *i.e.* of initiation factors and ribosomal proteins (phosphorylation, dephosphorylation), resulting in preferential translation of viral mRNA (Helentjaris and Ehrenfeld, 1978; Baglioni *et al.*, 1979; Trachsel *et al.*, 1980).

e) The interaction of viral proteins with cellular components involved in protein synthesis, which would directly or indirectly interfere with the translation of host mRNAs (Wright and Cooper, 1974; Racevskis *et al.*, 1976).

Finally, we suggest that the cell membrane may play a central role in mediating and regulating host cell protein synthesis.

The shut-off induced by EMC virus in L cells—although similar in kinetics to that induced by poliovirus in HeLa cells—differs in several aspects (Jen *et al.*, 1980), notably in that the activity of cap binding proteins (CBP), is not destroyed but a change in the subcellular distribution of several initiation factors occurs (Jen and Thach, 1982). In addition, the rapid shut-off induced in L cells by EMC infection is not induced in at least four other cell types by EMC virus indicating that the mechanisms operating in shut-off not only vary within the picornavirus group, but also depend upon the particular cell type employed (Jen and Thach, 1982).

a) The Activation of an Inherent Host-Cell Regulatory Mechanism

The poliovirus induced shut-off in many aspects resembles the reduction of protein synthesis which occurs in uninfected cells in response to a variety of environmental factors (the so-called negative pleiotropic response) (see Table 52, p. 329). Nutrient starvation, serum depletion or elevated osmolarity of the culture medium, high cell densities, and poliovirus infection all lead to a specific block in the initiation phase of protein synthesis, reflected in the breakdown of polysomes and the accumulation of 80S monosomes. In all cases the pattern of newly synthesized proteins is shifted in a characteristic manner, reflecting the competition of mRNAs with different translational efficiencies for the now-limited fraction of active initiation components (see Table 49, p. 319).

Poliovirus infection may trigger a normal "built-in" regulatory mechanism of the host cell for selective blockage of polypeptide chain initiation (Koch *et al.*, 1976, 1980a). Indeed, when we analyzed the particular changes induced in the pattern of proteins synthesized by an elevation of the osmolarity of the culture medium (hypertonic initiation block) and compared them to those induced in the same cells upon infection with vesicular stomatitis virus (myeloma cells) (Nuss and Koch, 1976; Koch *et al.*, 1976) or poliovirus (HeLa cells) (unpublished observations and Oppermann and Koch, 1976; Koch *et al.*, 1980a), we found the changes to be remarkably similar.

The notion that a common mechanism is involved is supported further by the finding that the effect of the conditions is additive. The severity of poliovirus induced shut-off is dependent on the nutritional state of the cell. In the absence of serum for example or under conditions of elevated medium osmolarity, shut-off occurs faster and is more severe. Conversely, addition of 5 % serum to infected cells (Schaerli and Koch, unpublished) results in a delay of both shut-off and virus replication by approximately 30 minutes, suggesting that efficient virus replication depends on interference with macromolecular synthesis of the host cell. Many other similarities could be found in the negative pleiotropic response of the host cell to environmental factors and the shut-off phenomenon including a reduction in the transport of hexose, uridine, Na^+, and amino acids, and changes in the phosphorylation state of ribosomal proteins and initiation factors (Koch *et al.*, 1982b).

There are a number of arguments which imply that the cell membrane plays an important role in mediating both the negative pleiotropic response and the virus-induced shut-off phenomenon (Koch *et al.*, 1976, 1979, 1980a, 1982b). The membrane functions as a barrier between the cell and its environment, thus environmental perturbations are necessarily communicated to the cell interior via the membrane. For example, poliovirus adsorption to cell surface receptors causes conformational alterations in the cell membrane, as manifested by changes in membrane fluidity and permeability to cations and other low molecular weight substances (Carrasco and Smith, 1976; Carrasco, 1977; Kohn, 1979; Schaefer *et al.*, 1983a, b) (see Chapters 6 and 7). Furthermore, as discussed above (section I B 3), a wide variety of substances inhibit protein synthesis in intact cells but not in cell-free extracts from these cells, implying a mediatory role of the cell membrane in the inhibitory process (see Table 53, p. 330) (Koch *et al.*, 1976).

The mechanism by which such apparent membrane-mediated inhibition of protein synthesis is affected, is still uncertain. This process may involve the release or activation of low molecular weight mediators which bind to ribosomes, such that the binding affinity of mRNA for ribosomes and/or initiation factors is reduced (see section I B 3). It is conceivable that the first interactions of poliovirus with the host cell, which take place at the cell membrane, trigger this inherent control mechanism. Indeed, the concentration of small molecular weight protein synthesis inhibiting factors is markedly enhanced in extracts of infected cells as compared to uninfected cells (our unpublished observations).

Other modifications of membrane functions, in particular alterations in the permeability of the cell-membrane to ions and other substances, may also contribute to a reduction of protein synthesis (see Table 38, p. 263).

As mentioned above, one of the common aspects of conditions which inhibit protein synthesis at the level of initiation is, that they all lead to a reduced uptake of amino acids (Koch *et al.*, 1980a). After infection with poliovirus there is also a rapid decline in the uptake of amino acids in HeLa cells, which is partially reversed beginning at 90 min post infection. The inhibition of amino acid uptake is maximal at a time when protein synthesis in infected cells has reached a minimum, that is: the rate of amino acid uptake and the rate of protein synthesis decline concomitantly (Fig. 88) (Koch *et al.*, 1980a, 1982b). Presently we do not know whether this inhibition of amino acid transport is the cause (as in the case with amino acid starvation) or the consequence of reduced protein synthesis (via a feedback control of amino acid uptake) (Ring and Heinz, 1966). In this context, it is perhaps interesting to note that an increase in amino acid transport has been observed as an early event following the stimulation of cell proliferation in several

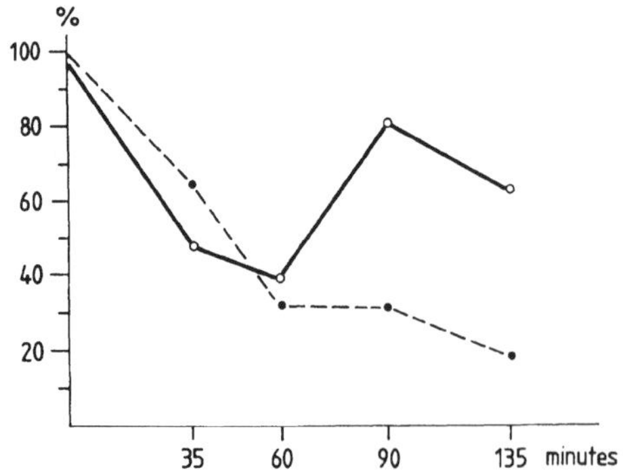

Fig. 88. Effect of poliovirus infection on amino acid uptake of HeLa cells
This figure illustrates the reduction of amino acid uptake (dashed line) accompanying the shut-off of protein synthesis (solid line) in poliovirus infected HeLa cells. HeLa cells were infected with 200 PFU/cell of poliovirus type I, Mahoney strain. Cells were labeled for 10 min with (^{35}S) methionine at the times indicated. In parallel, aliquots were used to determine the incorporation of label into TCA-precipitable material (total protein synthesis = solid line) and to determine the total cell-associated radiolabel (in newly synthesized protein and in the amino acid pool = dashed line). The data is presented as % of values in mock-infected cells. The solid line illustrates the rapid shut-off of cellular protein synthesis and the initiation of viral protein synthesis during the second hour post infection, the dashed line shows the parallel reduction in amino acid uptake. – Figure from Koch *et al.*, 1982 b [in: Protein Biosynthesis in Eukaryotes, p. 355 (1982)]

systems (Foster and Pardee, 1969; Isselbacher, 1972; Costlow and Baserga, 1973; Vaheri *et al.*, 1973; Villereal and Cook, 1977). Sander and Pardee (1972) also found cell cycle-related changes in amino acid transport. While neither the mechanism nor the significance of these changes is well understood, it has been proposed that alterations in membrane transport of small molecules may play a role in the control of cell proliferation (Pardee, 1964; Holley, 1972). Again, it is conceivable that the alterations in amino acid transport observed after poliovirus infection reflect the activation by poliovirus of certain inherent cellular control mechanisms of macromolecular synthesis.

b) Competition Between Viral and Host Cell mRNAs

As discussed above, a massive increase in the concentration of mRNAs above the saturation level for initiation components will lead to a competition between the mRNAs for the initiation components, and to the preferential translation of those mRNAs with relatively high translational efficiencies and a corresponding reduction in the translation of mRNAs with low efficiencies. While overall protein synthesis is reduced by inhibition of polypeptide chain initiation, those mRNAs with a high RTE are still able to participate in the formation of an initiation complex and thus are preferentially translated. Thus, given the relatively high translational efficiency of picornaviral mRNA, the massive production of viral mRNAs late in infection could by itself account for a significant reduction in the level of host protein synthesis. In cardiovirus infected tumor cells, this seems to be the principle mechanism for the relatively mild and late reduction of host protein synthesis (Lawrence and Thach, 1974). An mRNA-concentration dependent competition will certainly also contribute to the far more dramatic shut-off in poliovirus infected cells, particularly in the later stages of infection.

However, the shut-off of host protein synthesis in poliovirus infected cells sets in long before significant amounts of polioviral mRNAs are made and culminates rapidly in a dramatic event. Apparently, the predominant mechanism by which a competitive situation is achieved in poliovirus infected cells is the limitation of initiation components, which is a simple correlate to increasing mRNA concentrations. Evidently, initiation components are inactivated in poliovirus infected cells at a time when there are only low levels of poliovirus mRNA. Of course, it is precisely this limitation of initiation components that favors the translation of the highly efficient poliovirus mRNA and permits the direction of the available metabolic precursors (in particular amino acids and energy) away from host cell protein synthesis to viral protein synthesis (Nuss *et al.*, 1975).

mRNAs specified by cytopathic viruses in general and poliovirus in particular, are translated with high efficiency both *in vitro* and *in vivo.* They possess higher affinities for components of the protein synthesizing machinery of the cell. Picornavirus mRNAs for example outcompete cellular mRNAs in cell free systems when viral and cellular-mRNAs are added in identical amounts (Abreu and Lucas-Lenard, 1976). Golini *et al.* (1976) showed that the initiation step of translation is the site of competition. The efficiencies of binding of viral and cellular mRNAs to the 40S ribosomal subunits appear to be unequal for both mRNA classes, suggesting that the components responsible for competition may be the initiation factors themselves. This suggestion was supported by the observation that addition of excess crude initiation factors from reticulocytes or plasmacytoma cells could relieve the competition, thus allowing the translation of cellular mRNAs (10S RNA or globin mRNA) (Golini *et al.*, 1976).

Recent observations indicate that eIF-2 binds to different mRNAs with different affinities (Kaempfer, 1982). The translation of globin mRNA is progressively inhibited by addition of even low amounts of mengovirus RNA. This competition is relieved by addition of highly purified eIF-2 without concomitant stimulation of overall protein synthesis. In direct binding assays using isolated mRNA and eIF-2, mengovirus RNA exhibits a 30-fold higher affinity for eIF-2 than globin mRNA. In addition, the association of eIF-2 and mengovirus RNA is more resistant to high salt concentrations than is the complex with globin mRNA. These results suggest that the affinity of a given mRNA species for eIF-2 determines its translation, relative to that of other mRNA species. Messenger RNA competition for eIF-2 may contribute to the selective translation of viral RNA in infected cells.

Conditions which sensitize cells for infection by viruses and isolated viral RNA, such as treatment of cells with the hypertonic initiation block or exposure of cells to ethanol, cytochalasin B, DMSO, or DEAE-dextran (Hövel, 1973; Koch, 1973; Koch and Koch, 1974; Saborio *et al.*, 1975; England *et al.*, 1975) all inhibit protein synthesis at the level of initiation. These apparently paradoxical observations are explicable if one assumes that the indiscriminate reduction of polypeptide chain initiation provides the high efficiency message (the polioviral RNA) with a translational advantage over the low efficiency messages (host cell mRNAs).

c) The Role of Ionic Disturbances and Membrane Leakiness

When the effects of varying concentrations of NaCl on the *in vitro* translation of different mRNAs were studied, it was observed that each message appeared to have its own characteristic ionic optimum for efficient translation (Carrasco and Smith, 1976). In other words, the relative translational efficiency of an mRNA is dependent on ionic conditions. In particular, picornaviral mRNAs were found to be translated most efficiently at strikingly high NaCl concentrations, a condition under which translation of host mRNAs was inhibited. These observations led to the proposal that ionic disturbances also play an important role in the *in vivo* shut-off during poliovirus infection (Carrasco, 1977). Selective permeability changes to monovalent ions after infection of cells by animal viruses would lead to an increase in the intracellular concentration of Na^+ and a decrease in the concentration of K^+. It was further proposed that this change in the intracellular ionic environment results in a preferential synthesis of viral proteins while the synthesis of host proteins is supressed (Carrasco, 1977).

More detailed investigations of the changes in intracellular ionic environments after infection with different viruses showed, that whereas an ion-concentration mediated shut-off may indeed be operating in Sindbis-virus infected cells (Garry *et al.*, 1979), a comparable mechanism certainly does not play a significant role in the shut-off phenomenon in mengovirus infected tumor cells (Egberts *et al.*, 1977), or poliovirus infected HeLa or HEp-2 cells (Bossart and Bienz, 1981; Koch *et al.*, 1982b). Figure 89 shows the changes in Na^+ and K^+ concentrations during the course of poliovirus infection (Schaefer *et al.*, 1983). The Na^+ content per cell increases dramatically at 120 min post infection, long after the onset of the shut-off phenomenon. Of course, like the effect of increasing poliovirus

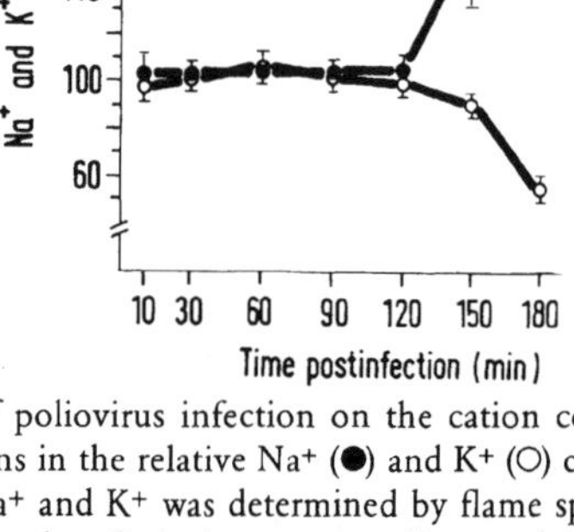

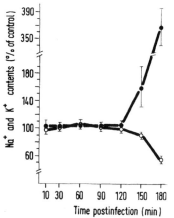

Fig. 89. The effect of poliovirus infection on the cation content of HeLa cells
This figure shows the alterations in the relative Na^+ (●) and K^+ (○) contents of poliovirus-infected HeLa cells. The content of Na^+ and K^+ was determined by flame spectrophotometry. Results are expressed as percentages of the values from the corresponding mock-infected cells and represent the means of three experiments ± standard error. Averages of the Na^+ and K^+ contents of control HeLa cells were 0.081 and 0.72 μmol/mg of protein, respectively. – Figure from Schaefer *et al.*, 1983 a [J. Virol. *44*, 446 (1983)]

mRNA concentrations late in infection, the late changes in ionic environments contribute to the reduction of host cell mRNA translation, but they are certainly not the principle cause for triggering of the shut-off.

On the other hand, injury to the cell membrane and an increased "leakiness" of the cell membrane after picornavirus infection are established phenomena (see Chapter 7) (Kohn, 1979; Impraim *et al.*, 1980). Changes in membrane fluidity after virus adsorption were observed by fluorescence depolarization of DPH (1.6 diphenyl 1, 3, 5 hexatriene) (Levanon, *et al.*, 1977; Lyles and Landsberger, 1977; Levanon and Kohn, 1978; Moore *et al.*, 1978; Schaefer *et al.*, 1983). These changes in membrane fluidity are accompanied by a redistribution of some membrane proteins (Bächi, *et al.*, 1973), and by capping of the adsorbed viruses (Gschwender and Traub, 1979). Viruses may interfere with interactions between the cytoskeleton and the membrane (Lyles and Landsberger, 1977) and lead to alterations in several membrane functions, notably the transport of metabolites and the permeability barrier for ions (Carrasco and Smith, 1976; Micklem and Pasternak, 1977). The transport of certain amino acids is reduced after poliovirus infection (Koch *et al.*, 1980a, see Fig. 88). Leakage of cellular constituents as a result of adsorption of different viruses (also UV inactivated polioviruses) to various host cells has also been demonstrated by gas chromatographic studies (Levanon *et al.*, 1977). The induction of membrane leakiness by many cytopathogenic viruses does not only result in loss of small molecules from the infected cells but might also allow the penetration of a number of specific inhibitors of protein synthesis selectively into the infected cells. Inhibitors, such as GppCH2p, edeine, blasticidin S, and various plant proteins normally bind only to isolated ribosomes and are excluded from uninfected cells (Contreras and Carrasco, 1979; Lacal *et al.*, 1980; Koch *et al.*, 1980a; Foà-Tomasi *et al.*, 1982, Lacal and Carrasco, 1982). Thus, permeability changes in the cell membrane are likely to affect several aspects of protein synthesis and may contribute to a reduction in host cell protein synthesis.

d) Alterations of Initiation Factors or Ribosomes

Ample evidence has been obtained in the past five years for significant and specific poliovirus induced modifications of initiation factors and ribosomes resulting in functional alterations which strongly favor translation of polioviral mRNA over host mRNA. The identity of the initiation factor(s) involved and the nature of the modification remains a matter of controversy (see Table 50, p. 323).

Crude initiation factor preparations from poliovirus infected cells are inactive in stimulating polypeptide chain initiation with cellular mRNAs although they do promote the initiation of viral mRNA translation (Kaufmann *et al.*, 1976; Helentjaris and Ehrenfeld, 1978; Helentjaris *et al.*, 1979). Addition of large quantities of crude initiation factors from uninfected cells could restore the translatability of host mRNAs. Attempts were made to identify the components responsible for the competitive disadvantage of host mRNAs either by partial purification of the inactivated factors or in reconstitution experiments with purified initiation factors from uninfected cells. In these studies eIF-3 (Helentjaris *et al.*, 1979), eIF-4B

(Golini *et al.*, 1976a+b; Rose *et al.*, 1978; Baglioni *et al.*, 1978), eIF-2 (Kaempfer *et al.*, 1978), and the cap binding protein (Trachsel *et al.*, 1980; Tahara *et al.*, 1981) were implicated as the affected factors.

Given the striking difference between poliovirus and host cell messages at the 5' terminal end, where host cell mRNAs are capped and poliovirus mRNA is not, the cap binding protein should be an especially suitable target for inactivation by poliovirus. Inactivation of the cap binding protein would provide the poliovirus message with a translational advantage and could explain the observed poliovirus induced shut-off. Support for this notion has come from studies on the influence of poliovirus infection on the translatability of other viral mRNAs which are capped, in particular vesicular stomatitis virus (VSV) mRNAs. Indeed, poliovirus superinfection of VSV-infected cells results in inhibition of VSV as well as host protein synthesis (Doyle and Holland, 1972; Ehrenfeld and Lund, 1977). It was predicted that initiation factor preparations from poliovirus infected cells would be unable to support translation of VSV mRNA and cellular mRNA, but would promote initiation of poliovirus mRNA translation. Initiation factors from uninfected cells should allow the translation of all mRNAs. This prediction was experimentally verified (Brown *et al.*,1980, 1982). Later on, Trachsel *et al.* (1980) showed that the initiation factor preparations from poliovirus infected cells lack a functional cap binding protein (CBP) which was previously characterized as a protein with a molecular weight of 24,000 (Sonenberg *et al.*, 1979). One step in the initiation process that is specifically inhibited by ribosomal salt washes or crude initiation factor preparations from poliovirus infected cells is the binding of host or VSV mRNAs to $40S^{met}$ tRNA complexes (Brown and Ehrenfeld, 1980). This finding is consistent with the notion of an inactivated cap binding protein. James and Tershak (1981) have recently shown that poliovirus infected HeLa and VERO cells contain several proteins including a 24 kd protein which are more highly phosphorylated than a comparable protein from uninfected cells. Further studies are in progress to show whether this 24 kd protein is identical with the cap binding protein mentioned above. It has been suggested that the earlier reports of other altered initiation factors (eIF-3, eIF4B) may simply reflect copurification of cap binding protein with these factors.

However, it is likely that more than one initiation factor is altered by poliovirus infection. In accordance with this view are recent observations on the association of CBP with eIF-3. This association is abolished in extracts of poliovirus infected cells (Hansen *et al.*, 1982). Using an antiserum against eIF-3, a protein of 220 kd was detected in uninfected cells which was absent in extracts from poliovirus infected cells. The latter extracts contained instead antigenically related polypeptides of 100 and 130 kd—presumably degradation products of the 220 kd protein (Etchison *et al.*, 1982). The authors postulate that the 220 kd protein is an essential component of the cap recognition complex and that its degradation in poliovirus infected cells results in the inhibition of host cell protein synthesis. Lee and Sonenberg (1982) reported that various CBP with 24, 28, 32, 50 and 80 kd from uninfected cells can be crosslinked to capped VSV mRNA, but that the corresponding proteins from poliovirus infected cells show only reduced crosslinking ability. In addition, initiation factor preparations from poliovirus infected cells rapidly inactivate CBP from uninfected cells by degradation or phosphorylation.

As discussed above (in section IB4), chemical modification of ribosomal proteins is another mechanism for affecting the control of translation. Polysome and ribosome preparations from poliovirus infected HeLa cells contain 10–14 fold more protein kinase activity in vitro than similar preparations from uninfected cells (Tershak, 1978a). And indeed, changes in the phosphorylation state of 60S ribosomal proteins are observed in cells 1 1/2 hour after poliovirus infection (Kruppa, 1979 and pers. communication). Whether the observed quantitative changes in the ribosomal protein phosphorylation are the cause or consequence of virus induced inhibition of protein synthesis remains to be determined.

Whatever the precise mechanism and site of inactivation may be, the effect is a reduction in the pool of active initiation components which favors the translation of poliovirus mRNA. Protein phosphorylation has been mentioned as a possibility (see above). In this context, it is of interest to note that a protein kinase activity is associated with the capsids of intact poliovirions (Schärli and Koch, 1984). Binding of small effector substances has also been discussed. Another mechanism which has been suggested is the direct interference with the translational machinery by poliovirus specific proteins (see below).

e) The Role of Virus-Specific Factors in Mediating the Shut-Off

A number of observations and arguments have been published that imply a direct involvement of poliovirus specific factors, particularly the capsid proteins, in the alteration of the host translational machinery during the shut-off.

Translation of the viral RNA is generally required to induce the shut-off (see below). Addition of an inhibitor of protein synthesis interrupts the process of shut-off. Upon removal of a reversible inhibitor, protein synthesis resumes at the level shown before it was blocked (Baltimore *et al.*, 1963b, Holland, 1964, Penman and Summers, 1965). These observations indicate that synthesis of viral protein is a prerequisite for the shut-off. In contrast, inhibition of viral RNA synthesis by guanidine slows but does not stop the shut-off (Tershak, 1982). Since viral RNA synthesis is not required for the shut-off, it is likely that the synthesis of double-stranded RNA plays no essential role for shut-off *in vivo*, although it is a potent inhibitor of *in vitro* protein synthesis (Ehrenfeld and Hunt, 1971).

A role of newly synthesized viral coat proteins in the shut-off is indicated by experiments with temperature-sensitive mutants of polioviruses defective in coat proteins. These mutants are unable to inhibit cellular protein synthesis at the non-permissive temperature (Steiner-Pryor and Cooper, 1973). Temperature-sensitive mutants with defects in the non-structural proteins, including those involved in RNA replication, were able to induce the shut-off. Further studies with defective interfering (DI) particles of polioviruses suggested that only part of the structural protein gene is required for shut-off. Cole and Baltimore (1973) have isolated DI particles of poliovirus with deletions in the 5' region of the viral RNA. These DI particles lack one third of the gene coding for the coat precursor protein, nevertheless they do elicit the shut-off phenomenon. The corresponding coat precursor protein, which is synthesized after infection by DI particles, is unstable and rapidly degraded. Further experiments are called for to determine whether the unstable two thirds of the capsid precursors, or part of them, persist long enough in the cell to induce the shut-off.

Matthews *et al.* (1973) and Wright and Cooper (1974) reported on the association of viral proteins with host ribosomes and proposed that this binding of viral proteins results in the suppression of host mRNA translation. An inhibition of initiation of protein synthesis was observed in reticulocyte lysates after addition of intact polioviruses (Racevskis *et al.*, 1976).

An intriguing question is whether or not the capsid proteins of the parental virion(s) could play a role in the shut-off. After the release of VP4 from intact polioviruses which is induced by the adsorption process, RNA and all of the viral proteins VP1-3 remain associated in particles sedimenting with 135S or 80S. A significant portion of these parental virion particles are found in association with the endoplasmic reticulum. Some of the capsid proteins of infecting particles are also found in association with virus-specific polysomes (Habermehl *et al.*, 1974). The fate of VP4 in infected cells is not known. However, it is recovered from infected cells in higher yields than the other three coat proteins (Habermehl *et al.*, 1974). It is tempting to speculate that VP4 or one of the other parental virion-derived capsid proteins associates with the endoplasmic reticulum or with polysomes in order to mitigate—perhaps only a "local"—shut-off. Another tempting speculation in this respect is to attribute a role to the protein kinase activity associated with the viral capsid in the phosphorylation mediated inactivation of initiation factors or ribosomal proteins. Alternatively, given the affinity of the capsid proteins for the translational machinery, the parental capsid proteins may just serve to "lead" the incoming genomic RNA to the correct site for its translation.

With some viruses (vaccinia), inhibition of cellular protein synthesis in infected cells occurs even when UV inactivated viruses interact with host cells and when the synthesis of viral mRNA is blocked by other means (Moss, 1968), indicating that parental viral proteins can trigger the shut-off and that their release from the virions is not altered by UV irradiation. However, infection of cells by UV inactivated poliovirions does not trigger the shut-off phenomenon. The UV mediated inactivation of the shut-off ability of polioviruses follows one hit kinetics indicating that one hit anywhere in the genome is sufficient to abolish the induction of the shut-off (Helentjaris and Ehrenfeld, 1977). However, UV irradiation of poliovirions also consistently results in covalent linkage of viral proteins, notably VP4 and VP1 to viral RNA (Wetz and Habermehl, 1981). Thus, for the case of poliovirus, it is difficult to say whether a defect in the RNA or in the capsid proteins is responsible for the failure of UV inactivated poliovirions to induce the shut-off. The UV induced linkage of viral proteins to RNA might prevent their function in the shut-off.

Further support for the notion that parental viral proteins function in the shut-off was obtained in studies on the infectivity of isolated polioviral RNA (see Chapter 7, Section V). Isolated poliovirus RNA has an infectivity which is several thousand fold lower than that of intact poliovirions (Koch, 1973). The greatly reduced infectivity of isolated RNA is due in part to the loss of RNase resistance and specific adsorbtion capacity which are important functions of the capsid proteins in intact virions (Crowell and Landau, 1982). The infectivity of isolated poliovirus RNA can be increased by incorporation of the RNA into liposome vesicles. This provides protection against RNase and circumvents the requirement of a specific receptor since liposomes fuse non-specifically with the cell membrane (Taber *et al.*, 1980). Still, the infectivity of liposome incorporated poliovirus RNA is far less than that of intact virions, which supports the hypothesis that the capsid proteins of parental virions also play a role in enhancing RNA infectivity after penetration of the RNA into the cell (Koch, 1973).

Treatment of poliovirions with diethylpyrocarbonate reduces infectivity to the level of isolated RNA, while RNase resistance and adsorbility are retained. The structure and function of the genomic RNA is not impaired. We suppose that such treatment either prevents the "correct" intracellular release of capsid proteins, or that it destroys the binding capacity of the capsid proteins to the translational machinery (Koch, unpublished observations), thereby abolishing the capacity of the parental capsid proteins to induce a local shut-off.

Interestingly, the variety of conditions which have been found to enhance the infectivity of isolated viral RNA, all have in common an inhibitory affect on the initiation of protein synthesis, which resembles in principle the shut-off observed during infection by intact virions. It seems plausible that these conditions are a substitute for a comparable viral RNA-translation enhancing function which is normally assumed by the parental capsid proteins during the early stages of poliovirus infection. The hypothesis of a viral protein mediated interference with translation of host mRNA is supported by observations with other picornaviruses. The association of an EMC virus capsid precursor protein with host mRNA (Skarlat and Kalinina, 1978), and of an EMC virus noncapsid polypeptide (corresponding to poliovirus protein NCVPX) with ribosomes (Medvedkina et al., 1974; Skarlat *et al.*, 1976) have been reported.

Finally, double stranded viral RNA which accumulates after the peak of viral replication may contribute to the decline of protein synthesis late in infection (Cordell-Stewart and Taylor, 1971, 1973; Hunt and Ehrenfeld, 1971; Celma and Ehrenfeld, 1974).

f) In Summary, a Concert of Mechanisms with a Purpose

By way of summary then, the shut-off of host protein synthesis in poliovirus infected cells appears to be the result of several different mechanisms which act in a concerted manner. In the past, most authors who proposed alternative explanations have stressed the dominant role of their respective hypotheses. Particular mechanisms indeed may predominate in different virus host cell systems, but usually different mechanisms act simultaneously, consecutively or in concert to induce the shut-off.

Thus, by its interaction with the cellular membrane early in infection, poliovirus may trigger an inherent cellular control mechanism in which the increased membrane fluidity, the transient increase in amino acid transport, and the release of small effector substance(s) may play an important mediating role. Parental viral proteins may induce a local shut-off effect. Freshly synthesized viral proteins may contribute to the shut-off in later stages of infection. Cellular mRNAs remain intact and translatable in *in-vitro* translation systems; they may be withdrawn from the translatable pool *in vivo* by binding to viral proteins or other factors, analogous to untranslated mRNPs in oocytes. In poliovirus infected cells, specific inactivation of the cap binding protein - and possibly other initiation factors and ribosomes - appears to assume a central role in the shut-off phenomenon. Inactivation of these factors may be mediated by phosphorylation, possibly directed by a virus-associated proteinkinase and/or a polysome-associated kinase activity. Other mechanisms may serve to amplify the shut-off late in infection: membrane leakiness can lead to ionic conditions unfavorable for cellular mRNA translation or to the entrance of protein synthesis inhibitors. Last but not least, the ever increasing concentrations of poliovirus mRNAs during the virus replication cycle can displace host mRNA from the translational machinery, thereby favoring the synthesis of viral proteins.

In any case, the ultimate purpose of the poliovirus induced shut-off appears to be the establishment of a competitive situation, where host and viral RNAs compete for a limited set of initiation components. The competitive situation favors the translation of the high efficiency poliovirus mRNA and thereby serves to direct the flow of metabolites essential for translation, *i.e.* of ATP, GTP, and amino acids away from host protein synthesis to the production of large amounts of poliovirus proteins.

2. *Non-Uniform Synthesis of Viral Proteins*

Several years ago, we analyzed the effect of medium hypertonicity on the synthesis of poliovirus specific proteins in infected HeLa cells (Nuss *et al.*, 1975). No significant differences were observed in the ratios of the individual viral proteins when their synthesis was analyzed at 3.5 hours post infection by pulse labelling with ^{35}S-methionine under isotonic and hypertonic conditions. We concluded that the hypertonic initiation block had no effect on the processing of viral proteins (Saborio *et al.*, 1974).

In a later study, we performed similar experiments with poliovirus infected BSC 1 cells (Koch *et al.*, 1980a). To our surprise the labelling pattern of viral proteins changed considerably when the osmolarity in the medium was increased. In BSC 1 cells the hypertonic initiation block (HIB) leads to a preferential accumulation of virus specified proteins coded by the 3' end of the genome. This is in contrast to the non-uniform synthesis in mengovirus infected cells at earlier times in the replication cycle, when the viral coat proteins appear to accumulate at a higher rate than the virus proteins coded by RNA regions near the 3' end (Paucha *et al.*, 1974). We, therefore, reexamined viral protein synthesis in HeLa cells incubated in media with different osmolarities (see details in Table 55). Under restrictive conditions of polypeptide chain initiation, the ratio of NCVP1a to other proteins changes considerably in favor of 1b, 2 and X by +87, 50 and 136 %, respectively. This result resembles our previous observation obtained with BSC 1 cells (Koch *et al.*, 1980a) and suggests either the existence of independent initia-

Table 55. *Differential accumulation of poliovirus proteins under normal and restricted conditions of polypeptide chain initiation in infected HeLa cells*

Labeling conditions	Viral proteins			
	NCVP1a	NCVP1b	NCVP2	NCVPX
Isotonic	100	100	100	100
Hypertonic	100	187	150	236

HeLa cell monolayers were infected with 300 PFU/cell of poliovirus type I. At 2.5 h.p.i., part of the cells were exposed to hypertonic medium (150 mM excess NaCl) for 25 min in order to halt initiation of protein synthesis. The tonicity of the medium was then reduced to 40 mM excess NaCl in order to permit reinitiation of protein synthesis. 6 min later, at 3 h.p.i., cells were pulsed-labeled with (^{35}S) methionine for 6 min and chased with a 1,000 fold excess of unlabeled methionine for 5 min. Cells were lysed and proteins separated by SDS-polyacrylamide gel electrophoresis on slab gels. The autoradiographs were scanned with a densitometer, and the peak areas of viral proteins labeled under isotonic and hypertonic conditions were compared. The densitometric tracings were adjusted to identical values for NCVP1a under isotonic and hypertonic conditions. The data is expressed as percent of peaks in isotonic conditions.

Table from Koch *et al.*, 1982 a [Protein Synthesis in Eukaryotes, p. 285 (1982)]

tion sites on viral mRNA(s), or a preferential degradation of NCVP1a when initiation of protein synthesis is inhibited by hypertonic medium, or the synthesis of viral mRNA with deletions in the coat region (DI-RNA), or a partial-degradation of mRNA at the 5'end.

The experiments described in Table 55 were performed at a time (2.5 hours p.i.) in the viral replication cycle when relatively large amounts of viral proteins are synthesized, the hypothetical existence of two independent initiation sites on viral mRNA(s) called for an analysis of viral protein synthesis at early times after infection (Koch *et al.*, 1982). Only two viral proteins were clearly detectable by this procedure in the infected cells at early times of infection: NCVP2 and NCVPX (Fig. 90). In addition two other viral proteins, NCVP1b and NCVP3a, might be present. Again these results suggested that NCVP1a was either not synthesized or preferentially degraded at early times after infection. Early in the replication cycle DI-RNA is synthesized which carries a deletion in the coat region (Etchison and Ehrenfeld, 1980). Such RNA might serve as mRNA and thereby give rise to defective coat protein precursors only. It is also conceivable that mRNA molecules are formed early after infection which miss all the coding capacity for coat proteins (see Chapter 9). A comparison of the two densitometer tracings of Figure 90 also reveaed the differential effect of the virus induced shut-off on the synthesis of individual host cell proteins.

To sum up, a non-uniform synthesis and/or accumulation of various viral proteins seems to take place early during infection, and under conditions of hypertonic shock. These findings indicate initiation of viral protein synthesis on more than one species of viral mRNA, or on internal initiation sites on the viral mRNA, or preferential degradation of proteins coded for by the 5' terminal region of the viral mRNA. Different species of viral mRNA could originate from fragmentation of viral mRNA (McClain *et al.*, 1981) or by synthesis of deletion-carrying RNA (Etchison and Ehrenfeld, 1980). Internal initiation of translation may be favored

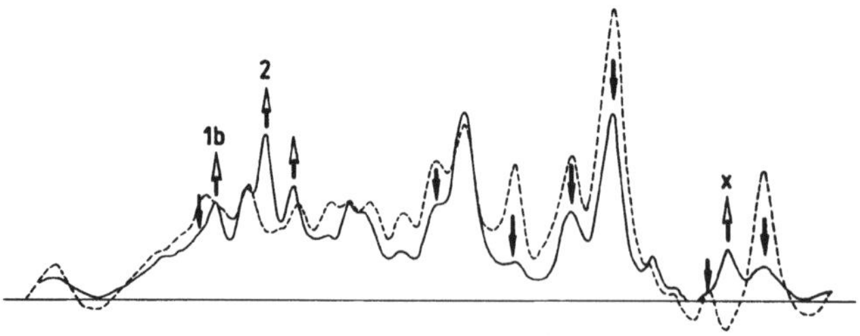

Fig. 90. Changes in the pattern of newly synthesized proteins following infection of HeLa cells by poliovirus

This figure shows the first appearance of viral non-capsid polypeptides in poliovirus-infected monolayer HeLa cells at 2 h.p.i. and the differential effect of the shut-off on individual cellular proteins. HeLa cells were infected with 300 PFU/cell. At 1 hour 45 min p.i., cells were incubated in medium lacking methionine, supplemented with 7.5% fetal calf serum. At 2 h.p.i., cells were pulse-labeled with (^{35}S) methionine for 5 min and chased for 4 min with a thousand-fold excess of unlabeled methionine. Cells were lysed and extracts subjected to SDS polyacrylamide gel electrophoresis. The figure shows the densitometer tracing of an autoradiograph of the gels, the dashed line shows the polypeptides synthesized in mock-infected cells, the solid line those synthesized in poliovirus infected cells. The open headed arrows mark the virus specific proteins appearing at this time. The proteins were identified by coelectrophoresis of polypeptides from late-stage infected cells. The solid arrows point out the differential effect of the shut-off on individual cellular proteins. Note that the kinetics of infection (and of appearance of viral polypeptides in particular) are markedly slower in monolayer cultures of HeLa cells (this figure) than in suspension culture HeLa cells (e.g. Figures 85 or 93). – Figure from Koch *et al.*, 1982 a [in: Protein Synthesis in Eukaryotes, p. 286 (1982)]

by certain structural features of the RNA, such as A-U-rich regions (Both *et al.*, 1976), and may be influenced by ionic conditions. Conformational changes in the RNA, possibly due to the binding of many ribosomes to the RNA strand may render the internal sites inactive soon after the onset of viral translation (Dorner *et al.*, 1984). The question is still unresolved and calls for further experiments to decide between these three alternatives. In any case, preferential synthesis of proteins from the 3' side of the genome may be of advantage early in the replication cycle, when viral protease and replicase are needed for efficient protein processing and RNA replication.

3. *The Role of Interaction Between Poliovirus Proteins and Intracellular Membranes*

Synthesis of a protein on membrane bound polyribosomes rather than on soluble ribosomes is usually taken as an indication for the destiny of the protein as some form of membrane associated or secreted protein. Poliovirus protein synthesis proceeds on membrane bound ribosomes. The newly synthesized viral proteins remain membrane associated even after processing into the primary and secondary cleavage products. It can be expected that some of the poliovirus proteins are integral membrane proteins or become translocated across the endoplasmic reticulum; however, it is not yet known which of the polio proteins actually participates in the membrane incorporation system of the host cell. There are several extended stretches of hydrophobicity within the poliovirus polyprotein, these are illustrated in Figure 91. These stretches are potential membrane anchoring sites. The N-terminal protein of the P-3 domain, NCVP9, has been shown to be an integral membrane protein, it contains a strong hydrophobic stretch of amino acids

(Takegami *et al.*, 1983a). Vesicle formation for secretory proteins in uninfected cells is induced by an accumulation of cleaved leader peptides in the wall of the endoplasmic reticulum. The formation of the specific poliovirus vesicles might be induced in a similar manner by one of the smaller poliovirus polyprotein cleavage products. Guanidine inhibits the migration of poliovirus induced membranes from the rough endoplasmic reticulum to the smooth membrane fraction (Caliguiri and Tamm, 1970; Mosser *et al.* 1972); this observation suggests that the guanidine sensitive P-2 proteins are responsible for the formation of the poliovirus specific vesicles (see Chapter 6).

The importance of membrane association for the functioning of poliovirus proteins is illustrated by a number of observations: Etchison and Ehrenfeld (1981) isolated the poliovirus RNA replication complex and studied RNA synthesis by this complex in vitro. The membrane bound replication complex yielded full length viral RNA. Removal of membranes resulted in the synthesis of shorter viral RNAs. Readdition of isolated virus specific membranes restored the capacity to synthesize full length viral RNA. A second example of a membrane dependent functioning of poliovirus proteins is the assembly of the viral capsid. Assembly of poliovirus capsid proteins in vitro in the absence of membranes yields stable capsids which are never detected as assembly intermediates in vivo; in the presence of membranes, shells resembling those formed in the infected cell are assembled (Palmenberg, 1981; Putnak and Phillips, 1981) (see Chapter 10).

Orderly assembly of the capsid proteins may require that they are either translocated into the lumen of the endoplasmic reticulum or that they are inserted into the membrane of the endoplasmic reticulum. The replicase proteins, at least those required for the initiation of vRNA synthesis may be integral membrane proteins also. The activity responsible for intiation of RNA synthesis is probably exposed on the cytoplasmic side of the endoplasmic reticulum, since the replicative intermediate evidently is also somehow associated to the cytoskeleton (Lenk and Penman, 1979). The actual polymerase activity responsible for RNA elongation (NCVP4, p63) is probably a soluble protein, possibly liberated from a more tightly associated complex after initiation of RNA synthesis (see Chapter 9).

Optimal protein processing may be dependent on the presence of membranes. The interaction of proteins and membranes might expose the proper cleavage site, or the proteases are membrane associated or membrane activated. The timely ordered cleavages might be an essential prerequisite for proper assembly (see Chapter 10).

4. Distribution of Viral Proteins

The different poliovirus proteins are not distributed uniformly within the infected cell. Some viral proteins are regularly found in larger aggregates (Korant, 1973). Most viral proteins are associated with specific organelles of the host cell, in particular with the rough and smooth endoplasmic reticulum (Caliguiri and Tamm, 1970 a, b). Approximately 5% of the viral proteins are found in the nucleus (Bienz *et al.*, 1982; Fernandez-Tomas, 1982). The cellular localization of a protein may be a hint as to its function.

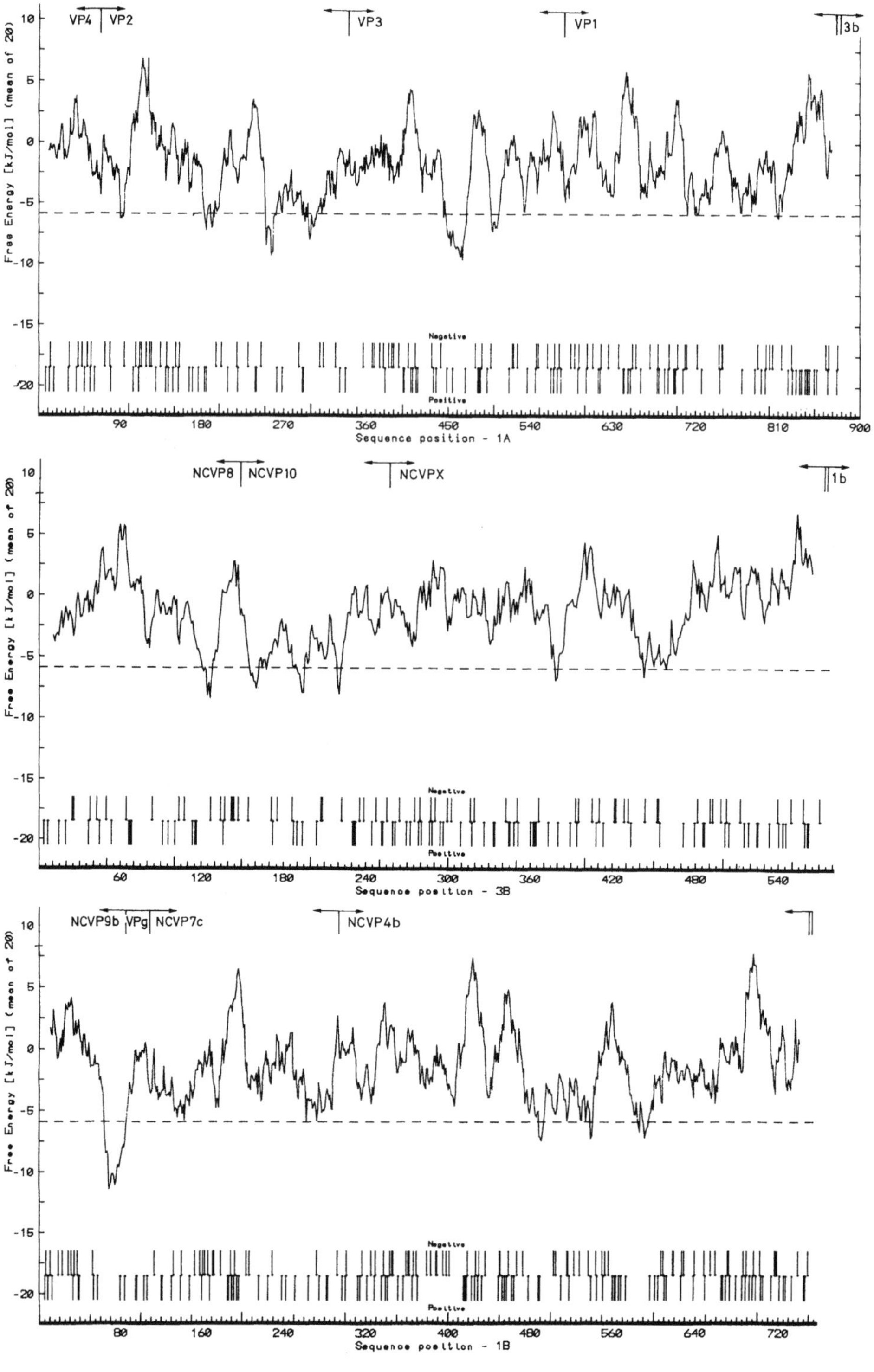
VP4
VP2
VP3
VP1
3b
Free Energy [kJ/mol] (mean of 20)
Negative
Positive
Sequence position - 1A
NCVP8
NCVP10
NCVPX
1b
Sequence position - 3B
NCVP9b
VPg
NCVP7c
NCVP4b
Sequence position - 1B

The distribution of viral proteins within the cell can be determined by fractionation of cytoplasmic extracts by differential centrifugation, density gradient centrifugation, or electrophoresis. Caliguiri and Tamm (1970 a+b) fractionated nondetergent treated lysates of poliovirus infected cells by centrifugation in sucrose density gradients. In short labelling pulses (5 min and less), radioactive amino acids were almost exclusively incorporated into nascent polypeptides on ribosomes attached to the rough endoplasmic reticulum (see Table 36 and Fig. 74, p. 246–247). In 10–30 min labelling pulses, part of the capsid proteins appeared in the smooth membrane fraction containing most of the polymerase activity. The association of the capsid proteins to this membrane fraction was only transient; after longer labelling periods, radioactivity in capsid proteins appeared in mature virions. The mature virions were associated again with the rough endoplasmic membrane fractions. Capsid proteins of parental and progeny virions also have been found in tight association to ribosomes (Fenwick and Wall, 1973; Wright and Cooper, 1976), where they may have shut-off functions. The smooth membrane fraction contained mainly replicase related proteins, in particular NCVP2, NCVP9, and NCVPX (Caliguiri and Mosser, 1971; Yin, 1977; Bienz *et al.*, 1983).

Specific migration of viral proteins into the host cell nucleus has recently been demonstrated. According to one report, several of the P-3 proteins, notably NCVP2, are found in the nuclei of infected HeLa cells (Fernandez-Tomas, 1982). These proteins accumulate at a rate which is directly proportional to the decrease in the rate of cellular RNA synthesis (see Chapter 9). No other viral proteins were detected in the nucleus in this study. In a similar study with poliovirus infected HEp-2 cells, many other viral proteins were detected in nuclei isolated from these cells (Bienz *et al.*, 1982). Quantitative determination revealed that the three primary cleavage products, NCVP1a, -3b, and -1b, as well as the viral proteins NCVP3a, -3c, -5b, -2, and VP1 were present in the nucleus at higher relative concentrations than in the cytoplasm. All of these proteins, except for VP_1, are cleavage intermediates. Since viral proteins are processed rapidly (see below), the high concentration of the primary and intermediate cleavage products in the nucleus suggests that their transport occurs shortly after their synthesis. Transport to the nucleus may require the release of these proteins from association to the intracellular membranes.

Fig. 91. Potential membrane insertion sites of the poliovirus proteins
This figure illustrates the relative distribution of hydrophobic and hydrophilic segments in the poliovirus primary cleavage products P-1, P-2, and P-3. The plots are obtained by averaging the hydrophobic constants of the amino acid side chains over stretches of 20 consecutive residues. For each consecutive stretch, the mean change in free energy required for insertion of the amino acid sequence into a hydrophobic environment was calculated. The "threshold" line at 5.9 kJ/mol indicates whether a certain segment of the protein could lie within a membrane (von Heijne, 1981). Valleys in the plot indicate relatively hydrophobic segments, peaks indicate hydrophilic segments. Strongly hydrophobic segments are potential membrane insertion sites. The bottom of each figure illustrates the distribution of negatively charged (top row) and positively charged (bottom row) amino acid residues along the polypeptide sequence. The localization of the stable protein products are indicated in the top of each figure. — Figures courtesy of J. Hoppe, Braunschweig

The viral RNA which is not incorporated into mature virions, is transported to the rough endoplasmic reticulum to serve as mRNA. The direction of transport of viral mRNA is opposite to that of the capsid proteins to their assembly sites. The bidirectional transport in the infected cell must be controlled by specific signals, which have not yet been elucidated. It is conceivable that transport of viral components occurs in specialized vesicles or along components of the cytoskeleton.

5. Protein Processing

a) The Role of Cleavage

Modification and/or processing of viral and cellular polypeptides have a profound effect on and may be required for their biological activity.

Studies of the picornaviruses led to the discovery of the important principles of nascent and post-translational processing of larger protein precursors (polyproteins) to functionally distinct secondary polypeptides (Holland and Kiehn, 1968; Jacobson and Baltimore, 1968; reviewed in Korant, 1975, 1979). Subsequently, different model systems have been developed which have led to the observations that processing and modification play important roles in many biological processes and often have important regulatory functions (activation, inactivation of enzymes and regulatory molecules, transport, compartmentalization and secretion of glycoproteins, and assembly of viruses and components of the cytoskeleton) (Laemmli, 1970; Doolittle, 1973; Bachrach and Benchetrit, 1974; Goldstein and Champe, 1974; Hershko and Fry, 1975; Cooper and Ziccardi, 1976; Schultz, 1976; Neurath and Walsh, 1976). For instance, processing of precursor proteins play an important role in the synthesis and maturation of the RNA tumor viruses (Yoshinaka *et al.*, 1980, Koch *et al.*,1982a).

b) Types of Cleavages

Functionally distinct types of proteolytic cleavage are involved in the processing of viral proteins. The formative cleavages serve to convert primary precursors to functional proteins. The second type of proteolytic processing are morphogenetic cleavages, which are intimately associated with the intermediary and final steps of virus assembly. These might serve to render the assembly reactions in which they participate irreversible. A third type of change may be involved in activating the viral replicase for initiation of RNA synthesis.

Straight forward pulse-chase studies clearly showed the flow of labelled amino acids from larger into smaller virus specific polypeptides (see Fig. 85, p. 339). Many subsequent experiments, including the biochemical mapping of precursors and products and inhibition of the processing reactions, confirmed the original experiments (see Figs. 86 and 87, p. 340). Cellular and viral proteases are responsible for the different cleavages. Figure 92 illustrates the processing pattern and kinetics of cleavage of poliovirus proteins. The nucleotide sequence of the poliovirus genome indicates that three different types of cleavage sites are involved: gln/gly sites are cleaved most often, two or three tyr/gly sites and one asn/ser site provide additional cleavage sites (see also Figs. 48, 49, p. 168–170).

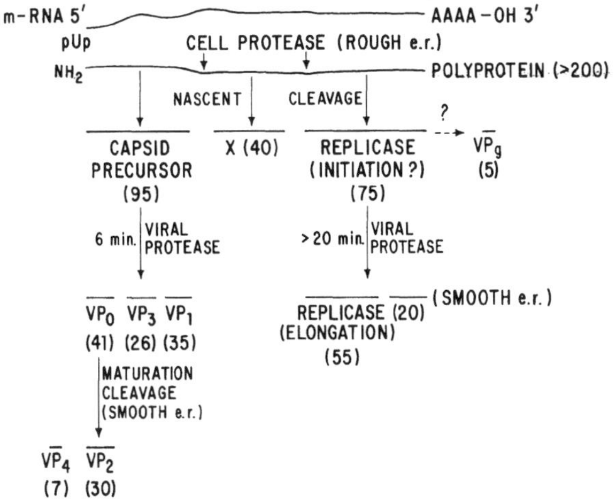

Fig. 92. Overview of proteolytic processing of poliovirus proteins
The numbers given refer to estimated molecular weights (in SDS-PAGE). – Figure from Korant *et al.*, 1980 [Ann. N.Y. Acad. Sci. *343*, 305 (1980)]

The processing of the viral coat protein precursor is associated with the assembly steps, *i.e.* these are morphogenetic cleavages. The capsid precursor protein NCVP1a (the protomer) is processed proteolytically and is then assembled into pentamers. The cleavage of NCVP1a into VP_0, VP_1, VP_3 appears to commit the protomers to assemble. The cleavage of VP_0 into VP_4 and VP_2, accompanies or follows the encapsidation of virion RNA and completes poliovirus assembly and maturation (for details see Chapter 10). Comparable morphogenetic cleavages of viral proteins were first discovered in phage infected bacteria (Laemmli, 1970) and later found in many viral systems.

c) Types of Proteases

Synthesis of prohormones, secretory proteins (Steiner *et al.*, 1974, 1980; Koch and Richter, 1980) and of picornavirus proteins takes place on membrane bound polysomes with an associated protein processing system (Caliguiri and Tamm, 1969; Levintow, 1974; Pérez-Bercoff, 1979). Picornaviruses, therefore, may utilize the same processing system as cellular proteins for some of their cleavages. Table 56 lists some of the evidence for involvement of cellular proteases in processing of viral proteins (Korant *et al.*, 1980).

Many viruses, notably reoviruses, the myxo- and paramyxo-viruses, rely on host cell proteases only for cleavage of virus specific proteins (Homma, 1971; Lazarowitz *et al.*, 1973; Scheid and Choppin, 1974; Ohuchi and Homma, 1976). These viruses have apparently evolved such that they can utilize proteolytic mechanisms already present in host cells, rather than investing their own limited genetic capacity for this purpose. Some viruses incorporate proteases in their particles during maturation and these proteases can be purified with the virus particles. Among the animal viruses the RNA tumor viruses are the best studied

Table 56. *Evidence for cellular proteases that may be involved in virus precursor polypeptide cleavage*

1. There are differences in proteolytic enzyme inhibitor sensitivities, depending on the cell line, rather than the infecting virus. This is commonly seen with chloromethyl ketones of phenylalanine or lysine.
2. Gel analyses indicate small but reliable differences in the size of cleavage products, depending on the host cell.
3. The proteases are associated with primary cleavages of nascent precursors, or removal of "signal-type" sequences, and are present in polyribosomes, or membranes to which the polysomes are bound.
4. The reactions are sensitive to diisopropyl-fluorophosphate, and by inference the proteases are of the serine active site type.
5. The proteases in cell-free extracts have a neutral to somewhat alkaline pH optimum. They are unlike the acidic lysosomal cathepsins.
6. The proteases are endoproteolytic, and are usually unstable with time, and difficult to purify.
7. They are present in several eukaryotic cells examined, including human, monkey, murine and avian cells.
8. In extracts, they produce primary cleavages on viral polyproteins, but cannot produce bona-fide, stable end products, except with the paramyxoviruses.

Table from B. D. Korant *et al.*, 1980 (Ann N.Y. Acad. Sci. *343,* 306).

example of virion associated proteases (v.d.Helm, 1977; Yoshinaka and Luftig, 1977; Korant, 1978). Other viruses may utilize their own proteases in addition to host protease systems. The picornaviruses, for example, utilize at least one distinct virus specified protease (Korant *et al.*, 1979; Palmenberg *et al.*, 1979).

Cellular protease can perform the primary cleavage of the polioviral polyprotein (Korant, 1972; Butterworth and Korant, 1974; Korant, 1975 a, b). The further specific processing of viral proteins depends on viral protease(s). Uninfected cells contain a ribosome associated protease with a minimum molecular weight of 12 kd which preferentially acts on polypeptides containing incorporated amino acid analogues and proteins "foreign" to the cell. This host protease may function as a restriction protease degrading false peptides at the site of synthesis. Recently it has been shown that a cellular protease itself is inactivated by proteolytic cleavage after poliovirus infection (Korant *et al.*, 1979). Total proteolytic activity in cell extracts decreases rapidly following infection with polioviruses concurrent with the loss of synthesis of cellular proteins (Korant *et al.*, 1979, 1980), probably reflecting the inactivation of cellular protease. It has been suggested that the virus thereby inactivates an inherent cellular defense mechanism for the degradation of foreign proteins (Korant *et al.*, 1979). With the onset of viral protein synthesis overall protease activity in cell extracts increases again, probably reflecting the synthesis of virus specific protease.

The properties of picornavirus protease are summarized in Table 57 (Korant, 1980). Direct evidence for virus-coded proteases of picornaviruses has been obtained in studies with cell-free systems after allowing viral mRNA to direct the synthesis of polyproteins in vitro (Shih *et al.*, 1979). Pelham (1978) showed that encephalomyocarditis (EMC) RNA can be translated completely in rabbit reticulocyte extracts. The translation of EMC RNA yielded an active proteolytic processing enzyme. Shih *et al.* (1978) reported that poliovirus RNA could be translated completely in rabbit reticulocyte lysates and that the products were processed extensively. Only when the translation was permitted to proceed past the coat protein region, a further processing of coat precursor proteins was observed. This kinetic analysis revealed that the gene for the putative protease must be located near the middle of the genome. In continuation of this study, Palmenberg *et al.* (1979) purified the protease synthesized in reticulocyte lysates in response to the addition of EMC RNA. The protease migrated on slab gels with an apparent molecular weight of 22 kd. By tryptic peptide analysis, p22 was identified as the complement to the viral replicase E (polio NCVP4b) in the replicase precursor D (polio NCVP2); i.e. as the peptide corresponding to polio NCVP7c. This conclusion was supported by the results of independent studies (Gorbalenya *et al.*, 1979; Svitkin *et al.*, 1979). Reports from other laboratories have assigned protease activity also to the coat protein VP3 of poliovirus (Lawrance and Thach, 1975). However, the viral protease P3-7c has a molecular weight almost identical with VP3. A contamination of viral coat proteins with this protease appears likely.

Viral protease activity has been ascribed also to the poliovirus protein NCVPX (Korant *et al.*, 1979). In order to determine which of the poliovirus proteins is responsible for cleavage in vitro, Hanecak *et al.* (1982) prepared monospecific antibodies against NCVP7c and NCVPX. The proteolytic activity responsible for the cleavage of gln-gly pairs was inactivated only by antibodies against NCVP7c and not by antibodies against NCVPX. The origin of enzyme(s) responsible for cleavage at tyr-gly sites is not known. It is conceivable that this peptide linkage is broken by a second viral protease or only by a cellular enzyme(s).

Only 2 out of 10 tyr-gly pairs but 9 out of 13 gln-gly pairs are utilized in processing (see Fig. 49, p. 170) (Larsen *et al.*, 1982, Emini *et al.*, 1982). The manner of selection of these cleavage sites and the sequence of processing are not well understood. The inhibition of processing at elevated temperatures and the stability of viral polyproteins at physiological temperatures subsequent to their synthe-

Table 57. *Properties of picornavirus protease*

1. Protease activity increases with time in the cytoplasm of infected cells.
2. The increase is virus multiplicity-dependent, and requires virus RNA and protein synthesis.
3. Protease is made after host protein synthesis is abolished.
4. The enzyme is specific for virus precursors. Cleavage generates new C-terminal leucine or glutamine residues.
5. The enzyme has a neutral pH optimum with virus precursor as substrate.
6. The protease copurifies with a virus non-structural polypeptide present in infected cytoplasm.
7. The protease is synthesized in cell-free extracts, programmed with viral RNA.
8. Virus mutants can be isolated that have altered proteases. This indicates the enzyme is specified by the virus genome.
9. Inhibitor studies indicate that the protease may be of the SH-active-site type.
10. There are pleiotropic effects of protease mutations on viral RNA synthesis.
11. The primary structure of the protease is highly conserved among the human picornaviruses.

Table from B. D. Korant, 1980 (Biochem. Soc. Transact. *8*, 418).

sis at elevated temperatures indicates an important role of the secondary and tertiary structure of the viral polypeptides for processing (Garfinkle and Tershak, 1971) and/or a necessary association of the peptides with ribosomes (nascent state) or membranes. The latter view is supported by the observation that the viral protease(s) is also associated with ribosomes (Shih *et al.*, 1982). In vitro viral precursor proteins can be processed after their release from ribosomes. The essential role of a given secondary and tertiary structure of the capsid precursor for orderly processing is well documented by recent studies with temperature sensitive mutants of EMC (Young and Radloff, 1982). In vitro synthesized capsid precursors of these mutants are cleaved by in vitro synthesized viral wildtype protease only at the permissive temperature indicating a temperature sensitive conformation. Comparison of the genome sequences of the three poliovirus serotypes has shown that the cleavage sites themselves have been conserved, whereas neighboring amino acid residues show a high degree of variation (Toyoda *et al.*, 1984), indicating that the conformation of the cleavage site is more important than the primary sequence of amino acid residues.

d) Role of Cleavage in RNA-Replication

The processing of the replicase precursor protein NCVP1b to NCVP9a + NCVP2, and of these intermediates to NCVP9b + VPg and NCVP7c + NCVP4b, respectively, may play a regulatory role during viral RNA replication. It has been proposed that the protease contained within the replicase precursor cleaves VPg (or NCVP9) and itself autocatalytically from the replicase enzyme (Rueckert, 1979; Rueckert *et al.*, 1980; Palmenberg and Rueckert, 1982), i.e. that the protease in NCVP1b may bite off its own tail. The cleavage of NCVP1b could serve to activate VPg as a primer for RNA synthesis and/or to activate the replicase enzyme (Nomoto *et al.*, 1977; Flanegan *et al.*, 1977). The synthesis of every new molecule of viral RNA (minus as well as plus strands) would then require a new molecule of NCVP1b. With the aid of anti-VPg antibodies (prepared against synthetic oligopeptides to all or part of the 22 amino-acid sequence of VPg) a number of potential primers for initiation of RNA replication have been identified, including polypeptide precursors to VPg (NCVP9a, NCVPX/9, NCVP3b/9, in addition to the replicase precursor NCVP1b) (Baron and Baltimore, 1982a, b; Semler *et al.*, 1982), as well as free VPg and a nucleotidyl peptide pUpU-VPg (Vartapetjan *et al.*, 1982; Crawford and Baltimore, 1983). Thus, the precise moment and functional role of cleavage for activation of RNA synthesis is still uncertain. An attractive model has been proposed that comprises the following steps: cleavage of NCVP9 from the replicase precursor and membrane-insertion of this peptide via its hydrophobic component, uridylation of a tyr residue near the carboxy terminus of the membrane bound peptide, cleavage of uridyl-VPg from NCVP9, hybridization of pUpU-VPg to template RNA, and RNA elongation (Takegami *et al.*, 1983b; Crawford and Baltimore, 1983) (see also Chapter 9).

e) Interference with Protein Processing

Interference with protein processing is possible in several independent ways: 1. by inhibition of protease, 2. by alteration of the primary, secondary or tertiary structures of the primary translation product or of the primary cleavage products, 3. by the non-coordinate transport of viral proteases and their substrates, *i.e.* interference with compartmentalization, 4. furthermore, protein processing is influenced by membrane association and membrane alterations might cause changes in the secondary or tertiary structure of proteins.

Incorporation of amino acid analogues into nascent polypeptides can prevent co- and post-translational cleavages (see Fig. 86, p. 339). This is probably not solely due to alterations in primary structure, but to induced changes in secondary and tertiary structure as well. Similarly, incubation of poliovirus infected cells at elevated temperatures leads to inhibition in processing and thereby to disturbances in the protein cleavage pattern (Garfinkle and Tershak, 1971; Schärli and Koch, 1982). Viral protein processing is also inhibited or disturbed in the presence of Zn^{++} or serine-protease inhibitors (Korant, 1981).

Inhibitors have been employed to obtain information on the details of the enzymatic mechanism and active sites in the poliovirus protease (Summers *et al.*, 1973; Korant, 1973, 1981; Korant and Butterworth, 1976; Pelham, 1978). The activity of the protease is not sensitive to phosphonofluoridate, and it is unlike the classical serine-type enzyme. Thiol-reactive compounds do not inhibit the activity, indicating–though not proving–that free thiol groups are not part of the catalytic site. Substituted leucine or glutamine compounds with alkylating groups, such as chloromethyl ketones, irreversibly inactivate the poliovirus protease. The treatment blocks virus production and affinity labels a unique site in the enzyme. A labelled tryptic peptide fragment of the enzyme has been isolated, and further studies on its composition may yield interesting information on the active site (Korant, 1980). Peptide fragments corresponding to the amino acid sequences at the cleavage sites in the polyprotein have been synthesized and tested for their ability to inhibit the protease; so far, it has not been possible to specifically inhibit the protease with such peptides (Korant, 1981).

f) The Effect of Guanidine on the Processing of Polioviral Proteins

Addition of 2mM guanidine to poliovirus-infected cells rapidly blocks viral RNA synthesis (Crowther and Melnick, 1961; Cords and Holland, 1964; Ikegami *et al.*, 1964; Caliguiri and Tamm, 1968; Jacobson and Baltimore, 1968a; Tershak *et al.*, 1982). The precise mechanism by which guanidine exerts these and other effects is not yet certain (see Table 64, p. 406).

We have obtained evidence suggesting that guanidine may also alter the primary cleavage pattern of poliovirus proteins (Koch *et al.*, 1980b, 1982a) (Fig. 93). Guanidine has no detectable effect on cellular protein synthesis in uninfected cells. When guanidine is present in infected cells from the onset of infection, the shut-off of cellular protein synthesis is less efficient and cellular proteins can still be produced at approximately 35% of the rate in uninfected cells. The cleavage pattern of poliovirus proteins is altered markedly under these conditions (Fig. 93A). Most notably, NCVPX, NCVP4b, and NCVP6b are made only in small amounts in the presence of guanidine. In turn, several protein bands appear in guanidine treated cells which normally are not formed in uninfected cells or

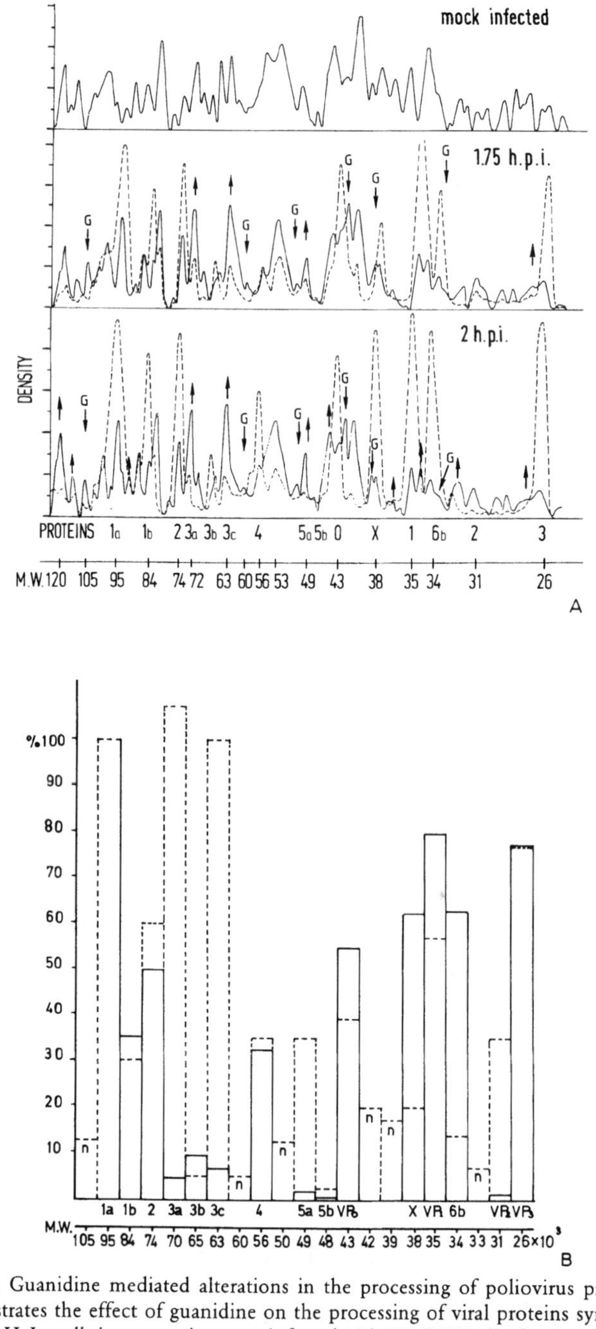

Fig. 93. Guanidine mediated alterations in the processing of poliovirus proteins
This figure illustrates the effect of guanidine on the processing of viral proteins synthesized from parental vRNA. HeLa cells in suspension were infected with 500 PFU/cell of poliovirus type I. Cells were incubated with or without 2 mM guanidine (this concentration completely blocks RNA replication). In order to decrease the background of cellular protein synthesis and to achieve a high incorporation of labeled amino acids, cells were subjected in addition to guanidine treatment to the follow-

untreated infected cells. These proteins (marked by the superscript G in Fig. 93A) migrate with apparent molecular weights of 105, 60, 50, 42, 39, and 33 kd in PAGE. Other protein bands are labeled to a higher extent in guanidine treated cells relative to untreated infected or uninfected cells. The guanidine induced changes can be quantitated tentatively by comparing the areas under individual peaks in the densitometer tracings as described in Fig. 93B. It is possible that the missing viral proteins are present in larger, new peaks, e.g. NCVPX in the 49 or 50 kd peaks. Of course, this remains a matter of speculation until peptide mapping experiments clarify the matter. Guanidine may exert the described effects on protein processing by inducing structural alterations in the viral proteins, by disturbing the interaction of viral proteins with cellular membranes, or by directly inhibiting the viral protease. In this respect, it is interesting to recall that the guanidine mediated inhibition of poliovirus replication can be reversed under certain conditions by some amino acids (Dinter and Bengtson, 1964), protease antagonists, and low temperature (Tershak, 1974).

III. Summary

Viral protein synthesis follows the same principles as cellular protein synthesis, albeit with some interesting modifications. 30–60 minutes post infection, host

---

ing procedure either at 1 hour 15 min (center scans) or at 1 hour 35 min (bottom scans): Cells were exposed to hypertonic medium for 25 min in order to block completely polypeptide chain initiation. Isotonicity was restored to permit resumption of protein synthesis, and 6 min later—at 1 hour 45 min and 2 hours 05 min respectively, cells were pulse labeled with (^{35}S) methionine for 4 min and chased with a 10,000 fold excess of unlabeled methionine. Cells were lysed and polypeptides analyzed by polyacrylamide gel electrophoresis, autoradiography and densitometer scanning at described in Fig. 85

A The profile on top illustrates the polypeptide pattern in uninfected cells, the superimposed scans in the lower part of the figure show the polypeptide patterns of infected cells in the presence (solid lines) and absence (dashed lines) of 2 mM guanidine. Note that the pattern of polypeptides synthesized in infected cells treated with guanidine differs markedly from that in uninfected cells and in infected cells without guanidine. The arrow ↑ indicates protein peaks which appear higher in guanidine-treated infected cells than in uninfected or in untreated infected cells. The arrow ↓ G marks polypeptide peaks which appear only in guanidine-treated infected cells

B In order to quantitate the guanidine-induced alterations, the relative amounts of polypeptides synthesized in infected cells in the absence and presence of guanidine were determined by comparing the areas of each individual peak of the densitometer scans. Since the shut-off of cellular protein synthesis is not complete (reduced to 35%) in guanidine treated infected cells, 35% of the corresponding host cell polypeptide peaks were subtracted first from the respective peaks in the guanidine induced pattern. The values obtained were related to those obtained for NCVP1 a which were arbitrarily set as 100%. Peaks marked n are polypeptides which appear only in guanidine-treated infected cells and not in uninfected or untreated infected cells. Notice in particular the marked increase in the polypeptide peaks comigrating with 3 a, 3 c, 5 a, and VP_2 and the appearance of new peaks with apparent molecular weights of 39, 42, 50, and 102 kilodaltons in guanidine-treated infected cells, and a corresponding decrease in the polypeptide peaks marked X and 6 b. – Figures from Koch *et al.*, 1982 a [in: Protein Synthesis in Eukaryotes, pp. 282–283 (1982)] cell protein synthesis is shut off; *i.e.* cellular polysomes disintegrate, indicating a block of initiation. A number of complementing mechanisms are utilized by the infecting virion to free the ribosomes from cellular mRNA. As a result of its interaction with the cell membrane after adsorption, the virions affect a number of cellular functions which influence protein synthesis also in uninfected cells. Membrane fluidity increases, amino acid transport is transiently enhanced, phenomena reminiscent of a mitogenic response. Small effector substances, possibly mediators in an inherent cellular control mechanism of initiation, are released as a response to virus infection. Parental capsid proteins, in particular VP4, may mediate a local shut-off phenomenon. Freshly synthesized viral proteins contribute to the shut-off at later stages of virus replication. Initiation factors and ribosomes are modified in their activities by phosphorylation. The virus exploits the natural difference in structure between cellular mRNAs and its own RNA—the lack of a 5' terminal cap—and inactivates the cap binding protein which cellular RNAs require for initiation. A specific increase in the permeability of the cell membrane for small monovalent cations at the peak of viral protein synthesis induces ionic conditions unfavorable for host mRNA translation. Compared to cellular mRNA, the viral genome is a very efficient messenger; the ever increasing concentrations of polioviral RNAs at the midpoint of infection displace cellular mRNA from ribosomes and favor translation of viral RNA. The cellular mRNA, however, is not destroyed after displacement from the ribosomes; it seems to form aggregates with other cellular and viral components, but it retains its integrity.

The shut-off of cellular protein synthesis is complete by 2.5—3 hours post infection. Meanwhile, ribosomes are reassembled into very large polysomes characteristic for translation of poliovirus RNA. Up to 35 ribosomes may be bound to the viral mRNA. The viral mRNA has some peculiar features not found in cellular RNAs: the lack of the cap at the 5' terminus, a very long untranslated 5' terminal stretch of 743 nucleotides containing 8 AUG initiating codons which are not recognized, in contrast a relatively short 3' untranslated region of only 70 nucleotides preceding the poly (A) tract, and it contains at least two internal ribosomal binding sites. It cannot be excluded that these binding sites are utilized for initiation at some early stage during viral replication. Finally, the parental viral mRNA is temporarily removed from the translational apparatus to be copied by viral RNA polymerase, a process which never occurs with cellular RNA.

Synthesis of poliovirus proteins occurs mainly on membrane bound ribosomes, many of the viral products become tightly associated to intracellular membranes. Indeed, many of the viral proteins depend on this association for efficient functioning. The replication complex synthesizes full length +RNA only when associated to membranes; the capsid proteins assemble into functional procapsids only in the presence of membranes; and protein processing may depend on membrane modified protein configurations. The poliovirus polyprotein contains several hydrophobic stretches which are potential membrane insertion sites. The relative localization of the viral proteins (cytoplasmic, membrane inserted, or luminal within endoplasmic reticulum and vesicles) is an important theme for future investigations.

The viral proteins are not distributed uniformly in the infected cell. The P-3 (replicase proteins) and P-2 proteins (X proteins) are concentrated on the virus induced vesicles which contain most of the viral RNA polymerase activity. Capsid proteins have been found in association with ribosomes. Capsid proteins also become transiently associated to the virus induced vesicles where, presumably, they encapsidate freshly synthesized RNA. The P-3 proteins and incompletely processed intermediates accumulate in the nucleus where they may modify nuclear functions, notably cellular hnRNA synthesis, or "borrow" cellular proteins required for efficient viral RNA synthesis.

During the peak of protein synthesis and at later stages of viral replication, the stable viral proteins are synthesized in equimolar ratios. In early stages of virus replication, however, there is nonuniform accumulation of viral proteins. P-2 and P-3 proteins appear preferentially, capsid proteins are either made at a very low rate or are degraded rapidly. This could reflect the activities of internal initiation sites, or the presence of defective or processed viral mRNA. The function of the early nonuniform synthesis is not clear, and further experiments are called for to decide between the alternative explanations.

Poliovirus proteins are processed proteolytically by cellular and viral proteases. A ribosome-associated cellular protease is utilized for primary cleavage of the poliovirus polyprotein. The cellular ribosomal protease activity, however, itself is inactivated by proteolytic cleavage, presumably by the accumulating viral protease. The viral protease, NCVP7c, is responsible for most of the formative cleavages and for the processing of capsid proteins prior to assembly of the protomers. NCVP7c is part of the replicase protein precursors, NCVP1b and NCVP2; it is active as part of the larger protein—where it may "bite off its own tail"—as well as in the stable processed form. NCVPX has been implicated as a protease, also, since it copurifies with protease activity in extracts from infected cells. Monoclonal antibodies against NCVPX, however, do not interfere with any of the formative cleavages in an in vitro system. This does not rule out a proteolytic activity for NCVPX, but makes it seem unlikely.

Three types of sites are cleaved in the poliovirus polyproteins: two tyr/gly sites may be cleaved by the cellular protease; nine gln/gly sites can be cleaved by NCVP7c; the asn/ser site in VPO is susceptible to cleavage only in the final stages of virion morphogenesis, the protease responsible for this cleavage still is not identified.

Proteolytic processing may have important regulatory functions for poliovirus replication, in particular for RNA synthesis. The small genome linked protein VPg is cleaved from its precursor during initiation of RNA synthesis, shortly before or after it is enzymatically linked to the first nucleotide of the progeny RNA strand. The initiation of RNA synthesis and the translocation of the nascent RNA to the capsid proteins for encapsidation are delicate processes which depend on the correct conformation of the participating factors. In this respect, it may be of interest that guanidine interferes with the processing of the P-2 proteins, and concomitantly blocks initiation of RNA synthesis, the association of procapsid and replicative intermediate in smooth membranes, and the transport of virus induced membranes from the rough endoplasmic reticulum to the virus specified smooth membrane vesicles.

9

Replication of the Viral RNA

I. Introduction

In the cytoplasm of an uninfected mammalian host cell, so far no enzymes have been detected that synthesize RNA on an RNA template. The incoming poliovirus contains a plus strand RNA but does not carry an RNA-replicase. Formation of replicase(s) is a prerequisite for the synthesis of viral progeny RNA. Therefore, the parental RNA which encodes for replicase protein(s) has first to function as mRNA. Replication of the viral RNA then proceeds in two distinct steps. First, the parental RNA serves as template for the synthesis of complementary RNA, then the minus strand serves as template for the synthesis of new viral RNA chains with the same polarity as the parental RNA (plus-strand RNA). This newly formed + RNA has three potential fates: a) it can again act as template for the synthesis of cRNA, b) it may function as mRNA, or c) it may become incorporated into progeny virus particles. Figure 94 (p. 374–375) shows a schematic overview of the principle steps in poliovirus RNA replication.

Double stranded (RF-RNA) and partially double stranded (RI-RNA) viral RNA species have been isolated from poliovirus infected cells. These have been implicated as intermediates in the two steps of viral RNA replication. The RF-RNA is thought to be the product of the first step in RNA synthesis. If the synthesis of complementary strand RNA is initiated only once on the parental RNA template, a double stranded RF-RNA-molecule is formed (Fig. 94, steps 2 and 3). Plus strand RNA synthesis, in turn, is initiated repeatedly on the minus strand in the RF-RNA, which automatically yields RI-RNA (Fig. 94, steps 4 and 5).

In principle, four different models of nucleic acid replication are conceivable (see Table 58). 1. As a single stranded template is copied, the newly synthesized chain may retain its association with the template strand, yielding a double stranded product. 2. If the newly synthesized chain is somehow prevented from extensive base pairing with the single stranded template, a single stranded progeny RNA is formed. 3. In analogy to mechanism 1, during replication on a double stranded template, the newly synthesized chain may form a stable duplex with the

Table 58. *Possible modes of RNA replication*

	examples
Type 1 repair synthesis ss template – ds product	repair synthesis on cellular DNA, *poliovirus cRNA synthesis*
Type 2 ss template copying ss template – ss product	some ssRNA phages (possibly poliovirus)
Type 3 semiconservative displacement synthesis ds template – ds product, displaced parental strand	*poliovirus mRNA and vRNA synthesis*
Type 4 conservative synthesis ds template – ss product, conserved parental template	cellular RNA synthesis, some dsRNA viruses

template strand concomitantly displacing the other of the parental strands from the preexisting duplex (semiconservative displacement synthesis). 4. Finally, in analogy to mechanism 2, if the nascent chain is prevented from base pairing with the template strand of a double stranded template, a single stranded progeny RNA strand is formed while the other parental strand reanneals to the template (since both parental chains are conserved in this mode, it has been termed conservative replication). Cellular RNA synthesis on the double stranded DNA template and the RNA synthesis of some double stranded RNA viruses (for example reovirus) proceed according to model 4. RNA replication of some single stranded RNA phages is believed to occur according to model 2. As indicated in Figure 94, and as discussed in this chapter we consider model 1 to best fit synthesis of poliovirus cRNA, and model 3 to best fit the mode of mRNA and vRNA synthesis of poliovirus (see, however, the discussion of model 2 on p. 384–386).

Different multi-component enzyme systems–the viral replication complexes–are involved in the two steps of viral RNA replication. A soluble complex consisting of host and viral proteins is responsible for the synthesis of minus strands. A larger membrane bound complex synthesizes progeny plus strand RNA. The actual polymerase activities–at least for RNA elongation–are due to the same viral coded protein (NCVP4b) in both steps. Nevertheless, there are important functional and corresponding structural differences between the two replication complexes responsible for the two steps. The templates for RNA synthesis–in particular the initiation sites–differ markedly, minus strand synthesis is initiated on a homopolymer (3' poly (A) of plus strand RNA); and plus strand synthesis is initiated on a heteropolymer (3'-end of minus strand RNA). The different tasks of the two phases are accomplished by different specificities of the replication complexes involved. The other viral and cellular subcomponents of the replication complexes may determine and modulate the specificity of the viral replicase. The conformation of the polymerase itself may depend on the state of the protein of which it is a part, as a precursor protein (as in NCVP1b and NCVP2) or cleavage product (NCVP4b, NCVP4a, and NCVP6b). In addition, modification of essential components may occur by association to different membranes and by

Fig. 94. Overview of poliovirus RNA replication
Steps in the replication of poliovirus RNA

1–7: Exponential phase of RNA synthesis

1 Parental virion RNA is first translated to produce sufficient quantities of virus-specific replicase
2 ssRNA of + polarity (parental virion RNA or mRNA) serves as template for cRNA synthesis. Initiation requires an oligo-U primer or a host-cell factor from ribosomes. RNA elongation can be accomplished by NCVP4 alone. (This step has been successfully performed *in vitro*)
3 The end product of + RNA copying is ds RNA or RF-RNA
4 The RF-RNA is probably short lived. Initiation of RNA synthesis on the RF template automatically results in the formation of RI-RNA. Initiation of RNA synthesis on this ds template is more complex than on the ss template in step 2, and has to date been achieved *in vitro* only with crude enzyme preparations. This initiation step presumably involves an enzyme system including a host cell factor (of nuclear origin?) and may require the association of the replication complex to cytoplasmic membranes
5 Repeated initiation as in step 4 produces the typical RI-RNA found in infected cells
6 Completed RNA strands of + polarity are released and then serve again as template for cRNA giving rise to RF-RNA and RI-RNA. Repeating steps 2–6 results in exponential replication of RNA
7 Alternatively, nascent or released ss + RNA may bind to ribosomes and function as mRNA (6 + 7 are the predominant fates of progeny RNA during early and intermediate phases of poliovirus replication)

5 and 8: Linear phase of RNA replication

8 RNA synthesis is tightly associated to newly formed virus-specific vesicles, and it is intimately coupled with virion assembly: during or shortly after initiation of RNA synthesis, the nascent RNA engages in the encapsidation process by preformed capsid precursors. Smaller amounts of ss progeny + RNA may not participate in encapsidation and are released into the cytoplasmic matrix to function as mRNA as in step 7)

9 and 10: Phase of declining RNA synthesis

9 When initiation of RNA synthesis is blocked either by specific inhibitors, such as guanidine, or as a result of energy or precursor depletion late in infection, elongation continues, and the last initiated nascent RNA chain remains hydrogen bound to the cRNA template RNA
10 The endproduct is RF-RNA
11 Upon removal of initiation inhibitors, RF-RNA may again serve as template as in step 4.

Notes: Steps 2 and 3, as well as elongation in step 5 are probably insensitive to guanidine. The guanidine sensitive step appears to be initiation of RNA synthesis in steps 4 and 5. Recombinational events occur during the exponential phase of RNA synthesis only (during steps 2–6). Defective interfering RNA is made also only during the exponential phase of RNA replication. Steps 2 and 3 are the probable sites of DI-RNA formation: secondary structure in the 5′-half of the + strand RNA or an occasional ribosome that may still be bound somewhere along the + strand template RNA may result in erroneous copying or "jumps" of the replicase, yielding deletions preferentially within the 5′-region of the newly synthesized + RNA.–Virus specific factors involved in the initiation of + strand RNA synthesis during steps 4 and 5 are the P-2 proteins NCVP-3 b, -5 b (and -8 and -10), the precise functions of which remain to be elucidated further; the P-3 proteins NCVP-9, which may act as a primer or as membrane fixation point for the replication complex and which contains VPg–to be linked to the 5 terminus of progeny RNA, NCVP-2 and its products the replicase activity NCVP-4 and the protease NCVP-7. In addition, some of the listed proteins may function to direct the progeny RNA to the capsid protein precursors for encapsidation (step 8)

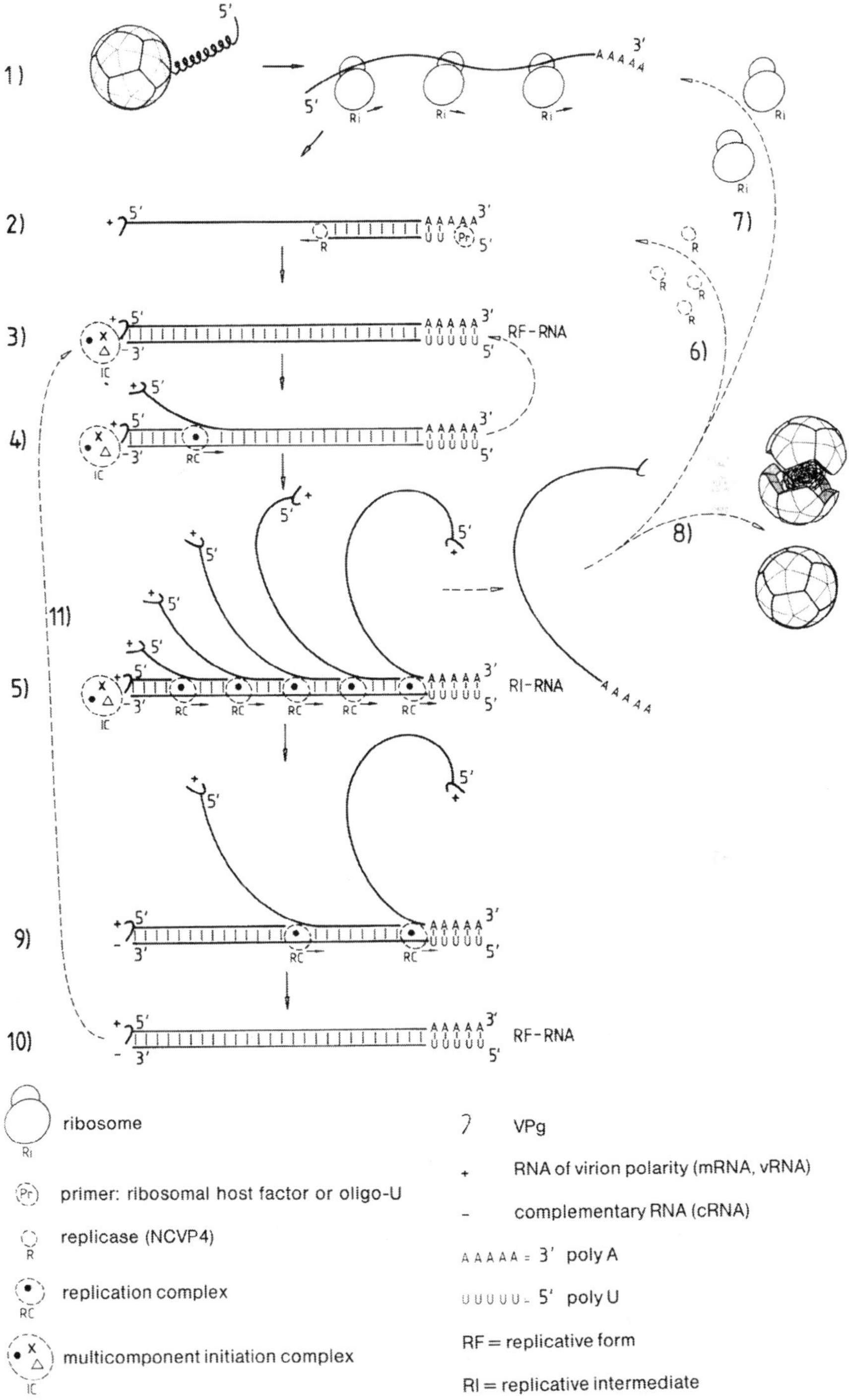
1)
2)
3)
4)
5)
6)
7)
8)
9)
10)
11)
RF-RNA
RI-RNA
RF-RNA
ribosome
primer: ribosomal host factor or oligo-U
replicase (NCVP4)
replication complex
multicomponent initiation complex
VPg
RNA of virion polarity (mRNA, vRNA)
complementary RNA (cRNA)
AAAAA = 3' poly A
UUUUU = 5' poly U
RF = replicative form
RI = replicative intermediate

compartmentalization. For example, plus strand RNA synthesis is membrane associated, minus strand synthesis seems to proceed on a soluble, not membrane associated, complex.

The fact that all cellular RNA synthesis is sensitive to inhibition by actinomycin D, whereas poliovirus RNA synthesis is resistant to this antibiotic, has often been exploited to facilitate analyses of poliovirus RNA replication. In the presence of 2 μg actinomycin D/ml, poliovirus specific RNAs in infected cells can be selectively labelled with radioactive precursors. The intracellular pools of nucleic acid precursors equilibrate at relatively slow rates, which has hampered pulse chase studies aimed at following the fate of freshly labelled species of RNA *in vivo*.

The kinetics of overall incorporation of radioactive precursors into poliovirus specific RNA are depicted in Figure 95. Replication of poliovirus RNA begins within the first hour post infection and then proceeds in three distinct phases: an early phase of exponential RNA synthesis (Fig. 95 and Fig. 94, steps 1-6), followed by an intermediate phase of linear RNA synthesis (Fig. 95 and Fig. 94, steps 5 and 8), and finally a late phase of declining RNA synthesis (Fig. 95 and Fig. 94, steps 9 and 10). The exponential phase is concerned mainly with the production of cRNA template and mRNA, the linear phase predominantly with the synthesis of virion RNA for encapsidation into progeny virions, the late phase follows the peak of virion production and signals the end of the replication cycle.

The different phases of viral RNA synthesis are controlled in several aspects: 1. initially in the switch from translation to replication of the parental RNA, 2. in the switch from the exponential to linear phase of RNA synthesis, 3. in determin-

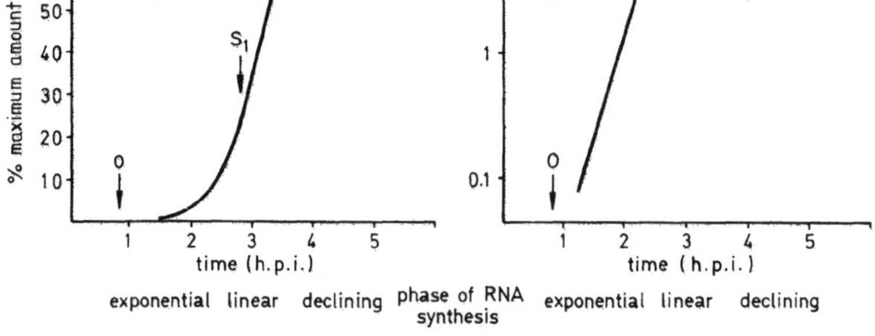

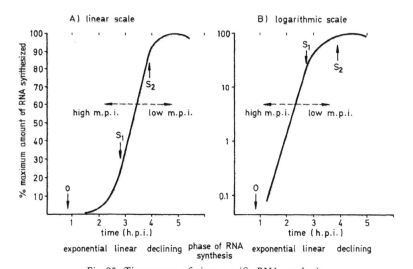

Fig. 95. Time-course of virus specific RNA synthesis

O onset of RNA synthesis; S_1 switch from exponential to linear rate of RNA synthesis; S_2 switch from linear to delining rate of RNA synthesis. The curve may be shifted forward or backward in time, depending on the multiplicity of infection, the overall pattern remains the same. – Figures redrawn from Baltimore *et al.*, 1966 [Virology *29*, 179–189 (1966)], and Oppermann and Koch, 1973 [Biochem. Biophys. Res. Commun. *52*, 635–640 (1973)] ing the fate of newly synthesized plus strand RNA as mRNA, template RNA or genomic RNA, and 4. in limiting the total amount of viral RNA synthesized. In addition, RNA replication in the infected cell is affected by other factors. Deletions and recombinational events can occur early during synthesis of poliovirus RNA. Infected cells contain a number of enzymatic activities, which are able to repair UV induced lesions in nucleic acids, and harbor RNA ligase activity (Yin, 1977b). Finally, the cell may be able to recognize unusual forms of RNA, such as the double and multi-stranded intermediates of poliovirus RNA replication, as foreign components and activate defense mechanisms against these molecules.

II. Isolation and Characterization of Virus Specific RNAs Isolated from Infected Cells

Extracts from picornavirus infected cells contain single-stranded, double-stranded, and multistranded virus specific RNA species. The presence of double-stranded RNA was first described by Montagnier and Sanders (1963) in encephalomyocarditis virus infected cells. A complete hybrid of plus and minus RNA was detected shortly thereafter in extracts from E. coli following infection by RNA phage M 12 (Amman *et al.*, 1964). This RNA was termed replicative form RNA (RF-RNA). Another characteristic RNA induced by infection with RNA viruses, a multistranded RNA, was first detected in RNA phage infected E. coli (Fenwick *et al.*, 1964; Erikson *et al.*, 1964). This RNA contains an RNase resistant "core" and attached—by base pairing—single-stranded RNAs of various lengths. This complex RNA was termed replicative intermediate RNA (RI-RNA). Table 59 provides an overview of the properties of the three corresponding poliovirus specific RNA species in infected cells (Bishop *et al.*, 1967a; Bishop and Levintow, 1971; Koch, 1973).

A. Preparation and Purification of Poliovirus Specific RNAs

Various isolation and purification procedures for poliovirus specific RNAs have been reported (for reviews see Bishop and Levintow, 1971; Koch, 1973). RNA can be isolated from infected cells by repeated extraction with phenol (to remove lipids and proteins) at room temperature or at 60° C. The hot phenol method is used to inactivate cellular RNase but has the disadvantage of yielding smaller amounts of RI and more RF-RNA because some of the ss branches of RI are released by high temperature. Phenol extraction at room temperature, in turn,—even in the presence of SDS—does not completely abolish degradation of viral RNA by cellular RNase. The content of RI-RNA or at least of single stranded branches of RI-RNA, therefore, is higher *in vivo* than determined *in vitro*. The residual phenol is removed from the RNA containing aqueous phase by repeated precipitation with ethanol at -20° C. In order to separate the different RNA species, the RNA is taken up in a concentrated form (0.5 mg/ml) in water, and NaCl or LiCl is added to a concentration of 1—2 M. Only RF-RNA and tRNA are so-

Table 59. *Properties of poliovirus specific RNAs in infected cells*

	ssRNA[a]	dsRNA (RF)[b]	RI[c]
Composition:			
of main fraction	ss + RNA	linear full length + RNA/-RNA	complete – RNA/incomplete + RNA tails
of minor fraction(s)[1]	(ss – RNA)	(circular molecules) (dimers)	ds core/+ RNA tails (complete + RNA/incomplete – RNA tails)
Sedimentation coefficient (S)	35–37	18–20	18–70 (18–40)[2]
S affected by ionic environment	+++	–	(+)
Buoyant density in Cs_2SO_4	1.68	1.60	1.61–1.65
Solubility in high salt (1M NaCl or 2M LiCl)	–	+	–
Elution from cellulose columns	15–20 % EtOH/saline	0 % EtOH/saline	0 % EtOH/saline
Sensitivity to RNase	completely degraded	resistant in 0.3M salt	ds-core resistant, ss-tails degraded
Melting transition	ø	sharp	biphasic
Migration in PAGE	fast	slow	very slow
Biological role in the infected cell	mRNA, genomic RNA in virions, template for cRNA synthesis	shortlived intermediate in early + RNA synthesis; "dead" endproduct	template for + RNA (mRNA and viral RNA) synthesis
Infectivity	++	+[3]	+
susceptibility to inhibition by actinomycin D, α-amanitin, and cordicepin	–	++	–?

[1] Forms listed in parentheses are hypothetical structures or structures detected so far in other picornavirus–but not in poliovirus–infected cells.
[2] Phenol extraction at 60 °C causes release of nascent chains and a corresponding reduction in S.
[3] Progeny virions carry the genetic information encoded by the + RNA strand of the infecting RF-RNA (Best *et al.*, 1972).
[a] Noble *et al.*, 1969; Roy and Bishop, 1970.
[b] Bishop and Koch, 1967; Bishop *et al.*, 1967, 1969.
[c] Katz and Penman, 1966; Baltimore, 1968; Bishop and Koch, 1969; Öberg and Philipson, 1971; Lundquist and Maizel, 1978; Meyer *et al.*, 1978.

luble in 1 M salt solutions whereas all single-stranded RNAs and single-strand containing RI-RNA are precipitated.

The salt precipitated viral RNA (ssRNA and RI-RNA) can be separated from cellular ribosomal RNA by agarose-gel-filtration (Koch *et al.*, 1969) or—less efficiently—by sucrose gradient centrifugation (Bishop *et al.*, 1969). The ssRNA and RI-RNA can be further purified by chromatography on cellulose. On cellulose, the behavior of RI is determined by the double-stranded portion of the molecule. Whereas ss RNA is eluted with 15–20% ethanol, all RNA containing as little as 15 % double-helical structure behave like double-stranded RNA, eluting only with water from the cellulose column. Thus, RI-RNA chromatographs in a manner identical to RF, allowing separation from ssRNA. Repeated chromatography is called for if a high degree of purity is required. RF-RNA and transfer RNA are easily separable by agarose-gel-filtration (Koch *et al.*, 1969) and also by chromatography on cellulose (Bishop and Koch, 1967; Bishop *et al.*, 1969). Other procedures that have been used to separate ssRNA, RF-RNA, and RI-RNA include sedimentation in sucrose gradients, and electrophoresis on agarose or polyacrylamide gels.

B. Properties of Poliovirus Specific RNAs

1. The Single-Stranded Viral RNA

The single-stranded RNA of positive polarity is by far the major molecular RNA species found in infected cells (Baltimore, 1969; Hewlett *et al.*, 1977) (Fig. 96). RNA isolated from intact virions and all single-stranded viral RNA isolated from infected cells show the same polarity (+) and identical base composition. cRNA (minus strand RNA) has never been detected in a single stranded form in poliovirus infected cells. Minus strands are always associated with plus strand RNA in double stranded and partially double stranded RNA forms (see below). ssRNA from infected cells reveals similar biological (infectivity and translatability) and

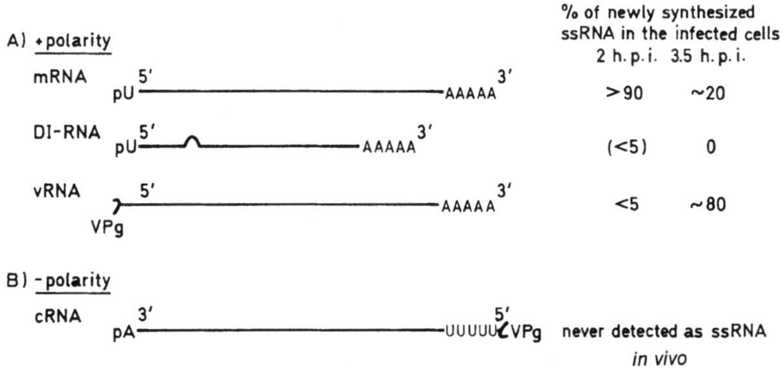

Fig. 96. Forms of single stranded poliovirus RNA

physicochemical properties as the RNA extracted from viruses. ssRNA sediments at 35S in sucrose gradients and has a buoyant density of 1.68 in Cs_2SO_4 gradients. However, the non-encapsidated, single-stranded viral RNA in infected cells does not contain VPg. This form of RNA serves mainly as mRNA for the synthesis of viral proteins. Early after infection, plus strand RNAs with deletions in the coat regions are detectable (especially after high multiplicities of infection). It is possible that this RNA—referred to as defectice interfering-DI-RNA (see p. 401, below)—also functions as mRNA.

2. *The Replicative Form-RNA*

From the onset of viral RNA synthesis a small but defined fraction of the virus specific RNAs in cell extracts is found in an RNase resistant state. At first this RNA accounts for 0.5 % of all virus specific RNAs but at the end of replication more than 10 % of the newly synthesized RNA is found in this RNase resistant form. The RF-RNA consists of an intact viral RNA chain hydrogen bonded to a complementary RNA minus strand with identical size (Baltimore, 1966; Bishop *et al.*, 1967b; Larsen *et al.*, 1980) (Fig. 97). Its molecular weight is 5×10^6. RF-RNA is as infectious as single stranded viral RNA (see below). The buoyant density in Cs_2SO_4 gradients is considerably lower than that of single-stranded RNA (Table

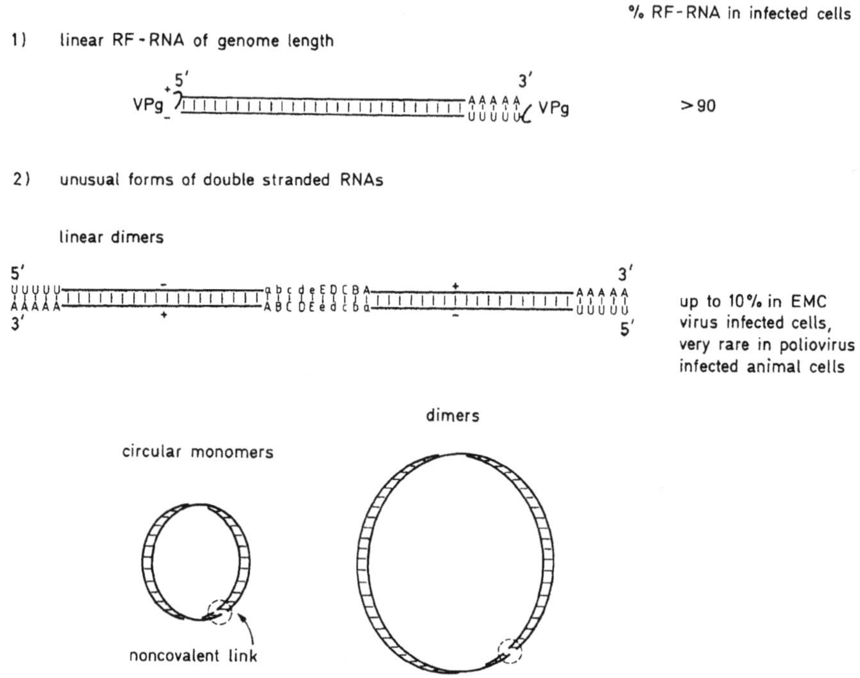

Fig. 97. Forms of replicative form RNA (RF-RNA)

59). Due to its rigid structure RF-RNA sediments relatively slow in sucrose gradient centrifugation with an S value of 18. As expected for a completely base paired double-stranded nucleic acid, RF-RNA shows a defined T_m with a sharp melting transition as revealed by UV-hyperchromicity (Amman et al., 1964; Bishop and Koch, 1967) and by the heat induced sensitivity to nucleases (Bishop *et al.*, 1965).

The base-paired double-stranded structure of RF-RNA is stabilized by cations. In high salt solutions, RF-RNA is completely resistant to digestion with RNase A or T_1: there is no conversion of incorporated radiolabeled nucleotides from an acid precipitable to an acid soluble state (Bishop and Koch, 1967; Mechali *et al.*, 1973). Biologically, however, RF-RNA is partly sensitive to RNase: a fraction of RF-RNA infectivity is lost by exposure to RNase A or T_1 (Mittelstaedt *et al.*, 1975). This latter observation is poorly understood but it may reflect the presence of unusual forms of RF-RNA (see below and Fig. 97). For example, dimeric and circular forms of RF-RNA have been described in EMC and mengo virus infected cells (Agol *et al.*, 1970, 1972, 1981; Romanova and Agol, 1979; Thornton *et al.*, 1981; Robberson *et al.*, 1982).

The significance of RF-RNA in the infected cell appears to be two-fold: 1) as a short-lived intermediate between cRNA and +RNA synthesis (Fig. 94, step 2), and 2) as endproduct of RNA synthesis after cessation of ssRNA synthesis (Fig. 94, step 10). In addition, RF-RNA is thought to exert adverse effects on host cell metabolism: addition of RF-RNA to an *in vitro* protein synthesizing system rapidly inhibits polypeptide chain initiation (Ehrenfeld and Hunt, 1971; Hunt and Ehrenfeld, 1971; Robertson and Mathews, 1973) and promotes the dissociation of the initiator tRNA-40S ribosomal subunit complex (Dambrought *et al.*, 1972). dsRNA activates a protein kinase which phosphorylates eIF-2 (Jackson, 1979). RF-RNA has thus been proposed as a mediator of poliovirus induced shut-off of host-cell protein synthesis (Hunt and Ehrenfeld, 1971). However, at the time of the shut-off, the intracellular concentration of RF-RNA is so low, that it cannot be responsible for this *in vitro* effect (see Chapter 8). The dsRNA accumulating in the infected cell after the peak of virion formation, however, may contribute to the decline of virus specific protein synthesis. Isolated dsRNA is cytotoxic when added to uninfected cells (Cordell, Stewart, and Taylor, 1973a, b). It is conceivable that dsRNA mediates the late cytopathic changes in the infected cell (see Chapter 6).

The infectivity of RF-RNA reveals some peculiarities. In contrast to infectivities of single stranded +RNA and RI-RNA, the infectivity of RF-RNA is sensitive to inhibition by actinomycin D (Koch *et al.*, 1967), and RF-RNA is not infectious in enucleated cells (Detjen *et al.*, 1978). These observations suggest that some host cell nuclear factors may be essential for the replication of RF-RNA. In E.coli and other cells, RF-RNA is converted to RI-RNA, suggesting that transcription preceeds translation after infection with RF-RNA (Koch *et al.*, 1967; Pérez-Bercoff *et al.*, 1979). On the other hand, it has been reported that RF-RNA can direct protein synthesis in cell-free extracts (Agol, 1981), suggesting that ribosomes may contain factors which unwind ds-RNA. Experimentally reconstituted hybrid RF-RNA molecules give rise only to progeny virus showing the characteristics encoded in the plus strand RNA (Best *et al.*, 1972). The issue of RF-RNA infectivity is discussed in more detail in section VID (p. 415), below.

3. The Replicative Intermediate – RNA

The major component released by deproteinization of active replication complexes isolated from infected cells is a multistranded RNA species–RI-RNA (Girard *et al.*, 1967) (Fig. 98 and Table 59). Incubation of infected cells with short pulses of radioactive nucleic acid precursors at the peak of RNA synthesis results predominantly in labelling of RI molecules (Baltimore and Girard, 1966; Öberg and Philipson, 1969). RI-RNA is partly sensitive to degradation by RNAse, indicating that RI contains single and double stranded parts. The double stranded core of RI remaining after RNase digestion behaves physically alike RF-RNA, although its composition differs somewhat (see below). Thermal denaturation of RI results in a biphasic hyperchromic shift in absorbance at 260 nm. At first, a gradual and continual rise is observed over a broad range of temperatures indicative of melting of double-stranded regions in single-stranded RNA. This is followed by an abrupt increase in optical density at a temperature typical for melting of double-stranded RNA. Due to its content of single-stranded RNA, RI-RNA–like single-stranded RNA–is precipitable in concentrated salt solutions. The buoyant densities of RI-RNAs, determined in Cs_2SO_4 gradients, are intermediate between those of RF and single-stranded RNA (Bishop and Koch, 1967, 1969; Bishop *et al.*, 1969). The sedimentation coefficient of RI ranges between 20 and 70S in sucrose gradients, the slower sedimenting fractions resemble RF-RNA molecules. The infectivity of RI-RNA on a molar basis is intermediate between the one of single-stranded RNA and the one of RF-RNA. The fastest sedimenting fractions of RI may contain up to 10 nascent single-stranded RNA strands in addition to the template RNA (Bishop *et al.*, 1969). Estimates of the content of single-stranded RNA have been obtained by analysis of nucleotide composition and of sensitivity towards RNase, by centrifugation through sucrose gradient containing 50 % DMSO (Bishop and Koch, 1969a), and by electron microscopy (Savage *et al.*, 1971). Average content of single-stranded RNA in RIs is approximately 6, but may vary from 1–10. It has been estimated that the poliovirus infected HeLa cell contains approximately 2,200 replicative intermediate molecules in replication complexes, and from 5,000 to 12,000 active replicase molecules (Lundquist *et al.*, 1974; Lundquist and Maizel, 1978 a+b).

Theoretically, several structurally different RI molecules could exist. The base paired nascent chains could be attached to one continuous template strand only (semiconservative RI) (Fig. 98A) or to a double stranded core consisting of a complete plus and minus strand RNA (conservative RI) (Fig. 98B). For both of these models the template RNA could be either of plus polarity and all single-stranded RNA tails of minus polarity ("negative" RI), or the template RNA is of minus polarity and the single-stranded tails of plus polarity ("plus" RI). The predominant form of RI in infected cells is semiconservative, "plus" RI.

Partial digestion of RI with RNase yields a double-stranded "core" which resembles RF in sedimentation coefficient and buoyant density. On denaturation of the "core" RI-RNA, half of the remaining RNA is found in the form of 35S RNA, the other half in fragments of various sizes. At least 90 % of the 35S RNA show minus polarity, less than 10 % are plus. Denaturation of "intact" RI yields

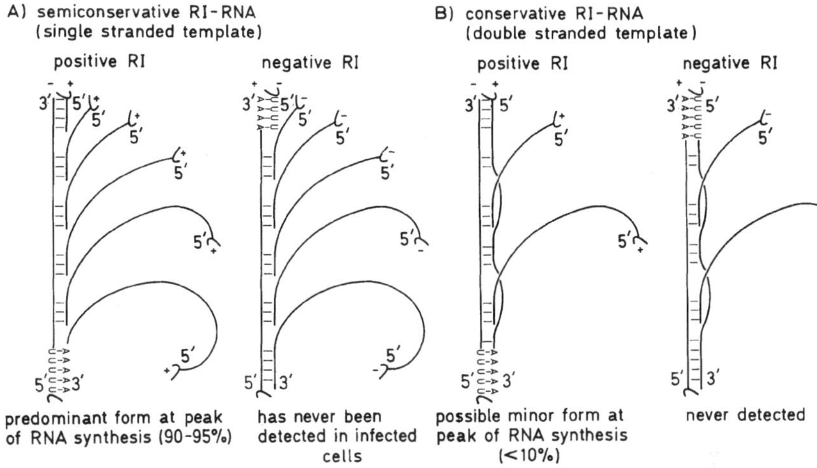

Fig. 98 A–B. Forms of replicative intermediate RNA (RI-RNA)

complete plus and minus strand RNA and a heterogeneous population of smaller molecules mainly or exclusively of plus strand polarity; indicating predominance of plus RI. The replicative intermediate RNA contains one poly (A) segment per molecule (Yogo and Wimmer, 1975). Labelling kinetics of RI *in vivo* and *in vitro* strongly suggest that RI is indeed the template on which +strand progeny RNA is synthesized during the peak of RNA replication (see Fig. 99, p. 387, and Fig. 100, p. 393).

The occurrence of conservative or negative RI in poliovirus infected cells has never been demonstrated directly. Nevertheless, the possible involvement of such structures in poliovirus replication has been inferred from a number of observations and discussed accordingly (see for example Bishop *et al.*, 1969; Bishop and Levintow, 1971). The detection of infectious RNA in RI preparations, which have been treated so as to inactivate the single stranded tails, indeed points to the presence of unusual forms of RI. cRNA alone is not infectious (Roy and Bishop, 1970), and the single intact plus strand in plus RI should be inactivated by these procedures. However, some 10% of the infectivity of RI preparations is resistant to inactivation by formaldehyde and a smaller fraction to inactivation by single strand specific RNase.

The significance of the minor fraction of RI which contains intact infectious plus-strand is not clear. One possible interpretation is the synthesis of viral RNA on a conserved double-stranded RNA. This form of RI-RNA may appear after infection of animal cells or bacteria by RF-RNA (Koch, 1973; Pérez-Bercoff, 1979). As mentioned above, RF-RNA is infectious. Studies with labeled RF-RNA revealed that input RF-RNA is rapidly converted into RI in tissue culture cells (Koch *et al.*, 1969; Pérez-Bercoff *et al.*, 1979) as well as in E. coli (Koch and Vollertsen, 1972a; Koch, 1973). An argument against conservative RI is the difficulty of envisioning a mechanism for its formation. In contrast to the well known conservative mechanism of RNA transcription from the double stranded DNA template of cellular genes, RNA transcription of poliovirus RNA begins at the very terminal

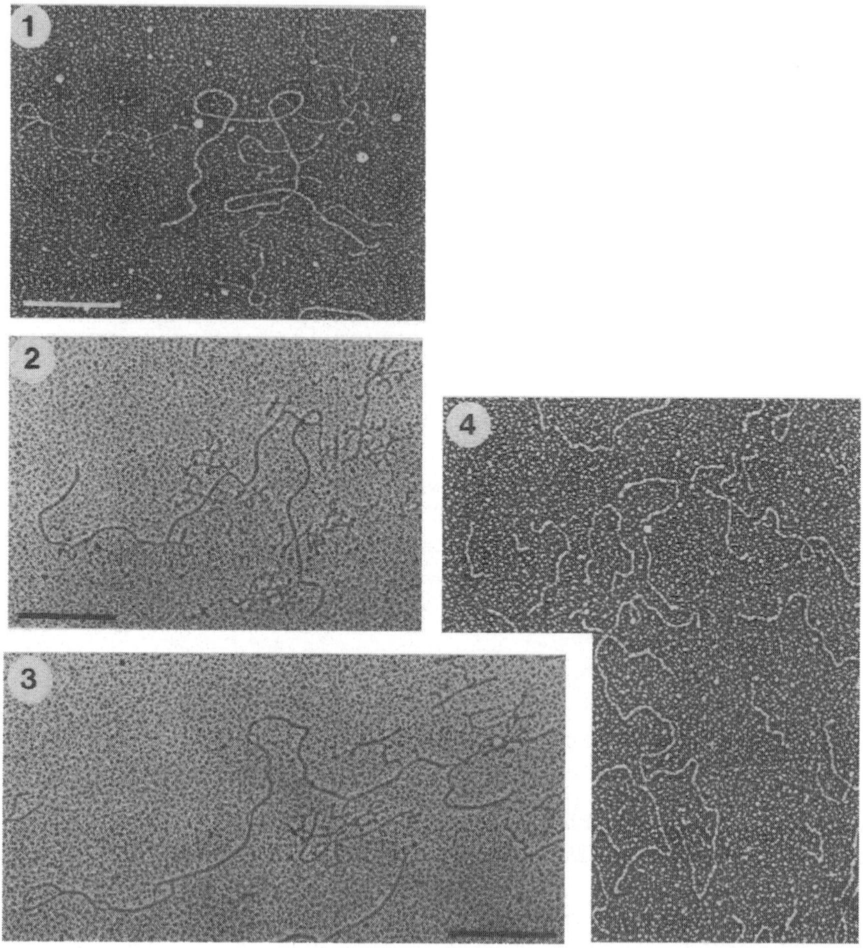

Fig. 98 C. Recent evidence to support the single-stranded backbone model of poliovirus replicative intermediate RNA *in vivo*

After the completion of this text, we were made aware of recent data supporting the single-stranded backbone model of poliovirus RI rather than the double-stranded model portrayed in this chapter. In order to determine the extent of base pairing in RI, RI molecules were cross-linked with psoralen derivatives either in the intact cell or *in vitro*. Any extensively hydrogen bonded regions in the replicating structures would be trapped by the cross-linking and become detectable by the acquired resistance to denaturation

1 Electron micrograph of uncross-linked, undenatured poliovirus RI. When isolated replicative intermediate (RI) RNA is spread for analysis in the electron microscope under non-denaturing conditions, classical RI molecules are observed–they exhibit a double stranded backbone of genome length with single stranded tails of varying lengths

2 Electron micrograph of purified poliovirus RI cross-linked with the psoralen derivative AMT (4'-aminomethyl-4, 5', 8-trimemthylpsoralen). When the same RI as in Fig. 1 is cross-linked after purification *in vitro* under non-denaturing conditions, essentially the same structures were observed, except that the single-stranded tails appear as "bushes" reflecting the cross-linkage of extensive secondary structure in the nascent chains

3 Electron micrograph of purified, cross-linked RI after denaturation. When the *in vitro* cross-linked end of the double stranded RF or RI template and not in the middle of a double stranded region. Initiation of RNA synthesis on a double stranded template necessitates first a dissociation of the strands at the 3' terminus of the actual template strand. Initiation of RNA synthesis involves the consecutive formation of hydrogen bonds between template strand and daughter strand. It is difficult to understand how the former +strand RNA could have a higher affinity for the cRNA strand than the freshly synthesized daughter RNA, which would be required if the parental strand were to replace the daughter strand again from the template RNA. One would need to postulate not only an additional unwinding activity behind the polymerase, but also a mechanism for reannealing the free tail of the former +RNA strand in favor of that from the daughter +RNA strand to the tail of the cRNA template.

A second interpretation is that cells infected with virus or single-stranded RNA might contain "negative RI", at least early in infection. Some 5—10 % of all virus specific RNA synthesized in the course of one infectious cycle consists of minus-strand RNA. Minus strand RNA is detectable, however, only as a constituent of RF or RI-RNA (Baltimore and Girard, 1966). The generation of minus-strands by a semiconservative mechanism comparable to that involved in the production of plus-strands would give rise to "cores" containing intact plus-strands and smaller molecules of segmented minus-strands. However, a careful search for small segments of minus-strands in RI-RNA and for free minus-strand RNA was not successful (Bishop, pers. commun.). Therefore, the questions of the existence of negative RI remains unresolved.

A third possible interpretation for the occurrence of infectious RNA in RI cores is one in terms of unusual forms of RNA intermediates from the early stages of RNA replication that may copurify with RI. Such unusual RNAs may play a role in the formation of defective RNA and as intermediates in recombinational events. Finally it has been shown for EMC virus, that an intact cRNA chain becomes infectious when hybridized to noninfectious fragments of vRNA (Béchet, 1972).

RI was denatured with dimethyl sulfoxide and glyoxal before spreading for electron microscopy, the single stranded tails retained their bushy appearance, and bubbles appeared in the double stranded backbone next to the sites of AMT-induced cross-linkages

4 Electron micrographs of denatured RI, cross-linked with AMT *in vivo*. In order to determine the conformation of RI *in vivo*, infected cells were treated with AMT at 3¼ h.p.i., when maximum levels of RI had accumulated. As expected, undenatured RI appeared similar to that observed for uncross-linked RI (not shown). When the RI was denatured before spreading for electron microscopy, however, a population of single strands of heterogeneous lengths without bushy appearances were observed (4). Accessibility of the cross-linking reagent to RI molecules was demonstrated by labelling of RI with radioactive AMT *in vivo*

The results suggest that RI molecules are comprised of a predominantly single-stranded backbone, attached to several nascent RNA chains, with few or no regions of extensive base pairing. The results suggest further that proteins in the replication complex or membrane attachment serve to extend the nascent single-stranded segments so as to reduce intrastrand cross-linking as well as base pairing with the template. The bar denotes one kilobase contour length. — Figures from Richards *et al.*, 1984 [J. Mol. Biol. 173, 235—240 (1984)]

4. Double Stranded Forms of Poliovirus RNA — Extraction Artefacts?

The actual occurrence of RF- and RI-RNA as such in virus infected cells has been questioned (Weissmann *et al.*, 1968). Some investigators maintained that replication of picornavirus genomes proceeds analogous to RNA replication in RNA phage systems (Rekosh, 1977; Luria *et al.*, 1978; Wimmer, 1979). Here, both minus and plus strand synthesis are believed to proceed on single-stranded templates, the nascent chains being separated from the template strand directly following polymerization (see Table 58, model 2, p. 373) (August *et al.*, 1968; Banerjee *et al.*, 1967, 1969). It was proposed that the complementary viral RNA strands were separated by proteins and that only after extraction of the proteins the complementary strands would anneal and form the double-stranded RNAs (Thach *et al.*, 1974). In apparent support of this view was the observation that only small amounts of RF-RNA were obtained from infected cells when the extraction of proteins by phenol was performed in the presence of diethylpyrocarbonate (DEP) (Öberg and Philipson, 1971). Later on, it was shown that DEP lowered the melting temperature of RF- and RI-RNA (Ehrenfeld, 1974) suggesting that the absence of RF-RNA (rather than the presence) was an artefact of the extraction procedure. The sensitivity of polymerase activity in isolated replication complexes to RNase also suggests that the only single stranded regions of the isolated replication complex are in nascent chains: The RNA template used for synthesis of viral RNA is unaffected by preincubation with the ssRNA-specific RNase A or RNase T_1, but it is sensitive to dsRNA- specific RNase III (Lundquist and Maizel, 1978 a+b). Electron microscopy of replication complexes indicated a double-stranded template molecule in the replication complex (Meyer *et al.*, 1978). Finally, recent studies on the treatment of cell lysates with RNase prior to RNA extraction support the conclusion that long dsRNA regions in RI- and RF-RNA are not artefacts of the extraction procedure (Koliais, 1981).

Although the issue of single-or double-stranded template in RI RNA is still not settled (even not for the intermediates of RNA phages), the available data on poliovirus RNA replication support the double-stranded model. The following discussion on possible modes of poliovirus RNA replication builds on this concept of a semiconservative displacement for the synthesis of poliovirus RNA (Baltimore, 1969; Levintow, 1974; Agol *et al.*, 1980). Very recently, a detailed electron-microscope analysis was reported which supports the single-stranded backbone model of poliovirus replicative intermediate rather than the double-stranded model portrayed in this chapter (see Fig. 98C, Richards *et al.*, 1984).

III. Time Course and Kinetics of Synthesis of Virus Specific RNAs

The time course of synthesis of viral RNA has been followed in two ways: by determining the rise in titer of infectious RNA isolated from infected cells at different times p. i. (Darnell *et al.*, 1961), and by measuring the incorporation of radioactive uridine into virus specific RNA (Scharff *et al.*, 1963). Within the first 30 min after infection, the recoverable infectivity in viral RNA decreases. This decrease is probably due to degradation of incoming parental viral RNA by cellular RNase and not caused by conversion of parental RNA into RF-RNA. Only a relatively small amount of infectious viral RNA can thus be detected at early times after infection. Synthesis of infectious viral RNA begins between 30 and 60 min p. i. Figure 94 (p. 374) provides an overview of the steps involved in poliovirus RNA replication, and Figure 95 (p. 376) shows that RNA synthesis in the infected cell proceeds in distinct phases with different kinetics.

The relative proportions of the different types of virus specific RNAs synthesized during these phases are illustrated in Table 60 and Figure 99. 80—85 % of all viral RNA synthesized is found in the form of single stranded plus RNA and 5—10 % in minus strand RNA, the latter exclusively as part of RF and RI-RNA. The additional viral RNA is plus strand RNA in RF and RI-RNA. During the exponential phase, most of the ss +RNA made is utilized as mRNA; later, more of the ss +RNA made is encapsidated as vRNA.

Figure 99 shows that the rate of incorporation of labeled nucleotides into RI-RNA is exponential until 2 h p. i., then linear, and highest between 2 and 4 hours after infection. RF-RNA accumulates linearly between 2 3/4 and 4 hours,

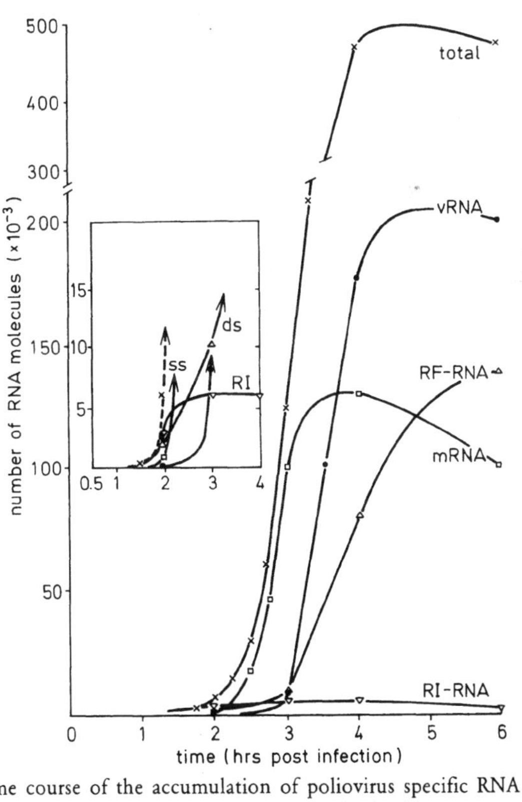

Fig. 99. Time course of the accumulation of poliovirus specific RNA molecules
The figure is compiled from the data in Table 61. Events of the first hours are magnified in the insert, ss indicates the sum of single stranded RNAs synthesized (mRNA + vRNA)

Table 60. *Time pattern of RNA species present in poliovirus infected cells*

Hrs p.i. / Phase	0.5 Exponential	1.0	1.5	2.0 Switch	3.0 Linear	4.0 Declining	6.0
Number of RNA molecules/ (equivalents of viral RNA)	10 (10)	10 (20)	200 (500)	3,000 (6,000)	117,000 (125,000)	420,000 (470,000)	410,000 (480,000)
RF - RNA functional	–	10 (20)	100 (200)	600 (1,200)	500 (1,000)	100 (200)	–
RI - RNA	–	–	100 (300)	1,000 (3,000)	1,500 (6,000)	2,000 (6,000)	1,200 (3,000)
mRNA	10 (10)	–	5 (5)	1,000	100,000	130,000	100,000
vRNA	–	–	–	5	8,000	180,000	200,000
+ RNA endproducts	–	–	–	50	2,000	70,000	40,000
RF - RNA endproducts	–	–	–	300 (600)	5,000 (10,000)	40,000 (80,000)	70,000 (140,000)
% RNA in ds	0	100	40	33	10	17	29
RF functional		100	100	66	10	1	0
RF endproduct		0	0	33	90	100	100
% RNA in RI	0	0	60	50	5	1	1
% RNA in ss	100	0	1	17	85	82	70
mRNA	100		100	95	91	33	29
vRNA	0		0	0	7	48	59
+ RNA	0		0	5	2	19	12

Compiled from data from Darnell *et al.*, 1961; Noble and Levintow, 1970; McDonnel and Levintow, 1970; Hecht and Summers, 1970; Oppermann and Koch, 1973; Mitchell and Tershak, 1973; Hewlett *et al.*, 1977; Lundquist and Maizel, 1978.

single stranded RNA accumulates exponentially until 3 1/4 hour p. i. and then linearly until 4 hours p. i. Late in infection from 4—5 hours p. i. there is a steep increase in the accumulation of RF-RNA. The phases of RNA synthesis are correlated to other events of poliovirus replication in Table 61.

A. The Onset of RNA Synthesis

The quantities of RNA synthesized during the early exponential phase are low, and hence RNA replication is difficult to study in this phase. But we can make some inferences about the beginning of RNA synthesis. In addition to serving as mRNA for the translation of the polyprotein, the viral RNA of an infecting picornavirus particle has to function as a template for RNA replication. The gene coding for the replicase protein lies at the 3' end of the viral RNA, and its translation is completed only 10 minutes after the attachment of the first ribosome to the 5' end. In this time interval, up to 35 ribosomes can consecutively attach to the 5' end of the viral mRNA. Several rounds of translation of the infecting viral RNA molecule are likely to occur before viral replication commences.

The switch from translation to transcription is faced with a dilemma. The ribosomes continually move down along the RNA from the 5' end. Any replicase daring an attempt to copy a polio RNA molecule involved in translation, will be confronted with considerable head-on traffic. One way to overcome this dilemma would be the ability of the replicase to "knock off" the ribosomes from the parental RNA as it travels down the RNA. This is not a desirable mechanism, since it would interfere with translation as replicase molecules accumulate. It would require some additional mechanism to prevent the replicase from binding to mRNA during the peak of viral protein synthesis. Therefore, it is more likely that the RNA-ribosome association is temporarily blocked. One possibility is that the initiation of protein synthesis is prevented by modification of the 5'end of the RNA or by somehow tying up ribosomes before or shortly after binding to RNA, another possibility is interference with polypeptide chain elongation, causing premature termination of translation and permitting the replicase to initiate 3'—5' transcription of the RNA. Whatever the mechanism, it is likely that this step is triggered by one of the viral products produced during the first rounds of replication. Among the possibilities that have been proposed are, binding of capsid proteins to ribosomes or the 5'end of the RNA (equestron hypothesis, Cooper *et al.*, 1973), stabilization of secondary structures in the 5' untranslated region of the viral RNA (Kitamura *et al.*, 1981), circularization of the RNA by proteins bound to the 3' terminal end with affinity also for the 5' end (this may be analogous to a circularization mechanism proposed to occur during the translation of cellular mRNA). It is also conceivable that the replicase protein, after attaching to the 3' end of the RNA, releases ribosomes from the RNA in concerted action with other proteins.

In analogy to the RNA phage system, it has been proposed that poliovirus replicase proteins have binding affinities for essential elongation factors. Binding of eEF-X by accumulating replicase, for example, may temporarily inhibit elongation

Table 61. *Phases of RNA synthesis in poliovirus infected cells—relation to other events in poliovirus replication*

	Early phase	Late phase	End of replication
Time (min)	30–120/150	150–240	240 ff.
Rate of RNA synthesis	exponential	*peak* constant (linear)	decreasing
Virion formation	none	at a constant rate	few–none
Types of RNA made	cRNA (in RF) → *RI* → mRNA → cRNA (RF) → *RI* → ···	RI → *vRNA*, (mRNA)	*RF*, (vRNA)
Localization of RNA synthesis	cytoplasmic–soluble?	membrane vesicles	diffuse, fading
Aim of RNA synthesis	production of cRNA (RI) and mRNA	production of vRNA (coupled to assembly)	–
Viral protein synthesis	increasing *peak*	levelling off	very little
Membrane formation	begins	*peak* (virus specific)	many autophagic vesicles
Lysosomal enzyme redistribution/ cytopathic lesions	none	none beginning	major event
Release of progeny	none	beginning *peak*	continuing
Alterations of host cell nucleus	beginning *peak*	continuing	degenerative changes
Sensitivity of RNA synthesis to inhibition by			
cycloheximide	++++	+ (after 30 min)	–
phleomycin	++++	–	–
guanidine	(+)	++++	(+)
Sensitivity of DNA shut off to inhibition by guanidine	++++	–	–

of translation and cause its premature termination (Agol, 1980). A different way to prevent elongation would be to tie up the docking protein required for the cotranslational insertion of membrane proteins into the membrane of the endoplasmic reticulum. It should be kept in mind that all of these mechanisms may also be localized phenomena, effective only within the close vicinity of the replication center initated by the infecting RNA.

Several lines of observations indicate that RNA replication in the ensuing early intermediate phase is different and unusual compared to that in later phases. P-2 and P-3 proteins are detectable first in infected cells, capsid proteins are conspicuously absent (Koch *et al.*, 1982) (see Chapter 8). The quantities of NCVPX and NCVP1b produced early in infection can be enhanced by incubation of cells in hypertonic medium (Koch *et al.*, unpublished). Defective interfering particles seem to originate during early stages of viral replication, and at high multiplicities of infection the yield of DI particles can be increased substantially (Cole and Baltimore, 1973). Recombinational events occur preferentially during the early phase of RNA synthesis (Cooper, 1972). And lastly, the specialized virus induced membranes, sites for RNA synthesis during the linear phase of replication, have not yet started to accumulate (Caliguiri and Tamm, 1970 a and b). In vitro studies have suggested that soluble forms of the viral replication complex do not efficiently synthesize full length copies of plus strand RNA, but instead produce defective RNAs reminiscent of those observed in DI particles, *i.e.* with deletions in the 5' terminal half of the RNA (Etchison and Ehrenfeld *et al.*, 1981). It is tempting to speculate on the basis of these observations that early in vivo viral RNA synthesis by a soluble replication complex is also deficient, producing mainly defective RNA during this phase of RNA replication. The defective RNA acting as mRNA would direct the synthesis of P-2 and P-3 proteins, *i.e.* replicase proteins and X proteins. Defective capsid proteins are presumably degraded. It should be possible to determine the types of RNA made early during RNA replication with probes of radioactive labeled poliovirus cDNA. Hypertonic medium would stabilize secondary structures in RNA and may enhance mistakes made by the replication complex. The accumulating P-2 and P-3 proteins may be required for the induction of the specialized membrane vesicles.

B. The Exponential Phase: cRNA → mRNA → cRNA ...

RNA synthesis, once initiated, proceeds exponentially at first. (Table 61). This implies that a significant fraction of the freshly synthesized RNA is utilized again as template for further RNA synthesis. The cRNA product of parental vRNA transcription serves as template for the synthesis of +RNA of the parental type, some of this RNA in turn, as template for the synthesis of more cRNA and so on. The precise structure of the templates involved in these early steps, again can only be inferred. A number of observations, however, support the model presented in Figure 95 (p. 375). 1. RF and RI are the major RNA species labeled at the earliest times, labelling of single stranded RNA occurs only shortly thereafter (Fig. 99, p. 387). 2. As discussed in section II above, neither single stranded cRNA nor negative RI have ever been detected in infected cells. 3. Viral RNA (+RNA) has been successfully transcribed by the soluble poliovirus replicase *in vitro,* the products is always RF-RNA. 4. Inhibition of RNA synthesis early in infection leads to the accumulation of label in RF-RNA; upon reversal of guanidine inhibition, the label in RF can be chased into RI-RNA and ss +RNA.

It seems that initiation of cRNA synthesis on the single-stranded +RNA template occurs only once and that reinitiation of cRNA synthesis on the now double stranded homopolymer (oligoA/oligoU) tail of RF or RI is not possible or very inefficient (single run hypothesis, Pérez-Bercoff, 1979). Initiation of +RNA synthesis on the opposite, heteropolymer tail of RF or RI, in contrast, appears to be quite efficient and leads to the formation of plus RI. The singlestranded +RNA product—at least initially—reparticipates in RNA replication (as template for cRNA/RF-RNA synthesis) for several turns of exponential RNA replication.

Soon after RNA replication has commenced part of the ss +RNA product leaves the replication complex to function as mRNA. The detection of RI-RNA in tight association with ribosomes during the early phase of viral RNA replication suggests that ribosomes may be able to bind to the 5' end of nascent RNA during ongoing RNA synthesis, at least early after infection. Only a thousand (or so) minus strand RNAs must be synthesized in order to provide sufficient templates for maximal RNA synthesis (see below). One hundred thousand viral RNAs, in contrast, serve as mRNA in polysomes and 2–4 x 10^5 viral RNAs are destined for genomic RNA in progeny viruses. None of the ss +RNA formed during the exponential phase is destined for virion formation, which begins only an hour or so later. Early ^{3}H labelled ss RNA thus is mainly found in polyribosomes (Baltimore *et al.*, 1966).

C. The Linear Phase: cRNA → vRNA, mRNA

When approximately 10–25 % of the total viral RNA has been produced—usually at about 3 h p.i.— an abrupt change in the kinetics of RNA synthesis is observed: from here on viral RNA continues to accumulate in the cell at a linear rate for about one hour. The time point for this change is dependent on the multiplicity of infection, the nutritional state of the cell, and is also influenced by stages in the cell cycle (Eremenko *et al.*, 1972 a, b). The curve of the kinetics of RNA synthesis may thus be shifted ahead or backward in time (see Fig. 95, p. 376). The pattern of kinetics, however, stays the same (Baltimore *et al.*, 1966). The time point of the switch from exponential to linear rates of RNA synthesis correlates with the peak of viral protein synthesis, the beginning of the formation of virus specific vesicles (see Table 61, p. 390), and the onset of virion formation.

At the time of maximal RNA synthesis (between 3 and 4 hours p. i.), RNA accumulates linearly, and the synthesis of a full length RNA molecule takes about one minute (Darnell *et al.*, 1967) or 45 seconds (Baltimore, 1969). At this time viral RNA accumulates in the cell at a rate of 3,000 (Baltimore *et al.*, 1966) to 6,000 (Oppermann and Koch, 1974) molecules per minute, that is 1.8 to 3.6 x 10^5 molecules/hour. The major portion of the plus strand RNA which is formed at this time only slightly preceeds the course of the appearance of infectious virus (Darnell *et al.*, 1961; Oppermann and Koch, 1973). Most of the viral RNA synthesized during the linear phase is destined to be incorporated into virions. The assembly process appears to begin with the nascent RNA, a conclusion which is supported by the fact that labelled RNA can be detected in mature virions within less than 5 minutes of its synthesis (Baltimore *et al.*, 1966).

The fate of radioactive precursors can be followed relatively easily during this phase of RNA replication. In short pulses, labelled nucleosides are most rapidly incorporated into RI-RNA, and the peak in the accumulation of RI molecules in infected cells is reached before the peak in single-stranded RNA as well as RF-RNA (Noble and Levintow, 1970). Longer labelling periods result in relatively higher portions of label in ss +RNA, late in infection also in dsRNA. A precursor product relationship between one RNA species and another is difficult to demonstrate in intact cells, since short pulses of radioactive label cannot be chased because the pools of intracellular nucleotides do not equilibrate fast enough. Such experiments, however, can be performed in vitro with crude cytoplasmic extracts from the infected cells and confirm the conclusion that RI is the functional intermediate in the synthesis of single-stranded viral RNA (Girard, 1969). Figure 100 shows that label is first incorporated into RI and later accumulates in ssRNA. About 90 % of the incorporated label from ^{3}H-uridine triphosphate is found in RI-RNA after 5 min at 37° C. Already after a 6 min chase, half of the label originally present in RI is now found in single-stranded viral RNA, and after a 20

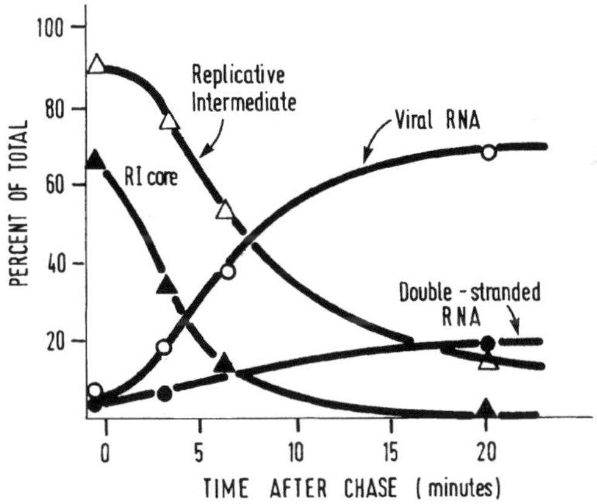

Fig. 100. The kinetics of RNA formation by poliovirus replication complex *in vitro*
The figure illustrates the repartitioning of label between the various species of poliovirus RNA during an *in vitro* pulse chase experiment. Crude RNA polymerase was incubated in a complete assay system with ^{3}H UTP. After 15 min of incubation, a 125 fold excess of unlabeled UTP was added to the reaction mixture and the samples chased for the times indicated. The samples were withdrawn and immediately deproteinized with SDS and precipitated with 2 M LiCl, and analyzed by sucrose density gradient centrifugation, exclusion chromatography. LiCl-soluble material (dsRNA) was precipitated with ethyl alcohol and analyzed by sucrose density gradient centrifugation. Radioactivity in 35 S viral RNA (○), double-stranded 18 S RNA (●), and ribonuclease resistant portion of replicative intermediate (▲) are plotted as percent of total in function of time after chase. Label is first incorporated into RI-RNA, and is chased into ssRNA and dsRNA. (Δ) total replicative intermediate. — Figure from M. Girard, 1969 [J. Virol. *3*, 382 (1969)]

min chase the RI-RNA contains only 20 % of the label. RF-RNA also accumulates as a final end product from the last round of RI transcription, when there is no reinitiation.

During a poliovirus replication cycle in HeLa cells, at least 2.4 x 10^7 potential polymerase molecules are synthesized (2 x 10^5 virus particles/cell each containing 60 copies of the four capsid proteins corresponding to 50 % of total viral protein synthesis). An infected cell makes only 4-6 x 10^5 copies of viral RNA (Hewlett *et al.*, 1977). It has been calculated that only 3,000 to 12,000 active polymerase molecules are present in the replication complexes isolated from infected cells (Lundquist *et al.*, 1974), that is only 0.1 % of potential replicases are employed in RNA synthesis. RI-RNA molecules contain 5 (to 10) replication sites, so only 600 to 2400 RI-RNA molecules are required for RNA synthesis. The time required for the synthesis of one full length RNA strand is less than one minute. Theoretically up to 12,000 RNA molecules could accumulate per minute under optimal conditions. This is near the observed average rate of 3,000–6,000 RNA molecules/min. The cRNA template of RI-RNA is replaced at a slow yet constant rate throughout the period of linear RNA synthesis (Baltimore, 1969).

There are several ways to explain the observed switch in the kinetics from exponential to linear. Withdrawal of template, decrease in replicase activity, and increased degradation of progeny RNA could all lead to an altered kinetics of RNA accumulation (Baltimore, 1969; Levintow, 1974). It is most likely that the altered kinetics reflect the withdrawal of template. Freshly synthesized RNA is used as mRNA; later on RNA is withdrawn from the replication pool by encapsidation into progeny virions. Compartmentalization may also play a role in the removal of template. In addition, the ions which start to accumulate in this phase may affect the state of the replicase activities.

If new template cRNA is furnished at a rate comparable to inactivation of template RNA—the constant level of RI RNA is indicative of such a process—then the steepnees of the linear rate of RNA synthesis must be determined by the activity of the replication complexes. This rate in turn may depend on the availability of modified membrane space, on the coupling to assembly, on the availability and fidelity of the proteins required for initiation of RNA synthesis (P-2, P-3, host factors), and on the availability and activity of the viral protease. The rate of viral RNA synthesis *in vivo* correlates with the level of viral RNA polymerase activity extractable from the infected cell, although some exceptions have been reported (Roeder and Koschel, 1974). However, the level of polymerase activity does not correlate with the total intracellular level of NCVP4b, the replicase responsible for elongation. As mentioned before, only 0.1% of the total potential viral replicase is present in a functional state in infected cells. The active replicase molecules in the membrane bound replication complexes seem to be the same proteins during most of the linear phase of RNA replication. They are proteins synthesized early during virus replication and do not seem to be replaced by the additionally accumulating replicase proteins (Lundquist *et al.*, 1974). The template for the production of viral RNA, on the other hand, is replaced at a slow but constant rate (Baltimore, 1969). After 10-20 rounds of initiation, the template RI seems to be inactivated and is either converted to RF-RNA or degraded by cellular enzymes.

D. Cessation of RNA Synthesis

The linear phase of RNA synthesis ceases late in infection (usually between 4 and 6 h.p.i.) when a rather constant number of RNA molecules (4–6 x 10^5) has been synthesized per infected cell (Levintow, 1974; Hewlett *et al.*, 1977). At the same time the altered permeability of the cell membrane allows nucleotides (ATP) to leave the cell (Egberts *et al.*, 1977) and thereby depletes the cell of essential precursors for RNA synthesis. Cessation of RNA synthesis could also be a simple consequence of the inhibition of protein synthesis.

Rapid cessation of viral RNA synthesis can occur at any time in the replication cycle by elevation of the temperature. In this instance nucleolytic degradation of viral RNA is observed. Activated (or lysosome released) enzymes of the host cell appear to be responsible (Killington *et al.*, 1977). At suboptimal temperatures (32-34°C) larger amounts of picornavirus specific RNA are synthesized in infected cells. Whether or not cellular RNases are more restricted (compartmentalized?) at suboptimal temperatures has not been studied. It is conceivable that other events regulating the quantity of viral RNA synthesis, such as an increased membrane permeability with loss of ATP, are temperature sensitive.

IV. The Sites of RNA Synthesis

As disscussed in Chapter 6 (see Figs. 66, 67, p. 232–234), formation of virus specific vesicles and compartmentalization in the infected cell cytoplasm play important roles in poliovirus RNA replication (Caliguiri and Tamm, 1970; Bienz *et al.*, 1980, 1983). From the onset of vesicle formation, RNA synthesis seems to be intimately coupled to smooth membranes (Table 62). Biochemical analyses showed that the vast amount of RI and RF labeled during the linear phase of RNA synthesis are associated to the virus induced smooth membrane fraction (fraction 2 of Mosser *et al.*, 1972a). Most of the replicase activity is found in the same fraction (Caliguiri and Compans, 1973; Yin, 1977).

Electron microscopic observations of autoradiographs from infected cells that had been labeled with radioactive nucleotide-precursors, also showed that RNA synthesis takes place almost exclusively on virus specific membranes once these are formed (Bienz *et al.*, 1980). Electron microscopic examinations of membrane fraction 2 (crude RC) have revealed the peculiar, sticky, fuzyy vesicular structures typical of poliovirus induced vesicles (Caliguiri and Tamm, 1971; Bienz *et al.*, 1983) (see also Figs. 71, 72, p. 242–243).

Synthesis of RNA on virus induced membrane enclosed vacuoles is a common process in picornavirus infected cells: FMDV and mengovirus RNA synthesis also occurs in close association to membrane enclosed vesicles (Polatnick and Wool, 1982). A superinfecting picornavirus is capable of utilizing vesicles induced by prior infection with a nonreplicating poliovirus, the normal eclipse phase of mengovirus replication thus being shortened by three hours

Table 62. *Distribution of virus specific RNAs in membrane fractions of poliovirus infected cells*

Fraction	RNA	Viral protein	Cellular location	E.M. appearance	Density (g/cm^3)
2	RI + RF	replicase	central	s.m.v.	1.12
3	RI + RF + SS	(capsid proteins)[1]		s.m.v.	1.18
4			not defined clearly	s.m.v. + r.m.v.	1.21
5	SS	all viral proteins		r.m.v.	1.25
6	SS	all viral proteins	periphery	r.m.v.	1.27
7	SS	mature virions			1.31

s.m.v. = smooth membrane vesicles.
r. m.v. = rough membrane vesicles.
[1] The capsid proteins are transiently associated to this fraction from 10–30 min after their synthesis.

Data from Caligiuri and Tamm, 1970 b.

(Zeichhardt *et al.*, 1982). Most viral RNA synthesis during the linear phase proceeds in conjunction with the newly formed virus specific vesicles, and all virion RNA destined for encapsidation is formed on these membranes.

In contrast, localization of RNA synthesis during the early exponential phase of RNA synthesis is more difficult. Microscopically detectable virus specific vesicles and biochemically detectable alterations of membranes in fraction 2, first occur near the midpoint of the replication cycle and thus correlate with the switch from exponential to linear RNA replication. Of course, it is possible that some virus specific membranes are formed already during the early exponential phase but are too low in quantity to be detected. The quantity of RNA synthesized during the exponential phase is also very low, and few·membranes might be sufficient. The possibility that exponential RNA replication occurs on a site different from linear phase RNA replication, however, deserves consideration. Replicase activity has been isolated from the soluble cytoplasmic fraction of infected cells, and it is possible that early RNA synthesis occurs in the soluble phase of the cytoplasm. RI-RNA of early RNA synthesis has been found in association with ribosomes, and soluble replicase has been found to be capable of synthesizing cRNA (RF-RNA) from +RNA in vitro (see below). This raises the interesting hypothesis that the formation of the vesicles is a prerequisite for the switch from exponential mRNA–cRNA synthesis to linear vRNA synthesis, and that the switch is coupled to formation of the vesicles.

It is further postulated that encapsidation of virion RNA by subunits of capsid proteins occurs only after binding of RI template to virus specific vesicles. In this way, vRNA synthesis may be separated spatially from the site of mRNA synthesis. The small genome protein VPg is automatically attached to the 5' end of nascent RNA chains in the replication complex. Failure of RNA to be incorporated into capsids may automatically result in the release of single stranded plus RNA into the soluble phase of the cytoplasm and render the VPg RNA linkage susceptible to cleavage by the specific cleavage enzyme which is abundant in the cytoplasm of HeLa cells, thereby giving rise to poliovirus mRNA. Less likely, presence or absence of VPg may determine the fate of the freshly synthesized RNA by influencing transport mechanisms of the RNA either to ribosomes or to assembly sites (see Fig. 78, p. 261).

V. The Viral RNA Polymerases

All cellular RNA synthesis proceeds on a DNA template, and can be inhibited by the antibiotic actinomycin. In contrast, viral RNA synthesis, occuring exclusively on RNA templates, is not inhibited by actinomycin (Reich *et al.*, 1961; Shatkin, 1962). On this basis, cellular and viral RNA synthesis can be conveniently separated.

Normal cells do not employ RNA dependent RNA polymerases. It was therefore assumed that the RNA polymerase(s) responsible for replication of picornavirus RNA is virus induced. Support for this view was first obtained in studies with

Table 63. *Poliovirus RNA replication complexes*

	Crude RC[a]	Large RC[b]	Small RC[c]	Soluble RC[d]	Replicase enzyme[a]
Isolation procedure	crude cytoplasmic extract, S10 (smooth membrane fraction 2)	from crude RC with 0.5 % DOC	from crude RC with 0.5 % DOC	from small RC with detergents	S30 (soluble fraction of extract); or from crude RC with SDS + DOC
Sedimentation in sucrose gradients	400S	100–300S	20–70S	20–26S	4S
Composition					
membranes	yes	some	no	no	no
template	RI	RI (8 % cRNA)	RI (60 % cRNA)	RI	none
viral proteins	procapsids, P-2 and P-3 proteins	several	few	NCVP 2 and 4	NCVP 4
Host proteins	several	several	few	none	none
In vitro products					
pulse	RI	RI	RI	elongates endogenous template	short cRNA fragments
chase	70 % ss (+) RNA, 20 % RF	50 % ss (+) RNA, 50 % RF	95 % RF, 5 % ss (+ DI) RNA	100 % RF	complete cRNA with primer (oligo U or host factor)
Dependence on endogeneous template	+	+	+	+	–
In vitro initiation	yes*	no	no	no	yes*
Sensitivity to					
guanidine	yes	yes	no	no	no
ds specific RNase	no		yes		
ss specific RNase	no		no	yes	yes

* blocked by antibody against host factor (Baron and Baltimore, 1982; Dasgupta *et al.*, 1982; Morrow and Dasgupta, 1983).

[a] Baltimore, 1964; Girard *et al.*, 1967; Girard, 1969; Caliguiri and Tamm, 1970; Caliguiri and Compans, 1973; Yin, 1977.

[b, c] Caliguiri and Tamm, 1970; Ehrenfeld *et al.*, 1970; Caliguiri and Mosser, 1971; Caliguiri and Compans, 1973; Caliguiri, 1974; Lundquist *et al.*, 1974; Butterworth *et al.*, 1976; Wright and Cooper, 1976; Meyer *et al.*, 1978; Etchison and Ehrenfeld, 1980, 1981.

[d, e] Mitchell and Tershak, 1973; Flanegan and Baltimore, 1977, 1979; Dasgupta *et al.*, 1979, 1980; van Dyke and Flanegan, 1980; van Dyke *et al.*, 1982; Tuschall *et al.*, 1982.

cells infected with mengovirus (Baltimore and Franklin, 1962) and poliovirus (Baltimore *et al.*, 1963). The first viral RNA polymerase preparations isolated from infected cells contained membrane associated enzyme activity, endogenous template, and several host and viral proteins including coat proteins. These ribonucleoprotein enzyme preparations were named replication complex (Girard *et al.*, 1967).

For some time, viral RNA was considered to be replicated by two distinct replicases (Penman *et al.*, 1970), one responsible for cRNA synthesis, the other for plus strand synthesis. Support for the two enzyme hypothesis came from genetic studies (e.g. Cooper, 1970): detailed studies were carried out with poliovirus temperature sensitive mutants which were defective in the synthesis of RNA at the restrictive temperature. The data indicated that the poliovirus genome codes for two replicase activities, one coded for by the 3' terminal end of the genome seemed to be responsible for the synthesis of doublestranded RNA; the other, encoded by more central parts of the genome, for the synthesis of singlestranded RNA. In accord with the two enzyme hypothesis was the ability of some drugs to preferentially inhibit the synthesis of singlestranded RNA (Dmitrieva and Agol, 1974; Tonew and Fahlbusch, 1977). The evidence available today suggests that two different states of the same polymerase molecule–modified by different associated factors–synthesize the RNA in both steps of replication.

Of particular interest are the host factors that may be involved in poliovirus RNA replication. The replicase of the small RNA phage of E. coli, Qß, provides an example of a host cell factor modified RNA polymerase (Kamen, 1975; Blumenthal and Carmichael, 1979). The Qß replicase consists of four subunits, one of which is virus coded, the others are host cell proteins. Two of these are elongation factors, the third is a component of the small ribosomal subunit. It is conceivable that poliovirus utilizes proteins of its host cells in a similar manner (Baron and Baltimore, 1982d, e; Dasgupta, 1983), although the identity of the host factors involved is still not known to date.

Over the years, different kinds of viral RNA polymerase preparations have been isolated from picornavirus infected cells. These have been characterized and distinguished on the basis of size, template dependence, composition, and activity. Table 63 provides an overview of the different replication complexes and replicases that have been described for poliovirus.

A. The Crude Replication Complex: Synthesis of Plus Strand RNA

Most polymerase preparations have been harvested from cells obtained at a time of maximal RNA synthesis. At these times most viral RNA synthesis proceeds in tight association with smooth cytoplasmic membranes (Penman *et al.*, 1964), and most of the singlestranded RNA product is rapidly encapsidated.

Cytoplasmic extracts prepared by Dounce homogenization contain a crude replication complex which sediments at 10,000 g (S10) and consists of smooth membrane vesicles with very low density (1.12 g/cm^3) corresponding to fraction 2 of the Caliguiri and Tamm (1970a) procedure (see Table 62, p. 396). This crude membrane associated replicase activity incorporates radioactive nucleotides into 35S single stranded RNA with plus polarity. In most such S10 preparations initiation of RNA synthesis does not occur, although some in vitro initiation has been achieved with careful preparation procedures (Tershak, 1982). All nascent RNA chains are completed and released by the crude RC. At the end of the reaction most remaining membrane associated RNA is found in an RNase resistant state (Girard, 1969). The radioactive precursors incorporated in vivo into RNA by the replication complex (RC) can be liberated from the complex by exposure to deoxycholate. RI-RNA is the predominant RNA species liberated (Girard *et al.*, 1967).

Several viral proteins—mainly gene products of the replicase gene and proteins coded for by the center of the genome (NCVPX)—and perhaps also some crucial host proteins are tightly associated with the crude viral replication complex (Caliguiri and Mosser, 1971; Caliguiri and Compans, 1973; Butterworth *et al.*, 1976; Wright and Cooper, 1976; Yin, 1977; Bienz *et al.*, 1983; Takegami *et al.*, 1983a). Virion-like particles and procapsids have been detected in the membrane fraction by electron microscopy and biochemical studies (Caliguiri and Compans, 1973; Yin, 1977), suggesting that the RNA is encapsidated as it is being synthesized on the membrane bound replication complex. This notion is supported by the finding that antiserum against the 14S assembly intermediates inhibits RNA synthesis in an *in vitro* system (Ghendon *et al.*, 1973). The membrane fraction is enriched in a protein fraction comigrating with VP_1 (Caliguiri and Mosser, 1971); the corresponding protein has been identified as NCVPX (Butterworth *et al.*, 1976; Yin, 1977; Takegami *et al.*, 1983a) or as NCVP6a (the alternative cleavage pathway precursor to the protease NCVP7c) (Bienz *et al.*, 1983). In addition, the membrane fraction contains relatively high amounts of the viral proteins P2-5b, P3-9, P3-4b, and P3-7c (Caliguiri and Compans, 1973; Yin, 1977; Bienz *et al.*, 1983). Ribosomes are not found in association with the replication complex, in the smooth cytoplasmic membrane fraction at the time of maximal viral RNA synthesis (Penman *et al.*, 1964; Girard *et al.*, 1967). Interestingly, the elongation activity during the peak of RNA synthesis seems to reside in a small number of NCVP4 molecules which have been synthesized during the early phases of the replication cycle (Lundquist and Maizel, 1978 a+b). A constant reproduction of RNA polymerase molecules thus does not seem to be essential for RNA elongation. For each newly initiated RNA molecule, however, synthesis of a polyprotein is required for the supply of one VPg molecule. The concerted action of host proteins, viral P-2 proteins (NCVP8, -10, -X), the membrane bound precursor to VPg—NCVP9, and one of the proteins with replicase activity (NCVP1b, - 2, -4, 6a), may be required for the initiation of plus strand synthesis by the crude replication complex.

Disruption of the membranes present in the crude RC by treatment with detergents markedly reduces the polymerase activity and enhances the proportion of double stranded RNA product. Reconstitution with membranes does not restore the lost activity or the synthesis of single stranded RNA (Caliguiri, 1974). The solubilized replication complex, as well as the predominant protein NCVPX bind spontaneously to synthetic lipid vesicles, indicating an affinity for hydrophobic components (Butterworth *et al.*, 1976). P3-9 and P3-7c, as well as a protein named NCVP3 (P2-3b?) are enriched in the lipoproteins released from crude RC by low concentrations of deoxycholate (Caliguiri and Mosser, 1971; Bienz *et al.*, 1983), indicating that these are integral membrane proteins.

Treatment of the S10 fraction with detergents liberates two distinct replication complexes sedimenting with 100–300S and about 70S. Procapsid proteins, NCVPX (or P3-6a), NCVP4, as well as a few minor proteins and some membrane components remain associated to the large replication complex (Caliguiri and Compans, 1973; Wright and Cooper, 1976; Yin, 1977; Bienz *et al.*, 1983). The large, fast sedimenting complex has retained most of the properties of the S10 preparation. It contains most of the acid precipitable label after short pulses with radioactive nucleosides. Earlier studies revealed that the two types of picornaviral RCs synthesize either preferentially single-stranded or double-stranded viral RNA species (Arlinghaus and Polatnick, 1969; Caliguiri, 1974). It was suggested that the smaller 70S RC which synthesizes mainly RF-RNA is engaged *in vivo* in the synthesis of – strand RNA while the larger RC, synthesizing single-stranded RNA, is the site of + strand RNA synthesis. However, more recent studies indicate that radioactive precursors are preferentially or exclusively incorporated into plus strand RNA by all membrane attached high molecular weight replication complexes *in vitro.*

Recently, the activities of the small replication complex derived from the S10 membrane fraction by a variety of detergent treatments have been reevaluated in some detail (Ehrenfeld *et al.*, 1978; Etchison and Ehrenfeld, 1980, 1981). It consists mainly of an intact minus strand RNA, a partially plus strand RNA and a tightly bound replicase which is able to elongate the plus strand RNA to completion. Neither initiation nor release of newly synthesized RNA does occur *in vitro.* The RNA synthesized on these 70S complexes is mainly plus strand RNA with a deletion in the coat-protein region comparable but not identical to the deletion in DI particles. These observations may be of interest to early RNA synthesis *in vivo,* before any of the virus specific smooth membrane vesicles have been formed: The first newly synthesized plus strands *in vivo* may carry a deletion in the coat protein region (see p. 356 ff., Chapter 8). The *in vitro* studies indicate that the deletion is due to a soluble (not membrane bound) state of the viral replicase. This could explain a number of puzzling observations early in infection, where RNA synthesis may occur by a soluble enzyme (see Section III, p. 391, above).

Virus specific polymerase molecules analogous to poliovirus NCVP4 have been described also in association with 70S and 250S replication complexes in other picornavirus infected cells (Loesch and Arlinghaus, 1975; Traub *et al.*, 1976; Polatnick and Wool, 1981). The membrane bound replication complex of foot and mouth disease virus also can be separated into two complexes exhibiting sedimentation rates of 20-70S and 100-300S, respectively (Arlinghaus and Polatnick, 1969). The 70S FMDV replication complex was further characterized to contain four host coded proteins in addition to viral protein p56 (corresponding to polio NCVP4). The five proteins of the replication complexes had the following molecular weights (in kd): 50, 56 (viral), 60, 70, and 74; they occurred in the replication complex with molar ratios of 1:2:2:1:1, respectively (Polatnick and Wool, 1981).

Electron microscopic examinations suggested a cartwheel shape for the replication complex, which dissociated upon addition of poly(A), similar to the viral/host factor replication complex of phage Qß. In addition to heterogeneous virus-induced RNAs, the 70S FMDV replication complex contained small cellular RNAs of 4.5 and 5 S, the latter presumably of ribosomal origin (Polatnick and Wool, 1981). Antibody previously shown to inhibit polymerase activity (Polatnick *et al.*, 1967), aggregated the cartwheel like complex.

The EMC-RNA replication complex is bound to smooth membranes in such a way that—after isolation—the proteins are protected against protease and the enzyme activity against nuclease, suggesting that the RC may be enclosed in membrane vesicles (Rosenberg *et al.*, 1979). Again, replicase proteins and the protein corresponding to NCVPX are present. RNA-associated enzyme activity of EMC virus can be separated from the membrane lipids by exposure to 0.05 % Triton X-100 and centrifugation through sucrose gradients containing also 0.05 % Triton X-100. The most prominent ^{35}S labeled virus specific peak before Triton treatment is the p38 (corresponding to poliovirus protein X) followed by p56 and p23. After Triton treatment the RC contains in decreasing amounts p75, p56, p65 and only trace amounts of p38 (Traub *et al.*, 1976; Rosenberg *et al.*, 1979). All of these observations suggest that viral protein NCVPX and undefined host cell proteins are important factors for viral RNA replication.

B. Soluble Replicase(s): Synthesis of Minus Strand RNA

Attempts to purify the proteins responsible for viral RNA synthesis have been carried out for several years (Ehrenfeld *et al.*, 1970; Mitchell and Tershak, 1973; Lundquist *et al.*, 1974; Flanegan and Baltimore, 1977, 1979). Further exposure of the 70S replicase to detergent yields a 25S replicase with retained replicase properties. The 25S replicase freed from its endogeneous template yields a 7S and a 4S template dependent replicase. The 7S replicase contains NCVP2 and -4 and the 4S only NCVP4 (see Table 63, p. 398). Only recently enzyme preparations were obtained which can initiate viral RNA synthesis in vitro. A soluble enzyme fraction was obtained from infected cells, which could synthesize cRNA in vitro on a single stranded +RNA template (Dasgupta *et al.*, 1979; Van Dyke and Flanegan, 1980).

Replicase activity from poliovirus infected cells was detected and purified using synthetic poly (A) oligo (U) as template-primer (Flanegan and Baltimore, 1977). The solubilized and partially purified enzyme showed preference for poliovirus RNA (plus strand) as template over other RNAs. Among homopolymers only poly (A) oligo (U) was copied. Replicase and poly (U) polymerase co-purified through several steps. The soluble replicase contained mainly the viral proteins NCVP2 and p63 (probably equivalent to NCVP4 or NCVP6b)(Flanegan and Baltimore, 1979).

Further purification of the replicase on poly (U) agarose resulted in a complete loss of replicase activity, but the poly (U) polymerase activity was retained (Dasgupta *et al.*, 1980). Only virus specific protein p 63 was present in the purified preparation of soluble polymerase (Flanegan and Van Dyke, 1979; Van Dyke and

Flanegan, 1980). The activity of the purified inactive replicase could be restored by addition of a host protein factor isolated from a 0.5 M KCl wash of rabbit reticulocyte ribosomes. The restoring activity was purified 50fold and was shown to be correlated with a 72,000 dalton protein. A comparable host factor was obtained from a ribosomal salt wash from uninfected HeLa cells (Baron and Baltimore, 1982; Dasgupta, 1983a, b). The factor was identified as a single 67,000 dalton protein and was shown to physically interact with virus specific polymerase. This protein is not part of or identical with any of the known eukaryotic initiation factors. The activity of the host factor could be mimicked by addition of oligo (U). This also restores the activity of the inactive replicase, but only to the extent of poly (U) synthesis, whereas the host factor allows the transcription of the complete viral RNA. Oligo(U) was maximally active at a ratio of 20-40 moles oligo (U) to one mole poliovirus RNA and was inhibitory at higher concentrations. The replicase-host factor combination showed an even higher selectivity for poliovirus RNA as a template than the partially purified replicase. Poly (A) on the 3'end is required since its removal by polynucleotide phosphorylase destroys the template activity of the RNA.

A partially purified host factor preparation, containing the 60 and 80kd proteins as well as a few minor components, was used to inoculate animals for the production of anti-host factor antibodies (Dasgupta *et al.*, 1982). The anti-host factor antibodies completely inhibited initiation of cRNA synthesis by the soluble replicase (viral polymerase plus host factor), but interfered neither with elongation of already initiated cRNA synthesis by the soluble replicase, nor with elongation on the endogenous template by the membrane derived replication complex (Dasgupta *et al.*, 1982). This indicates that the host factor associated with the soluble replicase is required for the initiation of cRNA synthesis. Whether identical or different factors are required also for initiation of plus strand RNA synthesis on the dsRNA template is not yet known. It will be of interest to determine whether the same host factors are present in the membrane bound replication complex in which RNA synthesis can be initiated in vitro. Antibodies against synthetic peptides of VPg were also shown to inhibit the host factor stimulated, but not the oligo (U) primed transcription of poliovirus RNA (Baron and Baltimore, 1982; Morrow and Dasgupta, 1983). These observations indicate that contaminating VPg containing peptides in the polymerase fraction are involved in host factor stimulated initiation of polivirus RNA synthesis.

The *in vitro* synthesis of a complete minus-strand copy of poliovirus RNA was reported by Flanegan and coworkers (Van Dyke *et al.*, 1982; Tuschall *et al.*, 1982). With viral plus-strand RNA as template, oligo (U) as primer, and in the presence of all four ribonucleoside triphosphates and Mg^{++} the purified polymerase synthesizes genome sized RNA of negative polarity. Under optimal conditions the polymerase synthesized RNA at a rate of 1,200 nucleotides per min or one RNA molecule per 6 minutes. The purified RNA polymerase was able to copy a variety of polyadenylated RNAs in addition to poliovirus RNA when oligo(U) was used as a primer, including HeLa cell mRNA and different viral RNAs. In the presence of the host factors described above, the poliovirus polymerase could even copy viral RNAs known to lack 3'terminal poly(A) (brome mosaic virus

RNA 3 and tobacco mosaic virus RNA), but which contain 3'terminal hairpin structures similar to those found in tRNA (Hall, 1979).

It is important to keep in mind, that the soluble replicase preparations described above use viral plus strand RNA as template for the synthesis of complementary RNA, the product always being dsRNA. This is in contrast to the crude enzyme preparations from infected cells, discussed in Section V A which preferentially yield single stranded viral plus RNAs as new products.

The observations discussed so far tend to support the model of poliovirus RNA replication presented in Figure 94, p. 375. During the earliest stages of RNA replication in the infected cell, a soluble polymerase complex, consisting of freshly synthesized viral replicase p63 (NCVP4 or NCVP6b) and a ribosome derived host factor, transcribes poliovirus RNA into a double stranded RF-RNA product, a process which is readily copied under *in vitro* conditions. An intriguing question arises: How is the early replicase complex prevented from copying host mRNA—which it does in vitro—and how is its specificity for the parental poliovirus RNA determined? A larger, more complex multicomponent enzyme system appears to be required for the synthesis of single stranded +RNA from this early template. In a soluble state, this larger complex may synthesize efficiently only defective + RNA, presumably missing a segment in the 5' terminal half. Once associated with specialized, virus induced membranes, the fidelity of the large replication complex appears to be enhanced; it now successfully synthesizes full length progeny plus strands by a displacement type of RNA synthesis. Additional host factors appear to be required for the synthesis of +RNA. The origin of these host factors—ribosomal or nuclear—remains a matter of speculation at present; recent evidence has revitalized the idea of an involvement of nuclear factors in poliovirus RNA synthesis (see Section VII C, below).

VI. The Effects of Guanidine on Poliovirus Replication

Guanidine is the most extensively studied inhibitor of poliovirus replication (for reviews see Sergiescu *et al.*, 1972; Caliguiri and Tamm, 1973; Tershak *et al.*, 1982), yet its precise mode(s) of action is still uncertain. In light of the increasing evidence that the locus for guanidine sensitivity resides in the P-2 proteins (see Chapter 4, p. 180 and 197), proteins which may be important for membrane associated RNA synthesis, it is of interest to review the effects of guanidine here.

Guanidine is a basic component with a pKa of about 12.5. At physiological pH the molecule gains a proton and is positively charged. The protonated molecule is probably much smaller than hydrated sodium or potassium ions (Davidoff, 1973). The guanidino group is a component of arginine, creatine, creatinine, methylguanidine, dimethylguanidine, guanidinosuccinic acid and guanidinoacetic acid, metabolites commonly present in animal tissue. Guanidine at molar concentrations is used to dissociate proteins and to isolate nucleic acids from cells. However, its antiviral activity is shown at thousand-fold lower concentrations and it is not yet known to what extent guanidine can alter the structure and conformation of proteins at mM concentrations. The effects of guanidine on guanidine sensitive and resistant mutants can be reversed by certain amino acids (met, val, leu, thr in decreasing order of effectiveness), membrane precursors (choline, dimethylpropanolamine, ethanolamine), and protease inhibitors (Dinter and Bengston, 1964; Lwoff, 1965, 1966; Philipson *et al.*, 1966; Mosser *et al.*, 1971; Tershak, 1974).

Since the discovery of the antiviral activity of 1–2 mM guanidine in the early sixties (Rightsel *et al.*, 1961; Crowther and Melnick, 1961; Tamm and Eggers, 1962), the compound has been shown to have a wide variety of effects on poliovirus replication. Even at 10 mM concentration, in contrast, guanidine does not inhibit the synthesis of cellular RNA and protein (Loddo *et al.*, 1962; Baltimore *et al.*, 1963; Bablanian *et al.*, 1965). Table 64 provides an overview of the reported effects of guanidine on poliovirus replication. Guanidine neither inactivates extracellular virions nor interferes with adsorption, penetration, and uncoating (Loddo *et al.*, 1961; Pringle, 1964; Carp, 1964; Eggers *et al.*, 1965, 1981). The early events of the growth cycle up to and including the shut off of cellular protein synthesis proceed in the presence of guanidine (Penman and Summers, 1965), and early RNA synthesis is relatively resistant to inhibition by guanidine (Koch *et al.*, 1974).

The accumulated evidence indicates that guanidine has two main sites of action: 1. it blocks the initiation of membrane dependent (+)-strand 35S RNA synthesis, and 2. it interferes with the formation and function of virus specific vesicles. In addition, guanidine may interfere with protein processing (see Chapter 8, p. 367).

As a consequence of 1) the coupled process of virion assembly and vRNA synthesis is stopped (14S subunits and procapsids accumulate, and cleavage of VP_0 in assembled particles is blocked) (Jacobson and Baltimore, 1968; Ghendon *et al.*, 1973; Wright and Cooper, 1974). The initiation and synthesis of cRNA is not affected as strongly: when guanidine is added early in infection, the content of cRNA and dsRNA still increases in the presence of guanidine (Caliguiri and Tamm, 1968; Huang and Baltimore, 1970; Koschel and Wecker, 1971; Koch *et al.*, 1974). Elongation of viral RNA synthesis by the membrane bound replication complex *in vitro* is not affected even in 50–100 fold higher concentrations than employed for *in vivo* inhibition (Baltimore *et al.*, 1963; Baltimore, 1964; Koschel, 1969) but *in vitro* initiation is dramatically inhibited (Tershak, 1982).

As a consequence of 2) choline incorporation is blocked (Penman and Summers, 1965), vesicle formation is prevented in the presence of guanidine (Bienz *et al.*, 1980), the movement of membranes from fraction 5 (rough ER) to fraction 2 (virus specific membranes) is prevented (Caliguiri and Tamm, 1973), and the association of capsid proteins with the replication complex in the virus induced membranes is inhibited (Yin, 1977).

Guanidine sensitivity is dominant in mixed infections of guanidine sensitive and resistant strains, indicating that the viral protein made by guanidine sensitive strains alters all replication centers equally. Early genetic studies indicated that the site of guanidine sensitivity lies near, but to the 5' side of genome (Cooper, 1969; Cooper *et al.*, 1970). This was originally interpreted as evidence for the capsid proteins as the site of guanidine action (Cooper, 1977). In apparent support of this conclusion, mutants with altered guanidine sensitivities were characterized

Table 64. *The effects of guanidine on poliovirus replication*

A. Effects of guanidine on different phases of poliovirus replication

Time post infection = addition of guanidine	1	2	3	4	5	6 hours
Adsorption Penetration Uncoating	not affected by guanidine					
Synthesis of viral RNA	+	+	+	(+)	(–)	(–)
Synthesis of viral protein	+	(+)	(–)	–	–	–
Shut off of host protein synthesis	(+)	(+)	–	–	–	–
Shut off of host DNA synthesis	+	+	–	–	–	–
Choline incorporation	+	+	(–)	–	–	–
Vesicle formation	+	(–)	(–)	–	–	–
Movements of membranes from rER to sER	+	+	+	–	–	–
Late cytopathic lesions	+	+	(+)	(+)	–	–
Assembly of virions	+	+	+	+	+	+
Processing of VP 0	+	+	+	+	+	+
Processing of other viral proteins	+	+	(?)	(?)	(?)	(?)
Cytoskeletal alterations	(+)	(+)				

\+ Blocked by guanidine.
– Proceeds in the presence of guanidine.

B. Effects of guanidine on virus-specific processes in infected cells

I. Inhibition of Viral RNA Synthesis
- blocks development of UV resistance by infectious centers (1)
- synthesis of ssRNA is more sensitive than production of RF- or RI-RNA (2)
- synthesis of +RNA more sensitive than cRNA (2)
- initiation of viral RNA synthesis is the sensitive step *in vitro,* elongation and release of completed chains are not affected significantly (3)

II. Inhibition of Membrane Formation
- reduces incorporation of choline into membranes (4)
- inhibits vesicle formation (5)
- prevents the movement of membranes from rER (fraction 5) to virus-specific sER (fraction 2) (6)
- delays cytopathic lesions (7)
- reduces virus-induced alterations of cytoskeleton (8)
- blocks accumulation of sodium ions in late stages of infection (9)
- effect antagonized by choline, ethanolamine (10)

III. Inhibition of Membrane Functions
- prevents association of viral capsid polypeptides with smooth membrane fraction and with replication complex (11)
- blocks initiation of vRNA synthesis (see above): limits availability of RNA for maturation process (12)
- blocks virion assembly: 14 S subunits and 80 S procapsids accumulate (12)
- blocks release of RNA from the replication complex *in vivo* (12 a)

IV. Interference with Processing of Virion Proteins
- processing of VPO is blocked (result of inhibited encapsidation) (12)
- processing of capsid proteins is delayed in FMDV (13)
- processing of P-2 proteins is altered in early stages of replication (14)
- when added for 15 minutes between 1 and 2 h.p.i., virus growth is delayed by 1.5 to 2 hours, TPCK and TLCK shorten this lag period (15)
- effect reversed by amino acids (16)

V. Interference with Aggregation of Virion Proteins
- 5mM guanidine dissociates poliovirus polypeptide aggregate (NCVP1 a, NCVP2, NCVP-X) of guanidine sensitive, but not of guanidine resistant strains *in vitro* (17)

VI. Interference with Virus Induced Shut-off
- delays shut-off of host protein synthesis at low m.o.i. of less than 10 PFU/cell (result of inhibition of RNA synthesis) (18)
- prevents poliovirus induced shut-off of host DNA synthesis only if added before 2 h.p.i. (effect reversed if removed for as little as 10 minutes at 4 hours p.i. even though no virions formed) (19)

References

1. Eggers *et al.,* 1965.
2. Crowther and Melnick, 1961; Loddo *et al.,* 1962; Tamm and Eggers, 1963; Summers *et al.,* 1965; Caliguiri and Tamm, 1968; Koschel and Wecker, 1971.
3. Baltimore *et al.,* 1963; Baltimore, 1964; Koschel, 1969; Noble and Levintow, 1970; Tershak, 1982.
4. Penman and Summers, 1965.
5. Bienz *et al.,* personal communication.
6. Caliguiri and Tamm, 1973.
7. Bablanian *et al.,* 1965; Zeichhardt *et al.,* 1982.
8. Lenk and Penman, 1979.
9. Nair *et al.,* 1979.

(Continuation see p. 408)

10. Lwoff, 1965; Philipson *et al.*, 1966; Mosser *et al.*, 1971.
11. Caliguiri and Compans, 1973; Yin and Knight, 1975; Yin, 1977.
12. Halperen *et al.*, 1964; Jacobson and Baltimore, 1968; Ghendon *et al.*, 1972, 1973; Caliguiri and Tamm, 1973; Wright and Cooper, 1974.
12 a. Baltimore, 1968; Huang and Baltimore, 1970.
13. Vande Woude and Ascione, 1974.
14. Koch *et al.*, 1980 b, 1982 a.
15. Tershak, 1974.
16. Dinter and Bengston, 1964; Lwoff, 1965.
17. Korant, 1977.
18. Holland, 1964; Koschel and Wecker, 1971; Bablanian, 1972.
19. Powers *et al.*, 1969.

---

which showed altered electrophoretic and chromatographic properties of capsid proteins (Korant, 1977b). On the other hand, DI-particles lacking detectable amounts of capsid proteins were shown to retain guanidine sensitivity, indicating that the guanidine sensitivity is not associated with capsid proteins (Cole and Baltimore, 1973). An alignment of the genetic map with the published nucleotide sequence indicates that the site for guanidine sensitivity resides in the P-2 region of the genome (see Figs. 52, 58, p. 178 and p. 198). Recent biochemical characterization of genetic recombinants have confirmed the localization in P-2 (Tolskaya *et al.*, 1983). In addition, presence of guanidine from the beginning of the replication cycle alters the cleavage pattern of the P-2 proteins (Koch *et al.*, 1980b, 1982a). Recently, a careful characterization of the proteins of guanidine resistant or dependent mutants has revealed altered electrophoretic mobilities of the P-2 proteins (Anderson-Sillman *et al.*, 1983), and, finally, one of the P-2 proteins has been implicated as the inducer of virus specific vesicle formation (Bienz *et al.*, 1983).

One may speculate that guanidine causes minute alterations in the secondary and tertiary structure of proteins (Tershak *et al.*, 1982), thereby affecting the delicate function of the guanidine sensitive P-2 proteins. As a consequence, the specific association of viral proteins and RNA with membranes is disturbed, thus preventing first the induction of vesicle formation, and later the membrane dependent initiation of +strand RNA synthesis and cotranscriptional encapsidation of the RNA. In addition, guanidine appears to alter the activity of a protease (or its substrate) required in these steps. It may be of pertinence in this regard, that the cleavage sites between P-1 and P-2 (between VP1 and NCVP3b), and between P2-6a and P2-6b are tyr/gly sites, (several additional tyr/gly cleavage sites are located in the vicinity of the former site) and that the responsible protease for these cleavages has not yet been identified.

VII. Some Thoughts on the Mode of RNA Replication

Poliovirus specific RNA synthesis is unususal in several aspects. First, the process of RNA-directed RNA synthesis in itself is a biological activity that usually does not occur at all in eukaryotic cells. Then, both minus and plus strand synthesis are initiated with uridine, in contrast to all known cellular RNA synthesis which is always initiated with a purine nucleotide. Poliovirus RNA replication begins with the transcription of the poly A tail on the parental poliovirus mRNA, *i.e.* RNA synthesis is initiated on a single stranded homopolymer which is present in the cytoplasm of the infected cell in abundance also as tails on many host specific mRNAs. For unknown reasons, however, the latter do not seem to be efficient templates for the poliovirus replicase, even though the replicase can transcribe such templates with high efficiency *in vitro*. Finally, in contrast to all cellular RNA polymerases, the polioviral polymerase is primer dependent.

There appear to be similarities as well as fundamental differences in the initiation and elongation of minus strand and plus strand RNA synthesis. Similarities include use of pU as the initiating nucleotide, linkage of this pU to VPg, and the identity of the replicase activity responsible for chain elongation as NCVP4 or p63. Marked differences are 1) the nature of the template–single stranded RNA vs. a double stranded RNA double-helix, 2) the mechanism of initiation–on a single stranded homopolymer on the one hand, on a double stranded heteropolymer on the other, 3) the localization of RNA synthesis in the soluble fraction or in membrane association, and 4) the involvement of different additional virus-specific and host cell proteins.

Initiation of RNA synthesis directed by RNA phages (which do not possess poly (A) at the 3' end of the +RNA) is believed to use a common recognition signal for both plus and minus strand RNA synthesis. In poliovirus RNAs, however, there are no common sequences on the 3' ends of plus and minus strand RNA.

A. Initiation of Viral RNA Synthesis

There are some fundamental conceptual problems pertaining to the initiation of RNA replication on poliovirus RNA, be it on a single or double stranded template RNA. Since the entire length of the template is copied, initiation of RNA synthesis must begin at the very 3'terminal end of the template. For cellular RNA synthesis on a dsDNA template, the binding and orientation sites for the RNA polymerase usually lie several nucleotides ahead of the stretch of DNA that is to be copied into RNA. For poliovirus RNA templates, polymerase recognition and orientation sites must lie within the region that is then copied by the replicase.

Since no DNA polymerases can start a de novo DNA chain, initiation of DNA synthesis by DNA polymerase always involves a primer, although this may be a segment of the parental DNA as in "rolling circle" or "hairpin" primed synthesis. RNA polymerases can start de novo RNA synthesis–however, transcription of the very terminal ends of eukaryotic DNA into RNA usually does not occur. A number of mechanisms have evolved in the cell for the copying of terminal DNA regions, and it is concievable that one of these, in principle, applies also to the initiation of poliovirus RNA synthesis. A segment of the parental DNA itself may serve as primer for initiation of replication on terminal ends, as in the "rolling circle" concatemer or hairpin models of DNA synthesis (Mitra, 1980; Challberg and Kelley, 1982).

Initiation of synthesis according to the hairpin primed mode (Cavalier-Smith, 1974) is used during replication of some single stranded DNA viruses (the parvoviruses): The ends of these genomes are palindromic so that the individual strands can form hairpins, yielding a free 3' OH primer for elongation of RNA synthesis. Elongation results in the formation of an RF molecule in which the two strands are covalently linked at one end by the hairpin. This hairpin is "resolved" when the palindromic sequence is subsequently cleaved at its 5' end by a sequence specific nuclease so that the palindrome end is now attached to the freshly synthesized progeny strand in inverted sequence. The palindromic end is copied, using the 3' OH nick of the parental strand as primer. If the palindrome is not "perfect", *i.e.* exactly base paired, the terminal ends of genome strands will exist in two—inverted—configurations, termed "flip" and "flop". Initiation of replication on the double stranded RF-RNA intermediate uses the same principles: after strand separation at the terminal regions, the ends of the individual strands can again form hairpins to serve as primers. If replication of the hairpin-linked RF would proceed through the terminal hairpin prior to its resolution, longer molecular forms—concatemers—of RF-RNA would be produced, in which RF dimers were linked in an end to end fashion. Even though the poliovirus RF-RNA contains palindromic sequences at the 5' end of the plus strand and corresponding 3' end of the minus strand, it is unlikely that the hairpin thus formed participates in initiation of RNA synthesis. The poliovirus hairpin is quite imperfect in comparison to those in parvovirus DNAs, for example (Rhode and Klaassen, 1982). The flip-flop configurations that would then be required by the Cavalier-Smith model have never been detected in poliovirus infected cells.

Models similar in principle to hairpin primed initiation of replication are rolling circle models (Tattersall and Ward, 1976). In such models the template nucleic acid is pictured to rearrange into a circular form, its ends either becoming linked by hydrogen bonding between complementary sequences at the opposing ends of the molecule, or mediated by some protein. Circular and concatemer forms of RF-RNA have not been detected in poliovirus infected cells. But in cells infected by EMC virus and mengovirus, close relatives of poliovirus, such structures have regularly been observed (Agol *et al.* 1970, 1972; Robberson *et al.*, 1982). At present it is still impossible to say whether any circular or concatemer RNA molecules are involved in initiation and transcription of picornavirus RNA synthesis.

A third possibility for initiation of polynucleotide synthesis on terminal ends of duplex templates may be a protein-primed mechanism. All picornaviruses, the dsDNA adenoviruses, and even the ssDNA parvoviruses which initiate replication by the hairpin transfer model, carry a protein covalently linked to the 5' termini of both complementary strands in RF and all nascent chains in RI-RNA.

Similarly, all newly synthesized molecules of polioviral RNA—at least during the late linear phase of RNA synthesis—carry the 20 amino acid long VPg covalently attached to their 5'ends. The cRNA template as well as even the shortest single stranded nascent RNA chains on the RI-RNA are linked to VPg (Flanegan *et al.*, 1977; Nomoto *et al.*, 1977a; Peterson *et al.*, 1978; Wu *et al.*, 1978). It has been suggested, that VPg serves as a primer for both the synthesis of plus and minus strand RNA. The molecular mechanism by which VPg participates in the initiation of RNA synthesis is still unknown. Covalent linkage of VPg to the 5' phosphate of UTP may suffice to provide the 3'-OH of the uridine in a form suitable for chain elongation by the replicase complex. Alternatively VPg might be covalently bound to a de novo synthesized oligonucleotide primer analogous to the cap on some mRNAs, thereby providing the primer necessary for chain elongation.

VPg is part of the replicase precursor protein NCVP-1b, *in vivo* VPg is usually found as the C-terminal tail of NCVP-9. It is not yet known wheter VPg is linked to UTP while still part of a larger precursor or only after its cleavage therefrom. With antibodies against VPg, several VPg containing viral proteins have been detected (Baron and Baltimore, 1982a, b; Semler *et al.*, 1982). A 12,000 dalton protein–NCVP-9–is found in a membrane attached state (Emini *et al.*, 1982). VPg is probably directly cleaved from this precursor either shortly before or after it is attached to nascent RNA (Rueckert *et al.*, 1979). Cleavage of NCVP-1b generates both the VPg precursor NCVP-9 and NCVP-2 (Dasgupta *et al.*, 1980), providing thereby a potential primer (VPg) for new RNA synthesis as well as a native active replicase or replicase precursor. Recently, free VPg and a nucleotidyl peptide pUpU-VPg have been detected in infected cells (Vartapetjan *et al.*, 1982; Takegami *et al.*, 1983b; Crawford and Baltimore, 1983), increasing the list of potential primers for RNA synthesis.

It is conceivable that NCVP-1 first binds to template RNA, the replicase fragment perhaps recognizing and binding to a specific sequence. Membrane insertion of the hydrophobic NCVP9 containing tail and cleavage of this peptide from the replicase precursor may activate the uridylation sensitive tyr residue in VPg. Enzymatic linkage of the VPg containing fragment (tail) to UTP (or a short oligonucleotide) would then provide the primer for RNA elongation. Cleavage of the

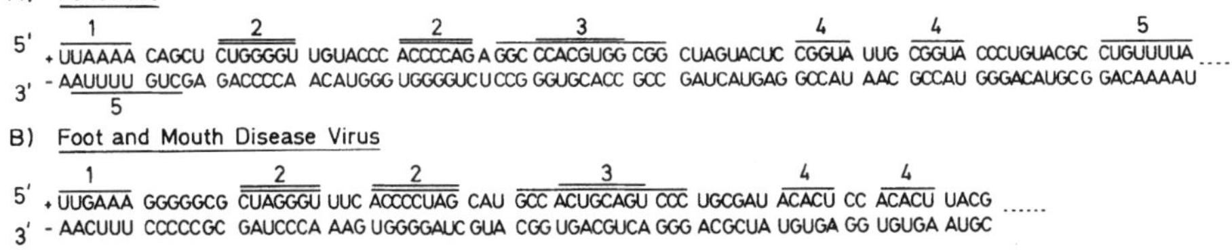

Fig. 101. The terminal sequences of poliovirus and foot and mouth disease virus replicative form RNAs

The figure illustrates some similarities in the terminal sequences of poliovirus and foot and mouth disease virus RF-RNAs. Some peculiar features, palindrome sequences, tandem repeats, inverted sequences, are expressed in both viruses. These structures seem to have been conserved in the evolution of these relatively distantly related picornaviruses. This may indicate participation of these structures in binding to RNA polymerase, RNA-binding proteins, strand-separation proteins, VPg linking enzyme, and other factors involved in the replication of these RNAs.

1 terminal A-U rich segment; *2* palindrome regions separated by several asymmetric base pairs; *3* palindrome bordered on each side by three C-G pairs; *4* tandem repeat of 5 base pair sequence; *5* inversion of sequence at terminal end. – Figure compiled from data from Harris, 1979 [Nucl. Acid. Res. *7*, 1765–1785 (1979)], and Hewlett and Florkiewicz, 1980 [Proc. Nat. Acad. Sci. *77*, 303–307 (1980)]

VPg-U (oliogonucleotide) primer from the precursor and hybridization to template RNA may provide the signal for chain elongation. Cleavage of VPg occurs at a gln/gly site by NCVP-7 which is also contained within the replicase precursor. For EMC virus it has been demonstrated that this protease is active in the precursor molecule and is capable of cleaving off the VPg containing tail (Palmenberg and Rueckert, 1982). The mechanism of forming the unique tyrosine nucleoside linkage to pU at the 5'end of viral RNA has not been elucidated. The responsible enzyme has not been identified. These are themes for further investigations.

An interesting pattern of features emerges when one compares the structures of the initiating end of RF-RNAs of poliovirus and of foot and mouth disease Virus, a very distantly related picornavirus (see Figure 101). It is possible that these features play a role in initiation of plus strand synthesis and have thus been conserved in evolution. Similarities between the RFs of these two viruses include: a 6-nucleotide long AU-rich stretch at the very terminal end, several base pairs onward a palindromic region separated by several nonpairing sets of bases. Shortly thereafter there is a second 8-nucleotide long palindrome bordered on each side by three CG pairs. A few nucleotides further, a 5 base pair sequence follows in tandem repeat. In the case of poliovirus this tandem repeat is quickly followed by an inversion of the 8 terminal nucleotide sequence stretch.

B. Elongation of Viral RNA Synthesis

The process of nucleic acid synthesis involves the stepwise addition of a nucleotide to a primer or oligonucleotide on a single- or double-stranded nucleic acid template. Synthesis on a single-stranded template generally results in the continous stable base pairing of the newly synthsized chain to its template resulting in the formation of a duplex. Synthesis of poliovirus minus strand RNA seems to follow this pattern.

Initiation and elongation of minus strand RNA synthesis evidently is less complex than that of plus strand synthesis. *In vitro*, the former requires only template RNA, poly U or a host cell protein as primer, and poliovirus RNA replicase (p63 or NCVP-4b). Linkage to VPg may be involved in host-factor primed cRNA synthesis *in vitro* (Morrow and Dasgupta, 1983). Elongation of minus strand RNA synthesis begins with the transcription of the poly A stretch at the 3'end of the plus strand. Elongation of minus strand RNA probably is accomplished by NCVP-4b alone. At most, an additional factor may be required for unwinding of secondary structures in the single stranded template RNA; a host cell elongation factor—eEF-X—could accomplish this task (Agol, 1982). cRNA synthesis does not require association of template or replicase proteins to intracellular membranes, although some cRNA synthesis might nevertheless proceed membrane-associated. The product of cRNA synthesis is double stranded RF-RNA.

Elongation of plus strand RNA synthesis is far more complex, since the template is a double stranded heteropolymer (RF-or RI-RNA). As discussed, most experimental evidence is in accord with the model in which newly synthesized

RNA displaces one of the parental RNA strands in RF-RNA. In general (or exclusively) only the minus strand of RF-RNA serves as a template. The enzymatic activity responsible for elongation of poliovirus RNA also appears to be NCVP-4.

In order to copy the minus strand of ds RF or RI-RNA, the ends of the two strands must first be separated and the double helix ahead of the replicase complex must be unwound. Most polymerases are unable to unwind duplex DNA and thus can only utilize single stranded templates. For cellular DNA and RNA syntheses, different enzymes are required for strand separation and unwinding of the double stranded helix.

An important class of oligomeric proteins—the DNA binding proteins—acts in concert with RNA polymerases during the synthesis of DNA or RNA on a dsDNA template by stabilizing single stranded regions of DNA. These proteins—which usually are present in larger number—bind cooperatively to ssDNA with a preference for AT-rich regions, thereby facilitating the melting of duplex DNA required for replication and transcription. DNA helicases can unwind the double helix of the template. They have a DNA-dependent ATPase activity, which provides the energy required for unwinding by hydrolysis of two ATP molecules for each base pair unwound. A DNA-helicase of E. coli—the rep protein—is required for the replication of RF- DNA of ssDNA coliphages. In the presence of DNA binding proteins, rep protein unwinds duplex DNA containing a single stranded leader region, processively and catalytically along the 3'—5' direction of the transcribed DNA strand, while the 5'end of the other strand is displaced concomitantly.

Some of these functions—strand separation, stabilization of single stranded RNA, duplex unwinding—may also be accomplished by ribosomal factors, which normally are responsible for unwinding shorter stretches of dsRNA in mRNA molecules to permit translation.

Some of these protein functions may be required for the copying of the cRNA strand in poliovirus RF- or RI-RNA. The poliovirus genome certainly does not carry enough genetic information to code for such proteins, some of the host cell proteins found associated to the poliovirus replication complex may serve these functions. It is likely that host cell proteins do not have a high affinity for dsRNA as such molecules usually do not occur in the cell. Perhaps only the concerted action of virus and host cell proteins provides the complex with the required efficiency.

Another class of proteins—the DNA topoisomerases—play a vital role in DNA replication by interconverting topological forms of DNA by concerted breakage and rejoining of DNA backbones. These enzymes are needed for removing positive supercoils which are generated as a result of duplex unwinding ahead of the fork during DNA replication. Since poliovirus RF with its length of 7,500 base pairs is relatively short, it probably does not require such enzymes.

Unwinding of the poliovirus duplex replication intermediate and the associated displacement of the plus strand poses other thermodynamic and geometric problems: The replicative form contains 750 complete helical turns of 10 base pairs each. If the template remains stationary relative to the replication proteins during RNA elongation, the displaced plus strand must rotate about the template strand as it is being displaced to avoid a tangled-up mess of RNA strands. Taking the observed rate of RNA synthesis of 45 sec—1 min per progeny strand, this would require the revolution of the displaced strand about the template at a rate of 10—15 turns per sec. As the displaced strand grows in length, its drag will increase, an effect which would be dramatically augmented if capsid assembly began about this strand as it is being displaced. The second possibility—the template rotating relative to stationary replication complexes and encapsidation sites—seems more feasible on thermodynamic considerations. This would imply a template rotating at a speed of 10—15 rotations per sec or 750—1,000 rpm (corresponding to the rate of centrifugation required for the sedimentation of cells). Assuming that replication complex and assembly site are relatively fixed in cytoplasmic space (by attachment to vesicle membranes or cytoskeletal components), this would further imply a movement of the template along these sites as it is being replicated. The terminal end of an RI template revolving at a rate of 750 rpm may pose a problem for recognition and binding of the proteins responsible for strand separation at the terminal end and for initiation of RNA synthesis by the next replication complex. Perhaps these proteins and possibly the entire initiation complex itself—are structured somewhat like a nut for the corresponding screw (rotating template), recognizing the revolving template on one side, while expulsing the displaced strand and template-progeny duplex at the other end.

C. Inhibition of Host Nuclear Functions

The nucleus plays a dominant role in the regulation of the cell metabolism. The principle steps of poliovirus replication, translation, RNA replication, and virion morphogenesis are cytoplasmic. Poliovirus replication can proceed in enucleated cells (Crocker *et al.*, 1964; Pollack and Goldmann, 1973; Follet *et al.*, 1975). Enucleated cells, however, give rise to lower yields of progeny virions (Bossart and Bienz, 1979). These observations have generally been taken as an indication that poliovirus RNA replication does not depend on any nuclear factors of the host cell.

Following poliovirus infection, host cell RNA and DNA synthesis are blocked by a virus specific product, peculiar alterations of nuclear morphology are observed, and some poliovirus proteins appear to migrate to and accumulate specifically in the nucleoplasm (see Chapters 6 and 8). Such observations in turn, may indicate involvement of some nuclear factors in poliovirus replication—factors which might also be present or synthesized in the cytoplasm of enucleated cells.

Nuclear functions, notably RNA synthesis, are rapidly inhibited in poliovirus infected cells, provided that the synthesis of viral proteins is not blocked (Contreras *et al.*, 1973). This led to the proposal that viral proteins are responsible for the inhibition of host RNA synthesis. In accord with this view are recent data (Bossart *et al.*, 1982) demonstrating that RNA synthesis in nuclei isolated from uninfected cells is inhibited when incubated together with cytoplasmic extracts from poliovirus infected cells.

Autoradiographic EM studies indicated a specific effect on hnRNA synthesis, ribosomal RNA synthesis in the nucleolus was not affected (Bienz *et al.*, 1978).

Biochemical studies have confirmed that poliovirus infection leads to the inhibition of hnRNA synthesis by prevention of initiation of transcription by RNA polymerase II (Flores-Otero *et al.*, 1982). The isolation of RNA polymerasse II from infected and uninfected cells yields enzyme preparations with identical activities, suggesting that inhibition of host RNA polymerase activity in infected cells is due to a transient effect of a viral or virus induced protein (Crawford *et al.*, 1981). A possible candidate is the viral protein NCVP-2 which accumulates in the nucleus of infected HeLa cells at a rate proportional to the decrease in host RNA synthesis (Fernandez-Tomas, 1982).

The presence of viral proteins in the nucleus could simply reflect their involvement in the inhibition of cellular hnRNA synthesis. It is tempting to speculate that the interference is triggered by the adherance of a viral protein to an essential component of RNA polymerase II, and further, that this component is carried by the viral protein back to the cytoplasm where it may serve to stimulate viral RNA synthesis.

D. On the Infectivity of RF-RNA

All of the rather theoretical considerations discussed above are important themes for future investigations. They may be of relevance for the interpretation of a peculiar phenomen observed in studies on the infectivity of poliovirus specific RNAs. Poliovirus ssRNA of virion polarity and the positive RI-RNA are infectious, cRNA is not infectious. Poliovirus RF-RNA is also infectious; the RNAs of the progeny virions after infection with heteroduplex RF-RNA have the same polarity as the plus strand of the infecting RF-RNA molecule. Plus strand virion RNA and RI-RNA are infectious even in enucleated cells or in cells in which host cell RNA synthesis has been interrupted by actinomycin D. Surprisingly however, poliovirus RF-RNA is not infectious in enucleated cells or in cells treated with actinomycin D (Koch *et al.*, 1967; Koch and Bishop, 1968; Detjen *et al.*, 1978).

Single stranded vRNA and RI-RNA provide single stranded plus RNA which can function as mRNA in the synthesis of poliovirus specific proteins, including replication proteins—the mechanism of their infectivity appears to be analogous to that of intact poliovirus, which can also proceed in enucleated cells or in the presence of actinomycin D.

The molecular mechanism of RF-RNA infectivity is much more difficult to explain. dsRNA is a very potent inhibitor of protein synthesis. To date there is no convincing evidence for the capacity of ribosomes to dissociate dsRNA and translate its plus strand. In a report claiming translation of dsRNA by an *in vitro* protein synthesizing system, it was proposed that the dissociation of the duplex is mediated by an elongation factor (Agol *et al.*, 1981). Translation of dsRNA would then initiate infection by the classical mechanism, which does not offer an explanation for the sensitivity of RF-RNA infectivity to inhibition by enucleation or actinomycin treatment. The proposed involvement of host cell nuclear factors in the replication of RF-RNA, e.g. in strand separation at the terminal end and duplex unwinding, however, could provide a plausible explanation: In the absence of poliovirus specific replication proteins, the RF-RNA is a poorer template for these enzymes than in the presence of poliovirus proteins. If initiation of RNA synthesis on the RF-template by host cell enzymes alone were relatively inefficient, but nevertheless occuring at a significant rate, the result would be initiation of a replication cycle by the displaced plus strand of the RF molecule. The freshly synthesized plus strand would remain bound to the minus strand RNA in RF. Enucleation and actinomycin treatment would prevent this displacement replication of the RF-RNA. This model also accounts for the observed plus strand polarity in the progeny after infection by heteroduplex RF-RNA: Initiation on RF-RNA at a relatively slow rate would first provide a larger number of free plus strands (those displaced from the infecting RF-RNA). The displaced strands should be free to initiate replication by association with ribosomes. Statistically, the slowly increasing number of plus strand RNA molecules complementary to the cRNA strand of the infecting RF-RNA would then have little chance of "catching up" with the plus strands released during the first few replication events.

E. Regulation of Viral RNA Synthesis

The orderly sequence of events in the poliovirus replication cycle is probably controled by several mechanisms. RNA synthesis, in particular, proceeds in distinct phases: switch from translation to transcription of parental RNA, exponential phase of template replication and mRNA production, linear phase of virion RNA production, and final phase of RF-RNA accumulation. Changing concentrations of intermediates and products may exert non-specific regulatory effects (Baltimore, 1969) and determine the sequence of the phases. Specific regulatory mechanisms may involve viral proteins, protein processing, and protein or RNA compartmentalization (Cooper, 1977; Korant, 1977b; Yin, 1977a; Bienz *et al.*, 1978; Agol, 1980). Many of the virus specific processes may acquire different characteristics when occuring in association with intracellular membranes: the replication complex gains fidelity, and protease activity or specificity may be altered. Table 65 summarizes some ideas and hypotheses on possible regulatory mechanisms during the different stages of poliovirus RNA synthesis.

VIII. Summary

The poliovirus genome is a single-stranded RNA molecule of mRNA polyarity. Neither the host cell nor the invading parental virions carry RNA polymerizing enzymes which can initiate RNA synthesis on single-stranded RNA templates. Therefore, translation of the parental RNA is required in order to produce virus-specific RNA polymerase before the viral RNA can be replicated.

The replication of poliovirus RNA proceeds in distinct steps and phases. Virus specific RNA synthesis is initiated when the viral replicase (or a precursor) binds to the '3 poly(A) tail of the parental RNA, and an efficient primer is present. *In*

Table 65. *Regulation of RNA replication*

I. Exponential phase: cRNA–mRNA synthesis

A. Early Phase

– no virus modified intracellular membranes

1. very early protein synthesis proceeds on parental RNA on free ribosomes
2. viral RNA which has served as mRNA has to be freed from ribosomes to act as template for the synthesis of minus strand RNA: ribosomes or RNA are transiently inactivated by binding of capsid proteins (equestron hypothesis), by interaction with replicase, circularization of RNA, or by elongation block–removal of eEF-X.
3. RNA synthesis:
   – RNA synthesis is not yet membrane associated and proceeds on a soluble replicase complex resulting in the production of minus strand RNA and thereby of RF-RNA
   – formation of RF-RNA prevents attachment of replicase on 3′ end of parental RNA
   – Soluble replicase attaches to RF-RNA and synthesized plus strand RNA, the ss-product of soluble replicase is mostly DI-RNA
4. protein synthesis:
   – DI like RNA serves as mRNA, NCVPX and NCVP 1 b are the first detectable proteins, defective capsid proteins are degraded.

B. Intermediate Phase

– poliovirus induced membranes are formed but no vesicles are detectable (possible role for P-2 genomic region: NCVP 8 + 10, NCVPX may induce membrane synthesis and/or bind the replication complex to membrane)

1. RNA synthesis:
   – RNA synthesis occurs by membrane bound replication complex
   – intact VPg–RNA is synthesized and continues to function as template for cRNA synthesis or is released into cytoplasmic matrix where VPg is cleaved off to yield mRNA
2. protein synthesis:
   – protein synthesis directed by newly synthesized mRNA of genome size
   – proteins including the capsid proteins are formed in equimolar amounts
   – most viral proteins are found membrane associated.

II. Linear Phase: vRNA synthesis

– virus induced membranes combine into membrane bound vesicles, compartmentalization

1. RNA synthesis:
   – RNA synthesis on membrane bound vesicles, virion formation begins
   – VPg-RNA is directly inserted during synthesis into membrane bound procapsids or is translocated into vesicles where RNA is protected from VPg cleaving enzyme, the RNA is condensed by hypertonicity and is rapidly packaged into viral capsid precursors.
   – other membrane associated replicase complexes continue to synthesize cRNA and mRNA as in phase II.

2. Protein:
   – meanwhile capsid proteins have begun the assembly process–they move to the cell center and are rapidly assembled with RNA to mature virions.

Prediction:

Inhibition of membrane synthesis should prolong phase I. Synthesis of virus specific membranes by one virus shortens lag phase of replication of a superinfecting picornavirus.

vitro, short stretches of oligo(U) and a host factor, present in a ribosomal salt wash, act as efficient primers. These factors or VPg may serve as a primer for cRNA synthesis *in vivo*.

The parental RNA is then copied by the viral replicase (NCVP-4 or p63) in the 3'-5' direction. The nascent RNA retains its association to the complementary template RNA, yielding a double-stranded product RF-RNA. Reinitiation of cRNA synthesis on the oligo(A)-oligo(U) tail of RF-RNA does not occur or is very inefficient. Initiation of plus strand RNA synthesis on the opposite, heteropolymer end of RF-RNA (5' end of plus strand, 3' end of minus strand), in contrast, is possible. Additional viral and host cell factors–constituing the RNA initiation complex–are required for this step. The process of initiation on the double-stranded heteropolymer tail of RF-RNA is more complex than the initiation of cRNA synthesis. It appears to occur with high efficiency only in the presence of virus induced membranes, it has to date not been possible to reconstitute initiation of plus strand synthesis *in vitro*.

Elongation of plus strand RNA synthesis requires the unwinding of the double-stranded template. Whereas cellular factors may be recruited for this task and for the initiation step, the actual polymerization of nucleotides into the nascent plus strand RNA is accompanied by the viral replicase NCVP-4 as in cRNA synthesis. The parental plus strand is displaced from the ds template as the nascent RNA chain is elongated. The nascent RNA retains its association to the cRNA strand of the template. Repeated reinitiation on the ds template yields a multistranded form of RNA–RI-RNA consisting of a complete strand of cRNA (the functional template) and several plus stranded tails of nascent RNA of different lengths. The nascent chains of RNA are associated to the cRNA by base-pairing in a doublehelix over several hundred nucleotides at the leading (3') edge of the polymerizing chains. A model which holds that the polymerizing nascent chains are prevented from extensive base pairing to the cRNA yielding a RI-RNA mainly composed of single stranded RNAs however is supported by recent data from an electron microscopic study of RI-RNA.

The progeny plus strand of RNA has one of three possible fates: 1. It may again function as template for the synthesis of cRNA (RF-RNA), 2. it may serve as mRNA, and 3. it may associate with capsid proteins in the morphogenesis of progeny virions. In the infected cell there appears to be a temporal control over the relative use of these three fates. Sequential kinetic phases of RNA accumulation are discernable and can be correlated in part to the fates of the freshly synthesized plus strand RNAs. The recirculation of plus strand RNA as template in the pool of replicating RNAs in the early phase of RNA synthesis leads to an exponential rate of RNA accumulation (mainly RI-RNA) (fate 1). As increasing quantities of nascent plus strand RNA are withdrawn from the replicating pool to function as mRNA in translation, the kinetics of RNA accumulation change from exponential to linear (fate 2). In direct consequence of the increasing translation of viral messages, capsid proteins and other viral products accumulate, inducing the formation of virus-specific smooth membranes. These developments, in turn, provide the basis for virion morphogenesis: plus strand RNA synthesis becomes associated to the transformed smooth membranes and directly coupled to the assembly of the viral capsid (fate 3). At the peak of viral RNA synthesis, the rate of RNA accumulation is linear, most RNA synthesis occurs in a membrane-bound form, and the progeny RNA strands become encapsidated as they are being synthesized. Finally, RNA synthesis and virion formation cease—presumably as a result of energy and substrate depletion, breakdown of the intracellular ionic environment, and the activation of autolytic host mechanisms. The active RI-RNAs and some of the single-stranded plus RNAs are converted to ds RF-RNA, much of the ssRNA is digested by cellular nucleases.

The phases of RNA synthesis may be shifted foward or backward in time (post infection) depending on the culture conditions of the host cell or the stage of the cell cycle at the time of infection; however, the relative pattern of RNA synthesis and the total amount of RNA synthesized per cell are relatively constant.

At the peak of RNA synthesis there are approximately 1,000–2,000 RI-RNA templates per cell—containing 5–10 active replication complexes each. Most of these RI-RNAs were formed during the early exponential phase of RNA synthesis; a slow but constant turnover of RI-templates occurs throughout the peak of RNA synthesis (RI molecules are converted to RF-RNA when reinitiation stops; new RI-RNAs are formed by the synthesis of cRNA on ss plus RNA). Approximately 100,000–150,000 molecules of mRNA are synthesized per cell during the switch from the exponential to the linear phase of RNA synthesis and during the early parts of the linear phase. Approximately 200,000 molecules of virion RNA are encapsidated as they are synthesized during the rest of the linear phase.

The relative roles of viral and cellular factors in the different phases of RNA replication are still largely unknown. The actual nucleotide polymerizing activity resides in the viral protein NCVP-4, the very C-terminal portion of NCVP-1b and of the polyprotein. Interestingly, most of the 5,000–20,000 active NCVP-4 replicases at the peak of RNA synthesis themselves were synthesized during the early phases of replication; the other hundredthousands of potential replicase proteins, which are synthesized later on in replication, evidently never participate in RNA replication. The other parts of NCVP-1b are certainly also involved in RNA synthesis, at least in initiation. The N-terminal NCVP-9 contains the small peptide VPg as well as a potential membrane insertion site. VPg is linked to the 5' end of all nascent RNA chains and may serve as a primer for RNA synthesis. An internal part of NCVP-1b is NCVP-7c, the viral gln/gly protease. This protease cleaves the viral precursor proteins including NCVP-1b and may thereby control the orderly sequential activation of initiation and elongation of RNA synthesis. The P2 proteins, in particular NCVPX, seem to be involved in the assembly-coupled synthesis of viral RNA. A cellular, evidently ribosomal, factor is required for initiation of cRNA synthesis on the polyA tail of ss plus RNA. Intriguingly, the cellular factor appears to confer a high specificity to viral RNA templates of the viral replicase, which *in vitro*—but not *in vivo*—copies also cellular mRNAs. Additional cellular factors are required for initiation and elongation of plus strand RNA synthesis on the double-stranded template. Studies on the infectivity of isolated RF-RNA as well as the specific alterations of host cell nuclear functions provide ground for speculations on the involvement of nuclear factors in the synthesis of plus strand RNA.

Further investigations may shed light on these and other still puzzling aspects of poliovirus RNA replication: synthesis of defective RNA by a soluble replicase early after infection; the switch from translation to RNA replication of the parental RNA; mechanisms of genetic recombination; the appearance of circular and concatameric forms of RNA; the relative localization of RNA synthesis and virion assembly with respect to the virus induced membranes; possible mechanisms of transport of RNA across these membranes; possible connections of viral RNA or of the replication complex to the cytoskeleton; the relative rotation of the replicase and template; and the effects of regulatory factors on the different phases and destinies of viral RNAs.

10

Assembly of the Virion

I. The Cytoplasmic Sites of Assembly — Virus-Induced Intracellular Membranes

A. Electron Microscopic Observations

The first complete virus particles that can be detected within the infected cell by electron microscopy or biological assay appear near the midpoint of the replication cycle: At 2 1/2 to 4 hours post infection, depending on the multiplicity of infection and on the nutritional state of the cell (Kallmann *et al.*, 1958; Dales *et al.*, 1965; Eremenko *et al.*, 1972 a, b). The appearance of the first complete virus particles thus correlates with the time of maximal rates of viral protein and RNA synthesis (Darnell and Levintow, 1960; Darnell *et al.*, 1961) (see Table 32, p. 209).

At this time, marked structural reorganization of cellular organelles and of the cell-cytoskeleton are in progress (Horne and Nagington, 1959; Dales *et al.*, 1965; Mattern and Daniel, 1965; Heding and Wolff, 1973; Bienz *et al.*, 1973; Lenk and Penman, 1979): Concurrent with a dramatic increase in the synthesis of intracellular membranes, an entire system of peculiar membrane enclosed vesicles grows in the cell center in close proximity to the deformed nucleus ("centrospheric region") (see also Chapter 6, Figs. 66–68, 71–72). These vesicles measure 70–200 nm in diameter and are bound by a single smooth membrane. At earlier times (3 h. p. i.), they are observed in small groups in the perinuclear zone, but soon the entire cell center is composed of this system of vesicles. Viral RNA synthesis is structurally tightly connected with the onset and increase of the formation of these vesicles (Bienz *et al.*, 1980) which exhibit a fast membrane turnover (Mosser *et al.*, 1972 a, b). Viral proteins radiolabeled early or midway in the infectious cycle are seen to move preferentially to the vesicles and into the nucleus (Fig. 102 B) (Bienz, 1981). Cell ribosomes are conspicuously absent from the central region. Examinations of the cytoskeletal framework of infected cells reveal that the central region contains a pattern of intermediate filaments unique to infected cells. This presumably reflects a substantial rearrangement of the zone rich in intermediate filaments which is next to the nucleus in infected cells (Lenk and Penman, 1979) (see Fig. 75, p. 248).

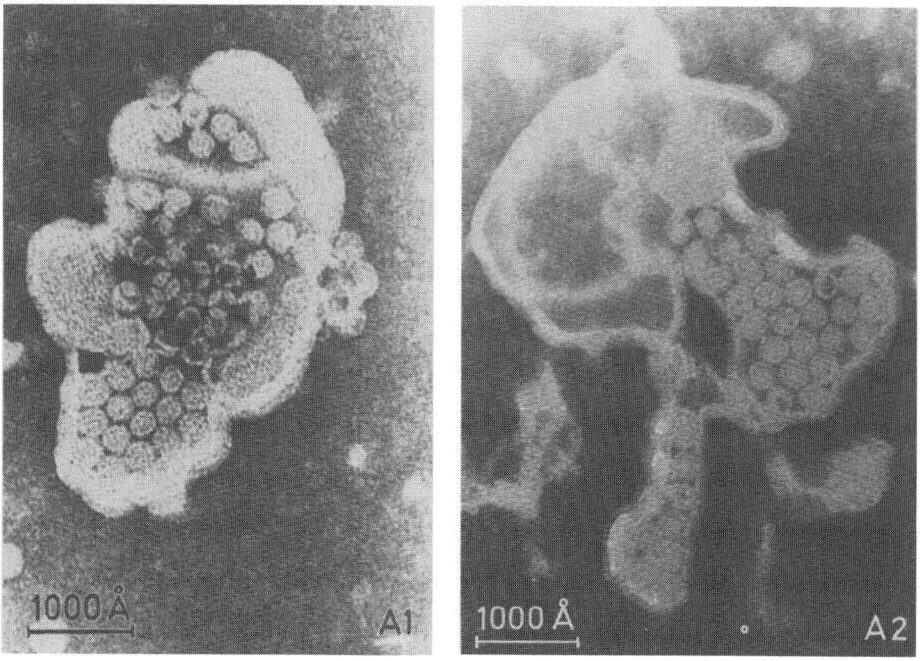

Fig. 102. Electronmicrographs of polioviruses in assembly

Fig. 102 A. Assembly of particles in membrane bound bodies from poliovirus infected HeLa cells
These figures show polioviruses in various stages of assembly in membraneous bodies found in cell fragments produced by a single cycle of freezing and thawing of poliovirus infected HeLa cell monolayers at 4 h.p.i. *(1)* and $5\frac{1}{2}$ h.p.i. *(2)*
Recognizable virus particles were frequently observed in these well-defined membrane bound bodies which measured from 30 to 1,000 nm. Subunits are visible in both complete and incomplete particles. The central region of Fig. *(1)* shows virus particles overlapping where they are more than one layer deep. Fig. *(2)* shows a small canicular appendage to the membrane bound body containing a single layer of particles. A higher magnification of Fig. *(2)* was presented in Fig. 9 C (p. 37). – Figures from Horne and Nagington, 1959 [J. Mol. Biol. *1*, 333–338 (1959)]

Fig. 102 B. Association of viral proteins and RNA synthesis with virus-induced membraneous vesicles
These figures illustrate the association of viral proteins *(1)* and viral RNA synthesis *(2)* with several clusters of virus induced vesicles in poliovirus infected HEp-2 cells
(1) shows an EM-autoradiograph of an infected cell, pulse labeled with (^{3}H)-leucine for 15 min at 2 hours 20 min p.i. and chased with cold leucine until 3 hours 30 min p.i. Silver grains representing viral proteins tend to be associated with the clusters of virus induced vesicles. Inset: The first virus-induced vesicles are found on the trans side of the Golgi complex
(2) shows an EM-autoradiograph of an infected cell, pulse labeled for 10 min with (^{3}H) uridine at 3 h.p.i. Silver grains demonstrate viral RNA synthesis in connection with several individual clusters of newly formed vesicles. In order to block the contribution of cellular protein or RNA synthesis, cells were pulsed under high salt conditions in *(1)* and treated with actinomycin in *(2)*. – Figures from (1): Bienz *et al.*, 1983 [Virology *131*, 41 (1983)]; (2): Bienz *et al.*, 1980 [Virology *100*, 390 (1980)]

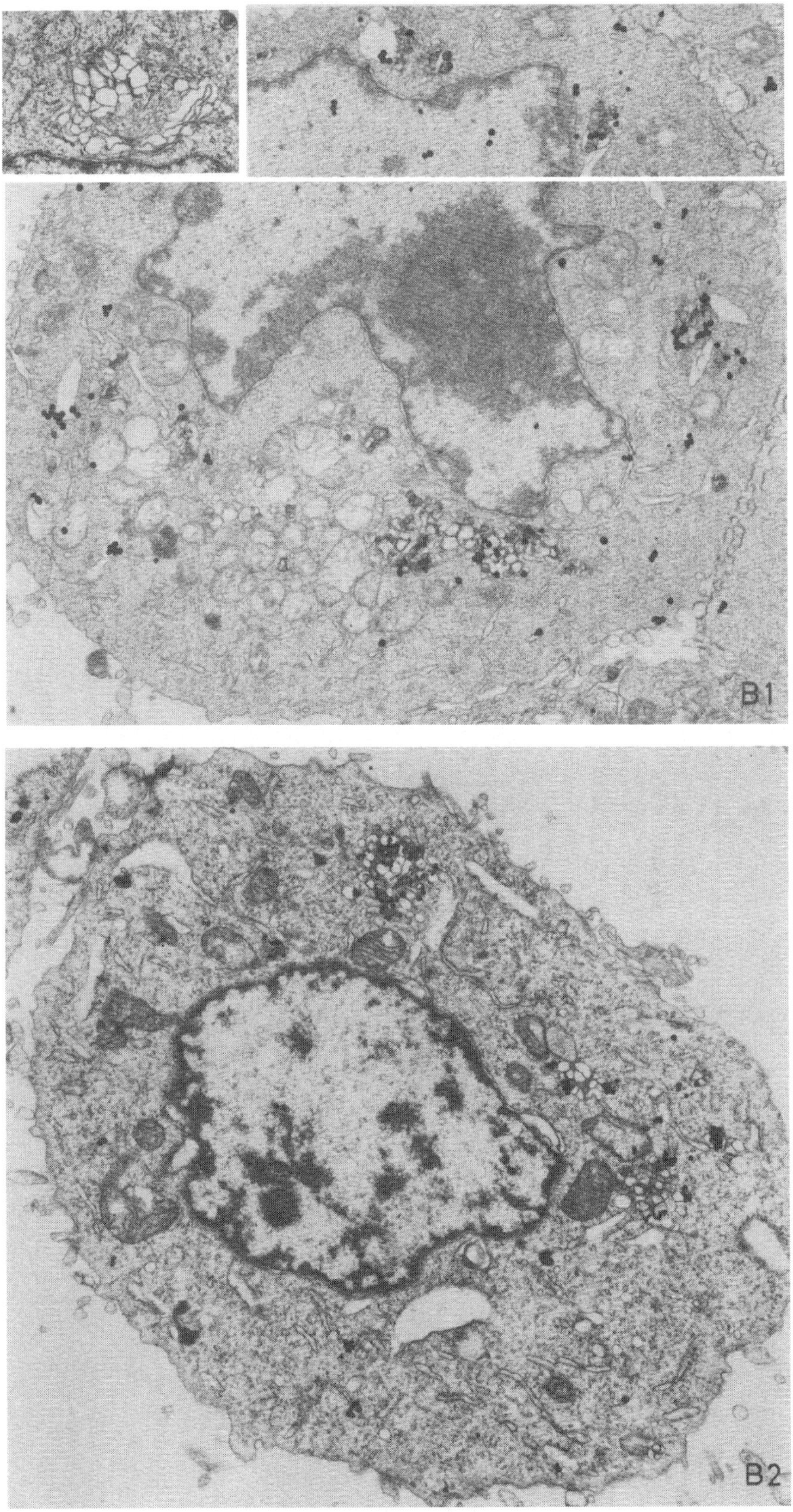
B1
B2

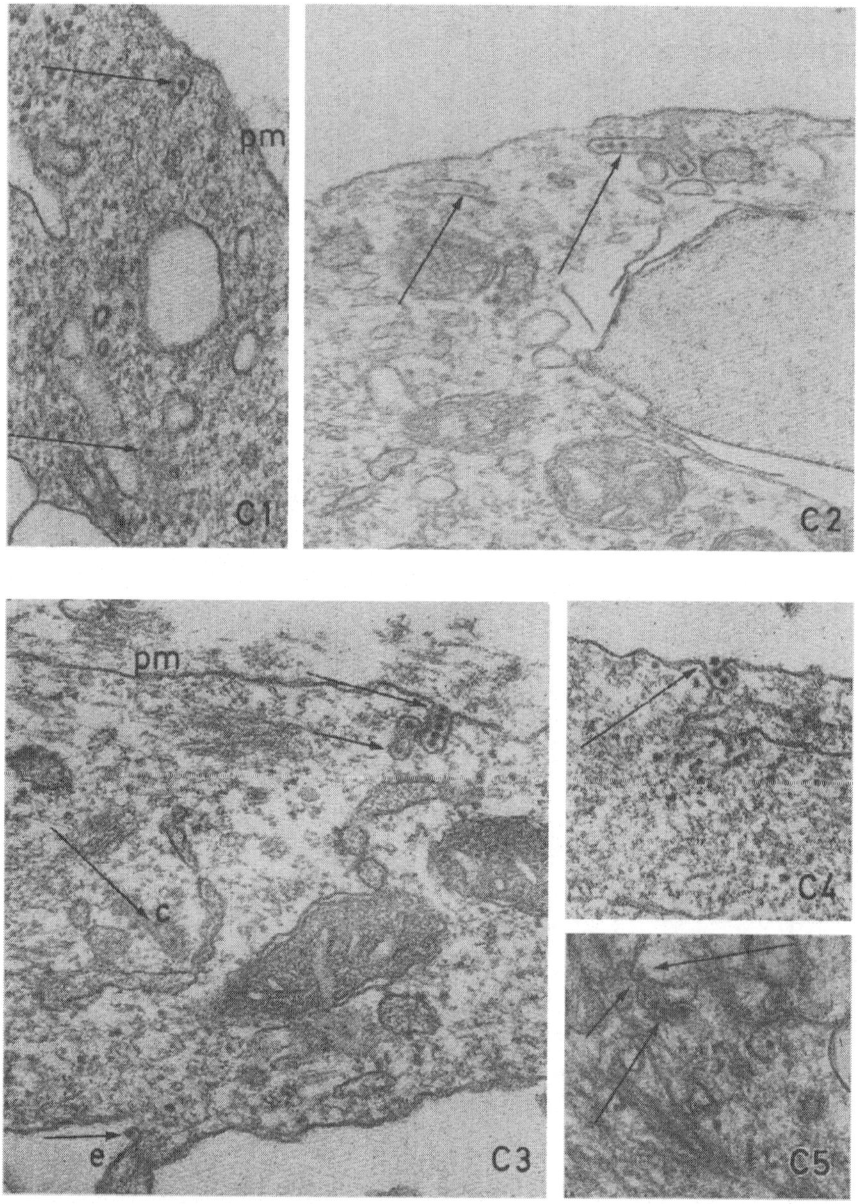

Fig. 102 C. Accumulation of progeny virions within cisternae of endoplasmic reticulum and release of virions from cisternae by fusion with the plasma membrane

Electron micrographs *(1)* and *(2)* show progeny polio-virions within cisternae of endoplasmic reticulum at various depths in the cytoplasm of infected HeLa cells, 8 h.p.i. *(1)* and of infected human chorion cells, 20 h.p.i. *(2)*. Electron micrographs *(3–5)* illustrate the release of progeny virions from cisternae *(c)* by fusion with the plasma membrane *(pm)* in poliovirus infected human chorion cells, 20 h.p.i. *(3)* and infected HeLa cells, 8 h.p.i. *(4)* and 12 h.p.i. *(5)*. – Figures from Dunnebacke *et al.*, 1969 [J. Virol. *4*, 511, 512 (1969)]

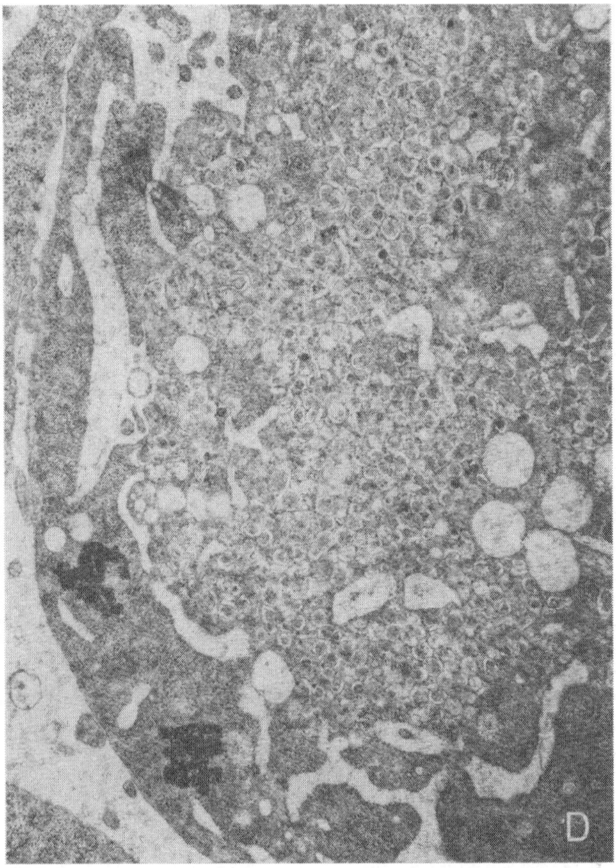

Fig. 102 D. Crystals of poliovirus in the cytoplasmic periphery
This electron micrograph shows a portion of the centrosphere region and peripheral cytoplasm of a poliovirus infected HeLa cell, 7 h.p.i. The membrane-enclosed small bodies are very numerous and occupy the whole central region of the cell. Two crystals of virus particles are present in the matrix near the surface of the cell. – Figure from Dales *et al.*, 1965 [Virology *26*, 379–389 (1965)]

Early in infection, units the size of capsomers (6 to 8 nm) are observed in the vesicles (Horne and Nagington, 1959) (Fig. 102 A). Later, dark staining, irregularly shaped objects that appear to be virus particles in various states of assembly are also clustered in the centro-sphere region. These particles are approximately 20 nm in diameter, and – in contrast to mature virions – they display a high affinity for the cytoskeletal framework (Lenk and Penman, 1979).

The first viral progeny particles are localized predominantly in the central perinuclear region (Dales *et al.*, 1965; Mattern and Daniel, 1965). Midway in infection, virus particles are often seen "trapped" within the vesicles or dispersed in the intervesicular space (Fig. 102 C, D). Empty capsids are occasionally seen in the same location. In the subsequent course of infection, virus particles steadily accumulate—either free in the peripheral cytoplasm or in membrane enclosed sacs with a density of the rough endoplasmatic reticulum (Caligiuri and Tamm, 1970 a+b). Late in infection, virus crystals may form. Such crystals are usually located in the cytoplasmic periphery below the cell membrane (Fig. 102D) but are also observed within the endoplasmic reticulum (Blinzinger *et al.*, 1969). Under certain conditions, viral crystals may also be formed within the nucleus (Anzai and Ozaki, 1969; Kawanishi, 1978). Initially, infectious virus is released through vacuoles which fuse with the plasma membrane. After several hours, virions escape from the cell in a burst, when host cells lyse and die (Dunnebacke *et al.*, 1961, 1969). These observations indicate that the intracellular membranes that are formed in the course of poliovirus infection and the resulting compartmentalization of the cell cytoplasm play a central role in *in vivo* viral morphogenesis. It seems reasonable to consider the system of membraneous vesicles in the cell center to function in part as virus-assembly factories. Virion assembly thus evidently proceeds in close proximity to viral RNA synthesis, that has also been localized to the central region (see Fig. 78, p. 261).

B. Biochemical Approaches

Biochemical approaches to the detection of the cytoplasmic sites of virion assembly usually involve the fractionation of the crude cytoplasmic extract from an infected cell by sedimentation on a discontinous gradient, followed by biochemical analyses of the components associated with the different fractions (*i.e.* by assay of polymerase or other enzymes, PAGE etc.) (Caligiuri and Tamm, 1969, 1970 a, b; Yin, 1977). Already in 1964, such studies led to the conclusion that poliovirus is made on membraneous structures rather than in the cytoplasmic matrix (Becker *et al.*, 1963; Penman *et al.*, 1964).

The capsid proteins — as all poliovirus proteins — are synthesized on polyribosomes associated with the rough endoplasmic reticulum in the peripheral cytoplasm (Baltimore *et al.*, 1963a; Roumiantzeff *et al.*, 1971 a, b). The capsid proteins themselves are always found in association with intracellular membranes, hardly any capsid related proteins are ever found in a soluble state in the cytoplasmic matrix (Caligiuri and Tamm, 1970 a, b; Korant, 1973). The capsid proteins are first associated to the rough endoplasmic reticulum near the site of their synthesis. An activity that facilitates the in vitro assembly of the second-order assembly intermediates (the 14S subunits) into 80S shells is also found in association with the rough membrane fraction (see below) (Perlin and Phillipps, 1973).

Newly formed capsid proteins and related subviral particles are found in transient association with the smooth membrane fraction between 10 and 30 minutes after the time of capsid protein synthesis (Girard *et al.*, 1967; Caligiuri and Compans, 1973; Yin, 1977). The smooth membrane fraction contains about 50% of the RNA replication complex as measured by the in vitro polymerase activity and, evidently is a major site of viral RNA synthesis (Caligiuri and Tamm, 1970 b; Butterworth *et al.*, 1976) (Fig. 74, p. 247). Freshly synthesized viral RNA can be detected in complete virions within 3—10 minutes after addition of radiolabeled nucleosides, labeled capsid proteins, in contrast, appear in complete virions only after a 20–30 minute chase (Baltimore *et al.*, 1966; Oppermann and Koch, 1973). Correspondingly, after a short pulse with labeled uridine, virus particles associated with the RNA replication complex in the smooth membrane fraction were found to have a three-to eightfold higher ratio of radioactivity to infectivity than virus particles from other cell fractions (Caliguiri and Compans, 1973). For these reasons and in accord with the electron microscopic observations it was postulated that virus RNA replication and virion formation are coupled processes which occur in association with smooth membranes (Caliguiri and Mosser, 1971; Caliguiri and Compans, 1973). A potential intermediate structure that may reflect this coupling of processes has been characterized: the procapsid in the smooth membrane fraction is associated with the viral RNA replication complex, presumably physically linked to the nascent viral RNA of the replicative intermediate (Yin, 1977).

Complete progeny virions are rarely found in the smooth membrane fraction. Instead they usually cosediment with the rough microsomal fraction (Caliguiri and Tamm, 1970a) supporting the view that completed virions are released into the peripheral cytoplasmic matrix, perhaps enclosed within sacs of endoplasmic reticulum.

Taken together, the electron microscopical and biochemical observations implicate the poliovirus induced intracellular membranes – in particular the perinuclear membrane enclosed vesicles – as the cytoplasmic sites that control the orderly assembly of virions. The peripheral rough endoplasmic reticulum evidently serves as the site of capsid protein synthesis, and in addition, contains factors that can support and enhance some of the early steps of capsid formation. The final formation of the viral capsid and its association with the viral RNA seems to occur on the membranes of the perinuclear smooth membrane vesicles in close proximity to ongoing viral RNA synthesis. Freshly formed virions probably are released mainly into sacs of endoplasmic reticulum, and are transported back to the cytoplasmic periphery (see Fig. 78, p. 261).

II. Subviral Particles in the Infected Cell – Potential Assembly Intermediates

A. Overview

In the search for potential assembly intermediates, a number of capsid related particles have been isolated from extracts of infected cells and biochemically characterized (for reviews see Phillips, 1972; Casjens and King, 1975; Rueckert, 1976; Putnak and Phillips, 1981). The different forms of capsid and capsid precursors that can be distinguished in poliovirus infected cells are listed in Table 66. The classic distinguishing feature is the sedimentation coefficient which reflects mass and size of the particles (Watanabe *et al.*, 1962; Phillips *et al.*, 1968) (Fig. 103). In addition to the complete virions, which sediment at 156S, and consist of the final capsid proteins VP_1, VP_2, VP_3 and VP_4, the putative intermediates include:

Table 66. *Subviral particles in the infected cell - potential assembly intermediates*

	Sedimentation coefficient ($S_{20,w}$)	Composition: protein	Composition: RNA	Sensitivity to RNase: −EDTA	Sensitivity to RNase: +EDTA	Sensitivity to disruption by 0.5% SDS	Cell-attachment capacity	Protease sensitivity
Capsid protein precursor	5	NCVP 1a	–				no	
Protomer	5	VP_1, VP_3, VP_0	–			yes	no	
Pentamer	14	$(VP_1, VP_3, VP_0)_5$	–			yes	no	
55S particle	55	$(\text{pentamer})_{5\ (\text{or } 6?)}$	–			yes	no	
Procapsid A	65	$(\text{pentamer})_?$		–	–	yes	yes	no
Procapsid B	80	$(\text{pentamer})_{12}$		–	–	yes	no	yes
80S RNP	60–80	$(\text{pentamer})_{10-12}$	RNA/VPg	n.d.	yes	yes	n.d.	n.d.
Provirion	125	$(VP_1, VP_3, VP_0)_{60}$	RNA/VPg	no	yes	yes	no	n.d.
Immature virions	150	$(VP_1, VP_3, VP_0/VP_2 + VP_4)_{60}$	RNA/VPg	no	yes	yes	yes	n.d.
Virion	155	$(VP_1, VP_3, VP_0)_2 + (VP_1, VP_3, VP_2, VP_4)_{58}$	RNA/VPg	no	no	no	yes	no

1. "Light" proteins or protein aggregates (sedimenting at 5-10S) (Watanabe *et al.*, 1962; Scharff *et al.*, 1964; Ghendon *et al.*, 1972). These are the initial translation product of the 5' (P1) genome region, a 100,000 dalton polyprotein, the capsid precursor protein NCVP1a (Summers *et al.*, 1965; Jacobson and Baltimore, 1968a); or aggregates of a single copy of each of the three cleavage products of the polyprotein (VP_0, VP_1 and VP_3) (Korant, 1973), the "protomer" (Rueckert, 1976);

2. Aggregates of VP_0, VP_1 and VP_3 that sediment at 14S and are thought to be pentamers of the 5S protomer (Phillips *et al.*, 1968);

3. Aggregates of the 14S pentamers sedimenting at 55-63S possibly representing a pentamer of the 14S subunits (Su and Taylor, 1976; Lee *et al.*, 1978; Lee and Colter, 1979; Delgadillo and Vanden Berghe, 1979; La Colla *et al.*, 1981);

4. A protein shell, sedimenting at 65-80S, thought to represent a complete capsid, *i.e.* consisting of 60 copies each of VP_0, VP_1 and VP_3, lacking the genomic RNA and termed the procapsid (Hummeler *et al.*, 1962; Scharff and Levintow, 1963; Maizel *et al.*, 1967; Jacobson and Baltimore, 1968b);

5. Ribonucleoprotein particles (RNPs) of intermediate sedimentation coefficients (at or near 80S) (Agol *et al.*, 1970a; Marongiu *et al.*, 1981a);

6. A particle that has viral RNA plus the procapsid proteins, the "provirion", found to sediment at 125S (Fernandez-Tomas and Baltimore, 1973);

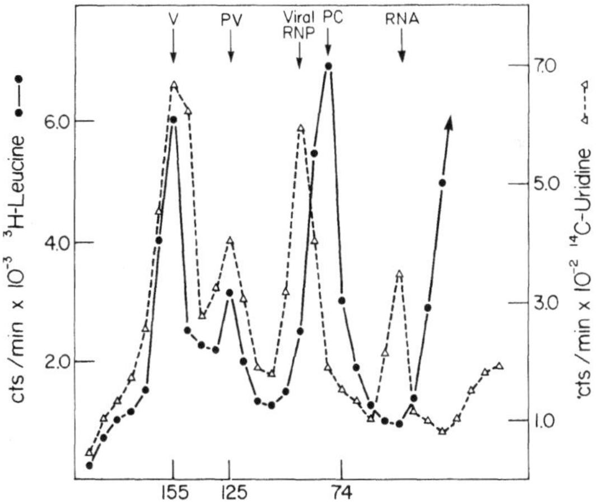

Fig. 103. Characterization of potential poliovirus assembly intermediates by sucrose density centrifugation

This figure illustrates the separation of potential poliovirus assembly intermediates by sucrose density centrifugation of cytoplasmic extracts from infected HeLa cells. Infected cells were pulse labeled with (^{3}H)-leucine at 3 h.p.i. for 30 min and with (^{14}C)-uridine for 15 min at 3.25 h.p.i. Cells were chased with a 100 fold excess of unlabeled leucine for 5 min. Cytoplasmic extracts were prepared and centrifuged on a linear 15–35% sucrose gradient. Radioactivity in protein (solid line) and RNA (dashed line) was determined in fractions from the gradient. Arrows indicate the positions of 156 S virion (V), 125 S provirion (PV), 80 S viral RNP, 76 S procapsid (PC), and 35 S viral RNA (RNA). Potential 55 S assembly intermediates may be hidden in the right hand shoulder of the 80 S procapsid peak; 14 S and smaller capsid precursors are at the top of the gradient. — Figure from Fernandez-Tomás and Baltimore, 1973 [J. Virol. *12*, 1122–1130 (1973)]

7. Immature virions sedimenting at 150S that are enriched in VP_0 and have some provirion and virion like properties (Guttmann and Baltimore, 1977a).

All of these structures have been implicated in one way or the other in the assembly process, although the precise role of some structures (in particular of the 80S particles) is still controversial (see below).

Whereas the relative stoichiometry of the different capsid proteins remains constant during the association into larger structures, marked changes in conformation seem to occur. Such conformational changes in viral structures in the process of morphogenesis are reflected by changes in the antigenicity of these structures. The three major poliovirus-specific structures produced in infected HeLa cells (5S, 14S, 80S) apparently show little antigenic relationship (Roizmann et. al., 1958; Scharff *et al.*, 1964; Ghendon and Yacobson, 1971). The smaller 5S subunits carry antigenic determinants also present on the individual capsid proteins after guanidine dissociation; these determinants have been named S (Scharff *et al.*, 1964). The subviral particles are considered to be poor immunogens (Hummeler and Tumilowicz, 1960; Ghendon and Yacobson, 1971): antisera obtained after inoculation of mice with isolated 5S, 14S or 80S particles had no significant or only little neutralizing capacity (Hinuma *et al.*, 1970; Ghendon and Yacobson, 1971). On the other hand, a recently obtained monoclonal neutralizing antibody for poliovirus can also aggregate naturally occurring 80S particles very efficiently, 14S pentamers less efficiently, but not the isolated capsid proteins, the 5S protomer, or artificial empty capsids (lacking VP_4) (Icenogle *et al.*, 1981) (see also Chapter 3, section VIII). The unstable *in vivo* capsids actually seem to have N antigenicity *in vivo* which is readily converted to the stable C antigenic state (Rombaut *et al.*, 1982) (see p. 433 ff.).

Conformational reorientations within the subviral structures may be necessary for a sequential exposure of new bonding domains. This would allow an orderly step by step assembly of the viral components (Rueckert, 1971, 1976). Factors that may be involved in regulating sequential conformational shifts of the capsid proteins are: 1. proteolytic cleavages of the capsid proteins (NCVP1a to VP_0-VP_1-VP_3, and VP_0 to VP_4-VP_2), 2. the interplay of hydrophobic and hydrophilic interactions between the different capsid proteins and between these and cellular components, in particular the intracellular membranes, and 3. interaction of capsid fragments or "procapsids" with viral RNA.

Before we discuss the dynamic and regulatory aspects of assembly, however, we will summarize the properties of the assembly intermediates in the following sections.

B. NCVP1a and the 5S Protomer

The coding region of polio RNA for coat protein (P-1) is translated into a large, 95,000 molecular weight precursor protein, NCVP1a. NCVP1a is derived by cleavage at a tyr/gly site from a larger polio polypeptide chain during ongoing translation, presumably by a host cell protease. NCVP1a has never been found in a soluble state in the cytoplasm (Korant, 1973). It becomes rapidly associated to intracellular membranes, perhaps already during its synthesis. NCVP1a itself is rapidly cleaved at two specific gln/gly sites by viral protease NCVP7 into VP$_0$, VP$_1$ and VP$_3$. *In vivo,* VP$_0$, VP$_1$ and VP$_3$ are always found associated to one another and to cellular membranes. Application of the helical hairpin hypothesis for insertion of proteins into membranes (Engelman and Steitz, 1981) and the hydrophobicity profile (von Heijne, 1981) reveal several possible membrane insertion sites in NCVP1a (see Fig. 15, p. 57).

The isolated capsid precursor protein NCVP1a as well as the polypeptide aggregates composed of one of each of VP$_0$, VP$_1$ and VP$_3$ sediment at 5S (Phillips *et al.*, 1968; Korant, 1973). It is not certain whether these represent consecutive intermediates or alternative starting points of the regular assembly process. Since aggregates of NCVP1a have never been found in poliovirus infected cells, it is more likely that the 5S VP$_{0\text{-}3\text{-}1}$ aggregate is the starting unit for further assembly. In any case, 5S particles that contain the complete sequence of one of each of the final capsid proteins are the most fundamental building blocks for the viral capsid: 60 of these so-called protomers are eventually associated to form the capsid (see Section IV, below).

C. The 14S Pentamer

The subviral particles that sediment as a distinct peak in sucrose gradients at 14S (Watanabe *et al.*, 1965; Phillips *et al.*, 1968) have been implicated as the fundamental second order assembly intermediate by a variety of experimental studies (see p. 443 ff.). Comparable particles are found for all picornaviruses. Analysis of the composition of 14S particles isolated by PAGE revealed that they contain mainly VP$_0$, VP$_1$, and VP$_3$ (Phillips, 1971; Phillips and Fennell, 1973), as well as trace amounts of NCVP1b, NCVP2, NCVP3, and NCVP4. The relative proportion of the main components VP$_0$, VP$_1$ and VP$_3$ of the 14S particles are similar to those observed in the 80S empty capsids or in complete virions (*i.e.* in roughly equimolar amounts). Nonetheless, some variations in the relative contents of VP$_0$, VP$_1$, and VP$_3$ were observed among different 14S particle preparations (Phillips and Wiemert, 1978). The significance of the other viral proteins, traces of which cosediment at 14S, is not clear. They might have some as of yet unknown function in the assembly process although they probably do not represent integral components of the 14S particles. The same can be said for the host protein component cosedimenting at 14S which was estimated to comprise at least 80% of total protein in the 14S region (similar proteins were found in the 14S region of mock-infected cells) (Phillips and Wiemert, 1978). Observations from other picornavirus systems suggest that the 14S particles are composed of 5 copies each of VP$_0$, VP$_1$, and VP$_3$, most likely resulting from the association of 5 protomers (Rueckert, 1976). The expected molecular weight of $4.5–5 \times 10^5$ for a pentamer of protomers is consistent with the 14S sedimentation velocity. The VP$_0$ and VP$_1$ components of the 14S particle are sensitive to proteolytic cleavages by trypsin and chymotrypsin (Putnak and Phillips, 1982).

Three functionally distinct types of 14S subunits have been recognized in 14S particle preparations from infected HeLa cells. 14S* mature particles which are able to self-assemble *in vitro* into 80S shells, 14S which can be assembled only when they have been "activated", and defective particles which cannot be activated (Perlin and Phillips, 1975). The functional distinctions of the different types of 14S subunits could not be correlated with any compositional differences. Isolated 14S subunits are assembled into 80S shells upon in vitro incubation at 37–40°C only at low salt concentrations and at neutral pH (Phillips *et al.*, 1968; Putnak and Phillips, 1981).

A particle comparable to the 13S particle observed in cells infected by other picornaviruses such as EMC (a cardiovirus), composed of a pentamer of the precursor protein (McGregor *et al.*, 1975, 1977) (NCVP1a for polio), has not been detected in polio-infected cells.

D. The 55S-Particle

Early reports on the purification and crystallization of Coxsackie virus (Briefs *et al.*, 1952; Mattern, 1958) showed the presence of material sedimenting at about 80S and 40S, but such material was then regarded as contamination. Subviral particles sedimenting at 55S–63S have rarely been noted in the traditional cell systems used for studies of poliovirus infection. The attention for the existence of such a structure and a possible central role in the assembly process instead has come first from studies on other picornaviruses, in particular from close relatives of poliovirus in the family of enteroviruses: a) A particle sedimenting at about 45S was isolated from bovine enterovirus lysates (Su and Taylor, 1976). The particle is composed of VP_0, VP_1 and VP_3. The particles are unstable at high ionic strength (150 mM NaCl) even at 4°C, where they are converted to 80S particles. b) Virus-related particles isolated from cell culture harvests of swine vesicular disease virus, a porcine enterovirus, included 148S virions, 81S RNA-free empty capsids, and a third 49S particle with a polypeptide composition typical of procapsids (*i.e.* VP_0, VP_1, VP_3), and antigenicity similar to virions (Moore, 1977). The 49S particle reveals a buoyant density alike that of the 81S particles at 1.28 g/ml (148S virions = 1.35 g/ml). In contrast to poliovirus infected cells, however, no empty particles could be detected in infected cells with the EM. The 49S particles were lost when incubated in serum or at high temperatures, probably by being converted to 81S particles.

Recently, a 50–55S particle has also been found in mengovirus (a cardiovirus) infected cells (Lee *et al.*, 1978; Lee and Colter, 1979). The particle is also composed of VP_0, VP_1 and VP_3. It can be disassembled into 14S subunits by prolonged incubation at room temperature. It is converted into a 75S particle by increasing the ionic strength of the suspending buffer or by banding in CsCl (75S particles usually cannot be detected in extracts of mengovirus infected cells). 53S particles band sharply at a buoyant density of 1.296 g/cm^3 like empty capsids in CsCl gradients. When such particles were recovered from CsCl gradients they sedimented at 75S in sucrose gradients. Careful molecular weight determinations suggest that the molecular composition of the 53S and 75S particles is (VP_0, VP_1,

$VP_3)_{25}$ and (VP_0, VP_1, $VP_3)_{50}$, respectively. The 53S particle thus represents an incomplete halfshell composed of five 14S subunits (instead of six in a complete halfshell). If the five 14S subunits are fit into an equivalent and symmetric arrangement in the 53S particles, the resulting structure would be a ring with a central "hole" (see Fig. 111, p. 61).

A 50—60S particle is also occasionally seen in poliovirus infected cells as a small peak (3% of total radioactivity) next to the 80S peak or as a shoulder on the side of the 80S peak. Such a small peak can be seen in some figures of the early publications of assembly studies (for example Jacobson and Baltimore, 1968a; Agol *et al.*, 1970) (see Figs. 103 and 104, p. 429 and p. 438). Perhaps due to the relative smallness of the peak, the particle was at the time not implicated as an assembly intermediate. It is the main subviral particle in infected cells, however, shortly after the release from treatment with the assembly inhibitor Py 11 (see p. 449) (La Colla *et al.*, 1981). The relative proportion of the 55S particle also increases in polio infected MiO cells when the cells are grown under suboptimal culture conditions, i.e. in suspension instead of monolayers (Ghendon *et al.*, 1972). A particle reported to sediment at 63S was formed during *in vitro* assembly of isolated 14S subunits under rate limiting conditions (Phillips, 1971). The 65S N antigenic particle (Rombaut *et al.*, 1982), recently isolated from poliovirus infected cells, presumably is equivalent to the *in vivo* procapsid (see below).

E. The 80S Shell

Particles sedimenting in the range of 70 - 80S consisting of 30 nm shells lacking RNA, have long been known to occur during the course of poliovirus multiplication (Hummeler *et al.*, 1962; Scharff and Levintow, 1963). These particles, which display a lighter density in CsCl gradients (1.29g/cm^3, typical of proteinacious structures) than virus (1.34g/cm^3) have been called *natural t*op *c*omponent (NTC) or procapsids. They are composed primarily of VP_0, VP_1 and VP_3 (Maizel *et al.*, 1967; Jacobson and Baltimore, 1968a), although NTCs with varying amounts of VP_2 have also been isolated from infected cells (Maizel *et al.*, 1967; Agol *et al.*, 1970; Vanden Berghe and Boeye, 1973a; Phillips and Fenell, 1973). Differences in polypeptide composition of NTC preparations may reflect different origins of the empty shells (see below).

Recently, the existence of two different functional states of procapsids have been reported by several authors (Marongiu *et al.*, 1981b; Putnak and Phillips, 1981; Rombaut *et al.*, 1982). The properties of the two particles are compared in Table 67. Different authors have reported different estimates of the sedimentation velocities of unstable NTCs (65S, 75S, 80S). It is likely, though not yet proven, that these are indeed identical particles in different conformational states. The "unstable" procapsid can be isolated from extracts of poliovirus infected cells under careful conditions. This unstable particle can be disassembled into 14S subunits; it can be rapidly and irreversibly converted into a stable 80S particle under a number of conditions that are generally used during common isolation procedures. The naturally occuring procapsids resemble the shells of virions in

Table 67. *Properties of stable and unstable 80S particles*

	Unstable particles[a]	Stable particles[b]
Isolation conditions	pH 7.8, 4 °C	neutral pH, isolation or sedimentation at 20 °C–37 °C, *in vitro* assembly
Treatment with pH 8.4	disassemble into 14S subunits (reversible)	stable
0.1 % SDS (pH 7.4)	disassemble into 5S subunits	stable
Incubation or sedimentation at 20 °C or 37 °C	conversion to stable 80S particles	stable
Formation during *in vitro* assembly of 14S subunits	no	yes
Presence *in vivo*		
at neutral pH 7.8	yes	no
at alkaline pH (8.4)	no	no
In MiO cells	no	no
Sedimentation coefficient	80S	80S
Composition	VP_0, VP_1, VP_3	VP_0, VP_1, VP_3
Significance	present in the cytoplasm of infected cells, possible assembly intermediate, storage of 14S subunits	artefactually made from unstable 80S particles during isolation procedures, probably never present in this form *in vivo*
Antigenicity	D/N	C/H
Isoelectric point	6.8	5.0

[a] Marongiu *et al.*, 1981; Putnak and Phillips, 1981, 1982; Rombaut *et al.*, 1982.
[b] Hummeler *et al.*, 1962; Scharff and Levintow, 1963; Maizel *et al.*, 1967.

many aspects: they carry D/N antigenicity, have a buoyant density of 1.31g/cm^3, an isoelectric point of 6.8, are insensitive to protease attack, and harbor tyrosine residues on VP_1 that are sensitive to gentle iodination (Putnak and Phillips, 1982; Rombaut *et al.*, 1982). Interestingly, the VP_0 component of in vivo procapsids becomes sensitive to cleavage by trypsin or chymotrypsin only after its conversion to the stable C antigenic, pI 5.0 form. The "unstable" 80S particle is probably the only form of procapsid occuring in vivo, and therefore its characteristics are the ones relevant to possible models for virion assembly. The stable 80S particle is apparently a normal byproduct of isolation procedures or conditions employed for in vitro studies of the assembly process. A capsid conformation resembling that of stable procapsids nevertheless may be important during the provirion stage of assembly to permit cleavage of VP_0.

Recently, two different forms of 80S shells – analogous to those obtained from infected cells – were reported to arise also from the in vitro assembly of 14S particles, depending on the reaction conditions (Putnak and Phillips, 1981; Putnak and Phillips, 1982). The isoelectric point of 80S shells formed by self-assembly of 14S subunits is 5.0, whereas that of 80S shells formed in an assembly reaction mediated by infected cell extracts is 6.8, similar to that of 80S particles isolated from infected cells. Such differences in pH probably reflect different conformational states of 80S shells similar to those observed for intact virions (see Chapter 3, Fig. 32, p. 110), or they may reflect the association of some other components to the shell, for example association of lipid or assembly-enhancing factors.

The characteristics of the stable empty shells, in particular their relatively rigid structural stability, had for a long time provided difficulties for envisioning the role of such particles in the subsequent assembly process. The demonstration of unstable 80S particles *in vivo* has revitalized a thermodynamically more favorable concept on the interaction of viral RNA with a capsid that exhibits some inherent "flexibility" (for a more detailed discussion, see section IVD, below). Since ribonucleoprotein particles of polioviruses (see below) are quite unstable, it is possible that the empty shells detected in infected cells may arise by two different pathways: 1. by assembly from 14S subunits, and 2. from ribonucleoprotein particles that have lost their RNA. 80S particles, however, are not found in all poliovirus infected cell systems: f. e. no such particles could be detected in poliovirus infected MiO cells, a continuous line of tonsil cells of Macaca rhesus monkeys which are highly sensitive to poliovirus under the normal monolayer culture growth conditions. When infected MiO cells are taken off the glass and are grown in suspension, however, 74S (and 55S) particles can be detected (Ghendon *et al.*, 1972). Also, the 14S particles isolated from infected MiO cells, self-assemble *in vitro* into stable 73S particles.

It should be noted that comparable particles (empty capsids) have only rarely or not at all been found in cells infected with other members of the picornavirus family, f. e. mengovirus (Lee and Colter, 1979), although they are often observed for such viruses as products of *in vitro* self assembly processes. The 75S particle described for mengovirus, which is formed during *in vitro* assembly reactions, but not found *in vivo*, has been carefully analyzed for composition, molecular weight, and stokes radii: the data indicate an incomplete shell composed of $(VP_0, VP_1, VP_3)_{50}$. Apparently, the 73S shells of mengovirus are "missing" two 14S subunits. The relative location of the corresponding "holes" in the capsid are not known. The corresponding shells of poliovirus have never been analyzed with sufficient care to permit an answer to the question whether they are also incomplete shells or rather complete $(VP_O, VP_1, VP_3)_{60}$ shells.

F. Ribonucleoprotein Particles

At some intermediate stage of poliovirus morphogenesis, the genomic RNA must interact with the viral capsid proteins if the RNA is to be encapsidated successfully by the viral capsid. It is difficult to envision a simple one-step interaction between the RNA and a complete capsid to a completed virus particle. Rather, the encapsidation of a relatively large molecule of RNA by a proteinacious capsid into a compact virion is likely to involve a series of steps via structurally and perhaps compositionally different ribonucleoprotein particles (RNPs). It should be noted that in the cytoplasm cellular and viral RNA is always present in association with proteins (Baltimore and Huang, 1968, 1969). Either such proteins are removed from viral RNA during assembly or the newly synthesized RNA is directly associated with viral capsid proteins during or immediately following its synthesis.

It is probable that the late steps of virus morphogenesis proceed through a series of RNPs no matter whether the encapsidation of the RNA occurs as a) an insertion of a long single stranded RNA molecule in the course of its synthesis into some "procapsid" – this would require some time consuming mechanism for the compact folding of RNA inside the procapsid – or as b) a completely synthesized RNA of some globular secondary structure (with 60–80 % double stranded regions) that is surrounded in sequence by a set of defined subcapsid components (see section IV below).

1. The Slow-Sedimenting (80S) RNPs

The occurrence of poliovirus-related RNPs—sedimenting as a broad peak at about 80S—in poliovirus infected cells was demonstrated more than 10 years ago (Agol *et al.*, 1970). Yet these relatively slow sedimenting RNPs were neglected for a long time, and the study of these RNPs has been intensified only recently (Marongiu *et al.*, 1981a, and La Colla, personal communication, Table 68). When comparable particles were observed in other studies, they were generally disregarded or considered to be breakdown products of polyribosomes or replication complexes (e. g. Fernandez-Tomas *et al.*, 1973; Yin, 1977). Perhaps the relative instability of such RNPs to some procedures employed for the isolation and characterization of assembly intermediates from infected cells may explain why they have evaded attention for so long.

60-80S subviral particles usually reveal a sharp peak in CsCl gradients at a buoyant density of 1.29 typical for empty capsids containing only protein and no RNA (Jamison and Mayor, 1966). In contrast, when 70-80S particles are treated with formaldehyde, which is known to stabilize ribonucleoproteins (Spirin *et al.*,

Table 68. *Comparative properties of poliovirus RNPs*

	In vivo[a, b] RNPs	Provirions[d]	*In vitro*[c] (virion heated for 2 min at 56 °C)	Virions
Sedimentation velocity (S° 20, W)				
– isotonic conditions	60–80	125 or 155	80	155
– low salt	n.d.	n.d.	decreased	unchanged
– plus 200 μg/ml DEAE-dextran	n.d.	n.d.	increased	unchanged
Buoyant density in CsCl (g/cm³)				
– without H_2CO	release of RNA	release of RNA	release of RNA	1.34
– after treatment				
with 2 % H_2CO	n.d.	n.d.	release of RNA	n.d.
with 4 % H_2CO	1.37–1.42	n.d.	1.32 (empty capsids)	1.44–1.45
RNase sensitivity	n.d.	no	yes	no
Incubation at 37 °C	converted to 125–155S RNPs	n.d.	stable	stable
Treatment with				
– pH 8.5	stable	n.d.	n.d.	stable
– 20 mM EDTA	stable	release of RNA	n.d.	stable
– SDS 0.1 %	release of RNA	n.d.	n.d.	stable
1 %	n.d.	release of RNA	release of RNA	stable
Composition	(presumably VP_0, VP_1, VP_3 + 35S RNA)	VP_0, VP_1, VP_3 + 35S RNA	VP_2, VP_1, VP_3 plus 35S RNA	VP_2, VP_1, VP_3, VP_4, VP_0 plus 35S RNA
Antigenicity	n.d.	n.d.	C	D

[a] Agol *et al.*, 1970.
[b] Marongiu *et al.*, 1981a.
[c] Breindl, 1971.
[d] Fernandez-Tomas and Baltimore, 1973.
n.d. = not determined.

1965b), prior to centrifugation in CsCl gradients, a second peak in addition to the empty capsid peak is observed (Agol *et al.*, 1970) (Fig. 104). This is a broad heterogeneous peak with a buoyant density of 1.37–1.42 g/cm^3 indicative of a protein-RNA complex (recall that the density of poliovirus RNA in CsCl is 1.9 g/cm^3, of native virion 1.34 g/cm^3, and of formaldehyde treated virion like that of dense particles 1.44-1.45 g/cm^3).

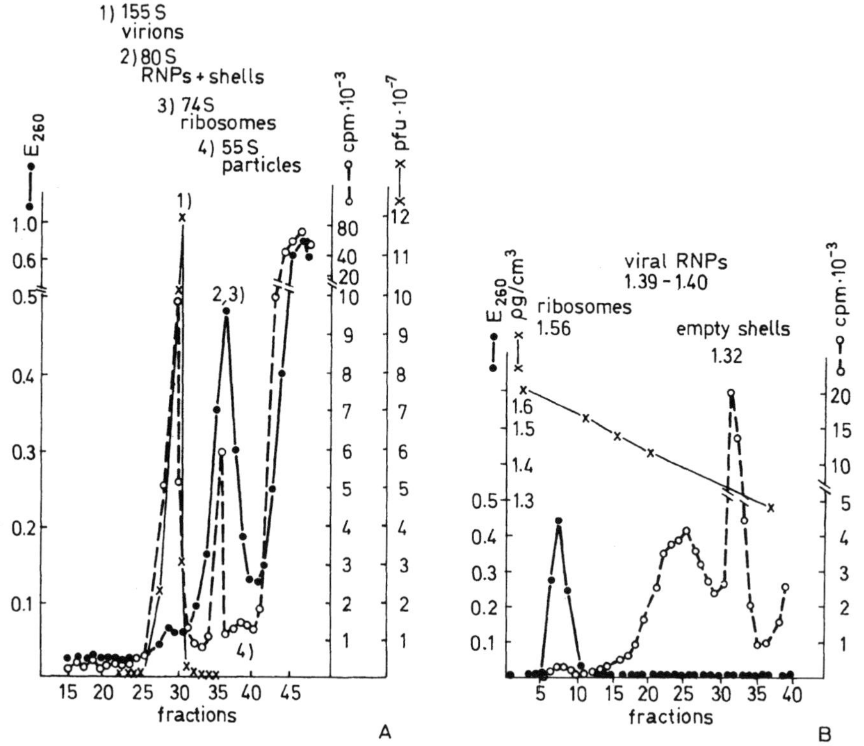

Fig. 104. Separation of poliovirus 80 S RNPs from empty capsids and ribosomes

A This figure illustrates the separation of poliovirus assembly intermediates by fractionation of cytoplasmic extracts from infected cells on sucrose density gradients (see also Fig. 103). HeLa cells in monolayer cultures were infected with the Neva strain of poliovirus. Actinomycin C was present in the medium after termination of the adsorption period in order to block cellular RNA synthesis. Cells were labeled with (^{14}C)-leucine and (^{14}C)-lysine at 4 h.p.i. and incubation was continued under normal osmotic pressure until 8 h.p.i. The cytoplasmic extract was prepared and centrifuged in a sucrose density gradient. Fractions were assayed for optical density at 260 nm (RNA), radioactivity (viral protein), and infectivity (progeny virions). 80 S viral RNPs, empty shells, and ribosomes comigrate under these conditions (peak 2/3)

B This figure shows the separation of ribosomes, viral RNPs, and empty shells from the 70–80 S peak of a sucrose gradient by centrifugation in CsCl density gradients. Cells were infected and treated as in A. The fractions from a sucrose gradient containing the peak of radioactivity sedimenting in the areas of ribosomes (peak 2/3 of A) were collected and treated with 4% formaldehyde in order to prevent dissociation of RNPs. The sample was then subjected to centrifugation in a CsCl density gradient. Fractions were assayed for optical density at 260 nm (solid line = ribosomal RNA) and radioactivity (dashed line = viral protein). Under these conditions, ribosomes, viral RNPs, and empty shells are clearly separated. – Figures from Agol *et al.*, 1970 b [Virology *41*, 533–540 (1970)]

The intermediate density of such RNPs could be explained by postulating a virion like particle containing less RNA than intact virion or – more likely – a virion like particle containing genome sized RNA of lesser density (*i.e.* more "unravelled" RNA). The sedimentation coefficient of 60-80S is also consistent with the interpretation of an RNA-containing particle that is larger (more "swollen") or more elipsoidal in shape (f. e. due to protruding RNA portions) than the native virion. Evidently the RNPs are quite unstable particles, which–unless stabilized by formaldehyde treatment–are dissociated into empty shells and free RNA by centrifugation in CsCl density gradients. Formaldehyde interacts in a number of ways with poliovirus: 1. Inactivation of RNA and stabilization of RNA structure by formation of crosslinkages; 2. loosening of capsid structure–increase in the permeability to Cs^+ ions and conversion to dense particles; 3. stabilization of capsid structure by crosslinks. Thus, the formaldehyde, so to speak, "traps" the RNA in a stable form in the RNPs. It is unlikely that these RNPs represent viral RNA associated to ribosomes, since ribosome-bound RNA exhibits a much higher density in CsCl gradients (between 1.56g/cm^3 for ribosomes and 1.9 g/cm^3 for viral RNA). It cannot be excluded, that they represent viral RNA associated with cellular proteins into 80S complexes (Huang and Baltimore, 1970). This is unlikely, however, since the proteins in the RNPs are synthesized (labeled with radioactive amino acids) even 4 hours after viral infection, at a time when cellular protein synthesis is blocked (Fig. 104B).

Interestingly, the 80S–90S ribonucleoprotein particle induced by mild heat dissociation of native poliovirus (electron microscopic appearance is similar to that of "coreless" virions) is similary unstable, being dissociated to empty capsids and RNA by centrifugation in CsCl gradients (Breindl, 1971a). The properties of *in vivo* and *in vitro* poliovirus RNPs are summarized in Table 68.

Recently, larger quantities of ^{14}C amino acid and ^3H uridine labeled 80S RNPs were isolated with careful procedures from poliovirus infected HeLa cells at 4.5 h p. i. and used for biochemical analysis (Marongiu *et al.*, 1981a, and La Colla, peronal communication, see Table 68). The particles are stable at pH 8.5 or in the presence of 20 mM EDTA (a chelator of divalent cations). The RNA within the particles, however, becomes sensitive to RNase in the presence of EDTA. Treatment of the RNPs with 0.1 % SDS results in the release of genome-sized 35S RNA. Prolonged incubation of isolated 90S RNPs at 37° C leads to their conversion to faster sedimenting RNPs with a broad range of sedimentation velocities between 125–155S, possibly representing provirions (see below).

The effects of treatment with low or high salt concentrations on these particles has not yet been elucidated. The decreased sedimentation velocity of *in vitro* heat induced 80S RNPs in gradient buffers of low ionic strength (Breindl, 1971a, Table 68), may reflect an osmotically induced swelling of these particles. The converse, an osmotically induced condensation of viron RNPs by high ionic strength is postulated to occur in the final stages of virion morphogenesis (see section IV).

The answer to the question, wether the VP_0 component is susceptible to proteolytic cleavage in the 80S RNPs and in the faster sedimenting RNPs that are produced in vitro by incubation or possibly also by high salt treatment of the 80S

RNP is of pertinence to concepts of virion morphogenesis. To date, it has not been possible to induce *in vitro* the cleavage of isolated VP_0 or of the VP_0 component in empty capsids to VP_2 and VP_4. It is probable that VP_0 acquires the conformation susceptible to protease only after interaction of the viral RNA with the capsid proteins. The elucidation of the stage at which VP_0 becomes susceptible to cleavage will provide an answer to the functional role of this cleavage: *i.e.* whether it serves as an aid in the condensation of RNPs or merely to stabilize the final condensed structure. With a suitable substrate, identification of the responsible protease should also be possible.

It would also be of interest to determine the susceptibility of the RNA component of *in vivo* RNPs to cleavage by RNase or to interaction with DEAE-dextran. This information is particularly important for concepts of viral RNA encapsidation. Susceptibility to RNase or interactions with DEAE-dextran (a multivalent cation that increases the sedimentation rate of viral RNA and of heat induced *in vitro* 80S RNPs but not that of empty shells, Breindl, 1971a, Table 68) would indicate that part of the RNA of the slow sedimenting RNPs, is exposed on the surface of the virion. Controlled RNase treatment might reveal the number of susceptible sites. The RNA of the next-step intermediate, the provirion, is resistant to RNase (see below).

Some further insights regarding the in vivo function of these RNPs, should come from the elucidation of their composition, cytoplasmic localization (e. g. possible membrane association), and interaction with other viral or host cell components.

2. *The 125S and 150S Provirion(s)*

A 125S ribonucleoprotein particle was first detected within poliovirus infected HeLa cells following reversal of a guanidine block (Fernandez-Tomas and Baltimore, 1973a). Its formation was also observed in extracts of infected cells (Fernandez-Tomas *et al.*, 1973b) (see Fig. 103, p. 429, and Tab. 68, p. 437). The particle has been called the provirion as it contains 35S RNA in an "immature" shell (*i.e.* composed exclusively of VP_0, VP_1, and VP_3). A similar particle has been described for bovine enterovirus; it contains viral RNA and polypeptides VP_0, VP_1, VP_2, VP_3, and VP_4 (Hoey & Martin, 1974).

The RNA in provirions is resistant to pancreatic ribonuclease, *i.e.* it appears to be burried within the protein shell. However, by treatment of provirion with EDTA, an agent which chelates divalent cations, the RNA can readily be released again, and the 125S particle is converted to an 80S empty shell. In contrast, the RNA of 80S RNPs or of complete virions cannot be released by a comparable treatment with EDTA.

Like the 80S RNPs, the provirion is unstable in CsCl gradients, where it is dissociated into empty shells and free RNA. The provirion is also sensitive to disruption by 1% SDS at neutral pH.

The provirion was originally reported to sediment at 125S (Fernandez-Tomas *et al.*, 1973). In subsequent studies from the same laboratory, the authors reported difficulties in detecting a 125S particle and instead demonstrated the occurrence of 150S "immature virions" with similar properties (composition enriched in VP_0, disruptable by EDTA and SDS) (Guttmann and Baltimore, 1977a). The VP_0 of the "immature" 150S virions could be chased into VP_2 "mature" 150S virions in the presence of guanidine. In contrast to the 125S provirion, 150S-VP_0 enriched particles could adsorb to host cells, and could be eluted in an altered state (130S, lacking VP_4 but not VP_0) like mature virions. In vitro formation of provirion or immature virions (Fernandez-Tomas *et al.*, 1973) could not be detected under the new experimental conditions (Guttman and Baltimore, 1977a).

It is unlikely that the 125S and 150S particles, both of which have been termed provirions, are indeed identical particles. The slower sedimentation coefficient of 125S could reflect either compositional difference (for example an incomplete particle still lacking one or two of the 14S building units), and/or conformational differences (a particle of identical composition but with a more "swollen" or elipsoidal conformation will sediment slower in sucrose gradients).

3. *Association of RNPs with the Replication Complex in Smooth Membranes*

When the smooth membrane fraction of polio-infected HeLa cells (fraction 2, Caliguiri and Tamm, 1970b) is gently solubilized in buffers of low ionic strength by the detergent Triton X-100, a heterogeneous set of poliovirus related structures (labeled only with radioactive amino acids) with a broad range of sedimentation velocities—from less than 80S to greater than 155S—is released (Yin, 1977) (Fig. 105 A). Unfortunately, the content of poliovirus RNA was not determined directly for any of these structures. In vitro RNA polymerase activity is associated with structures sedimenting faster than 80S, mainly in a broad peak centered about 160S. A small fraction of this peak which sedimented slightly faster than 155S marker virion and which showed a high polymerase activity was examined further with respect to polypeptide composition by PAGE: next to small peaks of radioactivity migrating like the polymerase NCVP4 and some smaller proteins it contained greater equimolar amounts of the capsid proteins VP_0, VP_1, and VP_3, and of the noncapsid protein NCVPX (Fig. 105 A). Treatment of the solubilized smooth membrane fraction with high ionic strength (700 mM KCl) (Fig. 105 B) or RNase (Fig. 105 C) results in a conversion of the very broad range of structures to particles sedimenting at a distinct 80S peak with a protein composition: VP_0, VP_1, and VP_3 (RNA content not determined). After treatment with RNase or high salt the polymerase activity that originally migrated in the broad peak at 160S now sediments as a distinct peak with 28S. The data of the original experiment are illustrated in Figure 105.

These observations indicate that the procapsid proteins are closely associated with the replication complex in smooth membranes, perhaps involving a physical linkage via NCVPX or viral RNA. In this context it is interesting to recall that guanidine prevents this association and that NCVPX or one of the smaller P-2 proteins NCVP-8 or-10 are implicated as the guanidine sensitive proteins from genetic studies (see Chapters 4 and 9, section VI). The broad range of viral struc-

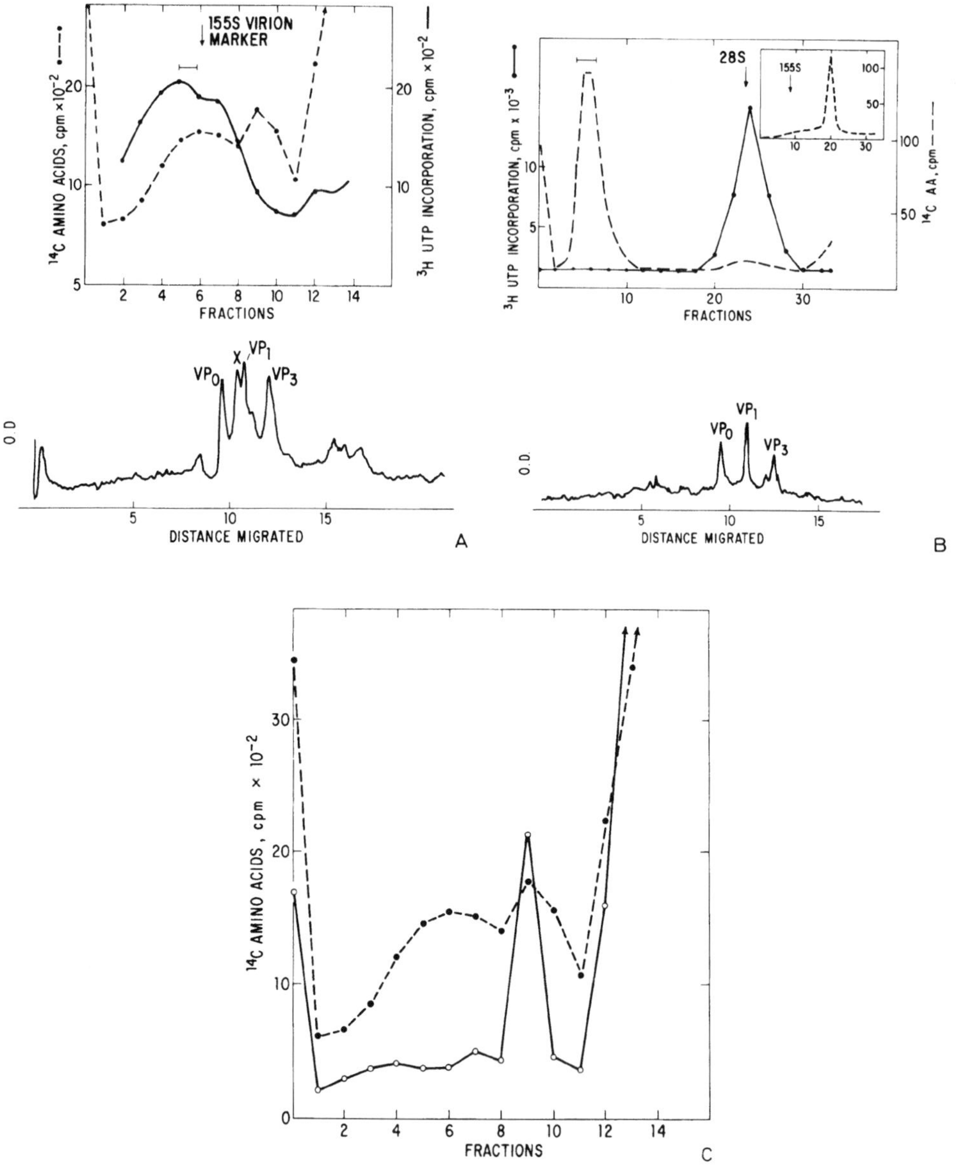

Fig. 105. Encapsidation of viral RNA on smooth membranes

A This figure illustrates the association of viral capsid proteins and NCVPX to the virus-induced smooth membrane fraction containing the active replication complex. Suspended HeLa cells were infected with poliovirus type 2 and labeled with (^{14}C)-amino acids from 2.5 to 3.5 h.p.i. The smooth membrane fraction was prepared by fractionation of the crude cytoplasmic extract on a discontinuous isopycnic sucrose gradient. The smooth membrane fraction was concentrated by centrifugation, solubilized with the detergent Triton X 100 (1%), and fractionated by centrifugation in 15—35% sucrose gradients (upper figure). *In vitro* polymerase activity (solid line) and radioactivity in viral protein (dashed line) were assayed. The fractions with the peak polymerase activity (5 and 6 = ca. 160 S) were analyzed further by SDS polyacrylamide gel electrophoresis on slab gels. The autoradiograph tures that are associated with the smooth membrane may reflect a series of intermediates in the process of RNA encapsidation. It would be highly interesting to localize and to trace the fate of viral RNA in these structures.

III. Assembly Kinetics

A good deal of evidence has been obtained in support of the concept that the subviral particles described above indeed represent consecutive intermediates in a stepwise assembly process of the poliovirion.

A. Chasing of Radioactive Precursors: Nucleosides and Amino Acids

When radioactive labeled uridine is added to poliovirus infected HeLa cells late in the replication cycle, the fate of newly synthesized RNA can be traced; it rapidly moves into virions, appearing there 2–10 min after addition of label (the time of synthesis of single stranded virion RNA from complementary (-) strand RNA is estimated to be 1 min) (Baltimore *et al.*, 1966). RNA is synthesized on smooth membranes at the site of virus assembly (Caliguiri and Tamm 1970 a, b; Caliguiri and Compans, 1973).

The flow of radioactive amino acids into virions is substantially slower. Kinetic analysis of the flow of radioactive amino acids into the various subviral particles revealed the following correlation: label in NCVP1a within 3–5 min., in the 14S subunit within 5 to 10 min., in the 80S shell within 15 minutes, and in 155S virion within 20–60 minutes (Penman *et al.*, 1964; Scharff *et al.*, 1964; Phillips *et al.*,

---

was analyzed by densitometer tracing (lower figure). The results show that Triton X 100 treatment solubilized the smooth membrane fraction to a heterogeneous set of structures sedimenting near 160 S, but did not separate the procapsid proteins from NCVPX and the RNA replication complex
B This figure illustrates the dissociation of procapsids from RNA replication complex by treatment of the solubilized smooth membrane fraction with high salt. Cells were infected and labeled as in A. The solubilization of the smooth membrane fraction by 1% Triton X 100 was carried out in the presence of 0.7 M KCl. The samples were fractionated by centrifugation in 15–35% sucrose gradients (upper figure) and assayed for *in vitro* polymerase activity (solid line) and radioactivity (dashed line). The peak of (^{14}C)-labeled viral protein was calculated to be 80 S from a separate experiment (inset). This fraction was shown to contain the viral procapsid proteins by polyacrylamide gel electrophoresis as in A (lower figure). The results suggest that the procapsid is tightly associated with the RNA replication complex in the smooth membrane fraction
C This figure illustrates the dissociation of procapsids from the RNA replication complex by treatment of the smooth membrane fraction with RNase. Cells were infected and labeled as in A. The Triton X 100 solubilized smooth membrane fraction was divided into two portions; one portion was digested with RNase (solid line), the other was left untreated (dashed line). Both fractions were centrifuged on a 15–35% sucrose gradient and analyzed for radioactivity. The results show that RNase digestion decreased the broad peak of 160 S in the Triton X 100 solubilized sample into an 80 S peak which represented procapsids composed of VP_0, VP_1, and VP_2 (slab-gel pattern not shown). The results indicate that the linkage between procapsid and the RNA replication complex in the smooth membrane fraction occurs through viral RNA. – Figures from Yin, 1977 [Virology *82*, 302, 303 (1977)]

1968; Jacobson *et al.*, 1970; Oppermann & Koch, 1973; Saborio *et al.*, 1974). Corresponding experiments in infected MiO cells (Ghendon *et al.*, 1972) showed that within 5–7 min after the addition of radiolabeled amino acids, radioactivity was found only in 5S virus-specific structures; the label is chased into the 14S area within 10–20 minutes, and within 20 to 30 minutes also into 150S virions. Fig. 106 illustrates the consecutive flow of radiolabeled amino acids – added at 1.5 h post infection and chased thereafter by addition of a large excess of unlabeled amino acids at 3.25 h – via 14S (S-antigen) and 73S (C-antigen) particles into complete virions (D-antigen) (Scharff *et al.*, 1964). RNPs or provirions were not considered in these studies. The described kinetics are consistent with precursor-product relationships of the subviral particles in poliovirus assembly.

The relatively rapid appearance of newly synthesized RNA in complete virions probably reflects the proximity of the sites of RNA synthesis and virion

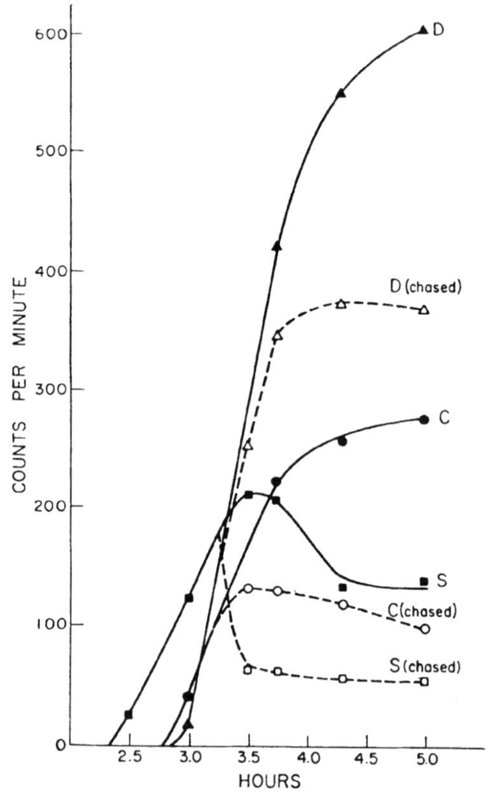

Fig. 106. The assembly of capsid related particles in poliovirus infected HeLa cells
This figure illustrates the formation and fate of poliovirus antigens in infected HeLa cells. Cells were labeled with (^{14}C)-threonine at 1.5 h.p.i. and the culture was divided in half. Samples from one half were withdrawn and frozen at the indicated intervals, and the amounts of labeled D, C, and S antigen were determined in duplicate. The other half of the culture was chased with a 20 fold excess of unlabeled threonine beginning at 3.25 h.p.i. Samples were withdrawn at the indicated times, and the amounts of label in D, S, and C antigen determined in duplicate. – Figure from Scharff *et al.*, 1964 [Proc. Nat. Acad. Sci. U.S.A. *51*, 333 (1964)]

assembly: namely, smooth membranes, probably those of the perinuclear system of channels and vesicles. The RNA kinetics might also reflect a close temporal relationship of virion RNA synthesis and encapsidation (see below).

The rather long lag period between the time of capsid protein synthesis and appearance of capsid protein in complete virions may reflect a time consuming transport system for the capsid proteins from their site of synthesis (peripheral rough endoplasmic reticulum) to the site of assembly, the central vesicular region. A movement of freshly synthesized poliovirus proteins from the rough to the smooth membrane fraction can be detected during the first 10 to 15 minutes of a chase after a 3-minute pulse labelling with radioactive amino acids. After a 30minute chase, the label reappears in the rough membrane fraction in the form of complete virions (Caliguiri & Tamm, 1970 a, b). The time course observed in these studies parallels the kinetics of passage of label through the subviral particles and supports the concept of a transport of capsid proteins to the site of assembly.

B. Assembly of Isolated Subviral Particles *in vitro*

Some of the proposed poliovirus assembly intermediates have been isolated from infected cells and characterized in terms of their assembly activity upon in vitro incubation in the absence or presence of extracts from infected cells. Table 69 summarizes the results of some of these experiments.

Table 69. *In vitro assembly processes of isolated poliovirus assembly intermediates*

Particle	Products after incubation of isolated particles at 37–40 °C	Products of infected cell-extract mediated incubation
5S	5S	broad spectrum of larger aggregates, 80S particles[f]
14S	80S (pI 5.0)[a, b]	80S (pI 6.8)[a, b]
Stable + unstable 80S shells	stable 80S[c]	stable 80S[c]
80S RNP	125–155S[d]	135S provirion[e]

[a] Phillips, 1969; Ghendon *et al.*, 1972.
[b] Putnak and Phillips, 1981, 1982.
[c] Marongiu *et al.*, 1981b.
[d] Marongiu *et al.*, 1981a.
[e] Yin, 1977.
[f] Palmenberg, 1982.

1. *Self Assembly of Isolated Subunits*

The primary building unit, the 5S protomers isolated from poliovirus infected cells could not self-assemble into larger structures *in vitro* (Ghendon *et al.*, 1972; Phillips, 1968). When 5S particles are added to cytoplasmic extracts from infected cells, however, some conversion to a broad spectrum of large aggregates is observed, although no well-defined subviral structures are formed (Phillips *et al.*, 1968).

Purified 14S subunits from infected HeLa cells or MiO cells, however, are capable of self assembly into stable 80S empty procapsids in the absence of cell free extracts (Phillips, 1969; Ghendon *et al.*, 1972). Here, assembly is dependent upon the concentration of the subunits (Phillips, 1969, 1971): 14S particle preparations show different efficiencies of self-assembly and, at a sufficient dilution, the capacity to self-assemble is lost. Preparations of 14S particles also vary in the extent of self-assembly: the process usually stops eventually and a certain proportion of 14S particles (usually about 20%) remain that appear to be "resistant" to assembly (Phillips, 1969; Phillips & Wiemert, 1978). Self assembled capsids are C-antigenic and resemble the stable procapsids derived from the *in vivo* procapsids; when extracts from infected cells are added to the cell free assembly system, D-antigenic particles resembling *in vivo* procapsids are formed (Putnak and Phillips, 1982, see below). These *in vitro* capsids, however, exhibit a slightly lighter density in CsCl gradients: 1.29 g/cm^3 compared to 1.31 g/cm^3.

Within the infected cell, the polymerization of 14S subunits must involve some slightly different, more intricate mechanisms than the *in vitro* self-assembly reactions that lead to stable 80S particles. Firstly, *in vivo* 80S particles are in ready equilibrium with assembly active 14S subunits; secondly, the entire *in vivo* assembly reaction involves RNA encapsidation and occurs membrane associated, probably on or within the membrane enclosed vesicles in the cell center.

Within the cell, a 55S particle, that most likely corresponds to a pentamer of the 14S component, is the next putative intermediate. Self-assembly capacity of isolated 55S poliovirus particles has not yet been determined, in part due to the difficulties involved in identifying and isolating these structures. As mentioned above, comparable particles obtained from other enteroviruses, however, are readily converted to 80S particles by incubation at high temperatures (37° C) (swine vesicular disease virus) (Moore, 1977) or by increase of the ionic strength (bovine enterovirus and also mengovirus, a cardiovirus) (Su and Taylor, 1976; Lee *et al.*, 1978). In contrast to the *in vitro* assembled mengovirus 80S particles, which are very stable, mengovirus 53S particles can be disassembled into 14S particles by prolonged incubation at room temperature.

Incubation of isolated poliovirus 80S ribonucleoproteins at 37° C induces the conversion of these particles to a broad heterogeneous set of faster sedimenting ribonucleoprotein particles (125—155S) (La Colla, personal communication). The specific properties of these products have not yet been examined sufficiently to warrant any conclusions to their nature, for example as provirion or virion-like particles.

So far, it has not been possible to promote the association of any isolated subviral capsid-related structures with RNA to complete virions or provirions *in vitro.* Yet, the replication complex associated procapsids can be chased into complete virions *in vitro* under careful conditions (Yin, 1977).

Recently a very efficient and complete translation of virion RNA from EMC virus (a cardiovirus) was achieved in reticulocyte lysates (Palmenberg, 1982). After long incubation times (15 hours), and stimulation by the addition of membranes,

5S protomers, 14S pentamers (containing only the VP_0, VP_3, VP_1 cleavage products), and 73S empty capsids were formed in this system.

2. *Assembly-Enhancing Activity in Extracts of Infected Cells*

Addition of extracts from infected cells to purified 14S preparations enhances their self-assembly activity, promotes the further assembly of the "resistant" fraction, and restores the self-assembly activity of diluted preparations (Phillips, 1969; Perlin and Phillips, 1975). The properties of the assembly enhancing activity are summarized in Table 70. Extracts from non-infected cells do not have any assembly enhancing effect. Some prior viral specific RNA and protein synthesis is required for the appearance of the activity.

Table 70. *Properties of the assembly–enhancing activity from infected cell extracts*

1. Promotes the assembly of isolated 14S subunits into 80S shells;
2. optimal reaction conditions at neutral pH, 37–40 °C, low salt concentration, does not require ATP;
3. present only in poliovirus infected cells;
4. requires prior viral RNA and protein synthesis (ongoing protein synthesis is not required);
5. associated with the rough endoplasmic reticulum;
6. involves a membrane-associated protein component (sensitive to trypsin, SDS, 1% DOC, but insensitive to RNase);
7. is not dependent on the presence of assembly-active 14S subunits: preincubation to deplete endogenous 14S subunits does not eliminate the activity;
   mutants exist that synthesize defective 14S particles (and altered VP_0 or VP_1) but retain the activity;
   mutants exist that synthesize acitve 14S subunits but show a greatly reduced activity;
8. is either induced by capsid related proteins or involves these proteins directly: the activity is not found in DI particle infected cells. These cells contain complete P 2 (NCVPX) and P 3 (NCVP 2, 4 and 7) proteins, but only part of the P1 proteins.

High concentrations of puromycin or cycloheximide had no effect on the assembly enhancing activity of infected cell extracts, indicating that ongoing protein synthesis is not required. Metabolic poisons such as cyanide and fluoride also had little – if any – effect on the assembly-enhancing activity indicating that it is not an energy requiring process (Phillips, 1969). Both, the extract-mediated reaction and the self assembly reaction are highly temperature dependent (optimal temperature 37 to 40°C) and occur only at neutral pH and low salt concentrations (Phillips *et al.*, 1968; Putnak and Phillips, 1981).

The assembly-promoting activity is destroyed by treatment of extracts with the protease trypsin or the membrane-disrupting detergent DOC (0.1–0.5%), but not by RNase treatment. This indicates that virus-induced membranes or membrane associated proteins (but not viral RNA) play a role in promoting assembly (Phillips, 1969). The assembly-enhancing activity is associated with membranes of the rough endoplasmic reticulum fraction isolated from infected cells (Perlin and Phillips, 1973, 1975). It was suggested that the function of viral-modified membranes is to adsorb and concentrate 14S subunits and thereby facilitate their assembly.

The assembly-promoting activity of infected-cell extracts is not reduced by preincubation which is sufficient to assemble the endogenous 14S particles of the extracts, indicating that it is not merely a concentration effect of the endogenous 14S particles. Two poliovirus mutants were recently characterized that retained assembly enhancing-activity, but contained altered VP_0 or VP_1 and defective 14S particles (Drescher-Lincoln *et al.*, 1983). Other mutants that showed greatly decreased assembly promoting activity also showed a reduced rate of processing of NCVP1a; the 14S particles formed by these mutants, however, had *in vitro* self-assembly capacity (Mikhejeva *et al.*, 1973; Putnak and Phillips, 1981). In sum, these observations show that the assembly-enhancing activity does not reside in active 14S subunits.

Nonetheless, some capsid precursor protein-related structures appear to be involved in the assembly promoting activity: the assembly factor which enhances or promotes the transition from inactive 14S to active $14S^*$ is present in extracts from infected cells, but is not present in extracts from uninfected cells or in extracts from cells infected with a defective interfering particle of poliovirus (Phillips *et al.*, 1980). Since DI particles are unable to synthesize the viral capsid precursor protein NCVP1a (Cole and Baltimore, 1973; Cole, 1975), it was suggested that the assembly factor resides in the capsid precursor protein or polypeptides derived therefrom (Phillips *et al.*, 1980). Alternatively, NCVP1a might serve as a trigger for the synthesis of polio-specific membranes and these are missing in DI-infected cells. NCVPX and a processable NCVP1b are produced in DI-infected cells, which would seem to rule out that either of these proteins or their cleavage products act as the assembly factor. Perhaps, NCVP1a or one of its products becomes associated with and promotes the modification of membranes in virus-infected cells to some type of assembly "factory". In this respect, it is of interest to note, that alternate cleavage products of the capsid precursor protein NCVP1a are regularly observed in infected cells, which are never detected, however, in any potential capsid related assembly intermediates. These products, NCVP3a (equivalent to VP_0 + VP_3) and NCVP3c (equivalent to VP_3 + VP_1), have not been assigned any function to date. As NCVP1a products they are potential candidates for the assembly-enhancing activity.

C. Studies with Inhibitors of Assembly

A number of inhibitors of poliovirus replication also directly or indirectly interfere with capsid morphogenesis (Table 71). Typically, the radioactivity of labeled amino acids added in the presence of such inhibitors accumulates in one of the intermediate subviral particles, depending on the particular site and mechanism of inhibition. Label accumulates in the capsid precursor protein NCVP1a, for example, in the presence of the nucleoside analogue Py-11 (2-amino-4,6-dichloropyrimidine), a potent specific inhibitor of virus assembly (La Colla *et al.*, 1976, 1981). Treatment with guanidine hydrochloride leads to the accumulation of label in 14S subunits in poliovirus infected MiO cells (Ghendon *et al.*, 1972) or in the accumulation of 80S procapsids in infected HeLa cells (Jacobson and Baltimore, 1968a). Treatment of infected Vero cells with glucosamine leads to the accumulation of label in 50S particles (Delgadillo and Vanden Berghe, 1980). Hypotonic growth conditions block assembly at the 80S ribonucleoprotein particle stage (Agol *et al.*, 1970). Upon removal of the inhibitory substance, the radioactivity can usually be chased through the subsequent intermediates into complete virions. Such studies provide more detailed insight into the individual steps of the assembly process.

Table 71. *Inhibitors of poliovirus assembly*

Affected stage of assembly	Accumulating viral intermediate	Inhibitor	Reversible	Reference
Cleavage of NCVP 1a	NCVP 1a	Py-11	yes	La Colla *et al.*, 1976, 1981
Association of vRNA with capsid proteins	14S subunits (MiO-cells)	guanidine	yes	Ghendon *et al.*, 1972
	80S shells (HeLa cells)			Jacobson and Baltimore, 1968
Condensation of 80S RNPs	80S RNP	hypotonic growth conditions	yes	Agol *et al.*, 1970
Assembly of 80S shells	50S	glucosamine		Delgadillo and Vanden Berghe, 1979

1. Reversible Inhibition of Assembly by Py-11

The pyrimidine analogue Py-11 (2-amino-4,6-dichloropyrimidine) is a potent reversible inhibitor of poliovirus growth (La Colla *et al.*, 1976). Py-11 specifically interferes with viral assembly: the production of 155S virus particles as well as intermediate 55S and 80S structures is completely suppressed and that of 14S protomers markedly reduced (La Colla *et al.*, 1981). All other viral functions appear to be nearly unaffected: the extents of viral protein and RNA synthesis may be slightly reduced, the rate of viral RNA synthesis, however, was identical to that determined in untreated cells; the RNA is fully infectious and detectable in the same amount as in untreated cells; the virus induced shut-off of host protein synthesis is also unaffected. PAGE reveals an accumulation of the capsid precursor NCVP1a with a comparable reduction of VP_0, VP_1 and VP_3 (by more than 50 %) and complete lack of VP_2 and VP_4; otherwise the pattern of proteins synthesized is normal. Thus Py-11 specifically blocks one of the first steps of assembly; apparently, the cleavage of the capsid protein precursor NCVP1a is impaired, thus preventing the proper association of these subunits into 14S pentamers.

The effect of Py-11 is completely reversible simply by removing the drug (Table 71). The label accumulated in NCVP1a in the presence of Py-11 is now chased via 14S, 55S (the predominant form after 1 h of chase), and 80S particles into complete virions (most label is found here after a 2–3 hour chase).

The effect of Py-11 is antagonized by the amino acids glutamine and cysteine, but only if these are added together and as L isomers; SH inactivating agents on the other hand, tend to synergize the inhibitory effect of Py-11, indicating that the mechanism of action of Py-11 does not involve a nonspecific combination of the drug with SH- or S-S groups on viral proteins. 14S particles made in the presence of Py-11 can self assemble into 80S particles *in vitro*.

These observations suggest that Py-11 interacts directly with a very specific site (involving gln and cys) either on the NCVP1a precursor or on the viral induced protease. The authors of the Py-11 studies favor the latter mechanism (La Colla *et al.*, 1981). An alteration of the protease is indeed consistent with the described observations; the binding of Py-11 to the protease could be irreversible and the synthesis of new protease in the absence of the drug would suffice to cleave the accumulated precursors. As no other cleavages are affected this would imply at least two different viral proteases for cleavage of capsid and noncapsid precursor proteins (see also Chapter 8, p. 362 ff.).

However, a modification of NCVP1a is also consistent with the data if one assumes the binding of Py-11 to NCVP1a to be reversible. The decrease in the rate of assembly and the clear demonstration of a 55S intermediate upon return to Py-11 free medium could then be explained by the gradual "release" of Py-11 from NCVP1a or its products permitting their participation in assembly. The gln residues present at each of the two cleavage (gln/gly) sites of NCVP1a might be candidates for the "interaction" site, explaining the antagonizing effect of gln.

In any case, these studies provide further support for the concept of a stepwise assembly of the capsid via 14S, 55S, and 80S particles.

2. *Studies with the Assembly Inhibitor Guanidine*

Guanidine hydrochloride reversibly inhibits poliovirus RNA synthesis and virion assembly over a wide concentration range (1–10 mM), in which it has no known effect on macromolecular synthesis of the host cell (Tamm and Eggers, 1963). The precise mechanims by which guanidine exerts this effect is still not certain. Apparently, the factor responsible for associating the capsid precursors with smooth membranes or with viral RNA possesses some abnormal conformation in the presence of guanidine (Cooper *et al.*, 1970; Yin, 1977; Koch *et al.*, 1980b, 1982a) (see Chapter 9, section VI). This factor could be either misshaped capsid proteins themselves, or a separate protein, such as NCVP8, 10, or X. The locus for guanidine sensitivity has been shown to lie on the coding region for these proteins (Romanova *et al.*, 1980; Tolskaya *et al.*, 1983), and these proteins when synthesized in the presence of guanidine have been shown to migrate abnormally in PAGE (Koch *et al.*, 1980b, 1982a). The important point for studies of the assembly process is that the process is blocked at a certain stage under guanidine treatment, and that this block is readily reversed again upon removal of guanidine.

When cells are exposed to guanidine midway through the replication cycle, virion formation as measured by the flow of radiolabeled uridine into virions, ceases abruptly (Halperen *et al.*, 1964; Baltimore, 1969, 1971a). In the presence of the drug, there is an accumulation of 80S particles, a rapid decline of RNA synthesis ensues, and all of the RNA which is made is trapped in a large structure called the guanidon (Baltimore, 1969). Viral protein synthesis declines much more gradually, being 50 % of normal after 30 min. When radioactive amino acids are added to infected cells at the same time as, or shortly after (15 min) guanidine, label accumulates in the 80S particles, but not in virions (Jacobson and Baltimore, 1968b). When these inhibited cells are transferred to fresh medium lacking the drug and labeled amino acids, and containing excess unlabelled amino acids to block incorporation of any residual label, radioactivity continually disappears from the 80S fraction, and a roughly equivalent amount of radioactivity appears in 125S provirions and in completed virions (Fig. 107) (Fernandez-Tomas and Baltimore, 1973). The total amount of label in these three particles increases slightly over an 1 1/2 hour sampling period, presumably reflecting a continuous assembly of prelabeled polypeptides.

In another type of experiment, radiolabeled amino acids were added for a 20 min pulse midway through the viral replication cycle, and labeled virions, pro-

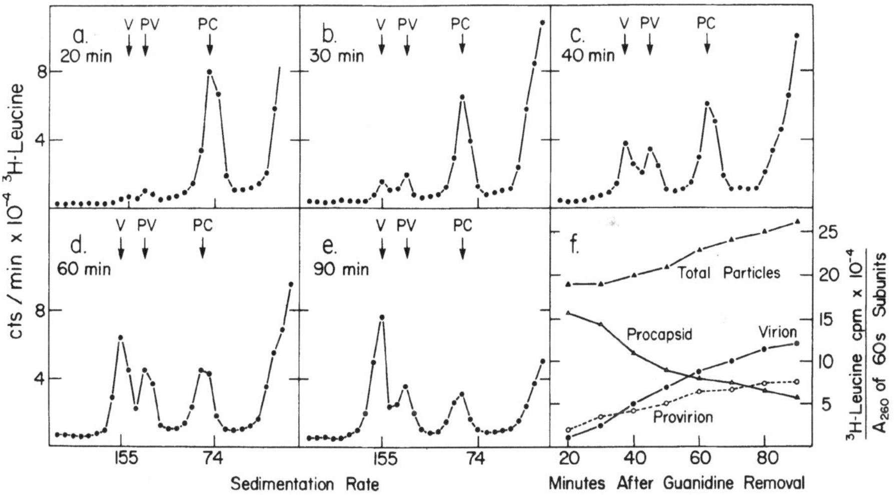

Fig. 107. The conversion of procapsids into provirion and virion after reversal of the guanidine-induced block in assembly

This figure illustrates the redistribution of labeled capsid proteins from procapsids (accumulated in the presence of guanidine) to provirion and virion after removal of guanidine. Poliovirus infected HeLa cells in suspension were exposed to 2 mM guanidine from 3–3.5 h.p.i. The cells were pulse labeled with (^{3}H)-leucine for 15 min from 3.25–3.5 h.p.i. and then chased in medium suplemented with a 100-fold excess of unlabeled leucine and lacking guanidine. At the indicated times after reversal of the guanidine block, cytoplasmic extracts were prepared and analyzed by sucrose gradient sedimentation. Samples were taken at 20 (a), 30 (b), 40 (c), 60 (d), and 90 (e) min of chase, respectively. Fractions were assayed for radioactivity. V = 156 S virion, PV = 125 S provirion, and PC = 76 S procapsid. f) illustrates the normalized amounts of radioactivity in the various particles. The amount of radioactivity in each peak was summed and the samples were normalized to the content of 60 S ribosomal subunits (measured by optical density at 260 nm). The total radioactivity in virus-specific particles was determined by summing the normalized values. — Figure from Fernandez-Tomás and Baltimore, 1973 [J. Virol. *12,* 1122–1130 (1973)] virions, and procapsids were allowed to be formed for 30 min (chase with excess unlabeled amino acids) before the addition of guanidine (Fernandez-Tomas and Baltimore, 1973). In the period after guanidine addition, radioactivity in provirions disappeared paralleled by an approximately equal increase of radioactivity in virions. Within the same time, radioactivity accumulated in procapsids in accord with the results described above. The data from these guanidine experiments suggest that guanidine interferes with the association of RNA to capsid proteins, but that once formed the provirions are converted into virions even in the presence of guanidine. The data also support the idea that the 80S particle and the provirion are consecutive intermediates in the biosynthesis of the poliovirion.

When corresponding guanidine inhibitor studies are carried out on polio infected MiO cells (where 73S particles cannot be detected under normal growth conditions) (Ghendon *et al.*, 1972), label is first detected in the 5S zone and then accumulates predominantly in the 14S region; 73S particles or complete virions are not detected. After removal of guanidine, the label appears predominantly in the 150S area, only very small amounts of label appear in the 55S and 73S regions. These results indicate that the formation of 5S aggregates and their subsequent association into 14S subunits are the earliest stages of poliovirus morphogenesis, and that – in MiO cells – assembly of capsid or virion-like structures are dependent on RNA synthesis.

3. *Inhibition of Poliovirus Maturation Under Hypotonic Culture Conditions*

Incubation of polio-infected HeLa cells under hypotonic culture conditions inhibits the formation of infectious virions, although synthesis of viral RNA continues at a considerable rate, and the patterns of viral protein synthesis and cleavage are not affected (Tolskaya *et al.*, 1966; Agol *et al.*, 1970) (Table 71). Virus assembly is blocked at the stage of 60-80S RNPs. 60-80S procapsids (1.32 g/cm^3 density) and 60-80S RNPs (1.37-1.42 g/cm^3 density)–the latter are discernible only after formaldehyde treatment–are formed in comparable ratios as in infected cells under isotonic conditions. The 60-80S particles appear to be somewhat more heterogeneous under the hypotonic culture conditions. The conversion to mature virions or virus specific particles sedimenting at the rate of virions is blocked under hypotonic culture conditions. In Figure 108, the results of the original study are shown.

A mutant of poliovirus was found, that is capable of partial reproduction under these conditions and that induces formation of 60-80S structures as well as mature virions (Tolskya *et al.*, 1966). This shows that the hypotonic conditions used are, in principle, compatible with the formation of mature virions. Unfortunately, the characteristics of this interesting mutant have not been studied in more detail.

4. *Assembly Defective Mutants*

Two temperature sensitive mutants of echovirus-12 (an enterovirus) have been obtained after mutagenization with 5-fluoruracil, that do synthesize considerable

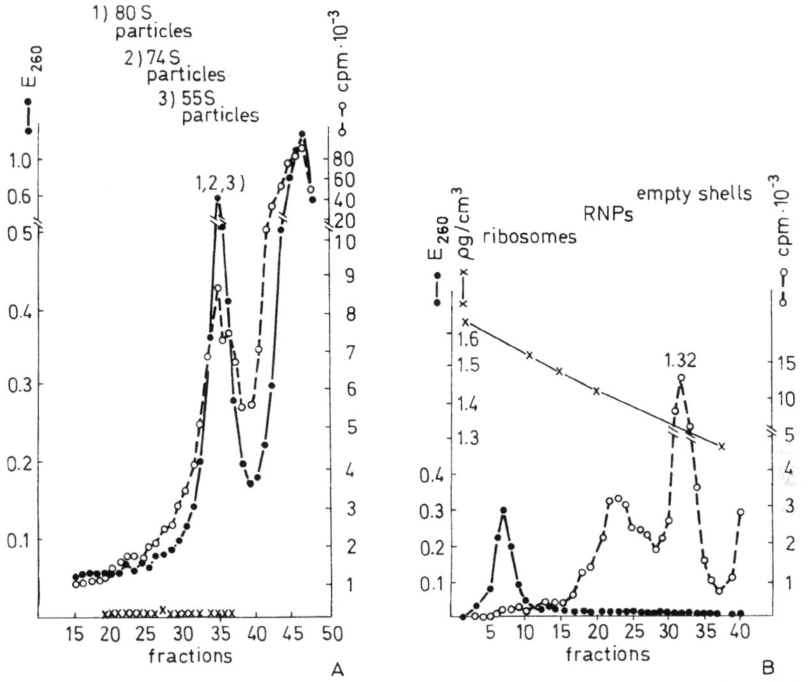

Fig. 108. Inhibition of poliovirus formation from 60—80 S RNPs under hypotonic growth conditions This figure shows that empty shells and 80 S RNPs may be assembled under hypotonic growth conditions, but that maturation to complete virions is blocked under these conditions. HeLa cell monolayer cultures were infected with the Neva strain of poliovirus and incubation carried out in hypotonic medium until 11 h.p.i. Cells were labeled with (^{14}C)-amino acids from 4 h.p.i., cytoplasmic extracts were prepared and analyzed by sucrose gradient sedimentation as in Fig. 104. Fractions were assayed for optical density (solid line), radioactivity (dashed line) and infectivity (xxx) (A). No infective virion is formed under these conditions, but label accumulates in the 80 S peak of ribosomes, RNPs, and procapsids. In order to separate the components in the 80 S peak, the fractions containing the peak of radioactivity were collected, treated with 4% formaldehyde (to prevent dissociation of RNPs), and centrifuged in CsCl density gradients (B). The results show that empty shells and viral RNPs are formed under hypotonic growth conditions, but maturation to complete virions is blocked. — Figures from Agol *et al.*, 1970 b [Virology *41*, 533—540 (1970)]

amounts of viral RNA and capsid proteins at the non-permissive temperature, but do not form any viral or subviral particles (Adrian *et al.*, 1979). This implies that maturation is impaired at the level of activation of the first bonding domain at the non-permissive temperature in these mutants. Small but reproducible differences in the isoelectric point of the mutant particles, as compared to wild type, and characteristic differences in the tryptic fingerprints of the mutant structural proteins, show that a defect in structural protein impairs the maturation process of these mutant viruses at the non-permissive temperature. A more detailed analysis of the defect might provide some insight into the particular parts of structural protein involved in the activation and interaction of the first bonding domain leading to formation of the 14S subunit.

Poliovirus mutants that exhibit defects in the assembly process have been observed in recent studies (Drescher, 1979; Drescher and Phillips, 1979; Putnak and Phillips, 1981; Drescher-Lincoln *et al.*, 1983). Two temperature-sensitive mutants containing altered VP_0 or VP_1 polypeptides were isolated. Both synthesize 14S subunits which are defective in the ability to polymerize into procapsids. Other temperature-sensitive + RNA and − RNA mutants making 150S particles with a very low infectivity possessing primarily VP_0, VP_1 and VP_3 were also described. In earlier studies, ts poliovirus mutants had been described that synthesized 5S material and 14S particles, but could not assemble 75S and 150S particles at the nonpermissive temperature (Milchejeva *et al.*, 1970, 1973). Table 72 summarizes the present-day information on assembly-defective mutants. The more detailed study of these viral mutants with specific defects in the maturation process should be a fruitful approach to a further understanding of the details of the assembly process.

Table 72. *Assembly defective mutants of enteroviruses*

Impaired stage of assembly	Accumulated product at the non permissive temperature	Detectable alterations in capsid protein	virus	Reference
Formation of 14S subunits	capsid proteins	all capsid proteins	ECHO	Adrian *et al.*, 1979
Polymerization of 14S subunits	14S subunits	VP_0 VP_1	polio	Mikhejeva *et al.*, 1970, 1973; Putnak and Phillips, 1981; Drescher-Lincoln *et al.*, 1983
Cleavage of VP_0	150S "immature" virons	VP_0	polio	Putnak and Phillips, 1981

IV. The Individual Steps of Assembly

A. Principles of Assembly

The process of poliovirus assembly evidently closely follows the cornerstone principles of virus structure (discussed in Chapter 3), in particular, the concept that viral capsids are constructed by the repeated use of equivalent sets of bonding contacts between identical subunits located in equivalent environments (Crick and Watson, 1956; Caspar and Klug, 1962). The following principles seem to be employed:

1. equivalent subunits are orderly assembled to larger intermediates, followed by further assembly of the intermediates and encapsidation of vRNA;

2. at defined stages in the assembly process, conformational changes of the subviral particles serve to activate the bonding domain that is required for the next step of assembly;

3. interaction of subunits with intracellular membranes and the location of the assembly process on such membranes play important regulatory functions;

4. linkage of the RNA to the small VPg may determine whether or not viral RNA is suited for encapsidation;

5. different steps in the assembly process may require different concentrations of salt, and intracellular compartmentalization may provide the required differences in ionic environments;

6. a stepwise condensation of the viral RNA in RNPs and provirion by the sequestering of cations—a salting out of the RNA—allows the formation of a compact virion;

7. specific morphogenetic cleavages of the capsid proteins drive the assembly reaction in the direction of completion and eventually render it irreversible.

Concerning the bonds required for association of the capsid, the stepwise expression of two bonding domains encoded within each protomer was proposed as the simplest principle to govern virion assembly (McGregor *et al.*, 1975; Rueckert, 1971 and 1976) (see also Fig. 11, p. 43). Such a sequence would be generated by the successive activation of a pair of bonding domains; the first domain fusing protomers into an intermediate subunit, and the second domain fusing the intermediate subunits into a 60-protomer capsid. The assembly process must proceed through one of three possible oligomers: a dimer, a trimer or a pentamer (Fig. 109, Rueckert, 1971). Two is the minimum number of bonding types required for the coherence of an icosahedral shell constructed from 60 identical building blocks, (see Chapter 3) if the process occurs in two discrete stages. A two-step bonding sequence requires that at least one of the 5:3:2 symmetry axes in the capsid is built into the intermediate at the first step. The finding that only 14S subunits (pentamers) and not, say 8S (dimers) or 10S (trimers) are found in infected cell extracts suggests that the two bonding domains are not used at random (e. g. type 1 domains forming pentamers; type 2 domains forming trimers), but rather that the domains are activated in a specific order. Such a tandem 1,2 activation sequence might be controlled by a delay in formation of the sterospecific complementary surface configurations required for the second bonding step until the product of the first step is complete (Rueckert, 1976). The information necessary for construction of the intermediate and final complexes (*i.e.*, the bonding domains) is built into the protomer. This mechanism has the advantage that control can be achieved at each level of organization: Any incorrectly constructed subunits are rejected automatically in the assembly process. Thus efficiency and accuracy of assembly is assured (see Fig. 109).

The possible role of virus-induced intracellular membranes in mediating conformational rearrangements of capsid-related structures during poliovirus assembly is still largely a matter of speculation. Tight association of virus components with cytoplasmic membranes during assembly and maturation has been observed in many different virus-host cell systems, including poliovirus (Caliguiri and Tamm, 1970b; Korant, 1973; Lonberg-Holm *et al.*, 1976). That membranes

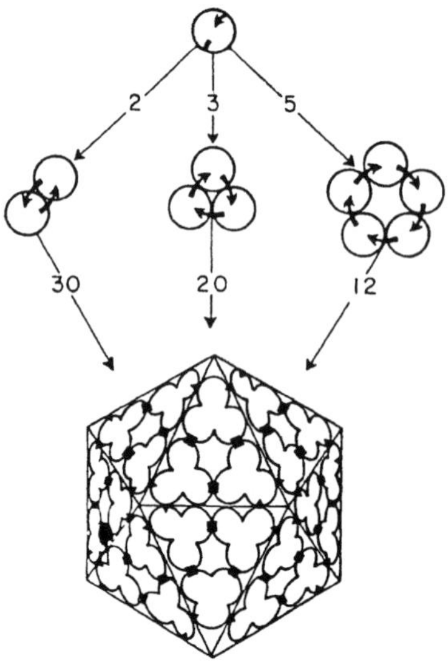

Fig. 109. Schematic diagram of possible pathways for the assembly of a 60-subunit capsid This figure illustrates schematically the three possible pathways by which the assembly of a 60-subunit capsid might take place by the stepwise activation of a paired set of two different bonding domains encoded into the identical subunit. Repeated use of the first set of bonds, consisting of complementary donor and acceptor surfaces (represented by arrows, top), leads to formation of a dimer, trimer, or a pentamer, thereby establishing the first of the three symmetry axes eventually appearing in the capsid. Completion of the intermediate is then followed, after activation of a previously latent second set of binding domains, by assembly of one of these sets of intermediates to form the 60-subunit capsid. The fact that only two types of bonds (▲, ■; bottom) are essential for coherence of a surface lattice constructed from an equilateral-triangular plane net was pointed out in 1965 by Caspar. — Figure from Rueckert, 1976 [in: Comprehensive Virology, pp. 131–213 (1976)]

can indeed mediate large structural transformations of protein complexes has been clearly demonstrated in studies with bacteriophage M13 (Wickner, 1979, 1980; Griffith *et al.*, 1981). The filamentous bacteriophage M13 contracts into a spheroid with a diameter of 0.04 μm upon exposure to a chloroform-water phase. The authors propose that the chloroform water interface mimicks properties of bacterial membranes and that a similar change in phage coat structure takes place during the entry and uncoating step in a normal M13 infection. Ideas on the involvement of membranes in poliovirus morphogenesis are discussed below.

One fundamental question remaining concerns the association of the genomic RNA with the viral capsid proteins. Two principally different concepts are considered: One concept, that was inspired by the elucidation of the assembly process of bacteriophages, envisions an RNA independent stepwise assembly of a complete or near complete proteinacious capsid into which the nascent RNA is inserted as one of the final steps of the assembly process. An alternative concept envisions virion assembly around some kind of assembly "nucleus" composed of the genomic RNA more or less specifically associated to some viral capsid protein, virion assembly would then proceed by a stepwise assembly of capsid subunits around the assembly nucleus.

Figure 110 presents an overview of the probable steps and possible pathways in the assembly of poliovirus.

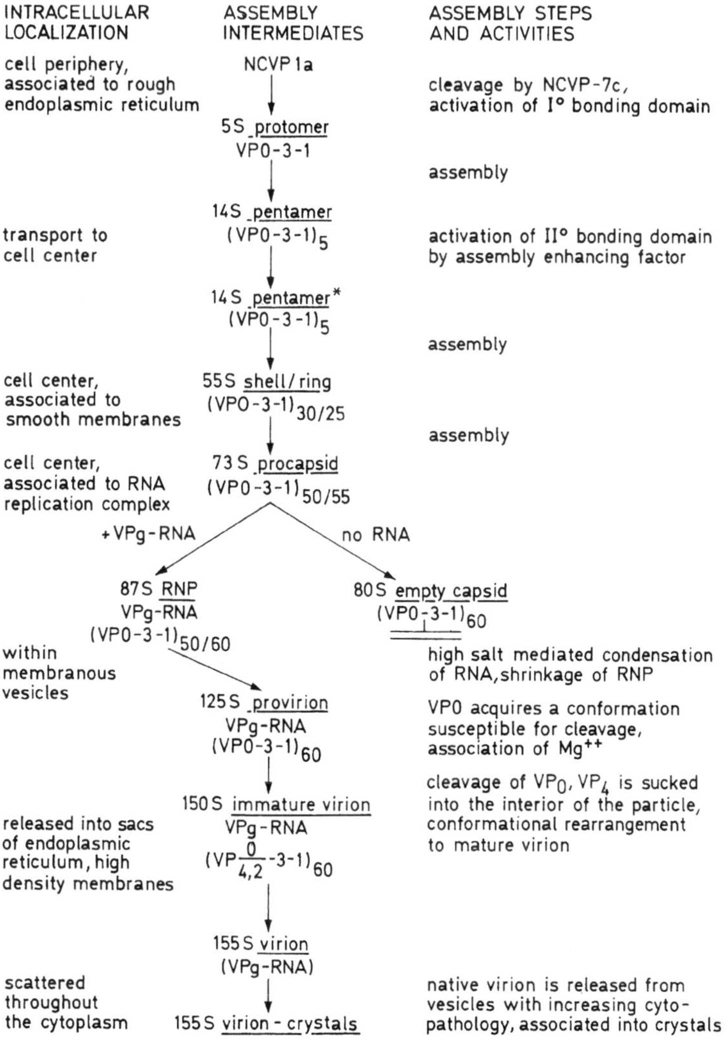

Fig. 110. Overview of poliovirus morphogenesis

B. Formation, Activation, and Assembly of the 5S Protomer

Capsid proteins are synthesized—like all viral proteins—on membrane-bound polysomes of the endoplasmic reticulum near the cell periphery. The capsid precursor protein NCVP1a (molecular weight 97 kd) is one of the three primary cleavage products of the nascent polypeptide chain during ongoing translation of the polio mRNA. NCVP1a soon becomes associated with membranes. We suspect that it becomes physically inserted into the membrane of the rough endoplasmatic reticulum as it is being synthesized. The physical insertion of virus coat proteins into membranes as a preparatory step for capsid assembly has been well documented for the bacteriophage M13 (Wickner 1976, 1979; Wickner *et al.*, 1978). Interestingly the precursor for M13 coat protein requires an electrochemical potential for its insertion into the membrane (Date *et al.*, 1980). An electrochemical potential across the intracellular membranes of poliovirus infected cells is also predicted (see below).

NCVP1a is cleaved at two points before initiation of assembly. The enzyme responsible for this cleavage appears to be NCVP7, a viral protein encoded by the P-3 replicase region (Korant, 1972, 1979; Korant *et al.*, 1979; Palmenberg *et al.*, 1979). The three cleavage products are of roughly equivalent size. VP_0-VP_3-VP_1 remain associated to one another as an aggregate (Korant, 1973) – the "protomer"—containing a single copy of each. The protomer remains bound to intracellular membranes, and may be transported to the cell center for further assembly. NCVP1a as well as the $VP_{0\text{-}1\text{-}3}$ aggregate sediment at 5S. Occasionally, NCVP1a is cleaved at only one of its two primary cleavage sites: this leads to the formation of the proteins NCVP3a (equivalent to VP_0 and VP_3) and NCVP3c (equivalent to VP_3 and VP_1). Neither one of these larger products are ever found in the larger capsid-related structures. It is not known whether NCVP3a and NCVP3c merely represent "waste products" of incomplete proteolysis or short lived intermediates for further cleavage, or whether they are of any functional significance. They are observed in the same membrane fractions as the assembly intermediates, and thus they are potential candidates for membrane-associated assembly-enhancing factors.

It is not known how the protomer is "activated" so as to associate with other protomers to form the 14S pentamer. This activation must involve the exposure of the first essential bonding domain in the "correct" configuration. It is conceivable that the cleavage of NCVP1a induces a shift in conformation so as to expose this bonding domain.

Evidently, cleavage of poliovirus NCVP1a is required before association to 14S subunits is possible. NCVP1a is never found in these subunits in an uncleaved form. In contrast, in EMC virus (a cardiovirus) infected cells, a 13S subunit consisting of a pentamer of the uncleaved capsid precursor polypeptide has been found (McGregor *et al.*, 1975; McGregor and Rueckert, 1977). This 13S particle undergoes proteolytic cleavage and conformational change to a 14S state before it can be further assembled. In a reticulocyte cell-free system for the translation of EMC virion RNA, however, the case is similar to that of poliovirus: The capsid precursor is cleaved to a 5S protomer *before* association to a 14S pentamer (Palmenberg, 1982).

It is difficult to ascertain whether the cleavage of NCVP1a alone suffices to promote the association into 14S pentamers. Isolated 5S poliovirus protomers do not self-assemble into 14S subunits *in vitro* (Phillips *et al.*, 1968; Ghendon *et al.*, 1972). In the infected cell, this process occurs in close association with intracellular membranes. Certainly a protein aggregate anchored by its hydrophobic regions to a membrane will assume a different conformation in an aqueous environment. If the 5S polio protomer is indeed physically inserted into intracellular membranes by some hydrophobic regions, and if these regions are involved in the primary bonding domain, the failure of 5S particle to associate in vitro in the absence of membranes is readily explained.

It is tempting to propose the following hypothetical scheme of protomer activation. NCVP1a is inserted into and anchored to intracellular membranes by hydrophobic regions during its synthesis. Cleavage of NCVP1a at two specific sites exposed to the aqueous environment of the cell cytoplasm induces a conformational shift that converts the primary bonding domain into an active configuration. These bonding domains, part or all of which presumably are located in the hydrophobic membrane associated region, promote the association of the 5S protomers into pentameric clusters. Other membrane associated factors – viral or host cell proteins, elements of the cytoskeleton, etc. – could in principle also be involved in bringing 5S protomers to sufficient spatial proximity or in promoting their association.

The pentameric 14S clusters thus are held together by bonds across the 5-fold axis. They are associated to the intracellular membrane by the future 5-fold apices of the viral capsid—at least during the time of their formation. The potential secondary bonding domains for association of the 14S pentamers across the 2- or 3-fold axes are exposed symmetrically to the aqueous phase of the cytoplasm. In reference to the poliovirus model presented in Chapter 3, the 14S pentamers are bound to the intracellular membranes predominantly by VP_0, and exposing VP_3 and VP_1 to the cytoplasmic matrix for further bonding.

C. Activation and Assembly of the 14S Pentamer

The next membrane mediated step is the activation of the 14S pentamer into an assembly active state, 14S*. Activation of the 14S pentamer exposes the second essential assembly-bonding domain that is responsible for the formation of the interpentamer bonds across the 2- or 3-fold axes. This is potentially a highly reactive bonding domain, leading to strong bonds: isolated 14S pentamers can be associated into stable empty shells in vitro. The energy released in this reaction step must be quite high, since the so-formed 80S shells are very stable.

The most simple possibility for activation of this bonding domain would be a conformational rearrangement brought about by the formation of the bonds between the 5 component protomers. Morphogenetic proteolytic cleavages are not involved in this step. Certainly, conformational rearrangements within the 14S pentamer are possible and perhaps are required in order to activate the interpentamer bonding domain. This is evidenced by the finding of assembly-active and assembly-inactive forms of poliovirus 14S subunits, that do not reveal any significant compositional differences (Perlin and Phillips, 1975; Phillips and Wiemert, 1978).

Preparations of purified 14S particles evidently contain a mixture of 14S subunits in different conformational states and in association with different amounts of other proteins. Part of the 14S particles may be in a state similar to that of intact virions: a monoclonal neutralizing antibody has been produced that also interacts efficiently with 14S particles (20 % compared to virion or 80S particles) (Icenogle *et al.*, 1981). The relative proportion of the conformational states varies in different preparations (Phillips and Wiemert, 1978), and certainly is different from that within the infected cells where other components may influence the equilibrium between the states in one direction or the other. Nevertheless, some insights into these steps may be gained from the *in vitro* studies: Isolated assembly inactive subunits can be activated to an assembly active state by some kind of "assembly factor". The activation of 14S particles *in vitro* is promoted by virus modified membranes from the rough microsomal fraction (Perlin and Phillips, 1973) and seems to involve the capsid protein precursor NCVP1a or one of its products (Phillips *et al.*, 1980).

Evidently, integral membrane association is not as important for the activity of the inter-(14S) pentamer bonding domain as it is for the activity of the inter-(5S) protomer bonding domain. Even purified 14S subunits, released from the intracellular membranes, may be in an assembly active conformation, and membranes alone do not promote assembly. It is difficult to say, whether in the infected cell the 14S pentamers are partly released from integral association in the intracellular membranes prior to the formation of interpentamer bonds or as a consequence thereof.

Insertion of 14S subunits into — or association with — the membrane by a defined polypeptide portion, for example by a proposed hydrophobic 5-fold axis, may pose a geometric problem for further assembly. The correct arrangement of 14S subunits for the formation of any larger capsid-like structures would be in a halfshell or sphere-like manner; such arrangement could be acquired readily in solution at a sufficient concentration of subunits, but it is more difficult to envision for a set of subunits bound by the same sites to a membrane structure.

Perhaps, most of the 14S subunits are released from integral membrane insertion after their formation and become instead more loosely associated to the membrane via other membrane components — for example by the assembly enhancing factor. Likely candidates for such factor(s) are membrane-anchored NCVP1a-related products: NCVP1a, other aggregates of VP_0, VP_1, and VP_3, or the alternative cleavage products NCVP3a and NCVP3c — these products are missing in the membrane fractions of DI-infected cells, and by their nature as products related to capsid protein, they inherently have — perhaps weak — 14S binding potential.

A tempting, albeit highly speculative idea, that could account for the next larger intermediates, the 55S and 80S particles, is the following: "released" 14S particles are associated by one of their five equivalent "edges" or "corners" (the future participants in 2-fold or 3-fold symmetry localities) to a corresponding binding site of an integral capsid-related membrane component (alternatively, such association might promote the "release" of the 14S subunit from its integral membrane insertion). These 14S particles are thereby "dangling" from the membrane, in an opportune arrangement for the formation of interpentameric bonds (Fig. 111). The "dangling" 14S subunits are in the activated conformation; they

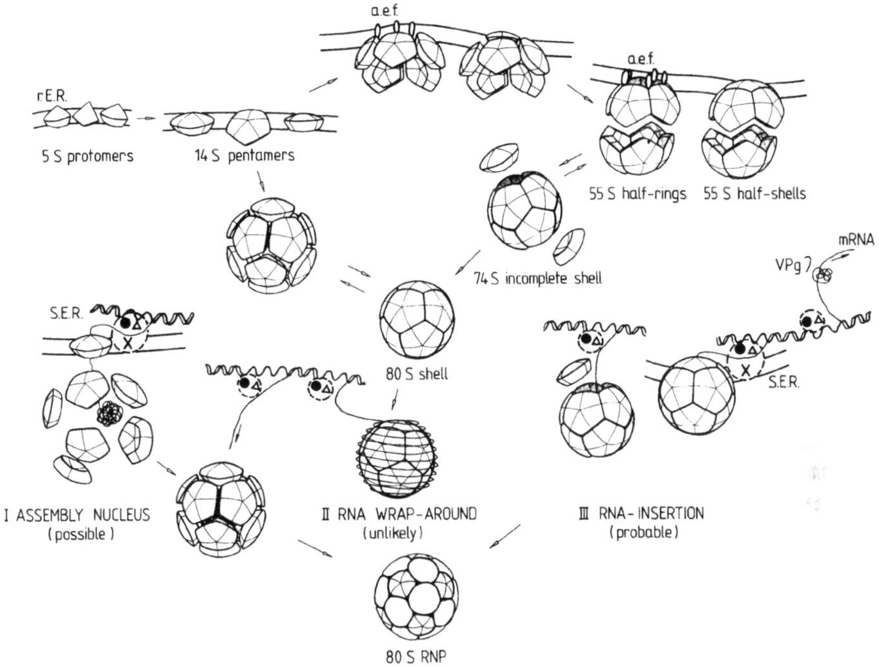

Fig. 111. Models of *in vivo* assembly pathways of 14 S subunits and initial interaction with viral RNA This figure illustrates schematically some of the proposed mechanisms for the assembly of the virion capsid and the encapsidation of the viral RNA. The 5 S protomer is synthesized on and becomes associated to membranes of the rough endoplasmic reticulum (rER). Assembly into 14 S pentamers seems to be an intramembrane process. 14 S pentamers are activated for the next step in assembly by an assembly enhancing factor (a.e.f.) present also in rER. It is not clear in what manner the membrane association of 14 S pentamers is altered by activation. The 14 S pentamers may assemble directly to complete 80 S empty shells or via 55 S (half-shell or half ring) and 74 S (incomplete shell) intermediates. Three different models have been proposed for the initial interaction of vRNA with capsid precursors, which evidently occurs during ongoing synthesis of the vRNA on membranes of the smooth endoplasmic reticulum (sER). One model envisions a catalytic type of assembly of 14 S subunits around a core of RNA (assembly-nucleus model); this model cannot be excluded on the basis of available evidence. A second model pictures the winding of RNA into grooves on the external face of a complete procapsid, followed by engulfment of the RNA during a conformational transition of the entire capsid (RNA wrap around model); this model is not supported by available evidence. A third, and most probable model on the basis of available evidence proposes the insertion of RNA into a complete or near-complete procapsid, driven by ongoing RNA synthesis (RNA insertion model). VPg, or a specialized capsid protein, or additional factors may mediate the initial interaction of the RNA with the capsid precursor. The endproduct of this phase of virion assembly appears to be an 80 S viral RNP in any of the possible models

can be dissociated from the membrane component in an active or inactive state. Five "dangling" active 14S particles could readily associate to form a ring-like structure (the 55S particle). Two such structures in turn could rapidly associate to form an incomplete capsid with two opposing "holes" (the 74S particle?). Indeed, 53S $(pentamer)_5$ and 73S $(pentamer)_{10}$ structures have been implicated in mengovirus assembly (Lee and Colter, 1979), but it is not known whether the 55S and 74S particles of poliovirus correspond to such structures as well.

The dimerization of a 55S ring like structure could occur either after release from the "loose" association to the membrane or still in some membrane-associated form when two opposing membranes approach each other (Fig. 111). For poliovirus, it is impossible to say when – *i.e.*, before or after dimerization – and how the two holes are filled by the two "missing" 14S subunits. It is tempting to speculate that the uncleaved VP_0 residues regularly found in completed virions arise from the two 14S structure units that complete the encapsidation of the viral genome.

An incomplete 74S capsid is in some way a more credible potential intermediate in the assembly process, as it is certainly easier to envision how a viral genome could be inserted into an incomplete capsid than into a complete one (Lee and Colter, 1979). To gain more confidence in the validity of the incomplete capsid hypothesis proposed for mengovirus and its applicability to the poliovirus system it would certainly be fruitful to carefully reexamine the molecular weights and protein compositions of the poliovirus subviral particles. Is the 55S particle observed after release of the Py-11 block a $(\text{pentamer})_5$ structure, and is it a ring like structure? What is the nature of the 63S poliovirus particle reported to occur during the first minutes of the in vitro assembly reaction of 14S particles (Phillips, 1971), and under conditions where the self assembly reaction was severely inhibited (Phillips and Wiemert, 1978)? Are the 74S unstable empty capsids and the 80S RNPs indeed incomplete capsids made up of 10 pentamers, instead of 12? Where are the two "holes" in the capsid localized, next to each other or opposite?

If there were indeed something special about one or both of the hole-filling 14S subunits, some interesting possibilities would follow: held inserted in the membrane by some component (NCVPX ?) such 14S subunits could provide a membrane binding site for the rest of the capsid on one hand, and could provide a defined site for the initial interaction of viral RNA with the capsid. Indeed, procapsids are found associated to the RNA replication complex and to large amounts of NCVPX in smooth membranes; an association that is broken by RNase or membrane disrupting detergents (Yin, 1977). The possible role of the viral RNA in these late steps of virion assembly is discussed in the next section. Figure 111 illustrates some of the proposed mechanisms for *in vivo* 14S pentamer assembly.

D. Encapsidation of the Viral RNA

The mechanism by which the viral RNA is encapsidated within the proteinacious capsid is still one of the most puzzling steps of poliovirus assembly. Direct experimental data on this step is not abundant.

The series of ribonucleoprotein particles containing RNA and the procapsid proteins show that extensive interaction between the viral RNA and capsid proteins occur *prior* to cleavage of VP_0 to VP_2 and VP_4 (in contrast to an earlier concept of RNA encapsidation, Baltimore, 1971a). In other words, the main steps of RNA encapsidation evidently occur before the final morphogenetic cleavage of VP_0 which locks the virion into its native state (see below).

The encapsidated poliovirus RNA is physically linked to the small genomic protein VPg, in contrast to the non-encapsidated viral mRNA (Fernandez-Munoz and Lavi, 1977; Lee *et al.*, 1977). VPg could conceivably play a role in some kind of recognition mechanism between vRNA and capsid proteins.

1. Condensation of the Viral RNA

Spatial requirements of the capsid core (considering a core diameter of 18 nm, the volume of the core would be 3050 nm^3) show that a compact condensation of the RNA is required to allow its packaging within the capsid. The density of poliovirus RNA within an 18 nm core would be 1.48 g/cm^3 (considering a calculated weight of 4.5 x 10^{18} g for the RNA-K^+/Na^+ salt).

Such compact condensation of RNA requires neutralization of the negatively charged phosphate groups predominantly by cations (most likely potassium or sodium) or by basic amino acid residues of proteins. As discussed in chapter 3, the virus eventually binds 4,900 K^+ ions, 900 Na^+ ions, some polyamines and 120 Mg^{++} ions (Mapoles, 1980). Approximately 6.000 charges on the RNA are neutralized by sodium and potassium, the remaining 1.500 by basic amino acids.

Neutralization and condensation of virion RNA requires a high concentration of cations. 6000 cations per 18 nm core correspond to a molarity of 3.3 M, 6000 cations per 20 nm core correspond to a molarity of 2.3 M. This corresponds to a 10 - to 15-fold increase in concentration of cations with respect to the normal tonicity of the cytoplasm.

Indeed, a virus induced accumulation of monovalent cations—notably Na^+— is observed in poliovirus infected HeLa cells beginning at approximately 2 h.p.i., reaching a maximum at 3.5–4 h.p.i. (Nair *et al.*, 1979; Nair, 1981; Schaefer *et al.*, 1983a and b), exactly paralleling the time course of virus maturation. Notice that the 5:1 ratio of K^+ to Na^+ ions in the virion is half that observed in uninfected cells. Some of the accumulated cations in infected cells are in some way bound and are only slowly released (Nair, 1981).

From these considerations and the evidence of electron microscopic studies we propose that the membrane-enclosed vesicles near the cell center represent encapsidation factories. Virion RNA is synthesized in close association with these membranes and is either inserted or synthesized directly into these vesicles, or into some form of procapsid associated to these membranes (see Fig. 78, p. 260). The vesicles contain a high concentration of cations necessary and useful for the neutralization of the negative groups on the RNA and for the condensation of the RNA into a compact structure.

2. Formation of the RNP

Principally two concepts are conceivable for the initial interaction of viral RNA with capsid proteins (see Fig. 111 above): 1. assembly of capsid subunits about an RNA-core, 2. insertion of viral RNA into a more or less completed capsid during ongoing synthesis of viral RNA. A third concept that envisioned the wrapping of viral RNA in grooves around a complete 80S procapsid, followed by a conformational shift of the capsid to "engulf" the RNA (Jacobson and Baltimore, 1968a; Phillips, 1972; Rueckert, 1976) is highly improbable on thermodynamic grounds.

A strong argument for this "procapsid" model was the observed stability of 80S shells that seemed to rule out an equilibrium between the 80S shells and 14S subunits (Rueckert, 1976). The recent demonstration that the *in vivo* 80S procapsids are not so stable after all and may be in an equilibrium with the 14S subunits (Marongiu *et al.*, 1981) opens the way for other more plausible models. In addition, an incomplete 50 subunit shell $(\text{pentamer})_{10}$, first proposed as an intermediate in mengovirus assembly (Lee and Colter, 1979) also provides for a more credible intermediate capsid structure to interact with viral RNA. A complete 60 subunit shell containing no RNA should be regarded either as a potential storage for 14S building blocks or as a flexible shell, possibly capable of pulling in an RNA molecule during ongoing RNA synthesis through some kind of "hole".

The involvement of a series of ribonucleoprotein particles including "immature" 150S virions that are still composed of the procapsid proteins VP_0, VP_3, VP_1 and RNase resistant RNA seems to rule out an earlier concept that held cleavage of VP_0 to VP_2 and VP_4 to be the driving force of viral RNA encapsidation (Baltimore, 1971a).

a) Assembly Around an RNP Core

Completed single stranded RNA is likely to assume a globular conformation with extensive secondary structure as dictated by complementary sequences unless it is stabilized in an extended form by bound proteins. 14S pentamers or even (14S $\text{pentamer})_n$ aggregates could assemble around an RNA core. This notion is supported by the finding that antisera against 14S virus-specific structures inhibited RNA synthesis in an *in vitro* system (Ghendon *et al.*, 1973). Polypeptide stretches containing clusters of basic amino acids may "stick out" as arms from the 14S subunits to interact with the negatively charged phosphate groups of the RNA. Such arms have been observed in the icosahedral plant picornaviruses (see Chapter 3, and Fig. 18, p. 64).

Synthesis within or secretion of the RNA into the hypertonic environment of the central vesicles would commence the condensation of the RNA. Highly reactive, 14S assembly intermediates might be associated to the inner membrane "waiting" for the RNA. The assembly intermediates may—reversibly—associate to 55S and 80S particles when RNA is not available. The exposed positively charged regions of the 14S pentamers are likely to react with the negatively charged phosphate groups of the RNA. A number of 14S subunits may interact with the RNA and assemble around the RNA either individually in an orderly fashion or via 55S pentameric groups (rings?), eventually producing 60-80S ribonucleoprotein particles (see Fig. 111 above).

b) Insertion of RNA into a Procapsid

In the course of its synthesis the viral RNA is produced as a single-stranded molecule. If this molecule were guided to a "sensitive" procapsid before the RNA assumes secondary structure, it could conceivably be "threaded through" a "hole" into the interior of the capsid. (This is reminiscent of bacterial conjugation where single stranded DNA is inserted into the receiving bacterium, driven by the ongoing synthesis of the DNA strand.) The 160S complex integrally associated with smooth membranes, consisting of small proteins, NCVPX, procapsid, and the replicative complex (Yin, 1977) could well represent a corresponding encapsidation intermediate (see Fig. 105, p. 442).

Again, the proposed function of the central membrane-enclosed bodies as encapsidation compartments leads to some compelling ideas: We might speculate that the replicase and its template RI-RNA are located on the outer side of the vesicular membrane, the procapsid on the internal or external side (see Fig. 78, p. 260). If procapsids were located inside the vesicles, viral RNA may be "captured" as it is being synthesized by a membrane component (e. g., NCVPX) and guided to a "channel" for passage through the membrane. This channel could be constituted by capsid related proteins as integral membrane component or by one of the 14S pentameric units in its transmembrane state (at the 5-fold apix). Symmetric oligomeric protein complexes have indeed been implicated as membrane channels (Klingenberg, 1981). Such oligomeric capsid-related proteins (for example a specialized 14S subunit) or also viral proteins that traverse the membrane could also function as channels for ions that may be necessary for concentrating cations within the vesicles.

The capsid proteins have a total net charge on the order of + 650 at neutral pH. The procapsids have a concentration of negative charge on the outside, thereby turning to the shell-interior the positive charges for potential interaction with the genomic RNA. The concentration of positive charge in the interior of the procapsid may provide part of the driving force for „sucking in" the viral RNA. Alternatively, the ongoing synthesis of the RNA molecule drives the encapsidation process.

A nice aspect of this model is the built-in regulatory mechanism for separating the RNA that is to function as virion RNA from that required as mRNA. The RNA that is not "captured" and directed to the procapsids is presumably bound to the cytoskeleton of the cytoplasmic matrix and transported to the ribosomes in the cellular periphery. The model also provides for a function of VPg as a recognition site for the RNA to the "channel" and insertion into procapsid: Some hypothetical mechanism might be postulated that selects VPg-associated virion RNA for secretion into the membrane-bound vesicles or directly into the procapsids, while releasing VPg-free RNA into the cytoplasmic matrix to function as mRNA. An alternative postulate might be that the enzyme responsible for the cleavage of VPg from the RNA (Ambros and Baltimore, 1980) is only present in the cytoplasmic matrix and not in the vesicles. A cytoplasmic VPg cleaving enzyme would cleave VPg only from mRNA not from virion RNA that was injected into the vesicles. If VPg had indeed some recognition site for capsid related proteins, the abundant, cleaved VPgs may represent reversible plugs of the proposed RNA channels or ionophores. Lastly, the model provides an explanation for the action of guanidine: the capture of the 5' end of the RNA and its guidance to a hole in the procapsid can be expected to be a step, sensitive to even minute alterations of conformation in the responsible proteins as may be induced by guanidine.

3. Stepwise Condensation of the RNPs

Whatever the mechanism of initial encapsidation, the product in all cases appears to be a swollen or awkwardly shaped 80S RNP (see Fig. 111, p. 461).

The problem facing the unfinished virus particle at this point is the sufficient condensation of the RNA and the "squeezing" away of the water. This is an important prerequisite for an optimal compact packaging of the virus, but it is not an easy task. Water has a high affinity for itself, for ionic groups and for polar amino acid side chains (ser, thr, etc.) with which the RNA and the capsid precursor particles, especially on their "interior" surface, are loaded. The sufficient condensation of RNA may be aided by cations that are concentrated in the central vesicles. This seems to be the step that is sensitive to hypotonic culture conditions: Hypotonic growth conditions which would antagonize any cation concentrating mechanism inside the cell prevent the conversion of 60-80S RNPs to mature virions.

The hypothesized reduction in size and "squeezing out" of water is consistent with the increase in sedimentation velocity of the RNPs from 60 -80S via 125S and 150S provirions to 155S native virions. It is also consistent with the increase in density from 1.37–1.42 g/cm^3 to 1.45 g/cm^3 found accompanying the transition from 80S RNPs to mature virions. (The densities were obtained after formaldehyde fixation, see Table 68, p. 437). Fig. 112 illustrates these final steps in virion morphogenesis.

4. The Possible Role of Mg^{++}

In tight association with native virions are 120 Mg^{++} ions (Mapoles, 1980). These ions evidently contribute significantly to the stability of complete virus particles (see Figs. 24, 32, p. 80 and p. 110). In the conversion of provirion to virion, these Mg^{++} ions move from a site sensitive to chelating agents to localities in which they are resistant to chelation. It is possible that these Mg^{++} bonds are stabilized by cleavage of VP_0. It is not yet known, which of the two steps occurs first or wether they occur simultaneously.

E. The Final Morphogenetic Cleavage

The final step in virus maturation – that renders the process irreversible – involves the cleavage of VP_0 to VP_2 and VP_4 (Holland and Kiehn, 1968; Jacobson *et al.,* 1970). The protease that is responsible for this cleavage still needs to be identified. The pair of amino acids at the VP_2–VP_4 cleavage site is unusual for poliovirus precursor proteins: asn/ser instead of the common gln/gly (Kitamura *et al.*, 1981). It is not known whether the protease for VP_0 cleavage is of viral or cellular origin and whether or not it is confined to a cellular compartment(s). The cleavage site on VP_0 seems to aquire the conformation that is susceptible to protease only after encapsidation of RNA into an RNP. Guanidine, which prevents the association of viral RNA and capsid proteins also blocks the cleavage of VP_0 (Jacobson and Baltimore, 1968a; Baltimore, 1971a). VP_0 is not cleaved in an in vitro translation system of EMC-RNA in which all other cleavages proceed

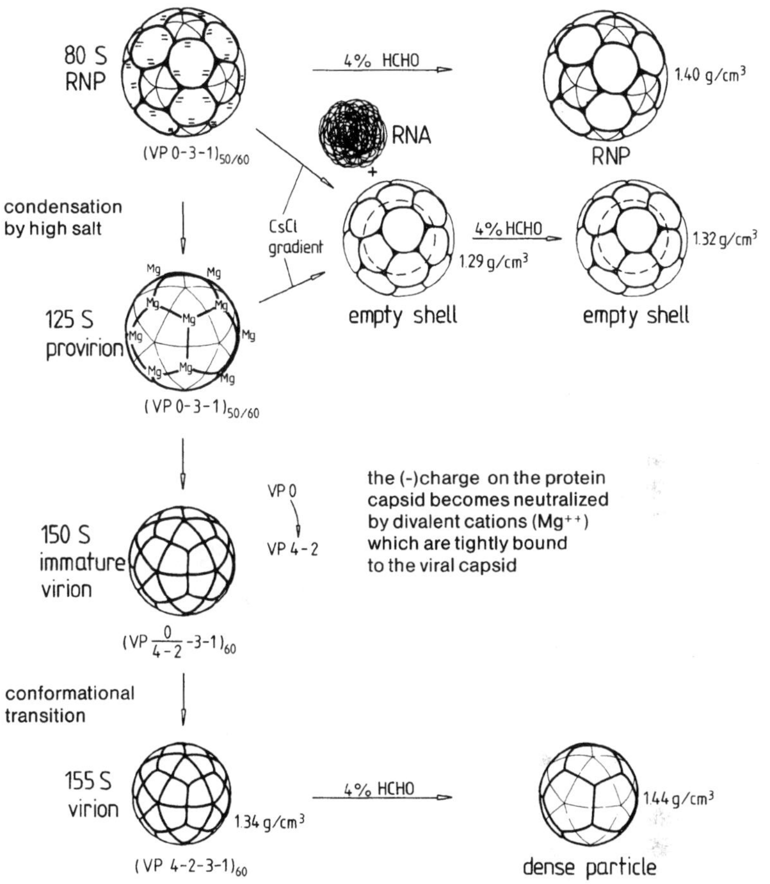

Fig. 112. Model for the condensation of poliovirus ribonucleoprotein particles
This figure illustrates schematically the hypothetical final steps of virion morphogenesis from the 80 S RNP to mature virions. Also illustrated are the behavior of particles upon centrifugation in CsCl density gradients. 80 S RNP and 125 S provirion are dissociated into empty shells and RNA upon density gradient centrifugation in CsCl unless they were fixed beforehand with 4% formaldehyde. After fixation with formaldehyde, 80 S RNPs have an apparent buoyant density of 1.40 g/cm^3 (see also figures 104 and 108), compared to apparent buoyant densities of 1.32 g/cm^3 and 1.44 g/cm^3 for empty shells and intact virions, respectively. Condensation of the ribonucleoprotein particle may be mediated by high salt concentrations in the corresponding cellular compartment. The 120 Mg^{++} ions per particle are localized in chelation-sensitive sites in provirions, and move to chelation-insensitive sites during the conformational rearrangement of the virion capsid which accompanies the cleavage of VP_0 to VP_4 and VP_2. Binding of Mg^{++} ions may be responsible for the transition in isoelectric point from pH 4.5 in 80 S RNPs to pH 7.2 in progeny virions. The conformational rearrangement accompanying cleavage of VP_0 leads to the full exposition of receptor recognition sites and D antigenic determinants on the virion surface

normally and in which formation of 74S empty shells can be observed (Palmenberg, 1982). Cleaved VP_0 occasionally observed in NTC preparations is probably due to the dissociation of RNPs in which the cleavage of VP_0 had already commenced.

The cleavage of VP_0 is accompanied by one last conformational transition of the viral capsid that finally locks the virion into the stable state of native virion. Indicators of this conformational shift are:

1. the acquired stability to chelating agents of divalent cations – which may also reflect a "disappearance" of negative charge from the capsid surface;

2. changes in the chemical reactivity of certain amino acid side chains of the capsid proteins, implying in particular the "burial" of portions of VP_2, and

3. the "disappearance" of the small cleavage product VP_4 from the capsid surface to an internal position where it comes to be in close spatial association with the RNA;

4. the aquisition of receptor binding capacity, and possibly also

5. change in antigenicity from the C to D antigenicity;

6. change from the isoelectric pI 4.5 form to the isoelectric pI 7.2 form which must involve quite some charge transfer from the outside to inside or vice versa.

Some negatively charged amino acid residues must "disappear" from the capsid surface and these are likely to carry their counter ions along with them to more internal positions of the capsid to keep a balanced charge environment internally. The involved cations are probably Mg^{++} and/or Ca^{++} which are small and tight and would not occupy much space and contribute only insignificantly to the molecular weight or density of the virion.

It is not clear whether the observed movement of completed virions to the cytoplasmic periphery and appearance there within the rough microsomal sacs precedes or follows the cleavage of VP_0. It is also not known whether the few molecules of VP_0 which remain uncleaved in the native virion are of functional significance in this stage of the virus growth cycle.

In any case, the main function of the final morphogenetic cleavage may be to render the process irreversible. Concomitantly, a potential starting point for the next infectious cycle has been "built in" ingeniously: the cleaved off little protein VP_4 seems to function as a potential uncoating plug (see Chapter 7). Thus, the circle of the virus "life cycle" has been closed.

V. Summary

1. A dramatic reorganization of cell structure proceeds under the direction of the virus, resulting in compartmentalization of the cell cytoplasm, and in functional differentiation of intracellular membranes and cytoskeleton. These structural and functional reorganizations reflect requirements for successful virion morphogenesis.

2. The synthesis of virion components proceeds in close association with virus-induced and virus-modified intracellular membranes and cytoskeleton: a) Viral capsid proteins are synthesized as a large common precursor protein NCVP1a. The synthesis takes place on large polyribosomes that are attached to the endoplasmatic reticulum in the cell periphery. b) Viral RNA is synthesized in association with smooth membranes in the central perinuclear region of the cell.

3. Morphogenesis of the virion from its component parts takes place in newly formed virus assembly factories in the central perinuclear region of the cell, *i.e.*, close to the site of RNA synthesis. The marked virus directed structural reorganization of the central region includes the construction of an intricate system of small vesicles and channels, and a reorganization of the associated cytoskeleton, in particular the long perinuclear filaments.

4. In contrast to the virion RNA which is encapsidated at the site ot its synthesis, the capsid proteins must be transported from the site of their synthesis to the "assembly factories". This transport presumably occurs on the inside or along the membranes of the endoplasmic reticulum – perhaps in association with the movement of freshly synthesized membranes from the rough to the smooth fraction. NCVP1a becomes attached to membranes of the endoplasmic reticulum during (or immediately following) its synthesis: NCVP1a is probably physically inserted into the membrane by hydrophobic polypeptide stretches–presumably in the VP_0 region. If the ensuing RNA encapsidation occurs within the virus induced vesicles, it is predicted that the main polypeptide regions that protrude from the membrane, presumably mainly the VP_1 and VP_3 regions, lie on the internal side of the endoplasmic reticulum, *i.e.*, on the side opposite to the ribosomes.

5. The capsid proteins undergo a series of conformational transitions, in the course of which two main bonding domains become exposed and activated in sequence. This serves to regulate the stepwise, orderly assembly of the viral capsid.

a) The first bonding domain is activated upon a morphogenetic cleavage of the capsid precursor protein by the viral protease NCVP7c, a part of the replicase P-3 region. NCVP1a is cleaved at two specific gln/gly sites into the three similarly sized procapsid proteins: VP_0, VP_3 and VP_1. The latter remain associated to the membrane and to one another in a 5S aggregate. This VP_0-VP_3-VP_1 complex has been termed the protomer–60 protomers eventually associate into an icosahedral virion capsid. Cleavage of NCVP1a, followed by some kind of conformational shift, may suffice to activate the interprotomer bond. The protomers then associate into a 14S pentameric subunit. NCVP1a is also occasionally cleaved at only one of its two cleavage sites producing NCVP3a and NCVP3c. These proteins or various aggregates of capsid related proteins (NCVP1a derived cleavage products) may serve some other functions which do not involve their direct use as capsid components. They may serve as membrane-integrated assembly-enhancing factors that are lacking in cells infected by DI-particles.

b) Activation of the second bonding domain–the interpentamer bonding domain–is not induced by a morphogenetic cleavage. Rather it seems to involve a conformational shift that is brought about either 1. by the formation of the interprotomer bonds in the 14S subunit itself, 2. by the activation by other membrane components ("assembly enhancing factor") that either are induced by NCVP1a or one of its products or represent such products themselves, or 3. by release of the subunit from physical membrane integration to a "loose association" via NCVP1a related membrane components.

6. The order of steps ensuing after the activation of the interpentamer bonding domain in the 14S particles remain obscure. It is proposed that they take

place within the membrane enclosed vesicles—the "encapsidation factories"—under the control of the vesicular membranes. Eventually the first ribonucleoprotein particle is formed.

a) The 14S subunits may associate first into 55S substructures. Presumably the 55S structure represents either a ring like structure composed of five 14S subunits or a "half-shell" composed of six 14S subunits. By dimerization the 55S subunits rapidly associate into 74S particles. It is not known whether this latter structure is a complete—yet still unstable—12 pentamer shell, or an "incomplete" 10-pentamer shell with two "holes".

There may be something special about one or two of the 14S pentamers, that form the final "caps" to the virion shell. They could provide a binding site of the shell to the vesicular membranes on the one hand and a potential recognition site for the initiation of RNA encapsidation (see below).

b) At some point during or shortly after the assembly of the 14S subunits into a capsid-like structure, these interact with viral RNA. This interaction presumably proceeds by means of ionic and hydrophilic interactions between the negatively charged phosphate groups of the RNA and basic or polar amino acid side chains. Hydrophobic interactions are also possible between the hydrophobic planes of the nucleotide bases and aromatic amino acids. Polypeptide arms specialized for interaction with RNA may stretch away from the shell-forming regions of the 14S subunits towards the future core of the virion.

Two possibilities for the time point and mechanism of initial RNA-capsid interaction are considered:

1. The viral RNA is synthesized into the vesicles and condensed into a globular structure—encapsidation ensues by the orderly assembly of 14S or dimerization of 55S particles around the RNA core.

2. During the course of its synthesis the viral RNA is directed to and injected into an unstable (near-) complete shell that is held attached to the vesicular membrane.

In either case, it is predicted that the replication complex is associated to the outer side of the vesicular membrane. The virion RNA may utilize a channel built from an integral membrane component (a 14S subunit ?) to traverse the vesicular membrane. A recognition mechanism could involve the small protein VPg that is attached to the 5' end of the viral RNA. RNA lacking VPg would remain "outside" the vesicles, become associated to the cytoskeleton and thus be "free" to move to the cytoplasmic periphery to function as mRNA.

7. Whatever the precise mechanism of encapsidation, the product is a swollen, unfinished virion like ribonucleoprotein. This structure is condensed in a stepwise fashion—reflected in increases in density and sedimentation velocity. It is proposed that this condensation is mediated by a high concentration of cations present in the encapsidation vesicles. 6,000 cations, mainly K^+ ions, are encapsidated within the virion as a neutralizing agent for the RNA. This corresponds to a molarity of 2–4 M. In a process akin to salting out, the virion RNA – and thereby the entire particle – is condensed into a compact state, water is "squeezed" out of the particle in the process. Mg^{++} ions essential for final capsid stablity – 2 per protomer – are transferred from a locality sensitive to chelating agents of divalent

cations in provirions to a position in which they are resistant to chelators. It is possible that part of this process takes place in conjunction with the final conformational reorganization of the viral capsid which is initiated by the cleavage of VP_0.

8. Finally, one last morphogenetic cleavage of VP_0 to VP_2 and VP_4 renders the process of virion assembly irreversible. The identity and localization of the responsible protease are not known. The cleavage site is formed by an unusual amino acid pair. It is in a sensitive conformation only in condensed ribonucleoprotein particles. The particle undergoes a conformational transition that is reflected in a change of antigenicity from the C to the D state and in a change of the isoelectric point from the pI 4.5 to the pI 7 form. The small cleavage product VP_4 is "sucked" into the virus particle to come to lie close to the RNA.

The virion is locked in its native, stable conformation and is released from membrane association. The freshly born virions probably first accumulate in the vesicles where they were formed. Later they are seen to move – still enclosed in membranous sacs – to the cytoplasmic periphery. Here they may be released from the cell by fusion of the membranous sacs with the cell membrane. Alternatively, virus particles may aggregate in large crystals below the cytoplasmic membrane, to be released when the cell dies and lyses.

Ingeniously, the cleavage of VP_4 from VP_0 confers to the virion an inherent mechanism for its decapsidation. This is a prerequisite for the initiation of any infection in the future. When the virion will encounter a susceptible host cell, it will utilize the cleaved VP_4 as an "uncoating plug" to initiate the liberation of its genome into the cytoplasmic matrix of its host cell where it may begin a new "living phase" of poliovirus biology.

11
Conclusions

The virus of poliomyelitis is one of the smallest and best studied animal viruses. Its apparent simplicity is striking: the virion is small, containing only a single strand of RNA and four species of protein. The RNA itself is small and seems simple, containing sufficient information only for about 10 average size proteins. This has furnished hopes and expectations that the functions of this virus some day can be completely described, and that the full understanding of poliovirus biology may provide models for studies of more complex systems. Its apparent simplicity and the fact that it is the agent of an emotionally provocative epidemic disease were important stimuli for the study of poliovirus. Many other diseases are caused by close relatives of the polioviruses: the common cold, hepatitis A, certain types of enteritis in humans, and the foot and mouth disease of live stock. It is expected that attempts to control these diseases should profit also from information obtained in studies on the poliovirus.

A very large body of information, indeed, has accumulated over the past three decades, concerning the molecular biology of poliovirus. More than 1,500 references were considered in this review and it is very likely that we have overlooked important contributions or simply lost some. Dozens of new publications on picornaviruses appear each month, and it is virtually impossible for any person to keep track of them all, let alone to gather them into a coherent whole. Many important insights were gained into the molecular biology of poliovirus. It has become clear that the apparent simplicity of poliovirus is deceiving – that we are still far from a complete understanding of poliovirus – and that by no means poliovirus research is "dead" (Baltimore, 1971a). Future work will certainly resolve some of the many unanswered questions.

Regarding the structure of the virion:

What exactly is the relative arrangement of the viral capsid proteins?

What is the three dimensional structure of the protomers, capsomers, and individual capsid proteins within the capsid?

Which are the specific parts of the viral surface designed for interaction with the receptor on the host cell membrane?

Are any particular types of surface protrusions involved?

What is the relative localization of antigenic determinants on the viral surface, which viral capsid proteins are involved in their specification?

What determines wether an antigenic site triggers the synthesis of antibodies which neutralize virions?

What is the nature of the bonds that hold the capsid together?

What determines the capacity of the capsid to assume different overall conformations?

In which part of the capsid does the proteinkinase activity reside? What is its function?

What is the nature of the interaction between capsid proteins and viral RNA?

Are there specific binding sites for the RNA on the capsid proteins?

Which capsid proteins are involved?

What portions of the RNA are recognized?

What is the relative arrangement of viral RNA and capsid protein in the interior of the particle?

How is the viral RNA uncoated?

What is the role of VP4 in the process? Where is VP4 located? Does the release of VP4 create "holes" in the capsid? How many, and where? What is the significance of the one or two uncleaved VP_Os?

What determines the lability of the capsid to disaggregation by heat, by urea, by other agents?

What is the role of the cysteine residues in VP_2, VP_3 and VP_1?

What is the role of Mg^{++} ions (or other divalent cations) in capsid stability and in the conformational shift (expansion of the capsid?).

How do the amino acid substitutions in the capsid proteins of vaccine strains and other structural mutants affect conformation, stability, antigenicity of the capsid?

What is the molecular basis of resistance to dextrane sulfate, to inhibitors in serum, to other inhibitors?

What part of the capsid constitutes the receptor recognition site? Is the receptor recognition site potentially antigenic? Will this site be useful for the construction of synthetic vaccines?

How do the capsid proteins mediate the shut-off of host protein synthesis?

Regarding genome structure and genetics:

What is the role of the unusually long untranslated segment of the genome at the 5' end?

Can the map derived from recombination studies with poliovirus mutants be correlated to the map derived from the combination of nucleotide sequence and amino acid analyses?

What are the functions of the genome products?

Do the precursors and processing intermediates carry out any functions distinct from those of their products?

What is the molecular basis of attenuation?

Which of the many mutations decribed in the vaccine strains are responsible for the alterations of its phenotype?

What is the molecular mechanism of recombination?

What inferences can be made regarding the evolution of the three poliovirus serotypes and other picornavirus relatives from a comparison of their nucleotide sequences?

Which segments of the genome are conserved, because they encode essential functions; which segments are more variable and what are the encoded functions?

Regarding poliovirus replication in the infected host cell:

What is the identity of the host cell receptor and the other host factors involved in the attachment, modification and uptake of the infecting virion?

Does the poliovirus receptor carry out any essential function in the uninfected cell?

What is the molecular basis of the far more common abortive infection pathway?

How does binding of poliovirus to the plasma membrane and its subsequent penetration effect the functions of the plasma membrane?

How is the Na^+/K^+ pump inhibited by the infecting poliovirus?

What causes the disturbance in intracellular ion concentrations, the specific increase of permeability for cations?

How is the rearrangement of the host cell cytoskeleton mediated?

What controls the formation of poliovirus specific smooth membrane vesicles?

What brings about the unusual appearance and configuration of the membrane clusters?

What is the molecular basis of interference with host cell protein, RNA and DNA synthesis?

What is the role of the nucleus, if any, in virus replication?

What regulates the switch from translation of the parental virion RNA to its transcription?

What is the identity of the host cell factors involved in the initiation of RNA synthesis?

Are any of the cellular polymerase activities utilized by the infecting virus for the replication of its RNA?

What is the identity and function of the host ribosomal factor(s) found in the replication complex?

Do any of the poliovirus proteins become cotranslationally inserted into or transported across intracellular membranes?

Where are the putative start-transfer and stop-transfer signals in the poliovirus proteins located?

What is the relative localization of viral RNA synthesis and capsid assembly with respect to intracellular membranes?

What determines the specificity of the poliovirus specific protease?

Is an additional host protease required for processing?

How are the proteolytic activities regulated?

How is the synthesis of vRNA initiated on the double stranded RF or RI template?

What is the role of VPg?

What is the significance of minor forms of viral RNAs, such as circular or dimeric forms of RF, conservative RI?

Is the poliovirus capsid assembled in the soluble phase of the cytoplasm, in the plane of intracellular membranes, or within preformed vesicles?

Is the RNA inserted into a preformed capsid, or is the capsid assembled around a core of RNA?

How are the processes of RNA synthesis and virion formation coupled?

What is the molecular basis of inhibition of virus replication by guanidine, HBB, and other inhibitors?

Indeed, poliovirus research should be very much alive for some time to come!

Appendix I: Laboratories Engaged in Poliovirus Research

The following is a list of laboratories presently engaged in research on the molecular biology of poliovirus. The brief descriptions of the field of research are based upon statements from members of the laboratories or on publications that have appeared within the last two years. It is also possible that some of these laboratories have since stopped working on poliovirus or that we have failed to include laboratories that only recently have begun work in this field.

This list is intended to give the reader an impression of the intensity and directions of present day polioresearch. We also hope that the listing will be of value for young researches intending to engage in research, as well as in stimulating communication between those researches already studying poliovirus.

In our search, we have come across quite a number of institutions engaged in analysis of polioviruses in sewage, food, waste-water, etc. as well as institutions engaged in administration and survey of vaccination programs, or institutions studying the clinical aspects of paralytic poliomyelitis disease and the rehabilitation of patients. Since these topics have not been discussed in our book, we have not included these institutions in the listing. We would greatly appreciate receiving information that we have missed or correction of mistakes made in the listing.

AUSTRALIA

The John Curtin School of Medical Research The Australian National University Canberra City A. C. T. 2601 P. Cooper	No active poliovirus research; library of ts mutants

AUSTRIA

Sandoz-Research Institut Brunnerstrasse 59 A-1235 Wien B. Rosenwirth	Inhibitors of the poliovirus protease

BELGIUM

Laboratory for Microbiology Antwerp University Universiteitsplein 1 B-2610 Wilrijk D. Vanden Berghe	Antiviral compounds from an African Euphorbia species
Vrije Universiteit Brussel Department of Micobiology and Hygiene Laaberbeklaan 103 B-1090 Brussels R. Vrijsen, A. Boeye	Amino acid sequencing of capsid proteins; virion conformation; monoclonal antibodies against poliovirus; morphogenesis

BERLIN

Freie Universität Berlin Institut für Experimentelle und Klinische Virologie Hindenburgdamm 27 D-1000 Berlin 45 K. Habermehl	Poliovirus structure; virus neutralization; lysosome mediated uptake of poliovirus; membrane alterations

BULGARIA

Medical Academy Institute of Infectious and Parasitic Diseases Viral Inhibitors and Interferon Laboratory 8, Belo More Street Sofia-1527 A. S. Galabor	Mechanism of action of inhibitors of poliovirus replication

ENGLAND

National Institute for Biological Standards and Control Holly Hill, Hampstedt London NW3 6RB P. D. Minor	Biochemical characterization of poliovirus strains; monoclonal antibodies; oligonucleotide mapping; genome sequencing
University of Leicester Department of Biochemistry Leicester J. Almond	Molecular basis for the antigenicity and virulence of poliovirus type 3

FEDERAL REPUBLIC OF GERMANY

Universität Hamburg Heinrich Pette Institut für Experimentelle Virologie und Immunologie Martinistrasse 25 D-2000 Hamburg 20 O. Drees, R. Dernick	Poliovirus structure; characterization of capsid proteins
Universität Hamburg Abteilung Molekularbiologie Grindelallee 117/IV D-2000 Hamburg G. Koch	Host cell receptor; Na^+/K^+ pump; poliovirus structure; shut-off of host protein synthesis
Universität Köln Institut für Virologie D-5000 Köln 41 H. J. Eggers	Antiviral drugs; cross-reacting enterovirus antigens

FRANCE

Institut Pasteur Unité de Virologie Médicale 1528 rue du Dr. Raux F-75724 Paris, Cedex M. Girard, F. Horodniceau	Genetic organization of poliovirus RNA; molecular genetics. Organisation of poliovirus proteins and antigenic function

ISRAEL

Israel Institute for Biological Research P.O.B. 19 Ness-Ziona A. Kohn	Poliovirus as model for naked viruses

ITALY

Università di Cagliari Istituto di Microbiologia 1 Via G. T. Procell 12 Cagliari P. La Colla	Morphogenesis of poliovirus; guanidine - mutants
University of Rome Cattedra di Virologica Via di Porta Tiburtina 28 Rome R. Pérez-Bercoff	RNA-replication; initiation sites on viral RNA

JAPAN

Sapporo Medical College Department of Hygiene and Epidemiology Minami 1, Nishi 17 Sapporo 060 S. Urasawa	Interactions of polioviruses and their antibodies
Kitasato University School of Pharmaceutical Sciences Shirokane, Minato-ku Tokyo, 108 A. Nomoto	Genome structures of the Sabin vaccine strains of poliovirus types 1, 2 and 3 and their biological significance
National Institute of Health Department of Enteroviruses Nakato, Musashimurayama Tokyo, 190–12 A. Totsuka, I. Tagaya	Development of new marker tests for characterization of live-vaccine strains

MEXICO

Centro de Investigación y de Estudios Avanzados des IPN Dept. de Genética y Biología Molecular Apartado Postal 14-740 Mexico City C. Fernández-Tomás	Mechanisms of inhibition of host cell RNA-synthesis

NETHERLANDS

Rijksinstitut voor de Volksgezondheit Postbus 1 NL-3720 BA Bilthoven T. van Wezel	Genetic engineering of polio virus cDNA

NORWAY

Norsk Hydro's Institute for Cancer Research Montebello Oslo 3 S. Olsnes, K. Sandvig	Mechanism of entry of protein toxins and poliovirus into the cytosol

SPAIN

Universidad Autónoma de Madrid Centro de Biologia Molecular Canto Blanco Madrid - 34 L. Carrasco	Virus induced changes in permeability of plasma membrane; inhibition of host cell nucleic acid synthesis

SWITZERLAND

Universität Basel Institut für Mikrobiologie und Hygiene Petersplatz 10 CH-4003 Basel K. Bienz	Morphological alterations of poliovirus infected cells; intracellular distribution and function of poliovirus proteins; virus induced vesicles

U.S.A.

University of Arizona Department of Cellular and Developmental Biology Tucson, AZ 85721 M. J. Hewlett	Interaction of RNA-binding proteins with terminal genome sequences of poliovirus RNA; ts mutants
University of California Department of Biological Chemistry Davis, CA 95616 J. W. B. Hershey	Shut-off of host protein synthesis; initiation factors
Scripps Clinic and Research Foundation Department of Molecular Biology La Jolla, CA 92037 J. M. Hogle	High resolution X-ray crystallography of poliovirus crystals
UCLA School of Medicine Department of Microbiology and Immunology Los Angeles, CA 90024 A. Dasgupta	Characterization of the polioviral replicase and related host factors
Central Research and Developmental Department E.I. du Pont de Nemours & Co. Wilmington, DE 19898 B. Korant	Processing of poliovirus proteins; viral and cellular proteases

University of Florida J. Hillis Miller Health Center Department of Immunology and Medical Microbiology Gainesville, FL 32610 J. B. Flanagan	RNA-replication
Department of Health, Education and Welfare Viral Exanthems Branch and WHO Collaborating Center for Smallpox and Other Poxvirus Infections Atlanta, GA 30333 J. H. Nakano	Molecular characterization of poliovirus strains
Medical College of Georgia Department of Cell and Molecular Biology Augusta, GA 30912 C. N. Nair	Monovalent cation metabolism in poliovirus infected cells
National Institute of Health Laboratory of Molecular Genetics Bethesda, MD 20205 J. Maizel	Structural organization of the poliovirus genome
Massachusetts Institute of Technology Center for Cancer Research Cambridge, MA 02138 D. Baltimore	Poliovirus RNA-replicating enzymes; molecular clones of viral cDNA
Washington University School of Medicine Department of Microbiology and Immunology St. Louis, MI 63112 M. McClure	Characterization of the deletions in defective interfering particles
University of Nebraska Department of Medical Microbiology Omaha, NE 68105 G. R. Dubes	Infection of cells by isolated RNA; inactivation of poliovirus RNA

Albany Medical College Department of Microbiology and Immunology Neil Hellman Medical Research Building Albany, NY 12208 L. A. Caliguiri	Uncoating; ts mutants; poliovirus structure
State University of New York at Stony Brook Department of Microbiology Stony Brook, NY 11794 E. Wimmer	Genome structure; characterization of the viral protease; neutralization of virions
The Public Health Research Institute of the City of New York Department of Virology New York, NY 10016 B. Mandel	Uncoating of poliovirus; interaction of poliovirus with neutralizating antibody
Columbia University Department of Microbiology New York, NY 10032 V. Racaniello	Genetic engineering of poliovirus cDNA
Pennsylvania State University Department of Microbiology and Cell Biology University Park, PA 16802 D. Tershak	Protein kinases in poliovirus infected cells; production of defective interfering particles; guanidine mutants
University of Pittsburgh Medical School Department of Microbiology Pittsburgh, PA 15261 B. Phillips	Viral protein synthesis; in vivo and in vitro morphogenesis
University of Utah Department of Biochemistry and Cellular, Viral and Molecular Biology Salt Lake City, UT 84132 E. Ehrenfeld	Inhibition of host cell macromolecular synthesis; mechanism of poliovirus RNA replication
Washington State University Biochemistry/Biophysics Program Pullman, WA 99164 K. Dunker	Structure of poliovirus; antigenic determinants and host cell receptor site

University of Wisconsin-Madison
Biophysics Laboratory
Madison, WI 53706
R. Rückert

Comparative approach to the structure and molecular biology of picornaviruses; characterization of the viral protease, neutralization of virions;

A. Palmenberg

comparative analysis of genome sequences of picornaviruses

USSR

Academy of Medical Sciences
Institute of Poliomyelitis and Viral Encephalitis
Moscow Region 142782
V. Agol, L. Romanova

Poliovirus ts mutants; host factors in RNA-replication; genetic recombination

Appendix II: Poliovirus Models

A. A Paper-Model of a Prototype Picornavirus (Fig. 113)

*a) Assembly Instructions**

1. Cut along the dashed lines.

2. Fold flaps A under—along the icosahedral edges—and glue them together (only in the triangular portions that extend outward beyond the icosahedral border) so that little "pockets" are formed.

3. Insert dashed areas B in the direction of the arrow exactly under the corresponding adjacent area and glue together.

4. For II only: fold the little triangular areas C inward.

5. Stick the two halves together by inserting the triangular extensions of I into the corresponding "pockets" above the extensions of II.

b) Explanation of the Model

Each half represents six building block pentamers (corresponding to the 14 S assembly intermediates).

The four capsid proteins are represented by roughly globular domains and are named according to the enterovirus terminology (genome order VP4-VP2-VP3-VP1, order of increasing size: 4-3-2-1).

In I, the outlines of potential protomers (VP4-3-2-1) are emphasized. (It is as of yet not possible to ascertain which group of 4 proteins actually corresponds to the set derived from a common precursor. For this model, the most globularly shaped set was chosen arbitrarily. The protomer could also have a more triangular, elongated or curved shape).

In II, the resulting morphological clusters are outlined: 1) a total of twelve VP2-VP4 pentamers about the icosahedral apices (5-fold vertices); 2) a total of twenty VP1-VP3 trimers about the icosahedral facets (3-fold vertices). Both clusters are of similar size (about 185,000 daltons); the lines in II most closely approximate the actual "grooves" in the virion capsid.

Note that the morphological clusters in the virion are distributed "more spherically", *i.e.* the VP1-VP3 trimeric clusters will be elevated somewhat in the virion in contrast to the relatively flat representation of the regular icosahedron.

---

* Assembly is facilitated when the drawing is enlarged by photocopy.

glue together
cut away
fold
cut
d
glue under d'
A
B
C
VP4-VP2-VP3-VP1 Protomer
= one of 60 identical crystallographic structure units
VP2-VP4 Pentamer
= icosahedral apex (containing the VP4 "uncoating plug", and the D antigenic site)
VP1-VP3 Trimer
= icosahedral facet (probable location of major inter-14 S bonding domain, C antigenic site)

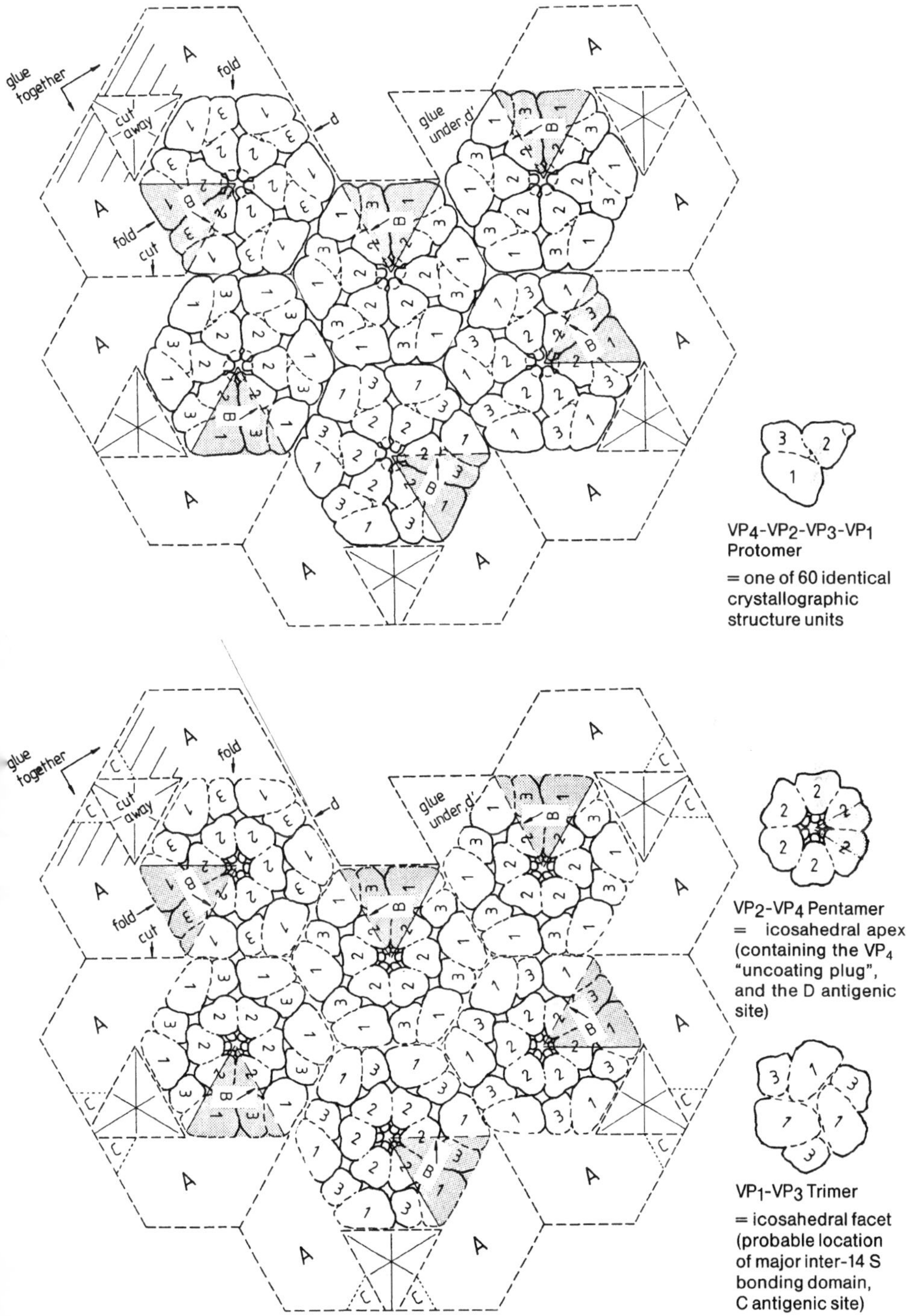

Fig. 113. A paper-model of a prototype picornavirus

The VP4 pentamers at the twelve 5-fold vertices are proposed to function as uncoating plugs.

The main bonding domains responsible for holding the 14 S pentamers together probably lie within the VP3-VP1 trimers, *i.e.* across the 3-fold axis of symmetry.

The model does not show any potential surface projections or internal polypeptide arms as are known to exist for some small plant viruses. Nonetheless there are striking similarities in construction principles to the unrelated small plant viruses.

B. The Apple Model of Poliovirus (Fig. 114)

1. Pick a fairly round apple. Start out by drawing a regular pentagon about the stem. (A ballpoint pen should draw well on an apple.) The size of the pentagon should be about 1/12th of the apple's surface. After one or two practice trials, you should become fairly efficient at drawing the correct size. (If you practice on an orange, you can discard the peals of poor trials and eat the orange).

2. Draw a second regular pentagon on the other end of the apple. The two pentagons should be arranged so that the corners of each point to the center of the sides of the other pentagon.

3. Draw lines from each of the corners of the two pentagons in the direction of the other. These lines are the continuation of the lines bisecting the angles of the pentagon. These lines should be the same lengths as the edges of the two original pentagons.

4. Connect the ends of the lines just drawn by a jig-jaw line, thus creating a ring of 10 additional pentamers. You have now constructed a regular dodecahedron onto a spherical surface and have outlined the positions of the twelve 14 S pentamers of your polioviurs-model (Fig. A).

Note: If you wish to disassemble the virion now into its component parts, proceed to steps 6—11 below. If you wish to complete visualizing the 60 protomers of the model, proceed to step 5 first. It is easier to disassemble and reassemble the model correctly if you start with steps 6—11 before completing the icosahedron. Or repeat steps 1—4 on a second apple and then proceed to the steps below.

5. From the center of each pentagon draw lines to each of the corners of the pentagon. Alternatively, you can draw lines from the center of each pentagon to the centers of each edge. In either case, you have constructed a regular icosahedron onto the model's surface. In the former case, you have concomitantly outlined a pentakis dodecahedron, in the latter a trapezoidal hexecontrahedron. You can now readily visualize the duality of dodecahedron and icosahedron (Fig. B). In principle, you could design an infinite number of regular subdivisions of the regular pentagons into five equivalent areas. All of these subdivisions yield 60 subunits in equivalent environments on a spherical surface.

6. Now take a sharp knife and stab it into each of the 10 lines of the long jigjaw line that you drew in step 4. Point the knife at the center of the apple, and try to stab roughly perpendicular to the surface of the apple. The apple should then fall apart into two halves (Fig. C).

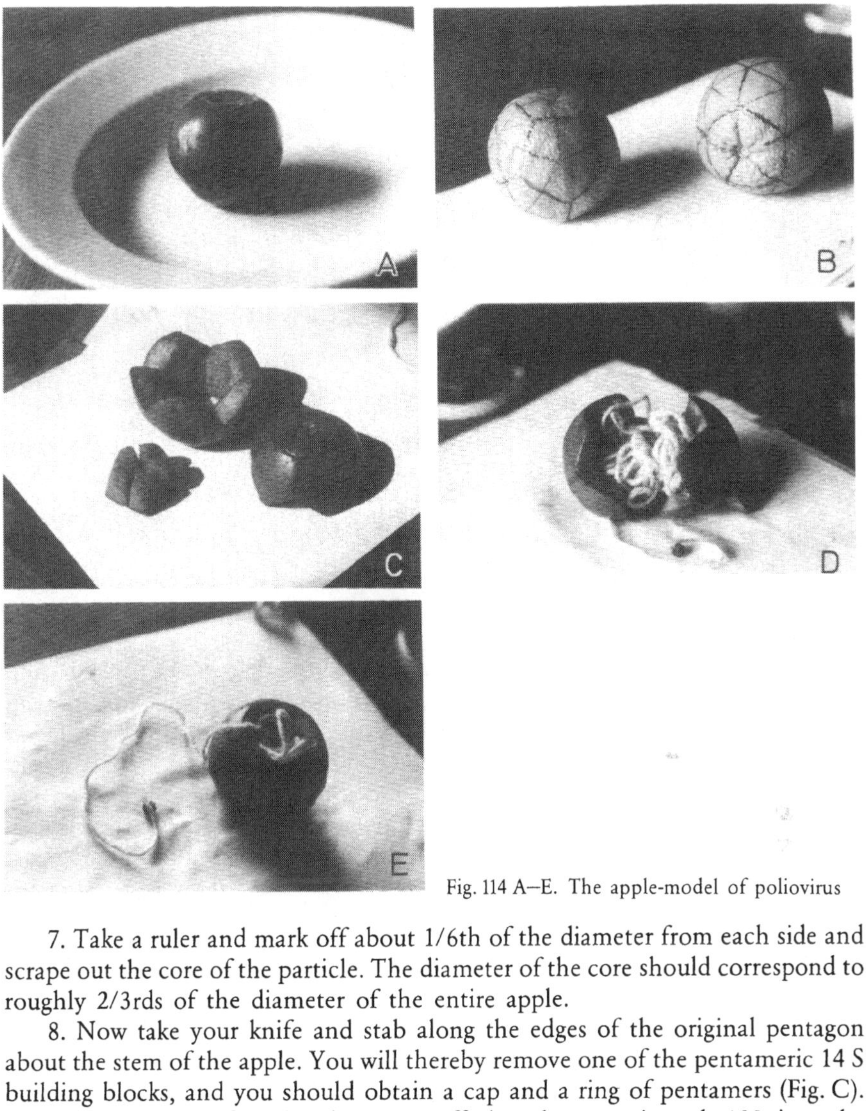

Fig. 114 A–E. The apple-model of poliovirus

7. Take a ruler and mark off about 1/6th of the diameter from each side and scrape out the core of the particle. The diameter of the core should correspond to roughly 2/3rds of the diameter of the entire apple.

8. Now take your knife and stab along the edges of the original pentagon about the stem of the apple. You will thereby remove one of the pentameric 14 S building blocks, and you should obtain a cap and a ring of pentamers (Fig. C).

9. Take a string of wool and measure off a length, approximately 100 times the diameter of the apple. This is your poliovirus RNA molecule. Try sticking it into the core and close the cap. This is your native virion (Fig. D, E). A relatively thick string of wool will illustrate better the forces required for tight packing of the virion.

10. If you wish, you can now easily cut along the remaining lines, obtaining 12 pentameric 14 S subunits; dividing each of these in 5 equivalent pieces, will yield the protomers (Fig. C). Each of these contains one molecule of each of the capsid proteins, VP_{4-1}. This is your disassembled virion.

11. Enjoy reassembling your model, playing with it, or eating it.

Appendix III: The Geometry of Isometric Polyhedra

A. The Platonic Polyhedra

a) Models

Models of the Platonic polyhedra can be constructed from planar net diagrams of regular polygons (Albrecht Dürer, 1525; Caspar and Klug, 1962), a procedure which helps to illustrate the geometric beauty of the design. The planar equilateral triangular net can be folded into a convex surface if 5, 4 or 3 of the triangular facets join at a polyhedron vertex instead of 6 as in the plane. If polyhedral vertices of only one kind are introduced to form the closed surface, then regular polyhedra will be formed, namely the icosahedron with twelve 5-vertices, the octahedron with six 4-vertices, and the tetrahedron with four 3-vertices, respectively. When planar quadratic or pentagonal net diagrams are folded similarly into convex surfaces, the cube and dodecahedron are formed respectively. (Hexagons or larger regular polygons cannot be joined to produce regular convex polyhedra, explaining why there are only 5 Platonic polyhedra, Fig. 115). The bonding pattern is not disturbed geometrically by the folding procedure. Appendix IIA (p. 484–485, Fig. 113) illustrates the folding procedure for the construction of an icosahedron. Asymmetric units also may be arranged in the planar equilateral triangular net, prior to folding. Although the asymmetric units are in six different orientations in space (in five different orientations after folding into an icosahedron), they all have exactly equivalent relations. Again the pattern of relations and bonds is not disturbed geometrically by the folding procedure (Caspar and Klug, 1962).

b) Characteristics

The characteristics of the Platonic polyhedra are summarized in Figure 115. The Platonic polyhedra are each constructed from identical faces—regular, plane, convex polygons with straight sides. The same number of faces meet at each of the vertices. The dihedral angle between the adjacent faces of a given Platonic

DATA FOR THE PLATONIC POLYHEDRA

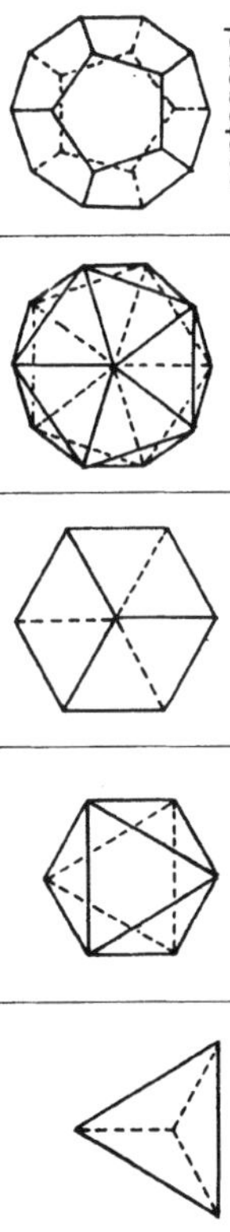

	tetrahedron	octahedron	cube	icosahedron	pentagonal dodecahedron
faces	4 triangles	8 triangles	6 squares	20 triangles	12 pentagons
vertices	4	6	8	12	20
edges	6	12	12	30	30
radius of circumsphere	·612 4	·7071	·8660	·9511	1·4013
radius of intersphere	·3536	·5000	·7071	·8090	1·3090
radius of insphere	·2041	·4082	·5000	·7558	1·1135
dihedral angle	70°32′	109°28′	90°0′	138°11′	116°34′

Radii of circumspheres, interspheres and inspheres are given (to four decimal places) in terms of the edge lengths of the polyhedra. Dihedral angles are given to the nearest minute.

Fig. 115. The five Platonic polyhedra. – Figure from Pugh, 1976 [in: Polyhedra–a Visual Approach, p. 14 (1976)]

polyhedron is constant. All of the vertices, all of the face centers, and all of the edge midpoints of a Platonic polyhedron are each at a constant distance from the center of the polyhedron. Consequentially, each Platonic polyhedron can be surrounded by a sphere that touches each of its vertices, the circumsphere; similarly, a slightly smaller sphere can be constructed that touches the midpoints of each edge, the intersphere; lastly, the insphere touches all of the face centers. Figure 116A shows the corresponding spheres for the icosahedron. The vertex figures – defined as the polygons that are formed by joining the midpoints of the edges that meet at a common vertex–of a Platonic polyhedron are congruent regular polygons (Figure 116B).

The circumsphere touches all of the icosahedral vertices, the vertices are all at a constant distance from the center of the icosahedron

The intersphere touches all of the edge-midpoints

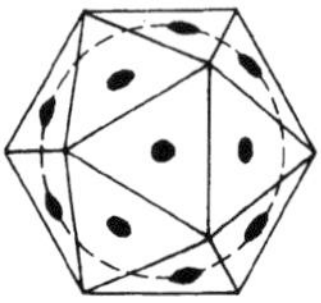

The insphere touches each of the facet centers

Fig. 116. Some geometric relationships of the icosahedron
Every corresponding (i.e. equivalently situated) point on the icosahedron is at a constant distance from the center of the icosahedron. This is readily evident for the 12 vertices, 30 edges and 20 facets from the above illustrations. Groups of three equivalently situated points can be established on every facet, and of two equivalently situated points on each edge. Both procedures yield a total of 60 equivalently situated points at a constant distance from the center of the icosahedron (see Fig. 120, p. 495)

Fig. 116 A. The circum-, inter- and inspheres of the icosahedron

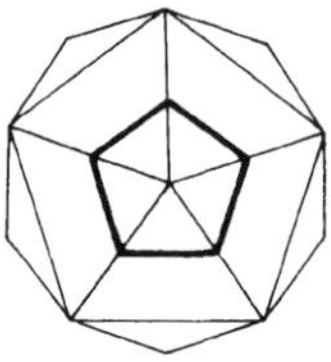

The vertex figure of a polygon is formed by connecting the midpoints of adjacent edges which meet at a common vertex. The vertex figure of the icosahedron is a regular pentagon

Fig. 116 B. The vertex figure of the icosahedron. – Figures from Pugh, 1976 [in: Polyhedra–A Visual Approach, pp. 4, 8 (1976)]

c) Duality

The symmetry-relations of the Platonic polyhedra were described above and illustrated in Figure 10 (p. 40). Since each of the Platonic polyhedra shares symmetries with each of the other four, the figures can each be arranged inside the others in a symmetrical manner creating some fascinating structures. Pairs of Platonic polyhedra with the same number of edges are "duals" of one another (the cube and octahedron, two tetrahedra, and the dodecahedron and icosahedron). The duals can be arranged about a common intersphere so that their edges cross at right angles and the vertices of each figure lie exactly outside the facet centers of the other figure. Figure 117 illustrates the duality of the dodecahedron and icosahedron.

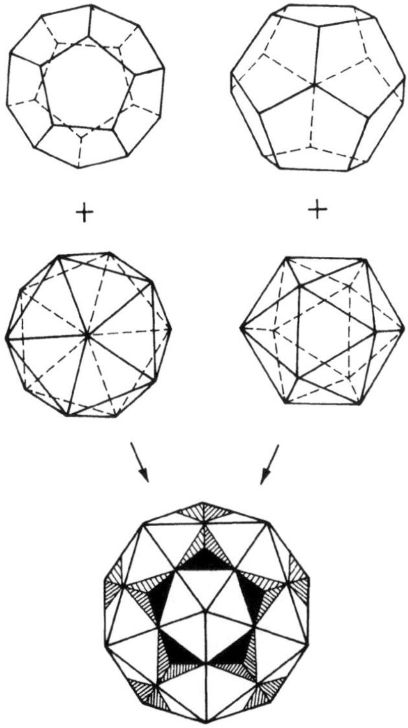

Fig. 117. Dualities of the Platonic polyhedra

Fig. 117 A. The dodecahedron and icosahedron can be placed about a common intersphere so that their edges cross at right angles and touch the intersphere at these thirty common points. Each vertex lies outside, and directly above, the facet center of the other figure. The two figures are said to be the duals of one another

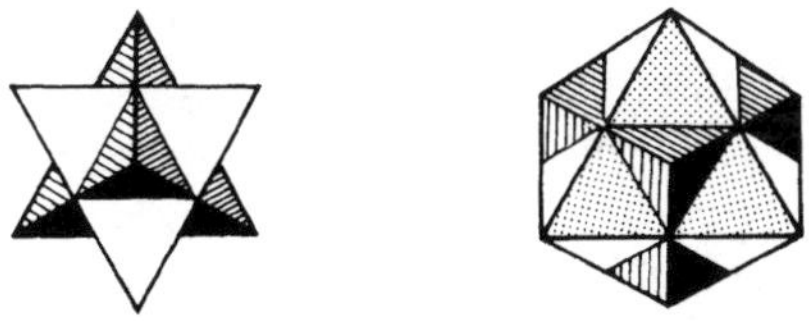

Fig. 117 B. Analogous principles of duality hold for the cube and octahedron and for two tetrahedra. – Figures from Pugh, 1976 [in: Polyhedra–A Visual Approach, p. 7 (1976)]

d) *The Golden Proportion*

The Platonic polyhedra are full of the seemingly mystical golden proportion relationship. When a line is divided into two parts so that the whole is to the larger part as the larger part is to the smaller part, the two parts are said to be in the golden proportion. The golden proportion is an irrational number which can also be derived from the Fibonacci Series of numbers—1, 1, 2, 3, 5, 8, 13, 21 . . .—produced by adding the last two numbers in the series to give the next one in the series: if two adjacent numbers of this series are expressed as a ratio, the larger the numbers are, the closer the ratio will approach the value of the golden proportion = 1:1.6180.

The golden proportion has often been used by artists and architects especially in Ancient Greece and during the Renaissance—as a basis for harmonious relationships. Figure 118 illustrates the appearance of the golden proportion in the golden rectangle (Fig. 118A), the regular pentagon and pentagram (Fig. 118B), and in the subdivision of a circle into a regular decagon (Fig. 118C). The golden proportion appears in the icosahedron in a variety of ways of which just a few examples are illustrated in Figure 119.

e) *Geometric Restriction of the Maximal Number of Subunits*

An important inherent geometric property of the Platonic polyhedra and all related structures bearing cubic symmetry is that they provide restrictions for the number of identical asymmetric structural units that can be fit into equivalent environments in a spherical lattice: namely, 12 units for tetrahedral structures, 24 for octahedral, and 60 for icosahedral structures. This property is illustrated in Figures 120 and 121 for the dodecahedron and icosahedron.

As can be easily seen in drawings (e.g. Fig. 120, p. 495) or models (Appendix II) of icosahedrons, the regular icosahedron has 12 vertices—the centers of 5-fold symmetry, 20 equilateral triangular facets—the centers of 3-fold symmetry, and 30 edges—the centers of 2-fold symmetry. The icosahedron thereby allows the arrangement of 12, 20 or 30 identical symmetric structure units into equivalent environments about the corresponding geometric points.

The surface of the icosahedron, however, can be subdivided in a way that fits 60 identical symmetric or asymmetric structure units in equivalent environments within the icosahedral lattice (Figs. 120, 121). An infinite number of sets of 60 equivalently situated points is defined by the icosahedron and dodecahedron (see Fig. 120). By connecting corresponding points on the icosahedron or dodecahedron in an identical manner sets of 60 identical areas are defined which are all equivalently situated in a spherical lattice (Fig. 121A, B). The spherical symmetry is more easily visualized when an intersphere is superimposed onto the icosahedron. Each facet of an icosahedron can also be divided in a regular manner into three equivalent areas. (This is the same as dividing its vertices into 5 equivalent areas each, and analogous to dividing the facets of the dodecahedron into 5 equivalent areas.) This procedure also provides for the arrangement of 60 identical structure units into equivalent environments in a spherical lattice (Fig. 121A).

In the construction of an icosahedral lattice from 60 identical subunits, the

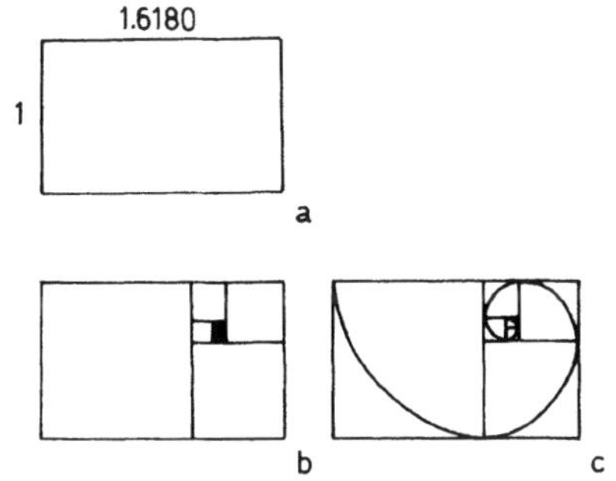

Fig. 118. The golden proportion in some geometric figures

Fig. 118 A. The golden rectangle
Adjacent sides of the golden rectangle are in the golden proportion *(a)*. When a square is divided off from a golden rectangle, the sides of the remaining rectangle will again be in the golden proportion, a procedure which can be carried on ad infinitum *(b, c)*

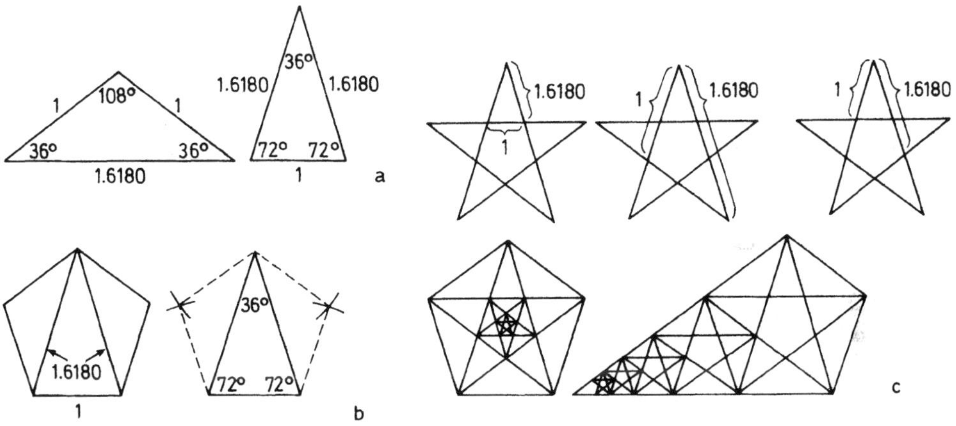

Fig. 118 B. The regular pentagon and pentagram
The sides of the 36° 72° 72° and 36° 36° 108° isocoles triangles are in the golden proportion *(a)*. Consequently the sides and diagonals of the regular pentagon are also in the golden proportion *(b)*. The golden proportions of the regular pentagram are presented in *(c)*

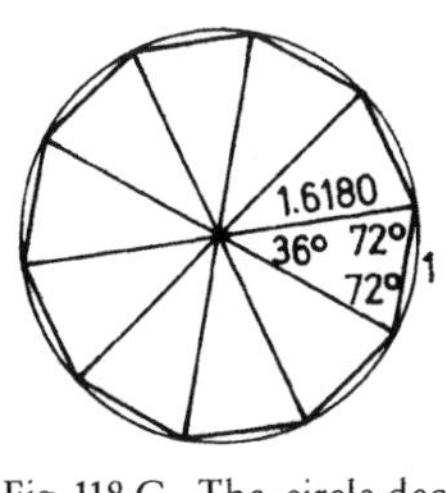

Fig. 118 C. The circle-decagon
When a circle is divided into 10 equal sections, the corresponding radii and chords will form 36° 72° 72° isocoles triangles. In other words the sides of a regular decagon and the radius of its circumscribing circle are in the golden proportion. – Figures redrawn from Pugh, 1976 [in: Polyhedra–A Visual Approach, p. 9 (1976)]

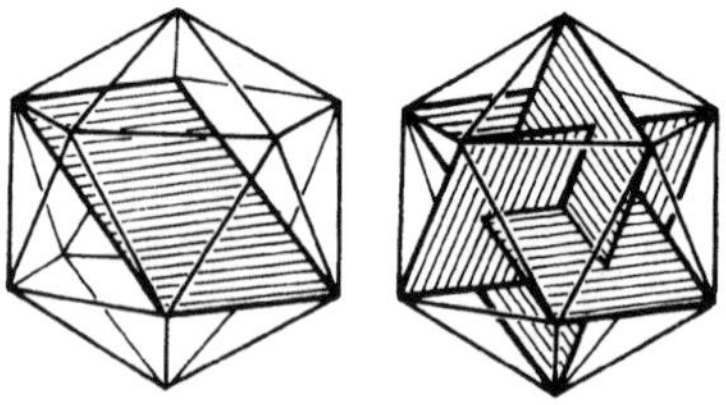

Fig. 119. The golden proportion in the icosahedron

Fig. 119 A. Golden rectangles in the icosahedron
Each pair of opposing edges, when connected by straight lines, forms a golden rectangle (since there are 30 edges in the icosahedron there are 15 such golden rectangles). These golden rectangles pass through the center of the icosahedron and lie in those bisecting planes of the icosahedron that pass through (are defined by) the icosahedral vertices (four vertices in each bisecting plane). In other words, the length of an icosahedral edge and the distance between opposing edges is in the golden proportion

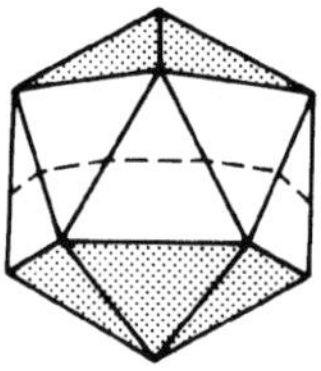

Fig. 119 B. The decagon-circle in the icosahedron
When opposing "caps" are removed from the icosahedron, a ring of ten triangles remains. Bisection of the ring defines a regular decagon and its circumscribing circle. There are six such golden rings in the icosahedron. These golden decagon/circles also pass through the center of the icosahedron and lie on those bisecting planes of the icosahedron (and its intersphere) that pass through (are defined by) the edge midpoints (ten in each bisecting plane). Or, to put it in another way, the distance between the centers of two adjacent icosahedral edges and the distances between these edge centers and the center of the icosahedron are in the golden proportion

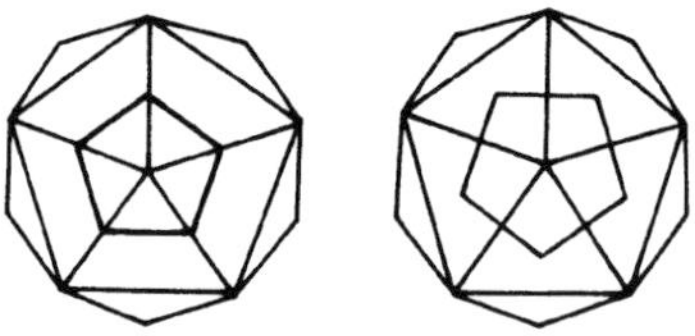

Fig. 119 C. Regular pentagons in the icosahedron
Since the arrangement of the facets that meet at a common vertex is that of a regular pentagon, the distance between any corresponding points on adjacent facets and the distance between those on nonadjacent facets (at a common vertex) are in the golden proportion

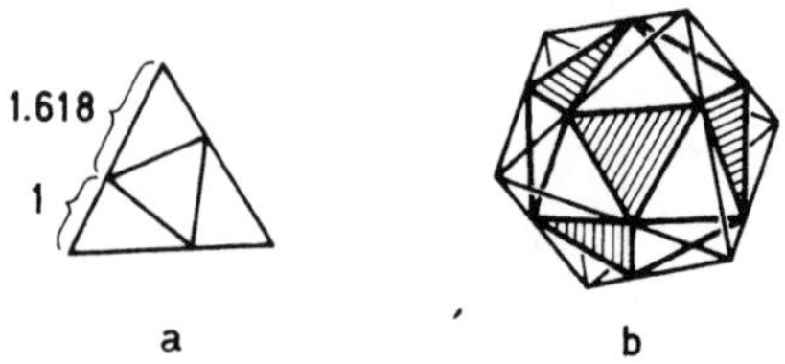

Fig. 119 D. The golden octahedron-icosahedron relationship
When the sides of an octahedron are divided in the golden proportion and the points so formed are connected by straight lines, an icosahedron is created. The octahedron and corresponding icosahedron share 8 common facial planes, i.e. 8 of the 20 icosahedral facets lie exactly on the facets of the octahedron. – Figures A and D redrawn from Pugh, 1976 [in: Polyhedra–A Visual Approach, p. 10 (1976)]

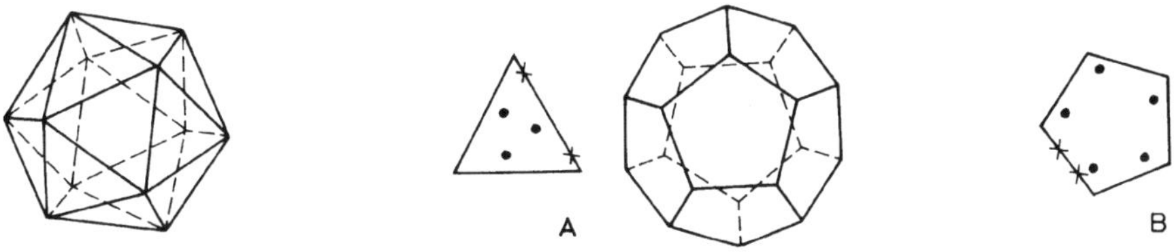

Fig. 120. An illustration of the phenomenon that the icosahedral lattice provides for no more than 60 equivalently situated subunits

A Restriction of the number of identically situated points in the icosahedron to sets of 12, 20, 30, or 60

Every point on the icosahedral facet (except for the facet center) has 2 corresponding points on the same facet. Each of the twenty identical facets has corresponding sets of three equivalently situated points yielding a total of 60 such points. Every point on the icosahedral edge (except of the edge midpoint) has another corresponding point on the same edge; there are corresponding sets of points on each of the 30 edges again yielding a total of 60 equivalently situated points. There are thus an infinite number of sets of 60 equivalently situated points on the icosahedron. In addition there are unique sets of 12 (at the vertices), 20 (at the facet centers), and 30 (at the edge midpoints) equivalently situated points. By connecting corresponding sets of points on the icosahedron by straight or curved lines, an infinite variety of sets of 60 equivalent areas are defined which are situated in strictly equivalent environments in a spherical frame work (see Fig. 121)

B Restriction of the number of identically situated points in the dodecahedron to sets of 12, 20, 30, or 60

Similar sets of equivalent points and areas are defined by the dodecahedron: every point on the dodecahedral facet (except for the facet center) has four corresponding points on the same facet. Each of the twelve identical facets has equivalent sets of five corresponding points. Every point (except for the edge mid- and endpoints) on a dodecahedral edge has another corresponding point on the same edge and corresponding sets of points on each of the 30 edges. There are thus an infinite number of sets of 60 equivalent situated points on the dodecahedron. In addition there are unique sets of 12 (at the facet centers), 20 (at the vertices), and 30 (at the edge midpoints) equivalently situated points

shape of the individual subunit needs not to be as rigidly symmetric as is implied by the subdivisions of Figure 120 and 121. As illustrated for the construction of an icosahedral model in Appendix II (p. 484), an infinite variety of 60 identical, awkwardly shaped subunits can be superimposed onto the subdivided lattice of the icosahedron. Corresponding points need only be connected in the same manner on all facets of the icosahedron or dodecahedron by curved lines.

There is no way of dividing the icosahedral surface into more than 60 equivalent areas so that all subunits are still arranged equivalently. For example, if the facets of the icosahedron were subdivided into four identical triangles each, there would still only be 60 equivalently situated triangles—the ones touching an icosahedral 5-fold vertex. The remaining 20 triangles at the facet centers—though all equivalently situated in respect to each other—are situated in different environments from those at the apices (Fig. 121B).

Dividing the facets of the dodecahedron equivalently also leads to a maximum of 60 identical and equivalently arranged structure units (Fig. 120B). As stated above, maximal subdivision of the other Platonic polyhedra leads to fewer, 12 or 24, equivalent structure units. Since there exist only these five spherically symmetric, principle polyhedra, and since all other isometric spherical structural

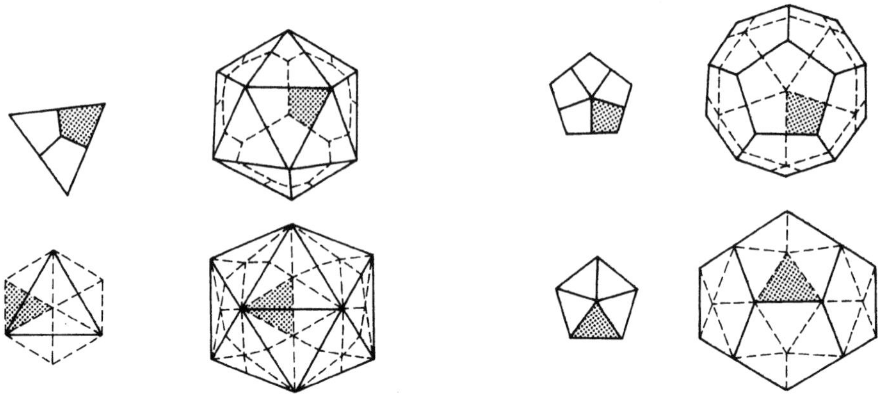

Fig. 121. Construction of an icosahedral shell from 60 identical structure units

Fig. 121 A. Regular subdivisions (shaded areas) of the icosahedron and dodecahedron that create 60 equivalently situated structure units

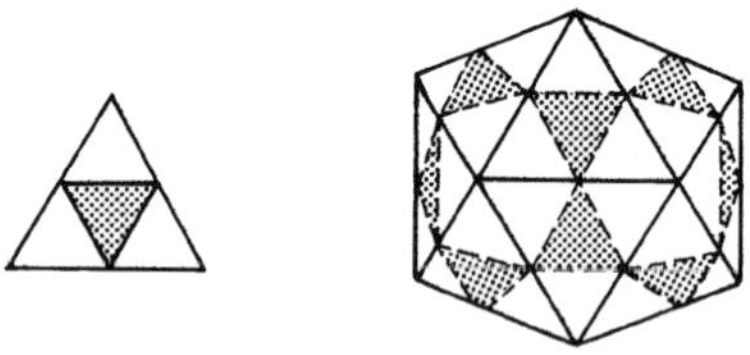

Fig. 121 B. Example of a regular subdivision of the icosahedron into more than 60 congruent areas There are still 60 areas that are equivalently situated (unshaded triangles). The twenty triangles in the facet centers (shaded triangles) are situated in different environments from those at the vertices

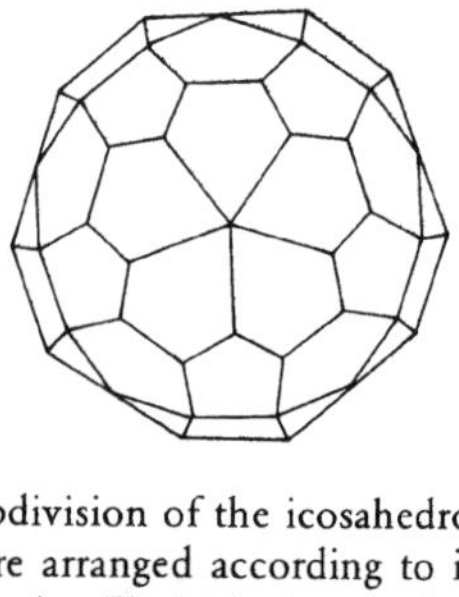

Fig. 121 C. Example of a regular subdivision of the icosahedron into 60 congruent structure units The 60 identical polygonal units are arranged according to icosahedral symmetry. This figure is named pentagonal hexecontrahedron (see Fig. 122). An edge between two adjacent polygons may be regarded as the contact between units. Each unit makes identical contact with its neighbors, i.e. the units are situated in strictly equivalent environments. – Figure C redrawn from Caspar and Klug, 1961 [Cold Spring Harbor Symposia on Quantitative Biology *27*, 8 (1961)]

lattices are related to these polyhedra, it is impossible to arrange more than 60 identical structural units on the surface of a sphere in a way that each is situated in an identical environment.

For the discussion of poliovirion structure it is important to keep in mind that these geometric considerations say nothing about the nature or composition of the biological "structural unit" in viral capsids with cubic symmetry. Such a structural unit may be a single protein molecule (e. g. calciviruses, satellite tobacco necrosis virus), or a set of different proteins (poliovirus and all other picornaviruses), or even a set of identical proteins in different conformational states (small insect and plant viruses such as the tomato bushy stunt virus (TBSV), or southern bean mosaic virus (SBMV) (see Fig. 13, p. 48; Fig. 18, p. 64, Chapter 3).

B. Other Icosahedron — Related Polyhedra

a) Characteristics

Next to the Platonic dodecahedron and icosahedron described above, other regular icosahedral lattices are found among the so called Archimedean polyhedra and their corresponding duals. (There are 13 figures in each group.) The general features of these three groups of polyhedra are summarized and compared in Tables 73 and 74 and Figure 122.

The Archimedean polyhedra were described in detail already in 1619 by Johannes Kepler. The icosahedral symmetry bearing Archimedean polyhedra (Fig. 122A) can be constructed by dividing the edges of the icosahedron or dodecahedron into equal parts and connecting the derived points with a series of lines to provide the edges of the new facets. According to the number of equal subdivisions, two-frequency, three-frequency, etc. figures are produced. These new figures are characterized in that they each are composed of two or more types of regular polygons which are associated in equivalent arrangements about each of the vertices. The edges of an Archimedean polyhedron are all equal in length,

Table 73. *Comparison of the properties of the three main groups of regular polyhedra*

	Platonic polyhedra	Archimedean polyhedra	Archimedean duals
Facets	one type regular, congruent	more than one type, regular	one type, non regular but congruent
Edges	all the same length	alle the same length	different lengths
Vertices	identical arrangement of only one type of polygon	identical arrangements of more than one polygon	more than one type of arrangement of identical polygons
Vertex figures	one type of regular congruent polygon	one type of non regular congruent polygon	more than one type of regular congruent polygons
Dihedral angle	constant	one or more than one	constant
Circumsphere	+	+	∅
Intersphere	+	+	+
Insphere	+	∅	+

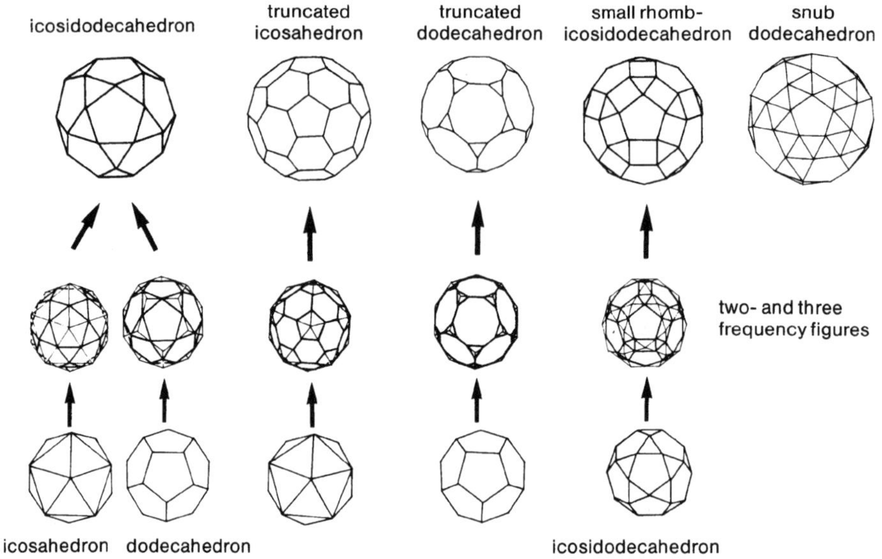

Fig. 122. Other icosahedron-related regular polyhedra which are used occasionally a models for spherical viruses

Fig. 122 A. The Archimedean polyhedra with icosahedral symmetry
The five Archimedian polyhedra with icosahedral symmetries are illustrated in the top row. The figures below illustrate how the Archimedian polyhedra are derived from the icosahedron and dodecahedron

Fig. 122 B. The duals of the Archimedian polyhedra with icosahedral symmetry
Each of the respective duals of the Archimedian polyhedra is drawn beneath the Archimedian polyhedron from which it was derived (top row)

the dihedral angles between adjacent facets, however, are usually of more than one type (Table 73 and Fig. 122A).

Each Archimedean polyhedron can be circumscribed by one of the Platonic polyhedra so that all of its vertices lie evenly arranged on the faces or edges of the circumscribing figure. Correspondingly, all of the vertices and all of the edge-midpoints are at equal distances from the center of the figure, so that each Archimedean polyhedron has a circumsphere touching all of its vertices and an intersphere touching the midpoints of each of the edges. Since they are constructed from different types of facets, however, the face centers are not at a constant distance from the edges or vertices, and hence are not at an equal distance from the figure center, so that no insphere can be constructed to touch every face of the Archimedean polyhedron.

The two-frequency figures of the dodecahedron and icosahedron are identical, the icosadodecahedron. The three-frequency figures derived from the dodecahedron and icosahedron appear as though the vertices have been cut away from the parental figures, hence they are named truncated dodecahedron and truncated icosahedron respectively. Of potential interest for poliovirus are also the small rhombicosidodecahedron (the four-frequency figure of either dodecahedron or icosahedron) and the snub dodecahedron, an intriguing figure, which—in contrast to all other Archimedean polyhedra—exists in two mirror image forms.

The duals of the Archimedean polyhedra are an entirely new set of polyhedra (Fig. 122B). The respective duals can be arranged about a common intersphere that touches each of their edges. The edges of the duals are perpendicular to each other at these points, the vertices of one lie exactly above the facet centers of the other (see also Fig. 117, p. 490). The faces of the Archimedean duals are not regular but they are all of only one type, *i.e.*, congruent. The Archimedean duals have been named vertically regular polyhedra, since their vertex figures are regular polygons (corresponding to the faces of the Archimedean polyhedra). Whereas the Archimedean polyhedra had the same number of edges meeting at each vertex and usually more than one dihedral angle between their facets, the vertically regular polyhedra have only one dihedral angle but do not have the same number of edges meeting in each vertex (Table 73).

Note that although there may be more than a total of 60 facets, vertices or edges there are never more than 60 equivalent structure units in any of the polyhedra (Table 74). Still the edges (and vertices) of these models also do not necessarily correspond to borders of the structure units in the virus capsid. For example, a vertex or edge might result from a "corner" or "bend" within the structure unit itself or neighboring structure units may form contacts so that the adjacent regions form a flat surface (corresponding to a facet in the model). In the latter case, (some of) the borders of the structure units would pass through the facet in the skeleton model, as for example in a rhombic triacontrahedron constructed from 60 structure units.

b) The Triangulation Number-Classification of Icosahedral Lattices

There is a trick employed by some viruses to increase the number of identical structure units employed for the construction of an icosahedral lattice. Though not of relevance for poliovirus, it is mentioned here, since it is often employed as a basis for the classification of viruses. The facets of any principle polyhedron can be subdivided in a regular manner. The most common procedure is to subdivide each triangular facet into a series of smaller triangles (by lines running parallel to the edges of the original triangle, this procedure is termed the Alternative Method). All of the new vertices created by this procedure will be formed from 6 equivalent triangles, whereas the 12 original vertices will always be formed from such 5 triangles. No matter how many 6-triangle vertices are created by the sub-

Table 74. *Characterization of the icosahedral polyhedra their vertices, facets, edges and dihedral angels*

Platonic	Archimedean	Archimedean duals
Equivalent vertices equivalent facets one dihedral angle	equivalent vertices two dihedral angles	 equivalent facets one dihedral angle
Icosahedron (138° 11′)* 12V, 20F, 30E (all V_5) (all F_3)	icosidodecahedron 30V, 32F, 60E ($20F_3 + 12\ F_5$)	rhombic triacontrahedron (144°) 32V, 30F, 60E ($12V_5 + 20V_3$)
	truncated icosahedron 60V, 32F, 90E ($12F_3 + 20F_6$)	pentacis dodecahedron (156° 43′) 32V, 60F, 90E ($12VP_5 + 20V_6$)
Dodecahedron (116° 34′) 20V, 12F, 20E (all V_3) (all F_5)	truncated dodecahedron 60V, 32F, 90E ($20F_3 + 12F_{10}$)	triakis icosahedron (160° 37′) 32V, 60F, 90E ($12V_{10} + 20V_3$)
	snub dodecahedron 60V, 92F, 150E ($60F_{3a} + 20\ F_{3b} + 12F_5$)	pentagonal hexecontra- hedron (153° 11′) 92V, 60F, 150E ($12V_5 + 20\ V_{3a} + 60V_{3b}$)
	small rhombicosido- decahedron 60V, 62F, 120E ($20F_5 + 30\ F_4 + 12F_5$)	trapezoidal hexecontra- hedron (154 ° 7′) 62V, 60F, 120E ($12V_5 + 30\ V_4 + 20V_3$)

* = dihedral angle.
V = vertices.
F = facets.
E = edges.
V_n—n = number of edges meeting at a vertex.
F_n—n = number of facial edges, f.e. F_3 = triangular facets, F_5 = pentagonal facets.
Compiled with data from A. Pugh, 1976.

division procedure, twelve 5-triangle vertices must always be present at the original icosahedral 5-fold vertices. (It is impossible to design a convex polygon which has six triangles meeting at every vertex, see p. 492—496.) The strains exerted on the bonding types between the subunits to from a 6-subunit vertex instead of a 5-subunit vertex may be small, and a slightly altered bonding environment may suffice to permit the assembly of a thus-expanded icosahedral shell. The subunits in the 5- and 6-fold vertices are then said to be situated "quasi-equivalently".

Two other methods for a regular subdivision of the icosahedral facets are possible, which all lead to principally identical situations but with different sets of additional 6-fold vertices. In the Triacon Method, the triangular facets are divided by lines parallel to the lines bisecting their angles. In this series, a number of unit triangles is shared equivalently by adjacent icosahedral facets. The third series is generated by subdividing the triangular facets with lines parallel to the lines trisecting their angles.

The structures that are thus derived from the icosahedron are called deltahedra, and each is characterized by the number of unit sub-triangles composing a single facet, the number being called triangulation number, T. Any icosahedral polyhedron thus has 20 T facets. T is given by the rule: $T = Pf^2$, where P can be a number of the series 1, 3, 7, 13, 19 etc. ($= h^2 + kh + k^2$ for all pairs of integers h and k having no common factors and f is an integer). The correspondence of such deltahedra to various icosahedral viruses was worked out by Caspar and Klug (1962, 1965) and was summarized well by Mattern (1969, 1971). The simplest icosahedral lattice with triangulation number of 1, for example, is the one employed by poliovirus.

Spherical structures can be generated from the triangulated deltahedra by projecting the newly created vertices from the center of the polydron onto the surface of the circumsphere (*i.e.*, the sphere that passes through the twelve original icosahedral vertices). The structures so formed are called "geodesic polyhedra". The procedure was worked out and elaborated by the architect Buckminister Fuller as the basis for his very successful geodesic domes, designs which apparently provided part of the inspiration for the work of Caspar and Klug on spherical viruses.

Appendix IV

Complete Nucleotide and Amino Acid Sequences of Poliovirus Type 1, 2 and 3*

```
                                                                                                       PV1(Sab) VPg-
                                                                                                       PV2(Sab) VPg-
                                                                                                       PV3(Sab) VPg-

                                                                                                                 100
UUAAAACAGC UCUGGGGUUG CACCCACCCC AGAGGCCCAC GUGGCGGCUA GUACUCCGGU AUUGCGGUAC CCUUGUACGC CUGUUUUAUA CUCCCU*UCC
---------- ---------- U------U-- ---------- ---------- -C-------- ---A------ -U----G--- ---------- -----*C---
---------- ---------- UU-------- ---------- ---------- ----A-U--- --CA------ -U-------- ---------- ------CC--

                                                                                                                 200
CGUAACUUAG ACGCACA*AA ACCAAGUUCA AUAGAAGGGG GUACAAACCA GUACCACCAC GAACAAGCAC UUCUGUUUCC CCGGUGAUGU UGUAUAGACU
--C------- -A----G--- ---------- ---------- ---------- -------U-- ---------- ---------- -------CA- --C-------
--C------- -A---U-C-- UU-------- ----G----- --G---G--- -CG--U--GU -GG------- -A-------- -------G-C C-C-------

                                                                                                                 300
GCUUGCGUGG UUGAAAGCGA CGGAUCCGUU AUCCGCUUAU GUACUUCGAG AAGCCCAGUA CCACCUCGGA AUCUUCGAUG CGUUGCGCUC AGCACUCAAC
---CA--C-- -------U-- UC-------- -C------G- ---------A -----U---- U-G---U--- --------C- ---------- -----C-G--
-U-CC-AC-- -------U-U -C-------- -----ACC-- ---------- -----U---- --G-UCU--- --------C- ---------- ----------

                                                                                                                 400
CCCAGAGUGU AGCUUAGGCU GAUGAGUCUG GACAUCCCUC ACCGGUGACG GUGGUCUAGG CUGCGUUGGC GGCC*U*ACC UAUGGCU*AA CGCCAUGGGA
---G-G---- G--------- ---------- -----U---- ---------- ---------- ---------- ---------- ---------- ------A---
---G------ -----G---C ---------- ----GU--C- --U--C---A ----CUCCA- -----C---- ----GCG--- -G----CC-- A----C----

                                                                                                                 500
CGCUAGUUGU GAACAAGGUG UGAAGAGCCU AUUGAGCUAC AU*AAGAAUC CUCCGGCCCC UGAAUGCGGC UAAUCCCAAC CUCGGGGCAG GUGGUCACAA
--U---A--- ---------- ---------- ---------- -------G-- ---------- ---------- ------U--- -A---AA--- -C----G-G-
---------- -----G---- ---------- ---------- --G----G*- ---------- ---------- ----U-U--- -AU--A---- -CA-CUG---

                                                                                                                 600
ACCAGUGAUU GGCCUGUCGU AACGCGCAAG UCCGUGGCGG AACCGACUAC UUUGGGUGUC CGUGUUUCCU UUUAUUUU*A UUGUGGCUGC UUAUGGUGAC
--------C- ---U------ ---------- --U------- ---------- ---------- ---------- G-------U- -CA------- ----------
C----CAGCC A--------- ---------- ---------- ---------- ---------- ---------- ------C-UG A*A------- ----------

                                                                                                                 700
AAUCACAGAU UGUUAUCAUA AAGCGAAUUG GAUUGGCCAU CCGGUGAAAG UGA*GAUUCA UUAUCUAUCU GUUUGCUGGA UUCGCUCCAU UGAG*UGUGU
-----G---- ---------- ---------- ---------- -------GU- -UGU-*-CAG G---AC-A-- -----U---- AC-A--GUG- -*---C***-
-----U---- ---------- -----G---- ---------- --A---UG-A -C------A- ---CUCCCU- -----U---- -C-A----*C *--AAC--*-

                                                                                                                 800
UUACU*CUAA G*UA*C*AAU U**UCAACAG UUAUUUC*A* *AUCAGACAA UU*C*UAUCA UAAUGGUGC  UCAGGUUUCA UCACAGAAAG UGGCCGCACA   1a
-----U--C- UU--A-C--- -AA----*-A AC-A-A-G-G G--A-A---- **CAA--CU- C-----C--  C--A------ ---------- -U--A--C--
-------*C* *U--A--**- -A--UG-*-A --G---*C-A G-CAG--U** --CAG-G--- C-----A--  ---A--A--- --C--A---- -A-----U--
                                                                     VP4
P1                                                                                                               900
UGAAAACUCA AAUAGAGCGU AUGGUGGUUC UACCAUUAAU UACACCACCA UUAAUUAUUA UAGAGAUUCA GCUAGUAACG CGGCUUCGAA ACAGGACUUC
C-----U--- --C-----C- ----C--G-- C-----C--- -----U--A- -C-----C-- ---G--C--U --A--C--U- -A--AAGC-- G--A--U--U
C--G--U--U --CC----C- -C-------- ---G--C--C --------A- ---------- --A-----C --A-----U- ----G--C-- ---A--U-A-

                                                                                                                1000
UCUCAAGACC CUUCCAAGUU CACCGAGCCC AUCAAGGAUG UCCUGAUAAA AACAUCCCCA AUGCUAAACU CGCCAAACAU AGAGGCUUGC GGGUAUAGCG
G-A-----U- -G-------- ------A--- --U-----C- ----U--U-- G--CG-U--C ---------  -C-------- U-----G--U --U-----U-
--A--G--U- -A--A--A-- ---------A C-A-----C- -G--C----- ----G-U--- GCA--C--U  -A-----UG- G--A--G--U --------U-
                                                                                    VP2
                                                                                                                1100
AUAGAGUACU GCAAUUAACA CUGGGAAACU CCACUAUAAC CACACAGGAG GCGGCUAAUU CAGUAGUCGC UUAUGGGCGU UGGCCUGAAU AUCUGAGGGA
-C--G---A- ---GC----U -----C--U- -A--G--C-- ---C--A--A -----C---- -U--U--U-- C--C--UA-A ---------- -CA-C--A--
-------GU- ----C-C--U U-A--C--U- -------U-- U--------- --A--A---- -------G-- ---C--A--- --------G- U-A-U--A--

                                                                                                                1200
CAGCGAAGCC AAUCCAGUGG ACCAGCCGAC AGAACCAGAC GUCGCUGCAU GCAGGUUUUA UACGCUAGAC ACCGUGUCUU GGACGAAAGA GUCGCGAGGG
U-C---G--A -----U--A- ----A--A-- C--G--C--U --A--C--G- -------C-- C--AU----U -----CA--- --CGC--G-- ---CA-----
UGA------A --C--G---- ----A--A-- U-------U --G---A--- ----A--C-- C--A------ --U--AAUG- --GCU--G-- ----AA---C

                                                                                                                1300
UGGUGGUGGA AGUUGCCUGA UGCACUGCGG GACAUGGGAC UCUUUGGCCA AAAUAUGUAC UACCACUACC UAGGUAGGUC CGGGUACACC GUGCAUGUAC
---------- -AC-A--A-- C--UU-AAAA --------GU -A-----U-- ---C----UU --U-----U- -U--G---G- U--C-----A -----C----
---------- ----A----- C------A-A --------U- -G-----A-- ---C-----U ---------- ----A--A-- ---------U -----C--G-
```

* Figure courtesy of A. Nomoto, Tokyo [see Toyoda *et al.*, J. Mol. Biol. 174: 561–585 (1984)].

```
                                                                                                         1400
AGUGUAACGC CUCCAAAUUC CACCAGGGGG CACUAGGGGU AUUCGCCGUA CCAGAGAUGU GUCUGGCCGG GGAUAGCAAC ACCACUACCA UGCACACCAG
----C--U-- U-U---G--U --U--A--A- -U-------- G--U--A--U -----A---- --U-A--U-- U------*** --A---CA-- --UU---A-A
-------U-- A--------U -----A--U- ----C----- G--U--GA-U --U---UAU- -------G-- U--C--UG-- ***-AGCAA- G-U----U--

                                                                                                         1500
CUAUCAAAAU GCCAAUCCUG GCGAGAAAGG AGGCACUUUC ACGGGUACGU UCACUCCUGA CGACAACCAG ACAUCACCUG CCCGUAGGUU CUGCCCGGUG
G--CG-G--- --G-----A- ----A----- ---UGAA--- -AA--G-GU- ----C-U--- UAC----GCC --UAAC---- -A--G-AC-- ------A--U
U---GC---- --G-----A- -U--A-G--- G--A-AA--U UACUCCCAA- ---ACAAG-- UA--GCAGUA -----C--AA AAA-AGA--- ------A---

                                                                                                         1600
GAUUACCUCU UUGGAAAUGG CACGUUAUUG GGGAAUGCCU UUGUGUUCCC GCACCAGAUA AUAAACCUAC GGACCAACAA CUGUGCUACA CUGGUACUCC
---------- -C--G-G--- AGU-C-GG-A --------A- ----U-AU-- A--U--A--- --------G- -C--U----- ---------G --A---U-G-
-----U---C -G---UG--- GGU----C-- --A------- ----A-A--- A--U--A--C --U--U--GA ---------- -A-C--A--U A-U--C--A-

                                                                                                         1700
CUUACGUGAA CUCCCUCUCG AUAGAUAGUA UGGUAAAGCA CAAUAAUUGG GGAAUUGCAA UAUUACCAUU GGCCCCAUUA AAUUUUGCUA GUGAGUCCUC
-C--U--A-- ---A-----A --------C- --AC------ ---C--C--- --G--C--U- -CC-C--CC- ---G---C-- G-C----UC- C---A--U--
-A--U----- UG-UU-GG-C --U---UCA- ----U--A-- ---C--C--- --C-----C- -UC-G--C-- AU-A--GC-G G--------C AA--U--A--

                                                                                                         1800
CCCAGAGAUU CCAAUCACCU UGACCAUAGC CCCUAUGUGC UGUGAGUUCA AUGGAUUAAG AAACAUUACC CUGCCACGCU UACAGGGCCU GCCGGUCAUG
-A-U-----A --C--U--AC -------U-- U--C------ --C--A---- ----U---C- C-----C--U G-----A-AA CC--A--AU- A--A---C--
AGUU--A--- -----U--UG ----A--U-- ---A-----U A-C------- -C--CC-UC- C---G-G--U GCA--UAAA- -U--A--A-- A--A--GU--
                                                                                           VP3   1900
AACACCCCUG GUAGCAAUCA AUAUCUUACU GCAGACAACU UCCAGUCACC GUGUGCGCUG CCUGAAUUUG AUGUGACCCC ACCUAUUGAC AUACCCGGUG
-----U--A- -G--U--C-- G--C--G--C --------U- A------U-- -------A-A -----G---- ----C--U-- ---C--A--- -----A--G-
-----U---- ----U--C-- G--C--G--G U--------C A---A----- A--C--AA-C --A------- ----C--U-- G--------U --C--A----

                                                                                                         2000
AAGUUAAGAA CAUGAUGGAA UUGGCAGAAA UCGACACCAU GAUUCCCUUU GACUUAAGUG CAAAAAAAAA GAACACCAUG GAAAUGUAUA GGGUUCGGUU
-G--GCGC-- ---------- -----G---- -A-------- ---A---C-C A----G-CAA GUC--CGC-G A-----A--- --C------- -A--CGA---
-G-----A-- ---------G C-C--C--G- -A-------- ------UC-C A-U--GGAGA GC-CC--G-G A-----A--- --C-----C- -A---ACUC-

                                                                                                         2100
AAGUGACAAA CCACAUACAG ACGAUCCCAU ACUCUGCCUG UCACUCUCUC CAGCUUCAGA UCCUAGGUUG UCACAUACUA UGCUUGGAGA AAUCCUAAAU
G--C----CG G-U--CU-U- --ACG--G-- CU-G--U--C --GU-G--C- -C-------- C--C--A--- G----C---- --U-G--U-- G--AU-----
G--C----GU G-CG--CU-U CGC-A--A-- UU-G---U-- -----A--C- ----A-UU-- ---GC-C--- -----C--C- -------G-- -G-A--G--C

                                                                                                         2200
UACUACACAC ACUGGGCAGG AUCCCUGAAG UUCACGUUUC UGUUCUGUGG AUCCAUGAUG GCAACUGGCA AACUGUUGGU GUCAUACGCG CCUCCUGGAG
---------- ---------- G---U----A --U--C---- -C--U--C-- C--A------ --C--C--A- -GU-A----- U--U-----A --A--C----
--U--U--U- -U-----C-- G---U----A --U--C--C- ---------- U--A------ --U--G--G- --A-CC-A-- -G-C--U--A --A--A--U-

                                                                                                         2300
CCGACCCACC AAAGAAGCGU AAGGAGGCGA UGUUGGGAAC ACAUGUGAUC UGGGACAUAG GACUGCAGUC CUCAUGUACU AUGGUAGUGC CAUGGAUUAG
-A--GG-C-- C----GU--C --A--A--A- --C-U--G-- ---------A --------U- -GU------- U-----C--- -----G--A- -U-----C--
-AC-A--C-- C-CC-GC--- --------U- -------C-- ------C--U -----UC-U- -C-----A-- A--U------ -----G---- -G--------

                                                                                                         2400
CAACACCACG UAUCGGCAAA CCAUAGAUGA UAGUUUCACC GAAGGCGGAU ACAUCAGCGU CUUCUACCAA ACCAGAAUAG UCGUCCCUCU UUCGACACCC
U--U-----A --CA-A---- ----CA-C-- ---------A -----U--C- ----U---A- G-----U--- --U--GG-U- -U-----GU- G--C------
U--UGUG--A --CA-A--G- -U-C-C-A-- ---------U --G------- -U------A- G--------- --A-----U- -G--G--A-- G--C--C--U

                                                                                                         2500
AGAGAGAUGG ACAUCCUUGG UUUUGUGUCA GCGUGUAAUG ACUUCAGCGU GCGCUUGAUG CGAGAUACCA CACAUAUAGA GCAAAAAGCG CUAGCACAGG
---A------ -------G-- ---------- --U--C---- -------U-- ------AC-- --------A- ----C--UAG U---G-G--U A-GC----A-
-AGAGU---A G---G--G-- G--------- --C------- -U-----U-- ---A---C-- -----C---- -U--C--UUC A---UCU--- --UC------

                                                                                                         2600
GGUUAGGUCA GAUGCUUGAA AGCAUGAUUG ACAACACAGU CCGUGAAACG GUGGGGGCGG CAACGUCUAG AGACGCUCUC CCAAACACUG AAGCCAGUGG
-AA-U---G- C---A----- G-GGCCG--- -AGGG-UUAC UAAAA-UG-A U---UUC-CC -G--U--C-C CA-UAGC--G --UGGACACA -GC-G--C--
-UA-U-AAG- UU--A--*** UCUGAAG--* **GCACAG-G -GCCCU---U U--UCACUCC -G-A-CAACA G--UAGCU-A --UG-U---A -G--------
VP1
                                                                                                         2700
ACCAGCACAC UCCAAGGAAA UUCCGGCACU CACCGCAGUG GAAACUGGGG CCACAAAUCC ACUAGUCCCU UCUGAUACAG UGCAAACCAG ACAUGUUGUA
U-----C--- --------G- -A--U---U- G--A--C--- --G--A---- -U--C----- GU-G--G--- --G--C--C- -------GC- C-----CA-C
C--G--G--U --------GG -A--U----- ---U-----C --G-----A- ----C----- U--G-CA--A --C--C---- -U-----GC- C--C--A--C

                                                                                                         2800
CAACAUAGGU CAAGGUCAGA GUCUAGCAUA GAGUCUUUCU UCGCGCGGGG UGCAUGCGUG GCCAUUAUAA CCGUGGAUAA CUCAGCUUCC ACCAAGAAUA
--GAGAC-AA -GC-A----- ---C-CGG-U -----A---- -U--AA-A-- G--U------ --U--C--UG AG-----C-- UGAU--AC-G -A---CGCG
----GAC-CA GC-------- ---C-CA--- --A--A---- ----A--C-- G--G-----C --U-----UG AG-----C-- UGA-CAAC-A ----CCCGGG

                                                                                                         2900
AGGAUAAGCU AUUUACAGUG UGGAAGAUCA CUUAUAAAGA UACUGUCCAG UUACGGAGGA AAUUGGAGUU CUUCACCUAU UCUAGAUUUG AUAUGGAAUU
CCAGC-GAU- G---U-G--U -----A--A- ----C----- ------U--A C-GA-AC-C- --C----A-- U-----A--- --G------- -C-----G--
CAC-G--A-- ----G-CA-- ---CGC--U- -A--C----- ---A--G--- --G--CC-U- -G-------- U-----A--C ---C-U---- -C--------

                                                                                                         3000
UACCUUUGUG GUUACUGCAA AUUUCACUGA GACUAACAAU GGGCAUGCCU UAAAUCAAGU GUACCAAAUU AUGUACGUAC CACCAGGCGC UCCAGUGCCC
C--U------ --C--CU--- -C-A--U--- UG-A--U--C --A-----A- -G--C----- U--U--G--A -----UA--- ----C--A-- A--UA-C--U
C-----C--- --A--C--C- -C-----CA- CG----U--- --------AC -C--C--G-- ------G--A ------A-C- -C-----G-- A--CACA--A

                                                                                                         3100
GAGAAAUGGG ACGACUACAC AUGGCAAACC UCAUCAAAUC CAUCAAUCUU UUACACCUAC GGAACAGCUC CAGCCCGGAU CUCGGUACCG UAUGUUGGUA
-GU------A -U-----U-- G-----G--G --C--U--C- -G--GG-G-- ---------U --GG-GC-C- ----AA-A-- A--A--G--C --C--G--A-
A--UC----- ---------- U--------A --U--C--C- -G--C--A-- ---------U --GG-U--C- -G--G--A-- ---A--G--A --C--G--GU
```

```
                                                                                                         3200
UUUCGAACGC CUAUUCACAC UUUUACGACG GUUUUUCCAA AGUACCACUG AAGGACCAG* **UCGGCAGC ACUAGGUGAC UCCCUCUAUG GUGCAGCAUC
--G-U--U-- G-----C--- -----U--U- -G---G-A-- ---------A GC--GU--AG CC--AA-U-A -***--C--U --GU-G--C- ----U--C--
-AG-C--U-- U--C--G--- ---------- -C--CG---- G--G---U-- ---ACAG-UG CCAAU-ACCA GA-U-----U ---U-G--CA -C--CAUGA-

                                                                                                         3300
UCUAAAUGAC UUCGGUAUUU UGGCUGUUAG AGUAGUCAAU GAUCACAACC CGACCAAGGU CACCUCCAAA AUCAGAGUGU AUCUAAAACC CAAACACAUC
A--G-----U --U--AUCAC --------C- C--G--A--- ---------- -C--GCG-C- ---------G ---------- -CA-G--G-- A--G--UG--
AG-UG----- --U---G-A- ----A---C- U--U------ ---------- -C--U--A-- A--------- G--C-CA-U- -CA-G----- -------G-A

                                                                                                         3400
AGAGUCUGGU GCCCGCGUCC ACCGAGGGCA GUGGCGUACU ACGGCCCUGG AGUGGAUUAC AAGGAUGGUA CGCUUACACC CCUCUCCACC AAGGAUCUGA
---------- ----A--A-- U--AC-A--- --CC-A---- U---A--A-- U--U-----U --A-----G* **--C--C-- A--AC-AGAA ----GAU-A-
C-U------- ----UA-A-- G---C-C--G --AC-U--U- -U--A--A-- G-----C--U -G-A-C***- ACU-GGAC-- -U-A--UGAG --A-G-U---

                                                                                                         3500
CCACAUAUGG AUUCGGACAC CAAAACAAAG CGGUGUACAC UGCAGGUUAC AAAAUUUGCA ACUACCAUUU GGCCACUCAG GAAGAUUUGC AAAACGCAGU   3b
-G--U----- ---U------ ---------- -U-------- A--U--C--- ---------- -U-----CC- A--U--A--A -----C---- ----U--C--
---------- C--U--G--U --G--U---- -U-------- ---U------ --G--C---- --------C- C------A-- --G-----A- ----U--U--

P2                                                                                                       3600
GAACGUCAUG UGGAAUAGAG ACCUCUUAGU CACAGAAUCA AGAGCCCAGG GCACCGAUUC AAUCGCAAGG UGCAAUUGCA ACGCAGGGGU GUACUACUGC
--GU------ -----C---- ---------- GG-U------ --G----UU- -------C-- G--------- ----GC--U- --A-G--U-- ---------U
A-G-A----- ---------- -------G-- UGUU------ -A---U--A- -U-----C-- ---A------ ---------- -U--G----- ------U--U

                                                                                                         3700
GAGUCUAGAA GGAAAUACUA CCCAGUAUCC UUCGUUGGCC CAACGUUCCA GUACAUGGAG GCUAAUAACU AUUACCCAGC UAGGUACCAG UCCCAUAUGC
--A--C--G- -A-----U-- U-----U--U ---A----G- -C--C----- A--------A --C---G-A- -------G-- ---A--U--A --A--C----
-----C---- ---------- ---U--G--G --U--G--A- -C--C----- A--------- ------G--- -C-------- ---A-----A -----C---U

                                                                                                         3800
UCAUUGGCCA UGGAUUCGCA UCUCCAGGGG AUUGUGGUGG CAUACUCAGA UGUCACCACG GGGUGAUAGG GAUCAUUACU GCUGGUGGAG AAGGGUUGGU
-U-----U-- ---G--U--- --A--G--U- ---------- ------U--- -----A---- ---------- A--A--C--- --------G- ----C-----
-A--C--G-- C--C--U--C --A-----U- -C-------- U--C--U--G -----A--U- -C--C--C-- A---G-G--A ---------- -G--A--A--

                                                                                                         3900
UGCAUUUACA GACAUUAGAG ACUUGUAUGC CUACGAAGAA GAAGCCAUGG AACAAGGCAU CACCAAUUAC AUAGAGUCAC UUGGGGCCGC AUUUGGAAGU
-------U-- ---------- --C------- U--U--G--G -----U---- -G--G----- UU----C--U --U------- ----U--U-- ------U---
C-----CU-U -----A--G- ---------- U-----G--- --G------- -G--G----- UU-A--C--U --U------- -C--U--U-- G--C--U---
                                                          5b
                                                                                                         4000
GGAUUUACUC AGCAGAUUGG AGACAAAAUA ACAGAGUUGA CUAAUAUGGU GACCAGUACC AUCACUGAAA AGCUACUUAA GAACUUGAUC AAGAUCAUAU
-----C---- -A--A----- U--U---G-U U-C---C-A- -C-GC----- A--U--C--- --U--A--G- --U-G----- A-----A--- --A--U--C-
--G--C---- ----A--A-- G--U--G--- U----AC-A- -C-GC----- ------C--G --U--A--G- ---------- A---C-A--- --A--U--U-

                                                                                                         4100
CCUCACUAGU UAUUAUAACU AGGAAUUAUG AAGACACCAC AACAGUGCUC GCUACCCUGG CCCUUCUUGG GUGUGAUGCU UCACCAUGGC AGUGGCUUAG
-A-----U-- G--C--U--- ---------- -G-----U-- C--------U --C-----C- ----C----- ---C--CAUC -----G---- -------A-A
-A--U--G-- G-----C--- --A-----C- ----U----- C--------- --C--U--A- -U-------- --------U- -----G---- -A-----G-A

                                                                                                         4200
AAAGAAAGCA UGCGAUGUUC UGGAGAUACC UUAUGUCACC AAGCAAGGUG ACAGUUGGUU GAAGAAGUUU ACUGAAGCAU GCAACGCAGC UAAGGGACUG
G-----G--- --U--CA-C- ----A--U-- A--C-C--U- G-A-----A- -U-------- ------A--C G---G----- -U--U--U-- A---------
G--------- --U--CAC-U -------U-- C-----U-UU -GA--G---- -U-------- ---A--A--- -----G--G- ---------- ------GU--
                                                 X
                                                                                                         4300
GAGUGGGUGU CAAACAAAAU CUCAAAAUUC AUUGAUUGGC UCAAGGAGAA AAUUAUCCCA CAAGCUAGAG AUAAGUUGGA AUUUGUAACA AAACUUAGAC
---------- -C--U--G-- A--C------ ---AG----U -GC----U-- ---C------ -----G---- -C--A--A-- G-----C--U -----A-AG-
-GA------- -C-------- ---------U -----C---U -G-GA--A-G ---C------ -----C--G- -C---C-U-- G--------C ---U-G-A--

                                                                                                         4400
AACUAGAAAU GCUGGAAAAC CAAAUCUCAA CUAUACACCA AUCAUGCCCU AGUCAGCAAC ACCAGCAAAU UCUAUUCAAU AAUGUCAGAU GGUUAUCCAU
--U------- ---------U --G--U--C- -C-------- ---U--U--A -----A---- -U-----G-- CU-------C -----GC-G- --C----U--
-GU-G----- ---A--G--U --G--A--C- -A-------- ---U--U--A ---------- ---------- -U-G-----C -----AC-C- ----G-----

                                                                                                         4500
CCAGUCUAAG AGGUUUGCCC CUCUUUACGC AGUGGAAGCC AAAAGAAUAC AGAAACUAGA GCAUACCAUU AACAACUACA UACAGUUCAA GAGCAAACAC
------C--- --------A- -A--A--U-- -CAU-----U -----G--U- -A--G--G-- ---------A --U--U---G ---------- ------G---
U--A--C--- --A--C--U- -AU-G----- -C-U--G--- --G------- -A--GU-G-- A--C------ --U--U---- ---------- ----------

                                                                                                         4600
CGUAUUGAAC CAGUAUGUUU GCUAGUACAU GGCAGCCCCG GAACAGGUAA AUCUGUAGCA ACCAACCUGA UUGCUAGAGC CAUAGCUGAA AGAGAAAACA
--------G- ---------- -U-------- -----U--A- -G-----A-- ---A--U--- -----U--A- ---------- A-----C--G -A---G----
--------G- ---------- -U----G--- --G-----A- -U-----A-- ---A--U--G --U-----A- ---------- ---------G -A---G----

                                                                                                         4700
CGUCCACGUA CUCGCUACCC CCGGAUCCAU CACACUUCGA CGGAUACAAA CAACAGGGAG UGGUGAUUAU GGACGACCUG AAUCAAAACC CAGAUGGUGC
-C-----A-- ---A--G--A --U-----G- -U-----U-- U--C-----G --------U- ----U----- ---U-----A --C-----U- ----C--A--
-C-----C-- ---------A -----C--G- -U-----U-- U--------- -----A--U- ----U--C-- ---------A --C------- -G-----G--
```

```
                                                                                                            4800
GGACAUGAAG CUGUUCUGUC AGAUGGUAUC AACAGUGGAG UUUAUACCAC CCAUGGCAUC CCUGGAGGAG AAAGGAAUCC UGUUUACUUC AAAUUACGUU
A--------A --U--U---- -A-----G-- C--U------ -----U---- -G-----C-- G--A--A--- -----C--UU ----C--A-- U---------
A--U------ --C--U---- -A-----G-- C--U------ -----C---- -U-----C-- G-----A--- -----C--U- ----C--A-- C--C--U---

                                                                                                            4900
CUAGCAUCCA CGAACUCAAG CAGAAUUUCC CCCCCCACUG UGGCACACAG UGAUGCAUUA GCCAGGCGCU UUGCGUUCGA CAUGGACAUU CAGGUCAUGA
U----C---- -C-----C-- UC-G--CA-A --A-----G- ----U----- ------GC-G ------A-A- -C--A--U-- -G-------A --A-------
U----C---- -C-----C-- UC-C--CA-A --A--U--A- -A--C----- ---C--UC-G ------A-G- -C--U----- U-----U--- --A--G---G

                                                                                                            5000
AUGAGUAUUC UAGAGAUGGG AAAUUGAACA UGGCCAUGGC UACUGAAAUG UGUAAGAACU GUCACCAACC AGCAAACUUU AAGAGAUGCU GUCCUUUAGU
GC-----C-- C-----C--A --GC-C---- ----A----- ---------- --C--A---- ----U----- ---------C --A-----U- ----------
GC-----C-- C--------U ---C-C---- ----A----- --------C- --C---G--- -C----G--- ---------C --A------- ----------

                                                                                                            5100
GUGUGGUAAG GCAAUUCAAU UAAUGGACAA AUCUUCCAGA GUUAGAUACA GUAUUGACCA GAUCACUACA AUGAUUAUCA AUGAGAGAAA CAGAAGAUCC
------C--- --------G- -------U-- ---------G ---------- -C-----U-- ----------------G-U- ---------- ---------A
---------- --------G- ---------- ---------- -----G---- --G------- ---U------ ---------- -C-------- ---------U

                                                                                                            5200
AACAUUGGCA AUUGUAUGGA GGCUUUGUUC CAAGGACCAC UCCAGUAUAA AGACUUGAAG AUUGACAUCA AGACGAGUCC CCCUCCUGAA UGUAUCAAUG   1b
--------U- ----C----- A---C-A--- --G------- -G-------- ---UC-A--A --A--UG-U- ----C----- ------G--G --------C-
---------- ----C----- ---------- ---------- -------C-- ----C----A ---------- -------G-- ---C------ --C-------
                                       → 9
P3                                                                                                          5300
ACUUGCUCCA AGCAGUUGAC UCCCAGGAGG UGAGAGAUUA CUGUGAGAAG AAGGGUUGGA UAGUCAACAU CACCAGCCAG GUUCAAACAG AAAGGAACAU
-U-------- G--------U --------A- ---------- ------A--- --A--C---- -U--U----- U-----U--- ---------- -G--------
-UC----U-- ---------- --------A- ----G----- U-----A--- --A--A---- -C-------- ---U-----A ---------- -G--A-----

                                                                                                            5400
CAACAGGGCA AUGACAAUUC UACAAGCGGU GACAACCUUC GCCGCAGUGG CUGGAGUUGU CUAUGUCAUG UAUAAACUGU UUGCUGGACA CCAGGGAGCA
----C----G -----U--C- -------A-- A--U--U--- --U-----A- -C--U--C-- G--C--U--- --C--G---- -C-----G-- -------U---
U---C-A--- -----C---U -G--G--A-- ------U--- ---------- ----U--C-- G--C------ --C--GU-A- -C-------- ----------
                                                                                                             → VPg
                                                                                                            5500
UACACUGGUU UACCAAACAA AAAACCCAAC GUGCCCACCA UUAGGACAGC AAAGGUACAA GGCCCAGGGU UCGAUUACGC AGUGGCUAUG GCUAAAAGAA
---------- -G-----U-- -CG------U --A-----U- -C-------- ---A--G--- --C--U---- -U-------- ------C--- ----------
---------C -G-------- --G------U ---------- ----AG---- ---A--G--- -----U---- -U--C--U-- ---------- ----------
                                                                  → 2;7c;6a
                                                                                                            5600
ACAUUGUUAC AGCAACUACU AGCAAGGGAG AGUUCACUAU GUUAGGAGUC CACGACAACG UGGCUAUUUU ACCAACCCAC GCUUCACCUG GUGAAAGCAU
---------- ------C--C -----A--G- ----U--G-- ---G------ U-U--U--U- ----C--C-- G--------- --C------- ----------
---------- ---------- -----A--G- -------A-- -C-------- ---------- ----C----- ------U--U --C------- ----G--U--

                                                                                                            5700
UGUGAUCGAU GGCAAAGAAG UGGAGAUCUU GGAUGCCAAA GCGCUCGAAG AUCAAGCAGG AACCAAUCUU GAAAUCACUA UAAUCACUCU AAAGAGAAAU
--C------C --U-----G- ----A--UC- U--C------ --C--U---- ----G----- ---U------ -----U--C- ----U--A-- ------G--C
---A--U--- --------G- -U--A---C- A--C--U--- --C------- ----G----- C--U-----G --------C- ----A--C-- C--A------

                                                                                                            5800
GAAAAGUUCA GAGACAUUAG ACCACAUAUA CCUACUCAAA UCACUGAGAC AAAUGAUGGA GUCUUGAUCG UGAACACUAG CAAGUACCCC AAUAUGUAUG
--G------- ----U--C-- G-----C--U --C--C---- ----C--A-- ---------- --U------- ---------- U--------- --C-------
---------- ----U--C-- --A---C--- --C------- ----C----- G--------- --UC----U- ---------- U--------- --C-------

                                                                                                            5900
UUCCUGUCGG UGCUGUGACU GAACAGGGAU AUCUAAAUCU CGGUGGGCGC CAAACUGCUC GUACUCUAAU GUACAACUUU CCAACCAGAG CAGGACAGUG
----C--U-- ---------C -----A--G- ----U----- ------A--A -----C---- ----G----- ---------- -----U---- ----U-----
---------- ---------- --G------- -C-------- ---------- --G------- ---U------ ---------- ---------- -U--U-----
                            → 4a
                                                                                                            6000
UGGUGGAGUC AUCACAUGUA CUGGGAAAGU CAUCGGGAUG CAUGUUGGUG GGAACGGUUC ACACGGGUUU GCAGCGGCCC UGAAGCGAUC AUACUUCACU
------U--- -----G--C- ----U----- ---U------ ---------- ---------- ---U-----C --G------- -A-----G-- ----------
---------- --------C- ---------- ---------- --C------- ----U----- ---U------ ---------- -------G-- ----------

                                                                                                            6100
CAGAGUCAAG GUGAAAUCCA GUGGAUGAGA CCUUCGAAGG AAGUGGGAUA UCCAAUCAUA AAUGCCCCGU CCAAAACCAA GCUUGAACCC AGUGCUUUCC
----U----- ----G--U-- A-------A- --A--A--A- ---------- C--G------ -----U---- ---------- A--------- --C-----U-
-----C---- ---------- ---------- --A--A---- -G-CA--G-- ------U--- --C-----AA ----G----- ---C--G--- ----------
          → 4b
                                                                                                            6200
ACUAUGUGUU UGAAGGGGUG AAGGAACCAG CAGUCCUCAC UAAAAACGAU CCCAGGCUUA AGACAAACUU UGAGGAGGCA AUUUUCUCCA AGUACGUGGG
---------- ---------- ---------- -------U-- C-----U--- --------C- G----G---- ---A--A--- --A-----U- ----U--A--
---------- ------A--A ---------- ---------- A--G--U--- -----A--C- -A---G---- ---A--A--- --C-----U- ----U--A--

                                                                                                            6300
UAACAAAAUU ACUGAAGUGG AUGAGCACAU GAAAGAGGCA GUAGACCACU AUGCUGGCCA GCUCAUGUCA CUAGACAUCA ACACAGAACA AAUGUGCUUG
C-----G--C -----U---- -----U---- ---------- --G--U--U- -C-------- A--------U ---------- ---------- ----------
G-----G--C -----G---- -----U---- ---------- --G-----U- -------A-- A--U-----G --G--U---- G------G-- ------UC-A
```

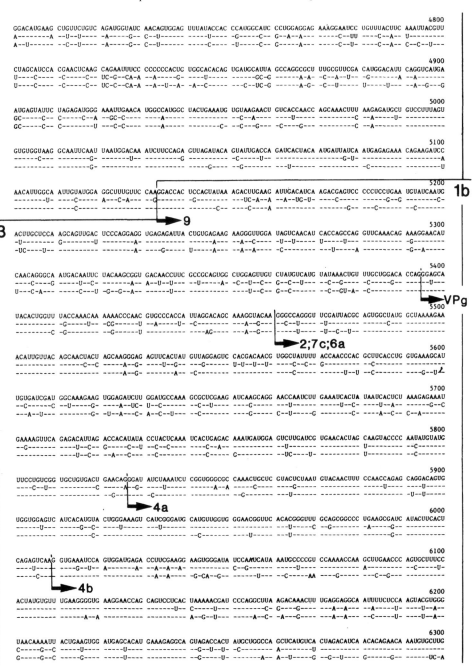

```
                                                                                                          6400
GAGGAUGCCA UGUAUGGCAC UGAUGGUCUA GAAGCACUUG AUUUGUCCAC CAGUGCUGGC UACCCUUAUG UAGCAAUGGG AAAGAAGAAG AGAGAUAUCU
-----C---- ----C----- C-----C--G ---------- -C---A---- U--------A ---------- ---------- ------A--- -----C----
--A--C---- -------U-- ---------G --G--G--A- --C----U-- ------C--G -----C--C- -G-------- G--------- ---------C

                                                                                                          6500
UGAACAAACA AACCAGAGAC ACUAAGGAAA UGCAAAAACU GCUCGACACA UAUGCAAUCA ACCUCCCACU GGUGACUUAU GUAAAGGAUG AACUUAGAUC
----U--G-- G--U------ --C------- ---GG-G--- CU-A--U--U -------U- --U-A--G-- U--A--A--- --U--A---- ----A--G--
-A-----G-- ---------- --C--A---- ------G--- UU-G---G-U --C------- ----A---U- A-----A--- --C-----C- -G--G--G--
                                                        → 6b
                                                                                                          6600
CAAAACAAAG GUUGAGCAGG GGAAAUCCAG AUUAAUUGAA GCUUCUAGUU UGAAUGACUC AGUGGCAAUG AGAAUGGCUU UUGGGAACCU AUAUGCUGCU
A-----U--- --G------- -A-------- ---G------ --C--C---- -------U-- ------C--- --------A- ----A--U-- C-----A--A
---------A --G--A---- -A-------- -C-G------ -----C---C -A-------- ------C--- --------A- ----A----- U-----A--A

                                                                                                          6700
UUUCACAAAA ACCCAGGAGU GAUAACAGGU UCAGCAGUAG GGUGCGAUCC AGAUUUGUUU UGGAGCAAAA UUCCGGUAUU GAUGGAAGAG AAGCUGUUUG
---------- ---------- UG-C--U--C AGU-----U- -U--U----- ----C-A--- --------G- -C--A--GC- A--------- -----C----
--C----GG- -U-----G-- CG-C--U--- AGU-----U- -A-------- ---CC-A--C --------G- -C--A--G-- ---------A -----A----

                                                                                                          6800
CCUUUGACUA CACAGGGUAU GAUGCAUCUC UCAGCCCUGC UUGGUUCGAG GCACUAGAGA UGGUGCUUGA GAAAAUCGGA UUCGGAGACA GAGUUGACUA
-U-------- ------U--- --------A- -------G-- C-----U--- -----CA-A- -------C-- ---------- --U--G---- -G--G--U--
-------U-- ------A--C --C-----A- -U-----A-- ------U--- -----CA--- -----U-A-- ------U--U --U-----U- ----G--U--

                                                                                                          6900
CAUCGACUAC CUAAACCACU CACACCACCU GUACAAGAAU AAAACAUACU GUGUCAAGGG CGGUAUGCCA UCUGGUUGCU CAGGCACUUC AAUUUUUAAC
U--U--U--- --C-----U- -C-------- ------A--C -----U--U- -C--A--A-- ---C------ -----C---- -------A-- ----------
---A------ --U-----U- --------U- ------A--C --G-U---U- ----U----- ---C------ -----C---- -C-------- ---------U

                                                                                                          7000
UCAAUGAUUA ACAACUUGAU UAUCAGGACA CUCUUACUGA AAACCUACAA GGGCAUAGAU UUAGACCACC UAAAAAUGAU UGCCUAUGGU GAUGAUGUAA
---------- -------A-- C--U------ ---C------ ---------- ---------- -----U---- ----G----- ---------- ----------
---------- ----U----- C--U-----G --U------- ---------- ---------- --G------U ---------- ---------- --C-------

                                                                                                          7100
UUGCUUCCUA CCCCCAUGAA GUUGACGCUA GUCUCCUAGC CCAAUCAGGA AAAGACUAUG GACUAACUAU GACUCCAGCU GACAAAUCAG CUAUAUUUGA
---------- ---------G -----U---- ---------- ---------- ---------- -------C-- ---------A -----G---- ---CC-----
-A-------- U--------G ---------- ---------- ---------- ---------- -------C-- ------G--A --U-----U- -C-CU--C--

                                                                                                          7200
AACAGUCACA UGGGAGAAUG UAACAUUCUU GAAGAGAUUC UUCAGGGCAG ACGAGAAAUA CCCAUUUCUU AUUCAUCCAG UAAUGCCAAU GAAGGAAAUU
---------- ---------- ---------- ---A------ --U--A--G- -U-----G-- U--C--C--C --A------- ---------- ------G---
G--------- ---------- ----U----- ---A------ -----A---- -U-------- ---C--C--C --A------- ---------- ----------

                                                                                                          7300
CAUGAAUCAA UUAGAUGGAC AAAAGAUCCU AGGAACACUC AGGAUCACGU UCGCUCUCUG UGCCUAUUAG CUUGGCACAA UGGCGAAGAA GAAUAUAACA
---------- ---------- --G----C-- --A----A-- ---------- C-----AU-- --------G- -C-------- C--------- -----C----
---------- -C-------- ---------- C----U--G- ----C--U-- A-----CU-- --U-----G- ---------- CA-G------ -----C----

                                                                                                          7400
AAUUCCUAGC UAAAAUCAGG AGUGUGCCAA UUGGAAGAGC UUUAUUGCUC CCAGAGUACU CAACAUUGUA CCGCCGUUGG CUUGACUCAU UUUAGUAACC
-G---U---- ---------- ---------- ----G----- ---------- --G------- -U-------- ---------- --C-----U- ----------
----UU---- ------U--- ---------- -C-------- ---G------ ---------- ---------- ---------- ---------- ----------

CUACCUCAGU CGAAUUGGAU UGGGUCAUAC UGCUGUAGGG GUAAAUUUUU CUUUAAUUCG GAGG-poly(A)
---------- ---------- --------G- --U------- ---------- ---------- -*---poly(A)
---------- ---------- ---------- --U------- ---------- ---------- -----poly(A)
```

```
                                                                                                                      100
V1(Sab) MGAQVSSQKV GAHENSNRAY GGSTINYTTI NYYRDSASNA ASKQDFSQDP SKFTEPIKDV LIKTSPMLNS PNIEACGYSD RVLQLTLGNS TITTQEAANS  1a
V2(Sab) ---------- ---------- ---------- ---------- ------A--- ---------- ----A----- ---------- --M------- ----------
V3(Sab) ---------- ---------- ---------- ---K------ -----Y---- ------L--- ----A-A--- --V------- ---------- ----------
        VP4                                                                      VP2
                                                                                                                      200
P1      VVAYGRWPEY LRDSEANPVD QPTEPDVAAC RFYTLDTVSW TKESRGWWWK LPDALRDMGL FGQNMYYHYL GRSGYTVHVQ CNASKFHQGA LGVFAVPEMC
        ---------- I--T------ ---------- --------T- R--------- -----K---- -----F---- --A------- ---F------ ----------
        ---------F I--D------ --------T- --------M- G---K----- ---------- ---------- ---------- ---------- -----I--Y-
                                                                                                                      300
        LAGDSNTTTM HTSYQNANPG EKGGTFTGTF TPDDNQTSPA RRFCPVDYLF GNGTLLGNAF VFPHQIINLR TNNCATLVLP YVNSLSIDSM VKHNNWGIAI
        -----*--H- F-K-E----- ----E-K-S- -L-T-A-N-- -N-------- -S-V-V---- -Y-------- ---------- ---------- T---------
        -----D*KQR Y---A----- -R--K-YSQ- NK-NAV---K -E-------L -C-V------ -Y-------- ---S--I--- ---A-A---- ----------
                                                                                                                      400
        LPLAPLNFAS ESSPEIPITL TIAPMCCEFN GLRNITLPRL QGLPVMNTPG SNQYLTADNF QSPCALPEFD VTPPIDIPGE VKNMMELAEI DTMIPFDLSA
        ------D-VT ---T------ ---------- ------V--T ----L----- ---------Y -----I---- ---------- -R-------- -----LN-TS
        ---S--D--Q D--V-----V ------S--- ----V-A-KF ----L----- ------S--H -----I---- ---------- ---------- -----LN-ES
                                                  VP3
                                                                                                                      500
        KKKNTMEMYR VRLSDKPHTD DPILCLSLSP ASDPRLSHTM LGEILNYYTH WAGSLKFTFL FCGSMMATGK LLVSYAPPGA DPPKKRKEAM LGTHVIWDIG
        QRR---D--- -E---TA-S- T--------- ------A--- ---------- ---------- ---------- ---------- EA--S----- ----------
        T-R---D--- -T---SADLS Q--------- -F-------- ---V------ ---------- ---------- I--A------ Q--TS----- --------L-
                                                                                                                      600
        LQSSCTMVVP WISNTTYRQT IDDSFTEGGY ISVFYQTRIV VPLSTPREMD ILGFVSACND FSVRLMRDTT HIEQKALAQG LGQMLESMID NTVRETVGAA
        ---------- ---------- -N-------- --M-----V- -------K-- ---------- -----L---- --S-E-MP-- I-D-I-GAVE GITKNALVPP
        ---------- ----V----- TQ-------- --M------- ------KS-S M--------- -----L---- --S-S--P-- IEDLI*-EV* AQGAL-LSLP
                                                                                                VP1
                                                                                                                      700
        TSRDALPNTE ASGPAHSKEI PALTAVETGA TNPLVPSDTV QTRHVVQHRS RSESSIESFF ARGACVAIIT VDNSASTKNK DKLFTVWKIT YKDTVQLRRK
        --TNS--GHK P--------- ---------- ---------- -----I-R-T ----TV---- ---------E ---D-P--RA SR--S----- ----------
        KQQ-S--D-K ---------V ---------- ----A----- -------R-- ----T----- ---------E ---EQP-TRA Q---AM-R-- ----------
                                                                                                                      800
        LEFFTYSRFD MEFTFVVTAN FTETNNGHAL NQVYQIMYVP PGAPVPEKWD DYTWQTSSNP SIFYTYGTAP ARISVPYVGI SNAYSHFYDG FSKVPLKDQ*
        ---------- --------S- YIDA------ --------I- ----I-G--N ---------- -V-----AP- ---------- A--------- -A----AG-A
        ---------- ---------- --NA------ --------I- ----T-KS-- ---------- -------A-- ---------L A--------- -A-----TDA
                                                                                                                      900
        SAALGDSLYG AASLNDFGIL AVRVVNDHNP TKVTSKIRVY LKPKHIRVWC PRPPRAVAYY GPGVDYKDGT LTPLSTKDLT TYGFGHQNKA VYTAGYKICN  3b
        -TE*------ --------S- ---------- -RL------- M----V---- -------P-F ---------* ----PE-G-- ---------- ----------
        NDQI-----S -MTVD---V- ---------- ------V-I- M----V---- -------P-- ------RN*N -D---E-G-- ---------- ----------
                                                                                                                     1000
P2      YHLATQEDLQ NAVNVMWNRD LLVTESRAQG TDSIARCNCN AGVYYCESRR KYYPVSFVGP TFQYMEANNY YPARYQSHML IGHGFASPGD CGGILRCHHG
        ---------- ---S------ ---A----L- ---------- -------S-- T--------- -------I-- --------E- ---------- -------Q--
        -----K---- ---SI----- ---V--K--- ---------- ---------- ---------- ---------- --------D- ---------- -------Q--
                                                                                                                     1100
        VIGIITAGGE GLVAFTDIRD LYAYEEEAME QGITNYIESL GAAFGSGFTQ QIGDKITELT NMVTSTITEK LLKNLIKIIS SLVIITRNYE DTTTVLATLA
        ---------- -----S---- ---------- --S------- ---------- -----VS--- S--------- ---------- ---------- ----------
        ----V----- -----S---- ---------- --S------- ---------- -----S---- S--------- ---------- ---------- ----------
                                          5b
                                                                                                                     1200
        LLGCDASPWQ WLRKKACDVL EIPYVTKQGD SWLKKFTEAC NAAKGLEWVS NKISKFIDWL KEKIIPQARD KLEFVTKLRQ LEMLENQIST IHQSCPSQEH
        -----I---- --K-----I- ----AIE--- ------A--- ---------- -------S-- QD-------- --------K- ---------- ----------
        -----V---- --K-----T- -----IR--- ---------- ------G--- ---------- R-R------- --------K- ---------- ----------
                                        X
        QEILFNNVRW LSIQSKRFAP LYAVEAKRIQ KLEHTINNYI QFKSKHRIEP VCLLVHGSPG TGKSVATNLI ARAIAERENT STYSLPPDPS HFDGYKQQGV
        ---------- ---------- ---H------ ---------V ---------- ---------- ---------- ------K--- ---------- ----------
        ---------- ---------- ---L------ ---------- ---------- ---------- ---------- ------K--- ---------- ----------
                                                                                                                     1400
        VIMDDLNQNP DGADMKLFCQ MVSTVEFIPP MASLEEKGIL FTSNYVLAST NSSRISPPTV AHSDALARRF AFDMDIQVMN EYSRDGKLNM AMATEMCKNC
        ---------- ---------- ---------- ---------- ---------- -----T---- ---------- ---V-----S ---------- ----------
        ---------- ---------- ---------- ---------- ---------- -----T---- ---------- ---------G ---------- -----T--D-
                                                                                                                     1500
        HQPANFKRCC PLVCGKAIQL MDKSSRVRYS IDQITTMIIN ERNRRSNIGN CMEALFQGPL QYKDLKIDIK TSPPPECIND LLQAVDSQEV RDYCEKKGWI
        ---------- ---------- ---------- --------V- ---------- ---------- --------V- ---------- ---------- ----------  1b
        -R-------- ---------- ---------- V--------- ---------- ---------- ---------- -R-------- ---------- ----------
                                                                   9
                                                                                                                     1600
P3      VNITSQVQTE RNINRAMTIL QAVTTFAAVA GVVYVMYKLF AGHQGAYTGL PNKKPNVPTI RTAKVQGPGF DYAVAMAKRN IVTATTSKGE FTMLGVHDNV
        ---------- ---------- ---------- ---------- ---------- ---R------ ---------- ---------- ---------- ------Y---
        ---------- ---------- ---------- ---------- ---------- ---R------ -A-------- ---------- ---------- ----------
                                                       VPg              2;7c;6a
```

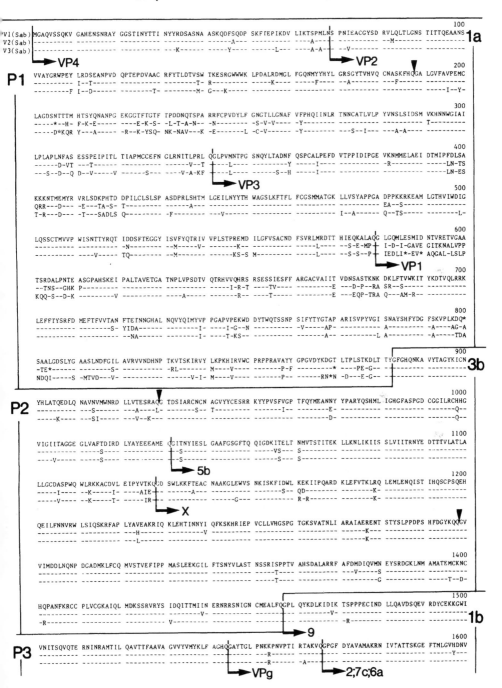

```
                                                                                                          1700
AILPTHASPG ESIVIDGKEV EILDAKALED QAGTNLEITI ITLKRNEKFR DIRPHIPTQI TETNDGVLIV NTSKYPNMYV PVGAVTEQGY LNLGGRQTAR
---------- ---A------ ---------- ---------- ---------- ---------- ---------- ---------- ---------- ----------
---------- ---------- ---------- ---------- ---------- ---Q------ ---------- ---------- ---------- ----------
                                                                                          4a
                                                                                                          1800
TLMYNFPTRA GQCGGVITCT GKVIGMHVGG NGSHGFAAAL KRSYFTQSQG EIQWMRPSKE VGYPIINAPS KTKLEPSAFH YVFEGVKEPA VLTKNDPRLK
---------- ---------- ---------- ---------- -------I-- -----K---- ---------- ---------- ---------- ---------R
I--------- ---------- ---------- ---------- ---------- ---------- A--------T ---------- ---------- ----------
                                                   4b
                                                                                                          1900
TNFEEAIFSK YVGNKITEVD EHMKEAVDHY AGQLMSLDIN TEQMCLEDAM YGTDGLEALD LSTSAGYPYV AMGKKKRDIL NKQTRDTKEM QKLLDTYGIN
-D-------- -------D-- -Y-------- ---------- ---------- ---------- -T-------- ---------- ---------- RR--------
-D-------- ---------- -Y-------- ---------S ---------- ---------- ---------- ---------- ---------- -R---A----
                                                                                                          2000
LPLVTYVKDE LRSKTKVEQG KSRLIEASSL NDSVAMRMAF GNLYAAFHKN PGVITGSAVG CDPDLFWSKI PVLMEEKLFA FDYTGYDASL SPAWFEALEM
---------- ---------- ---------- ---------- ---------- ---V------ ---------- ---------- ---------- --------K-
---------- ---------- ---------- ---------- --------R- ---V------ ---------- ---------- ---------- --------K-
6b
                                                                                                          2100
VLEKIGFGDR VDYIDYLNHS HHLYKNKTYC VKGGMPSGCS GTSIFNSMIN NLIIRTLLLK TYKGIDLDHL KMIAYGDDVI ASYPHEVDAS LLAQSGKDYG
---------- ---------- ---------- ---------- ---------- ---------- ---------- ---------- ---------- ----------
---------- ---------- -------I-- ---------- ---------- ---------- ---------- ---------- ---------- ----------
                                                                                                          2200
LTMTPADKSA IFETVTWENV TFLKRFFRAD EKYPFLIHPV MPMKEIHESI RWTKDPRNTQ DHVRSLCLLA WHNGEEEYNK FLAKIRSVPI GRALLLPEYS
---------- T--------- ---------- ---------- ---------- ---------- ---------- ---------- ---------- ----------
---------- T--------- ---------- ---------- ---------- ---------- ---------- ---R------ ---------- ----------

TLYRRWLDSF
----------
----------
```

References

Abad-Zapatero, C., Adbel-Meguid, S.A., Johnson, J.E., Leslie, A.G.W., Rayment, I., Rossman, M., Suck, D., Tsukihara, T.: Structure of Southern bean mosaic virus at 2.8 A resolution. Nature 286: 33-39 (1980).

Abraham, G., Cooper, P.D.: Poliovirus polypeptides examined in more detail. J. Gen. Virol. 29: 199-213 (1975a).

Abraham, G., Cooper, P.D.: Relations between poliovirus polypeptides as shown by tryptic peptide analysis. J. Gen. Virol. 29: 215-221 (1975b).

Abraham, G., Cooper, P.D. : Anomalous behavior of certain poliovirus polypeptides during SDS gel electrophoresis. Anal. Biochem. 73: 439-446 (1976).

Abreu, S., Lucas-Lenard, J.: Cellular protein synthesis shut-off by mengovirus: Translation of non-viral mRNAs in extracts from uninfected and infected Ehrlich ascites tumor cells. J. Virol. 18: 184-192 (1976).

Ackermann, W.W., Rabson, A., Kurtz, H.: Growth characteristics of poliomyelitis virus in HeLa cell cultures: Lack of parallelism in cellular injury and virus increase. J. Exp. Med. 100: 437 (1954).

Adesnik, M., Salditt, M., Thomas, W., Darnell, J.E.: Evidence that all messenger RNA molecules (except histone messenger RNA) contain poly(A) sequences and that poly(A) has a nuclear function. J. Mol. Biol. 71: 21-30 (1972).

Adler, R., Garfinkle, B. D., Mitchell, W., Tershak, D.R.: Degradation of poliovirus RNA in vivo. Canad. J. Microbiol. 19: 539 (1973).

Adrian, T., Rosenwirth, B., Eggers, H.J.: Isolation and characterization of temperature-sensitive mutants of echovirus 12. Virology 99: 229-239 (1979).

Ageeva, O.N., Gutkina, A.V., Ghendon, Y.Z.: Synthesis of ribonucleic acid and protein of poliomyelitis virus at low temperature. Vop. Virus. 2: 164 (1971).

Agol, V.I.: Structure, translation and replication of picornaviral genomes. Prog. Med. Virol. 26: 119-157 (1980).

Agol, V.I., Chumakova, M.Y.: An agar polysaccharide and d marker of poliovirus. Virol. 17: 221-223 (1962).

Agol, V.I., Chumakova, M.Y.: Effect of polyanions on the multiplication of two variants of poliovirus. Acta virol., Prague 7: 97-106 (1963).

Agol, V., Shirman, G.A.: Interaction of guanidine-sensitive and guanidine-dependent variants of poliovirus in mixedly infected cells. Biochem. Biophys. Res. Commun. 17: 28 (1964).

Agol, V.I., Shirman, G.A.: Formation of virus particles by means of enzyme systems and structural proteins induced by another „helper" virus. Vop. Virus. 10: 8 (1965).

Agol, V.I., Lipskaya, G.U., Tolskaya, E.A., Voroshilova, M.K., Romanova, L.I.: Defect in poliovirus maturation under hypotonic conditions. Virology 41: 533-540 (1970a).

Agol, V.I., Dryzin, Y., Romanova, L., Bogdanov, A.: Circular structure in preparation of replicative form of EMC virus RNA. FEBS Lett. 8: 13-16 (1970b).

Agol, V.I., Romanova, L.I., Chumakov, K.M., Dunaevskaya, L.D., Bogdanov, A.A.: Circularity and cross-linking in preparations of replicative form of encephalomyocarditis virus RNA. J. Mol. Biol. 72: 77-89 (1972).

Agol, V.I., Chumakov, K.M., Dmitrieva, T.M., Svitkin, Y.V.: Functional and structural studies on the genome of picornaviruses. Soviet Sci. Rev. Sect. D, Biology Reviews 1: 319-370 (1981).

Agrawal, H.O.: Studies on the structure of poliovirus. Arch. Gesamt Virusforsch. 19: 365-372 (1966).

Ahlquist, P., Kaesberg, P.: Determination of the length distribution of poly (A) at the 3' termini of the virion RNAs of EMC virus, poliovirus, rhinovirus, RAV-61 and CPMV, and mouse globin mRNA. Nucl. Acids Res. 7: 1195-1204 (1979).

Akihama, S., Okude, M., Sato, K., Iwabuchi, S.: Inhibitory effect of 1,2-bis(2-benzimidazolyl)1,2-ethanediol derivatives on poliovirus. Nature (London) 217: 526-563 (1968).

Alexander, H.E., Koch, G., Mountain, I.M., Van Damme, O.: Infectivity of ribunucleic acid from poliovirus in human cell monolayers. J. Exp. Med. 108: 493-506 (1958a).

Alexander, H.E., Koch, G., Mountain, I.M., Sprunt, K., Van Damme, O.: Infectivity of ribonucleic acid of poliovirus on HeLa cell monolayers. Virology 5: 172 (1958b).

Alonso, M.A., Carrasco, L.: Reversion by hypotonic medium of the shut-off of protein synthesis induced by encephalomyocarditis virus. J. Virol. 37: 535-540 (1981).

Alonso, M.A., Carrasco, L.: Selective inhibition of cellular protein synthesis by amphotericin B in EMC virus-infected cells. Virology 114: 247-251 (1982).

Amako, K., Dales, S.: Cytopathology of mengovirus infection. II. Proliferation of membranous cisternae. Virology 32: 201-215 (1967).

Ambros, V., Baltimore, D.: Protein is linked to the 5' end of poliovirus RNA by phosphodiester linkage to tyrosine. J. Biol. Chem. 253: 5263-5266 (1978).

Ambros, V., Baltimore, D.: Purification and properties of a HeLa cell enzyme able to remove the 5'-terminal protein from poliovirus RNA. J. Biol. Chem. (1980).

Ambros, V., Pettersson, R.F., Baltimore, D.: An enzymatic activity in uninfected cells that cleaves the linkage between poliovirion RNA and the 5' terminal protein. Cell 15: 1439-1446 (1978).

Ammann, J., Delius, H., Hofschneider, P.H.: Isolation and properties of an intact phage-specific replicative form of RNA phage M12. J. Mol. Biol. 10: 557 (1964).

Amos, L.A., Baker, T.S.: The three-dimensional structure of tubulin protofilaments. Nature 279: 607-612 (1979).

Amstey, M.S.: Enhancement of polio-RNA infectivity with dimethylsulfoxide. Fed. Proc. 25: 492 (1966).

Anderson-Sillman, K., Bartal, S., Tershak, D.R.: Guanidine-resistant poliovirus mutants produce modified 37-kilodalton proteins. J. Virol. 50: 922-928 (1984).

Anderton, B.H.: Intermediate filaments: a family of homologous structures. J. Muscle Res. Cell Motility 2: 141-166 (1981).

Anzai, T., Ozaki, Y.: Intranuclear crystal formation of poliovirus. Electron microscopic observations. Exp. Molec. Path. 10: 176-185 (1969).

Argos, P.: Secondary structure prediction of plant virus coat proteins. Virology 110: 55-62 (1981).

Argos, P., Schwarz, J., Schwarz, J.: An assessment of protein secondary structure prediction methods based on amino acid sequences. Biochim. Biophys. Acta 439: 261-273 (1976).

Arlinghaus, R.B., Polatnick, J.: The isolation of two enzyme-ribonucleic acid complexes involved in the synthesis of foot-and-mouth disease virus ribonucleic acid. Proc. Natl. Acad. Sci. USA 62: 821-828 (1969).

Armstrong, C.: The experimental transmission of poliomyelitis to the Eastern cotton rat, Sigmodon hispidus hispidus, Pub. Health Rep. 54: 1719-1721 (1939).

Armstrong, C.: Seasonal Distribution of Poliomyelitis. J. Pub. Health 40: 1296-1304 (1950).

Armstrong, J.A., Edmonds, M., Nagazato, H., Philips, B.A., Vaughan, M.H.: Polyadenylic acid sequences in the virion RNA of poliovirus and eastern equine encephalitis virus. Science 176: 526-528 (1972).

Attardi, G., Smith, J.: Virus specific protein and a ribonucleic acid associated with ribosomes in polio-virus infected HeLa cells. Cold Spring Harbor Symp. Quant. Biol. 27: 271 (1962).

August, J.T., Ortiz, P.J., Hurwitz, J.: Ribonucleic acid-dependent ribonucleotide incorporation: I. Purification and properties of the enzyme. J. Biol. Chem. 237: 3786 (1962).

August, J.T., Banerjee, A.K., Eoyang, L., Franze de Fernandez, M.T., Hori, K., Kuo, C.H., Rensing, U., Shapiro, L.: Synthesis of bacteriophage Qß RNA. Cold Spring Harbor Symp. Quant. Biol. 32: 73 (1968).

Austin, S.A., Clemens, M.J.: The regulation of protein synthesis in mammalian cells by amino acid supply. Biosci. Reports 1: 35-44 (1981).

Avery, O.T., MacLeod, C.M., McCarty, M.: Studies on the chemical nature of the substance inducing transformation of pneumococcal types. Induction of transformation by a desoxyribonucleic acid fraction isolated from pneumococcus type III. J. Exp. Med. 79: 137-158 (1944).

Axler, D.A., Crowell, R.L.: Effect of anticellular serum on the attachment of enteroviruses to HeLa cells, J. Virol. 2: 813-821 (1968).

Babich, A., Wimmer, E., Toyada, H., Nomoto, A.: The genome-linked protein of picornaviruses VI. The 5'-terminal protein of poliovirus type 2 RNA is covalently linked to a nonanucleotide identical to that of poliovirus type 1 RNA. Intervirology 13: 192-199 (1980).

Bablanian, R.: Depression of macromolecular synthesis in cells infected with guanidine-dependent poliovirus under restrictive conditions. Virology 47: 255-259 (1972).

Bablanian, R.: Structural and functional alterations in cultured cells infected with cytocidal viruses. Prog. Med. Virol. 9: 40-83 (1975).

Bablanian, R., Eggers, H.J., Tamm, I.: Studies on the mechanism of poliovirusinduced cell damage. Virology 26: 100-113 (1965a).

Bablanian, R., Eggers, H., Tamm, I.: Studies on the mechanism of poliovirusinduced cell damage. II. Relation between poliovirus growth and virus-induced morphological changes in cells. Virology 26: 114-121 (1965b).

Bachrach, H.L., Schwerdt, C.E.: Purification studies on Lansing poliomyelitis virus, pH stability, CNS extraction and butanol. Purification experiments. I. J. Immunol. 69: 551-560 (1952).

Bachrach, U., Benchetrit, L.: Studies on phage internal proteins. II. Cleavage of a precursor of internal proteins during the morphogenesis of bacteriophage T4. Virology 59: 51-58 (1974).

Bachthold, J.G., Bubel, H.C., Gebhardt, L.P.: The primary interaction of poliomyelitis virus with host cells of tissue culture origin. Virology 4: 582-589 (1957).

Bächi, T., Aguet, M., Howe, C.: Fusion of erythrocytes by Sendai virus studied by immuno-freeze-etching. J. Virol. 11: 1004-1012 (1973).

Baer, B.W., Kornberg, R.D.: Repeating structure of cytoplasmic poly(A) ribonucleoprotein. Proc. Natl. Acad. Sci. USA 77: 1890-1892 (1980).

Baglioni, C., Minks, M.A., Maroney, P.A.: Interferon action may be mediated by activation of a nuclease by pppA2'p5'A2'p5'A. Nature 273: 684-687 (1978a).

Baglioni, C., Simili, M., Shafritz, D.A.: Initiation activity of EMC virus RNA, binding to initiation factor eIF-4B and shut-off of host cell protein synthesis. Nature 275: 240-243 (1978b).

Baglioni, C., Maroney, P. A., Simili, M.: The role of initiation factors in the shut-off of protein synthesis. In: The Molecular Biology of Picornaviruses (Pérez-Bercoff, R., ed.), 101–111. New York: Plenum Press 1979.

Balayan, M.S., Tolskaya, E.A., Voroshilova, M.K., Yurovetskaya, A.L.: Study of intratypic differences between type 2 polioviruses. I. Relationship between neurovirulence, antigenicity and other properties determined in cells in vitro. Virology 23: 125-140 (1964).

Baltimore, D.: In vitro synthesis of viral RNA by the poliovirus RNA polymerase. Proc. Natl. Acad. Sci. USA 51: 450-456 (1964).

Baltimore, D.: Purification and properties of poliovirus double-stranded ribonucleic acid. J. Mol. Biol. 18: 421-428 (1966).

Baltimore, D.: Structure of the poliovirus replicative intermediate RNA. J. Mol. Biol. 32: 359-368 (1968).

Baltimore, D.: The replication of picornaviruses. In: The Biochemistry of Viruses (Levy, H.B., ed.), 103-176. New York: Marcel Dekker 1969.

Baltimore, D.: Polio is not dead. In: Perspectives in Virology. VII. From Molecules to Man (Pollard, M., ed.), 1-12. New York: Academic Press 1971a.

Baltimore, D.: Expression of animal virus genomes. Bacteriol. Rev. 35: 235-241 (1971b).

Baltimore, D.: Purification of a factor that restores translation of vesicular stomatitis virus mRNA in extracts from poliovirus-infected HeLa cells. Proc. Natl. Acad. Sci. USA 77: 770-774 (1980).
Baltimore, D., Franklin, R.M.: The effect of mengovirus infection on the activity of the DNA-dependent RNA polymerase of L-cells. Proc. Natl. Acad. Sci. USA 48: 1383-1390 (1962).
Baltimore, D., Franklin, R.M.: A new ribonucleic acid polymerase appearing after mengovirus infection of L-cells. J. Biol. Chem. 238: 3395-3400 (1963).
Baltimore, D., Girard, M.: An intermediate in the synthesis of poliovirus RNA. Proc. Natl. Acad. Sci. USA. 56: 741-748 (1966).
Baltimore, D., Huang, A.S.: Isopycnic separation of subcellular components from poliovirus-infected and normal HeLa cells. Science, New York 162: 572-574 (1968).
Baltimore, D., Huang, A.S.: Interaction of HeLa cell proteins with RNA. J. Mol. Biol. 47: 263-273 (1969).
Baltimore, D., Eggers, H., Tamm, I.: Altered location of protein synthesis in the cell after poliovirus infection. Biochim. Biophys. Acta 76: 644-646 (1963a).
Baltimore, D., Franklin, R.M., Callender, J.: Mengovirus induced inhibition of host ribunucleic acid and protein synthesis. Biochim. Biophys. Acta 76: 425-530 (1963b).
Baltimore, D., Franklin, R.M., Eggers, H.J., Tamm, I.: Poliovirus induced RNA polymerase and the effects of virus-specific inhibitors on its production. Proc. Natl. Acad. Sci. USA 49: 843-849 (1963c).
Baltimore, D., Girard, M., Darnell, J.E.: Aspects of the synthesis of poliovirus RNA and the formation of virus particles. Virology 29: 179-189 (1966).
Baltimore, D., Jacobson, M.F., Asso, J., Huang, A.: The formation of poliovirus proteins. Cold Spring Harbor Symp. Quant. Biol. 34: 741-746 (1969).
Banerjee, A.K., Eoyand, L., Hori, K., August, J.T.: Replication of RNA viruses. IV. Initiation of RNA synthesis by Qß RNA polymerase. Proc. Natl. Acad. Sci. USA 57: 986 (1967).
Banerjee, A.K., Kuo, C.H., August, J.T.: Replication of RNA viruses. VIII. Direction of chain growth in the Qß polymerase reaction. J. Mol. Biol. 40: 445 (1969a).
Banerjee, A.K., Rensing, U., August, J.T.: Replication of RNA viruses. IX. Isolation of a small self-replicating RNA from the Qß polymerase. J. Mol. Biol. 45: 181 (1969b).
Baron, M.H., Baltimore, D.: Antibodies against the chemically synthesized genome-linked protein of poliovirus react with native virus-specific proteins. Cell 28: 395-404 (1982a).
Baron, M.H., Baltimore, D.: Anti-VPg antibody inhibition of the poliovirus replication reaction and production of covalent complexes of VPg-related proteins and RNA. Cell 30: 745-752 (1982b).
Baron, M.H., Baltimore, D.: Antibodies against a synthetic peptide of the poliovirus replicase protein: reaction with native, virus-coded protein and inhibition of virus-coded polymerase activities in vitro. J. Virol. 43: 969-978 (1982c).
Baron, M.H., Baltimore, D.: Purification and properties of a host cell protein required for poliovirus replication in vitro. J. Biol. Chem. 257: 12351-12358 (1982d).
Baron, M.H., Baltimore, D.: In vitro copying of viral positive strand RNA by poliovirus replicase: characterization of the reaction and its products. J. Biol. Chem. 257: 12359-12366 (1983e).
Baron, S., Friedman, R.M., Buckler, C.E.: Properties of poliovirus inhibitor from monkey brain. Proc. Soc. Exp. Biol. Med. 113: 107-110 (1963).
Barrell, B.G., Air, G.M., Hutchiso, C.A.: Overlapping genes in bacteriophage ØX174. Nature 264: 34-41 (1976).
Barrieux, A., Ingraham, H.A., Nystul, S., Rosenfeld, M.G.: Biochemistry 15: 35233528 (1976).
Barski, G., Robineaux, R., Endo, M.: Evolution of the cellular lesion induced by poliomyelitis virus in vitro as studied with phase contrast microcinematography. Ann. N.Y. Acad. Sci. 61: 899 (1955).
Bartsch, H.D., Habermehl, K.O., Diefenthal, W.: Correlation between poliomyelitis virus reproductive cycle, chromosomal alterations and lysosomal enzymes. Arch. ges. Virusforsch. 27: 115-127 (1969).
Baxt, B., Bachrach, H.L.: Early interactions of foot-and-mouth-disease virus with cultured cells. Virology 104: 42-55 (1980).
Baxt, B., Grubman, M.J., Bachrach, H.L.: The relation of poly(A) length to specific infectivity of viral RNA: A comparison of different types of foot-and-mouth disease virus. Virology 98: 480-483 (1979).

Beatrice, S.T., Katze, M.G., Zajac, B.A., Crowell, R.L.: Induction of neutralizing antibodies by the coxsackievirus B3 virion polypeptide, VP2. Virology 104: 426-438 (1980).
Beaud, G., Dru, A.: Protein synthesis in vaccinia virus-infected cells in the presence of amino acid analogs: A translational control mechanism. Virology 100: 10-21 (1980).
Bechet, J.M.: Isolement de la cháine „moins" d'acide ribonucleique du virus de l'encephalomyocardite de la souris et étude de son infectivité. C. R. Acad. Sci. Ser. D 274: 1761 (1972).
Becker, Y., Penman, S., Darnell, J.E.: A cytoplasmic particulate involved in poliovirus synthesis. Virology 21: 274-276 (1963).
Beckman, L.D., Caliguiri, L.A., Lilly, L.S.: Cleavage site alterations in poliovirus-specific precursor proteins. Virology 73: 216-227 (1976).
Behrens, N.H.: Polyprenol sugars and glycoprotein synthesis. In: Biology and Chemistry of Eukaryotic Cell Surfaces (Less, E.Y.C., Smith, E.D., eds.), 159-180. New York: Academic Press 1974.
Beneke, T.W., Habermehl, K.O., Diefenthal, W., Buchholz, M.: Iodination of poliovirus capsid proteins. J. Gen. Virol. 34: 387-390 (1977).
Bengtsson, S.: Mechanism of dextran sulfate inhibition of attenuated poliovirus. Proc. Soc. Exp. Biol. Med. 118: 47-53 (1965).
Bengtsson, S.: Attempts to map the poliovirus genome by analysis of selected recombinants. Acta Path. Microbiol. Scan. 73: 592-604 (1968).
Bengtsson, S., Philipson, L.: Countercurrent distribution of poliovirus type 1. Virology 20: 176-184 (1963).
Bengtsson, S., Philipson, L., Persson, H., Laurent, T.C.: The basis for the interaction between attenuated poliovirus and polyanions. Virology 24: 617-625 (1964).
Benne, R., Edman, J., Traut, R.R., Hershey, J.W.B.: Phosphorylation of eukaryotic protein synthesis initiation factors. Proc. Natl. Acad. Sci. USA 75: 108-112 (1978).
Benyesh-Melnick, M., Melnick, J.L.: The use of in vitro markers and monkey neurovirulence tests to follow genetic changes in attenuated poliovirus multiplying in the human alimentary tract. In: Live poliovirus vaccines, Vol. 44, 179-202. 1st Int. Conf. on Live Poliovirus Vaccines, Washington, D.C. Pan Am. San. Bur. Publ. 1959.
Beremand, M.N., Blumenthal, T.: Overlapping genes in RNA phage: a new protein implicated in lysis. Cell 18: 257-266 (1979).
Bergen, L.G., Borisy, G.G.: Head to tail polymerization of microtubules in vitro: electron microscope analysis of seeded assembly. J. Cell Biol. 84: 141-150 (1980).
Berlin, R.D., Fera, J.P.: Changes in membrane microviscosity associated with phagocytosis: effects of colchicine. Proc. Natl. Acad. Sci. USA 74: 1072-1076 (1977).
Best, M., Evans, B., Bishop, J.M.: Double-stranded replicative form of poliovirus RNA: Phenotype of heterozygous molecules. Virology 47: 592-603 (1972).
Bhargava, P.M., Shanmugam, G.: Uptake of nonviral nucleic acids by mammalian cells. Prog. Nucleic Acid Res., Mol. Biol. 11: 103-159 (1971).
Bienz, K.: Reaktionen von Wirtszellen auf eine Virusinfektion. Path. Microbiol. 41: 143-150 (1974).
Bienz, K., Bienz-Isler, G., Egger, D., Weiss, M., Löffler, H.: Coxsackie virus infection in skeletal muscles of mice: an electron microscopic study. Arch. ges. Virusforsch. 31: 257-265 (1970).
Bienz, K., Egger, D., Wolff, D.A.: Virus replication, cytopathology, lysomal enzyme response of mitotic and interphase HEp-2 cells infected with poliovirus. J. Virol. 11: 565-574 (1973).
Bienz, K., Egger, D., Rasser, Y., Loeffler, H.: Differential inhibition of host cell RNA synthesis in several picornavirus-infected cell lines. Intervirology 10: 209-220 (1978).
Bienz, K., Egger, D., Rasser, Y., Bossart, W.: Kinetics and location of poliovirus macromolecular synthesis in correlation to virus-induced cytopathology. Virology 100: 390-399 (1980).
Bienz, K., Egger, D., Bossart, W.: Migration of polioviral proteins into different host cell organelles. 2nd Meeting of the European Study Group on the Molecular Biology of Picornaviruses, Hamburg (1981).
Bienz, K., Egger, D., Rasser, Y., Bossart, W.: Accumulation of poliovirus proteins in the host cell nucleus. Intervirology 18: 189-196 (1982).
Bienz, K., Egger, D., Rasser, Y., Bossart, W.: Intracellular distribution of poliovirus proteins and the induction of virus-specific cytoplasmic structures. Virology 131: 39-48 (1983).
Bilello, J.A., Warnecke, G., Koch, G.: Inhibition of polypeptide chain initiation by inducers of

erythroid differentiation in friend erythroleukemic cells. In: Modern Trends in Human Leukemia III (Neth, R., Gallo, R.C., Hofschneider, P.H., Mannweiler, K., eds.), 303-306. Berlin-Heidelberg-New York: Springer 1978.

Bilello, J.A., Colletta, G., Warnecke, G., Koch, G., Frisby, D., Pragnell, I.B., Ostertag, W.: Analysis of the expression of SFFV related RNA and gp55, a Friend and Rauscher virus specific protein. Virology 107: 331-344 (1980).

Bishop, J.M., Koch, G.: Purification and characterization of poliovirus induced infectious double-stranded RNA. J. Biol. Chem. 242: 1736-1743 (1967).

Bishop, J.M., Koch, G.: Infectious replicative intermediate of poliovirus: Purification and characterization. Virology 37: 521-534 (1969a).

Bishop, J.M., Koch, G.: Plaque assay for poliovirus and poliovirus specific RNAs. In: Fundamental Techniques in Virology, 131-145. New York: Academic Press 1969b.

Bishop, J.M., Levintow, L.: Nucleic acid as the carrier of viral activity. In: Progress in Medical Virology (Melnick, J.L., ed.), Vol. 13, 1-65. Basel: Karger 1971.

Bishop, J.M., Summers, D.F., Levintow, L.: Characterization of ribonuclease-resistant RNA from poliovirus-infected HeLa cells. Proc. Natl. Acad. Sci. USA 54: 1273-1281 (1965).

Bishop, J.M., Koch, G., Levintow, L.: Biological and physicochemical aspects of poliovirus induced double-stranded RNA. In: The Molecular Biology of Viruses, 355-374. New York: Academic Press 1967a.

Bishop, J.M., Quintrell, N., Koch, G.: Poliovirus double-stranded RNA: Inactivation by ultraviolet light. J. Mol. Biol. 24: 125-139 (1967b).

Bishop, J.M., Koch, G., Evans, B., Merriman, M.: Poliovirus replicative intermediate: Structural basis of infectivity. J. Mol. Biol. 46: 235-249 (1969).

Blinzinger, K., Simon, J., Magrath, D., Boulger, L.: Poliovirus crystals within the endoplasmic reticulum of endothelial and mononuclear cells in the monkey spinal cord. Science 163: 1336-1337 (1969).

Blobel, G., Dobberstein, B.: Transfers of proteins across membranes. I. Presence of proteolytically processed and unprocessed nascent immunoglobulin light chains on membrane bound ribosomes of murine myeloma. J. Cell. Biol. 67: 835-851 (1975a).

Blobel, G., Dobberstein, G.: Transfer of proteins across membranes. II. Reconstitution of functional rough microsomes from heterologues components. J. Cell. Biol. 67: 852-862 (1975b).

Blondel, B., Crainic, R.: A technique to detect the reaction between poliovirus structural polypeptides and neutralizing anti-polio IgG. Dev. Biol. Stand. 47: 335-337 (1980).

Blondel, B., Crainic, R., Akacem, O., Bruneau, P., Girard, M., Horodniceanu, F.: Evidence for common, intertypic antigenic determinants on poliovirus capsid polypeptides. Virology 123: 461-463 (1982a).

Blondel, B., Crainic, R., Horodniceanu, F.: Le polypeptide structural VP1 du poliovirus type 1 induit des anticorps neutralisants. C. R. Acad. Sci. 294: 91 (1982b).

Blondel, B., Akacem, O., Crainic, R., Couillin, P., Horodniceanu, F.: Detection by monoclonal antibodies of an antigenic determinant critical for poliovirus neutralization present on VP1 and on heat inactivated virions. Virology 126: 707-710 (1983).

Blumenthal, T., Carmichael, G.G.: RNA replication: function and structure of Qß-replicase. Ann. Rev. Biochem. 48: 525-548 (1979).

Bodian, D.: The virus, the nerve cell and paralysis. A study of experimental poliomyelitis in the spinal cord. Bull. Johns Hopkins Hosp. 83: 1 (1948).

Bodian, D., Horstmann, D.M.: Polioviruses. In: Viral and Rickettsial Infections of Man (Horsfall, F., Tamm, I., eds.), 4th ed., 430-473. Philadelphia: Lippincott 1965.

Bodian, D., Morgan, I.M., Howe, H.A.: Differentiation of types of poliomyelitis viruses. III. The grouping of fourteen strains into three basic immunological types. Am. J. Hyg. 49: 234-245 (1949).

Böttiger, M.: Studies on characteristics of poliovirus. I. d marker and temperature. Arch. ges. Virusforsch. 17: 119-126 (1966a).

Böttiger, M.: Studies on characteristics of poliovirus. III. Investigation of d, rct, h and n markers of poliovirus isolated after vaccination with the type I Chat strain. Arch. ges. Virusforsch. 17: 135-154 (1966b).

Boeÿe, A.: Induction of mutation in poliovirus by nitrous acid. Virology 9: 691-700 (1959).

Boeÿe, A., Van Elsen, A.: Alkaline disruption of poliovirus: Kinetics and purification of RNA-free particles. Virology 33: 335 (1967).

Bolton, A.E., Hunter, M.W.: The labelling of proteins to high specific radioactivities by conjugating to a ^{125}I-containing acylating agent. Biochemical J. 133: 529-539 (1973).

Bonner, W.M.: Protein migration and accumulation in nuclei. In: The Cell Nucleus 6 (Busch, H., ed.), 97-148. New York: Academic Press 1978.

Boothroyd, J.C., Highfield, P.E., Cross, G.A.M., Rowlands, D.J., Lowe, P.A., Brown, F., Harris, T.F.R.: Molecular cloning of foot and mouth disease virus genome and nucleotide sequences in the structural protein genes. Nature 290: 800 (1981).

Boothroyd, J.C., Harris, T.J., Rowlands, D.J., Lowe, P.A.: The nucleotide sequence of cDNA coding for the structural proteins of foot-and-mouth disease virus. Gene 17: 153-161 (1982).

Borgert, K., Koschel, K., Taeuber, H., Wecker, E.: Effect of inactivation by hydroxylamine on early functions of poliovirus. J. Virol. 8: 1-6 (1971).

Borriss, E.: Wechselbeziehungen zwischen ^{32}P-markierten Ribonucleinsäuren und suspendierten Gewebekulturzellen. III. Einfluß von Arsen, Mg^{2+} und Protamin. Z. Naturforsch. 20: 752 (1965).

Borriss, E., Koch, G.: Wechselbeziehungen zwischen ^{32}P-markierten Ribonucleinsäuren und suspendierten Gewebekulturzellen. I. Adsorption. Z. Naturforsch. 19b: 32 (1964a).

Borriss, E., Koch, G.: Wechselbeziehungen zwischen ^{32}P-markierten Ribonucleinsäuren und suspendierten Gewebekulturzellen. II. Übergang der RNS von einem RNase-sensitiven in einen RNase-resistenten Zustand. Z. Naturforsch. 19b: 688 (1964b).

Borriss, E., Koch, G.: Isolation of poliovirus-neutralizing antibodies by affinity chromatography. Virology 67: 356-364 (1975).

Borriss, E., Shu, H.L., Koch, G.: Wechselbeziehungen zwischen ^{32}P-markierten Ribonucleinsäuren und suspendierten Gewebekulturzellen. IV. Schicksal der RNS nach Adsorption und Penetration. Z. Naturforsch. 20b: 759 (1965).

Bosmann, H.B.: Cellular membranes. Membrane marker enzyme activities in synchronized mouse leukemic cells L5178Y. Biochim. Biophys. Acta 203: 256-260 (1970).

Bosmann, H.B., Hagopian, A., Eylar, E.H.: Cellular membranes: the isolation and characterization of the plasma and smooth membranes of HeLa cells. Arch. Biochem. 128: 51-69 (1968).

Bossart, W., Bienz, K.: Virus replication, cytopathology, and lysosomal enzyme response in enucleated HEp-2 cells infected with poliovirus. Virology 92: 331-339 (1979).

Bossart, W., Bienz, K.: Regulation of protein synthesis in HEp-2 cells and their cytoplasmic extracts after poliovirus infection. Virology 111: 555-567 (1981).

Bossart, W., Egger, D., Rasser, Y., Bienz, K.: Poliovirus-induced inhibition of host RNA synthesis studied in isolated HEp-2 cell nuclei. J. Gen. Virol. 63: 131-140 (1982).

Both, G.W., Banerjee, A.K., Shatkin, A.J.: Methylation-dependent translation of viral messenger RNAs in vitro. Proc. Natl. Acad. Sci. USA. 72: 1189-1193 (1975a).

Both, G.W., Moyer, S.A., Banerjee, A.K.: Translation and identification of the viral mRNA species isolated from subcellular fractions of vesciular stomatitis virusinfected cells. J. Virol. 15: 1012-1019 (1975b).

Both, G.W., Furuichi, Y., Muthukrishnan, S., Shatkin, A.J.: Ribosome binding to reovirus mRNA in protein synthesis requires 5'-terminal 7-methylguanosine. Cell 6: 185-195 (1975c).

Both, G.W., Furuichi, Y., Muthukrishnan, S., Shatkin, A.J.: Effect of 5'terminal structure and base composition on polyribonucleotide binding to ribosome. J. Mol. Biol. 104: 637-658 (1976).

Boublik, M., Drzeniek, R.: Demonstration of a core in poliovirus particles by electron microscopy. J. Gen. Virol. 31: 447-449 (1976).

Boublik, M., Drzeniek, R.: Structural subunits of poliovirus particles by electron microscopy. J. Gen. Virol. 37: 127-134 (1977).

Bowles, S.A., Tershak, D.R.: Proteolysis of noncapsid protein 2 of type 3 poliovirus at the restricitve temperature: Breakdown of non-capsid protein 2 correlates with loss of RNA synthesis. J. Virol. 27: 443-448 (1978).

Bradbeer, C., Woodrow, M.L., Khalifah, L.I.: Transport of vitamin B12 in E. coli: common receptor system for vitamin B12 and bacteriophage BF 23 on the outer membrane of the cell envelope. J. Bact. 125: 1032-1039 (1976).

Branton, D., Cohen, C.M., Tyler, J.: Interaction of cytoskeletal proteins on the human erythrocyte membrane. Cell 24: 24 (1981).

Braun, V., Hancock, R.E.W., Hantke, K., Hartmann, A.: Functional organization of the outer membrane of Escherichia coli: Phage and colicin receptors as components of iron uptake systems. J. Supramol. Struct. 5: 37 (1976).

Brawerman, G.: Eukaryotic messenger RNA. Ann. Rev. Biochem. 43: 621-642 (1974).

Breindl, M.: Zur Struktur hitzebehandelter Poliovirus-Partikel. Hoppe Seylers Z. Physiol. Chem. 350: 1165 (1969).

Breindl, M.: The structure of heated poliovirus particles. J. Gen. Virol. 11: 147-156 (1971a).

Breindl, M.: VP_4, the D-reactive part of poliovirus. Virology 46: 962-964 (1971b).

Breindl, M., Koch, G.: Competence of suspended HeLa cells for infection by inactivated poliovirus particles and by isolated viral RNA. Virology 48: 136-144 (1972).

Bresler, S., Katsushikina, N., Kolikov, V., Chumakov, M., Pervikov, J., Zhdanov, S.: Adsorption chromatography of polioviruses on porous glass. Virology 59: 36-39 (1974).

Bretscher, M.S., Raff, M.C.: Mammalian plasma membranes. Nature 258: 43-49 (1975).

Briefs, A., Breese, S.S., Jr., Warren, J., Huebner, R.J.: Physical properties of two group A Coxsackie (Herpangina) viruses when propagated in egg and mice as determined by ultracentrifugation and electron microscopy. J. Bact. 64: 237-246 (1952).

Brinkley, B.R., Fuller, G.M., Highfield, D.P.: Cytoplasmic microtubules in normal and transformed cells in culture: Analysis by tubulin antibody immunofluorescence. Proc. Natl. Acad. Sci. USA 72: 4981-4985 (1975).

Brionen, P., Sijens, R.J., Vrijsen, R., Rombaut, B., Thomas, A.A., Jackers, A., Boeye, A.: Hybridoma antibodies to poliovirus N and H antigens. Arch. Virol. 74: 325-330 (1982).

Brodie, M., Park, W.H.: Active immunization against poliomyelitis. Am. J. Pub. Health 26: 119-125 (1936).

Browerman, G.: Characteristics and significance of the polyadenlyate sequence in mammalian mRNA. Prog. Nucleic Acid Res. Mol. Biol. 17: 118-148 (1977).

Brown, B.A., Ehrenfeld, E.: Translation of poliovirus RNA in vitro: changes in cleavage pattern and initiation sites by ribosomal salt wash. Virology 97: 396-405 (1979).

Brown, B.A., Ehrenfeld, E.: Initiation factor preparations from poliovirus-infected cells restrict translation in reticulocyte lysates. Virology 103: 327-339 (1980).

Brown, D., Hansen, J., Ehrenfeld, E.: Specificity of initiation factor preparations from poliovirus-infected cells. J. Virol. 34: 573-575 (1980).

Brown, D., Jones, C.L., Ehrenfeld, E.: Translation of capped and uncapped VSV mRNAs in the presence of initiation factors from poliovirus-infected cells. Virology 123: 60-68 (1982).

Brown, F.: Structure function relationships in the picornaviruses. In: The Molecular Biology of Picornaviruses (Perez-Bercoff, R., ed.), 49—72. New York-London: Plenum Press 1979.

Brown, F.: Molecular basis of antigenic variation in the picornaviruses. Ann. N. Y. Acad. Sci. 354: 202-218 (1980).

Brown, F., Hull, R.: Comparative virology of the small RNA viruses. J. Gen. Virol. 20: 43-60 (1973).

Brown, F., Smale, C.J.: Demonstration of three specific sites on the surface of foot-and-mouth disease virus by antibody complexing. J. Gen. Virol. 7: 115-127 (1970).

Brown, F., Steward, D.L.: The influence of proflavine on the synthesis of foot-and mouth disease virus. Gen. Microbiol. 23: 369-379 (1960).

Brown, F., Newman, J., Stott, J., Porter, A., Frisby, D., Newton, C., Carey, N., Fellner, P.: Poly (C) in animal viral RNAs. Nature 251: 342-344 (1974).

Brunette, D.M., Till, J.E.: A rapid method for the isolation of L-cell surface membranes using an aqueous two-phase polymerase system. J. Membrane Biol. 5: 215-224 (1971).

Buckingham, M.E., Gros, F.: The use of metrizamide to separate cytoplasmic ribonucleoprotein particles in muscle cell cultures: a method for the isolation of messenger RNA, independent of its poly A content. FEBS Letts. 53: 355-359 (1975).

Buckingham, M.E., Caput, D., Cohen, A., Whalen, R.G., Gros, F.: The synthesis and stability of cytoplasmic messenger RNA during myoblast differentiation in culture. Proc. Natl. Acad. Sci. USA 71: 1466-1470 (1974).

Buckley, S.M.: Visualization of poliomyelitis virus by fluorescent antibody. Arch. ges. Virusforsch. 6: 388-400 (1956).

Buckley, S.M.: Cytopathology of poliomyelitis virus in tissue culture. Fluorescent antibody and tinctorial studies. Amer. J. Path. 33: 691 (1957).

Buhl, W.-J., Sarre, T.F., Hilse, K.: Characterization of a native mRNA containing preinitiation complex from rabbit reticulocytes: RNA and protein constituents. Biochem. Biophys. Res. Commun. 93: 979-987 (1980).

Buhl, W.-J., Ernst, H., Hilse, K.: Characterization of two different messenger ribonucleoprotein particles isolated from a postpolysomal fraction of the rabbit reticulocyte lysate. Biochem. Biophys. Res. Commun. 99: 1108-1116 (1981).

Burell, C.J., Cooper, P.O.: N-terminal asparate, glycine, serine in poliovirus capsid protein. J. Gen. Virol. 21: 443-451 (1973).

Burness, A.T.H., Clothier, F.W.: Particle weight and other biophysical properties of encephalomyocarditis virus. J. Gen. Virol. 6: 381-393 (1970).

Burness, A.T.H., Pardoe, I.U., Duffy, E.M., Bhalla, R.B., Goldstein, N.O.: The size and location of the poly (A) tract in EMC virus RNA. J. Gen. Virol. 34: 331-345 (1977).

Burroughs, N., Brown, F.: Presence of a covalently linked protein on calcivirus RNA. J. Gen. Virol. 41: 443-446 (1978).

Butterworth, B.E.: Proteolytic processing of animal virus proteins. Curr. Top. Microbiol. Immunol. 77: 1-41 (1971).

Butterworth, B.E.: A comparison of the virus-specific polypeptides of encephalomyocarditis virus, human rhinovirus-1A, poliovirus. Virology 56: 439-453 (1973).

Butterworth, B.E., Korant, B.D.: Characterization of the large picornaviral polypeptides produced in the presence of zinc ion. J. Virol. 14: 282-291 (1974).

Butterworth, B.E., Rueckert, R.R.: Gene order of encephalomyocarditis virus as determined by studies with pactamycin. J. Virol. 9: 823-828 (1972b).

Butterworth, B.E., Shimshick, E.J., Yin, F.H.: Association of the polioviral RNA complex with phospholipid membranes. J. Virol. 19: 457-466 (1976).

Bülow-Hansen, Harbitz, F.: Beitrag zur Lehre der acuten Poliomyelitis. Beiträge path. Anat. 25: 517-546 (1899).

Cairns, E., Gschwender, H.H., Primke, M., Yamakawa, M., Traub, P., Schweiger, H.G.: Translation of animal virus RNA in the cytoplasm of a plant cell. Proc. Natl. Acad. Sci. USA 75: 5557-5559 (1978).

Caliguiri, L.A.: Analysis of RNA associated with the poliovirus RNA replication complexes. Virology 58: 526-535 (1974).

Caliguiri, L.A., Compans, R.W.: The formation of poliovirus particles in association with the RNA replication complexes. J. Gen. Virol. 21: 99-108 (1973).

Caliguiri, L.A., McSharry, J.J., Lawrence, G.W.: Effect of arildone on modifications of poliovirus in vitro. Virology 105: 86-93 (1980).

Caliguiri, L.A., Mosser, A.G.: Proteins associated with the poliovirus RNA replication complex. Virology 46: 375-386 (1971).

Caliguiri, L.A., Tamm, I.: Action of guanidine on the replication of poliovirus RNA. Virology 35: 408-417 (1968).

Caliguiri, L.A., Tamm, I.: Membranous structures associated with translation and transcription of poliovirus RNA. Science 166: 885-886 (1969).

Caliguiri, L.A., Tamm, I.: The role of cytoplasmic membranes in poliovirus biosynthesis. Virology 42: 100-111 (1970a).

Caliguiri, L.A., Tamm, I.: Characterization of poliovirus-specific structures associated with cytoplasmic membranes. Virology 42: 112-122 (1970b).

Caliguiri, L.A., Tamm, I.: Guanidine and 2-(α-hydroxybenzyl)-benzimidazole (HBB): selective inhibitors of picornavirus multiplication. In: Selective Inhibitors of Viral Function (Carter, W., ed.), 257-294. Cleveland: CRC Press 1973.

Carp, R.I.: Studies on the guanidine character of poliovirus. Virology 22: 270-279 (1964).

Carrasco, L.: The inhibition of cell functions after viral infection: A proposed general mechanism. FEBS Lett. 76: 11-15 (1977).

Carrasco, L.: Membrane leakiness after viral infection and a new approach to the development of antiviral agents. Nature 272: 694-699 (1978).

Carrasco, L., Smith, A.E.: Sodium ions and the shut-off of host cell protein synthesis by picornaviruses. Nature 264: 807-809 (1976).

Carrasco, L., Smith, A.E.: Molecular biology of animal virus infection. Pharmacol. Ther. 9: 311-355 (1980).
Carthew, P.: The surface nature of proteins of a bovine enterovirus, before and after neutralization. J. Gen. Virol. 32: 17-23 (1976).
Carthew, P., Martin, S.J.: The iodination of bovine enterovirus particles. J. Gen. Virol. 24: 525-534 (1974).
Casjens, S., King, J.: Virus assembly. Ann. Rev. Biochem. 44: 555-611 (1975).
Caspar, D.L.D.: The design of icosahedral virus particles. In: Viral and Rickettsial Infections of Man (Horsfall, F., Tamm, I., eds.), 4th ed., 64—93. Philadelphia: Lippincott 1965.
Caspar, D.L.D., Klug, A.: Physical principles in the construction of regular viruses. Cold Spring Harbor Symp. Quant. Biol. 27: 1-24 (1962).
Catterall, J.F., O'Malley; B.W., Robertson, M.A., Staden, R., Tana-ka, Y. Brownlee, G.G.: Nucleotide sequence homology at 12 intron-exon junctions in the chick ovalbumin gene. Nature 257: 510-513 (1978).
Cavalier-Smith, T.: Palindromic base sequences and replication of eukaryotic chromosome ends. Nature 250: 467-470 (1974).
Cavanagh, D., Sangar, D.V., Rowlands, D.J., Brown, F.: Immunogenic and cell attachment sites of FMDV; further evidence for their location in a single capsid polypeptides. J. Gen. Virol. 35: 149-158 (1977).
Caverly, C.S.: Notes of an epidemic of acute anterior poliomyelitis. J. A. M.A. 26: 1-5 (1896).
Celma, M.L., Ehrenfeld, E.: Effect of poliovirus double-stranded RNA on viral and host-cell protein synthesis. Proc. Nat. Acad. Sci. USA. 71: 2440-2444 (1974).
Celma, M.L., Ehrenfeld, E.: Translation of poliovirus RNA in vitro: Detection of two different initiation sites. J. Mol. Biol. 98: 761-780 (1975a).
Celma, M.L., Ehrenfeld, E.: Detection of two initiation sites utilized during translation of poliovirus RNA in vitro. In: In Vitro Transcription and Translation of Viral Genomes (Haenni, A.L., Beaud, G., eds.). 1975b.
Cervera, M., Dreyfuss, G., Penman, S.: Messenger RNA is translated when associated with the cytoskeletal framework in normal and VSV-infected HeLa cells. Cell 23: 113-120 (1981).
Challberg, M.D., Kelly, T.J.: Eukaryotic DNA replication: viral and plasmid model systems. Ann. Rev. Biochem. 51: 901-934 (1982).
Chan, V.F., Black, F.L.: Uncoating of poliovirus by isolated plasma membranes. J. Virol. 5: 309-312 (1970).
Chanock, R.M.: New opportunities for development of safe, effective live virus vaccines. Yale J. Biol. Med. 55: 361-367 (1982).
Chanock, R.M., Lerner, R.A. (eds.): Modern Approaches to Vaccines. Molecular and Chemical Basis of Virus Virulence and Immunogenicity. Cold Spring Harbor Laboratory 1984.
Chatterjee, N.K., Koch, G., Weissbach, H.: Initiation of Protein Synthesis in vivo in poliovirus-infected HeLa cells. Arch. Biochem. Biophys. 154: 431-437 (1973).
Chatterjee, N.K., Bachrach, H.L., Polatnick, J.: Foot-and-mouth disease virus RNA. Presence of 3'-terminal polyriboadenylic acid and absence of amino acid binding ability. Virology 69: 369-377 (1976).
Chauvin, C., Witz, J., Jacrot, B.: The structure of tomato bushy stunt virus: A model for RNA-protein interaction. J. Mol. Biol. 124: 641-651 (1978).
Chow, M., Baltimore, D.: Isolated poliovirus capsid protein VP1 induces a neutralizing response in rats. Proc. Natl. Acad. Sci. USA 79: 7518-7521 (1982).
Chow, M., Bittle, J.L., Hogle, J., Baltimore, D.: Antigenic determinants in the poliovirus capsid protein VP1. In: Modern Approaches to Vaccines (Chanock, R.M., Lerner, R.A., eds.), 257-261. Cold Spring Harbor Laboratory 1984.
Choy, P.C., Paddon, H.B., Vance, D.E.: An increase in cytoplasmic CTP accelerates the reaction catalyzed by CTP: phosphocholine cytidylyltransferase in poliovirus-infected HeLa cells. J. Biol. Chem. 255: 1070-1073 (1980).
Christensen, H.N.: On the development of amino acid transport systems. Fed. Proc. 32: 19-28 (1973).
Christensen, H.N.: Biological Transport, 2nd ed. Reading, Pa.: Benjamin 1975.
Chumakov, M.P., Dzagurov, S.G., Gagarin, A.V., Lashkevich, V.A., Ralph, N.M., Mironova, L.L.,

Shmeleva, G.A., Khanina, M.K., Elbert, L.B., Chernyshev, V.I., Robinson, I.A., Ravkina, L.I., Tsypkin, L.B., Tufanov, A.V.: Problems of manufacture and control of live poliovirus vaccine from A. B. Sabin's attenuated strains. Oral Live Poliovirus Vaccine; 379-389 (1961a).

Chumakov, M.P., Voroshilova, M.K., Drozdov, S.G., Dzagurov, S.G., Lashkevich, V. A., Mironova, L.L., Ralph, N.M., Gargarina, A.V., Ashmarina, E.E., Shirman, G.A., Fleer, G.P., Tolskaya, E.A., Sokolova, I.S., Elbert, L.B., Sinyak, K.M.: Some results of the work on mass immuization in the Soviet Union with live poliovirus vaccine prepared from Sabin strains. Bull. WHO 25: 79-91 (1961b).

Chumakov, M.P., Rubin, S.G., Balayan, M.S., Savinskaya, S.S., Voroshilova, M.K., Zhevandrova, V.I., Shekoyan, L.A., Kazantseva, K.A.: Improvement of methods for intratypic differentiation of poliovirus. I. Agar gel diffusion precipitation test, with repeated addition of antigens for differentiation between wild and vaccine poliovirus types 1 and 2. Arch. ges. Virusforsch. 46: 61-65 (1974).

Civelli, O., Vincent, A., Buri, J.F., et al.: Evidence for a translational inhibitor linked to globin mRNA in untranslated free cytoplasmic messenger ribonucleoprotein complexes. FEBS Lett. 72: 71-76 (1976).

Clarkson, B., Baserga, R. (eds.): Control of Proliferation in Animal Cells, Cold Spring Harbor Conferences on Cell Proliferation, Vol. 1. Cold Spring Harbor Laboratory 1974.

Clements, J.B., Martin, S.J.: Evidence for large strands of ribonucleic acid induced by a bovine enterovirus. J. Gen. Virol. 12: 221-232 (1971).

Cohen, C.: Cell architecture and morphogenesis. I. The cytoskeletal proteins. Trends Biochem. Sci. 4: 73-77 (1979).

Colby, D.S., Finnerty, V., Lucas-Lenard, J.: Fate of mRNA of L-cells infected with mengovirus. J. Virol. 13: 858-869 (1974).

Cole, C.: Defective interfering (DI) particles of poliovirus. Prog. Med. Virol. 20: 180-207 (1975).

Cole, C.N., Baltimore, D.: Defective interfering particles of poliovirus. II. Nature of the defect. J. Mol. Biol. 76: 325-343 (1973).

Collins, F.D., Roberts, W.K.: Mechanism of mengovirus-induced cell injury in L cells: use of inhibitors of protein synthesis to dissociate virus-specific events. J. Virol. 10: 969-978 (1972).

Colonno, R.J., Cordova, A.A.: Cloning and nucleotide sequence determination of rhinovirus type 2 genome RNA. Second Meeting of the European Study Group on the Molecular Biology of Picornaviruses, Hamburg 1981.

Colter, J.S., Bird, H.H., Moyer, A.W., Brown, R.A.: Infectivity of ribonucleic acid isolated from virus infected tissues. J. Virol. 4: 522 (1957).

Committee on the Enteroviruses, National Foundation for Infantile Paralysis: The enteroviruses. Am. J. Pub. Health 47: 1556-1566 (1957).

Committee on the Enteroviruses: Classification of human enteroviruses. Virology 16: 501-504 (1962).

Committee on Typing of the National Foundation for Infantile paralysis: Immunologic classification of poliomyelitis viruses. I. A cooperative program for the typing of onehundred strains. Am. J. Hyg. 54: 191-204 (1951).

Content, J., Lebleu, B., Nudel, U., et al.: Blocks in elongation and initiation of protein synthesis induced by interferon treatment in mouse L cells. Eur. J. Biochem. 54: 1-10 (1975).

Contreras, A., Carrasco, L.: Selective inhibition of protein synthesis in virus-infected mammalian cells. J. Virol. 29: 114-122 (1979).

Contreras, G., Summers, D.F., Maizel, J.V., Ehrenfeld, E.: HeLa cell nucleolar RNA synthesis after poliovirus infection. Virology 53: 120-129 (1973).

Cooper, N.R., Ziccardi, R.J.: The nature and reactions of complement enzymes. In: Proteolysis and Physiological Regulation (Ribbons, D.W., Brew, K., eds.), Vol. 11, 167-187. New York: Academic Press 1976.

Cooper, P.D.: An improved agar cell-suspension plaque assay for poliovirus: Some factors affecting efficiency of plating. Virology 13: 153-157 (1961).

Cooper, P.D.: Studies on the structure and function of the poliovirion: Effect of concentrated urea solutions. Virology 16: 485-495 (1962).

Cooper, P.D.: The mutation of poliovirus by 5-fluorouracil. Virology 22: 186-192 (1964).

Cooper, P.D.: Rescue of one phenotype in mixed infections with heat defective mutants of type-1 poliovirus. Virology 25: 431-438 (1965).

Cooper, P.D.: A genetic map of poliovirus temperature-sensitive mutants. Virology 35: 584-596 (1968).

Cooper, P.D.: The genetic analysis of poliovirus. In: The Biochemistry of Viruses (Levy, H.B., ed.), 177-218. New York: Marcel Dekker 1969.

Cooper, P.D.: Genetics of picornaviruses. In: Comprehensive Virology (Fraenkel-Conrat, Wagner, eds.), Vol. 9, 133-207. New York: Plenum Press 1977.

Cooper, P.D., Bennett, D.J.: Genetic and structural implications of tryptic peptide analysis of poliovirus structural protein. J. Gen. Virol. 20: 151-160 (1973).

Cooper, P.D., Johnson, R.T., Garwes, D.J.: Physiological characterization of heat-defective (temperature-sensitive) poliovius mutants: Preliminary classification. Virology 30: 638-649 (1966).

Cooper, P.D., Stancek, D., Summers, D.F.: Synthesis of double-stranded RNA by poliovirus temperature-sensitive mutants. Virology 40: 971-977 (1970a).

Cooper, P.D., Summers, D.F., Maizel, J.V.: Evidence for ambiguity in the post-translational cleavage of poliovirus proteins. Virology 41: 408-418 (1970b).

Cooper, P.D., Wentworth, B.B., McCahon, D.: Guanidine inhibition of poliovirus: A dependence of viral RNA synthesis on the configuration of structural protein. Virology 40: 486-493 (1970c).

Cooper, P.D., Geissler, D., Scotti, P.D., Tannock, G.A.: Further characterization of the genetic map of poliovirus temperature-sensitive mutants. In: Ciba Symposium: The Strategy of the Viral Genome (Wolstenholme, O'Connor, eds.), 75-100. London: Churchill Livingstone 1971.

Cooper, P.D., Steiner-Pryor, A., Wright, P.J.: A proposed regulator for poliovirus: the equestron. Intervirology 1: 1-10 (1973).

Cooper, P.D., Steiner-Pryor, A., Scotti, P.D., Delong, D.: On the nature of poliovirus genetic recombinants. J. Gen. Virol. 23: 41 (1974).

Cooper, P.D., Geissler, E., Tannock, G.A.: Attempts to extend the genetic map of poliovirus temperature-sensitive mutants. J. Gen. Virol 29: 109-120 (1975).

Cooper, P.D., Agol, V.I., Bachrach, H.L., Brown, F., Ghendon, Y., Gibbs, A.J., Gellespie, J.H., Lonberg-Holm, K., Mandel, B., Melnick, J.L., Mohanty, S.B., Povey, R.C., Rueckert, R.R., Schaffer, F.C., Tyrell, D.A.: Picornaviridae: second report. Intervirology 10: 165-180 (1978).

Cordell-Stewart, B., Taylor, M.W.: Effect of double-stranded viral RNA on mammalian cells in culture. Proç. Natl. Acad. Sci. USA 68: 1326-1330 (1971).

Cordell-Stewart, B., Taylor, M.W.: Effect of viral double-stranded RNA on protein synthesis in intact cells. J. Virol. 11: 232-237 (1973a).

Cordell-Stewart, B., Taylor, M.W.: Effect of viral double-stranded RNA on mammalian cells in culture: Cytotoxicity under conditions preventing viral replication and protein synthesis. J. Virol. 12: 360 (1973b).

Cords, C.E., Holland, J.J.: Replication of poliovirus RNA induced by heterologous virus. Proc. Natl. Acad. Sci. USA 51: 1080-1082 (1964).

Cornatzer, W.E., Sandstrom, W., Fisher, R.G.: The effect of poliomyelitis virus Type 1 (Mahoney strain) on the phospholipid metabolism of the HeLa cell. Biochim. Biophys. Acta. 49: 414 (1961).

Costlow, M., Baserga, R.: Changes in membrane transport function in G_0 and G_1 cells. J. Cell Physiol. 82: 411 (1973).

Couillin, M.M., Boúe, A., Van-Cong, N., et al.: Confirmation de la localisation sur le chromosome humain F 19 d'un ĝene de structure des ŕecepteurs de poliovirus. C.R. Acad. Sci. (Paris) 281: 293-295 (1975).

Couillin, P., Fellous, M.: Recherche d'une correlation entre le systeme antigenique HLA et les recepteurs viraux de trois enterovirus a' l'aide de l'hybridation cellulaire. Comptes rendues des seances de la Societe de Biologie 168: 180-186 (1974).

Couillin, P., Bouye, A., Rebourcet, R., Van Cong, N.: Permissivity of mouse-man hybrid cell clones to three enterviruses: poliovirus II, coxsackie B3 and echovirus II. Role of human chromosome F 19. Path. Biol. (Paris) 24: 195-203 (1976a).

Couillin, P., Bouye, A., Van Cong, N., Weil, D., Rebourcet, R., Friezal, J.: Evidence for synteny between a polio receptor gene and glucose phosphate isomerase (GPI) by analysis of human-mouse hybrids. Cytogenet. Cell Genet. 16: 111-113 (1976b).

Cova, L., Aymard, M.: Comparative study of dense and standard echovirus 11 particles. The First Meeting of the European Study Group on the Molecular Biology of Picornaviruses, Enkhuizen 1979.

Cova, L., Aymard, M.: Isolation and characterization of non-hemagglutinating echovirus 11. J. Gen. Virol. 51: 219-222 (1980).
Cox, H.R., Cabasso, V.J., Markham, F.S., Moses, M.J., Moyer, A.W., Roca-Garcia, M., Ruegsegger, J.M.: Immunological response to trivalent oral poliomyelitis vaccine. Brit. M. J. 2: 591-597 (1959).
Craig, N.: Effect of reduced temperatures on protein synthesis in mouse L cells. Cell 4: 329-335 (1975).
Crainic, R.: The mode of action of inhibitors from normal horse sera on the multiplication of polioviruses (in Roumanian); Med. sci. D. thesis, Institute of Medicine, Bucharest (1970).
Crainic, R., Couillin, P., Cabau, N., Boue, A., Horodniceanu, F.: Determination of type 1 poliovirus subtype classes with neutralizing monoclonal antibodies. Dev. Biol. Stand. 50: 229-234 (1981).
Crainic, R., Couillin, P., Blondel, B., Cabau, N., Boue, A., Horodniceanu, F.: Natural variation of poliovirus epitopes. Infect. Immun. 41: 1217 (1983).
Crawford, N.M., Baltimore, D.: Genome-linked protein VPg of poliovirus is present as free VPg and VPg-pUpU in poliovirus-infected cells. Proc. Natl. Acad. Sci. USA 80: 7452-7455 (1983).
Crawford, N., Fire, A., Samuels, M., Sharp, P.A., Baltimore, D.: Inhibition of transcription factor by poliovirus. Cell 27: 555-562 (1981).
Crick, F.H.C., Watson, J.D.: Structure of small viruses. Nature (London) 177: 473-475 (1956).
Crocker, T.T., Pfendt, E., Spendlove, R.: Poliovirus: Growth in non-nucleate cytoplasm. Science 145: 401-403 (1964).
Crowell, R.L.: Specific cell-surface alteration by enteroviruses as reflected by viral-attachment interference. J. Bact. 91: 198-204 (1966).
Crowell, R.L.: Determination of the relative receptor concentration of HeLa cells for different enteroviruses by attachment kinetics. Bacteriol. Proc., 140 (1967).
Crowell, R.L.: Comparative generic characteristics of picornavirus-receptor interactions. In: Cell Membrane Receptors for Viruses, Antigens, and Antibodies, Polypeptide Hormones, and small Molecules (Beers, R.F. jr., Bassett, E.G., eds.), 179-202. New York: Raven Press 1976.
Crowell, R.L., Landau, B.J.: Picornaviridae: enterovirus-coxsackievirus. In: CRC Handbook in Clinical Laboratory Science (Hsiung, G.D., Green, R., eds.), Vol. I, 131-155. West Palm Beach: CRC Press 1978.
Crowell, R.L., Landau, B.J.: Receptors as determinants of cellular tropism in picornavirus infections. In: Receptors and Human Diseases (Bearn, A.G., Choppin, P. W., eds.), 1-33. New York: Jos. Macy Fdn. 1979.
Crowell, R.L., Landau, B.J.: Receptors in the initiation of picornavirus infections. In: Comprehensive Virology (Fraenkel-Conrat, H., Wagner, R.R., eds.), Vol. 18, 1-42. New York: Plenum Press 1983.
Crowell, R.L., Philipson, L.: Specific alterations of Coxackie virus B3 eluted from HeLa cells. J. Virol. 8: 509-515 (1971).
Crowell, R.L., Siak, J.-S.: Receptor for group B coxsackieviruses: characterization and extraction from HeLa cell membranes. In: Perspectives in Virology (Pollard, M., ed.), Vol. X, 39-53. New York: Raven Press 1978.
Crowther, D., Melnick, J.L.: The incorporation of neutral red and acridine orange into developing poliovirus particles making them photosensitive. Virology 14: 11-21 (1961a).
Crowther, D., Melnick, J.L.: Studies of the inhibitory action of guanidine on poliovirus multiplication in cell cultures. Virology 15: 65-74 (1961b).
Cumakov, I.M., Lipskaya, G.Y., Agol, V.I.: Comparative studies on the genomes of some picornaviruses: Denaturation mapping of replicative from RNA and electron microscopy of heteroduplex RNA. Virology 92: 259-270 (1979).

Dales, S.: Penetration of animal viruses into cells. Prog. Med. Virol. 7: 1-43 (1965).
Dales, S.: Role of lysosomes in cell-virus interactions. In: Lysosomes in Biology and Pathology 2 (Dingle, J.T., Fell, H.B., eds.). Amsterdam: North-Holland Publishing Co. 1969.
Dales, S.: The early events in cell-animal virus interactions. Bacteriol. Rev. 37: 103-135 (1973).
Dales, S., Eggers, H. J., Tamm, I., Palade, G. E.: Electron microscopic study of the formation of poliovirus. Virology 26: 379-389 (1965).
Darnbrough, C., Legon, S., Hunt, T., Jackson, R.J.: Initiation of protein synthesis: Evidence for mes-

senger RNA-independent binding of methionyl-transfer RNA to the 40 S ribosomal subunit. J. Mol. Biol. 76: 379-403 (1973).

Darnell, J.E.: Adsorption and maturation of poliovirus in single and multiply infected HeLa cells. J. Exp. Med. 107: 633-641 (1958).

Darnell, J.E., Jr.: Early events in poliovirus infection. Cold Spring Harbor Symp. Quant. Biol. 27: 149-158 (1962).

Darnell, J.E., Jr., Levintow, L.: Poliovirus protein: Source of amino acids and time course of synthesis. J. Biol. Chem. 235: 74-77 (1960).

Darnell, J.E., Sawyer, T.K.: The basis for variation in susceptibility to poliovirus in HeLa Cells. Virology 11: 665-675 (1960).

Darnell, J.E., Jr., Levintow, L., Thofen M.M., Hooper, J.L.: The time course of synthesis of poliovirus RNA. Virology 13: 271-279 (1961).

Darnell, J.E., Girard, M., Baltimore, D., Summers, D.F., Maizel, J.V.: The synthesis and translation of poliovirus RNA. In: The Molecular Biology of Viruses (Colter, J. S., Paranchych, W., eds.), 375-401. New York: Academic Press 1967.

Darzynkiewicz, E., Shatkin, A.J.: Assignment of reovirus mRNA ribosomes binding sites to virion genome segments by nucleotide sequence analyses. Nucl. Acids Res. 8: 337-350 (1980).

Dasgupta, A.: Purification of host factor required for in vitro transcription of poliovirus RNA. Virology 128: 245-251 (1983a).

Dasgupta, A.: Antibody to host factor precipitate poliovirus RNA polymerase from poliovirus-infected HeLa cells. Virology 128: 252-259 (1983b).

Dasgupta, A., Baron, M.H., Baltimore, D.: Poliovirus replicase: a soluble enzyme able to initate copying of poliovirus RNA. Proc. Natl. Acad. Sci. USA 76: 2679-2683 (1979).

Dasgupta, A., Zabel, P., Baltimore, D.: Dependence of the activity of the poliovirus replicase on a host cell protein. Cell 19: 423-429 (1980).

Dasgupta, A., Hollingshead, P., Baltimore, D.: Antibody to a host protein prevents initiation by the poliovirus replicase. J. Virol. 42: 1114-1117 (1982).

Date, T., Goodmann, T.M., Wickner, W.T.: Procoat, the precursor of M-13 coat protein, requires an electrochemical potential for membrane insertion. Proc. Natl. Acad. Sci. USA 77: 4669-4673 (1980).

Davidoff, F.: Guanidine derivatives in medicine. N. Eng. J. Med. 289: 141-146 (1973).

Davis, B.D., Dulbecco, R., Eisen, H.N., Ginsberg, H.S. (eds.): Microbiology Including Immunology and Molecular Genetics, 3rd ed. Hagerstown: Harper & Row 1980.

Dawson, W.O.: Guanidine inhibits tobacco mosaic virus RNA synthesis at two stages. Intervirology 6: 83-89 (1975).

De Meis, L., Vianna, A.L.: Energy interconversion by the Ca^{2+}-dependent ATPase of the sarcoplasmic reticulum. Ann. Rev. Biochem. 48: 275-292 (1979).

De Petris, S., Raft, M.C.: Normal distribution, patching and capping of lymphocyte surface immunoglobulin studied by electron microscopy. Nature New Biol. 241: 257-259 (1973).

De Sena, J., Jarvis, D.L.: Modification of the poliovirus capsid by ultraviolet light. Can. J. Microbiol. 27: 1185-1193 (1982).

De Sena, J., Mandel, B.: Studies on the in vitro uncoating of poliovirus. I. Characterization of the modifying factor and the modifying reaction. Virology 70: 470-483 (1976).

De Sena, J., Mandel, B.: Studies on the in vitro uncoating of poliovirus. II. Characterization of the membrane-modified particle. Virology 78: 554-566 (1977).

De Sena, J., Torian, B.: Studies on the in vitro uncoating of poliovirus. III. Roles of membrane-modifying and stabilizing factors in the generation of subviral particles. Virology 104: 149-163 (1980).

Deitch, A.D., Sawicki, S.G., Godman, G.C., Tanenbaum, S.W.: Enhancement of viral infectivity by cytochalasin. Virology 56: 417-428 (1973a).

Deitch, A.D., Godman, G.C., Tanenbaum, S.W., Rose, H.M.: Cytochalasin D enhances infectivity of poliovirus. Proc. Soc. Exp. Biol. Med. 144: 60-64 (1973b).

Dejong, W., Zweers, A., Cohen, L.H.: Influence of single amino acid substitutions on electrophoretic mobility of sodium dodecyl sulfate-protein complexes. Biochem. Biophys. Res. Commun. 82: 532-539 (1978).

Delgadillo, R.A., Vanden Berghe, D.A.: Glucosamine and poliovirus. The First Meeting of the European Study Group on the Molecular Biology of Picornaviruses, Enkhuizen 1979.

Denny, P.C., Tyler, A.: Activation of protein biosynthesis in non-nucleate fragments of sea urchin eggs. Biochem. Biophys. Res. Commun. 14: 245 (1964).

Dernick, R.: Antigenic structure of poliovirus. Dev. Biol. Stand. 47: 319-333 (1981).

Dernick, R., Heukeshoven, J., Higbrig, M.: Induction of neutralizing antibodies by all three structural poliovirus polypeptides. Virology 130: 243 (1983).

Detjen, B.M., Lucas, J.J., Wimmer, E.: Poliovirus single-stranded RNA and double-stranded RNA: Differential infectivity in enucleate cells. J. Virol. 27: 582-586 (1978).

DeWachter, R., Fiers, W.: Preparative two-dimensional polyacrylamide gel electrophoresis of ^{32}P-labeled RNA. Anal. Biochem. 49: 184-197 (1972).

Diana, G.D., Salvador, U.H., Jonson, D., Hinshaw, W.B., Lorenz, R.R., Thielking, W.H., Pancic, F.: Antiviral activity of some ß-diketones. 1. Aryl alkyl diketones. In vitro activity against both RNA and DNA viruses. J. Med. Chem. 20: 750-756 (1977a).

Diana, G.D., Salvador, U.H., Zalay, E.S., Carabateas, P.M., Williams, G.L., Collins, J.C., Pancic, F.: Antiviral activity of some ß-diketones. 2. Aryloxyl alkyl diketones. In vitro activity against both RNA and DNA viruses. J. Med. Chem. 20: 757-761 (1977b).

Dickson, C., Puma, J.P., Nandi, S.: Identification of a precursor protein to the major glycoproteins of mouse mammary tumor virus. J. Virol. 17: 275-282 (1976).

Diefenthal, W., Habermehl, K.-O., Lorenz, P.R., Beneke, T.: Virus-induced inhibition of host cell synthesis. Advances in Biosciences 11: 127-148 (1973).

Dimmock, N.J.: Differences between the thermal inactivation of picornaviruses at „high" and „low" temperatures. Virology 31: 338-353 (1967).

Dimmock, N.J., Harris, W.J.: Heterogeneity of apparently empty poliovirus particles with phosphotungstic acid after heating of the virus at „low" temperature. Virology 31: 715-719 (1967).

Dinter, Z., Bengston, Z.: Suppression of the inhibitory action of guanidine on virus multiplication by some amino acids. Virology 24: 254-261 (1964).

Di Segni, G., Rosen, H., Kaempfer, R.: Competition between alpha- and beta-globin messenger ribonucleic acids for eucaryotic initiation factor 2. Biochemistry 18: 2847-2854 (1979).

Diwan, A.R., Ozaki, Y., Fulton, F., Benyesh-Melnick, M.: Antigenic analysis of polioviruses employing the disc neutralization test. J. Immunol. 90: 280-287 (1963).

Dmitrieva, T.M., Agol, V.I.: Selective inhibition of the synthesis of single-stranded RNA of encephalomyocarditis virus by 2-(α-hydroxybenzyl)-benzimidazole in cell-free system. Arch. ges. Virusforsch. 45: 17-26 (1974).

Dmitrieva, T.M., Shcheglova, M.V., Agol, V.I.: Inhbition of activity of encephalomyocarditis virus-induced RNA polymerase by antibodies against cellular components. Virology 92: 271-277 (1979).

Dobos, P., Martin, E.M.: Virus-specific polypeptides in ascites cells infected with encephalomyocarditis virus. J. Gen. Virol. 17: 197-212 (1972).

Dobos, P., Plourde, J.Y.: Precursor-product relationship of encephalomyocarditis virus-specific polypeptides. Comparisons by tryptic peptide mapping. Eur. J. Biochem. 39: 463-469 (1973).

Dömök, I., Magrath, D.I.: Guide to poliovirus isolation and serological techniques for poliomyelitis surveillance. WHO Offset Publ. 46 (1979).

Doermann, A.H.: Lysis and lysis inhibition with Escherichia coli bacteriophage. J. Bact. 55: 257 (1948).

Doolittle, R.F.: Structural aspects of the fibrinogen to fibrin conversion. Advan. Protein Chem. 27: 1-109 (1973).

Dorner, A.J., Rothberg, P.G., Wimmer, E.: The fate of VPg during in vitro translation of poliovirus RNA. FEBS Lett. 132: 219-223 (1981).

Dorner, A.J., Dorner, L.F., Larsen, G.R., Wimmer, E., Anderson, C.W.: Identification of the initiation site of poliovirus polyprotein synthesis. J. Virol. 42: 1017-1028 (1982).

Dorsch-Haesler, K., Yogo, Y., Wimmer, E.: Replication of picornaviruses. I. Evidence from in vitro RNA synthesis that poly (A) of the poliovirus genome is genetically coded. J. Virol. 16: 1512-1527 (1975).

Dowers, C.D., Miller, B.A., Kurtz, H., Ackerman, W.W.: Specific effect of guanidine in the programming of poliovirus inhibition of deoxyribonucleic acid synthesis. J. Virol. 3: 337-342 (1969).

Doyle, S., Holland, J.: Virus-induced interference in heterologously infected HeLa cells. J. Virol. 9: 22-28 (1972).

Drake, N.L., Palmenberg, A.C., Gosh, A., Omilianowski, D.R., Kaesberg, P.: Identification of the polyprotein termination site on encephalomyocarditis viral RNA. J. Virol. 41: 726-729 (1982).

Drees, O.: Zur Problematik der Gewinnung reiner Enterovirus-Präparate durch isopyknische Gradientenzentrifugation in Cäsiumchlorid. Arch. ges. Virusforsch. 3: 344-355 (1965).

Drees, O., Borna, C.: Über die Spaltung physikalisch intakter Poliovirus-Teilchen in Nucleinsäure und leere Proteinhüllen durch Wärmebehandlung. Z. Naturforsch. 20b: 870 (1965).

Drescher, C.K.: Characterization of morphogenetic defects in temperature-sensitive mutants of poliovirus. Ph. D. thesis, University of Pittsburgh, Pittsburgh, Pa. (1979).

Drescher, C.K., Phillips, B.A.: Morphogenetic defects in temperature-sensitive mutants of poliovirus. Abstr. Ann. Meet. Am. Soc. Microbiol. S89: 254 (1979).

Drescher-Lincoln, C.K., Putnak, J.R., Phillips, B.A.: Use or temperature-sensitive mutants to study the morphogenesis of poliovirus. Virology 126: 301-316 (1983).

Drzeniek, R.: Dissociation and reassociation of poliovirus. I. Effect of urea on the virion. Z. Naturforsch. 30c: 523-531 (1975).

Drzeniek, R., Bilello, P.: Reconstitution of poliovirus. Biochem. Biophys. Res. Commun. 46: 719-724 (1972a).

Drzeniek, R., Bilello, P.: Dissociation and reassociation of infectious poliovirus particles. Nature 240: 118-122 (1972b).

Drzeniek, R., Bilello, P.: Absence of glycoproteins in poliovirus particles. J. Gen. Virol. 25: 125-132 (1974).

Dubes, G.R.: Methods for transfecting cells with nucleic acids of animal viruses: A review. (Experientia. Supplementum 16.) Basel: Birkhäuser 1971.

Dubes, G.R.: Protection of transfective poliovirion RNA by histidine. Biochim. Biophys. Acta 608: 368-377 (1980).

Dubes, G.R., Chapin, M.: Poliovirus mutants with altered response to cystine. J. Gen. Microbiol. 18: 320-329 (1958).

Dubes, G.R., Wenner, H.A.: Virulence of polioviruses in relation to various characteristics distinguishable on cells in vitro. Virology 4: 275 (1957).

Dubes, G.R., Wegrzyn, R.J.: Influence of polypeptide molecular and side chain lengths and of side chain steric location on the kinetics of basic polypeptide-induced sensitization of primate cells to transfection. Acta Virol. (Praha) 21: 15-24 (1977).

Dubes, G.R., Wegrzyn, R.J.: Rapid ephemeral cell sensitization as the mechanism of histone-induced and protamine-induced enhancement of transfection by poliovirus RNA. Protoplasma 96: 209-223 (1978).

Dubes, G.R., Archetti, I., Wenner, H.A.: Antigenic variation among type 3 polioviruses. Am. J. Hyg. 70: 91-105 (1959).

Dubes, G.R., Faas, F.H., Kelly, D.G., Chapin, M., Lamb, R.D., Lucas, T.A.: A study of the facilitations of infection with surviving poliovirus units. J. Infec. Dis. 114: 346 (1964).

Dubes, G.R., Wegrzyn, R.J., Masoud, A.N.: Inactivation of transfective poliovirion RNA by a product or products of the interaction of trace copper with an impurity or impurities in reagent-grade phenol. Arch. Virol. 66: 27-44 (1980).

Dulbecco, R.: Production of plaques in monolayer tissue cultures by single particles of an animal virus. Proc. Natl. Acad. Sci. USA 38: 747-752 (1952).

Dulbecco, R.: Interaction of viruses and animal cells. A study of facts and interpretations. Physiol. Rev. 35: 301-335 (1955).

Dulbecco, R.: The nature of viruses. In: Microbiology Including Immunology and Molecular Genetics, 3rd ed. (Davis et al., eds.), 853-884. Hagerstown: Harper & Row 1980.

Dulbecco, R., Vogt, M.: Plaque formation and isolation of pure lines with poliomyelitis viruses. J. Exp. Med. 99: 167-182 (1954).

Dulbecco, R., Vogt, M.: Biological properties of poliomyelitis viruses as studied by the plaque technique. Ann. N.Y. Acad. Sci. 61: 790-800 (1955).

Dulbecco, R., Vogt, M.: Studies on the induction of mutations in poliovirus by proflavine. Virology 5: 236-243 (1958).

Dulbecco, R., Vogt, M., Strickland, A.G.: A study of the basic aspects of neutralization of two animal viruses, western equine ecephalitis virus and poliomyelitis virus. Virology 2: 162-205 (1956).

Dunker, A.K.: Structure of isometric viruses containing nonidentical polypeptide chains. J. Virol. 14: 878-885 (1974).
Dunker, A.K.: The structure of picornaviruses: classification of the bonding networks. Virology 97: 141-150 (1979).
Dunker, A.K. Rueckert, R.R.: Fragments generated by pH dissociation of ME virus and their relation to the structure of the virion. J. Mol. Biol. 58: 217-235 (1971).
Dunn, J.J.: RNase III cleavage of single-stranded RNA. J. Biol. Chem. 251: 3807-3814 (1976).
Dunnebacke, T.H.: Correlation of the stage of cytopathic change with the release of poliomyelitis virus. Virology 2: 399 (1956).
Dunnebacke, T.H., Williams, R.C.: The maturation and relesae of infectious polio and coxsackie viruses in individual tissue cultured cells. Arch. ges. Virusforsch. 11: 583-591 (1961).
Dunnebacke, T.H., Levinthal, J.D., Williams, R.C.: Entry and release of poliovirus as observed by electron microscopy of cultured cells. J. Virol. 4: 505-513 (1969).
Dusing, S.K., Wolff, D.A.: Comparative biochemical and cytochemical examination of the relationship of lysosomal enzyme release to the development of diverse viral cytopathologies. Bacteriol. Proc. 187 (1969).

Eagle, H.: Nutrition needs of mammalian cells in tissue culture. Science 122: 501 (1955).
Eagle, H., Habel, K.: The nutritional requirements for the propagation of poliomyelitis virus by the HeLa cell. J. Exp. Med. 104: 271-287 (1956).
Egberts, E., Hackett, P.B., Traub, P.: Alteration of the intracellular energetic and ionic conditions by mengovirus infection of Ehrlich ascites tumor cells and its influence on protein synthesis in the midphase of infection. J. Virol. 22: 591-597 (1977).
Eggen, K.L., Shatkin A.J.: In vitro translation of cardiovirus ribonucleic acid by mammalian cell-free extracts. J. Virol. 9: 636-645 (1972).
Eggers, H.J.: Selective inhibition of uncoating of echovirus 12 by rhodanine. A study on early virus-cell interactions. Virology 78: 241-252 (1977).
Eggers, H.J.: Selective inhibition of virus replication by benzimidazoles in cell cultures and in the organism. In: Handbook of Experimental Pharmacology. Antiviral Chemotherapy (Caliguiri, L., Came, P., eds.). Berlin-Heidelberg-New York: Springer 1982.
Eggers, H.J., Tamm, I.: On the mechanism of selective inhibition of enterovirus multiplication by 2-(α-hydroxybenzyl)-benzimidazole. Virology 18: 426-438 (1962).
Eggers, H.J., Ikegami, N., Tamm, I.: Comparative studies with selective inhibitors of picornavirus reproduction. Ann. N.Y. Acad. Sci. 130: 267-281 (1965a).
Eggers, H.J., Ikegami, N., Tamm, I.: The development of ultraviolet-irradiation resistance by poliovirus infective centers and its inhibition by guanidine. Virology 25: 475-478 (1965b).
Eggers, H.J., Bode, B., Brown, D.T.: Cytoplasmic localization of the uncoating of picornaviruses. Virology 92: 211-218 (1979).
Ehrenfeld, E.: Interaction of diethylpyrocarbonate with poliovirus double-stranded RNA. Biochem. Biophys. Res. Commun. 56: 214-219 (1974).
Ehrenfeld, E.: In vitro translation of picornavirus RNA. In: The Molecular Biology of Picornaviruses (Perez-Bercoff, R., ed.), 175-190. New York-London: Plenum 1979.
Ehrenfeld, E., Hunt, T.: Double-stranded poliovirus RNA inhibits initiation of protein synthesis by reticulocyte lysates. Proc. Natl. Acad. Sci. USA 68: 1075-1078 (1971).
Ehrenfeld, E., Lund, H.: Untranslated vesicular stomatitis virus messenger RNA after poliovirus infection. Virology 80: 297-308 (1977).
Ehrenfeld, E., Manis, S.: Inhibition of 80S initiation complex formation by infection with poliovirus. J. Gen. Virol. 43: 441-445 (1978).
Ehrenfeld, E., Maizel, J.V., Summers, D.F.: Soluble RNA polymerase complex from poliovirus-infected HeLa cells. Virology 40: 840-846 (1970).
Eisen, H.N.: Immunology - an Introduction to Molecular and Cellular Principles of the Immune Responses. Hagerstown: Harper & Row 1974.
Ellem, K.A.O., Colter, J.S.: The interaction of infections ribonucleic acids with mammalian cells. Comparison of infection and RNA uptake in the HeLa cell-polio RNA and L cell-mengo RNA system. Virology 15: 113-126 (1961).
Emini, E.A., Elzinga, M., Wimmer, E.: Carboxy-terminal analysis of poliovirus proteins: termination

of poliovirus RNA translation and location of unique poliovirus polyprotein cleavage sites. J. Virol. 42: 194-199 (1982a).

Emini, E.A., Jameson, B.A., Lewis, A.J., Larsen, G.R., Wimmer, E.: Poliovirus neutralization epitopes: Analysis and localization with neutralizing monoclonal antibodies. J. Virol. 43: 997 (1982b).

Emini, E.A., Kao, S.-Y., Lewis, A.J., Crainic, R., Wimmer, E.: Functional basis of poliovirus neutralization determined with monospecific neutralizing antibodies. J. Virol. 46: 466 (1983a).

Emini, E.A., Ostapchuk, P., Wimmer, E.: Bivalent attachment of antibody onto poliovirus leads to conformational alteration and neutralization. J. Virol. 48: 547 (1983b).

Emini, E.A., Jameson, B.A., Wimmer, E.: Priming for and induction of antipoliovirus neutralizing antibodies by synthetic peptides. Nature 304: 699 (1983c).

Emini, E.A., Dorner, A.J., Dorner, L.F., Jameson, B.A., Wimmer, E.: Identification of a poliovirus neutralization epitope through use of neutralizing antiserum raised against a purified viral structural protein. Virology 124: 144 (1983d).

Emini, E.A., Jameson, B.A., Wimmer, E.: Identification of multiple neutralization antigenic sites on poliovirus type 1 and the priming of the immune response with synthetic peptides. In: Modern Approaches to Vaccines (Chanock, R.M., Lerner, R.A., eds.), 65-75. Cold Spring Harbor Laboratory 1984.

Enders, J.F.: General preface to studies on the cultivation of poliomyelitis viruses in tissue culture. J. Immunol. 69: 639-643 (1952).

Enders, J.F., Weller, T.H., Robbins, F.C.: Cultivation of the Lansing strain of poliomyelitis virus in cultures of various human embryonic tissues. Science 109: 85-87 (1949).

Engelhard, V.H., Guild, B.C., Helenius, A., Terhorst, C., Strominger, J.L.: Reconstitution of purified detergent-soluble HLA-A and HLA-B antigens into phospholipid vesicles. Proc. Natl. Acad. Sci. USA 75: 3230-3234 (1978).

Engelman, D.M., Steitz, T. A.: The spontaneous insertion of proteins into and across membranes: The helical hairpin hypothesis. Cell 23: 411-422 (1981).

Enger-Valk, B.E., Jore, J., Pouwels, P.H., Van der Marel, P., Van Wezel, T.L.: Expression in E. coli of capsid protein VP1 of poliovirus type 1. In: Modern Approaches to Vaccines (Chanock, R.M., Lerner, R.A., eds.), 173-177. Cold Spring Harbor Laboratory 1984.

England, J.M., Howett, M.K., Tan, K.B.: Effect of hypertonic conditions on protein synthesis in cells productively infected with Simian virus 40. J. Virol. 16: 1101-1107 (1975).

Eremenko, T., Benedetto, A., Volpe, P.: Virus infection as a function of the host cell life cycle. Replication of poliovirus RNA. J. Gen. Virol. 16: 61-68 (1972b).

Eremenko, T., Volpe, P., Benedetto, A.: Poliovirus replication during the HeLa cell life cycle. Nature New Biol. 237: 114-116 (1972a).

Erikson, R.L., Gordon, J.A.: Replication of bacteriophage RNA: purification of the replicative intermediate by agarose column chromatography. Biochem. Biophys. Res. Commun. 23: 422-428 (1966).

Erikson, R.L., Erikson, E., Gordon, J.A.: Structure and function of bacteriophage R 17 replicative intermediate RNA. I. Studies on sedimentation and infectivity. J. Mol. Biol. 22: 257-268 (1966).

Erlandson, R.A., Deharven, E.: The ultrastructure of synchronized HeLa cells. J. Cell Sci. 8: 353-397 (1971).

Esteban, M., Kerr, I.M.: The synthesis of EMC virus polypeptides in infected L cells and cell-free systems. Eur. J. Biochem. 45: 567-576 (1974).

Etchison, D., Ehrenfeld, E.: Viral polypeptides assicoated with the RNA replication complex in poliovirus-infected cells. Virology 107: 135-142 (1980).

Etchison, D., Ehrenfeld, E.: Comparison of replication complexes synthesizing poliovirus RNA. Virology 111: 33-46 (1981).

Etchison, D., Milburn, S.C., Edery, I., Sonenberg, N., Hershey, J.W.B.: Inhibition of HeLa cell protein synthesis following poliovirus infection correlates with the proteolysis of a 220,000 dalton polypeptide associated with eucaryotic initiation factor 3 and a cap binding protein complex. J. Biol. Chem. 257: 14806-14810 (1982).

Evans, D.M.A., Minor, P.D., Schild, G.C., Almond, J.W.: Critical role of an eight amino acid sequence of VP1 in neutralization of poliovirus type 3. Nature 304: 459 (1983).

Falcoff, R., Falcoff, E., Sanceau, J., Lewis, J.A.: Influence of pre-incubation on the development of the inhibition of protein synthesis in extracts from interferon-treated mouse L cells. Action on tRNA. Virology 86: 507-515 (1978).

Fan, H., Penman, S.: Regulation of protein synthesis in mammalian cells. I. Inhibition of protein synthesis at the level of initiation during mitosis. J. Mol. Biol. 50: 655-670 (1970).

Fanconi, G., Zellweger, H., Botsztejn, A.: Die Poliomyelitis und ihre Grenzgebiete. Basel: 1945.

Farquhar, M., Palade, G.: The Golgi apparatus (complex) - (1954-1981) from artifact to center stage. J. Cell Biol. 91: 77s-103s (1981).

Farrah, S.R., Shah, D.O., Ingram, L.O.: Effects of chaotropic and antichaotropic agents on elution of poliovirus adsorbed on membrane filters. Proc. Natl. Acad. Sci. USA 78: 1229-1232 (1981).

Fellner, P.: General organization and structure of the picornavirus genome. In: The Molecular Biology of Picornaviruses (Pérez-Bercoff R., ed.), 25-48. New York: Plenum 1979.

Fennell, R., Phillips, B.A.: Polypeptide composition of urea and heat-resistant mutants of poliovirus type 2. J. Virol. 14: 821-833 (1974).

Fenner, F.: Classification and nomenclature of viruses. Second report of the International Committee on Taxonomy of Viruses. Intervirology 7: 1-116 (1976).

Fenner, F., McAuslan, B.R., Mims, C.A., Sambrook, J., White, D.O. (eds.): The Biology of Animal Viruses, 2nd ed. New York: Academic Press 1974.

Fenwick, M.L.: The influence of poliovirus infection on RNA synthesis in mammalian cells. Virology 19: 241-249 (1963).

Fenwick, M.L.: The effect of ribonuclease on polysomes and ribosomes of bacteria and animal cells. Biochem. J. 107: 481-489 (1968).

Fenwick, M.L., Cooper, P.D.: Early interactions between poliovirus and ERK cells. Some observations on the nature and significance of the rejected particles. Virology 18: 212-223 (1962).

Fenwick, M.L., Wall, M.J.: The density of poliovirus-specific polysomes. J. Gen. Virol. 17: 143-146 (1972).

Fenwick, M.L., Wall, M.J.: Factors determining the site of synthesis of poliovirus proteins: the early attachment of virus particles to endoplasmic membranes. J. Cell Sci. 13: 403-413 (1973).

Fenwick, M.L., Erikson, R.L., Franklin, R.M.: Replication of the RNA of bacteriophage R 17: Science 146: 527-530 (1964).

Ferguson, M., Schild, G.C., Minor, P.D., Yates, P.J., Spitz, M.: A hybridoma cell line secreting antibody to poliovirus type 3 D-antigen: Detection in virus harvest of two D-antigen populations. J. Gen. Virol. 54: 437-442 (1981).

Ferguson, M., Minor, P.D., Magrath, D.I., Qui, Y.-H., Spitz, M., Schild, G.C.: Neutralization epitopes on poliovirus type 3 particles: An analysis using monoclonal antibodies. J. Gen. Virol. 65: 197-201 (1984).

Fernández-Munoz, R., Darnell, J.E.: Structural difference between the 5'-termini of viral and cellular mRNA in the poliovirus-infected cells: Possible basis for the inhibition of the host protein synthesis. J. Virol. 18: 719-726 (1976).

Fernández-Munoz, R., Lavi, U.: 5'-termini of poliovirus RNA: Difference between virion and non-encapsidated 35S RNA. J. Virol. 21: 820-824 (1977).

Fernández-Puentes, C., Carrasco, L.: Viral infection permeabilizes mammalian cells to protein toxins. Cell 20: 769-775 (1980).

Fernández-Tomás, C.: The presence of viral-induced proteins in nuclei from poliovirus-infected HeLa cells. Virology 116: 629-634 (1982).

Fernández-Tomás, C.B., Baltimore, D.: Morphogenesis of polio-virus. II. Demonstration of a new intermediate, the proviron. J. Virol. 12: 1122-1130 (1973a).

Fernández-Tomás, C.B., Guttmann, N., Baltimore, D.: Morphogenesis of polio-virus. III. Formation of provirion in cell-free extracts. J. Virol. 12: 1181-1183 (1973b).

Fillingame, R.: The proton-translocating pumps of oxidative phosphorylation. Ann. Rev. Biochem. 49: 1079-1114 (1980).

Finch, J.T., Klug, A.: Structure of poliomyelitis virus. Nature 183: 1709-1714 (1959).

Finch, J.T., Klug, A.: Arrangement of protein subunits and the distribution of nucleic acid in turnip yellow mosaic virus. II. Electron microscope studies. J. Mol. Biol. 15: 344-364 (1966).

Fiszman, M.Y., Bucchini, D., Girard, M., Lwoff, A.: Inhibition of poliovirus RNA synthesis by supraoptimal temperatures. J. Gen. Virol. 6: 293 (1970).

Fiszman, M., Reynier, M., Bucchini, D., Girard, M.: Thermosensitive block of the Sabin strain of poliovirus type I. J. Virol. 10: 1143-1151 (1972).
Fiszman, M., Reynier, M., Bucchini, D., Girard, M.: Retarded growth of poliovirus in contact inhibited cells. J. Gen. Virol. 23: 73-82 (1974).
Fitzgerald, M., Schenk, T.: The sequence 5'-AAUAAA-3' forms part of the recognition site for polyadenylation of late SV40 mRNAs. Cell 24: 251-260 (1981).
Flanagan, J.F.: Hydrolytic enzymes in KB cells infected with poliovirus and herpes simplex virus. J. Bact. 91: 789 (1966).
Flanegan, J.B., Baltimore, D.: Poliovirus-specific primer-dependent RNA polymerase able to copy poly(A). Proc. Natl. Acad. Sci. USA 74: 3677-3680 (1977).
Flanegan, J.B., Baltimore, D.: Poliovirus polyuridylic acid polymerase and RNA replicase have the same viral polypeptide. J. Virol. 29: 352-360 (1979).
Flanegan, J.B., Van Dyke, T.A.: Isolation of a soluble and template-dependent poliovirus RNA polymerase that copies virion RNA in vitro. J. Virol. 32: 155-161 (1979).
Flanegan, J.B., Pettersson, R.F., Ambros, V., Hewlett, M.J., Baltimore, D.: Covalent linkage of a protein to a defined nucleotide sequence at the 5'-terminus of virion and replicative intermediate RNAs of poliovirus. Proc. Natl. Acad. Sci. USA 74: 961-965 (1977).
Flexner, S., Lewis, P.A.: Experimental epidemic poliomyelitis in monkeys. Ascending nasal infection; characteristic alterations of the cerebrospinal fluid and its early infectivity; infection from human mesenteric lymph node. J. A. M. A. 54: 1140 (1910).
Flores-Otero, G., Fernández-Tomás, C., Gariglio-Vidal, P.: DNA-bound RNA polymerases during poliovirus infection: reduction in the number of form II enzyme molecules. Virology 116: 619-628 (1982).
Fòa-Tomasi, L., Campadelli-Fiume, G., Barbieri, L., Stripe, F.: Effect of ribosome-inactivating proteins on virus-infected cells. Inhibition of virus multiplication and of protein synthesis. Arch. Virol. 71: 323-332 (1982).
Fogel, A., Plotkin, S.A.: Genetic changes in attenuated poliovirus strains cultivated on human intestine in vitro. Amer. J. Epidemiol. 87: 385-395 (1969).
Fogh, J., Stuart, D.C., Jr.: Intracellular crystals of polioviruses in HeLa cells. Virology 11: 308-311 (1960a).
Fogh, J., Stuart, D.C., Jr.: Formation of polio and Coxsackie virus in FL cells. Fed. Proc. 19: 404 (1960b).
Folkman, J., Moscona, A.: Role of cell shape in growth control. Nature 273: 345 (1978).
Follett, E.A., Pringle, C.R., Pennington, T.H.: Virus development in enucleate cells: echovirus, poliovirus, pseudorabies virus, reovirus, respiratory syncytial virus and semliki forest virus. J. Gen. Virol. 26: 183-196 (1975).
Forchhammer, J., Nexo, B.A., Vaughan, H., Jr.: Cellular pools of viral proteins in 3T3 cells chronically infected with moloney-murine leukemia-virus. Virology 71: 134-142 (1976).
Fors, S., Schaller, H.: A tandem repeat gene in a picornavirus. Nucl. Acids Res. 10: 6441-50 (1982).
Foster, D.O., Pardee, A.B.: Transport of amino acids by confluent and nonconfluent 3T3 and polyoma virus transformed 3T3 cells growing on glass cover slips. J. Biol. Chem. 244: 2675 (1969).
Fraenkel-Conrat, H., Singer, B., Williams, R.C.: Infectivity of viral nucleic acid. Biochim. Biophys. Acta (Amst.) 25: 87 (1957).
Fraenkel-Conrat, H.: Descriptive Catalogue of Viruses „Viruses of Vertebrates and Insects". In: Comprehensive Virology (Fraenkel-Conrat, H., Wagner, R.R., eds.), Vol. 1, 38-41. Plenum Press 1974.
Franke, W.W., Schmid, E., Osborn, M., Weber, K.: Different intermediate-sized filaments distinguished by immunofluorescence microscopy. Proc. Natl. Acad. Sci. USA 75: 5034-5038 (1978).
Franklin, R.M., Baltimore, D.: Patterns of macromolecular synthesis in normal and virus-infected mammalian cells. Cold Spring Harbor Symp. Quant. Biol. 27: 175-198 (1962).
Friedmann, A., Lipton, H.L.: Replication of Theiler's murine encephalomyelitis viruses in BHK21 cells: An electron microscopy study. Virology 101: 389-398 (1980).
Friedman, R.M., Ramseur, J.M.: Mechanisms of persistent infections by cytopathic viruses in tissue culture. Arch. Virol. 60: 83-103 (1979).
Frisby, D.P., Eaton, M., Fellner, P.: Absence of the 5'-terminal capping group in encephalomyocarditis virus RNA. Nucl. Acids Res. 3: 2771-2788 (1976a).

Frisby, D.P., Newton, C., Carey, N.H., Fellner, P., Newman, J.F.E., Harris, T.J.R., Brown, F.: Oligonucleotide mapping of picornavirus RNAs by two dimensional electrophoresis. Virology 71: 379-388 (1976b).

Fujioka, R.S., Ackermann, W.W.: The inhibitory effects of $MgCl_2$ on the inactivation kinetics of poliovirus by urea. Proc. Soc. Exp. Biol. Med. 148: 1063-1069 (1975a).

Fujioka, R.S., Ackermann, W.W.: Evidence for conformational states of poliovirions: Effects of cations on reactivity of poliovirions to guanidine. Proc. Soc. Exp. Biol. Med. 148: 1070-1074 (1975b).

Fujioka, R., Kurtz, H., Ackermann, W.W.: Effects of cations and organic compounds on inactivation of poliovirus with urea, guanidine, and heat (34316). Proc. Soc. Exp. Biol. Med. 132: 825-829 (1969).

Fulton, A.B.: How do eucaryotic cells construct their cytoarchitecture? Cell 24: 4-5 (1981).

Fulton, A.B.: How crowded is the cytoplasm? Cell 30: 345-347 (1982).

Furesz, J., Armstrong, R.E., Moreau, P., Nagler, F.D.: Genetic markers of poliovirus strains isolated from paralytic patients prior to and after Sabin vaccination programs. II. Studies on type 3 strains. Am. J. Hyg. 80: 55-61 (1964).

Furesz, J., Armstrong, R.E., Contreras, G.: Viral and epidemiological links between poliomyelitis outbreaks in unprotected communities in Canada and the Netherlands. Lancet 2: 1248 (1978).

Furuichi, Y., Morgan, M.A., Shatkin, A.J.: Synthesis and translation of mRNA containing 5'-terminal 7-methylguanosine cap. J. Biol. Chem. 254: 6732-6738 (1979).

Gallwitz, D., Traub, U., Traub, P.: Fate of histone messenger RNA in mengovirus infected Ehrlich ascites tumor cells. Eur. J. Biochem. 81: 387-393 (1977).

Gard, S.: Immunological strain specificity within type 1 polioviruses. Bull. WHO 22: 235-242 (1960).

Garfinkle, B.D., Tershak, D.R.: Effect of temperature on the cleavage of polypeptides during growth of LSc poliovirus. J. Mol. Biol. 59: 537-541 (1971).

Garfinkle, B.D., Tershak, D.R.: Degradation of poliovirus polypeptides in vivo. Nature 238: 206-208 (1972).

Garoff, H., Schwarz, R.T.: Glycosylation is not necessary for membrane insertion and cleavage of Semliki Forest virus membrane proteins. Nature 274: 487-490 (1978).

Garry, R.F., Bishop, J.M., Parker, S., Westbrook, K., Lewis, G., Waite, M.R.: Na^+ and K^+ concentrations and the regulation of protein synthesis in Sindbis virus-infected chick cells. Virology 96: 108-120 (1979).

Garwes, D.J., Wright, P.J., Cooper, P.D.: Poliovirus temperature-sensitive mutants defective in cytopathic effects are also defective in synthesis of double-stranded RNA. J. Gen. Virol. 27: 45-59 (1975).

Geck, P., Pietrzyk, C., Burckhardt, B.C., Pfeiffer, B., Heinz, E.: Electrically silent cotransport of Na^+, K^+, and Cl^- in Ehrlich cells. Biochim. Biophys. Acta 600: 432-447 (1980).

Geoghegan, T., Cereghini, S., Brawerman, G.: Inactive mRNA-protein complexes from mouse sarcoma-180 ascites-cells. Proc. Natl. Acad. Sci. USA 76: 5587-5591 (1979).

Ghendon, Y.Z.: Mutations of virulent and attenuated poliovirus strains induced by nitrous acid. Acta Virol. 7: 16-24 (1963).

Ghendon, Y.Z.: Non-complementing rct_{40} - mutant of poliovirus. Acta. Virol. 10: 173 (1966).

Ghendon, Y.Z.: Conditional lethal mutants of animal viruses. Prog. Med. Virol. 14: 68-122 (1972).

Ghendon, Y.Z., Yakobson, E.A.: Antigenic specificity of poliovirus-related particles. J. Virol. 8: 589-590 (1971).

Ghendon, Y.Z., Yakobson, E.A., Mikhejeva, A.: Study of some stages of poliovirus morphogenesis in MiO cells. J. Virol. 10: 261-266 (1972).

Ghendon, Y.Z., Babushkina, L., Blagoveshienskaya, O.: Inhibition of poliovirus RNA synthesis in an in vitro system by antiserum against 14S virus-specific structures. Arch. ges. Virusforsch. 40: 47-51 (1973).

Gibbs, A.J., Harrison, B.D., Watson, D.H., Wildy, P.: What's in a virus name? Nature (London) 209: 450 (1966).

Gierer, A., Schramm, G.: Die Infektivität der Nucleinsäure aus Tabakmosaikvirus. Z. Naturforsch. 11b: 138 (1956).

Giloh, H., Mager, J.: Inhibition of peptide chain initiation in lysates from ATP depleted cells. I. Stages in the evolution of the lesions and its reversal by thiol compounds, cyclic AMP, or purine derivatives and phosphorylated sugars. Biochim. Biophys. Acta 414: 293–308 (1975).

Girard, M.: In vitro synthesis of poliovirus ribonucleic acid: Role of the replicative intermediate. J. Virol. 3: 376 (1969).

Girard, M., Baltimore, D., Darnell, J.E.: The poliovirus replication complex: Site for synthesis of poliovirus RNA. J. Mol. Biol. 24: 59-74 (1967).

Girard, M., Benichou, D., van der Werf, S., Kopecka, H.: Analysis by electron microscopy of recombinant plasmids carrying poliovirus type-1 cDNA inserts. Ann. Virol. (Inst. Pasteur) 132: 271-285 (1981).

Giron, M.L., Logeat, F., Hanania, N., Fossar, N., Huppert, J.: Size of the poly (A) sequences in encephalomyocarditis virus RNA. Intervirology 6: 367-371 (1975).

Goldstein, J., Champe, S.P.: T4-induced activity required for specific cleavage of a bacteriophage protein in vitro. J. Virol. 13: 419-427 (1974).

Goldstein, J., McCollough, J.E., Champe, S.P.: Identification of gene products required for in vitro formation of the internal peptides of bacteriophage T4. J. Virol. 18: 894-903 (1976).

Goldstein, N.O., Pardoe, I.U., Burness, A.T.H.: Requirement of an adenylic acid rich segment for the infectivity of EMC RNA. J. Gen. Virol. 31: 271-278 (1976).

Golini, F., Thach, S., Lawrence, C., Thach, R.: Regulation of protein synthesis in EMC virus-infected cells. In: Animal Virology, IV. ICN-ULCA Symposia on Molecular and Cellular Biology (Baltimore, D., Huang, A. S., Fox, D. F., eds.), 717-734. New York: Academic Press 1976a.

Golini, F., Thach, S.S., Birge, C.H., Safer, B., Merrick, W.C., Thach, R.E.: Competition between cellular and viral mRNAs in vitro is regulated by a messenger discriminatory initiation factor. Proc. Natl. Acad. Sci. USA 73: 3040-3044 (1976b).

Golini, F., Nomoto, A., Wimmer, E.: The genome-linked protein of picornaviruses. IV: Difference in the VPgs of encephalomyocarditis virus and poliovirus as evidence that the genome-linked proteins are virus-coded. Virology 89: 112-118 (1978).

Golini, F., Semler, B.L., Dorner, A.J., Wimmer, E.: Protein-linked RNA of poliovirus is competent to form an initiation complex of translation in vitro. Nature 287: 600-603 (1980).

Gomperts, B.D.: The Plasma Membrane: Models for Its Structure and Function. New York: Academic Press 1976.

Gorbalenya, A.E., Svitkin, Y.V., Kazachkov, Y.A., Agol, V.I.: Encephalomyocarditis virus-specific polypeptide p22 is involved in the processing of the viral precursor polypeptides. FEBS Lett. 108: 1-5 (1979).

Grado, C., Fischer, S., Contreras, G.: The inhibition by actinomycin D of poliovirus multiplication in HEp 2 cells. Virology 27: 623 (1965).

Grado, C., Friedlender, B., Ihl, M., Contreras, G.: Incorporation of methyl groups by viral and cellular RNA of HEp 2 cells after poliovirus infection. Virology 35: 339 (1968).

Graham, A.F.: Symposium on the Biology of cells modified by viruses or antigens: III. Physiological conditions for studies of viral biosynthesis in mammalian cells. Bacteriol. Rev. 23: 224 (1959).

Granboulan, N., Franklin, R.M.: Replication of bacteriophage ribonucleic acid: analysis of the ultrastructure of the replicative form and the replicative intermediate of bacteriophage R 17. J. Virol. 2: 129-148 (1968).

Granboulan, N., Girard, M.: Molecular weight of poliovirus RNA. J. Virol. 4: 475-479 (1969).

Grantham, R., Gautier, C., Gouy, M., Mercier, R., Pavé, A.: Codon catalog usage and the genome hypothesis. Nucl. Acids Res. 11: 49-62 (1980).

Griffin, M.J., Ber, R.: Cell cycle events in the hydrocortisone regulation of alkaline phosphatase in HeLa S3 cells. J. Cell Biol. 40: 297-304 (1969).

Griffith, J., Manning, M., Dunn, K.: Filamentous bacteriophages contract into hollow spherical particles upon exposure to a chloroform-water interface. Cell 23: 747-753 (1981).

Grimmel, M., Zibirre, R., Koch, G.: Fluorescence spectrophotometric study on structural alterations in the capsid of poliovirus. Arch. Virol. 78: 191-201 (1983).

Gross, P.R., Cousineau, G.H.: Macromolecule synthesis and the influence of actinomycin on early development. Exp. Cell Res. 33: 368-395 (1964).

Grulee, C.G. Jr., Eldridge, F.L., Ford, G.D.: A clinical study of phenosulfazole (darvisul) in acute poliomyelitis. Tex. Rep. Biol. Med. 8 (1950).

Grunstein, M., Hogness, D.: Colony hybridization: a method for the isolation of cloned DNAs that contain a specific gene. Proc. Natl. Acad. Sci. USA 72: 3961-3965 (1975).

Gschwender, H.H., Traub, P.: Mengovirus-induced capping of virus receptors on the plasma membrane of Ehrlich ascites tumor cells. J. Gen. Virol. 42: 439-442 (1979).

Gurd, F.R.N., Rothberg, T.M.: Motions in proteins. Adv. Protein Chem. 33: 73-165 (1979).

Guskey, L.E., Wolff, D.A.: The effects of actinomycin D on the replication and development of CPE in HEp-2 cells. J. Virol. 14: 1229-1234 (1971).

Guskey, L.E., Smith, P.C., Wolff, D.A.: Patterns of cytopathology and lysosomal enzyme release in polioinfected HEp-2 cells treated with either 2-(α-hydroxybenzyl)-benzimidazole or guanidine HCl. J. Gen. Virol. 6: 151-161 (1970).

Guttman, N., Baltimore, D.: Morphogenesis of poliovirus. IV. Existence of particles sedimenting at 150S and having the properties of provirion. J. Virol. 23: 363-367 (1977a).

Guttman, N., Baltimore, D.: A plasma membrane component able to bind and alter virions of poliovirus type 1: studies on cell free alteration using a simplified assay. Virology 82: 25-36 (1977b).

Habel, K., Hornibrook, J.W., Gregg, N.C., Silverberg, R.J., Takemoto, K.K.: The effect of anticellular sera on virus multiplication in tissue culture. Virology 5: 7-29 (1958).

Habermehl, K.O., Diefenthal, W., Buchholz, M.:Distribution of parental viral constituents in the course of polioinfection. In: Advances in the Biosciences (Raspé, G, ed.), Vol. 11, 41-63. Oxford: Pergamon Press/Vieweg 1974.

Hagenbüchle, O., Santer, M., Steitz, J.A., Mans, R.J.: Conservation of the primary structure at the 3'end of 18S rRNA from eucaryotic cells. Cell 13: 551-563 (1978).

Hahn, E.E.A.: Polioviruses. In: Strains of Human Viruses (Majer, M., Plotkin, S.A., eds.), 155-176. Karger 1972.

Hahn, E., Fogh, J.: Increased resistance of SV40 transformed human amnion cells to poliovirus infection. Arch. ges. Virusforsch. 29: 343-360 (1970).

Hahnemann, F., Siboni, K., von Magnus, H.: Serodifferentiation of type 1 polioviruses. Proceedings of the IXth Symposium of the European Association against Poliomyelitis and Allied Diseases, Stockholm, 306-310 (1963).

Hall, S., Crouch, R.J.: Isolation and characterizatin of two enzymatic activities from chicken embryo which degrade double-stranded RNA. J. Biol. Chem. 252: 4092-4097 (1977).

Hall, T.C.: Transfer RNA-like structures in viral genomes. Int. Rev. Cytol. 60: 1-26 (1979).

Halperen, S., Eggers, H.J., Tamm, L.: Evidence for uncoupled synthesis of viral RNA and viral capsids. Virology 24: 36-46 (1964).

Hamann, A., Wiegers, K.J., Drzeniek, R.: Isoelectric focusing and 2D-analysis of poliovirus proteins. Virology 78: 359-362 (1977).

Hamann, A., Reichel, C., Wiegers, K.J., Drzeniek, R.: Isoelectric points of polypeptides of standard poliovirus particles of different serological types and of empty capsids and dense particles of poliovirus type 1. J. Gen. Virol. 38: 567-570 (1978).

Hamlyn, P.H., Brownlee, G.G., Cheng, C.-C., Gait, M.J., Milstein, C.: Complete sequence of constant and 3' noncoding regions of an immunoglobulin mRNA using the dideoxynucleotide method of RNA sequencing. Cell 15: 1067-1075 (1978).

Hanecak, R., Semler, B.L., Anderson, C.W., Wimmer, E: Proteolytic processing of poliovirus polypeptides: antibodies to polypeptide P3-7c inhibit cleavage at glutamine-glycine pairs. Proc. Natl. Acad. Sci. USA 79: 3973-3977 (1982).

Hansen, J., Ehrenfeld, E.: Presence of the cap-binding protein in initiation factor preparations from poliovirus-infected HeLa cells. J. Virol. 38: 438-445 (1981).

Hansen, J., Etchison, D., Hershey, J.W.B., Ehrenfeld, E.: Association of cap-binding protein with eucaryotic initiation factor 3 in initiation factor preparations from uninfected and poliovirus-infected HeLa cells. J. Virol. 42: 200-207 (1982).

Hara, M., Saito, Y, Komatsu, T, Kodama H., Abo, W., Chiba, S., Nakao, T.: Antigenic analysis of polioviruses isolated from a child with a gammaglobulinemia and paralytic poliomyelitis after Sabin vaccine administration. Microbiol. Immunol. 25: 905-913 (1981).

Harbitz, F., Scheel, O.: Pathologisch-anatomische Untersuchungen über akute Poliomyelitis und verwandte Krankheiten. J. Dybwad, Christiania (1907).

Harris, T.J.R.: The nucleotide sequence at the 5' end of foot-and-mouth disease virus RNA. Nucl. Acids Res. 7: 1765-1785 (1979).

Harris, T.J.R., Dunn, J.J., Wimmer, E.: Identification of specific fragments containing the 5' end of poliovirus RNA after ribonuclease III digestion. Nucl. Acids Res. 5: 4039-4054 (1978).

Harrison, S., Olson, A.J., Schutt, E., Winkler, F.K., Bricogne, G.: Tomato bushy stunt virus at 2.8 A resolution. Nature 276: 368-373 (1978).

Harrison, S.C.: Protein interfaces and intersubunit bonding. The case of tomato bushy stunt virus. Biophys. J. 32: 139-153 (1980).

Harter, D.H., Choppin. P.W.: Adsorption of attenuated and neurovirulent poliovirus strains to central nervous system tissues of primates. J. Immunol. 95: 730-736 (1965).

Hashimoto, N.: Analysis of specificity of poliovirus inhibitors with inhibitor-resistant mutants of a strain of type 1 poliovirus. J. Immunol. 115: 569-574 (1975).

Hassell, J., Engelhardt, D.: The regulation of protein synthesis in animal cells by serum factors. Biochem. 15: 1375-1380 (1976).

Havliza, D., Koch, G.: Complex formation between poliovirus RNA and polycations. Arch. Biochem. Biophys. 147: 85-91 (1971).

Hecht, T.T., Summers, D.F.: The effect of phleomycin on poliovirus RNA replication. Virology 40: 441-447 (1970).

Heding, L.D., Wolff, D.A.: The cytochemical examination of poliovirus-induced cell damage. J. Cell Biol. 59: 530-536 (1973).

Heggeness, M.H., Wang, K., Singer, S.J.: Intracellular distribution of mechanochemical proteins in cultured fibroblasts. Proc. Natl. Acad. Sci. USA 74: 3883-3887 (1977).

Heine, J.: Beobachtungen über Lähmungszustände der unteren Extremitäten und deren Behandlung. Stuttgart: Köhler 1840.

Helenius, A., Simons, K.: Solubilization of membranes by detergents. Biochim. Biophys. Acta 415: 29 (1975).

Helenius, A., Kartenbeck, J., Simons, K., Fries, E.: On the entry of Semliki Forest virus into BHK-21 cells. J. Cell Biol. 84: 404-420 (1980).

Helentjaris, T., Ehrenfeld, E.: Inhibition of host cell protein synthesis by UV-inactivated poliovirus. J. Virol. 21: 259-267 (1977).

Helentjaris, T., Ehrenfeld, E.: Control of protein synthesis in extracts from poliovirus-infected cells. I. mRNA discrimination by crude initiation factors. J. Virol. 26: 510-521 (1978).

Helentjaris, T.G., Ehrenfeld, E., Brown-Lueidi, M.L., Hershey, J.W.B.:Alterations in initiation factor activity from poliovirus-infected HeLa-cells. J. Biol. Chem. 254: 10973-10978 (1979).

Hershey, A.D., Chase, M.: Independent functions of viral protein and nucleic acid in growth of bacteriophage. J. Gen. Physiol. 36: 39-56 (1952).

Hershey, J.W.B.: The initiation factors. In: Protein Biosynthesis in Eukaryotes (Perez-Bercoff, R., ed.), 97-117. New York: Plenum Press 1982a.

Hershey, J.W.B.: Messenger ribonucleoprotein particles. In: Protein Biosynthesis in Eukaryotes (Pérez-Bercoff, R., ed.), 157-166. New York: Plenum Press 1982b.

Hershko, A., Fry, M.: Posttranslational cleavage of polypeptide chains: Role in assembly. Ann. Rev. Biochem. 44: 775-797 (1975).

Hershko, A., Mamont, P., Shields, R., Tomkins, G.M.: Pleiotropic response. Nature 232: 206-211 (1971).

Hewlett, M.J., Florkiewicz, R.Z.: Sequence of picornavirus RNAs containing a radioiodinated 5'-linked peptide reveals a conserved 5' sequence. Proc. Natl. Acad. Sci. USA 77: 303-307 (1980).

Hewlett, M.J., Rose, J.K., Baltimore, D.: 5'-terminal structure of poliovirus polyribosomal RNA is pUp. Proc. Natl. Acad. Sci. USA 73: 327-330 (1976).

Hewlett, M.J., Rosenblatt, S., Ambros, V., Baltimore, D.: Separation and quantitation of intracellular forms of poliovirus RNA by agarose gel electrophoresis. Biochem. N. Y. 16: 2763-2767 (1977).

Hewlett, M.J., Axelrod, J.H., Antinoro, N., Feld, R.: Isolation and preliminary characterization of temperature-sensitive mutants of poliovirus type 1. J. Virol 41: 1089-1094 (1982).

Hickey, E.D., Weber, L.A., Baglioni, C.: Inhibition of initiation of protein synthesis by 7-methylguanosine-5'-monophosphate. Proc. Natl. Acad. Sci. USA 73: 19-23 (1976).

Hiller, R., Schaefer, A., Zibirre, R., Kaback, H.R., Koch, G.: Factors influencing the accumulation of tetraphenylphosphonium cation in HeLa cells. Mol. Cell. Biol. 4: 199-202 (1984).

Hinuma, Y., Katagiri, S., Fukuda, M., Fukushi, K., Watanabe, Y.: Kinetic studies on the thermal degradation of purified poliovirus. Biken Journal 8: 143 (1965).
Hinuma, Y., Katagiri, S., Aikawa, S.: Immune response to H particles of poliovirus. Virology 40: 773-776 (1970).
Hirsimäki, Y., Arstila, A.U., Trump, B.F.: Autophagocytosis: In vitro induction by microtubule poisons. Exp. Cell Res. 91: 11-14 (1975).
Hirst, G.K.: Genetic recombination with Newcastle disease virus, polioviruses and influenza. Cold Spring Harbor Symp. Quant. Biol. 27: 303-308 (1962).
Hirst, H.G.: Reversible inactivation of poliovirus through exposure to sulfhydryl compounds. 5th International Poliomyelitis Conference, Copenhagen, 1415 (1961).
Hodes, H.L., Zepp, H.D., Ainbender, E.: A physical property as a virus marker. Difference in avidity of cellulose resin for virulent (Mahoney) and attenuated (LSc, 2ab) strain of type 1 poliovirus. Virology 11: 306-308 (1960).
Hoey, E.M., Martin, S.J.: A comparison of urea and alkali degradation of full and empty bovine enterovirus particles. Microbios 10: 45-62 (1974a).
Hoey, E.M., Martin, S.J.: A possible precursor containing RNA of a bovine enterovirus: the provirion. II. J. Gen. Virol. 24: 515-524 (1974b).
Hogle, J.M.: Preliminary studies of crystals of poliovirus type I. J. Mol. Biol. 160: 663-668 (1982).
Hogle, J.M.: High-resolution structural studies of poliovirus. In: Modern Approaches to Vaccines (Chanock, R.M., Lerner, R.A., eds.), 7-11. Cold Spring Harbor Laboratory 1984.
Hogue, N.J., Mcallister, R., Green, A.S., Coriell, L.L.: The effect of poliomyelitis virus on human brain cells in tissue culture. J. Exp. Med. 102: 29 (1955).
Holland, J.J.: Receptor affinities as major determinants of enterovirus tissue tropisms in humans. Virology 15: 312-326 (1961).
Holland, J.J.: Irreversible eclipse of poliovirus by HeLa cells. Virology 16: 163-176 (1962a).
Holland, J.J.: Inhibition of DNA-primed RNA synthesis during poliovirus infection of human cells. Biochem. Biophys. Res. Commun. 9: 556-562 (1962b).
Holland, J.J.: Depression of host-controlled RNA synthesis in human cells during poliovirus infection. Proc. Natl. Acad. Sci. USA 49: 23-28 (1963).
Holland, J.J.: Enterovirus entrance into specific host cells and subsequent alterations of cell protein and nucleic acid synthesis. Bacteriol. Rev. 28: 3 (1964a).
Holland, J.J.: Inhibition of host cell macromolecular synthesis by high multiplicities of poliovirus under conditions preventing virus synthesis. J. Mol. Biol. 8: 574-581 (1964b).
Holland, J.J., Cords, C.E.: Maturation of poliovirus RNA with capsid protein coded by heterologous enteroviruses. Proc. Natl. Acad. Sci. USA 51: 1082-1085 (1964).
Holland, J.J., Hoyer, B.H.: Early stages of enterovirus infection. Cold Spring Harbor Symp. Quant. Biol. 27: 101-111 (1962).
Holland, J.J., Kiehn, E.D.: Specific cleavage of viral proteins as steps in the synthesis and maturation of enteroviruses. Proc. Natl. Acad. Sci. USA 60: 1015-1022 (1968).
Holland, J.J., McLaren, L.C.: The mammalian cell-virus relationship. II. Adsorption, reception, eclipse of poliovirus by HeLa cells. J. Exp. Med. 109: 487-504 (1959).
Holland, J.J., McLaren, L.C.: The location and nature of enterovirus receptors in susceptible cells. J. Exp. Med. 114: 161-171 (1961).
Holland, J.J., Peterson, J.A.: Nucleic acid and protein synthesis during poliovirus infection of human cells. J. Mol. Biol. 8: 556-573 (1964).
Holland, J.J., McLaren, L.C., Syverton, J.T.: Mammalian cell-virus relationship. III. Poliovirus production by non-primate cells exposed to poliovirus ribonucleic acid. Proc. Soc. Exp. Biol. Med. 100: 843 (1959a).
Holland, J.J., McLaren, L.C., Sysverton, J.T.: The mammalian cell-virus relationship. IV. Infection of naturally insusceptible cells with enterovirus ribonucleic acid. J. Exp. Med. 110: 65 (1959b).
Holland, J.J., McLaren, L.C., Hoyer, B.H., Syverton, J.T.: Enteroviral ribonucleic acid. II. Biological, physical and chemical studies. J. Exp. Med. 112: 841-864 (1960).
Holley, R.W.: A uniform hypothesis concerning the nature of malignant growth. Proc. Natl. Acad. Sci. USA 69: 2840 (1972).
Hollinshead, A.C.: Differences in chromatographic behavior on cellulose columns of virulent Maho-

ney and attenuated LSc strains of poliovirus and similarities of base ratios of their nucleic acids. Medna. exp. 2: 303-309 (1960).

Holzer, H., Wohlhueter, R.: (Glutamine synthetase) tyrosyl-O-adenylate: a new energy-rich phosphate bond. Adv. Enzyme Reg. 10: 121-132 (1972).

Homma, M.: Trypsin action on the growth of Sendai virus in tissue culture cells. I. Restoration of the infectivity for L cells by direct action of trypsin on L cell-borne Sendai virus. J. Virol. 8: 619-629 (1971).

Hordern, J.S., Leonard, J.D., Scraba, D.G.: Structure of the mengo virion. VI. Spatial relationship of the capsid polypeptides as determined by chemical crosslinking analysis. Virology 97: 131-140 (1979).

Horne, R.W.: The structure of viruses. Readings from Scientific American. The Molecular Basis of Life (1963).

Horne, R.W., Nagington, J.: Electron microscope studies of the development and structure of poliomyelitis virus. J. Mol. Biol. 1: 333-338 (1959).

Hotchkiss, R.D.: Towards a general theory of recombination in DNA. Adv. Genet. 16: 325 (1971).

Hövel, H.: Einfluß der Mediumosmolarität und Ionenstärke auf die Virusreproduktion in diploiden und heteroploiden Zellkulturen. Arch. ges. Virusforsch. 43: 200-212 (1973).

Howes, D.W.: The growth cycle of poliovirus in cultured cells. II. Maturation and release of virus in suspended cell populations. Virology 9: 96-109 (1959a).

Howes, D.W.: The growth cycle of poliovirus in cultured cells. III. The asynchronous response of HeLa cells multiply infected with type 1 poliovirus. Virology 9: 110-126 (1959b).

Howes, D.W., Melnick, J.L.: The growth cycle of poliovirus in monkey kidney cells. I. Maturation and release of virus in monolayer cultures. Virology 4: 97-108 (1957).

Hoyer, B., Bolton, E., Ritter, D., Ribi, E.: Simple method for preparation of purified radioactive poliovirus particles: electron micrograph. Virology 7: 462-464 (1959).

Hruby, D.E.: Effect of polyadenylic acid on the functional half-life of encephalomyocarditis virus RNA during translation. Biochem. Biophys. Res. Commun. 81: 1425-1434 (1978).

Hruby, D.E., Roberts, W.K.: Variations in polyadenylic acid content and biological activity. J. Virol. 19: 325-330 (1976).

Hruby, D.E., Roberts, W.K.: Encephalomyocarditis virus RNA. II. Polyadenylic acid requirement for efficient translation. J. Virol. 23: 338-344 (1977).

Hruby, D.E., Roberts, W.K.: Encephalomyocarditis virus RNA III. Presence of a genome-associated protein. J. Virol. 25: 413-415 (1978).

Hsiung, G.D., Melnick, J.L.: Adsorption, multiplication and cytopathogenicity of enteroviruses (poliomyelitis, coxsackie and ECHO groups) in susceptible and resistant monkey kidney cells. J. Immunol. 80: 45 (1958a).

Hsiung, G.D., Melnick, J.L.: Effect of sodium bicarbonate on plaque formation of virulent and attenuated polioviruses. J. Immunol. 80: 282-293 (1958b).

Huang, A.S., Baltimore, D.: Initiation of polyribosome formation in poliovirus-infected HeLa cells. J. Mol. Biol. 47: 275-291 (1970).

Huang, A., Baltimore, D.: Defective viral particles and viral disease processes. Nature 226: 325-327 (1976).

Hugentobler, A.L., Bienz, K.: Influence of poliovirus infection on S-phase and mitosis of the host cell. Arch. Virol. 64: 25-33 (1980).

Hull, R.: The grouping of small spherical plant viruses with single RNA components. J. Gen. Virol. 36: 289-295 (1977).

Hummeler, K., Hamparian, V.V.: Studies on complement fixing antigens of poliomyelitis. I. Demonstration of type and group-specific antigens in native and heated viral preparations. J. Immunol. 81: 499-505 (1958).

Hummeler, K., Tumilowicz, J.J.: Studies on the complement-fixing antigens of poliomyelitis. II. Preparation of type-specific anti-N and anti-H indicator sera. J. Immunol. 84: 630-634 (1960).

Hummeler, K.T., Anderson, T.F., Brown, R.A.: Identification of poliovirus particles of different antigenicity by specific agglutination as seen in the electron microscope. Virology 16: 84-90 (1962).

Humphrey, D.D., Kew, O.M., Feorino, P.M.: Monoclonal antibodies of four different specificities for neutralization of type 1 polioviruses. Infect. Immun. 36: 841-843 (1982).

Hunt, L.A.: In vitro translation of encephalomyocarditis viral RNA: Synthesis of capsid precursor-like polypeptides. Virology 70: 484-492 (1976).

Hunt, T., Ehrenfeld, E.: Cytoplasm from poliovirus-infected HeLa cells inhibits cell-free hemoglobin synthesis. Nature 230: 91-94 (1971).

Hunter, T.: Proteins phosphorylated by the RSV transforming function. Cell 22: 647-648 (1980).

Hunter, T.R., Hunt, R., Knowland,J., Zimmern, D.: Messenger RNA for the coat protein of tobacco mosaic virus. Nature 260: 759-764 (1976).

Icenogle, J., Gilbert, S.F., Grieves, J., Anderegg, J., Rueckert, R.: A neutralizing monoclonal antibody against poliovirus, its reaction with virus related antigens. Virology 115: 211-215 (1981).

Icenogle, J., Shiwen, H., Duke, G., Gilbert, S., Rueckert, R., Anderegg, J.: Neutralization of poliovirus by a monoclonal antibody: Kinetics and stoichiometry. Virology 127: 412 (1983).

Ikegami, N., Eggers, H.J., Tamm, I.: Rescue of drug-requiring and drug-inhibited enteroviruses. Proc. Natl. Acad. Sci. USA 52: 1419-1426 (1964).

Impraim, C.C., Foster, K.A., Micklem, K.J., Pasternak, C.A.: Nature of virally mediated changes in membrane permeability to small molecules. Biochem. J. 186: 847-860 (1980).

Isselbacher, K.J.: Increased uptake of amino acids and 2-deoxy-D-glucose by virus-transformed cells in culture. Proc. Natl. Acad. Sci. USA 69: 585 (1972).

Iwanowski, D.: Über die Mosaikkrankheit der Tabakspflanze. Bull. Acad. Imp. Sci. de St. Petersbourg, n. s. 3: 67-70 (1892).

Jackson, R.C., Blobel, G.: Post-translational cleavage of presecretory proteins with an extract of rough microsomes from dog pancreas containing signal peptidase activity. Proc. Natl. Acad. Sci. USA 74: 5598-5602 (1977).

Jackson, R.J.: The mechanism and cytoplasmic control of mammalian protein synthesis. In: The Molecular Biology of Picornaviruses (Pérez-Bercoff, R., ed.), 191-222. New York - London: Plenum Press 1979.

Jackson, R.J.: The control of initiation of protein synthesis in reticulocyte lysates. In: Protein Biosynthesis in Eukaryotes (Perez-Bercoff, R., ed.), 363-418. New York: Plenum Press 1982.

Jackson, R.J., Hunt, T.: Mechanisms on control of polypeptide-chain initiation in reticulocytes. Biochem. Soc. Trans. 8: 457-458 (1980).

Jacob, S.W., Bischel, M., Herschler, J.L.: Dimethylsulfoxide: Effects on the permeability of biologic membranes. Curr. Ther. Res. 6: 134-135 (1964).

Jacobson, M.F., Asso, J., Baltimore, D.: Further evidence on the formation of poliovirus proteins. J. Mol. Biol. 49: 657-669 (1970).

Jacobson, M.F., Baltimore, D.: Polypeptide cleavages in the formation of poliovirus proteins. Proc. Natl. Acad. Sci. USA 61: 77-84 (1968b).

Jacobson, M.F., Baltimore, D.: Morphogenesis of poliovirus. I. Association of viral RNA with coat protein. J. Mol. Biol. 33: 369-378 (1968a).

James, L.A., Thershak, D.R.: Protein phosphorylations in poliovirus infected cells. Can. J. Microbiol. 27: 28-35 (1981).

Jamieson, J.D., Palade, G.E.: Intracellular transport of secretory proteins in the pancreatic exocrine cell. J. Cell Biol. 34: 577-615 (1967).

Jamison, R.M.: Morphology of echovirus 22. J. Virol. 4: 904-906 (1969).

Jamison, R.M., Mayor, H.D.: Comparative study of seven picornaviruses of man. J. Bact. 91: 1971-1976 (1966).

Jen, G., Thach, R.E.: Inhibition of host translation in encephalomyocarditis virus-infected L cells: a novel mechanism. J. Virol. 43: 250-261 (1982).

Jen, G., Detjen, B.M., Thach, R.E.: Shut-off of HeLa cell protein synthesis by encephalomyocarditis virus and poliovirus, a comparative study. J. Virol. 35: 150-156 (1980).

Jense, H., Knauert, F., Ehrenfeld, E.: Two initiation sites for translation of poliovirus RNA in vitro: Comparison of LSc and Mahoney strains. J. Virol. 28: 387-394 (1978).

Johnson, G.L., Kaslow, H.R., Farfel, Z., Bourne, H.R.: Genetic analysis of hormone-sensitive adenylate cyclase. In: Advances in Cyclic Nucleotide Research (Greengard, P., Robison, G.A., eds.), Vol. 13, 1-38. New York: Raven 1980.

Johnson, T.C., Holland, J.J.: Ribonucleic acid and protein synthesis in mitotic HeLa cells. J. Cell Biol. 27: 565-574 (1965).
Johnston, M.D., Martin, S.J.: Capsid and procapsid proteins of a bovine enterovirus. J. Gen. Virol. 11: 71-79 (1971).
Johnston, R.E., Bose, H.R.: Correlation of messenger RNA function with adenylate-rich segments in the genomes of single-stranded RNA viruses. Proc. Natl. Acad. Sci. USA 69: 1514-1516 (1972).
Joklik, W.K., J.E. Darnell: The adsorption and early fate of purified poliovirus in HeLa cells. Virology 13: 439-447 (1961).
Jordan, L., Mayor, H.D.: Studies on the degradation of poliovirus by heat. Microbios 9: 51-60 (1974).
Jorgensen, A.O., Subrahmanyan, L., Turnbull, C., Kalnins, V.I.: Localization of the neurofilament protein in neuroblastoma cells by immunofluorescent staining. Proc. Natl. Acad. Sci. USA 73: 3192-3196 (1976).
Juliano, R.L.: Techniques for the analysis of membrane glycoproteins. Curr. Top. Memb. Transp. 11: 107-144 (1978).

Kabat, D., Chappell, M.: Competition between globin messenger ribonucleic acids for a discrimination initiation factor. J. Biol. Chem. 252: 2684-2690 (1977).
Kacian, D.L., Myers, J.C.: Synthesis of extensive, possibly complete DNA copies of poliovirus RNA in high yields and at high specific activities. Proc. Natl. Acad. Sci. USA 73: 2191-2195 (1976a).
Kacian, D.L., Myers, J.C.: Anticomplementary nature of smaller DNA produced during synthesis of extensive DNA copies of poliovirus RNA. Proc. Natl. Acad. Sci. USA 73: 3408-3412 (1976b).
Kaempfer, R.: Initiation factor/mRNA interactions and mRNA recognition. In: Protein Synthesis in Eukaryotes (Pérez-Bercoff, P., ed.), 223-243. New York: Plenum Press 1982a.
Kaempfer, R.: Messenger RNA competition. In: Protein Synthesis in Eukaryotes (Pérez-Bercoff, P., ed.), 441-457. New York: Plenum Press 1982b.
Kaempfer, R., Hollender, R., Abrams, W.R., Israeli, R.: Specific binding of messenger RNA and methionyl-tRNA-fMet by the same initiation factor for eukaryotic protein synthesis. Proc. Natl. Acad. Sci. USA 75: 209-213 (1978).
Kahn, C.R.: Membrane receptors for hormones and neurotransmitters. J. Cell Biol. 70: 261-286 (1976).
Kallmann, F., Williams, R.C., Dulbecco, R., Vogt, M.: Fine structure of changes produced in cultured cells sampled at specified intervals during a single growth cycle of poliovirus. J. Biophys. Biochem. Cytol. 4: 301-308 (1958).
Kalvelage, B., Koch, G.: Studies on HeLa cell surface glycopeptide alterations in membrane structure during the cell cycle. Eur. J. Cell Biol. 28: 233-237 (1982a).
Kalvelage, B., Koch, G.: Inhibition of mitogenic stimulation of human leukocytes by protease released membrane glycopeptides. Eur. J. Cell Biol. 28: 238-242 (1982b).
Kamen, R.I.: Structure and function of Qß RNA replicase. In: RNA Phages (Zinder, N., ed.), 203-234. Cold Spring Harbor Laboratory 1975.
Kanamitsu, M., Hashimoto, N., Urasawa, S., Chiba, S.: Studies on poliovirus inhibitors in sera of domestic animals. I. Distributions and properties of poliovirus inhibitors in bovine and equine sera. Jap. J. Med. Sci. Biol. 20: 471-482 (1967).
Kanda, Y., Melnick, J.L.: In vitro differentiation of virulent and attenuated polioviruses by their growth characteristics on MS cells. J. Exp. Med. 109: 9-24 (1959).
Kaper, J.M.: The chemical basis of virus structure, dissociation and reassembly. In: Frontiers of Biology 39 (Neuberger, Tatum, eds.), Amsterdam: North-Holland 1975.
Kaplan, A.S.: The susceptibility of monkey kidney cells to poliovirus in vivo and in vitro. Virology 1: 377-392 (1955).
Kaplan, J.G., Pasternak, C.A.: Fourth conference on macromolecular synthesis: membrane mediated controls and virus induced membrane alterations. Eur. J. Cell Biol. 29: 288-289 (1983).
Kassell, B., Kay, J.: Zymogens of proteolytic enzymes. Science 180: 1022-1027 (1973).
Katagiri, S, Aikawa, S., Hinuma, Y.: Stepwise degradation of poliovirus capsid by alkaline treatment. J. Gen. Virol. 13: 101-109 (1971).
Katagiri, S., Hinuma, Y., Ishida, N.: Biophysical properties of poliovirus particles irradiated with ultraviolet light. Virology 32: 337-343 (1967).

Katagiri, S. Hinuma, Y., Ishida, N.: Relation between the adsorption to cells and antigenic properties in poliovirus infection. Virology 34: 797-799 (1968).

Kates, J.: Transcription of the vaccinia virus genome and the occurrence of polyriboadenylic acid sequences in messenger RNA. Cold Spring Harbor Symp. Quant. Bio. 35: 743-752 (1970).

Katz, L., Penman, S.: The solvent denaturation of double-stranded RNA from poliovirus infected HeLa cells. Biochem. Biophys. Res. Commun. 23: 557 (1966).

Katze, M.G., Crowell, R.L.: Indirect enzyme-linked immunosorbent assay (ELISA) for the detection of coxsackievirus group B antibodies. J. Gen. Virol. 48: 225-229 (1980).

Kaufman, Y., Goldstein, E., Penman, S.: Poliovirus-induced inhibition of polypeptide initiation in vitro on native polyribosomes. Proc. Natl. Acad. Sci. USA 73: 1834-1838 (1976).

Kawanishi, M.: Intranuclear crystal formation in picornavirus infected cells. Arch. Virol. 57: 123-132 (1978).

Kazachkov, Y.A., Chernovskaya, T.V., Siyanova E.Y., Svitkin, Y.V., Ugarova, T.Y., Agol, V.I.: Leader polypeptides encoded in the 5'-region of the encephalomyocarditis virus genome. FEBS Lett. 141: 153-156 (1982).

Keller, R.: Reactivation by physical means of antibody-neutralized poliovirus. J. Immunol. 94: 143-149 (1965).

Keller, R.: Studies on the mechanisms of the enzymatic reactivation of antibody-neutralized poliovirus. J. Immunol. 100: 1071-1079 (1968).

Keller von Asten, H.: Wandlungen - Freundschaft mit den platonischen Körpern. Verlag Walter Keller 1980.

Kessel, J.F., Pait, C.F.: Differentiation of three groups of poliomyelitis virus. Proc. Soc. Exp. Biol. Med. 70: 315-316 (1949).

Kessler, S.W.: Rapid isolation of antigens from cells with a staphylococcal potein A-antibody adsorbent: Parameters of the interaction of antibody-antigen complexes with protein A. J. Immunol. 15: 1617-1624 (1977).

Kew, O.M., Nottay, B.K.: Evolution of the oral polio vaccine strains in human occurs by both mutation and intramolecular recombination. In: Modern Approaches to Vaccines (Chanock, R.M., Lerner, R.A., eds.), 357-361. Cold Spring Harbor Laboratory 1984.

Kew, O.M., Pallansch, M.A., Omilianowski, D.R., Rueckert, R.R.: Changes in three of the four coat proteins of oral polio vaccine strain derived from type 1 poliovirus. J. Virol. 33: 256-263 (1980).

Kew, O.M., Nottay, B.K., Hatch, M.H., Nakano, J.H., Obijeski J.F.: Multiple genetic changes can occur in the oral poliovaccines upon replication in humans. J. Gen. Virol. 56: 337-347 (1981).

Khan, S., Macnab, R.M.: Proton chemical potential, proton electrical potential and bacterial motility. J. Mol. Biol. 138: 599-614 (1980).

Kiehn, E.D., Holland, J.J.: Synthesis and cleavage of enterovirus polypeptides in mammalian cells. J. Virol. 5: 358-367 (1970).

Killington, R.A., Stott, E.J., Lee, D.: The effect of temperature on the synthesis of rhinovirus type 2 RNA. J. Gen. Virol. 36: 403-411 (1977).

King, A.M., McKahon, D., Slade, W.R., Newman, J.W.: Recombination in RNA. Cell 29: 921-928 (1982).

Kirschner, M.W.: Microtubules: assembly and nucleation. Int. Rev. Cytol. 54: 1-71 (1978).

Kitamura, N., Wimmer, E.: Sequence of 1060 3'-terminal nucleotides of poliovirus RNA as determined by a modification of the dideoxynucleotide method. Proc. Natl. Acad. Sci. USA 77: 3196-3200 (1980).

Kitamura, N., Adler, C., Wimmer, E.: Structure and expression of the picornavirus genome. In: Genetic Variation of Viruses. Ann. N.Y. Acad. Sci. (1980a).

Kitamura, N., Adler, C., Martinko, J., Nathenson, S.G., Wimmer, E.: The genome-linked protein of picornaviruses. VII. Genetic mapping of poliovirus VPg by protein and RNA sequence studies. Cell 21: 295-302 (1980b).

Kitamura, N., Semler, B.L., Rothberg, P.G., Larsen, G.R., Adler, C.J., Dorner, A. J., Emini, E.A., Hanecak, R., Lee, J.J., Van der Werf, S., Anderson, C.W., Wimmer, E.: Primary structure, gene organization, polypeptide expression of poliovirus RNA. Nature 291: 547-553 (1981).

Klein, R., Sergiescu, D., Teodorescu, M.: Induction with nitrous acid of dextran sulfate resistance in type-1 poliovirus. Virology 30: 145-147 (1966).

Klemperer, H.G.: Hemolysis and the release of potassium from cells by Newcastle disease virus (NDV). Virology 12: 540-552 (1960).

Kling, C., Levaditil, C., Lépine, P.: La pénetration du virus poliomyélitique à travers la muqueuse du tube digestif chez le singe et sa conservation dans l'eau. Bull. Acad. méd. (Paris) 102: 158-165 (1929).

Klingenberg, M.: Membrane protein oligomeric structure and transport function. Nature 290: 449-453 (1981).

Klingler, E.A., Jr., Chapin, M., Dubes, G.R.: Relationship between inactivation of poliovirus by phenol and appearance of ribonuclease-labile infectivity. Proc. Soc. Exp. Biol. N.Y. 101: 829 (1959).

Klug, A., Longley, W., Leberman, R.: Arrangement of protein subunits and the distribution uf nucleic acid in turnip yellow mosaic virus. I. X-ray diffraction studies. J. Mol. Biol. 15: 315-343 (1966).

Knauert, F., Ehrenfeld, E.: Translation of poliovirus RNA in vitro: studies on n-formylmethionine-labeled polypeptides initiated in cell-free extracts prepared from poliovirus infected HeLa cells. Virology 93: 537-546 (1979).

Koch, A.S., Feher, G.: The possible nature of the chance-event in initiation of virus infection at the cellular level. J. Gen. Virol. 18: 319-327 (1973).

Koch, A.S., Eremenko, T., Benedetto, A., Volpe, P.: A guanidine-sensitive step of the poliovirus RNA replication cycle. Intervirology 4: 221-225 (1974).

Koch, F.: Regulation of protein synthesis—a fine-tuning on the translational level. Honors Thesis, Wesleyan University, USA (1976).

Koch, F., Koch, G.: Reversible inhibition of macromolecular synthesis in HeLa cells by ethanol. Res. Commun. Chem. Path. Pharm. 9: 219-289 (1974).

Koch, F., Koch G., Kruppa, J.: Virus-induced shut-off of host specific protein synthesis. In: Protein Biosynthesis in Eukaryotes (Pérez-Bercoff, R., ed.), 339-361. New York - London: Plenum Press 1982.

Koch, G.: Influence of assay conditions on infectivity of heated poliovirus. J. Virol. 12: 601-603 (1960a).

Koch, G.: The kinetics of induction of plaque formation in cell monolayers by ribonucleic acid from poliovirus. Experientia 16: 490-493 (1960b).

Koch, G.: Zur unterschiedlichen Infektiosität von isolierter Ribonukleinsäure und intaktem Poliovirus in der Gewebekultur. Z. Naturforsch. 15B: 656-661 (1960c).

Koch, G.: Über die Hitzeinaktivierung von Polioviren und isolierter Poliovirusnukleinsäure. Zentralbl. Bakteriol. 184: 165-168 (1962).

Koch, G.: Zum Ablauf der Invasion von Virusnukleinsäure in Gewebekulturzellen. Z. Naturforschg. 18b: 899-902 (1963).

Koch, G.: Properties, functions and replication of viral RNA. In: Ribonucleic Acid - Structure and Function (Tuppey, H., ed.), 107-129. New York: Pergamon Press 1966.

Koch, G.: Differential effect of phleomycin on the infectivity of poliovirus and poliovirus-induced ribonucleic acids. J. Virology 8: 28-34 (1971a).

Koch, G.: Stability of polycation-induced cell competence for infection by viral ribonucleic acid. Virology 45: 841-843 (1971b).

Koch, G.: Interaction of poliovirus-specific RNAs with HeLa cells and E. coli. Curr. Top. Microbiol. Immunol. 62: 89-138 (1973).

Koch, G.: Reversible inhibition of protein synthesis in HeLa cells by exposure to proteolytic enzymmes. Biochem. Biophys. Res. Commun. 61: 817-824 (1974).

Koch, G., Bishop, J.M.: The effect of polycations on the interaction of viral RNA with mammalian cells: Studies on the infectivity of single- and double-stranded poliovirus RNA. Virology 35: 9-17 (1968).

Koch, G., Koch, F.: The use of cytochalasins in studies on the molecular biology of virus host cell interaction. In: Cytochalasins, Biochemical and Cell Biological Aspects (Tanenbaum, S.W., ed.), 475-498. Elsevier/North-Holland 1978.

Koch, G., Kubinski, H.: On the separation of poliovirus ribonucleic acid from cellular ribonucleic acids. Virology 17: 220-221 (1962).

Koch, G., Oppermann, H.: Sensitization of HeLa cells for viral RNA infection by cytochalasin B. Virology 63: 395-403 (1975).

Koch, G., Richter, D. (eds.): Regulation of Macromolecular Synthesis by Low Molecular Weight Mediators. New York: Academic Press 1979.

Koch, G., Richter, D. (eds.): Biosynthesis, Modification, and Processing of Cellular and Viral Polyproteins. New York: Academic Press 1980.

Koch, G., Vollertsen, I.: Interaction of viral RNA with E. coli: II. Synthesis of poliovirus specific RNA directed by poliovirus RF-RNA. Arch. Biochem. Biophys. 152: 475-487 (1972a).

Koch, G., Vollertsen, I.: Interaction of viral RNA with E. coli: III. Synthesis of poliovirus specific RNA directed by isolated poliovirus RNA. Arch. Biochem. Biophys. 153: 823-830 (1972b).

Koch, G., Koening, S., Alexander, H.E.: Quantitative studies on the infectivity of ribonucleic acid from partially purified and highly purified poliovirus preparations. Virology 10: 329-343 (1960).

Koch, G., Quintrell, N., Bishop, J.M.: An agar-cell suspension plaque assay for isolated viral RNA. Biochem. Biophys. Res. Commun. 24: 304-309 (1966).

Koch, G., Quintrell, N., Bishop, J.M.: Differential effect of actinomycin D on the infectivity of single- and double-stranded poliovirus RNA. Virology 31: 388-390 (1967).

Koch, G., Wiegers, K., Bishop, J.M., Lempidakis, G., von Bodelschwingh, H.: The replication of poliovirus RNA. XIIth European Symposium of Poliomyelitis and Allied Diseases, 429-439. Bucharest: Publishing House of the Academy of the Socialist Republic of Romania 1969.

Koch, G., Kubinski, H., Koch, F.: Inhibition of protein and RNA synthesis in HeLa cells by protease-released membrane glycopeptides. Hoppe Seylers Z. Physiol. Chem. 355: 1218 (1974).

Koch, G., Oppermann, H., Bilello, P., Nuss, D.: Control of peptide chain initiation in uninfected and virus infected cells by membrane mediated events. In: Modern Trends in Human Leukemia II (Neth, R., Gallo, R. C., Mannweiler, K., Moloney, W. C., eds.), 541-555. München: J.F. Lehmanns Verlag 1976.

Koch, G., Bilello, P., Kruppa, J.: Membrane mediated amplification of translational control in eukaryotes: A pleiotropic effect. In: Regulation of Macromolecular Syntheis by Low Molecular Weight Mediators (Koch, G., Richter, D., eds.), 273-290. Academic Press 1979.

Koch, G., Bilello, J.A., Kruppa, J., Koch, F., Oppermann, H.: Amplification of translational control by membrane-mediated events: A pleiotropic effect on cellular and viral gene expression. Ann. N.Y. Acad. Sci. 339: 280-306 (1980a).

Koch, G., Hiller, R., Schärli, C.: Influence of medium hyperosmolarity and of guanidine on the synthesis and processing of poliovirus proteins. In: Biosynthesis, Modification, Processing of Cellular and Viral Polyproteins, 249-262. New York: Academic Press 1980b.

Koch, G., Koch, F., Bilello, J.A., Hiller, E., Schärli, C., Warnecke, G., Weber, C.: Biosynthesis, modification and processing of viral polyproteins. In: Protein Biosynthesis in Eukaryotes (Pérez-Bercoff, R., ed.), 275-309. New York: Plenum Press 1982a.

Köhler, G., Milstein, C.: Derivation of specific antibody-producing tissue culture and tumour lines by cell fusion. Eur. J. Immunol. 6: 511-519 (1976).

Kohn, A.: Early interactions of viruses with cellular membranes. In: Advances in Virus Research, Vol. 24, 233-276. Academic Press 1979.

Koliais, S.I.: Presence of double-stranded regions of viral RNA in infected cells. Experientia 37: 971-972 (1981).

Kolmer, J.A., Klugh, G. Jr., Rule, A.M.: A successful method for vaccination against acute anterior poliomyelitis; Further report. J. A. M. A. 104: 456-460 (1935).

Kopecka, H., van der Werf, S. Bregegere, F., Kitamura, N., Dreano, M., Rothberg, P.G., Wimmer, E., Kourilsky, P., Girard, M.: Restriction map of poliovirus type 2 cDNA. Dev. Biol. Stand. 50: 301-309 (1981).

Koprowski, H., Norton, T.W., Jervis, G.A., Nelson, T.L., et al.: Clinical investigations of attenuated strains of poliomyelitis virus. Use as a method of immunization of children with living virus. J.A.M.A. 160: 954-966 (1956).

Korant, B.D.: Cleavage of viral precursor proteins in vivo and in vitro. J. Virol. 10: 751-759 (1972).

Korant, B.D.: Cleavage of poliovirus-specific polypeptides aggregates. J. Virol. 12: 556-563 (1973).

Korant, B.D.: Protein cleavage in picornavirus replication. In: In vitro Transcription and Translation of Viral Genomes (Haenni, A.L., Beaud, G., eds.), 273-279. Paris: INSERM 1975a.

Korant, B.D.: Regulation of animal virus replication by protein cleavage. In: Proteases and Biological Control (Reich, E., Rifkin D. B., Shaw, E., eds.), 621-644. Cold Spring Harbor Laboratory 1975b.

Korant, B.D.: Protein cleavage in virus-infected cells. Acta Biol. Med. Germ. 36: 1565-1573 (1977a).

Korant, B.D.: Poliovirus coat protein as the site of guanidine action. Virology 81: 17-28 (1977b).

Korant, B.D.: Protease activity associated with HeLa cell ribosomes. Biochem. Biophys. Res. Commun. 74: 926-933 (1977c).

Korant, B.D.: In: Molecular Basis of Biological Degradating Processes (Berlin, R., Herrmann, H., Lepow, I., Tanzer, G., eds), 171-220. New York: Academic Press 1978.

Korant, B.D.: Role of cellular and viral proteases in the processing of picornavirus proteins. In: The Molecular Biology of Picornaviruses (Pérez-Bercoff, R., ed.), 149-173. New York-London: Plenum Press 1979.

Korant, B.D.: Inhibition of viral protein cleavage. In: Antiviral Chemotherapy: Design of Inhibitors of Viral Functions (Gauri, K.K., ed.), 37-47. New York: Academic Press 1981.

Korant, B.D., Butterworth, B.E.: Inhibition by zinc of rhinovirus protein cleavage: Interaction of zinc with capsid polypeptides. J. Virol. 18: 298-306 (1976).

Korant, B.D., Lonberg- Holm, K.K.: Viral proteins and site-specific cleavage. Acta. Biol. Med. Ger. 40: 1481-1488 (1981).

Korant, B.D., Lonberg-Holm, K., Noble, J., Stasny, J.T.: Naturally occurring and artificially produced components of three rhinoviruses. Virology 48: 71-86 (1972).

Korant, B.D., Lonberg-Holm, K., Yin, F.H., Noble-Harvey, J.: Fractionation of biologically active and inactive populations of human rhinovirus type 2. Virology 63: 384-394 (1975).

Korant, B., Chow, D., Lively, M., Powers, J.: Virus-specified protease in poliovirus-infected HeLa cells. Proc. Natl. Acad. Sci. USA 76: 2992-2995 (1979).

Korant, B.D., Chow, N.L., Lively, M.O., Powers, J.C.: Proteolytic events in replication of animal viruses. Ann. N.Y. Acad. Sci. 343: 304-318 (1980).

Korn, E.D.: Actin polymerization and its regulation by proteins from nonmuscle cells. Physiol. Rev. 62: 672-737 (1982).

Koroleva, G.A., Lashkevich, V.A., Voroshilova, M.K.: Use of proflavine-sensitized poliovirus strains for determination of the range of susceptible animals. Acta Virol. 17: 310-318 (1973).

Koroleva, G.A., Lashkevich, V.A., Voroshilova, M.K.: Study of poliovirus multiplication in different animal species using photosensitized virus strains. Arch. ges. Virusforsch. 46: 11-28 (1974).

Koschel, K.: No effect of guanidine on an improved in vitro system for viral RNA polymerase. Biochem. Biophys. Res. Commun. 37: 938-944 (1969).

Koschel, K.: Freisetzung von ^{51}Chrom aus markierten HeLa-Zellen nach Poliovirus Infektion. Z. Naturforsch. 26: 929-933 (1971).

Koschel, K.: Poliovirus infection and poly(A) sequences of cytoplasmic cellular RNA. J. Virol. 13: 1061-1066 (1974).

Koschel, K., Wecker, E.: Early functions of poliovirus. III. The effect of guanidine on early functions. Z. Naturforsch. 26b: 940-944 (1971).

Koschel, K., Aus, H.M., Ter Meulen, V.: Lysosomal enzyme activity in poliovirus-infected HeLa cells and VSV-infected L-cells: biochemical and histochemical comparative analysis with computer-aided techniques. J. Gen. Virol. 25: 359-369 (1974).

Koschel, K., Täuber, H., Wecker, E.: Early functions of poliovirus I. The inhibition of cellular protein synthesis in HeLa cells after infection with active and inactivated poliovirus. Z. Naturforsch. 26: 798-803 (1971).

Koza, J.: Calcium phosphate adsorption patterns of virulent and avirulent strains of poliovirus. Virology 21: 477-481 (1963).

Koza, J.: Changes in secondary structure of poliovirus ribonucleic acid. Arch. Bioch. 170: 724-730 (1975).

Kozak, M.: Nucleotide sequences of 5'-terminal ribosome-protected initiation regions from two reovirus messages. Nature 269: 390-394 (1977).

Kozak, M.: How do eukaryotic ribosomes select initiation regions in messenger RNA? Cell 15: 1109-1123 (1978).

Kozak, M.: Migration of 40S ribosomal subunits on a messenger RNA when initations is perturbed by lowering magnesium or adding drugs. J. Biol. Chem. 254: 4731-4738 (1979).

Kozak, M.: Influence of mRNA secondary structure on binding and migration of 40S ribosomal subunits. Cell 19: 79-90 (1980).

Kozak, M.: Possible role of flanking nucleotides in recognition of the AUG initiator codon by eukaryotic ribosomes. Nucl. Acids Res. 9: 5233-5252 (1981).

Kozak, M.: Recognition of initiation sites. In: Protein Biosynthesis in Eukaryotes (Perez-Bercoff, R., ed.), 167-197. New York: Plenum Press 1982.
Kozak, M., Shatkin, A.J.: Migration of 40 S ribosomal subunits on messenger RNA in the presence of edeine. J. Biol. Chem. 253: 6568-6577 (1978).
Kozak, M., Shatkin, A.J.: Characterization of translational initiation regions from eukaryotic messenger RNAs. Methods Enzymol. 60: 360-375 (1979).
Krah, D.L., Crowell, R.L.: A solid-phase assay of solubilized HeLa cell membrane receptors for binding group B coxsackieviruses and polioviruses. Virology 118: 148156 (1982).
Krstic, R.V.: Ultrastructure of the Mammalian Cell. An Atlas. Berlin-Heidelberg-New York: Springer 1979.
Kruppa, J.: Membrangebundene Biosynthese zellulärer und viraler Proteine in Säugetierzellen. Habilitationsschrift, Hamburg University (1979).
Kruppa, J., Martini, O.H.W.: Dephosphorylation of one 40S ribosomal protein in MPC 11 cells induced by hypertonic medium. Biochem. Biophys. Res. Commun. 85: 428-435 (1978).
Krüse, J., Timmins, P.A., Witz, J.: A neutron scattering study of the structure of compact and swollen froms of Southern bean mosaic virus. Virology 119: 42-50 (1982).
Kunin, C.M.: Virus-tissue union and the pathogenesis of enterovirus infections. J. Immunol. 88: 556 (1962).
Kunin, C.M.: Cellular susceptibility to enteroviruses. Bacteriol. Rev. 28: 382-390 (1964).
Kunin, C.M., Jordan, W.S.: In vitro adsorption of poliovirus by noncultured tissues. Effect of species, age and malignancy. Am. J. Hyg. 73: 245-257 (1961).
Kuntz, I.D., Kauzmann, W.: Hydration of proteins and polypeptides. Adv. Protein Chem. 28: 239-345 (1974).
Kusano, T., Wang, R., Pollack, R., Green, H.: Human-mouse hybrid cell lines and susceptibility to poliovirus. II. Polio sensitivity and the chromosome constitution of the hybrids. J. Virol. 5: 682-685 (1970).
Kyte, J.: Molecular considerations relevant to the mechanism of active transport. Nature 292: 201-204 (1981).

Lacal, J.C., Carrasco, L.: Relationship between membrane integrity and the inhibition of host translation in virus-infected mammalian cells. Comparative studies between encephalomyocarditis virus and poliovirus. Eur. J. Biochem. 127: 359-366 (1982).
Lacal, J.C., Vázquez, D., Fernández-Sousa, J.M., Carrasco, L.: Antibiotics that specifically block translation in virus-infected cells. J. Antibiotics 33: 441-446 (1980).
La Colla, P., Marcialis, M.A., Mereu, G.P., Loddo, B.: Specific inhibition of poliovirus induced blockage of cell protein synthesis by a thiopyrimidine derivative. J. Gen. Virol. 17: 13-18 (1972).
La Colla, P., Marcialis, M.A., Flore, O., Firinu, A., Garzia, A., Loddo, B.: Bichlorinated pyrimidines as possible antiviral agents. Chemotherapy 6: 295 (1976).
La Colla, P., Corrias, M.V., Marongiu, M.E., Pani, A.: Dichloropyrimidines: inhibitors of viral morphogenesis. In: Design of Inhibitors of Viral Functions (Gauri, ed.), 21-35. 1981.
La Porte, J., Lenoir, G.: Structural proteins of foot-and-mouth disease virus. J. Gen. Virol. 20: 161-168 (1973).
Laemmli, U.K.: Cleavage of structural proteins during the assembly of the head of bacteriophage T4. Nature 227: 680-685 (1970).
Landsteiner, K., Popper, E.: Übertragung der Poliomyelitis acuta auf Affen. Z. Immunitätsforsch. Orig. 2: 377-390 (1909).
Larsen, G.R., Semler, B.L., Wimmer, E.: Stable hairpin structure within the 5'terminal 85 nucleotides of poliovirus RNA. J. Virol. 37: 328-335 (1981).
Larsen, G.R., Dorner, J.A., Harris, T.J.R., Wimmer, E.: The structure of poliovirus replicative form. Nucl. Acids Res. 8: 1217-1229 (1980).
Larsen, G.R., Anderson, C.W., Dorner, A.J., Semler, B.L., Wimmer, E.: Cleavage sites within the poliovirus capsid protein precursors. J. Virol. 41: 340-344 (1982).
Laskey, R.A., Mills, A.D., Gurdon, J.B., Partington, G.A.: Protein synthesis in oocytes of Xenopus laevis is not regulated by the supply of messenger RNA. Cell 11: 345-351 (1977).
Lawrence, G., Thach, R.E.: Encephalomyocarditis virus infection of mouse plasmacytoma cells. I. Inhibition of cellular protein synthesis. J. Virol. 14: 598-610 (1974).

Lawrence, C., Thach, R.E.: Identification of a viral protein involved in posttranslational maturation of the EMC virus capsid precursor. J. Virol. 15: 918-928 (1975).

Lazarides, E.: Actin, alpha-actinin, and tropomyosin interaction in the structural organization of actin filaments in non-muscle cells. J. Cell Biol. 68: 202-219 (1976).

Lazarides, E.: Intermediate filaments as mechanical integrators of cellular space. Nature 283: 249 (1980).

Lazarides, E.: Intermediate filaments - chemical heterogeneity in differentiation. Cell 23: 649-650 (1981).

Lazarides, E., Weber, K.: Actin antibody: the specific visualization of actin filaments in non-muscle cells. Proc. Natl. Acad. Sci. USA 71: 2268-2272 (1974).

Lazarowitz, S.G., Goldberg, A.R., Choppin, P.W.: Proteolytic cleavage by plasmin of the HA polypeptide of influenza virus: Host cell activation of serum plasminogen. Virology 56: 172-180 (1973).

Le Bouvier, G.L.: The modification of poliovirus antigens by heat and ultraviolet light. Lancet 2: 1013-1016 (1955).

Le Bouvier, G.L.: Poliovirus precipitins. A study by means of diffusion in agar. J. Exp. Med. 106: 661-676 (1957).

Le Bouvier, G.L.: Poliovirus D and C antigens; their differentiation and measurement by precipitation in agar. Brit. J. Exp. Path. 40: 452-463 (1959a).

Le Bouvier, G.L.: The D to C change in poliovirus particles. Brit. J. Exp. Path. 40: 605-620 (1959b).

Le Bouvier, G.L., Schwerdt, D.E., Schaffer, F.L.: Specific precipitates in agar with purified poliovirus. Virology 4: 590 (1957).

Leary, K., Blair, C.D.: Sequential events in the morphogenesis of Japanese encephalitis virus. J. Ultrastruc. Res. 72: 123-129 (1980).

Leavitt, R., Schlesinger, S., Kornfeld, S.: Tunicamycin inhibits glycosylation and multiplication of sindbis and vesicular stomatitis viruses. J. Virol. 21: 375-385 (1977).

Ledinko, N.: Genetic recombination with poliovirus type 1. Studies of crosses between a normal horse serum-resistant mutant and several guanidine-resistant mutants of the same strain. Virology 20: 107-119 (1963).

Ledinko, N., Hirst, G.K.: Mixed infection of HeLa cells with polioviruses types 1 and 2. Virology 14: 207-219 (1961).

Lee, K.A.W., Sonenberg, N.: Inactivation of cap-binding proteins accompanies the shut-off of host protein synthesis by poliovirus. Proc. Natl. Acad. Sci. USA 79: 3447-3451 (1982).

Lee, P.W.K., Colter, J.S.: Further characterization of mengo subviral particles: a new hypothesis for picornavirus assembly. Virology 97: 266-274 (1979).

Lee, P.W.K., Paucha, E., Colter, J.S.: Identification and partial characterization of a new (50S) subviral particle in mengovirus infected cells. Virology 85: 286-295 (1978).

Lee, Y.F., Wimmer, E.: "Fingerprinting" high molecular weight RNA by twodimensional gel electrophoresis: Application to poliovirus RNA. Nucl. Acids Res. 3: 1647-1658 (1976).

Lee, Y.F., Nomoto, A., Wimmer, E.: The genomes of poliovirus is an exceptional eukaryotic mRNA. Prog. Nucl. Acid Res. Mol. Biol. 19: 89-95 (1976).

Lee, Y.F., Nomoto, A., Detjen, B.M., Wimmer, E.: The genome-linked protein of picornaviruses. I. A protein covalently linked to poliovirus genome RNA. Proc. Natl. Acad. Sci. USA 74: 59-63 (1977).

Lee, Y.F., Kitamura, N., Nomoto, A., Wimmer, E.: Sequence studies of poliovirus RNA. IV. Nucleotide sequence complexities of poliovirus type 1, type 2 and two type 1 defective interfering particles RNA, fingerprints of the poliovirus type 3 genome. J. Gen. Virol. 44: 311-322 (1979).

Leibowitz, R., Penman, S.: Regulation of protein synthesis in HeLa cells. III. Inhibition during poliovirus infection. J. Virol. 8: 661-668 (1971).

Lempidakis, G.A., Koch, J.G.: Interaction of viral RNA with E. coli. I. Polycation augmented adsorption of poliovirus-induced double-stranded RNA. Arch. Biochem. Biophys. 151: 200-205 (1972).

Lenk, R., Penman, S.: The cytosketal framework and poliovirus metabolism. Cell 16: 289-301 (1979).

Lenk, R., Ransom, L., Kaufmann, Y., Penman, S.: A cytoskeletal structure with associated polyribosomes obtained from HeLa cells. Cell 10: 67-78 (1977).

Levanon, A., Kohn, A.: Changes in cell membrane microviscosity associated with adsorption of viruses. Differences between fusing and non-fusing viruses. FEBS Lett. 85: 245-248 (1978).

Levanon, A., Klibansky, Y., Kohn, A.: Picorna- and togavirus infection of cells detected by gas chromatography. J. Med. Virol. 1: 227-237 (1977a).

Levanon, A., Kohn, A., Inbar, M.: Increase in lipid fluidity of cellular membranes induced by adsorption of RNA and DNA virions. J. Virol. 22: 353-360 (1977b).

Levanon, A., Inbar, M., Kohn, A.: fluorescence polarization of DPH-labeled cells adsorbing viruses and its diagnostic potential. Arch. Virol. 59: 223-230 (1979).

Levine, E.M., Becker, Y., Boone, C.W., Eagle, H.: Contact inhibition, macromolecular synthesis, and polyribosomes in cultured human diploid fibroblasts. Proc. Natl. Acad. Sci. USA 53: 350-355 (1965).

Levine, R.A., Wolff, D.A.: Bovine enterovirus CPE at different multiplicities of infection in the absence of viral RNA synthesis. Intervirology 11: 255-260 (1979).

Levinthal, J.D., Dunnebacke, T.H., Williams, R.C.: Study of poliovirus infection of human and monkey cells by indirect immunoferritin technique. Virology 39: 211-233 (1969).

Levintow, L.: The reproduction of picornaviruses. In: Comprehensive Virology (Fraenkel-Conrat, H., Wagner, R.R., eds.), Vol. 2, 109-169. New York: Plenum Press 1974.

Levintow, L., Bishop, J.M.: Comparative aspects of poliovirus replication. J. Cell. Physiol. 76: 265-272 (1970).

Levintow, L., Darnell, J.E.: A simplified procedure for purification of large amounts of poliovirus: Characterization and amino acid analysis of tpye 1 poliovirus. J. Biol. Chem. 235: 70-73 (1960).

Levintow, L., Thoren, M.M., Darnell, J.E., Jr., Hooper, J.L.: Effect of p-fluorophenylalanine and puromycin on the replication of poliovirus. Virology 16: 220 (1962).

Levitt, N.H., Crowell, R.L.: Comparative studies of the regeneration of HeLa cell receptors for poliovirus T1 and coxsackievirus B3. J. Virol. 1: 693-700 (1967).

Levy, H.B.: Intracellular sites of poliovirus reproduction. Virology 15: 173-184 (1961).

Li, C.P., Habel, K.: Adaptation of Leon strain of poliomyelitis virus to mice. Proc. Soc. Exp. Biol. 78: 233-238 (1951).

Li, C.P., Jahnes, W.G.: Studies on variation in virulence of poliomyelitis virus. I. The loss and gain of virulence of the mouse-adapted type KKK virus. Virology 2: 828 (1956).

Liautard, J.-P., Egly, J.M.: In vitro translation studies of the cytoplasmic nonpolysomal particles containing messenger RNA. Nucl. Acids Res. 8: 1793-1804 (1980).

Liautard, J.-P., Jenteur, P.: Purification of histone messenger ribonucleoprotein particles from HeLa cell S-phase polysomes. Nucl. Acids Res. 7: 135-150 (1979).

Liautard, J.-P., Köhler, K.: A comparative study by sucrose gradient electrophoresis of messenger ribonucleoprotein complexes. Biochimie 58: 317-323 (1976).

Liautard, J.-P., Liautard, J.: Isolation of biologically active polysomal messenger ribonucleoprotein by centrifugation in cesium sulfate. Biochim. Biophys. Acta 474: 588-594 (1977).

Lichstein, D., Kaback, H.R., Blume, A.J.: Use of a lipophylic cation for determination of membrane potential in neuroblastoma-glioma hybrid cell suspension. Proc. Natl. Acad. Sci. USA 76: 650-654 (1979).

Lichstein, D., Dunlop, K., Kaback, H.R., Blume, A.J.: Mechanism of monensin induced hyperpolarization of neuroblastoma-glioma hybrid WG 108-15. Proc. Natl. Acad. Sci. USA 76: 2580-2584 (1979).

Lindberg, V., Sundquist, B.: Isolation of messenger ribonucleoproteins from mammalian cells. J. Mol. Biol. 86: 451-468 (1974).

Lipton, H.L.: Persistent Theiler's murine encephalomyelitis virus infection in mice depends on plaque size. J. Gen. Virol. 46: 169-177 (1980).

Littlefield, J.W., Gould, E.A.: The toxic effect of 5-bromodeoxyuridine on cultured epithelial cells. J. Biol. Chem. 235: 1129-1133 (1960).

Loddo, B., Ferrari, W., Brotzu, G., Spanedda, A.: In vitro inhibition of infectivity of poliovirus by guanidine. Nature 193: 97-98 (1962).

Loddo, B., Mutoni, S., Spanedda, A., Brotzu, G., Ferrari, W.: Guanidine conditional infectivity of ribonucleic acid extracted from a strain of guanidine-dependent polio-1 virus. Nature 197: 315 (1963).

Loddo, B., Gressa, G.L., Schivo, M.L., Spanedda, A., Brotzu, G., Ferrari, W.: Antagonism of the guanidine interference with poliovirus replication by simple methylated and ethylated compounds. Virology 28: 707-712 (1966).

Lodish, H.F.: Alpha and beta globin messenger ribonucleic acid. J. Biol. Chem. 246: 7131-7138 (1971).
Lodish, H.F.: Model for the regulation of mRNA translation applied to hemoglobin synthesis. Nature 251: 385-388 (1974).
Lodish, H.F.: Translational control of protein synthesis. Ann Rev. Biochem. 45: 39-72 (1976).
Lodish, H.F., Rose, K.M.: Relative importance of 7-methylguanosine in ribosome binding and translation of vesicular stomatitis virus mRNA in wheat germ and reticulocyte cell-free systems. J. Biol. Chem. 252: 1181-1188 (1977).
Loeffler, F., Frosch, P.: Berichte der Kommission zur Erforschung der Maul- und Klauenseuche bei dem Institut für Infektionskrankheiten in Berlin. Zbl. Bakter. Abt. I. Orig. 23: 371-391 (1898).
Loesch, W.T., Arlinghaus, R.B.: Stable polypeptides associated with the 250 S mengovirus-induced RNA polymerase structure. Arch. Virol. 47: 201-215 (1975).
LoGrippo, G.A., Earl, D.P. Jr., Brodie, B.B., Graef, I.P., Bowman, L., Ward, R.: Lack of effect of sodium phenosulfazole (Darvisul) on certain experimental virus infections. Proc. Soc. Exp. Biol. Med. 70: 528 (1949).
Lokteva, V.F., Chumakov, M.P., Zhevandrova, V.I., Shekoyan, L.A.: Differential density of virulent and attenuated strains of type 1 poliovirus to polyethylene glycol as a possible new genetic marker. Arch. ges. Virusforsch. 41: 155-159 (1973).
Lonberg-Holm, K.: The effects of concanavalin A on the early events of infection by rhinovirus type 2 and poliovirus type 2. J. Gen. Virol. 28: 313-327 (1975).
Lonberg-Holm, K.: Attachment of animal virus to cells, an introduction. In: Virus Receptors, Part 2. In: Receptors and Recognition (Lonberg-Holm, K., Philipson, L., eds.), Series B, Vol. 8. London: Chapman and Hall 1980.
Lonberg-Holm, K., Butterworth, B.: Investigation of the structure of polio and human rhinovirions through the use of selective chemical reactivity. Virology 71: 207-216 (1976).
Lonberg-Holm, K., Philipson, L.: Early interaction between animal viruses and cells. In: Monographs in Virology (Melnick, J.L., ed.), Vol. IX, 1-148. Basel: S. Karger 1974.
Lonberg-Holm, K., Philipson, L.: Molecular aspects of virus receptors and cell surfaces. In: Cell Membranes and Viral Envelopes (Blough, H.A., Tiffany, J.M., eds.), 789-848. London-New York: Academic Press 1980.
Lonberg-Holm, K., Philipson, L. (eds.): Virus Receptors, Part 2. In: Receptors and Recognition, Series B, Vol. 8. London: Chapman and Hall 1981.
Lonberg-Holm, K., Whiteley, N.M.: Physical and metabolic requirements for early interaction of poliovirus and human rhinovirus with HeLa cells. J. Virol. 19: 857-870 (1976).
Lonberg-Holm, K., Gosser, L.B., Kauer, J.C.: Early alteration of poliovirus in infected cells and its specific inhibition. J. Gen. Virol. 27: 329-342 (1975).
Lonberg-Holm, K., Gosser, L.B., Shimshick, E.J.: Interaction of liposomes with subviral particles of poliovirus type 2 and rhinovirus type 2. J. Virol. 19: 746-749 (1976a).
Lonberg-Holm, K., Crowell, R.L., Philipson, L.: Unrelated animal viruses share receptors. Nature 259: 679-681 (1976b).
Lowry, O.H., Rosebrough, N.J., Farr, A.L., Randall, R.J.: Protein measurement with the folin phenol reagent. J. Biol. Chem. 193: 265 (1951).
Lucas-Lenard, J.M.: Inhibition of cellular protein synthesis after virus infection. In: The Molecular Biology of Picornaviruses (Pérez-Bercoff, R., ed.), 73-93. New York London: Plenum Press 1979a.
Lucas-Lenard, J.M.: Virus-directed protein synthesis. In: The Molecular Biology of Picornaviruses (Pérez-Bercoff, R., ed.), 127-147. New York London: Plenum Press 1979b.
Ludwig, E.H., Smull, C.E.: Infectivity of histone-poliovirus ribonucleic acid preparations. J. Bact. 85: 1334 (1963).
Lukens, L.N.: Evidence for nature of precursor that is hydroxylated during biosynthesis of collagen as hydroxyproline. J. Biol. Chem. 240: 1661-1669 (1965).
Lund, G.A., Ziola, B.R., Salmi, A., Scraba, D.G.: Structure of the mengovirion. V. Distribution of the capsid polypeptides with respect to the surface of the virus particle. Virology 78: 35-44 (1977).
Lundquist, R., Maizel, J.V.: In vivo regulation of the poliovirus RNA polymerase. Virology 89: 484-493 (1978a).
Lundquist, R.E., Maizel, J.V.: Structural studies on the RNA component of the poliovirus replication complex. I. Purification and biochemical characterization. Virology 85: 434-444 (1978b).

Lundquist, R.E., Ehrenfeld, E., Maizel, J.V., Jr.: Isolation of a viral polypeptide associated with poliovirus RNA polymerase. Proc. Natl. Acad. Sci. USA 71: 4773-4777 (1974).

Lundquist, R.E., Sullivan, M., Maizel, J. V., Jr.: Characterization of a new isolate of poliovirus defective interfering particles. Cell 18: 759-769 (1979).

Luria, S.E., Darnell, J.E., Baltimore, D., Campbell, A. (eds.): General Virology, 3rd ed. New York: Wiley 1978.

Lwoff, A.: The specific effectors of viral development. Biochem. J. 96: 289-302 (1965).

Lwoff, A., Lwoff, M.: L'inhibition du développement du virus poliomyélitique à 39° et le probléme du role de l'hyperthermie dans l'évolution des infections virales. C. R. Hebd. Séanc. Acad. Sci. Paris 246: 190-192 (1958).

Lwoff, A., Dulbecco, R., Vogt, M., Lwoff, M.: Kinetics of the release of poliomyelitis virus from single cells. Virology 1: 128-139 (1955).

Lyles, D.S., Landsberger, F.R.: Sendai virus-induced hemolysis: Reduction in heterogeneity of erythrocyte lipid bilayer fluidity. Proc. Natl. Acad. Sci. USA 74: 1918-1922 (1977).

Maassab, H.F., Loh, P.C., Ackermann, W.W.: Growth characteristics of poliovirus in HeLa cells: Nucleic acid metabolism. J. Exp. Med. 106: 641-648 (1957).

Macieira-Coelho, A., Fernandes, M.V., Mellman, W.J.: Cortisone on in vitro infetion. II. Effect on distribution of cytopathic effect and acid phosphatase arrangement in human cells infected with poliovirus. Proc. Soc. Exp. Biol. Med. 119: 631 (1965).

Madshus, I.H., Olsnes, S., Sandvig, K.: Mechanism of entry into the cytosol of poliovirus type 1: Requirement for low pH. J. Cell Biol. 98: 1194-1200 (1984).

Maizel, J.V.: Evidence for multiple components in the structural protein of type I poliovirus. Biochem. Biophys. Res. Commun. 13: 483-489 (1963).

Maizel, J.V.: Preparative electrophoresis of proteins in acrylamide gels. In: New York Academy of Sciences Symposium on Gel Electrophoresis, 382-390. Albany: New York Academy of Sciences 1964.

Maizel, J.V.: Acrylamide gel electropherograms by mechanical fractionation. Radioactive adenovirus proteins. Science 151: 988-990 (1966).

Maizel, J. V. jr.: Polyacrylamide gel electrophoresis of viral proteins. In: Methods in Virology (Maramorosh, K., Koprowski, H., eds.), Vol. 5, 179-246. New York: Academic Press 1971.

Maizel, J.V., Summers, D.F.: Evidence for differences in size and composition of the poliovirus-specific polypeptides in infected HeLa cells. Virology 36: 48-54 (1968a).

Maizel, J.V., Summers, D.F.: Evidence for large precursor proteins in poliovirus synthesis. Proc. Natl. Acad. Sci. USA 59: 966-971 (1968b).

Maizel, J.V., Phillips, B.A., Summers, D.F.: Composition of artificially produced and naturally occurring empty capsids of poliovirus type 1. Virology 32: 692-699 (1967).

Mak, T.W., Colter, J.S., Scraba, D.G.: Structure of the mengo virion. II. Physicochemical and electron microscopic analysis of degraded virus. Virology 57: 543-553 (1974).

Mandel, B.: Studies on the interactions of poliomyelitis virus, antibody, and host cells in a tissue culture system. Virology 6: 424-447 (1958).

Mandel, B.: Reversibility of the reaction between poliovirus and neutralizing antibody of rabbit origin. Virology 14: 316-328 (1961).

Mandel, B.: The use of sodium dedecyl sulfate in studies on the interaction of poliovirus and HeLa cells. Virology 17: 288-294 (1962a).

Mandel, B.: Early stages of virus-cell interaction as studied by using antibody. Cold Spring Harbor Symp. Quant. Biol. 27: 123-136 (1962b).

Mandel, B.: The extraction of ribonucleic acid from poliovirus by treatment with sodium dodecyl sulfate. Virology 22: 360 (1964).

Mandel, B.: The fate of the inoculum in HeLa cells infected with poliovirus. Virology 25: 152-154 (1965).

Mandel, B.: The interaction of neutralized poliovirus with HeLa cells. I. Adsorption. Virology 31: 238-247 (1967a).

Mandel, B.: The relationship between penetration and uncoating of poliovirus in HeLa cells. Virology 31: 702-712 (1967b).

Mandel, B.: Characterization of type 1 poliovirus by electrophoretic analysis. Virology 44: 554-568 (1971).

Mandel, B.: An analysis of the physical and chemical factors involved in the reactivation of neutralized poliovirus by the method of freezing and thawing. Virology 51: 358-369 (1973).

Mandel, B.: Neutralization of poliovirus: A hypothesis to explain the mechanism and the one-hit character of the neutralization reaction. Virology 69: 500-510 (1976).

Mandel, B.: Interaction of viruses with neutralizing antibodies. In: Comprehensive Virology (Fraenkel-Conrat, H., Wagner, R.R., eds.), Vol. 15, 37. New York: Plenum Press 1979.

Mandel, B.: Uncoating - like modification of poliovirus capsid resulting from the cooperative effects of subfreezing temperature and submolar concentrations of urea. Arch. Virol. 71: 27-42 (1982).

Mapoles, J.E.: Interaction of cations with poliovirus. Dissertation, University of Wisconsin, USA (1980).

Mapoles, J.E., Anderegg, J.W., Rueckert, R.R.: Properties of poliovirus propagated in medium containing cesium chloride: Implications for picornaviral structure. Virology 90: 103-111 (1978).

Marchalonis, J.J.: An enzymatic method for the trace iodination of immunoglobulins and other proteins. Biochem. J. 113: 299-305 (1969).

Marcialis, M.A., Schivo, M.L., Atzeni, A., Garzia, A., Loddo, B.: On the mechanism of the inhibitory action of 2-amino-4, 6-dichloropyrimidine on poliovirus growth. Experientia 29: 1559-1561 (1973a).

Marcialis, M.A., Schivo, M.L., Cioglia, A.M., Atzeni, A., Loddo, B.: Irreversible impairment produced by guanidine on the functions of poliovirus proteins. Experientia 29: 1561-1563 (1973b).

Marcialis, M.A., Flore, O., Marongiu, M.E., Pompei, R., La Colla, P., Loddo, B.: Enhancement of virus growth produced by thiols and disulphides. In: Separatum Experientia, 33, 1044-1045. Basel: Birkhäuser 1977.

Marongiu, M.E., Caddia, M., La Colla, P.: Morphogenesis of picornaviruses. Second Meeting of the European Study Group on the Molecular Biology of Picornaviruses, Hamburg 1981a.

Marongiu, M.E., Pani, A., Corrias, M.V., Sau, M., La Colla, P.: Poliovirus morphogenesis. I. Identification of 80S dissociable particles and evidence for the artifactual production of procapsids. J. Virol. 39: 341-347 (1981b).

Marquardt, O.: Three strains of European foot-and-mouth disease virus are highly conserved in the 3'-termini and highly variable in the genes of two capsid proteins. J. Gen. Virol. 59: 283-94 (1982).

Marsh, M., Bolzau, E., Helenius, A.: Penetration of Semliki forest virus from acidic prelysosomal vacuoles. Cell 32: 931-940 (1983).

Martin, S.J., Johnston, M.D.: The selective release of proteins from a bovine enterovirus. J. Gen. Virol. 16: 115-125 (1972).

Martin, S.L., Zimmer, E.A., Kan, Y.W., Wilson, A.C.: Silent α-globin gene in Old World monkeys. Proc. Natl. Acad. Sci. USA 77: 3563-3566 (1980).

Martin, S.L., Zimmer, E.A., Davidson, W.S., Wilson, A.C.: The untranslated regions of β-globin mRNA evolve at a functional rate in higher primates. Cell 25: 737-741 (1981).

Martini, O.H.W., Kruppa, J.: Ribosomal phosphoproteins of mouse myeloma cells. Changes in the degree of phosphorylation induced by hypertonic initiation block. Eur. J. Biochem. 95: 349-358 (1979).

Matlin, K.S., Reggio, H., Helenius, A., Simons, K.: Pathway of vesicular stomatitis virus entry leading to infection. J. Mol. Biol. 156: 609-631 (1982).

Mattern, C.F.T.: In Poliomyelitis: from the 4th International Poliomyelitis Congress, 326-329; edited by the International Poliomyelitis Congress. Philadelphia: J.B. Lippincott 1958.

Mattern, C.F.T.: Virus architecture. In: The Biochemistry of Viruses (Levy, H.B., ed.), 55-100. New York: Marcel Dekker 1969.

Mattern, C.F.T., Daniel, W.A.: Replication of poliovirus in HeLa cells: Electron microscopic observations. Virology 26: 646-663 (1965).

Matthews, R.E.F.: Classification and nomenclature of viruses. Intervirology 12: 129147 (1979).

Matthews, T.J., Butterworth, B.E., Chaffin, L., Rueckert, R.R.: Encephalomyocarditis (EMC) virus and rhinovirus 1A (HRV-1A) peptides associated with the infected cell ribosomes. Fed. Proc. 32: 461 (1973).

Maundrell, K., Maxwell, E.S., Civelli, O., Vincent, A., Goldenberg, S., Buri, J.-F., Imaizumi-Scherrer, M.-T., Scherrer, K.: Messenger ribonucleoprotein complexes in avian erythroblasts: carriers of post-transcriptional regulation? Mol. Biol. Rep. 5: 43-51 (1979).

Maxam, A.M., Gilbert, W.: Sequencing end-labeled DNA with base-specific chemical cleavages. Meth. Enzymol. 65: 499-560 (1980).

Mayer, M.M., Rapp, H.J., Roizman, B., Klein, S.W., Cowan, K.M., Lukery, D., Schwerdt, C.E., Schaffer, F.L., Charney, J.J.: The purification of poliomyelitis virus as studied by complement fixation. J. Immunol. 78: 435-455 (1957).

Mayor, H.D.: Cytochemical and fluorescent antibody studies on growth of poliovirus in tissue culture. Tex. Rep. Biol. Med. 19: 106-122 (1961).

Mayor, H.D.: Picornavirus symmetry. Virology 22: 156-160 (1964).

Mayor, H.D., Diwan, A.R.: Studies on the acridine orange staining of two purified RNA viruses: poliovirus and tobacco mosaic virus. Virology 14: 74-82 (1961).

Mayor, H.D., Jordan, L.E.: Formation of poliovirus in monkey kidney tissue culture cells. Virology 16: 325-333 (1962).

Mayor, H.D., Melnick, J.L.: Icosahedral models and viruses: a critical evaluation. Science 137: 613-615 (1962).

McBride, W.D.: Antigenic analysis of polioviruses by kinetic studies of serum neutralization. Virology 7: 45-58 (1959).

McBride, W.D.: Biological significance of poliovirus mutants of altered cystine requirement. Virology 18: 118-130 (1962).

McCahon, D., Cooper, P.D.: Identification of poliovirus temperature-sensitive mutants having defects in virus structural protein. J. Gen. Virol. 6:51-62 (1970).

McClain, K., Stewart, M., Sullivan, M., Maizel, J. V., Jr.: Ribosomal binding sites on poliovirus RNA. Virology 113: 150-167 (1981).

McClure, M.A., Holland, J.J., Perrault, J.: Generation of defective interfering particles in picornaviruses. Virology 100: 408-418 (1980).

McCormick, W., Penman, S.: Regulation of protein synthesis in HeLa cells: Translation at elevated temperatures. J. Mol. Biol. 39: 315-333 (1969).

McDonnell, J.P., Levintow, L.: Kinetics of appearance of the products of poliovirus-induced RNA polymerase. Virology 42: 999-1006 (1970).

McGeady, M.L., Crowell, R.L.: Stabilization of "A" particles of coxsackievirus B3 by HeLa cell plasma membrane extract. J. Virol. 32: 790-795 (1979).

McGeady, M.L., Crowell, R.L.: Proteolytic cleavage of VP1 in "A" particles of coxsackievirus B3 does not appear to mediate virus uncoating by HeLa cells. J. Gen. Virol. 55: 439-450 (1981).

McGregor, S., Mayor, H.D.: Biophysical studies on rhinovirus and poliovirus. I. Morphology of viral nucleoprotein. J. Virol. 2: 149-154 (1968).

McGregor, S., Mayor, H.D.: Internal components released from rhinovirus and poliovirus by heat. J. Gen. Virol. 7: 41-56 (1971a).

McGregor, S., Mayor, H.D.: Biophysical studies on rhinovirus and poliovirus. II Chemical and hydrodynamic analysis of the rhinovirus. J. Virol. 7: 41-46 (1971b).

McGregor, S., Rueckert, R.R.: Picornaviral capsid assembly: Similarity of rhinovirus and enterovirus precursor subunits. J. Virol. 21: 548-553 (1977).

McGregor, S., Hall, L., Rueckert, R.: Evidence for the existence of protomers in the assembly of encephalomyocarditis virus. J. Virol. 15: 1107-1120 (1975).

McKeehan, W.L.: Regulation of hemoglobin synthesis - effect of concentration of messenger ribonucleic-acid, ribosome subunits, initiation-factors, and salts on ratio of alpha-chains and beta-chains synthesized in vitro. J. Biol. Chem. 249: 6517 (1974).

McLaren, L.C., Holland, J.J., Syverton, J.T.: The mammalian cell-virus relationship. I. Attachment of poliovirus to cultivated cells of primate and non-primate origin. J. Exp. Med. 109: 475-485 (1959).

McLaren, L.C., Scaletti, J.V., James, C.G.: Isolation and properties of enterovirus receptors. In: Biological Properties of the Mammalian Surface Membrane (Manson, ed.), 123-135. Philadelphia: Wistar Institute Press 1968.

McLean, C., Matthews, T.J., Rueckert, R.R.: Evidence of ambiguous processing and selective degradation in the non-capsid proteins of rhinovirus 1A. J. Virol. 19: 903-914 (1976).

McSharry, J.J., Caliguiri, L.A., Eggers, H: Inhibition of uncoating of poliovirus by arildone, a new antiviral drug. Virology 97: 307-315 (1979).

Means, G.E., Feeney, R.E.: Chemical Modification of Proteins, 1-254. San Francisco: Holden-Day 1971.

Mechali, M., Pérez-Bercoff, R., Carrara, G., Falcoff, E.: An improved methodology for preparation of virus-induced double-stranded RNA. Biochimie 55: 361-363 (1973).

Medin, O.: Über eine Epidemie von spinaler Kinderlähmung. Verhandl. X Internat. med. Kongr. 1890, Berlin, 2, Abt. 6: 37 (1891).

Medrano, L., Green, H.: Picornavirus receptors and picornavirus multiplication in human mouse hybrid cell lines. Virology 54: 515-525 (1973).

Medvedkina, O.A., Scarlat, I.V., Kalinia, N.O., Agol, V.I.: Virus-specific proteins associated with ribosomes of Krebs-II cells infected with encephalomyocarditis virus. FEBS Lett. 39: 4-8 (1974).

Melnick, J.L.: Attenuated poliovirus vaccine: virus stability. In: Poliomyelitis. Fifth International Poliomyelitis Conference, Copenhagen, 384-402. Philadelphia: Lippincott 1961.

Melnick, J.L.: Classification and nomenclature of viruses. In: Ultrastructure of Animal Viruses and Bacteriophages (Dalton, A.J., Haguenau, F., eds.), 1-6. New York - London: Academic Press 1973.

Melnick, J.L.: Enteroviruses. In: Viral Infections of Humans: Epidemiology and Control (Evans, ed.), 163-207. New York: Plenum 1976.

Melnick, J.L.: Classification of hepatitis A virus as enterovirus type 72 and of hepatitis B virus as hepadnavirus type 1. Intervirology 18: 105-106 (1982).

Melnick, J.L., Wallis, C.: Effect of pH on thermal stabilization of oral poliovirus vaccine by magnesium chloride. Proc. Soc. Exp. Biol. Med. 112: 894-897 (1963).

Melnick, J.L., Crowther, D., Barrera-Oro, J.: Rapid development of drug resistant mutants of poliovirus. Science 134: 557 (1961).

Melnick, J.L., Agol, V.I., Bachrach, H.L., Brown, F., Cooper, P.D., Fiers, W., Gard, S., Gear, J.H.S., Ghendon, Y., Kasza, L., LaPlaca, M., Mandel, B., McGregor, S., Mohanty, S.B., Plummer, G., Rueckert, R.R., Schaffer, F.L., Tagaya, I., Tyrrell, D. A.J., Voroshilova, M., Wenner, H.A.: Picornaviridae. Intervirology 4: 303-316 (1974).

Meloen, R.H., Rowlands, D.J., Brown, F.: Comparison of the antibodies elicited by the individual structural polypeptides of foot-and-mouth disease and polioviruses. J. Gen. Virol. 45: 761-763 (1979).

Meyer, D.I., Krause, E., Dobberstein, B.: Secretory protein translocation across membranes - the role of the "docking protein". Nature 297: 647-650 (1982).

Meyer, J., Lundquist, R.E., Maizel, J.V., Jr.: Structural studies of the RNA component of the poliovirus replication complex. Virology 85: 445-455 (1978).

Micklem, K.J., Pasternak, C.A.: Surface components involved in virally mediated membrane changes. Biochem. J. 162: 405-410 (1977).

Mietens, C., Koschel, K.: RNA Gehalt und Antigeneigenschaften von Polioviren vor und nach Hitzeinaktivierung. Z. Naturforsch. 26b: 945-950 (1971).

Mikhejeva, A., Yakobson, E., Soloviev, G.Y.: Characterization of some poliovirus temperature-sensitive mutants and poliovirus-related particle formation under nonpermissive conditions. J. Virol. 6: 188 (1970).

Mikhejava, A., Yakobson, E., Ghendon, Y.Z.: Studies on temperature sensitive events in synthesis of poliovirus temperature sensitive mutants. Arch. ges. Virusforsch. 43: 352 (1973).

Miller, A.O.: Study of RNA extracted from HeLa cell polysomes. Isopycnic centrifugation of cytoplasmic particles extracted from normal, poliovirus-infected HeLa cells and mouse A9 cells. Arch. Biochem. Biophys. 150: 282-295 (1972).

Miller, D.A., Miller, O.J., Dev, V.G., Hashmi, S., Tantravahi, R., Medrano, L., Green, H.: Human chromosome 19 carries a poliovirus receptor gene. Cell 1: 167-173 (1974).

Miller, J.R., Guntaka, R.V., Myers, J.C.: Amyotrophic lateral sclerosis: search for poliovirus by nucleic acid hybridization. Neurology 30: 884-886 (1980).

Miller, O.L.: The nucleolus, chromosomes, and visualization of genetic activity. J. Cell Biol. 91: 15s-27s (1981).

Milstein, C., Brownlee, G.G., Harrison, T.M., Methews, M.B.: A possible precursor of immunoglobin light chains. Nature New Biol. 239: 117-120 (1972).

Milstien, J.B., Walker, J.R., Eron, L.J.: Correlation of virus polypeptide structure with attenuation of poliovirus type 1. J. Gen. Virol. 23: 811-815 (1977).

Minor, P.D.: Characterization of strains of type 3 poliovirus by oligonucleotide mapping. J. Gen. Virol 59: 307-317 (1982).

Minor, P.D., Schild, G.C., Wood, J. M., Dandawate, C.N.: The preparation of specific immune sera

against type 3 poliovirus D-antigen and C-antigen, and the demonstration of two C-antigenic components in vaccine strain populations. J. Gen. Virol. 51: 147-156 (1980).
Minor, P.D., Kew, O., Schild, G.C.: Poliomyelitis-epidemiology, molecular biology and immunology. Nature 299: 109-110 (1982a).
Minor, P.D., Schild, G.C., Ferguson, M., Mackay, A., Magrath, D.I., John, A., Yates, J.P., Spitz, M.: Genetic and antigenetic variation in type 3 polioviruses: Characterization of strains by monoclonal antibodies and T1 oligonucleotide mapping. J. Gen. Virol. 61: 167-176 (1982b).
Minor, P.D., Schild, G., Bootman, J., Evans, D.M.A., Ferguson, M., Reeve, P., Spitz, M., Stanway, G., Cann, A.J., Hauptmann, R., Clarke, L.D., Mountford, R.C., Almond, J.W.: Location and primary structure of a major antigenic site for poliovirus neutralization. Nature 301: 674 (1983).
Miroff, G., Cornatzer, W.E., Fischer, R.G.: The effect of poliomyelitis virus type 1 (Mahoney strain) on the phosphorous metabolism of the HeLa cell. J. Biol. Chem. 228: 255-262 (1957).
Mitchell, W.R., Tershak, D.R.: Synthesis of complementary RNA by poliovirus polymerase in vitro. Virology 56: 386-389 (1973a).
Mitchell, W.R., Tershak, D.R.: The synthesis of complementary ribonucleic acid during infection with LSc poliovirus. Virology 54: 290-293 (1973b).
Mitra, S.: DNA replication in viruses. Ann. Rev. Genet 14: 347-397 (1980).
Mittelstaedt, R., Oppermann, H., Koch, G.: Poliovirus-induced infectious doublestranded RNA: Effect of RNA-degrading enzymes. Arch. Virol. 47: 381-392 (1975).
Montagnier, L., Sanders, F.K.: Sedimentation properties of infective ribonucleic acid extracted from encephalomyocarditis virus. Nature 197: 1178 (1963a).
Montagnier, L., Sanders, F.K.: Replicative form of encephalomyocarditis virus ribonucleic acid. Nature 199: 664 (1963b).
Montelaro, R.C., Rueckert, R.R.: Radiolabeling of proteins and viruses in vitro by acetylation with radioactive acetic anhydride. J. Biol. Chem. 250: 1413-1421 (1975).
Moolenaar, W.H., Mummery, C.L., van der Saag, P.T., de Laat, S.W.: Rapid ionic events and the initiation of growth in serum-stimulated neuroblastoma cells. Cell 23: 789-798 (1981).
Moore, D.M.: Characterization of three antigenic particles of swine vesicular disease virus. J. Gen. Virol. 34: 431-445 (1977).
Moore, D.M., Cowan, K.M.: Effect of trypsin and chymotrypsin on the polypeptides of large and small plaque variants of foot-and-mouth disease virus: relationship to specific antigenicity and infectivity. J. Gen. Virol. 41: 549-562 (1978).
Moore, N.F., Patzer, E.J., Shaw, J.M., Thompson, T.E., Wagner, R.R.: Interaction of vesicular stomatitis virus with lipid vesicles: depletion of cholesterol and effect on virion membrane fluidity and infectivity. J. Virol. 27: 320-329 (1978).
Moore, N.F., Tinsley, T.W.: The small RNA-viruses of insects. Arch. Virol. 72: 229-245 (1982).
Morrell, D., Brown, F.: The structure of foot-and-mouth disease virus. Second Meeting of the European Study Group on the Molecular Biology of Picornaviruses, Hamburg 1981.
Morrow, C.D., Dasgupta, A.: Antibody to a synthetic nonapeptide corresponding to the NH_2 terminus of poliovirus genome-linked protein VPg reacts with native VPg and inhibits in vitro replication of poliovirus RNA. J. Virol. 48: 429-439 (1983).
Moss, B.: Inhibition of HeLa cell protein synthesis by the vaccinia virion. J. Virol. 2: 1028-1037 (1968).
Mosser, A.G., Caligiuri, L.A., Tamm, I.: Blocking action of guanidine on poliovirus multiplication. Virology 45: 653-663 (1971).
Mosser, A.G., Caliguiri, L.A., Scheid, A.S., et al.: Chemical and enzymatic characteristics of cytoplasmic membranes of poliovirus-infected HeLa cells. Virology 47: 30-38 (1972a).
Mosser, A.G., Caliguiri, L.A., Tamm, I.: Incorporation of lipid precursors into cytoplasmic membranes of poliovirus-infected HeLa cells. Virology 47: 39-47 (1972b).
Moyer, A.W., Accorti, C., Cox, H.R.: Poliomyelitis. I. Propagation of MEF-1 strain of poliomyelitis virus in the suckling hamster. Proc. Soc. Exp. Biol. 81: 513-518 (1952).
Much, D.H., Zajac, I.: Homology of surface receptors for poliovirus on mammalian cell lines. J. Gen. Virol. 21: 385-390 (1973).
Much, D.H., Zajac, I.: The effect of an antireceptor serum on mammalian cell lines. J. Gen. Virol. 23: 205-208 (1974).
Munyon, W., Salzman, N.P.: The incorporation of 5-fluorouracil into poliovirus. Virology 18: 95-101 (1962).

Muthukrishnan, S., Both, G.W., Furuichi, Y., Shatkin, A.J.: 5'-terminal 7-methylguanosine in eukaryotic mRNA is required for translation. Nature 255: 33-37 (1975a).

Muthukrishnan, S., Filipowicz, W., Sierra, J. M., Both, G. W., Shatkin, A. J., Ochoa, S.: mRNA methylation and protein synthesis in extracts from embryos of brine shrimp, Artemia salina. J. Biol. Chem. 250: 9336-9341 (1975b).

Nair, C.N.: Monovalent cation metabolism and cytopathic effects of poliovirus-infected HeLa cells. J. Virol. 37: 268-273 (1981).

Nair, C.N., Panicali, D.L.: Polyadenylate sequence of human rhinovirus and poliovirus RNA and cordycepin sensitivity of virus replication. J. Virol. 20: 170-176 (1976).

Nair, C.N., Stowers, J.W., Singfield, B.: Guanidine-sensitive Na^+ accumulation by poliovirus-infected HeLa cells. J. Virol. 31: 184-189 (1979).

Nakano, J.H., Gelfand, H.M.: The use of a modified Wecker technique for the serodifferentiation of type 1 polioviruses related and unrelated to Sabin's vaccine strain. I. Standardization and evaluation of the test. Am. J. Hyg. 75: 363-376 (1962).

Nakano, J.H., Gelfand, H.M., Cole, J.T.: The use of a modified Wecker technique for the serodifferentiation of type 1 polioviruses related and unrelated to Sabin's vaccine strain. II. Antigenic segregation of isolates from specimens collected in field studies. Amer. J. Hyg. 78: 214-226 (1963a).

Nakano, M., Iwame, S., Tagaya, I.: A guanidine-dependent variant of poliovirus. Isolation of a variant and some of its biological properties. Virology 21: 264-266 (1963b).

Nakano, J.H., Hatch, M.H., Thieme, M.L., Nottay, B.: Parameters for differentiating vaccine-derived and wild poliovirus strains. Prog. Med. Virol. 24: 178-206 (1978)

Nakano, M., Matsukura, T., Yoshii, K., Komatsu, T., Mukoyama, A., Uchida, N., Kodama, H., Tagaya, I.: Genetical analysis of the stability of poliovirus type 3 Leon $12a_1b$. J. Biol. Stand. 7: 157-168 (1979).

Naso, R., Arcement, L.J., Karshin, W.L., Jamjoom, G.A., and Arlinghaus, R.B.: A fucose-deficient glycoprotein precursor to Rauscher Leukemia virus gp69/71. Proc. Natl. Acad. Sci. USA 73: 2326-2330 (1976).

Nathanson, N.: Eradication of poliomyelitis in the United States. Rev. Infect. Dis. 4: 940-950 (1982).

Neurath, H., Walsh, K.A.: Role of proteolytic enzymes in biological regulation (a review). Proc. Natl. Acad. Sci. USA 73: 3825-3832 (1976).

Newman, J.F.E., Rowlands, D.J., Brown, F.: A physicochemical sub-grouping of the mammalian picornaviruses. J. Gen. Virol. 18: 171-180 (1973).

Nobis, P., Meyer, A.G., Kühne, J., Koch, G.: The effect of the ionophore monensin on the replication of poliovirus. Eur. J. Cell Biol. Suppl. 2: 30 (1983).

Noble, J., Levintow, L.: Dynamics of poliovirus-specific RNA synthesis and the effects of inhibitors of virus replication. Virology 40: 634-642 (1970).

Noble, J., Lonberg-Holm, K.: Interactions of components of human rhinovirus type 2 with HeLa cells. Virology 51: 270-278 (1973).

Noble, J., Kass, S.J., Levintow, L.: Analysis of poliovirus-specific RNA in infected HeLa cells by polyacrylamide gel electrophoresis. Virology 37: 535-544 (1969).

Noble-Harvey, J., Lonberg-Holm, K.: Sequential steps in attachment of human rhinovirus type 2 to HeLa cells. J. Gen. Virol. 25: 83-91 (1974).

Nomoto, A., Lee, Y.J., Wimmer, E.: The 5'-end of poliovirus mRNA is not capped with $m^7G(5')$ ppp (5')Np. Proc. Natl. Acad. Sci. USA 73: 375-380 (1976).

Nomoto, A., Detjen, B., Pozzatti, R., Wimmer, E.: The location of the polio genome protein in viral RNAs and its implication for RNA synthesis. Nature (Lond.) 268: 208-213 (1977a).

Nomoto, A., Kitamura, N., Golini, F., Wimmer, E.: The 5'-terminal structures of poliovirion RNA and poliovirus mRNA differ only in the genome-linked protein VPg. Proc. Natl. Acad. Sci. USA 74: 5345-5349 (1977b).

Nomoto, A., Kajigaya, S., Suzuki, K., Imura, N.: Possible point mutation sites in LSc, 2ab poliovirus RNA and a protein covalently linked to the 5'-terminus. J. Gen. Virol. 45: 107-117 (1979a).

Nomoto, A., Jacobson, A., Lee, Y.F., Dunn, J., Wimmer, E.: Defective interfering particles of poliovirus: mapping of the deletion and evidence that the deletions in the genome of DI (1), (2), (3) are located in the same region. J. Mol. Biol. 128: 179-196 (1979b).

Nomoto, A., Kitamura, N., Lee, J.J., Rothberg, P.G., Imura, N., Wimmer, E.: Identification of point mutations in the genome of the poliovirus Sabin vaccine LSc 2ab, and catalogue of RNase T1- and RNAse A-resistant oligonucleotides of poliovirus type 1 (Mahoney) RNA. Virology 112: 217-227 (1981a).

Nomoto, A., Toyoda, H., Imura, N.: Comparative sequence analysis of the 5'-terminal noncoding regions of poliovirus vaccine strain Sabin 1, Sabin 2, and Sabin 3 genomes. Virology 113: 54-63 (1981b).

Nomoto, A., Toyoda, H., Imura, N., Noguchi, S., Sikiya, T.: Restriction map of double-stranded DNA copy synthesized from poliovirus Sabin 1 RNA. J. Biochem. 91: 1593-1600 (1982a).

Nomoto, A., Omata, T., Toyoda, H., Kuge, S., Horie, H., Kataoka, Y., Genba, Y., Nakano, Y., Imura, N.: Complete nucleotide sequence of the attenuated poliovirus Sabin 1 strain genome. Proc. Natl. Acad. Sci. USA 79: 5793-5797 (1982b).

Nomura, M., Yates, J.L., Dean, D., Post, L.E.: Feedback regulation of ribosomal protein gene expression in Escherichia coli: Structural homology of ribosomal RNA and ribosomal protein mRNA. Proc. Natl. Acad. Sci. USA 77: 7084-7088 (1980).

Nomura, S., Takemori, N.: Mutation of polioviruses with respect to size of plaques. I. Selection of minute plaque mutants of three types of polioviruses in tissue culture. Virology 12: 154-170 (1960).

Nottay, B.K., Kew, O.M., Hatch, M.H., Heyward, J.T., Obijeski, J.F.: Molecular variation of type 1 vaccine-related and wild polioviruses during replication in humans. Virology 108: 405-423 (1981).

Novikoff, A.: The endoplasmic reticulum: a cytochemist's view (a review). Proc. Natl. Acad. Sci. USA 73: 2781-2787 (1976).

Nuss, D.L., Koch, G.: Variation in the relative synthesis of IgG and non-IgG proteins in cultured MPC-11 cells with changes in the overall rate of polypeptide chain initiation and elongation. J. Mol. Biol. 102: 601-612 (1976).

Nuss, D.L., Oppermann, H., Koch, J.G.: Selective blockage of initiation of host protein synthesis in RNA-virus-infected cells. Proc. Natl. Acad. Sci. USA 72: 1258-1262 (1975a).

Nuss, D.L., Furuichi, Y., Koch, G., Shatkin, A.J.: Detection in HeLa cell extracts of a 7-methyl guanosine specific enzyme activity that cleaves -m^7GpppN^m. Cell 6: 21-27 (1975b).

Öberg, B.F.: Biochemical and biological characteristics of carbethoxylated poliovirus and viral RNA. Biochim. Biophys. Acta 204: 430-440 (1970).

Öberg, B.F., Philipson, L.: Replication of poliovirus RNA studied by gel filtration and electrophoresis. Eur. J. Biochem. 11: 305 (1969).

Öberg, B.F., Philipson, L: Replicative structures of poliovirus RNA in vivo. J. Mol. Biol. 58: 725-737 (1971).

Öberg, B.F., Shatkin, A.J.: Initiation of picornavirus protein synthesis in ascites cell extracts. Proc. Natl. Acad. Sci. USA 69: 3589-3593 (1972).

Ofengand, J.: Structure and function of tRNA and aminoacyl tRNA synthetases in eukaryotes. In: Protein Biosynthesis in Eukaryotes (Pérez-Bercoff, R., ed.), 1-67. New York: Plenum Press 1982.

Ohuchi, M., Homma, M.: Trypsin action on the growth of Sendai virus in tissue culture cells. IV. Evidence for activation of Sendai virus by cleavage of a glycoprotein. J. Virol. 18: 1147-1150 (1976).

Oliver, J.M., Berlin, R.D.: Mechanisms that regulate the structural and functional architecture of cell surfaces. Int. Rev. Cytol. 74: 55-94 (1982).

Olsnes, S., Sandvig, K.: Entry of toxic proteins into cells. In: Receptor-Mediated Endocytosis. Receptors and Recognition, Series B (Cuatrecasas, P., Roth, T.F., eds.), Vol. 15, 187-236. London: Chapman and Hall 1983.

Oppermann, H., Koch, J.G.: Kinetics of poliovirus replication in HeLa cells infected by isolated RNA. Biochem. Biophys. Res. Commun. 52: 635-640 (1973).

Oppermann, H., Koch, G.: On the regulation of protein synthesis in vaccinia virus infected cells. J. Gen. Virol. 32: 261-273 (1976a).

Oppermann, H., Koch, G.: Individual translational efficiencies of SV40 mRNAs in BSC-1 cells. Arch. Virol. 52: 123-134 (1976b).

Oppermann, H., Saborio, J., Zarucki, T., Koch, J.G.: Sensitization of cells for viral RNA infection by inhibition of macromolecular synthesis. Fed. Proc. 32: 531 A (1973).

Osborn, M., Weber, K.: Cytoplasmic microtubules in tissue culture cells appear to grow from an organizing structure towards the plasma membrane. Proc. Natl. Acad. Sci. USA 73: 867-871 (1976).
Osborn, M., Weber, K.: The detergent-resistant cytoskeleton of tissue culture cells includes the nucleus and the microfilament bundles. Exp. Cell Res. 106: 339-349 (1977).
Osborn, M., Weber, K.: Intermediate filaments: cell-type-specific markers in differentiation and pathology. Cell 31: 303-306 (1982).
Osterhaus, A.D., van Wezel, A.L., van Steenis, B., Drost, G.A., Hazendonk, T.G.: Monoclonal antibodies to polioviruses. Production of specific monoclonal antibodies to the Sabin vaccine strains. Intervirology 16: 218-224 (1981a).
Osterhaus, A.D., van Wezel, A.L., van Steenis, G., Hazendonk, A.G., Drost, G.: Production and potential use of monoclonal antibodies against polioviruses. Dev. Biol. Stand. 50: 221-228 (1981b).

Pagano, J.S.: Equine and bovine serum and dextran sulfate as genetic markers of poliovirus. Ann. N.Y. Acad. Sci. 130: 398-403 (1965).
Pagano, J.S.: Biologic activity of isolated viral nucleic acids. Prog. Med. Virol., Vol. 12: 1-48. Basel-München-New York: Karger 1970.
Pagano, J.S., Böttiger, M.: Studies of the vaccine virus isolated during a trial with an attenuated poliovirus vaccine prepared in a human diploid cell strain, including use of a strain-specific bovine serum inhibitor. Arch. ges. Virusforsch. 15: 19-34 (1964).
Pagano, J.S., Sedwick, W.D.: Identification of LSc2ab strain of type-1 poliovirus by inhibition with horse serum and dextran sulfate. Proc. Soc. Exp. Biol. Med. 122: 1232-1236 (1966).
Pagano, J.S., Gilden, R.W., Sedwick, W.D.: The specificity and interaction with poliovirus of an inhibitory bovine serum. J. Immunol. 95: 909-917 (1965).
Palade, G.: Intracellular aspects of the process of protein secretion. Science 189: 347 (1975).
Pallansch, M.A., Kew, O.M., Palmenberg, A.C., Golini, F., Wimmer, E., Rueckert, R.R.: Picornaviral VPg sequences are contained within the replicase gene. J. Virol. 35: 414-419 (1980).
Palmenberg, A.C.: In vitro synthesis and assembly of picornaviral capsid intermediate structures. J. Virol. 44: 900-906 (1982).
Palmenberg, A.C., Rueckert, R.R.: Evidence for intramolecular self-cleavage of picornaviral replicase precursors. J. Virol. 41: 244-249 (1982).
Palmenberg, A.C., Pallansch, M.A., Rueckert, R.R.: Protease required for processing picornaviral coat protein resides in the viral replicase gene. J. Virol. 32: 770-778 (1979).
Papaevangelou, G.J., Youngner, J.S.: Correlation between heat-restistance of poliovirus and other genetic markers. Proc. Soc. Exp. Biol. Med. 108: 505-507 (1961).
Pardee, A.B.: Cell division and a hypothesis of cancer. Nat. Cancer Inst. Monogr. 14: 7 (1964).
Pardee, A.B., Dubrow, R., Hamlin, J.L., Kletzien, R.F.: Animal cell cycle. Ann. Rev. Biochem. 47: 715-750 (1978).
Pasternak, C.A., Micklem, K.J.: Virally induced alterations in cellular permeability: a basis of cellular and physiological damage? Biosci. Rep. 1: 451-448 (1981).
Paucha, E., Colter, J.S.: Evidence for control of translation of the viral genome during replication of mengovirus and poliovirus. Virology 67: 300-305 (1975).
Paucha, E., Seehafer, J., Colter, J.S.: Synthesis of viral-specific polypeptides in mengo virus-infected L cells: evidence for asymmetric translation of the viral genome. Virology 61: 315-326 (1974).
Paul, J.R.: Observations on the history of poliomyelitis. Bull. Amer Clin. Lab., Penna. Hosp. 4: 21-33 (1951).
Paul, J.R.: A History of Poliomyelitis. New Haven: Yale University Press 1971.
Peabody, F.W., Draper, G., Dochez, A.R.: A clinical study of acute poliomyelitis, Monogr. 4, Rockefeller Inst. Med. Res. (1912).
Pearce, P., Pearce, S.: Experiments in Form. New York: Nostrand 1980.
Pearce, T.C., Rowe, A.J., Turnock, G.: Detrmination of molecular weights of RNAs by low-speed sedimentation equilibrium; 16S ribosomal RNA as a model compound. J. Mol. Biol. 97: 193-205 (1975).
Pelham, H.R.: Translation of encephalomyocarditis virus RNA in vitro yields an active proteolytic processing enzyme. Eur. J. Biochem. 85: 457-462 (1978).

Pelham, H.R.B., Jackson, R.J.: An efficient mRNA-dependent translation system from reticulocyte lysates. Eur. J. Biochem. 85: 457-462 (1978).
Penman, S.: Stimulation of the incorporation of choline in poliovirus-infected cells. Virology 25: 148 (1965).
Penman, S., Summers, D.: Effects on host cell metabolism following synchronous infection with poliovirus. Virology 27: 614-620 (1965).
Penman, S., Scherrer, K., Becker, Y., Darnell, J.E.: Polysomes in normal and poliovirus-infected HeLa cells and their relationship to messenger-RNA. Proc. Natl. Acad. Sci. USA 49: 654-661 (1963).
Penman, S., Becker, Y., Darnell, J.E.: A cytoplasmic structure involved in the synthesis and assembly of poliovirus components. J. Mol. Biol. 8: 541-555 (1964).
Penman, S., Fan, H., Perlman, S., Rosbash, M., Weinberg, R., Zylber, E.: Distinct RNA synthesis systems of the HeLa cell. Cold Spring Harbor Symp. Quant. Biol. 35: 561-575 (1970).
Penman, S., Fulton, A., Capco, D., Ben Zeev, A., Wittelsberger, S., Tse, C.F.: Cytoplasmic and nuclear architecture in cells and tissue: form, functions, and mode of assembly. Cold Spring Harbor Symp. Quant. Biol.: 1013-1028 (1982).
Pérez-Bercoff, R.: The mechanism of replication of picornavirus RNA. In: The Molecular Biology of Picornaviruses, (Pérez-Bercoff, R., ed.), 293-318. New York: Plenum Press 1979.
Pérez-Bercoff, R.: But is the 5' end of messenger RNA always involved in initiation? In: Protein Biosynthesis in Eukaryotes (Pérez-Bercoff R., ed.), 245-252. New York: Plenum Press 1982.
Pérez-Bercoff, R., Kaempfer, R.: Genomic RNA of mengovirus V. Recognition of common features by ribosomes and eucaryotic initiation factor 2. J. Virol. 41: 30-41 (1982).
Pérez-Bercoff, R., Cioe, L., Degemer, A.M., Mes, P., Rita, G.: Infectivity of mengovirus replicative form. IV. Intracellular conversion into replicative intermediate. Virology 96: 307-310 (1979).
Perlin, M., Phillips, B.A.: In vitro assembly of polioviruses. III. Assembly of 14S particles into empty capsids by poliovirus-infected HeLa cell membranes. Virology 53: 107-114 (1973).
Perlin, M., Phillips, B.A.: In vitro assembly of polioviruses. IV. Evidence for the existence of two assembly steps in the formation of empty capsids from 14S particles. Virology 63: 505-511 (1975).
Pernis, B., Forni, L. Amante, L.: Immunoglobin spots on the surface of rabbit lymphocytes. J. Exp. Med. 132: 1001-1018 (1970).
Perry, R.P.: RNA processing comes of age. J. Cell Biol. 91: 28s-38s (1981).
Perry, R.P.: Messenger RNA structure and biosynthesis. In: Protein Biosynthesis in Eukaryotes (Pérez-Bercoff, R., ed.), 119-135. New York - London: Plenum Press 1982.
Peters, K., Richards, F.M.: Chemical cross-linking reagents and problems in studies of membrane structure. Ann. Rev. Biochem. 46: 523-551 (1977).
Pettersson, R.F., Flanegan, J.B., Rose, J.K., Baltimore, D.: 5'-terminal nucleotide sequences of poliovirus polyribosomal RNA and virion RNA are identical. Nature 268: 270-272 (1977).
Petterson, R.F., Ambros, V., Baltimore, D.: Identification of a protein linked to nascent poliovirus RNA and to the polyuridylic acid of negative-strand RNA. J. Virol. 27: 357-365 (1978).
Philipson, L.: The early interaction of animal viruses and cells. Prog. Med. Virol. 5: 43-78 (1963).
Philipson, L., Choppin, P.W.: On the role of virus sulfhydryl groups in the attachment of enteroviruses to erythrocytes. J. Exp. Med. 112: 455 (1960).
Philipson, L., Bengtsson, S.: Interaction of enteroviruses with receptors from erythrocytes and host cells. Virology 18: 457-469 (1962).
Philipson, L., Choppin, P.W.: Inactivation of enteroviruses by 2,3-di-mercaptopropanol(BAL). Virology 16: 405-413 (1962).
Philipson, L., Lind, M.: Enterovirus eclipse in a cell-free system. Virology 23: 322 (1964).
Philipson, L., Bengtsson, S., Barberaoro, J.: The reversion of guanidine inhibition of poliovirus synthesis. Virology 29: 317-329 (1966).
Philipson, L., Lonberg-Holm, K., Petersson, U.: Virus receptor interaction in an adenovirus system. J. Virol. 2: 1064-1075 (1968).
Philipson, L., Beatrice, S.T., Crowell, R.I.: A structural model of picornaviruses as suggested from an analysis of urea-degraded virions and procapsids of coxsackievirus B3. Virology 54: 69-79 (1973).
Phillips, B.A.: In vitro assembly of polioviruses. I. Kinetics of the assembly of empty capsids and the role of extracts from infected cells. Virology 39: 811-821 (1969).
Phillips, B.A.: In vitro assembly of poliovirus. II. Evidence for the self-assembly of 14S particle into empty capsids. Virology 44: 307-316 (1971).

Phillips, B.A.: The morphogenesis of poliovirus. Curr. Top. Microbiol. and Immunol. 58: 156-174 (1972).

Phillips, B.A., Fennel, R.: Polypeptide composition of poliovirions, naturally occurring empty capsids, 14S precursor particles. J. Virol. 12: 291-299 (1973).

Phillips, B.A., Fennell, R.: Unusual attachment behavior exhibited by a urea-resistant mutant of poliovirus type 1. Virology 83: 295-304 (1978).

Phillips, B.A., Wiemert, S.: In vitro assembly of poliovirus. V. Evidence that the self-assembly of 14S particles is independent of extract assembly factor(s) and host proteins. Virology 88: 92-104 (1978).

Phillips, B.A., Summers, D.F., Maizel, J.V., Jr.: In vitro assembly of poliovirus related particles. Virology 35: 216 (1968).

Phillips, B.A., Lundquist, R.E., Maizel, J.V., Jr.: Absence of subviral particles and assembly activity in HeLa cells infected with defective-interfering (DI) particles of poliovirus. Virology 100: 116-124 (1980).

Phillips, S.K., Cramer, W.A.: Properties of the fluorescence probe response associated with the transmission mechanism of colicin E1. Biochemistry 12: 1170-1176 (1973).

Pohjanpelto, P.: Stabilization of poliovirus by cystine. Virology 6: 472-487 (1958).

Polatnick, J.: Isolation of a foot-and-mouth disease polyuridylic acid polymerase and its inhibition by antibody. J. Virol. 33: 774-779 (1980).

Polatnick, J., Arlinghaus, R.B.: Foot-and-mouth disease virus-induced ribonucleic acid polymerase in baby hamster kidney cells. Virology 31: 601-608 (1967).

Polatnick, J., Wool, S.H.: Characterization of a 70S polyuridylic acid polymerase isolated from foot-and-mouth disease virus-infected cells. J. Virol. 40: 881-889 (1981).

Polatnick, J., Wool, S.H.: A novel structure seen when foot-and-mouth disease virus-induced poly (U) polymerase acts in a cell-free system. J. Ultrastruct. Res. 80: 363-366 (1982a).

Polatnick, J., Wool, S.H.: Localization of foot-and-mouth disease - RNA synthesis on newly formed cellular smooth membranous vacuoles. Arch. Virol. 71: 207-215 (1982b).

Polatnick, J., Arlinghaus, R.B., Graves, J.H., Cowan, K.M.: Inhibition of cell-free foot-and-mouth disease virus RNA synthesis by antibody. Virology 31: 609-615 (1967).

Pollack, R., Goldman, R.: Synthesis of infective poliovirus in BSC-1 monkey cells enucleated with cytochalasin B. Science 179: 915-916 (1973).

Pollard, T.D., Craig, S.W.: Mechanism of actin polymerization. Trends Biochem. Sci. 7: 55-58 (1982).

Pong, S.S., Nuss, D.L., Koch, J.G.: Inhibition of initiation of protein synthesis in mammalian tissue culture cells by L-1-tosylamido-2-phenylethyl chloromethyl ketone. J. Biol. Chem. 250: 240-245 (1975).

Pons, M.: Stabilization of poliovirus by sodium tetrasulfide. Virology 22: 253-261 (1964a).

Pons, M.: Infectious double-stranded poliovirus RNA. Virology 24: 467-473 (1964b).

Porter, A., Carey, N., Fellner, P.: Presence of a large poly(C) tract within the RNA of encephalomyocarditis virus. Nature 248: 675-678 (1974).

Porter, A.G., Fellner, P., Balck, D.N., Brown, F.: 3'-terminal nucleotide sequences in the genome RNA of picornaviruses. Nature 276: 298-301 (1978).

Porter, A.G., Barber, C., Carey, N.H., Hallewell, R.A., Threlfall, G., Emtage, J.S.: Complete nucleotide sequence of an influenza virus haemagglutinin gene from cloned DNA. Nature 282: 471-477 (1979).

Porter, K.R.: Motility in cells - Introduction. In: Cell Motility (Goldman, R., Pollard, T., Rosenbaum, J., eds.), 1-28. New York: Cold Spring Harbor Laboratory 1976.

Poskanzer, D., Cantor, H., Kaplan, G.: The frequency of preceeding poliomyelitis in amyotrophic lateral sclerosis. In: Motor Neuron Diseases: Research on Amyotrophic Lateral Sclerosis and Related Disorders (Norris, F., Kurland, L., eds.), 286-289. New York: Grune & Stratton 1967.

Pouysségur, J., Chambard, J.C., Franchi, A., Paris, S., Van Obberghen-Schilling, E.: Growth factor activation of an amiloride-sensitive Na^+/H^+ exchange system in quiescent fibroblasts: Coupling to ribosomal protein S6 phosphorylation. Proc. Natl. Acad. Sci. USA 79: 3935-3939 (1982).

Powers, C.D., Miller, B.A., Kurtz, H., Ackermann, W.W.: Specific effect of guanidine in the programming of poliovirus inhibition of deoxyribonucleic acid synthesis. J. Virol. 3: 337-342 (1969).

Powers, J.C.: Reaction of serine proteases with halomethyl ketones. Methods Enzymol. 46: 197-208 (1977).

Price, D.L., Porter, K.R.: The response of ventral horn neurons to axonal transection. J. Cell Biol. 53: 24-37 (1972).

Priess, H., Eggers, H.J.: Synthesis and activity of RNA polymerase of a temperature sensitive poliovirus mutant at an elevated temperature. Nature 220: 1047 (1968).

Prokop, D.J., Kivirikko, K.I., Tuderman, L., Guzman, N.A.: The biosynthesis of collagen and its disorders. New Engl. J. Med. 301: 13-23, and 77-83 (1979).

Proudfoot, N.J.: Complete 3' noncoding region sequences of rabbit and human ß-globin messenger RNAs. Cell 10: 559-570 (1977).

Proudfoot, N.J., Brownlee, G.G.: 3' non-coding region sequences in eukaryotic messenger RNA. Nature 263: 211-214 (1976).

Puck, Marcus: Proc. Natl. Acad. Sci. USA 41: 432-437 (1955).

Pugh, A.: Polyhedra - A Visual Approach. Berkeley: University of California Press 1976.

Putnak, J.R., Phillips, B.A.: Differences between poliovirus empty capsids formed in vivo and those formed in vitro: a role for the morphopoetic factor. J. Virol. 40: 173-183 (1981a).

Putnak, J.R., Phillips, B.A.: Picornaviral structure and assembly. Microbiol. Rev. 45: 287-315 (1981b).

Putnak, J.R., Phillips, B.A.: Poliovirus empty capsid morphogenesis: evidence for conformational differences between self- and extract-assembled empty capsids. J. Virol. 41: 792-800 (1982).

Queen, C., Rosenberg, M.: Differential translation efficiency explains discoordinate expression of the galactose operon. Cell 25: 241-249 (1981).

Quersin-Thiry, L.: Action of anticellular sera on virus infections. I. Influence on homologous tissue cultures infected with various viruses. J. Immunol. 81: 253-260 (1958).

Quersin-Thiry, L.: Interaction between cellular extracts and animal viruses. I. Kinetic studies and some notes on the specificity of the interaction. Acta Virol. 5: 141-152 (1961).

Quersin-Thiry, L., Nihoul, E.: Interaction between cellular extracts and animal viruses. II. Evidence for the presence of different inactivators corresponding to different viruses. Acta Virol. 5: 283-293 (1961).

Quinn, A.J., Chapman, D.: The dynamics of membrane structure. CRC Crit. Rev. Biochem. 8: 1-117 (1980).

Racaniello, V.R., Baltimore, D.: Molecular cloning of poliovirus cDNA and determination of the complete nucleotide sequence of the viral genome. Proc. Natl. Acad. Sci. USA 78: 4887-4891 (1981a).

Racaniello, V.R., Baltimore, D.: Cloned poliovirus complementary DNA is infectious in mammalian cells. Science 214: 916-919 (1981b).

Racevskis, J., Kerwar, S.S., Koch, G.: Inhibition of protein synthesis in reticulocyte lysate by poliovirus. J. Gen. Virol. 31: 135-138 (1976).

Ramabhadran, T.V., Thach, R.E.: Specificity of protein synthesis inhibitors in the inhibition of encephalomyocarditis virus replication. J. Virol. 34: 293-296 (1980).

Rao, P.N., Engelberg, J.: HeLa cells: effects of temperature on the life cycle. Science 148: 1092-1094 (1965).

Ravin, A.W.: Bacterial genetics. Ann. Rev. Microbiol. 12: 309 (1958).

Reich, E., Franklin, R.M., Shatkin, A.J., Tatum, E.L.: Effect of actinomycin D on cellular nucleic acid synthesis and virus production. Science 124: 556 (1961).

Reich, E., Rifkin, D. Shaw, E. (eds.): Proteases and Biological Control. Cold Spring Harbor Laboratory, 1975.

Reissig, M., Howes, D.W., Melnick, J.L.: Sequence of morphological changes in epithelial cell cultures infected with poliovirus. J. Exp. Med. 104: 289-304 (1956).

Rekosh, D.: Gene order of the poliovirus capsid proteins. J. Virol. 9: 479-487 (1972).

Rekosh, D.M.K.: The molecular biology of picornaviruses. In: The Molecular Biology of Animal Viruses (Nayak, D.P., ed.), Vol. I, 63-110. New York: Dekker 1977.

Rekosh, D., Lodish, H.F., Baltimore, D.: Translation of poliovirus RNA by an E. Coli cell-free system. Cold Spring Harbor Symp. Quant. Biol. 34: 747-753 (1969).

Rekosh, D.M., Lodish, H.F., Baltimore, D.: Protein synthesis in Escherichia coli extracts programmed by poliovirus RNA. J. Mol. Biol. 54: 327 (1970).

Revel, M.: Interferon-induced translational regulation. Tex. Rep. Biol. Med. 35: 212-220 (1977).

Revel, M., Groner, Y.: Post-transcriptional and translational control of gene expression in eukaryotes. Ann. Rev. Biochem. 47: 1079-1126 (1978).

Revel, M., Content, J., Zilberstein, A., et al.: Control of mRNA translation by specific tRNAs in extracts from interferon treated mouse cells. In: In Vitro Transcription and Translation of Viral Genomes (Haenni, A.L., Beaud, G., eds.), 397-405. Paris: INSERM 1975.

Revie, D., Tseng, B., Grafstrom, R., Goulina, M.: Covalent association of protein with the replicative from DNA of parvovirus H-1. Proc. Natl. Acad. Sci. USA 11: 5539-5543 (1979).

Rhode, S.L., Klaassen, B.: DNA sequence of the 5' terminus containing the replication origin of parvovirus replicative form DNA. J. Virol. 41: 990-999 (1982).

Rice, J.M., Wolff, D.A.: Phospholipase in the lysosomes of HEp-2 cells and its release during poliovirus infection. Biochim. Biophys. Acta 381: 17-21 (1975).

Richards, O.C., Ehrenfeld, E., Manning, J.: Strand-specific attachment of avidin spheres to double-stranded poliovirus RNA. Proc. Natl. Acad. Sci. USA 76: 676-680 (1979).

Richards, O.C., Hey, T.D., Ehrenfeld, E.: Two forms of VPg on poliovirus RNAs. J. Virol. 38: 863-871 (1981).

Richter, J.D., Smith, D.: Differential capacity for translation and lack of competition between mRNAs that seregate to free and membrane-bound polysomes. Cell 27: 183-191 (1981).

Rifkind, R.A., Godman, G.C., Howe, C., Morgan, C., Rose, H.M.: Structure and development of viruses as observed in the electron microscope. VI. ECHO-virus, type 9. J. Exp. Med. 114: 1-12 (1961).

Righthand, V.F., Bagshaw, J.C.: Characterization of a regulator of host cell permissiveness to a specific enterovirus. Virology 86: 148-156 (1978).

Rightsel, W.A., Dice, J.R., McAlpine, R.J., Timm, E.A., McLean, I.W., Dixon, G.J., Schabel, F.M.: Antiviral effect of guanidine. Science 134: 558-559 (1961).

Ring, K., Heinz, E.: Active amino acid transport in streptomyces hydrogenans. I. Kinetics and uptake of α-aminoisobutyric acid. Biochem. Z. 344: 446 (1966).

Rissler, J.: Zur Kenntnis der Veränderungen des Nervensystems bei Poliomyelitis anterior acuta. Nord. med. Ark. 20, 22: 1-63 (1888).

Robberson, D.L., Thornton, G.B., Marshall, M.V., Arlinghaus, R.B.: Novel circular forms of mengovirus-specific double-stranded RNA detected by electron microscopy. Virology 116: 454-467 (1982).

Robbins, F.C., Enders, J.F.: Tissue culture techniques in study of animal viruses. Am. J. Med. Soc. 220: 316-338 (1950).

Robbins, F.C., Enders, J.F., Weller, T.H.: Cytopathogenic effect of poliomyelitis viruses in vitro on human embryonic tissues. Proc. Soc. Exp. Biol. Med. 75: 370-374 (1950).

Roberts, B.E., Paterson, B.M.: Efficient translation of tobacco mosaic virus RNA and rabbit globin 9S RNA in a cell-free system from commercial wheat germ. Proc. Natl. Acad. Sci. USA 70: 2230-2334 (1973).

Robertson, H.D., Mathews, M.D.: Double-stranded RNA as an inhibitor of protein synthesis and as a substrate for a nuclease in extracts of Krebs II ascites cells. Proc. Natl. Acad. Sci. USA 70: 225-229 (1973).

Robinson, I.K., Harrison, S.C.: The structure of the expanded state of tomato bushy stunt virus at 8 A resolution. Nature 297: 563-568 (1983).

Roca-Garcia, M., Jervis, G.A.: Experimentally produced poliomyelitis variant in chick embryo. Ann. N. Y. Acad. Sci. 61: 911-923 (1955).

Röder, A., Koschel, K.: Reversible inhibition of poliovirus RNA synthesis in vivo and in vitro by viral products. J. Virol. 14: 846-852 (1974).

Röder, A., Koschel, A.: Virus-specific proteins associated with the replication complex of poliovirus RNA. J. Gen. Virol. 28: 85-98 (1975).

Roesing, T.G., Toselli, P.A., Crowell, R.L.: Elution and uncoating of coxsackievirus B3 by isolated HeLa cell plasma membranes. J. Virol 15: 654-667 (1975).

Roizman, B, Mayer, M.M., Rapp, H.J.: Immunochemical studies of poliovirus particles produced in tissue culture. J. Immunol. 81: 419-425 (1958).

Roizman, B., Mayer, M.M., Roane, P.R.: Immunochemical studies of poliovirus. IV. Alteration of the immunologic specificity of purified poliovirus by heat and ultraviolet light. J. Immunol. 82: 19-25 (1959).

Romanova, L.I., Agol, V.I.: Interconversion of linear and circular forms of double-stranded RNA of encephalomyocarditis virus. Virology 93: 574-577 (1979).

Romanova, L.I., Tolskaya, E.A., Kolesnikova, M.S., Agol, V.I.: Biochemical evidence for intertypic genetic recombination of polioviruses. FEBS Lett. 118: 109-112 (1980).

Romanova, L.I., Tolskaya, E.A., Agol, V.I.: Antigenic and structural relatedness among non-capsid and capsid polypeptides of polioviruses belonging to different serotypes. J. Gen. Virol. 52: 279-289 (1981).

Rombaut, B., Vrijsen, R., Brioen, P., Boeye, A.: A pH-dependent antigenic conversion of empty capsids of poliovirus studied with the aid of monoclonal antibodies to N and H antigen. Virology 122: 215-218 (1982).

Rombaut, B., Vrijsen, R., Boeye, A.: Epitope evolution in poliovirus maturation. Arch. Virol. 76: 289-298 (1983).

Rose, J.K., Trachsel, H., Levy, K., Baltimore, D.: Inhibition of translation by poliovirus: Inactivation of a specific initiation factor. Proc. Natl. Acad. Sci. USA 75: 2732-2736 (1978).

Rose, K.M., Jacob, S.T., Kumar, A.: Poly (A) polymerase and poly (A) - specific mRNA binding protein are antigenically related. Nature 279: 260-262 (1979).

Rosenberg, H., Diskin, B., Kalmar, E., Traub, A.: The RNA-dependent RNA polymerase (replicase) of encephalomyocarditis virus. In: The Molecular Biology of Picornaviruses (Perez-Bercoff, R., ed.), 319-335. New York-London: Plenum Press 1979.

Rosenthal, E.T., Hunt, T., Ruderman, J.V.: Selective translation of mRNA controls the pattern of protein synthesis during early development of the surf clam, Spisula solidissima. Cell 20: 487-494 (1980).

Rosenwirth, B., Eggers, H.J.: Early processes of echovirus 12-infection: elution, penetration and uncoating under the influence of rhodamine. Virology 97: 241-255 (1979).

Rothberg, P.G., Wimmer, E: Monocucleotide and dinucleotide frequencies, and codon usage in poliovirion RNA. Nucl. Acids Res. 9: 6221-6229 (1981).

Rothberg, P.G., Harris, T.J.R., Nomoto, A., Wimmer, E.: The genome-linked protein of picornaviruses V. O4-(5'-Uridylyl)-tyrosine is the bond between the genome-linked protein and the RNA of poliovirus. Proc. Natl. Acad. Sci. USA 75: 4868-4872 (1978).

Rothman, J.: The Golgi apparatus: two organelles in tandem. Science 213: 1212-1219 (1981).

Rothman, J., Fine, R.: Coated vesicles transport newly synthesized membrane glycoproteins from endoplasmic reticulum to plasma membrane in two successive stages. Proc. Natl. Acad. Sci. USA 77: 780-784 (1980).

Rothman, J., Lenard, J.: Membrane asymmetry. Science 195: 743-753 (1977).

Rouhandeh, H., Chronister, R.R., Brinkman, M.L.: Inhibition of poliovirus minute plaque mutants and echoviruses by sulfated polysaccharides. Proc. Soc. Exp. Biol. Med. 118: 1118-1124 (1965).

Roumiantzeff, M., Maizel, J.V., Jr., Summers, D.F.: Comparison of polysomal structures of uninfected and poliovirus-infected HeLa cells. Virology 44: 239-248 (1971a).

Roumiantzeff, M., Summers, D.F., Maizel, J.V.: In vitro protein synthesis activity of membrane-bound poliovirus polyribosomes. Virology 44: 249-258 (1971b).

Rowlands, D.J., Sangar, D.V., Brown, F.: Relationship of the antigenic structure of foot-and-mouth disease virus to the process of infection. J. Gen. Virol. 13: 85-93 (1971a).

Rowlands, D.J., Sangar, D.V., Brown, F.: Buoyant density of picornaviruses in caesium salts. J. Gen. Virol. 13: 141-152 (1971b).

Rowlands, D.J., Sanger, D.V., Brown, F.: A comparative chemical and serological study of the full and empty particles of foot-and-mouth disease virus. J. Gen. Virol. 26: 227-238 (1975a).

Rowlands, D.J., Shirley, M.W., Sangar, D.V., Brown, F.: A high density component in several vertebrate enteroviruses. J. Gen. Virol. 29: 223-234 (1975b).

Roy, P., Bishop, D.H.L.: Isolation and properties of poliovirus minus strand-ribonucleic acid. J. Virol. 6: 604-609 (1970).

Rueckert, R.R.: Picornaviral architecture. In: Comparative Virology (Maramorosch, K., Kurstak, E., eds.), 255-306. New York: Academic Press 1971.

Rueckert, R.R.: On the structure and morphogenesis of picornaviruses. In: Comprehensive Virology (Frankel-Conrat, H., Wagner, R.R., eds.), Vol. 6, 131-213. New York: Plenum Press 1976.

Rueckert, R.R., Dunker, A.K., Stoltzfus, C.M.: The structure of Maus-Elberfeld virus: A model. Proc. Natl. Acad. Sci. USA 62: 912-919 (1969).

Rueckert, R.R., Mattews, T.J., Kew, O.M., Pallansch, M.A., Mclean, C. Omilianowski, D.: Synthesis and processing of picornaviral protein. In: The Molecular Biology of Picornaviruses (Perez-Bercoff, R. ed.), 113-125. New York: Plenum Press 1979.

Rueckert, R.R., Palmenberg, A.C., Pallansch, M.A.: Evidence for a self-cleaving precursor of virus-coded protease, RNA-replicase, and VPg. In: Biosynthesis, Modification, and Processing of Cellular and Viral Polyproteins (Koch, G., Richter, D., eds.), 263-275. New York: Academic Press 1980.

Rueckert, R., Icenogle, J., Anderegg, J.W., Gilbert, S., Grieves, J., Duke, G.: Neutralization of poliovirus by monoclonal antibody. Second Meeting of the European Study Group on the Molecular Biology of Picornaviruses, Hamburg (1981).

Ryser, H.J.: Studies on protein uptake by isolated tumor cells. III. Apparent stimulation due to pH, hypertonicity, polycations, or dehydration and their relation to the enhanced penetration of infectious nucleic acids. J. Cell Biol. 32: 737-750 (1967).

Ryser, H.J., Hancock, R.: Histones and basic polyamino acids stimulate the uptake of albumin by tumor cells in culture. Science 150: 501-503 (1965).

Sabatini, D.D., Kreibich, G., Morimoto, T., Adesnik, M.: Mechanisms for the incorporation of proteins in membranes and organelles. J. Cell Biol. 92: 1-22 (1982).

Sabin, A.B.: Epidemiologic patterns of poliomyelitis in different parts of the world. Papers & Disc. Internat. Polio. Conf. 1: 3 (1949).

Sabin, A.B.: Characteristics and genetic potentialities of experimentally produced and naturally occurring variants of poliomyelitis virus. Ann. N. Y. Acad. Sci. 61: 924-938 (1955).

Sabin, A.B.: Properties of attenuated polioviruses and their behavior in human beings. Spec. Pub. of the N.Y. Acad. of Sci. 5: 113-127 (1957).

Sabin, A.B.: Reproductive capacity of polioviruses of diverse origin at various temperatures. In: Perspectives in Virology II (Pollard, ed.), 90-110. The Gustav Stern Symposium 1961.

Sabin, A.B.: Vaccine control of poliomyelitis in the 1980s. Yale J. Biol. Med. 55: 383-389 (1982).

Sabin, A.B., Boulger, L.R.: History of Sabin attenuated poliovirus oral live vaccine strains. J. Biol. Stand. 1: 115-118 (1973).

Sabin, A.B., Lwoff, A.: Relation between reproductive capacity of poliovirus at different temperatures in tissue culture and neurovirulence. Science 129: 1287 (1959).

Sabin, A.S., Ward, R.: Nature of non-paralytic and transitory paralytic poliomyelitis in rhesus monkeys inocculated with human virus. J. Exp. Med. 73: 757-770 (1941a).

Sabin, A.B., Ward, R.: The natural history of poliomyelitis. I. Distribution of virus in nervous and non-nervous tissues. J. Exp. Med. 73: 771-793 (1941b).

Sabin, A.B., Hennessen, W.A., Winsser, J.: Studies on variants of poliomyelitis virus. I. Experimental segregation and properties of avirulent variants of three immunologic types. J. Exp. Med. 99: 551 (1954).

Saborio, J.L., Koch, G.: Reversible inhibition of protein synthesis in HeLa cells by dimethylsulfoxide. J. Biol. Chem. 248: 8343-8347 (1973).

Saborio, J.L., Pong, S.S., Koch, G.: Selective and reversible inhibition of initiation of protein synthesis in mammalian cells. J. Mol. Biol. 85: 195-21 (1974).

Saborio, J.L., Wiegers, K.J., Koch, G.: Effect of DEAE-dextran on protein synthesis in HeLa cells. Arch. Virol. 49: 81-87 (1975).

Safer, B., Adams, S.L., Kemper, W.M., Berry, K.W., Lloyd, M., Merrick, W.C.: Purification and characterization of two initiation factors required for maximal activity of a highly fractionated globin mRNA translation system. Proc. Natl. Acad. Sci. USA 73: 2584-2588 (1976).

Salk, J.E.: Principles of immunization as applied to poliomyelitis and influenza. Am. J. Pub. Health 43: 1384-1398 (1953).

Salk, J.E., Krech, U., Younger, J.S., Bennett, B.L., Lewis, L.J., Bazeley, P.L.: Formaldehyde treatment and safety testing of experimental poliomyelitis vaccines. Am. J. Pub. Health 44: 563 (1954).

Salzmann, N.P., Lockart, R.Z., Jr., Sebring, E.D.: Alterations in HeLa cell metabolism resulting from poliovirus infection. Virology 9: 244-259 (1959).

Samarina, O.P., Krichevskaya, A.A., Georgiev, G.P.: Nuclear ribonucleoprotein containing messenger ribonucleic acid. Nature (London) 210: 1319-1322 (1966).

Sander, G., Pardee, A.B.: Transport changes in synchronously growing CHO and L cells. J. Cell Physiol. 80: 267-271 (1972).

Sangar, D.V.: The replication of picornaviruses. J. Gen. Virol 45: 1-13 (1979).

Sangar, D.V., Black, D.N., Rowlands, D.J., Harris, T.J.R., Brown, F.: Location of the initiation site for protein synthesis on foot-and-mouth disease virus RNA by in vitro translation of defined fragments of the RNA. J. Virol. 33: 59-68 (1980).

Sangar, D.V., Bryant, J., Harris, T.J.R., Brown, F., Browlands, D.J.: Removal of the genome-linked protein of foot-and-mouth disease virus by rabbit reticulocyte lysate. J. Virol. 39: 67-74 (1981).

Sanger, F., Air, G.M., Barrel, B.G., Brown, N.L., Coulson, A.R., Fiddes, J.C., Hutchinson, C.A., Slogombe, P.M., Smith, M.: Nucleotide sequence of bacteriophage ∅X174 DNA. Nature 265: 687-695 (1977).

Saunders, K., King, A.M.: Guanidine-risistant mutants of aphthovirus induce the synthesis of an altered nonstructural polypeptide, P34. J. Virol. 42: 389-394 (1982).

Savage, T., Granboulan, N., Girard, M.: Architecture of the poliovirus replicative intermediate RNA. Biochimie 53: 533-543 (1971).

Savinskaya, S.S., Shekoyan, L.A., Shirman, G.A.: Demonstration of antigenic differences between strains of poliovirus type 1 by agar diffusion precipitation reaction. Acta. Virol. 23: 183-188 (1979).

Sawicki, S., Jelinic, W., Darnell, J.E.: 3'-Terminal addition to HeLa cell nuclear and cytoplasmic poly(A). J. Mol. Biol. 113: 219-235 (1977).

Schaefer, A., Kuehne, J., Zibirre, R., Koch, G.: Poliovirus-induced alterations in HeLa cell membrane functions. J. Virol. 44: 445-449 (1983a).

Schaefer, A., Zibirre, R., Kabus, P., Kuehne, J., Koch, G.: Alterations in plasma-membrane functions after poliovirus infection. Biosci. Rep. 2: 613-615 (1983b).

Schärli, C.E.: Post translational protein modifications. Thesis, Univ. Basel, 1982.

Schärli, C.E., Koch, G.: Protein kinase activity in purified particles and empty viral capsid preparations. J. Gen. Virol. 65: 129-139 (1984).

Schaffer, F.L.: Binding of proflavine and photoinactivation of poliovirus propagated in the presence of the dye. Virology 18: 412-425 (1962).

Schaffer, F.L., Frommhagen, L.H.: Similarities of biophysical properties of several human enteroviruses as shown by density gradient ultracentrifugation of mixtures of the viruses. Virology 25: 662-664 (1965).

Schaffer, F.L., Gordon, M.: Differential inhibitory effects of actinomycin D among strains of poliovirus. J. Bact. 91: 2309-2316 (1966).

Schaffer, F.L., Hackett, A.J.: Early events in poliovirus-HeLa cell interaction: Acridine orange photosensitization and detergent extraction. Virology 21: 124 (1963).

Schaffer, F.L., Schwerdt, C.E.: Crystallization of purified MEF-I poliomyelitis virus particles. Proc. Natl. Acad. Sci. USA 41: 1020-1023 (1955).

Schaffer, F.L., Schwerdt, C.E.: Purification and properties of poliovirus. Adv. Virus Res. 6: 159-204 (1959).

Schaffer, F.L., Moore, H.F., Schwerdt, C.E.: Base composition of the ribonucleic acids of the three types of poliovirus. Virology 10: 530 (1960).

Scharff, M.D., Levintow, L.: Quantitative study of the formation of poliovirus antigens in infected HeLa cells. Virology 19: 491-500 (1963).

Scharff, M.D., Shatkin, A.J., Levintow, L.: Association of newly formed viral protein with specific polyribosomes. Proc. Natl. Acad. Sci. USA 50: 686-694 (1963).

Scharff, M.D., Maizel, J.V., Levintow, L.: Physical and immunological properties of a soluble precursor of the poliovirus capsid. Proc. Natl. Acad. Sci. USA 51: 329-337 (1964).

Scharff, M.D., Summers, D.F., Levintow, L.: Further studies on the effect of pfluorophenylalanine and puromycin on poliovirus replication. Ann. N.Y. Acad. Sci. 130: 282-290 (1965).

Scheid, A., Choppin, P.W.: Identification of biological activities of paramyxovirus glycoproteins.

Activation of cell fusion, hemolysis, and infectivity by proteolytic cleavage of an inactive precursor protein of Sendai Virus. Virology 57: 475-490 (1974).

Scherrer, K.: In: Eukaryotic Gene Regulation (Kolodney, G., ed.), Florida: CRC Press 1979.

Scherrer, K., Imaizumi-Scherrer, M.T., Reynaud, C.A., Therwath, A.: On pre-messenger RNA and transcriptons. A review. Molec. Biol. Rep. Vol. 5, 1-2: 5-28 (1979).

Schild, G.C., Minor, P.D., Evans, D.M.A., Ferguson, M., Stanway, G., Almond, J.W.: Molecular basis for the antigenicity and virulence of poliovirus type 3. In: Modern Approaches to Vaccines (Chanock, R.M., Lerner, R.A., eds.), 27-35. Cold Spring Harbor Laboratory 1984.

Schliwa, M., van Blerkom, J.: Structural interaction of cytoskeletal components. J. Cell Biol. 90: 222-235 (1981).

Schmidt, N.J., Lennette, E.H.: Modification of the homotypic specificity of poliomyelitis complement-fixing antigens by heat. J. Exp. Med. 104: 99-120 (1956).

Schreier, M.H., Erni, B., Staehelin, T.: Initiation of mammalian protein synthesis. I. Purification and characterization of 7 initiation factors. J. Mol. Biol. 116: 727-753 (1977).

Schrom, M., Bablanian, R.: Altered cellular morphology resulting from cytocidal virus infection. Arch. Virol. 70: 173-187 (1981).

Schrom, M., Laffin, J.A., Evans, B., McSharry, J.J., Caliguiri, L.A.: Isolation of poliovirus variants resistant to and dependent on arildone. Virology 122: 492-497 (1982).

Schultz, E.W., Gebhardt, L.P.: Observations on the intranasal route of infection in experimental poliomyelitis. Proc. Soc. Exp. Biol. Med. 33: 1010-1012 (1933).

Schultz, E.W., Gebhardt, L.P.: Mechanism of zinc sulphate prophylaxis in experimental poliomyelitis. Proc. Soc. Exp. Biol. Med. 38: 603-605 (1938).

Schultz, M., Crowell, R.L.: Acquisition of suspeptibility to coxsackie-virus A2 by the rat L8 cell line during myogenic differentiation. J. Gen. Virol. 46: 39-49 (1980).

Schultz, R.D.: Biological functions of activated complement proteins in normal and disease states. In: Proteolysis and Physiological Regulation (Ribbons, D. W., Brew, K. eds.), Vol. 11, 143-167. New York: Academic Press 1976.

Schwartz, H., Darnell, J.E.: The association of protein with the polyadenylic acid of HeLa cell messenger RNA: evidence for a „transport" role of a 75,000 molecular weight polypeptide. J. Mol. Biol. 104: 833-851 (1976).

Schweiger, M., Wagner, E.F., Hirsch-Kaufmann, E.F., Ponta, H., Herrlich, P.: Biochemistry of development of E. coli virus T7 and T1. In: Gene Expression (Clark, B. F. C., Klenow, H., Zeuthen, J., eds.), 171. New York: Pergamon Press 1978.

Schweiger, M., Wagner, E.F.: Nucleosidetriphosphate mediated discrimination of gene expression in T1-infected E. coli. In: Regulation of Macromolecular Synthesis by Low Molecular Weight Mediators (Koch, G., Richter, D., eds.), 249-262. Academic Press 1979.

Schwerdt, C.E., Fogh, J.: The ratio of physical particles per infectious unit observed for poliomyelitis viruses. Virology 4: 41-52 (1957).

Schwerdt, C.E., Schaffer, F.L.: Some physical and chemical properties of purified poliomyelitis virus preparations. Ann. N. Y. Acad. Sci. 61: 740-753 (1955).

Schwerdt, C.E., Schaffer F.L.: Purification of poliomyelitis viruses propagated in tissue culture. Virology 2: 665-678 (1956).

Scraba, D.G.: The picornavirion: Structure and assembly. In: The Molecular Biology of Picornaviruses (Perez-Bercoff, R., ed.), 1-23. New York: Plenum 1979.

Seeburg, P.H., Shine, J., Martial, J.A., Baxter, J.D., Goodman, H.M.: Nucleotide sequence and amplification in bacteria of structural gene for rat growth hormone. Nature 270: 486-494 (1977).

Sefton, B.M., Gaffneys, B.J.: Effect of the viral proteins on the fluidity of the membrane lipids in Sindbis virus. J. Mol. Biol. 90: 343-358 (1974).

Segawa, K., Yamaguchi, N., Oda, K.: Simian virus 40 gene A regulates the association between a highly phosphorylated protein and chromatin and ribosomes in Simian virus 40-transformed cells. J. Virol. 22: 679-693 (1977).

Semler, B.L., Anderson, C.W., Kitamura, N., Rothberg, P.G., Wishart, W.L., Wimmer, E.: Poliovirus replication proteins: RNA sequence encoding P3-1b and the site of porteolytic processing. Proc. Natl. Acad. Sci. USA 78: 3464-3468 (1981a).

Semler, B.L., Hanecak, R., Anderson, C.W., Wimmer, E.: Cleavage sites in the polypeptide precursors of poliovirus protein P2-X. Virology 114: 589-594 (1981b).

Semler, B.L., Anderson, C.W., Hanecak, R., Dorner, L.F., Wimmer, E.: A membrane-associated precursor to poliovirus VPg identified by immuno-precipitation with antibodies directed against a synthetic heptapeptide. Cell 28: 405-412 (1982).

Semler, B.L., Hanecak, R., Dorner, L.F., Anderson, C.W., Wimmer, E.: Poliovirus RNA synthesis in vitro: Structural elements and antibody inhibition. Virology 126: 624-635 (1983).

Sen, A., Williams, W.P., Brain, A.P.R., Dickens, M.J., Quinn, P.J.: Formation of inverted micelles in dispersions of mixed galactolipids. Nature 293: 488-490 (1981).

Sen, G.C., Gupta, S.L., Brown, G.E., Lebleu, B., Rebello, M.A., Langyel, P.: Interferon treatment of Ehrlich ascites tumor cells: Effects on exogenous mRNA translation and tRNA inactivation in the cell extract. J. Virol. 17: 191-203 (1976).

Senkevich, T.G., Cumakov, I.M., Lipskaya, G.Y., Agol, V.I.: Palindrome-like dimers of double-stranded RNA of encephalomyocarditis virus. Virology 102: 339-348 (1980).

Sergiescu, D., Aubert-Combiescu, A.: Differential sensitivity to gliotoxin of virulent and attenuated poliovirus strains and its possible use as a genetic marker. Arch. ges. Virusforsch. 27: 268-281 (1969).

Sergiescu, D., Horodniceanu, F., Crainic, R.: Mutation towards dextran sulfate resistance in type-1 poliovirus. Nature 215: 313-315 (1967).

Sergiescu, D., Aubert-Combiescu, A., Crainic, R.: Recombination between guanidine resistant and dextran sulfate resistant mutants of type-1 poliovirus. J. Virol. 3: 326-330 (1969).

Sergiescu, D., Horodniceanu, F., Aubert-Combiescu, A.: The use of inhibitors in the study of picornavirus genetics. Prog. Med. Virol. 14: 123-199 (1972).

Shapiro, A.L., Vinuela, E., Maizel, J.V.: Molecular weight estimation of polypeptide chains by electrophoresis in SDS polyacrylamide gels. Biochem. Biophys. Res. Commun. 28: 815-820 (1967).

Shapiro, S.Z., Strand, M., August, J.T.: High molecular weight precursor polypeptides to structural proteins of Rauscher murine leukemia virus. J. Mol. Biol. 107: 459-477 (1976).

Sharp, P.: Speculations on RNA splicing. Cell 23: 643-646 (1981).

Shatkin, A.J.: Actinomycin inhibition of ribonucleic acid synthesis and poliovirus infection of HeLa cells. Biochim. Biophys. Acta 61: 310-313 (1962).

Shatkin, A.J.: Capping of eukaryotic mRNAs. Cell 9: 645-653 (1976).

Shatkin, A.J.: A closer look at the 5'-end of mRNA in relation to initiation. In: Protein Biosynthesis in Eukaryotes (Perez-Bercoff, R., ed.), 199-221. New York: Plenum Press 1982.

Shea, M.A., Plagemann, P.G.W.: Effects of elevated temperatures on mengovirus ribonucleic acid synthesis and virus production in Novikoff rat hepatoma cells. J. Virol. 7: 144 (1971).

Shih, C.-Y.T., Shih, D.S.: Cleavage of the capsid protein precursors of encephalomyocarditis virus in rabbit reticulocyte lysates. J. Virol. 40: 942-945 (1981).

Shih, D.S., Shih, C.T., Kew, O., Pallansch, M., Rueckert, R., Kaesberg, P.: Cell-free synthesis and processing of the proteins of poliovirus. Proc. Natl. Acad. Sci. USA 75: 5807-5811 (1978).

Shih, D.S., Shih, C.T., Zimmern, D., Rueckert, R.R., Kaesberg, P.: Translation of encephalomyeocarditis virus in reticulocyste lysates: kinetic analysis of the formation of virion proteins and a protein required for processing. J. Virol. 30: 472-480 (1979).

Shih, C.Y., Naseer, N., Shih, D.S.: Rapid method for the preparation of encephalomyocarditis virus protease from rabbit reticulocyte lysates. J. Virol. 42: 1127-1130 (1982).

Shine, J., Dalgarno, L.: The 3'-terminal sequence of Escheria coli 16S ribosomal RNA: complementary to nonsense triplets and ribosome binding sites. Proc. Natl. Acad. Sci. USA 71: 1342-1346 (1974).

Shine, J., Seeburg, P.H., Martial, J.A., Baxter, J.D., Goodman, H.M.: Construction and analysis of recombinant DNA for human chorionic somatomammotropin. Nature 270: 494-499 (1977).

Siak, J.-S., McGeady, M.L., Crowell, R.L.: In: Abstracts of the IV International Congress on Virology, 182 (1978).

Siegl, G., Frösner, G.G.: Characterization and classification of virus particles associated with hepatitis A. I. Size, density, sedimentation. J. Virol. 26: 40-47 (1978a).

Siegl, G., Frösner, G.G.: Characterization and classification of virus particles associated with hepatitis A. II. Type and configuration of nucleic acid. J. Virol. 26: 48-53 (1978b).

Siegl, G., Tratschin, J. D., Frösner, G. G., Gauss-Müller, V., Scheid, R., Deinhardt, F.: Hepatitis A virus, a picornavirus with some uncommon characteristics. Second Meeting of the European Study Group on the Molecular Biology of Picornaviruses, Hamburg 1981a.

Siegl, G., Frösner, G.G., Gauss-Müller, V., Tratschin, J.D., Deinhardt, F.: The physicochemical properties of infectious hepatitis A virions. J. Gen. Virol. 57: 331-341 (1981b).

Silverman, R., Atherley, A., Richter, D.: Guanosine 3', 5'-bis (diphosphate) search in eukaryotes. In: Regulation of Macromolecular Synthesis by Low Molecular Weight Mediators (Koch, G., Richter, D., eds.), 115-125. New York: Academic Press 1979.

Singer, S.J., Nicolson, G.L.: The fluid mosaic model of the structure of cell membranes. Science 175: 720-731 (1972).

Sjöstrand, F.S., Polson, A.: Macrocrystalline patterns of closely packed poliovirus particles in ultrathin sections. J. Ultrastruc. Res. 1: 365-374 (1958).

Skarlat, I.V., Kalinina, N.O.: (RS) Protein, detected in cell messenger RNA-containing structure, after infection with encephalomyocarditis virus. Dokl. Akad. Nauk. SSSR 243: 797-800 (1978).

Skarlat, I.V., Kalinina, N.O., Medvedki, O.A., Yatsina, A.A.: Virus-specific proteins bound to components of protein-synthesizing system in Krebs II ascites carcinoma cells infected with encephalomyocarditis virus. Mol. Biol. (Mosk) 10: 931-939 (1976).

Skinner, M.S., Halperen, S., Harkin, J.C.: Cytoplasmic membrane-bound vesicles in Echovirus 12-infected cells. Virology 36: 241-253 (1968).

Skulachev, V.P.: Transmembrane electrochemical H^+-potential as a convertible energy source for the living cell. FEBS Lett. 74: 1-9 (1977).

Smit, G.L., Wilterdink, J.B.: Intratypic serodifferentiation of polioviruses by comparison of neutralization indices and correlation of the antigenic marker, temperature marker and monkey neurovirulence. Arch. ges. Virusforsch. 18: 261-275 (1966).

Smith, A.E.: Control of translation of animal virus messenger RNA. In: Control Processes in Virus Multiplication (Burke, Russel, eds.), 183-224. Cambridge University Press 1975.

Soloviev, V.D., Krispin. T.I., Zaslavsky, V.G., Agol, V.I.: Mechanism of resistance to enteroviruses of some primate cells in tissue culture. J. Virol. 2: 553-557 (1968).

Sonenberg, N., Rupprecht, K.M., Hecht, S.M., Shatkin, A.J.: Eukaryotic mRNA cap binding protein: Purification by affinity chromatography on sepharose-coupled m^7 GDP. Proc. Natl. Acad. Sci. USA 76: 4345-4349 (1979).

Sonenberg, N., Guertin, D., Cleveland, D.: Probing the function of the eucaryotic 5' cap structure by using a monoclonal antibody directed against cap-binding proteins. Cell 27: 563-572 (1981).

Spector, D.H., Baltimore, D.: Requirement of 3'-terminal poly (adenylic acid) for the infectivity of poliovirus RNA. Proc. Natl. Acad. Sci. USA 71: 2983-2987 (1974).

Spector, D.H., Baltimore, D.: Polyadenylic acid on poliovirus RNA. II. Poly (A) on intracellular RNAs. J. Virol. 15: 1418-1431 (1975a).

Spector, D.H., Baltimore, D.: Polyadenylic acid on poliovirus RNA. III. In vitro addition of polyadenylic acid to poliovirus RNAs. J. Virol. 15: 1432-1439 (1975b).

Spector, D.H., Baltimore, D.: Polyadenylic acid on poliovirus RNA. IV. Poly (U) in replicative intermediate and double-stranded RNA. Virology 67: 498-505 (1975c).

Spector, D.H., Baltimore, D.: Poly (A) on mengovirus RNA. J. Virol. 16: 1081-1084 (1975d).

Spector, D.H., Villa-Komaroff, L., Baltimore, D.: Studies on the function of polyadenylic acid on poliovirus RNA. Cell 6: 41-44 (1975).

Spirin, A.S., Nemer, M.: Messenger RNA in early sea-urchin embryos -cystoplasmic particles. Science 150: 214 (1965).

Spirin, A.S., Belitsina, N.V., Ajtkhozhin, M.A.: Informatsionnye RNK v rannem émbriogeneze. Zhurnal Obshchei Biologii (Moscow) 25: 321-338 (1964). [English translation in Fed. Proc. 24, T907 (1965a).]

Spirin, A.S., Belitsina, N.V., Lerman, M.J.: Use of formaldehyde fixation for studies of ribonucleoprotein particles by caesium chloride density gradient centrifugation. J. Mol. Biol. 14: 611-615 (1965b).

Stanway, G., Cann, A.J., Hauptmann, R., Hughes, P., Clarke, L.D., Mountford, R.C., Minor, P.D., Schild, G.C., Almond, J.W.: The nucleotide sequence of poliovirus type 3 Leon 12 A, 43B: Comparison with poliovirus type 3: Nucl. Acids Res. 11: 5629 (1983).

Steere, R.L., Schaffer, F.L.: The structure of crystals of purified Mahoney poliovirus. Biochim. Biophys. Acta 28: 241-246 (1958).

Steiner, D.F., Kemmler, W., Tager, H.S., Rubenstein, A.H., Lernmark, A., Zühlke, H.: Proteolytic mechanism in the biosynthesis of polypeptide hormones. In: Proteases and Biological Control (Reich, E., Rifkin, D., Shaw, E., eds.), 531-549. Cold Spring Harbor Laboratory 1974.

Steiner, D.F., Quinn, P.S., Chan, S.J., Marsh, J., Tager, H.S.: Processing mechanisms in the biosynthesis of proteins. Ann. N.Y. Acad. Sci. 343: 1-16 (1980).

Steiner-Pryor, A., Cooper, P.D.: Temperature-sensitive poliovirus mutants defective in repression of host protein synthesis are also defective in structural protein. J. Gen. Virol. 21: 215-225 (1973).

Steitz, J.A., Jakes, K.: How ribosomes select initiator regions in mRNA: Base pair formation between the 3' terminus of 16S rRNA and the mRNA during initiation of protein synthesis in Escheria coli. Proc. Natl. Acad. Sci. USA 72: 4734-4738 (1975).

Stewart, M.L., Crouch, R.J., Maizel, J. V. Jr.: A high resolution oligonucleotide map generated by restriction of poliovirus type 1 genomic RNA by ribonuclease III. Virology 104: 375-397 (1980).

Stoner, G.D., Williams, B., Kniazeff, A., Shimkin, M.B.: Effect of neuraminidase pretreatment on the susceptibility of normal and transformed mammalian cells to bovine enterovirus 261. Nature 245: 319-320 (1973).

Stuart, D.C., Jr., Fogh, J.: Electron microscopic demonstration of intracellular poliovirus crystals. Exp. Cell Res. 18: 378-381 (1959).

Stuart, D.C., Jr., Fogh, J.: Micromorphology of FL cells infected with polio and coxsackie viruses. Virology 13: 177-190 (1961).

Su, R.T., Taylor, M.W.: Morphogenesis of picornavirus: Characterization and assembly of bovine enterovirus subviral particles. J. Gen. Virol. 30: 317-328 (1976).

Summers, D.F., Levintow, L.: Constitution and function of polyribosomes of poliovirus-infected HeLa cells. Virology 27: 44-53 (1965).

Summers, D.F., Maizel, J.V., Jr.: Disaggreagation of HeLa cell polysomes after infection with poliovirus. Virology 31: 550 (1967).

Summers, D.F., Maizel, J.V., Jr.: Evidence for large precursor proteins in poliovirus synthesis. Proc. Natl. Acad. Sci. USA 59: 966-971 (1968).

Summers, D.F., Maizel, J.V., Jr.: Determination of the gene sequence of poliovirus with pactamycin. Proc. Natl. Acad. Sci. USA 68: 2852-2856 (1971).

Summers, D.F., McElvain, N.F., Thoren, M.M., Levintow, L.: Incorporation of amino acids into polyribosome-associated protein in cytoplasmic extracts of poliovirus-infected HeLa cells. Biochem. Biophys. Res. Commun. 3: 290-295 (1964).

Summers, D.F., Maizel, J.V., Darnell, J.E.: Evidence for virus-specific noncapsid proteins in poliovirus-infected HeLa cells. Biochemistry 54: 505-513 (1965).

Summers, D.F., Maizel, J.V., Jr., Darnell, J.E.: The decrease in size and synthetic activity of poliovirus polysomes late in the infectious cycle. Virology 31: 427 (1967).

Summers, D.F., Shaw, E.N., Stewart, M.L., Maizel, J.V. Jr.: Inhibition of cleavage of large poliovirus-specific precursor proteins in infected HeLa cells by inhibitors of proteolytic enzyme. J. Virol. 10: 880-884 (1972).

Svehag, S., Mandel, B.: The formation and properties of poliovirus-neutralizing antibody. J. Exp. Med. 119: 1-19 (1964).

Svitkin, Y.V., Gorbalenya, A.E., Kazachkov, Y.A., Agol, V.I.: Encephalomyocarditis virus-specific polypeptide p22 possessing a proteolytic activity. FEBS 108: 6-9 (1979).

Sweadner, K.J., Goldin, S.M.: Active transport of sodium and potassium ions: mechanism, function and regulation. N. Engl. J. Med. 302: 777-783 (1980).

Taber, R., Rekosh, D., Baltimore, D.: Effect of pactamycin on synthesis of poliovirus proteins, a method for genetic mapping. J. Virol. 8: 395-401 (1971).

Taber, R., Wilson, T., Papahadjopoulos, D.: Use of large unilamellar vesicles (liposomes) for the introduction of poliovirus and poliovirus RNA into cells. In: Introduction of Macromolecules into Viable Mammalian Cells, 221-237. New York: Alan R. Liss 1980.

Tahara, S.M., Morgan, M.A., Shatkin, A.J.: Two forms of purified m^7G-cap binding protein with different effects on capped mRNA translation in extracts of uninfected and poliovirus-infected HeLa cells. J. Biol. Chem. 256: 7691-7694 (1981).

Takeda, N., Miyamura, K., Kono, R., Yamazaki, S.: Characterization of a temperature-sensitive defect of enterovirus type 70. J. Virol. 44: 98-106 (1982).
Takegami, T., Semler, B.L., Anderson, C.W., Wimmer, E.: Membrane fractions active in poliovirus RNA replication contain VPg precursor polypeptides. Virology 128: 33-47 (1983a).
Takegami, T., Kuhn, R.J., Anderson, C.W., Wimmer, E.: Membrane-dependent uridylylation of the genome-linked protein VPg of poliovirus. Proc. Natl. Acad. Sci. USA 80: 7447-7451 (1983b).
Takemori, N., Nomura, S., Nakano, M., Morioka, Y., Henmi, M., Kitaoka, M.: A new mutation in polioviruses. Science 125: 1196-1198 (1957a).
Takemori, N., Nomura, S., Morioka, Y., Nakano, M., Kitaoka, M.: A minute-plaque mutant (m) of type-2 poliovirus. Science 126: 924-925 (1957b).
Takemori, N., Nomura, S., Nakano, M., Moriaka, M., Henmi, M., Kitaoka, M.: Mutation of polioviruses to resistance to neutralizing substances in normal bovine sera. Virology 5: 30-55 (1958).
Takemoto, K.K.: Plaque mutants of animal viruses. Prog. Med. Virol. 8: 314-348 (1966).
Takemoto, K.K., Habel, K.: Sensitivity and resistance of type-1 polioviruses to an inhibitor in certain host sera. Virology 9: 228-243 (1959).
Takemoto, K.K., Kirschstein, R.L.: Dextran sulfate plaque variants of attenuated type-1 poliovirus. Relationship to in vitro and in vivo markers. J. Immunol. 42: 329-333 (1964).
Takemoto, K.K., Liebhaber: Virus-polysaccharide interactions. II. Enhancement of plaque formation and the detection of variants of poliovirus with dextran sulfate. Virology 17: 499-501 (1962).
Takemoto, K.K., Spicer, S.S.: Effects of natural and synthetic sulfated polysaccharides on viruses and cells. Ann. N.Y. Acad. Sci. 130: 365-373 (1965).
Talbot, P., Brown, F.: A model for foot-and-mouth disease virus. J. Gen. Virol. 15: 163 (1972).
Tamm, I., Eggers, H.J.: Specific inhibition of replication of animal viruses. Science 142: 24-33 (1963).
Tannock, G.A., Gibbs, A.I.,, Cooper, P.: A reexamination of the molecular weight of poliovirus RNA. Biochem. Biophys. Res. Commun. 38: 298-304 (1970).
Tattersall, P., Ward, D.C.: Rolling hairpin model for replication of parvovirus and linear chromosomal DNA. Nature (London) 263: 106-109 (1976).
Taylor, J., Graham, A.F.: Studies on ^{32}P-labeled poliovirus. Trans. N.Y. Acad. Sci. 21: 242-248 (1959).
Terasima, T., Tolmach, L.J.: Growth and nucleic acid synthesis in synchronously dividing populations of HeLa cells. Exp. Cell Res. 30: 344-362 (1963).
Tershak, D.R.: Synthesis of RNA in cells infected with LSc poliovirus at elevated temperatures. J. Virol. 3: 297 (1969).
Tershak, D.R.: Guanidine inhibition of poliovirus growth. Partial elimination by protease antagonists and low temperature. Can. J. Microbiol. 20: 817-824 (1974).
Tershak, D.R.: Effect of hypertonic medium on protein synthesis of Mahoney and LSc poliovirus. J. Virol. 20: 597-603 (1976).
Tershak, D.R.: Protein kinase activity of polysome-ribosome preparations from poliovirus infected cells. Biochem. Biophys. Res. Commun. 80: 283-289 (1978a).
Tershak, D.R.: Peptide-chain initiation with LSc poliovirus is intrinsically more resistant to hypertonic environment than is peptide-chain initiation with Mahoney virus and deletion mutants of Mahoney virus. J. Virol. 28: 1006-1010 (1978b).
Tershak, D.R.: Inhibition of poliovirus polymerase by guanidine in vitro. J. Virol. 41: 313-318 (1982).
Tershak, D.R., Yin, F.H., Korant, B.D.: Guanidine. In: Chemotherapy of Virus Infections (Caliguiri, L., Came, P., eds.). Berlin-Heidelberg-New York: Springer 1982.
Thach, S.S., Thach, R.E.: Mechanism of viral replication: I, Structure of replication complexes of R17 bacteriophage. J. Mol. Biol. 81: 367-380 (1973).
Thach, S.S., Dobbertin, D., Lawrence, C., Golini, F., Thach, R.E.: The mechanism of viral replication. Structure of the replication complexes of encephalomyocarditis virus. Proc. Natl. Acad. Sci. USA 71: 2549-2553 (1974).
Thacore, H., Wolff, D.A.: Activation of lysosomes by poliovirus-infected cell extracts. Nature 218: 1063-1064 (1968).
Thomsson, R., Koehne-Bartleben, G., Schober, A.: Determination of equine serum inhibitors for poliovirus by the gel-adsorption technique. Arch. ges. Virusforsch. 19: 415-434 (1966).
Thorne, H.V.: Electrophoretic study of the interaction of radioactive poliovirus with components of cultured cells. J. Bact. 185: 1247-1255 (1963).

Thornton, G.B., Robberson, D.L., Arlinghaus, R.B.: Attachment of avidin-coupled spheres to linear and circular forms of mengovirus double-stranded RNA. J. Virol. 39: 229-237 (1981).

Thorpe, R., Minor, P.D., Machay, A., Schild, G.C., Spitz, M.: Immunochemical studies of polioviruses: Identification of immunoreactive virus capsid polypeptides. J. Gen. Virol. 63: 487-492 (1982).

Thucydides: The Complete Writings of Thucydides: The Peloponesian War, 112 (Transl. J.H. Finley, Jr.). New York: The Modern Library 1951.

Tobia, A.M., Schildkraut, C.L., Maio, J.J.: Deoxyribonucleic acid replication in synchronized cultured mammalian cells. I. Time of synthesis of molecules of different average guanine and cytosine content. J. Mol. Biol. 54: 499-515 (1970).

Tolbert, O., Weaver, B., Engler, R.: Synthesis of uncoated viral RNA following picornavirus infection. Arch. ges. Virusforsch. 19: 221-229 (1966).

Tolskaya, E.A., Agol, V.I., Voroshilova, M.K., Lipskaya, G.Y.: The osmotic pressure of the maintenance medium and reproduction of poliovirus. Virology 29: 613-621 (1966).

Tolskaya, E.A., Romanova, L.A., Kolesnikova, M.S., Agol, V.I.: Intertypic recombination in poliovirus: genetic and biochemical studies. Virology 124: 121-132 (1983).

Tonew, E., Fahlbusch, B.: Effects of pyrimidine derivatives on RNA-dependent RNA polymerase of mengovirus-infected Fogh and Lund (FL) cells. J. Gen. Virol. 34: 37-45 (1977).

Totsuka, A., Ohtaki, K., Tagaya, I.: Aggregation of enterovirus small plaque variants and polioviruses under low ionic strength conditions. J. Gen. Virol. 38: 519-533 (1978).

Tovell, D.R., Colter, J.S.: Observations on the assay of infectious viral ribonucleic acid: Effect of DMSO and DEAE dextran. Virology 32: 84-92 (1967).

Toyoda, H., Michinori, K., Kataoka, Y., Suganama, T., Omata, T., Imura, N., Nomoto, A.: Complete nucleotide sequence of all three poliovirus serotype genomes: Implications for genetic relationship, gene function, and antigenic determinants. J. Mol. Biol. 174: 561-585 (1984).

Trachsel, H., Erni, B., Schreier, M.H., Staehelin, T.: Initiation of mammalian protein synthesis. II. The assembly of the initiation complex with purified initiation factors. J. Mol. Biol. 116: 755-767 (1977).

Trachsel, H., Sonenberg, N., Shatkin, A.J., Rose, J.K., Leong, K., Bergmann, J.E., Gordon, J., Baltimore, D.: Purification of a factor that restores translation of vesicular stomatitis virus mRNA in extracts from poliovirus-infected HeLa cells. Proc. Natl. Acad. Sci. USA 77: 770-774 (1980).

Trachsel, H., Stahli, C., Sonenberg, N., Staeheli, T., Fessler, R., Kuster, H., Zumbe, A.: Preparation and characterization of a monoclonal antibody directed against eukaryotic messenger-RNA 5' cap binding proteins. Mol. Biol. Rep. 7: 189 (1981).

Tracy, S., Smith R.A.: A comparison of the genomes of polioviruses by cDNA: RNA hybridization. J. Gen. Virol. 55: 193-199 (1981).

Trask, J.D., Vignec, A.J., Paul, J.R.: Poliomyelitis virus in human stools. J. A. M. A. 111: 6-11 (1938).

Traub, A., Diskin, B., Rosenberg, H., Kalmar, E.: Isolation and properties of the replicase of encephalomyocarditis virus. J. Virol. 18: 375-382 (1976).

Tucker, P.W., Marcu, K.B., Slightom, J.L., Blattner, R.R.: Structure of the constant and 3' untranslated regions of the murine γ 2b heavy chain messenger RNA. Science 206: 1299-1303 (1979).

Tuschall, D.M., Hiebert, E., Flanegan, J.B.: Poliovirus RNA-dependent RNA polymerase synthesizes full-length copies of poliovirion RNA, cellular mRNA, and several plant virus RNAs in vitro. J. Virol. 44: 209-216 (1982).

Udenfriend, S.: Fluorescence assay in biology and medicine. In: Molecular Biology (Horecker, B., Kaplan, N.O., Marmur, J., Scheraga, H.A., eds.), Vol. II. New York: Academic Press 1969.

Ullmann, A., Joseph, E., Danchin, A.: Cyclic AMP as a modulator of polarity in polycistronic transcriptional units. Proc. Natl. Acad. Sci. USA 76: 3194-3197 (1979).

Underwood, M.: A treatise on the diseases of children, ed. 2. London: 1789.

Urasawa, S., Urasawa, T., Chiba, S., Kanamitsu, M.: Studies on poliovirus inhibitors in sera of domestic animals. II. Determination of poliovirus inhibitors by the precipitation in agar-gel method. Jap. J. med. Sci. Biol. 21: 155-166 (1968a).

Urasawa, S., Urasawa, T., Chiba, S., Kanamitsu, M.: Studies on poliovirus inhibitors in sera of domestic animals. III. A comparison of physicochemical properties of poliovirus inhibitors and specific inhibitors. Jap. J. med. Sci. Biol. 21: 173-183 (1968b).

Urasawa, S., Urasawa, T., Kanamitsu, M.: Radioimmunoelectrophoretic identification of poliovirus inhibitors and their characteristic mode of action. Arch. ges. Virusforsch. 33: 113-125 (1971).

Urasawa, S., Urasawa, T., Kanamitsu, M.: Analysis of specificity of antibodies contained in antiserum against poliovirus with mutants resistant to inhibitors in equine serum. J. Immunol. 113: 537-542 (1974).

Urasawa, S., Urasawa, T., Kanamitsu, M.: Further studies of specificity of antibodies contained in antiserum against poliovirus. Japan J. Microbiol. 1: 11-16 (1976).

Urasawa, S., Ishizawa, F., Urasawa, T.: Antigenic variation of poliovirus caused by antibody components with different specificity. Microbiol. Immunol. 21: 299-307 (1977).

Urasawa, T., Urasawa, S., Taniguchi, K.: Reactivity of neutralizing antibodies with different specifities to H particles of poliovirus. Microbiol. Immunol. 7: 651-657 (1979).

Vaheri, A., Pagano, J.S.: Infectious poliovirus RNA: A sensitive method of assay. Virology 27: 434 (1965).

Vaheri, A., Ruoslahti, E., Hovi, T., Nordling, S.: Stimulation of density inhibited cultures by insulin. J. Cell. Physiol. 81: 355 (1973).

Vallbracht, A., Hofmann, L., Wurster, K.G., Flehmig, B.: Persistent infection of human fibroblasts by hepatitis A virus. J. Gen. Virol. 65: 609-615 (1984).

Van der Werf, S., Bŕegeġere, F., Kopecka, H., Kitamura, N., Rothberg, P.G., Kourilsky, P., Wimmer, E., Girard, M.: Molecular cloning of the genome of poliovirus type 1. Proc. Natl. Acad. Sci. USA 78: 5983-5987 (1981).

Van der Werf, S., Wychowski, C., Bruneau, P., Blondel, B., Crainic, R., Horodniceanu, F., Girard, M.: Location of a poliovirus type 1 neutralization epitope in viral capsid polypeptide VP1. Proc. Natl. Acad. Sci. USA 80: 5080 (1983).

Van Dyke, T.A., Flanegan, J.B.: Identification of poliovirus polypeptide p63 as a soluble RNA-dependent RNA polymerase. J. Virol. 35: 732-740 (1980).

Van Dyke, T.A., Rickles, R.J., Flanegan, J.B.: Genome-length copies of poliovirion RNA are synthesized in vitro by the poliovirus RNA dependent RNA polymerase. J. Biol. Chem. 257: 4610-4617 (1982).

Van Elsen, A., Boeÿe, A.: Disruption of type 1 poliovirus under alkaline conditions: Role of pH, temperature, and sodium dodecyl sulfate (SDS). Virology 28: 481 (1966).

Van Elsen, A., Boeye, A., Teuchy, H.: Formation of fibrillar structures from poliovirus by alkaline disruption and other treatments. Virology 36: 511-514 (1968).

Van Loon, L.C.: Induction by 2-chloroethyl-phosphonic acid of viral-like lesions, associated proteins, and systemic resistance in tobacco. Virology 80: 417-420 (1977).

Van Venrooij, W.J.W., Henshaw, E.C., Hirsch, C.A.: Effects of deprival of glucose or individual amino acids on polyribosome distribution and rate of protein synthesis in cultured mammalian cells. Biochim. Biophys. Acta 259: 127-137 (1972).

Van Wezel, A.L., Hazendonk, A.G.: Intratypic serodifferentiation of poliomyelitis virus by strain-specific antisera. Intervirology 11: 2-8 (1979).

Van Wezel, A.L., Van der Marel, P., Van Steenis, G., de Vries, F.A.J.: Studies on the antigenicity and immunogenicity of the structural polypeptides of poliovirus. Second Meeting of the European Study Group on the Molecular Biology of Picornaviruses, Hamburg 1981.

Van Wezel, A.L., Van der Marel, P., Hazendonk, T.G., Boer-Bak, V., Henneke, M.A.C.: Antigenicity and immunogenicity of poliovirus capsid proteins. Dev. Biol. Stand. (in press, 1984).

Vance, D.E., Trip, E.M., Paddon, H.B.: Poliovirus increases phosphatidylcholine biosynthesis in HeLa cells by stimulation of the rate-limiting reaction catalyzed by CTP: phosphocholine cytidylyltransferase. J. Biol. Chem. 255: 1064-1069 (1980).

Vande Woude, G., Ascione, R.: Translation products of foot-and-mouth disease virus-infected baby hamster kidney cells. Arch. ges. Virusforsch. 45: 259-271 (1974).

Vande Woude, G.F., Bachrach, H.L.: Number and molecular weights of foot-and-mouth disease virus capsid proteins and the effects of maleylation. J. Virol. 7: 250-259 (1971).

Vanden Berghe, D., Boeye, A.: New polypeptides in poliovirus. Virology 48: 605-606 (972).

Vanden Berghe, D., Boeye, A.: A new species of poliovirus top component. Arch. ges. Virusforsch. 41: 138-142 (1973a).

Vanden Berghe, D., Boeye, A.: Stepwise degradation of poliovirus and top component by concentrated urea. Arch. ges. Virusforsch. 41: 216-228 (1973b).

Vanden Berghe, D., Boeye A.: In situ fragments of RNA in poliovirus. Arch. ges. Virusforsch. 40: 215-222 (1973c).

Vanden Berghe, D., Neetens, A., Van de Sompel, W., Delgadillo, R.: Glucosamine therapy: a preliminary note on a new approach in treatment of herpetic keratitis. Bull. Soc. Belge Ophtalmol. 187: 41-44 (1980).

Vartapetjan, A.B., Drygin, Yu.-F., Chumakov, K.M.: The structure of the covalent linkage between proteins and RNA in encephalomyocarditis virus. Bioorganicheskaya Kimia 5: 1876-1878 (1979).

Vartapetjan, A.B., Kunin, E.B., Chumakov, K.M., Bogdanov, A.A., Agol, V.I.: Dokl. Akad. Nauk SSSR 267: 963-965 (1982).

Vaughan, M., Pawlowski, P., Forchhammer, J.: Regulation of protein synthesis initiation in HeLa cells deprived of single essential amino acids. Proc. Natl. Acad. Sci. USA 68: 2057-2061 (1971).

Villa-Komaroff, L., McDowell, M., Baltimore, D., Lodish, H.: Translation of reovirus mRNA, poliovirus RNA, and bacteriophage QB RNA in cell-free extracts of mammalian cells. Methods Enzymol. 30F: 709-723 (1974).

Villa-Komaroff, L., Guttmann, N., Baltimore, D., Lodish, H.F.: Complete translation of poliovirus RNA in a eukaryotic cell-free system. Proc. Natl. Acad. Sci. USA 72: 4157-4161 (1975).

Villereal, M.J., Cook, J.S.: Role of the membrane potential in serum-stimulated uptake of amino acids in a diploid human fibroblast. J. Supramol. Struct. 6: 179 (1977).

Vogt, M., Dulbecco, R.: Properties of a HeLa cell culture with increased resistance to poliomyelitis virus. Virology 5: 425 (1958).

Vogt, M., Dulbecco, R., Wenner, H.A.: Mutants of poliomyelitis viruses with reduced efficiency of plating in acid medium and reduced neurophathogenicity. Virology 4: 141-155 (1957).

von Heijne, G.: Membrane proteins: the amino acid composition of membrane-penetrating segments. Eur. J. Biochem. 120: 275-278 (1981).

von der Helm, K.: Cleavage of Rous sarcoma viral polypeptide precursor into internal structural proteins in vitro involves viral protein p15. Proc. Natl. Acad. Sci. USA 74: 911-915 (1977).

Vonka, V., Janda, Z., Tuckova, E.: Serological properties of antigenically different derivates of LSc-2ab poliovirus. Arch. ges. Virusforsch. 12: 3 (1962).

Voss, H.: Über den Einfluß von Dextransulfat auf die oralen Poliomyelitis-Impfstämme nach Sabin. Z. Naturforsch. 19: 226-229 (1964).

Vrijsen, R., Boeye, A.: Gel electophoresis of protein-dodecyl sulfate complexes in a pH gradient and improved resolution of poliovirus polypeptides. Anal. Biochem. 85: 355-366 (1978).

Vrijsen, R., Wouters, M., Boeye, A.: Resolution of the major capsid polypeptides into doublets. Virology 86: 546-555 (1978).

Vrijsen, R., Brioen, P., Boeye, A.: Identification of poliovirus precursor proteins by immunoprecipitation. Virology 107: 567-569 (1980).

Vrijsen, R., Rombaut, B., Boeye, A.: Intertypic crossreactions of nonneutralizing, monoclonal poliovirus antibodies. J. Virol. 49: 1002-1004 (1984).

Wallach, D.F.H.: Disposition of proteins in plasma-membranes of animal cells - analytical approaches using controlled peptidolysis and protein labels. Biochim. Biophys. Acta 265: 61 (1972).

Wallis, C., Melnick, J.L.: Stabilization of poliovirus by cations. Tex. Rep. Biol. Med. 19: 409-683 (1961).

Wallis, C., Melnick, J.L.: Cationic stabilization - a new property of enteroviruses. Virology 16: 504-506 (1962a).

Wallis, C., Melnick, J.L.: Magnesium chloride enhancement of cell susceptibility to poliovirus. Virology 16: 122-132 (1962b).

Wallis, C., Melnick, J.L.: Photodynamic inactivation of poliovirus. Virology 21: 332-341 (1963).

Wallis, C., Melnick, J.L.: Mechanism of plaque inhibition of polioviruses possessing the d marker. J. Gen. Virol. 3: 349-358 (1968a).

Wallis, C., Melnick, J.L.: Mechanism of enhancement of virus plaques by cationic polymers. J. Virol. 2: 267-274 (1968b).

Wallis, C., Melnick, J.L., Ferry, G.D., Wimberly, I.L.: An aluminum marker for the differentiation and separation of virulent and attenuated polioviruses. J. Exp. Med. 115: 763-775 (1962).
Wallis, C., Shirley, A., Melnick, J.L.: Total recovery of infectious virus from noninfectious type 1 poliovirus-antibody complex by heating in salts. Intervirology 1: 41-47 (1973).
Walter, P., Blobel, G.: Translocation of proteins across the endoplasmic reticulum. III. Signal recognition protein (SRP) causes signal sequence-dependent and site-specific arrest of chain elongation that is released by microsomal membranes. J. Cell Biol. 91: 557-561 (1981).
Warner, J., Madden, M.J., Darnell, J.E.: The interaction poliovirus RNA with Escherichia coli ribosomes. Virology 19: 393-399 (1963).
Watanabe, Y., Watanabe, K., Hinuma, Y.: Synthesis of poliovirus-specific proteins in HeLa cells. Biochim. Biophys. Acta 61: 976-977 (1962).
Watanabe, Y., Watanabe, K., Katagiri, K., Hinuma, Y.: Virus-specific proteins produced in HeLa cells infected with poliovirus: Charaterization of subunit-like protein. J. Biochem. 57: 733-741 (1965).
Weber, K., Osborn, M.: Intracellular display of microtubular structures revealed by indirect immunofluorescence microscopy. In: Microtubules (Roberts, K., Hyams, J. S., eds.), 279-313. New York: Academic Press 1979.
Weber, L.A., Hickey, E.D., Nuss, D.L., Baglioni, C.: 5'-terminal 7-methylguanosine and mRNA function: Influence of potassium concentration on translation in vitro. Proc. Natl. Acad. Sci. USA 74: 3254-3258 (1977).
Wecker, E.: A simple test for serodifferentiation of poliovirus strains within the same type. Virology 10: 376-379 (1960).
Wecker, E.: Virus and nucleic acid. Ergebn. Mikrobiol. 35: 1 (1962).
Wecker, E., Lederhilger, G.: Curtailment of the latent period by double infection with polioviruses. Proc. Natl. Acad. Sci. USA 52: 24-251 (1964a).
Wecker, E., Lederhilger, G.: Genomic masking produced by double infection of HeLa cells with heterotypic polioviruses. Proc. Natl. Acad. Sci. USA 52: 705-709 (1964b).
Weeds, A.: Actin-binding proteins - regulators of cell architecture and motility. Nature 296: 811-816 (1982).
Weinberg, R.A.: Nuclear RNA metabolism. Ann. Rev. Biochem. 42: 329-354 (1973).
Weissmann, C., Feix, G., Slor, H.: In vitro synthesis of phage RNA: the nature of the intermediates. Cold Spring Harbor Symp. Quant. Biol. 35: 83-100 (1968).
Wengler, G., Wengler, G.: Medium hypertonicity and polyribosome structure in HeLa cells. Eur. J. Biochem. 27: 162-173 (1972).
Wenner, H.: Some comparative observations on the behaviour of poliomyelitis viruses in animals and in tissue culture. Ann. N. Y. Acad. Sci. 61: 840 (1955).
Wenner, H.A.: The enteroviruses: recent advances. Yale J. Biol. Med. 55: 277-282 (1982).
Wenner, H.A., Archetti, I., Dubes, G.R.: Antigenic variations among type 1 polioviruses. A study of 16 wild type strains and 5 variants. Am. J. Hyg. 20: 66-90 (1959).
Wentworth, B.B., Mccahon, D., Cooper, P.D.: Production of infectious RNA and serum-blocking antigen by poliovirus temperature-sensitive mutants. J. Gen. Virol. 2: 297-307 (1968).
Wentzky, P., Koch, G.: Influence of polycations on the interaction between poliovirus multistranded ribonucleic acid and HeLa cells. Virology 35: 35 (1971).
Wetz, K., Habermehl, K.O.: Topographical studies on poliovirus capsid proteins by chemical modification and cross-linking with bifunctional reagents. J. Gen. Virol. 44: 525-534 (1979).
Wetz, K., Habermehl, K.-O.: Topographical studies on poliovirus: Cross-linking of the virus by UV irradiation. Second Meeting of the European Study Group on the Molecular Biology of Picornaviruses, Hamburg (1981).
Wetz, K., Habermehl, K.O.: Specific cross-linking of capsid proteins to virus RNA by ultraviolet irradiation of poliovirus. J. Gen. Virol. 59: 397-401 (1982).
White, J., Helenius, A.: pH-dependent fusion between the Semliki Forest virus membrane and liposomes. Proc. Natl. Acad. Sci. USA 77: 3273-3277 (1980).
White, J., Matlin, K., Helenius, A.: Cell fusion by Semliki Forest, influenza and vesicular stomatitis virus. J. Cell. Biol. 89: 674-679 (1981).
Wickmann, J.: Studien über Poliomyelitis acuta; zugleich ein Beitrag zur Kenntnis der Myelitis acuta. Eng. Trans. Nerv., Ment. Dis. Monog. Ser., No. 16: 1913 (1905). Berlin: S. Karger.

Wickmann, J.: Beiträge zur Kenntnis der Heine-Medinschen Krankheit. In: Die akute Poliomyelitis. Handbuch der Neurologie, Bd. 2 (1907). Berlin: Karger 1911.

Wickner, W.: Asymmetric orientation of phage M13 coat protein in Escherichia coli cytoplasmic membranes and in synthetic lipid vesicles. Proc. Natl. Acad. Sci. USA 73: 1159-1163 (1976).

Wickner, W.: The assembly of proteins into biological membranes: The membrane trigger hypothesis. Ann. Rev. Biochem. 48: 23-45 (1979).

Wickner, W.: Assembly of proteins into membranes. Science 210: 861-868 (1980).

Wickner, W., Mandel, G., Zwizinski, C., Bates, M., Killick, T.: Synthesis of phage M13 coat protein and its assembly into membranes in vitro. Proc. Natl. Acad. Sci. USA 75: 1754-1748 (1978).

Wiegers, K.J., Dernick, R.: Immunoprecipitation reactions of antisera raised with purified poliovirus capsid proteins isolated by isoelectric focusing in urea. Second Meeting of the European Study Group on the Molecular Biology of Picronaviruses. Hamburg, 1981a.

Wiegers, K.J., Dernick, R.: Peptide maps of labeled poliovirus proteins after two-dimensional analysis by limited proteolysis and electrophoresis in sodium dodecyl sulfate. Electrophoresis 2: 98-103 (1981b).

Wiegers, K.J., Drzeniek, R.: Preparative isoelectric focusing of poliovirus peptides in urea-sucrose gradients. J. Gen. Virol. 47: 423-430 (1980).

Wiegers, K.J., Koch, G.: Interaction of poliovirus-induced double-stranded RNA with HeLa-cells. Arch. Biochem. Biophys. 148: 89-96 (1972).

Wiegers, K.J., Yamaguchi-Koll, U., Drzeniek, R.: A complex between poliovirus RNA and the structural polypeptide VP1. Biochem. Biophys. Res. Commun. 71: 1308-1312 (1976).

Wiegers, K.J., Yanaguchi-Koll, U., Drzeniek, R.: Differences in the physical properties of dense and standard poliovirus particles. J. Gen. Virol. 34: 465-473 (1977).

Wiegers, K.J., Gschwender, H.H., Dernick, R.: The effect of cesium salts on dense poliovirus particles. Intervirology 10: 329-337 (1978).

Wild, T.F., Brown, F.: Nature of the inactivating action of trypsin on foot-and-mouth disease virus. J. Gen. Virol. 1: 247-250 (1967).

Wildy, P.: Classification and nomenclature of viruses. Monographs in Virology, Vol. 5 (1971).

Willems, M., Penman, S.: The mechanism of host cell protein synthesis inhibition by poliovirus. Virol. 30: 355-367 (1966).

Willis, D.B., Goorha, R., Miles, M., Hranoff, A. : Macromolecular synthesis in cells infected by frog virus 3. VIII. Transcriptional and posttranscriptional regulation of viral gene expression. J. Virol. 24: 326-342 (1977).

Wilson, D.B.: Cellular transport mechanisms. Ann. Rev. Biochem. 47: 933-965 (1978).

Wilson, J.N., Cooper, P.D.: Aspects of the growth of poliovirus as revealed by the photodynamic effects of neutral red and acridine orange. Virology 21: 135-145 (1963).

Wilson, J.T., De Riel, J.K., Forget, B.G., Marotta, C.A., Weissman, S.M.: Nucleotide sequence of 3' untranslated portion of human alpha globin mRNA. Nucl. Acids Res. 4: 2353-2368 (1977).

Wilson, T., Paphadjopoulos, D., Taber, R.: Biological properties of poliovirus encapsulated in lipid vesicles: antibody resistance and infectivity in virus-resistant cells. Proc. Natl. Acad. Sci. 74: 3471-3475 (1977).

Wilson, T., Paphadjopoulos, D., Taber, R.: The introduction of poliovirus RNA into cells via lipid vesicles (liposomes). Cell 17: 77-84 (1979).

Wimmer, E.: Sequence studies of poliovirus RNA. I. Characterization of the 5'-terminus. J. Mol. Biol. 68: 537-540 (1972).

Wimmer, E.: The genome-linked protein of picornaviruses: Discovery, properties and possible functions. In: The Molecular Biology of Picornaviruses (Perez-Bercoff, R., ed.), 175-189. New York - London: Plenum Press 1979.

Wolff, D.A., Bubel, H.C.: The disposition of lysosomal enzymes as related to secific viral cytopathic effects. Virology 24: 502 (1964).

Wolosewick, J.J., Porter, K.R.: Microtrabecular lattice of the cytoplasmic ground substance: artifact or reality. J. Cell Biol. 82: 114-139 (1979).

Woods, W.A., Robbins, F.C.: The elution properties of type 1 polioviruses from $Al(OH)_3$ gel. A possible attribute. Proc. Natl. Acad. Sci. USA 47: 1501-1507 (1961).

Wool, I.G.: The structure and function of eukaryotic ribosomes. Ann. Rev. Biochem. 48: 719-754 (1979).

Wool, I.G.: The structure of eukyryotic ribosomes. In: Protein Biosynthesis in Eukaryotes (Pérez-Bercoff R., ed.), 69-95. New York: Plenum Press 1982.

Wouters, M., Van Der Kerckhove, J.: Amino acid composition of the poliovirus capsid polypeptides isolated as fluorescamine conjugates. J. Gen. Virol. 33: 529-533 (1976).

Wouters, M., Miller, A.O.A., Fenwick, M.L.: Distortion of poliovirus particles by fixation with formaldehyde. J. Gen. Virol. 18: 211-214 (1973a).

Wouters, M., Vanden Berghe, D., Boeye, A.: Composition of poliovirus fibrils. Arch. ges. Virusforsch. 43: 24-33 (1973b).

Wright, P.J., Cooper, P.D.: Poliovirus proteins associated with ribosomal structures in infected cells. Virology 59: 1-20 (1974a).

Wright, P.J., Cooper, P.D.: Isolation of cold-sensitive mutants of poliovirus. Intervirology 2: 20 (1974b).

Wright, P.J., Cooper, P.D.: Poliovirus proteins associated with the replication complex in infected cells. J. Gen. Virol. 30: 63-71 (1976).

Wu, M., Davidson, N., Wimmer, E.: An electron microscope study of proteins attached to poliovirus RNA and its replicative form (RF). Nucl. Acids Res. 5: 4711-4723 (1978).

Wychowski, C., Van der Werf, S., Siffert, O., Crainic, R., Bruneau, P., Girard, M.: A poliovirus type 1 neutralization epitope is located within amino acid residues 93 to 104 of viral capsid polypeptide VP1. EMBO J. 2: 2019 (1983).

Yamaguchi-Koll, U., Wiegers, K.J., Drzeniek, R.: Isolation and characterization of „dense particles" from poliovirus-infected HeLa cells. J. Gen. Virol. 26: 307-319 (1975).

Yamaguchi-Koll, U., Wiegers, K.J., Dernick, R.: Dissociation and reassociation of poliovirus. II. Protein components obtained by urea treatment of the virus particle. Z. Naturforsch. 32c: 632-636 (1977).

Yamazi, Y., Takahashi, M., Harasawa, I., Tate, H., Yamamoto, M.: Chemotherapy 14: 439 (1966).

Yamazi, Y., Takahashi, M., Todome, Y.: Inhibition of poliovirus by effect of a methylthiopyrimidine derivative (34542). Proc. Soc. Exp. Bio. Med. 133: 674-677 (1970).

Yilma, T., McVicar, J.W., Breese, S.S.: Prelytic release of foot-and-mouth disease virus in cytoplasmic blebs. J. Gen. Virol. 41: 105-114 (1978).

Yin, F.H.: Involvement of viral procapsid in the RNA synthesis and maturation of poliovirus. Virology 82: 299-307 (1977a).

Yin, F.H.: Possible in vitro repair of viral RNA by ligase-like enzyme(s) in polio-infected cells. J. Virol. 21: 61-68 (1977b).

Yogo, Y., Wimmer, E.: Polyadenylic acid at the 3'-terminus of poliovirus RNA. Proc. Natl. Acad. Sci. USA 69: 1877-1882 (1972).

Yogo, Y., Wimmer, E.: Poly (A) and poly (U) in poliovirus double-stranded RNA. Nature New Biol. 242: 171-174 (1973).

Yogo, Y., Wimmer, E.: Sequence studies of poliovirus RNA. III. Polyuridylic acid and polyadenylic acid as components of the purified poliovirus replicative intermediate. J. Mol. Biol. 92: 467-477 (1975).

Yogo, Y., Teng, M.H., Wimmer, E.: Poly (U) in poliovirus minus RNA is 5'-terminal. Biochem. Biophys. Res. Commun. 61: 1101-1109 (1974).

Yoneyama, T., Hagiwara, A., Hara, M., Shimojo, H.: Alteration in oligonucleotide fingerprint patterns of the viral genome in poliovirus type 2 isolated from paralytic patients. Infect. Immun. 37: 46-53 (1982).

Yoshinaka, Y., Luftig, R.B.: Partial characterization of a p70 proteolytic factor that is present in purified virions of Rauscher leukemia virus (RLV). Bioch. Biophys. Res. Commun. 76: 54-63 (1977).

Yoshinaka, Y., Ishigame, K., Ohno, T., Kageyama, S., Shibata, K., Luftig, R.B.: Preparations enriched for „immature" Murine leukemia virus particles that remain in tissue culture fluids are deficient in $Pr65^{gag}$ proteolytic activity. Virology 100: 130-140 (1980).

Young, N.A., Hoyer, B.H., Martin, N.A.: Polynucleotide sequence homologies among polioviruses. Proc. Natl. Acad. Sci. USA 61: 548-555 (1968).

Young, N.A.: Polioviruses, coxsackieviruses, echoviruses: comparison of the genomes by RNA hybridization. J. Virol. 11: 832-839 (1973a).

Young, N.A.: Size of gene sequences shared by polioviruses types 1, 2, 3. Virol. 56: 400-403 (1973b).
Young, S.A., Radloff, R.J.: Functional characterization of cleavage-defective mutants of encephalomyocarditis virus. J. Virol. 42: 814-824 (1982).
Youngner, J.S.: Thermal inactivation studies with different strains of poliovirus. J. Immunol. 78: 282-290 (1957).

Zabel, P., Moerman, M., van Straaten, F., Goldbach, R., van Kannem, A.: Antibodies against the genome-linked protein VPg of cowpea mosaic virus recognize a 60,000-dalton precursor polypeptide. J. Virol. 41: 1083-1088 (1982).
Zähringer, J., Baliga, B.S., Munro, H.N.: Novel mechanism for translational control in regulation of ferritin synthesis by iron. Proc. Natl. Acad. Sci. USA 73: 857-861 (1976).
Zajac, I., Crowell, R.L.: Effect of enzymes on the interaction of enteroviruses with living HeLa cells. J. Bact. 89: 574 (1965a).
Zajac, I., Crowell, R.L.: Location and regeneration of enterovirus receptors of HeLa cells. J. Bact. 89: 1097 (1965b).
Zajac, I., Crowell, R.L.: Differential inhibition of attachment and eclipse activities of HeLa cells for enteroviruses. J. Virol. 3: 422-428 (1969).
Zeichhardt, H., Schlehofer, J.R., Hampl, H., Wetz, K., Habermehl, K.-O.: Comparative SEM and TEM studies on the cytopathic effect in HEp-2 cells induced by different picornaviruses. Electron Microscopy 1980. Proceedings of the Seventh European Congress on Electron Microscopy, The Hague, The Netherlands. In: Biol. 2: 358-359 (1980).
Zeichhardt, H., Habermehl, K.-O., Diefenthal, W.: Modification and exploitation of a poliovirus-induced membrane complex by superinfecting ME virus. J. Gen. Virol. 55: 265-274 (1981).
Zeichhardt, H., Schlehofer, J.R., Wetz, K., Hampl, H., Habermehl, K.-O.: Mouse Elberfeld (ME) virus determines the cell surface alterations when mixedly infecting poliovirus-infected cells. J. Gen. Virol. 58: 417-428 (1982).
Zeichhardt, H., Wetz, K., Habermehl, K.-O.: Intracellular transport of picornaviruses as investigated by ionophores and weak lysosomotropic bases. Zentralbl. Bakteriol. 255: 195 (1983).
Zibirre, R., Koch, G.: Influence of serum and poliovirus infection on the transport and accumulation of AIB in HeLa cells. Biosci. Rep. 2: 625- 630 (1983).
Zilberstein, A., Dudock, B., Berissi, H., et al.: Control of messenger RNA translation by minor species of leu-transfer RNA in extracts from interferon-treated L cells. J. Mol. Biol. 108: 43-54 (1976).
Zimmermann, E.F., Heeter, M., Darnell, J.E.: RNA synthesis in poliovirus-infected cells. Virology 19: 400 (1963).
Zimmern, D., Kaesberg, P.: 3'-Terminal nucleotide sequence of encephalomyocarditis virus RNA determined by reverse transcriptase and chain-terminating inhibitors. Proc. Natl. Acad. Sci. USA 75: 4257-4261 (1978).
Zolotor, L., Engler, R.: Equilibrium centrifugation of DNA, RNA and poliovirus in density gradients of cesium oxalate and cesium acetate. Biochim. Biophys. Acta 145: 52-59 (1967).

Subject Index*

* Page numbers in bold face refer to a major text discussion of the entry.

Assembly of Enveloped RNA Viruses

By **Monique Dubois-Dalcq,** M. D., Laboratory of Molecular Genetics, National Institute of Neurological and Communicative Disorders and Stroke, National Institutes of Health, Bethesda, Maryland,

Kathryn V. Holmes, M. D., Department of Pathology, Uniformed Services University of the Health Sciences, Bethesda, Maryland, and

Bernard Rentier, M. D., Laboratoire de Microbiologie Générale et Médicale, Université de Liège, Liège, Belgique

Editorial Assistance: **David W. Kingsbury,** M. D., St. Jude Children's Research Hospital, Memphis, Tenn.

1984. 94 partly coloured figures. XVI, 235 pages.
ISBN 3-211-81802-2

This book describes in a most comprehensive manner the assembly of nine families of enveloped RNA viruses which have two features in common: the stable repository of genetic information in each virus is RNA, and each virus modifies and appropriates a particular patch of the eukaryotic cell membrane system to complete its structure. It takes the reader from the level of virus genome structure and expression through the quaternary interactions between virus-specified elements and cellular components that cooperate to produce virus particles. There are spectacular illustrations in this volume, but it is much more than a picture gallery. New techniques such as high resolution scanning electron microscopy, stereo and electron microscopy of platinum replicas, and numerous immunofluorescence pictures obtained with monoclonal antibodies against viral proteins were used in gathering a great variety of documents illustrating these processes. Reading widely in this book is an effective antidote to overspecialization; the reader will discover illuminating parallels between diverse virus families and will come away with a sharpened awareness of important things that are still to be learned.

Springer-Verlag Wien NewYork